Applied Anatomy & Physiology for Manual Therapists

Pat Archer, MS, ATC, LMP
Director of Instruction
Discoverypoint School of Massage
Seattle, Washington

Lisa Nelson, BA, AT/R, LMP
Director of Education
Discoverypoint School of Massage
Seattle, Washington

Acquisitions Editor: Kelley Squazzo
Product Manager: Linda G. Francis
Manufacturing Manager: Margie Orzech
Marketing Manager: Shauna Kelley
Development Editor: Tanya M. Martin
Design Coordinatior: Joan Wendt
Production Service: SPi Global

Two Commerce Square
2001 Market Street
Philadelphia, PA 19103 USA
LWW.com

Printed in China

Library of Congress Cataloging-in-Publication Data
Archer, Patricia A.
Applied anatomy & physiology for manual therapists / Pat Archer, Lisa A. Nelson. — 1st ed.
p. ; cm.
Applied anatomy and physiology for manual therapists
Includes bibliographical references and index.
ISBN 978-1-60547-655-1
I. Nelson, Lisa A. II. Title. III. Title: Applied anatomy and physiology for manual therapists.
[DNLM: 1. Musculoskeletal Manipulations—methods—Problems and Exercises. 2. Musculoskeletal Physiological Phenomena—Problems and Exercises. 3. Musculoskeletal System—anatomy & histology—Problems and Exercises.
WB 18.2]
LC classification not assigned
616'70076—dc23

2011030176

---

10 9 8 7 6 5 4 3 2

To all of the students and teachers who balance the wonder and magic of the manual therapies with the science of anatomy and physiology.

About the Authors

Pat Archer, MS, ATC, LMP

Director of Instruction, Discoverypoint School of Massage, Seattle, Washington

Pat has been an educator and health care specialist for over 35 years, beginning her career as a high school health and physical education teacher in 1973. She received her MS degree with a specialty in exercise physiology and became a certified athletic trainer in 1977. That same year, she was appointed the head women's athletic trainer at the University of Montana. Pat added licensed massage practitioner (WA) to her credentials in 1985 and began teaching anatomy and physiology, kinesiology, pathology, and advanced massage techniques at the Brenneke School of Massage in Seattle early in 1986, where she happily remained through 2008.

Pat's combined knowledge as both a Certified Athletic Trainer and licensed massage therapist provides her with a unique blend of advanced manual therapy assessment and treatment skills that guide her in a successful private practice called Seattle Somatics. Pat has been honored to serve as the lead massage therapist for the Seattle Sonics (NBA), the WNBA Seattle Storm, and the University of Washington swim teams. She was chosen to be Director of Massage Services at the 1997 USA Track and Field National Championships and served as both Athletic Trainer and Massage Therapist at the 1996 Summer Olympics. She has several published workbooks and journal articles to her credit, as well as another textbook with Lippincott Williams & Wilkins, *Therapeutic Massage in Athletics*, 2007. Pat is a popular and highly regarded speaker who has presented at state, regional, and national conferences for both AMTA and NATA, as well as multiple CE courses for private schools and businesses. She is also cofounder and owner of Discoverypoint School of Massage in Seattle, where she serves as the Director of Instruction and core faculty.

Lisa Nelson, BA, AT/R, LMP

Director of Education, Discoverypoint School of Massage, Seattle, Washington

After receiving her degree in Biology and English Literature, Lisa began her professional career in 1987 as Head Athletic Trainer at North Park College in Chicago, Illinois. In charge of providing medical care for athletes participating in 14 intercollegiate varsity sports, she also developed and implemented North Park's initial Exercise Science major, which included an Athletic Training track. In 1990, Lisa joined the faculty at the Brenneke School of Massage and in 1991 became a licensed massage practitioner.

While she maintained a small private practice, Lisa began to focus her professional efforts toward program development and learning methodology, and in 1999, she transitioned from regular classroom teaching to being a Faculty Mentor and Learning Strategist. In this role, she designed and implemented a formal faculty development and training program, while she supported students through learning skills classes, individual tutoring, and academic counseling. Over the past 25 years, she has served a wide variety of education communities as a teacher, curriculum consultant, administrator, faculty mentor, and learning strategist. Lisa has been privileged to speak at numerous teaching conferences throughout the United States, published articles for vocational educators, and contributed to several textbooks in sports medicine, sports massage, and muscle anatomy. Lisa is a cofounder and owner of Discoverypoint School of Massage in Seattle, where she also serves as the Director of Education and core faculty member.

Preface

If you can't explain it simply, you don't understand it well enough.

Albert Einstein

Throughout our careers, we have used many texts to support the teaching of anatomy and physiology (A&P) to students of manual therapy. Most of these standard A&P texts have been written for a broad audience that includes students in a wide variety of health care programs such as nursing, dentistry, and physical and occupational therapy. These texts generally offer lots of specific information that, while essential for other scientific and professional endeavors, has minimal relevance to the practice of manual therapy. For example, most provide detailed facts on such things as chemical bonding, the structure of DNA, reading an EKG, and the steps of meiosis; details that are not necessary in order for manual therapists to make safe and informed therapeutic choices. At the same time, these standard texts lack adequate information about the intricacies of the myofascial network; important functional connections between the muscular, nervous, and fascial systems; and details of the lymphatic system's structure and function as it relates to fluid return. For manual therapists, all of these details help provide essential rationale that explains many of the benefits and effects of their form of therapy. We've consistently heard from students and colleagues alike that a different type of A&P book is needed: one that emphasizes the A&P concepts that manual therapists really need to know, and links that science to their hands-on work.

Applied Anatomy and Physiology for Manual Therapists is a clear, accurate, simple, and comprehensive A&P textbook that addresses the needs of students in manual therapy education programs. This is a *focused* text that deliberately emphasizes the information manual therapists need to be familiar with in order to explain the benefits, effects, indications, and contraindications of their specific form of manual therapy. The text includes detailed information not covered in standard A&P texts, adding an entire chapter on neuromuscular and myofascial connections (Chapter 8), and separating the structure and function of the lymphatic system (Chapter 11) from immunity and healing (Chapter 12). This, along with chapter features such as *Manual Therapy Applications, Pathology Alerts*, and *What Do You Think?* questions, and *By the Way* features, helps readers build bridges between the scientific facts and the application of that information to their therapeutic practice.

ORGANIZATION AND APPROACH

The table of contents is organized into six units to help readers grasp the connections between various body systems. With the exception of the first two chapters that introduce basic concepts and terminology, the unit headings help students establish a functional link between the individual systems of each unit. The order in which the body systems are presented is fairly standard. The choice to first explore the skin, skeletal, and muscular systems is designed to support a manual therapy program in which the study of A&P coincides with the development of students' manual skills. By exploring these systems first, students learn about the tissues they are working on, with, and through.

We have taken a *layered approach* to the presentation of content within chapters. For each topic, we establish a general understanding of and context for foundational information before delving into all the details. Using this method, a key term or concept is initially explained in basic terms the first time it is encountered, while a more detailed analysis of the information is saved until there is an opportunity to make a meaningful and useful connection to a manual therapy application. For example, the types of connective tissue are listed and defined in Chapter 3, Chemistry, Cells, and Tissues, but the detailed explanations of fascia and the molecular structure of collagen do not appear until Chapter 8, Neuromuscular and Myofascial Connections.

CHAPTER LEARNING FEATURES

Through 60 combined years of teaching in settings ranging from high school to vocational and collegiate classrooms, we have found that programs for manual therapists tend to enroll a large number of kinesthetic learners. Too often, these students experience A&P coursework as a grueling and cryptic academic pursuit. They find it easier to understand information when practical applications are emphasized and active learning exercises are included. The unique style and features of this text along with student resources, especially the *Study and Review Guide*, are designed to shift the student experience toward a more practical and accessible approach to the science of A&P.

Each chapter opens with art and an intriguing quotation to spark the imagination and interest of readers. Aristotle wrote that "wonder implies the desire to learn," and Abraham Heschel, rabbi and civil rights activist, echoed this when he wrote that "wonder rather than doubt is the root of knowledge." Without inviting curiosity and connecting the knowledge directly to what they do on a daily basis, the study of A&P can indeed be grueling and uninteresting for any student.

At the beginning of each chapter, learning objectives, key terms with pronunciations, and bulleted lists of the primary system components and functions provide learners with an overview that creates a foundation of key information about the basic workings of each system. The deeper and more comprehensive explanations of the A&P of each system follow. Within each chapter, learning features include

- ***Manual Therapy Applications.*** These sections, identified with a special icon, help students appreciate the physiologic processes that support the therapeutic benefits and effects of their techniques. For example, Chapter 9, The Endocrine System, includes a discussion of the proven benefits of massage in reducing anxiety and the negative effects of long-term stress.
- ***What Do You Think?*** Open-ended questions that invite creative thinking appear at the end of major topic sections within each chapter. These critical thinking questions serve two purposes: (1) they help break the chapter content into smaller bites and (2) they challenge learners to move beyond the simple declarative information by asking them to think about the information in the context of manual therapy, or to apply the information to everyday items or activities.
- ***By the Way.*** Interesting anatomic or physiologic facts, explanations, or reminders are showcased in this boxed feature. These small bites of interesting information help clarify the narrative or provide a "Wow!" factor to keep readers hooked.
- ***Pathology Alert.*** Sometimes a particular structure or function is best understood by exploring how damage to that structure or disruption of that function affects general health. While this text does not have comprehensive explanations on all system pathologies, it provides simple discussions of the pathophysiology, indications, and contraindications for many pathologies commonly seen by manual therapists.

Each chapter concludes with a bulleted summary of the key A&P concepts of that chapter. Review questions that provide students with an objective check of their declarative knowledge are also included. These questions are written in standard testing styles—multiple choice, short answers, and matching—that are appropriate for self-study or assignments. Answers to these and *What Do You Think?* questions may be found in Appendix C.

STUDY AND REVIEW GUIDE

The *Study and Review Guide for Applied Anatomy and Physiology for Manual Therapists* is a multipurpose resource that helps students focus on essential information. Although it is primarily intended to support this textbook, its design also allows it to be used as a learning tool for those using other textbooks or as a stand-alone A&P review guide for practicing therapists. Each study guide chapter opens with the text's learning objectives and a cross-reference to the specific learning activities that address each objective. This will help readers organize their study and review time by allowing them to choose activities according to their specific needs or learning style, or to work through all the exercises as listed. Activities offer learners multiple ways to work with the information. Each chapter offers new and different ways of looking at the system information, rather than simply repeating all of the textbook content. Group activities provide creative study options for student study groups or classroom exercises for instructors. Additionally, the *Study and Review Guide includes* short multiple-choice quizzes organized by unit rather than by chapter. This encourages students to consolidate and integrate their learning and helps prepare them for comprehensive testing.

ONLINE STUDENT RESOURCES

Online student resources that accompany this text include

- Audio pronunciations of key terms
- Animations
- Electronic flash cards
- Additional practice quiz and chapter review questions

INSTRUCTOR RESOURCES

Online instructor resources include

- Lesson plans
- Acland video clips
- Image bank
- Test bank with test generator software
- PowerPoint slides
- Sample syllabus and course outline for 50- and 100-hour courses

CONCLUSION

We, like many manual therapists, approach client care from a more holistic intention than found in standard allopathic medicine: one that includes consideration of the amazing and complex interplay between physical, emotional, mental, and metaphysical aspects of the human being. Additionally, we recognize that there are manual therapies based on very different cultural traditions and perspectives on how the body is organized, as well as how it functions in health and disease. While we embrace and respect the validity of these approaches and are excited to point out some of the interesting parallels between these alternative practices, it is simply not possible to catalog or explain all of the possible viewpoints and connections.

Ultimately, we know that a textbook is simply a tool. We hope that this tool will allow teachers and students of manual therapy to move beyond the basic facts and information of A&P to focus on how the information is important to, and supportive of, what we do. We hope that from the opening to the final chapters you will be fascinated and excited, intrigued and interested in learning more about the human body. Finally, we trust that our book will inspire teachers and students to find a multitude of wondrous links and applications to support and explain the value and benefits of manual therapy. We invite you to share these with us.

Pat Archer and Lisa Nelson
AppliedAP@yahoo.com

User's Guide

Applied Anatomy and Physiology for Manual Therapists provides students with a focused, comprehensive look at anatomy and physiology, emphasizing the information that you need to know in order to understand the benefits, effects, indications, and contraindications of manual therapy. This focused approach, combined with the book's features and robust ancillaries, aims at helping you understand how what you learn in class will apply in clinic. We've provided this User's Guide to help you put the book's features to work for you!

CHAPTER OPENERS

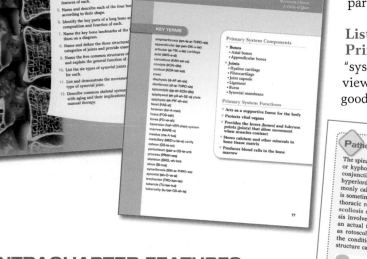

Learning Objectives offer a preview of each chapter, and what key points you will learn within

Key Terms with pronunciations introduce you to new language from the chapter that will become a part of your professional vocabulary

Lists of Primary System Components and Primary System Functions appear in all "system" chapters in the book. This "big picture" view of a system is a helpful preview tool, as well as good study tool!

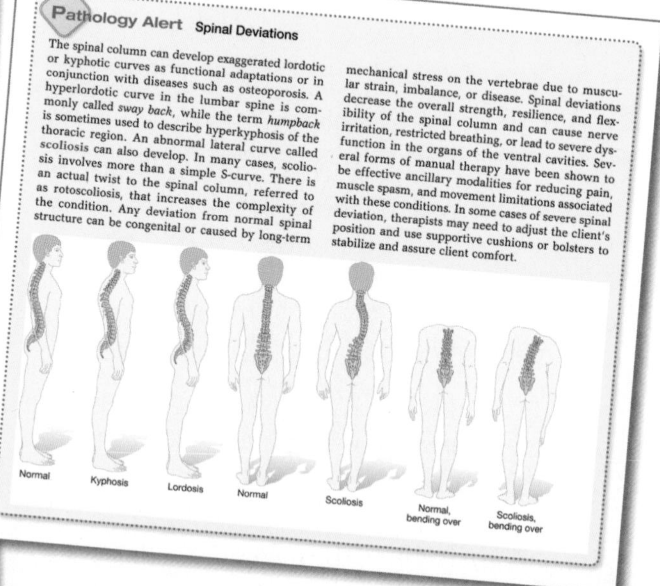

Pathology Alert Spinal Deviations

The spinal column can develop exaggerated lordotic or kyphotic curves as functional adaptations or in conjunction with diseases such as osteoporosis. A hyperlordotic curve in the lumbar spine is commonly called *sway back*, while the term *humpback* is sometimes used to describe hyperkyphosis of the thoracic region. An abnormal lateral curve called scoliosis can also develop. In many cases, scoliosis involves more than a simple S-curve. There is an actual twist to the spinal column, referred to as rotoscoliosis, that increases the complexity of the condition. Any deviation from normal spinal structure can be congenital or caused by long-term mechanical stress on the vertebrae due to muscular strain, imbalance, or disease. Spinal deviations decrease the overall strength, resilience, and flexibility of the spinal column and can cause nerve irritation, restricted breathing, or lead to severe dysfunction in the organs of the ventral cavities. Several forms of manual therapy have been shown to be effective ancillary modalities for reducing pain, muscle spasm, and movement limitations associated with these conditions. In some cases of severe spinal deviation, therapists may need to adjust the client's position and use supportive cushions or bolsters to stabilize and assure client comfort.

INTRACHAPTER FEATURES

Manual Therapy Applications, marked by the "DaVinci" icon, highlights material of particular relevance to manual therapists so you can become familiar with the topics that will be most crucial to you in practice.

Manual therapists may need to take extra precautions to ensure the safety and comfort of older clients. For example, clients with extreme kyphosis may require special bolstering, or they may not be able to lie prone at all. Some geriatric clients may require assistance getting on or off a treatment table, and caution should always be used with any deep tissue techniques, range of motion exercises, or stretching.

Pathology Alerts are brief, simple discussions of the pathophysiology, indications, and contraindications for many pathologies commonly seen by manual therapists to help you begin to understand conditions you will often encounter as a professional.

What Do You Think? 5.1

- What overall structural advantage does the medullary cavity provide for the skeleton?
- Knowing the dynamic activity that occurs at the epiphyseal plate, how much truth do you think there might be in the concept of "growing pains"?

What Do You Think? boxes present open-ended questions placed throughout the text invite you to think critically and creatively about what you're reading.

By the Way

The female pelvic girdle is flatter, broader, and more shallow than that of a male. This shape makes the pelvic inlet wider to accommodate a baby's head during childbirth.

By the Way boxes share fascinating A&P facts, explanations, and reminders that highlight just some of the most intriguing aspects of the human body!

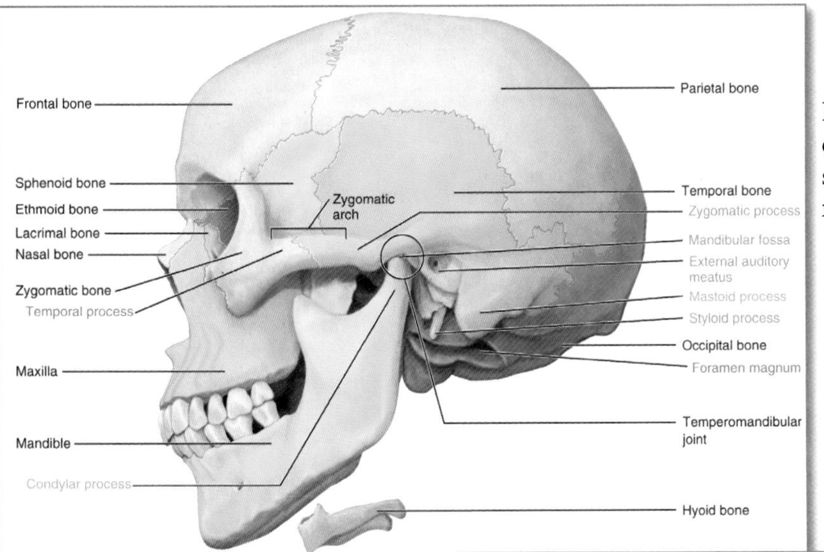

Fresh, colorful **illustrations** and tables explain complex concepts in a clear, simplified manner to provide visual reinforcement for what you learn.

Table 5-3 Joint Classifications

Structural	Functional	Examples
Fibrous	Synarthrotic (immovable)	Sutures of the skull Tibiofibular articulations
Cartilaginous	Amphiarthrotic (slightly movable)	Pubic symphysis Intervertebral joints
Synovial	Diarthrotic (freely movable)	Elbow (humeroulnar joint) Knee (tibiofemoral joint) Hip (iliofemoral joint)

Brightly colored **tables** organize important information into a manageable format.

Important terms are highlighted in **purple** in the text.

END-OF-CHAPTER FEATURES

Summary of Key Points gives you a thorough review of the chapter so you can ensure that you covered each of the chapter's Learning Objectives.

SUMMARY OF KEY POINTS

- The skeletal system provides framework, protection, levers for movement, calcium and other mineral storage, and is the site for blood cell production in the body.
- The bones of the skeleton are organized into two major divisions: axial and appendicular. See TABLE 5-1 to review the axial and appendicular bones.
- There are four types of bone cells: osteoblasts that build up bone, osteoclasts that break down bone, osteocytes that carry out nutrient–waste exchange to keep bones alive, and osteogenic cells that are the only bone cells capable of mitosis (developing into osteoblasts).
- There are two types of bone tissue: compact (dense), also called cortical, and spongy (cancellous). Compact bone forms the shaft of long bones and the outer layers of all bones. It is made up of osteons, or haversian systems, a group of concentric rings of bone tissue called lamellae. Spongy bone is found in the ends of long bones and throughout the flat bones. It is composed of a lattice work of bone tissue called trabeculae. Osteocytes that keep bone tissue alive reside in the tiny spaces of both types of bone tissue, connected by canaliculi, a network of canals.
- Bones are classified by shape as flat, long, short, or irregular. The majority of appendicular bones are long bones; carpals and tarsals are short or cuboid; the sternum, scapula, and bones in the skull and pelvic girdle are flat bones; and irregular bones include the vertebrae, some facial bones, and the sesamoid bones such as the patella.
- The five common parts of long bones are
  - Epiphysis—the end of the bone; contains red marrow
  - Diaphysis—the shaft
  - Medullary cavity—cavity in the middle of the shaft; contains yellow marrow
  - Articular or hyaline cartilage—covers the articular surfaces of the bone
  - Periosteum—outer connective tissue covering
  - Endosteum—lining of the medullary cavity
- Each bone has a variety of depressions, bumps, grooves, and projections called bone landmarks that serve as attachments for other bones, muscles, and ligaments, or as passageways for nerves and blood vessels. See TABLE 5-2 to review these landmarks.
- Joints (articulations) can be classified by structure and function. The structural classification is based mostly on the connective tissues of the joint and the manner in which they stabilize the bone ends. The structural classification of joints is as fibrous, cartilaginous, or synovial. The functional method of classification is based on the movements allowed at the joint. Fibrous joints are called synarthroses and are immovable, and cartilaginous joints are amphiarthrotic; allow partial movement. All synovial joints are diarthrotic (freely movable) joints.
- All synovial joints have four common structural features:
  - A stabilizing fibrous capsule
  - Stabilizing ligaments; can be inside or outside the joint capsule
  - A synovial membrane that secretes synovial fluid into the joint cavity
  - Articular or hyaline cartilage to protect the bone ends and stabilize the joint
- Synovial joints are classified according to shape and their movement capability as hinge, condyloid, pivot, ball-and-socket, gliding, or saddle joints.
- Common changes in the skeletal system related to aging include wearing away of the articular cartilage, decreased range of motion, decreased density of bone tissue matrix, and decreased production of blood cells.
- Manual therapists use their knowledge of the names and locations of bones, bone landmarks, types of joints and their movement capabilities to assess and treat clients.

Review Questions are a mini-test that includes short-answer, multiple-choice, and true-false questions. Answers are in Appendix C, so you can immediately assess your progress.

REVIEW QUESTIONS

Short Answer

1. Name and explain the five functions of the skeletal system.
2. List the four types of bones.
3. Name and define the three structural classifications of joints and describe the movement capability of each.
4. Name and explain the movement capability of the six types of synovial joints.

APPENDICES

This text has three valuable appendices that provide useful background and help students engage further:

Appendix A: Benefits and physiologic Effects of Swedish Massage

Appendix B: Innervation of Major Skeletal Muscles

Appendix C: Answer Key includes answers to "What Do You Think?" and Review Questions.

A full **glossary** with pronunciations and definitions is provided in this text.

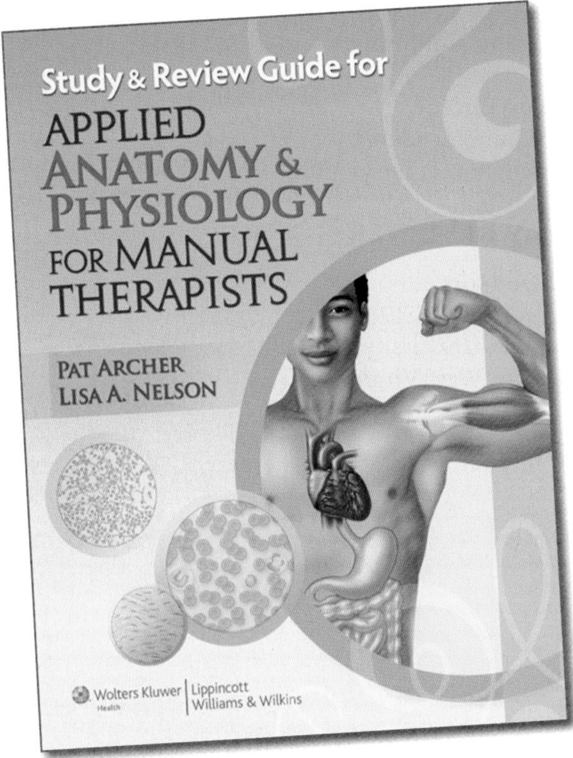

ALSO!

Study and Review Guide for Applied Anatomy & Physiology for Manual Therapists

- **Table** at beginning of each chapter links Chapter Objectives from the text to specific exercises in the Guide
- **Crossword puzzles,** at least 1 per chapter, reinforce vocabulary terms and definitions
- **Coloring** and **labeling exercises** provide a fun way to engage with complex anatomic structures and functions
- **Mnemonic exercises** give students a tool to help them remember important information
- **Quiz show-formatted activities** challenge students to work together and support each other
- **Questions in Fill-in-the-Blank and Matching** format encourage a close engagement with the material
- **Practice exams** organized by unit provide students with valuable practice taking tests
- **Answer Key** in the back of the Guide provides immediate feedback on all exercises and quizzes.

Online Resources:
The following resources are available at http://thePoint.lww.com/Archer-Nelson:

INSTRUCTOR RESOURCES

Case Studies
PowerPoint Slides
Lesson Plans
Image Bank
Acland Video Clips
Test Generator

STUDENT RESOURCES

Labeling Exercises
Audio Glossary
Animations
Electronic Flash Cards
Quiz Bank

Reviewers

John Butler
Cleveland Chiropractic College, Los Angeles
Los Angeles, California

Carol Carlson
Omaha School of Massage Therapy—Herzing University
Omaha, Nebraska

Thomas J. Farruggella
Hudson Valley School of Massage Therapy
West Park, New York

Jessica Keller
Shenandoah Valley School of Therapeutic Massage
Strasburg, Virginia

Johnathan Lambert
Western Career College
Citrus Heights, California

Maryganes Luczak
Career Training Academy
New Kensington, Pennsylvania

ANATOMY REVIEWER

Sean C. Dodson
Third Year Medical Student
West Virginia University School of Medicine
Morgantown, West Virginia

Acknowledgments

I was successful because you believed in me.

Ulysses S. Grant to Abraham Lincoln

Writing a book is a wondrously strange, arduous, and all-encompassing process that in some way seems to involve everything you've ever learned from everyone in your life. While it is impossible to list and acknowledge everyone, there have been some key people who helped bring this project to fruition by believing in us. Thank you to the team at LWW for their guidance, support, and long hours of work. Many thanks to our editors Linda Francis and Tanya Martin who worked with us to refine our words and ideas in order to clearly articulate our vision, and cheered us on. Also, thanks to all the reviewers for catching mistakes in our initial draft, and providing essential feedback that helped us add, delete, or rethink our content.

Special thanks go to Heida Brenneke, founder and owner of the Brenneke School of Massage, as well as all of our students and colleagues over the years. You formed the unique and vibrant educational community that truly forged us as teachers and practitioners. The lessons you taught us both in and out of the classroom provided the inspiration, intent, shape, and tone of this book. To Julie Darrah, our dear friend, business partner, and Executive Director at the Discoverypoint School of Massage: a huge thank you for picking up the slack when our attention was focused on finishing the book. Your smile, patience, support, sense of humor, and tireless commitment ensured the completion of this project and a successful launch of Discoverypoint.

From Lisa, more gratitude than can be expressed goes to my incredible husband Mark, and kids Karianne, Zach, and Tyler. You all endured my mental distraction, never-ending hours at the computer, and many moments of frustration. Thanks for seldom complaining about the dust, laundry, dishes, homework projects done without my help, and the multiple weeklong writing retreats that allowed Pat and me to get so much work done. Without your love, support, and sacrifice, the whole project would have fallen apart. And to Pat, the best coauthor I could ever ask for, you are amazing! Thank you, my treasured friend.

From Pat, my deepest thanks to Lisa for taking on the extra work of being the project manager for this text. Without your organizational and communication skills, I might have given up on this. The depth and breadth of your knowledge and the clarity of how you express those thoughts and ideas inspire me. You truly are the best teacher I have ever known, and I am honored and grateful to have you as my coauthor, business partner, and dear friend.

Contents

Unit III • Framework and Movement

Applied Anatomy & Physiology for Manual Therapists

1 Applying Anatomy and Physiology to the Practice of Manual Therapy

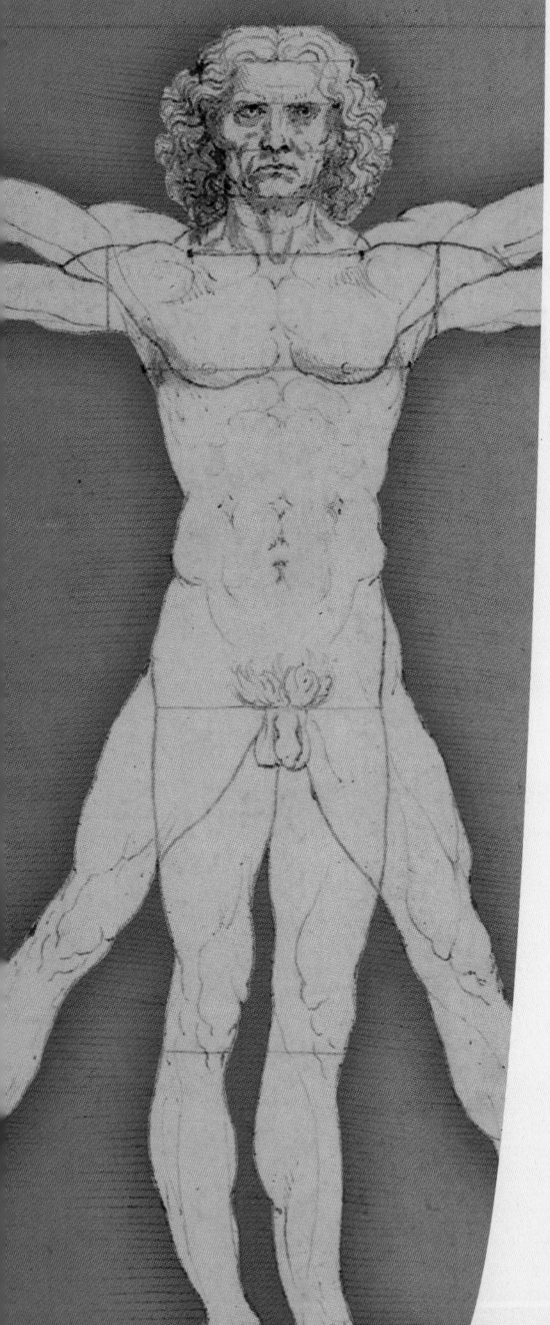

LEARNING OBJECTIVES

Upon completion of this chapter, you will be able to:

1. Discuss the importance of the study of anatomy and physiology for manual therapy practitioners.
2. List and define the levels of complexity for body organization and provide an example of each.
3. Define homeostasis and give a few examples of common homeostatic changes.
4. Explain the difference between the negative and positive feedback mechanisms used to maintain homeostasis.
5. Compare and contrast the benefits versus physiologic effects of manual therapy.
6. Distinguish between the structural and systemic effects of manual therapy.
7. Based on therapeutic intent, list and explain seven general manual therapy categories.
8. Name several forms of manual therapy that fall into each of the seven categories.
9. Identify the 11 body systems, including the primary components and general functions of each system.

The human body is a marvel of engineering. From the muscular and skeletal systems of the hand working in concert to allow us to type, eat, and caress to the circadian rhythms of the heart and digestive system keeping things moving despite our consciousness being elsewhere, our bodies are far more complex and awe-inspiring than any man-made creation."

ALEXANDER TSAIRAS AND BARRY WERTH

The Architecture and Design of Man and Woman

KEY TERMS

anatomy (ah-NAT-o-me)
cardiovascular (kar-de-o-VAS-ku-lar) **system**
cell (SEL)
digestive (di-JES-tiv) **system**
effector (e-FEK-tor)
endocrine (EN-do-krin) **system**
feedback system
holistic (ho-LIS-tik) **approach**
homeostasis (ho-me-o-STA-sis)
integumentary (in-teg-u-MEN-tah-re) **system**
lymphatic (lim-FAT-ik) **system**
manual therapy
muscular (MUS-ku-lar) **system**
negative feedback
nervous (NER-vus) **system**
organ (OR-gan)
organism (OR-gah-nizm)
physiologic effect (fiz-e-o-LOJ-ik e-FEKT)
physiology (fiz-e-OL-o-je)
positive feedback
receptor (re-SEP-tor)
reproductive (re-pro-DUK-tiv) **system**
respiratory (RES-peh-rah-tor-e) **system**
skeletal (SKEL-eh-tal) **system**
stimulus (STIM-u-lus)
structural (STRUK-chur-al) **effect**
system (SIS-tem)
systemic (sis-TEM-ik) **effect**
tissue (TISH-u)
urinary (YER-eh-nar-e) **system**

The study of the structure of an organism such as the human body is called **anatomy**. It involves how the body is organized, including its component parts, their locations, and how they are positioned relative to one another. **Physiology** is the study of the function of these parts. It explains mechanical functions such as how we chew and chemical processes such as how we break down food. Anatomy and physiology (A&P) are commonly studied together because structure and function are intricately linked. For example, the size and shape of blood vessels influences both the volume and pressure of blood within them, and the multiple folds within the small intestines increase the surface area of the digestive tract, allowing for greater absorption of nutrients from digested food.

The study of A&P helps us understand the physiologic shifts and changes that occur continually in our bodies. It explains how we grow and why our hearts sometimes race or we experience moments of breathlessness. It explains why our stomachs grumble, our skin feels itchy, and how we suffer discomfort from stubbed toes and bad colds. For manual therapists, understanding the structures and functions of the body also provides essential information and rationales for the safe and effective application of therapeutic techniques.

This chapter provides basic information to help you gain a general understanding of the body and the interdependent functions of its systems. It discusses the different levels of organization within the body, the physiologic mechanism that balances all bodily processes to sustain life, and the general role and primary structures of each body system. This chapter also provides a definition of manual therapy, categorizes various forms and methods, and sets out clear terminology to describe therapeutic benefits and effects. This overview of A&P and manual therapy provides the fundamental framework that will help you connect the study of the human body to your practice as a manual therapist.

By the Way

The human body is a highly organized community of approximately 100 trillion cells.

LEVELS OF ORGANIZATION

An **organism** is any living thing that functions as a whole. Like all organisms, the human body is able to use nutrients, excrete wastes, reproduce, move, grow, and respond or adapt to changes in its environment. These physiologic processes require a coordinated effort between multiple systems and levels of structural organization in the body. The five levels of organization for the body, moving from least to most complex (FIGURE 1.1) are:

- ***Cellular***— While all components of the human body are formed by microscopic particles called atoms and molecules, the smallest unit of life capable of existing on its own is the **cell**. A cell is considered to be the most basic building block of the body and therefore the simplest level of organization.
- ***Tissue***— A **tissue** is a group of similar cells that work together to carry out one or more specific functions. Fat is an example of a tissue. Fat cells are bound together to form a tissue that insulates us from the cold, cushions delicate structures, and serves as a source of energy when needed.
- ***Organ***—An **organ** is a well-defined and organized group of tissues working together to accomplish a specific set of tasks for the body. For example, the heart has several different types of tissue that function together to pump blood throughout the body.
- ***System***—A **system** is a group of interrelated and interdependent organs that work together to accomplish a specific function or set of functions. For example, the digestive system is made up of the mouth, esophagus, stomach, intestine, and other organs that work together to convert the nutrients in food into forms the body can use for energy, repair, and renewal. Each system plays an important role in creating and maintaining the health and well-being of the entire human organism.
- ***Organism (the body as whole)*** —The entire human organism, the highest level of complexity, is an integrated structure made up of the various systems, organs, tissues, and cells. Each level of organization is functionally linked to the next, making each essential to our health and well-being. No single change occurs in the body without impacting several other processes or structures.

By the Way

D'Arcy Thompson, a Scottish professor of biology in the late 1800s, emphasized the link between A&P by teaching that living forms are inseparable from their function.

Analogies for building from a simple unit to a complex organization can be found everywhere. For example, if we think of a person as representing the smallest building block or cell, increasing levels of complexity could be described as: a person is a member of a family; families make up a neighborhood; neighborhoods make up a city; and cities make up a state. Similarly, cells combine to make up a tissue; tissues working together make up an organ; organs function together to form a system;

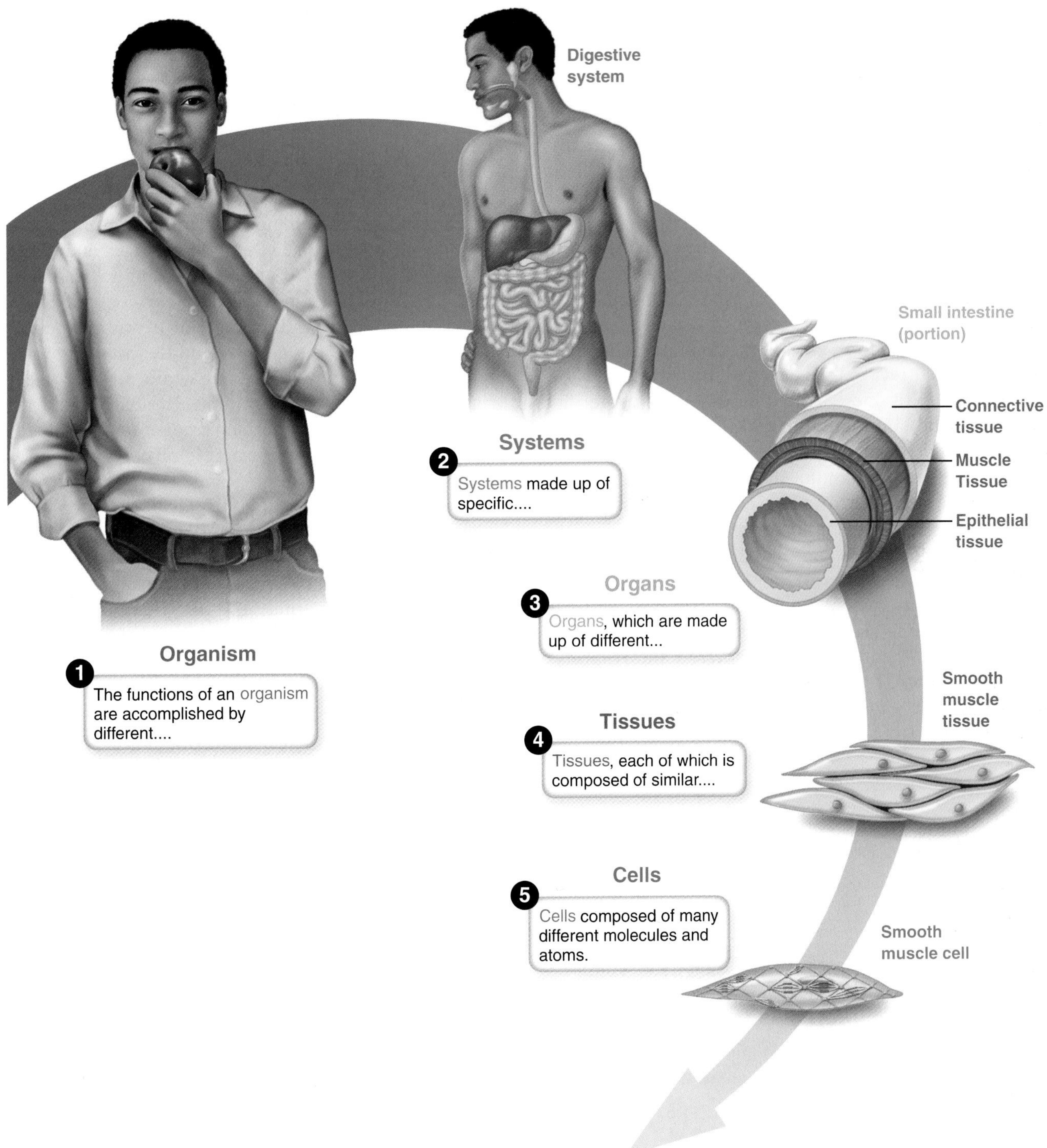

FIGURE 1.1 ▸ Levels of organization. The entire human organism represents the highest level of complexity, and the least complex level is the cell, the basic building block of life.

and systems function together to form the entire human organism and sustain life.

HOMEOSTASIS

Conditions both inside and outside our bodies are constantly changing. In order to survive in changing conditions, the human body must maintain a level of internal stability or balance, known as **homeostasis**. Rather than a static point of balance, homeostasis is a *range* of stability. For instance, we do not maintain a normal body temperature of precisely 98.6°F at every moment. Each day, our temperature fluctuates slightly above and below this point as the body responds to changes in the environment. Our temperature will go down slightly if a cool breeze blows through an open window as we sleep.

It will rise when we run on a treadmill or bicycle up a hill. Thus, homeostasis is a *dynamic* state of equilibrium.

When a change to the internal or external environment, called a **stimulus**, occurs, the body must respond quickly to restore conditions to the preferred range of balance. These changes are sensed by special organs called **receptors** that are sensitive to only one specific type of stimulus. Once the receptor is stimulated, the integration center, either the nervous system or an endocrine gland, causes specific cells, tissues, or organs to respond. The responding cell, tissue, or organ is called the **effector** (FIGURE 1.2). For example, when you drink a lot of water all at once, specific receptors sense an increase in fluid volume, and homeostatic mechanisms in the kidneys (effectors) are engaged to flush extra water out of the body by increasing the amount of urine that is produced and eliminated. In addition to body temperature and fluid balance, other examples of physiologic functions regulated by homeostatic mechanisms include growth, blood pressure, and heart and respiratory rates.

The body's homeostatic processes are dependent on two different types of feedback loops or mechanisms. A **feedback system**, a term used to describe the relationship between the original stimulus and the body's response, is classified as either negative or positive. A negative feedback system is when the response counteracts the original stimulus, while a positive feedback loop reinforces it.

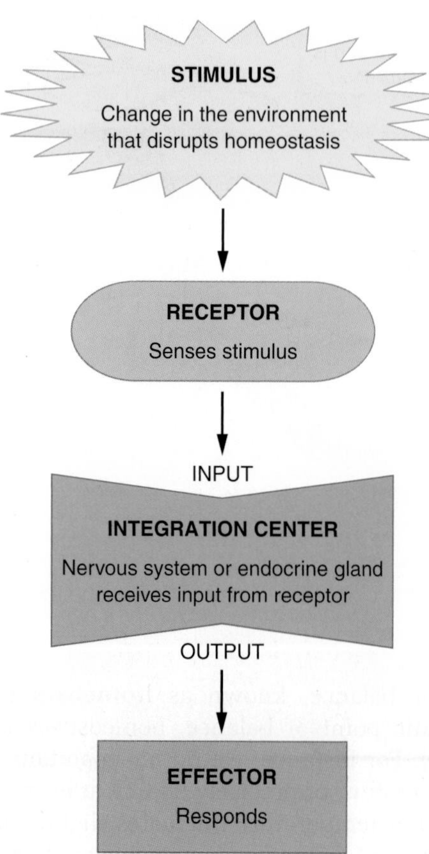

FIGURE 1.2 Homeostasis. The homeostatic mechanism proceeds from stimulus, through the integration center, to effector.

Negative Feedback

The most common type of homeostatic mechanism is a **negative feedback** loop. In this situation, the receptor senses a change and causes a response in the effector that reverses the change that initially stimulated the receptor (FIGURE 1.3). Let's say a teenager is due home each weeknight by 11:00 p.m. according to his parents' rules. One day, he comes in after curfew, and his parents respond by grounding him for a week. Their purpose in restricting his activity for the week is to stop their son from coming home late again. In this case, the original stimulus (returning home after curfew) creates a response (being grounded) that is meant to counteract the original behavior. A good example of negative feedback in the human body is the regulation of glucose levels in the blood. After a meal, blood levels of the simple sugar known as glucose rise. This stimulus triggers receptors in special cells in the pancreas. As the integration center, the pancreas produces a chemical called insulin that causes body cells to increase their usage of glucose. As the cells use more glucose for production of energy, the level of glucose in the blood decreases. Eventually, the amount of glucose in the blood returns to normal homeostatic levels, insulin production ceases, and the cells stop using glucose. In other words, the result (lowered glucose levels due to increased usage by body cells) counteracts the original stimulus of increased glucose levels.

Positive Feedback

In contrast to negative feedback, **positive feedback** mechanisms do not counteract the original stimulus, rather, they *reinforce* it. Instead of returning the body to homeostatic balance, they promote the continuation of some kind of temporary or special circumstance. For example, in childbirth, the body releases a hormone called oxytocin that stimulates powerful contractions of the uterus. These contractions push the baby against the opening to the birth canal, causing it to stretch. This stretching stimulates the continued release of oxytocin to sustain uterine contractions. This self-reinforcing cycle is an example of positive feedback that ensures the continuation of uterine contractions until the birth is completed.

Holistic View of Homeostasis

The standard view of homeostasis is generally limited to a narrow definition: maintenance of the internal physical environment of the body within tolerable physiologic limits. However, people are more than the sum of their parts and physiological processes. Many manual therapists interact and work together with their clients from a **holistic approach**, one that is guided by the principle

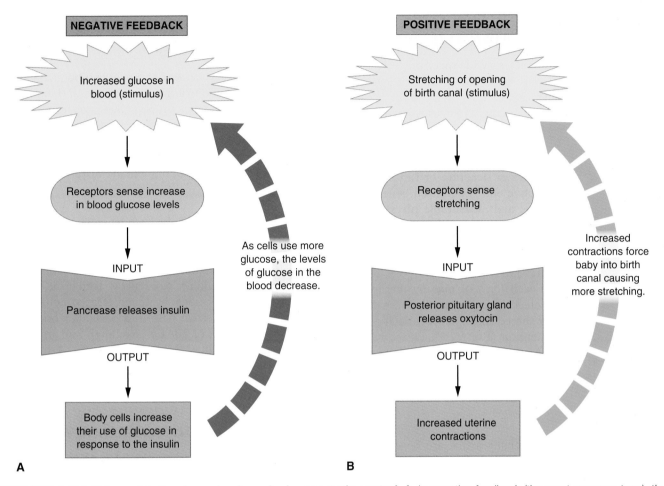

NEGATIVE FEEDBACK
Increased glucose in blood (stimulus)
Receptors sense increase in blood glucose levels
INPUT
Pancreas releases insulin
OUTPUT
Body cells increase their use of glucose in response to the insulin
As cells use more glucose, the levels of glucose in the blood decrease.
A

POSITIVE FEEDBACK
Stretching of opening of birth canal (stimulus)
Receptors sense stretching
INPUT
Posterior pituitary gland releases oxytocin
OUTPUT
Increased uterine contractions
Increased contractions force baby into birth canal causing more stretching.
B

FIGURE 1.3 ▸ Primary feedback mechanisms for homeostatic control. A. In negative feedback (the most common type), the response of the effector counteracts the original stimulus. **B**. Positive feedback loops are used to sustain special physiologic needs.

that the physical body, cognitive processes (mind), and emotional or spiritual aspect are inseparable parts of a whole and integrated person. From a holistic perspective, we can broaden the definition of homeostasis to also include support of the balance between body, mind, and spirit. This view acknowledges the dynamic equilibrium that exists and needs to be supported to maintain optimal health and well-being. For most manual therapists, this holistic approach is implicit in our approach to the clients with whom we work. It is reflected in the simple understanding that we cannot move or touch the body of another human being without also affecting their mind and spirit, whether consciously or not (▸ FIGURE 1.4).

What Do You Think? 1.1

- What everyday examples or analogies can you give for the five levels of body organization?
- What examples of homeostatic changes can you name based on your current understanding of how the body works?

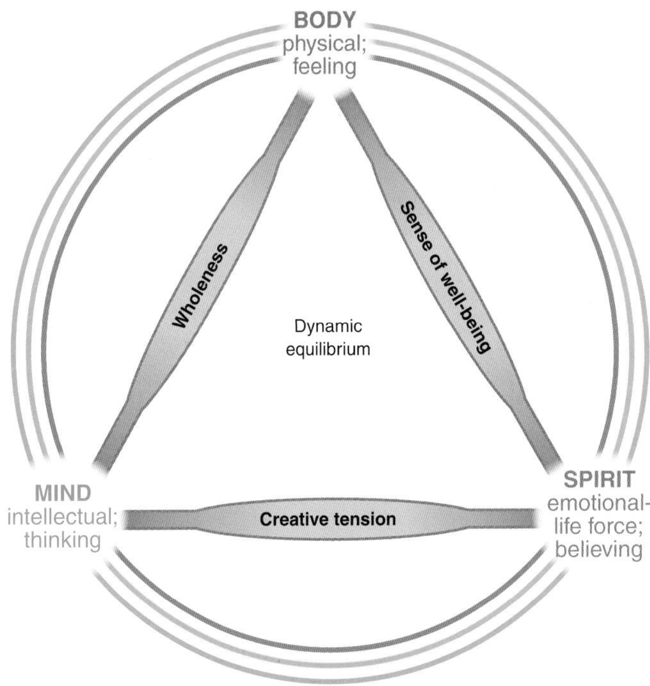

FIGURE 1.4 ▸ Holistic representation of homeostasis. This model of homeostasis depicts the dynamic equilibrium between body, mind, and spirit.

BENEFITS AND PHYSIOLOGIC EFFECTS OF MANUAL THERAPY

Manual therapy encompasses a wide spectrum of unique practices that utilize patterned and purposeful touch and/or movement with therapeutic intent. ◗ TABLE 1-1 provides an overview of various categories of manual therapies. Each discipline includes techniques designed to promote healing or overall health and well-being by creating changes in the body, mind, or spirit that supports homeostasis. These changes in structure or function are classified as either subjective and generalized benefits of manual therapy or specific and measurable **physiologic effects** of a particular discipline or technique. Although all manual therapy is beneficial, different forms and techniques produce a wide array of specific and differing physiologic results because they differ in purpose, style, and application. While each form of manual therapy is unique, it is helpful to categorize and describe these disciplines according to their shared concepts, vocabulary, and primary therapeutic intention (TABLE 1-1). Please recognize that this neither effectively represents the full range of manual therapies nor does it completely describe the unique perspective of any single therapeutic method.

Benefits

The term *benefit* refers to changes that are beneficial to health and well-being regardless of the particular form of manual therapy applied. Benefits are general and subjective, so how well they are achieved depends on the individual therapist's skills; the overall quality of touch and/or movement; and the unique medical history, attitudes, and experiences of the recipient. Therefore, all manual therapy is beneficial, but each individual can experience different levels of benefit from a manual therapy session. Some well-documented benefits of manual therapy include

- Decreased anxiety
- Improved mental focus
- Improved resilience to stress
- Enhanced sense of well-being
- Improved sleep patterns[1–8]

Ultimately, the benefits of manual therapy cannot be attributed to any specific method or technique. Generally, a healthy body will respond to the positive therapeutic relationship between practitioner and client and the skilled application of patterned and purposeful touch and/or movement.

Physiologic Effects

In contrast to benefits, **physiologic effects** of manual therapy are specific, objective, and quantifiable tissue, organ, or system changes created by a particular form or technique. Physiologic effects provide the rationale for the use of a particular technique and can be divided into two categories: structural and systemic. **Structural effects** are physical changes such as stretching, loosening, broadening, or unwinding that occur in the muscles and connective tissues of the body.[1,2,6,9–15] In contrast, **systemic effects** are regional or body-wide responses that are mediated by cellular, circulatory, endocrine, and/or nervous system processes. Systemic effects include outcomes such as decreased pain in a particular body area or region, reduction of edema (swelling), and improved local fluid movement.[1,2,4,6–10,13,16–21] Appendix A lists and describes the key benefits and effects of Swedish massage.

By the Way

Pain reduction is somewhat subjective because it is based on an individual's perceptions and experiences. However, since an objective scale can be used to document and track a client's report of the frequency, intensity, and duration of pain, pain reduction can be considered a physiologic effect, rather than a benefit, of manual therapy.

What Do You Think? 1.2

- Based on the categories shown in TABLE 1-1, how would you categorize any other forms of manual therapy that you've heard of or experienced?
- What are some benefits of manual therapy for you?
- What structural and/or systemic effects of manual therapy have you personally experienced?

SYSTEMS OF THE BODY

The body is an integrated organism made up of 11 different organ systems. While each system carries out a specific set of tasks, life is sustained only when all systems operate in sync with each other. Therefore, it is essential to explore each system individually and to keep in mind that the body's systems are highly interdependent. Just as an auto mechanic must learn about the locations, roles, and interplay between all the different parts and systems of a car, manual therapists must take a detailed look at each body system's structural components and the functional contributions it makes to the integrated whole. The following sections will help you establish a basic understanding of body systems and their general function prior to the detailed exploration of each in the chapters that follow. In addition, general responses of each system to manual therapy are described to spark your interest and provide some insight into the connection between A&P and professional therapeutic practice.

Table 1-1 Manual Therapy Categories

Category	Description	Primary Therapeutic Intention	Common Names
Swedish or relaxation	Sliding/gliding style of work that uses a lubricant for a full-body session	Relaxation Stress reduction Reconnecting Grounding	Relaxation massage Spa massage Stress reduction massage Traditional massage Wellness/health maintenance massage
Myofascial	Any technique focused on stretching, loosening, or broadening fascia and other connective tissues. Usually done with no lubricant or with minimal amounts of lubricant	Improve structural alignment Improve range of motion Decrease pain Improve local fluid flow	Active Release Technique (ART™) Aston Patterning® Cranio-sacral Hellerwork® Joint mobilization Myofascial Release (MFR) Rolfing® Soft tissue release (STR) Structural integration
Neuromuscular	Any technique that reduces resting muscle tension. Includes trigger point, tender point, and proprioceptive techniques.	Normalize muscle tension Decrease pain Improve range of motion Loosen myofascial components	Active Release Technique (ART™) Muscle Energy Technique® (MET) Myotherapy Neurokinetics Neuromuscular Technique (NMT) Positional release Proprioceptive Neuromuscular Facilitation (PNF) Strain Counterstrain® Tender point Trigger point
Lymphatic	Any system of light to moderate depth strokes based on A&P of the lymphatic system	Stimulate edema uptake, lymph flow, and other lymphatic processes	Comprehensive Decongestive Therapy (CDT) Lymphatic massage Lymphatic techniques Lymphedema techniques Manual Lymphatic Drainage (MLD)®
Movement therapies	Focused, patterned, conscious movement and/or positioning	Achieve pain-free movement General relaxation and body-awareness Decrease muscle tension Release holding patterns (emotional or physical)	Alexander technique Feldenkrais® Hakomi Qigong Rosenwork Tai chi Trager®
Reflexive/zone therapies	The use of light or deep pressure to stimulate defined energy zones, dermatomes, or points	Improve systemic and organ functions Decrease pain Enhance general well-being and energy flow	Acupressure Acupuncture Amma Bindegewebmassage (CTM) Reflexology Shiatsu Therapeutic Touch®
Energy techniques	Touching/holding/stroking (with or without physical contact) of chakras, energy zones, or chi points	Decongest, balance, or improve activity of energy (chi, prana, Qi, ki, life-force)	Aura techniques Ayurvedic Chakra balancing Polarity® Qigong Reiki® Therapeutic Touch® Touch For Health

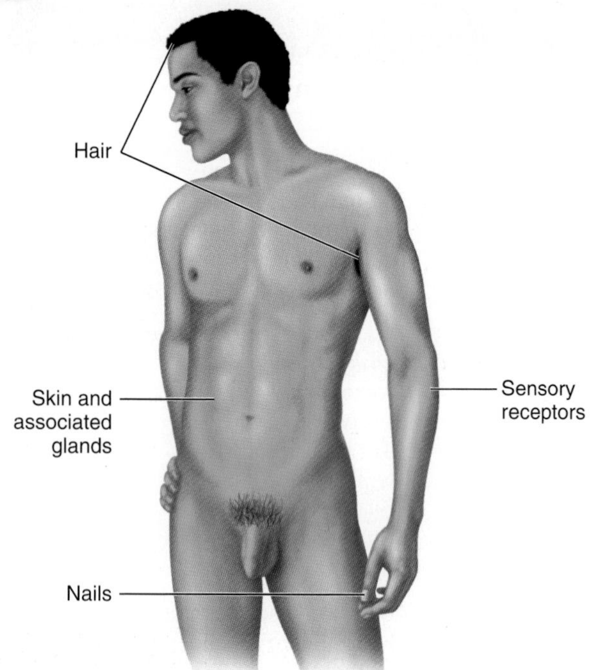

The integumentary system. The skin forms the body's protective outer covering, making it an important part of body defenses. Containing thousands of general sensory receptors, the skin serves as the body's largest sensory organ and also plays a key role in the regulation of body temperature. The color, texture, and temperature of the skin provide manual therapists with important information on the status of underlying tissues and systems, and serves as the therapeutic interface for all touch therapies.

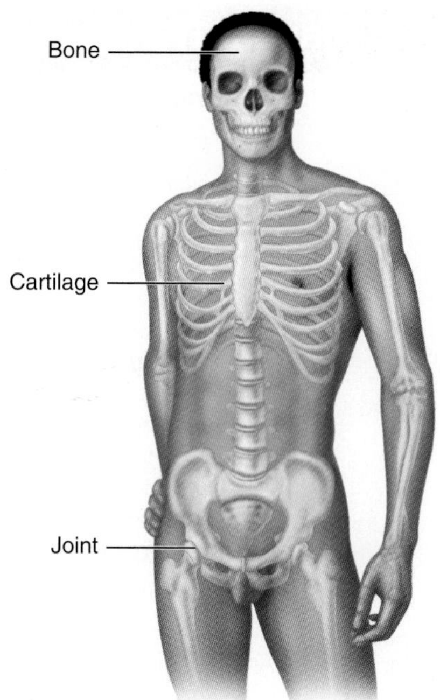

The skeletal system. This system provides the structure and framework for the body. This framework protects vital organs, while individual bones serve as levers used by skeletal muscles to move body parts. Bones are the site of blood cell production, and they store calcium and other important minerals. Many of the structural effects of manual therapy are directed toward the connective tissues and movement receptors of the joints in the skeletal system. Bone landmarks provide therapists with important information on structural alignment and postural adaptations.

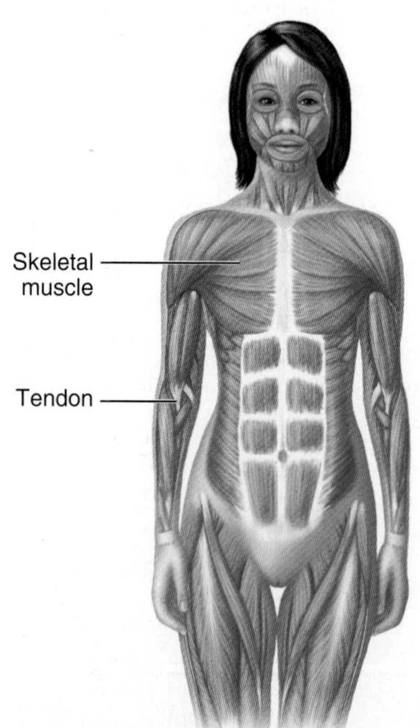

The muscular system. Muscles have one single function—to contract. Skeletal muscle contractions apply tension to the skeleton to maintain an upright posture and create movement of body parts. Since muscle contractions generate heat as a primary byproduct, heat generation is another important function of this system. Most structural effects of manual therapy are derived from changes that occur in skeletal muscles and connective tissue.

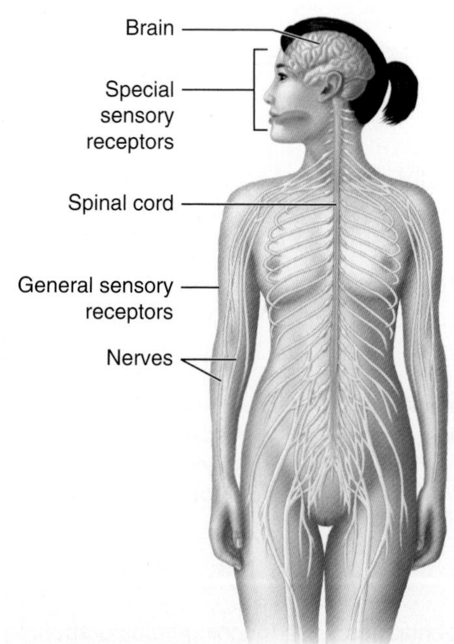

The nervous system. Acting as the body's communication and control center, the nervous system transmits information in the form of nerve impulses. Its complex functions can be summarized as communication, coordination, and control of all other body systems. Virtually all benefits and systemic physiologic effects of touch and movement therapies are mediated in some way by the nervous system.

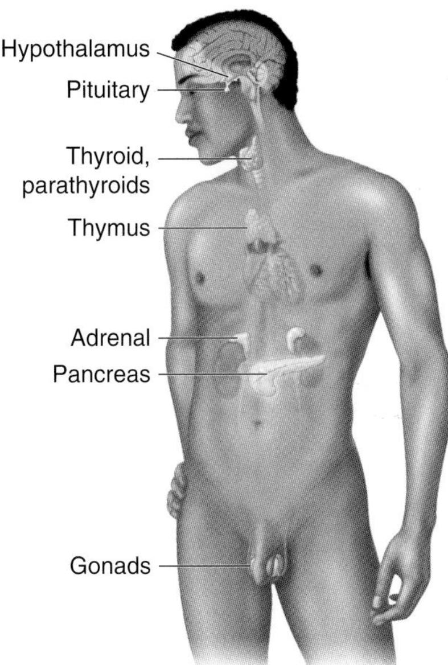

The endocrine system. This system works in conjunction with the nervous system in coordinating and controlling other body systems. It functions at a slower pace than the nervous system to initiate and regulate long-term physiologic processes such as growth and sexual development, and regulates sleep and reproductive cycles. Most manual therapies affect some change in endocrine function that results in benefits such as stress reduction, improved mental focus, and restorative sleep.

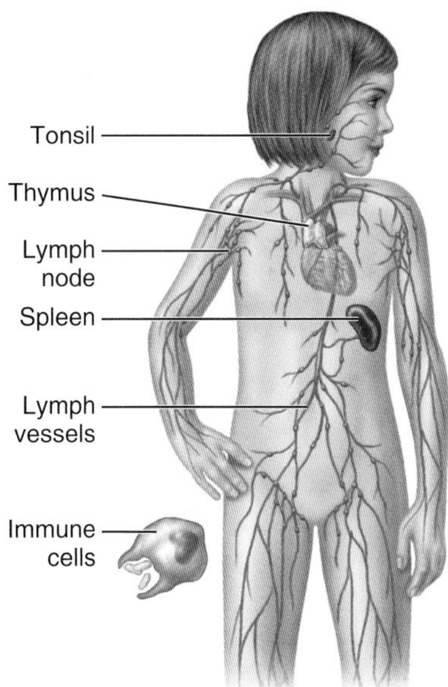

The lymphatic system. This system employs a variety of vital defense mechanisms that create resistance to disease and play a role in healing from injury or illness. It also returns proteins and fluids to the bloodstream, which is essential to maintaining proper fluid balance in the body. Specialty manual therapy techniques such as Manual Lymph Drainage® and Lymphatic Facilitation are used to boost fluid return. Benefits of manual therapy such as decreased stress, relaxation, and an enhanced sense of well-being often lead to improved immune responses.

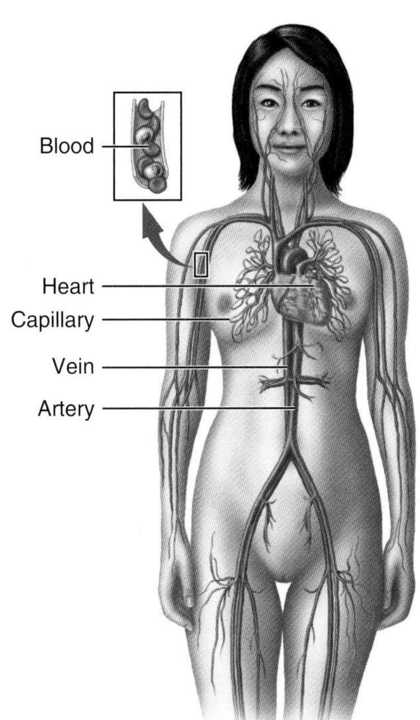

The cardiovascular system. The primary function of the cardiovascular system is to pump blood throughout the body via a complex system of blood vessels. Blood carries oxygen and other nutrients to cells and tissues and transports wastes for excretion. Blood also plays an important role in immune responses and transports hormones secreted by the endocrine glands. Some of the primary systemic effects of manual therapy on the cardiovascular system include improved local blood flow and decreased blood pressure.

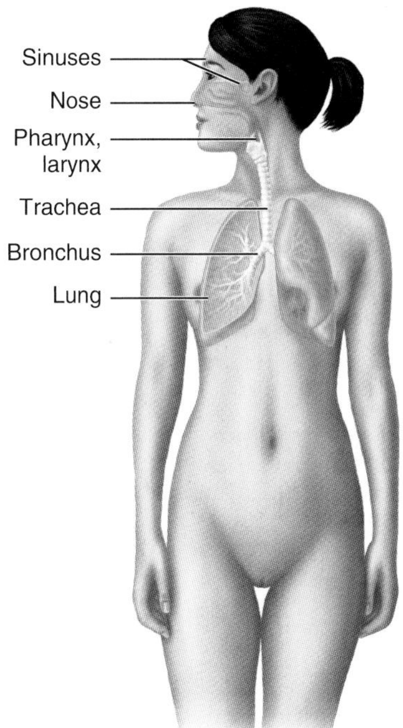

The respiratory system. This system regulates the exchange of oxygen and carbon dioxide in the body. When air from the external environment enters the lungs, carbon dioxide in the blood passes into the lungs and oxygen in the "new" air moves into the blood. Vocalization (speaking) is a secondary function of this system; the passage of air over the vocal cords results in the vibrations that create voice. Structural effects of manual therapy on the muscles of respiration can result in improved efficiency and ease of breathing.

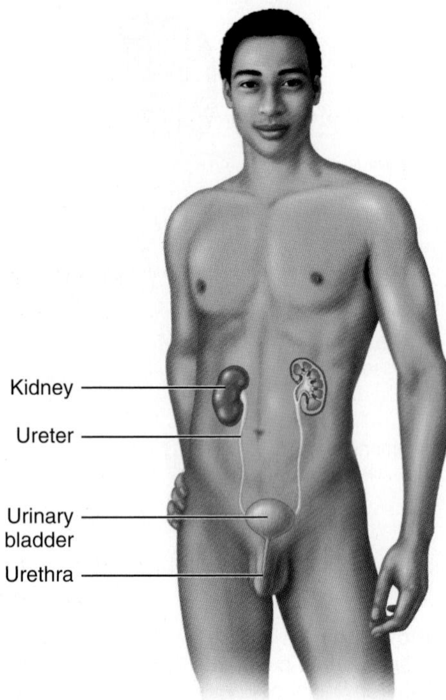

The urinary system. This system produces, stores, and eliminates urine, a process that helps the body regulate pH and electrolyte balances and plays a major role in regulating fluid volumes. Manual therapy techniques that have the most direct impact on the urinary system include reflex and zone therapies, as well as lymphatic techniques that improve fluid return.

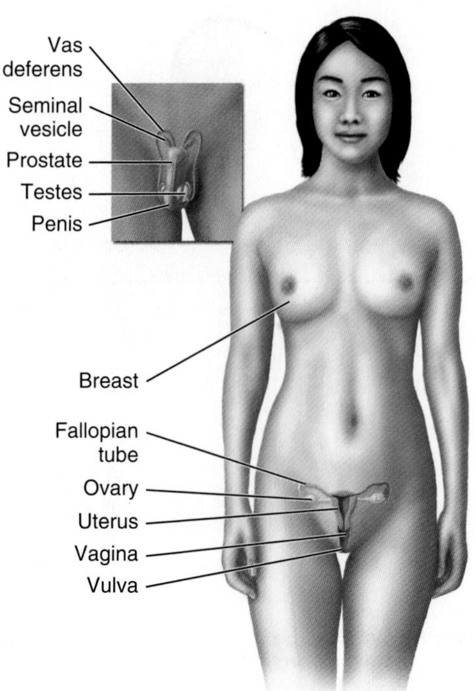

The reproductive system. While male and female structures differ, the general purpose of the reproductive system is the same—to regulate and control the reproductive process. This system is more dependent on the endocrine system than any other body system since hormones regulate sexual maturity, reproductive cycles, pregnancy, and birth. Manual therapy can affect fertility and the reproductive cycle through its impact on the endocrine system, and relaxation and stress reduction can ease labor.

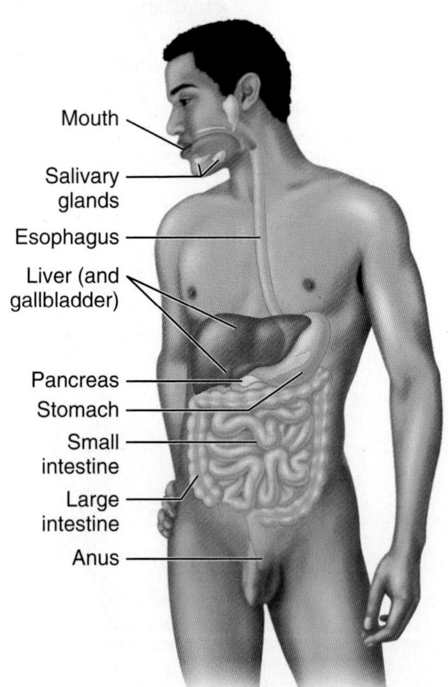

The digestive system. The process of digestion involves breaking down food into usable nutrients, which are then absorbed into the blood for transportation to the cells and tissues. The digestive system also eliminates solid waste. The relaxation and stress reduction effects of manual therapy may improve the efficiency of the digestive system. Reflex and zone therapies can be used to target specific organs and improve overall digestive function.

SUMMARY OF KEY POINTS

- From least to most complex, the body's levels of organization are
  - Cells—considered the basic building block of the body
  - Tissues—group of like cells working together
  - Organs—group of tissues working together
  - Systems—group of organs working together
  - Organism—the body as a whole
- Homeostasis is a state of internal stability or balance between all bodily processes. It is a dynamic equilibrium between all systems and processes designed to maintain optimal functioning of the body. Any change in the internal or external environment, or stimulus, requires these processes to shift and rebalance to maintain homeostasis.
- Changes in homeostasis occur through one of two mechanisms: negative feedback, in which the effector's response counteracts the original stimulus, or positive feedback, in which the effector's response reinforces the original stimulus.
- Manual therapy is defined as any style, discipline, or method in which patterned and purposeful touch and/or movement is applied with therapeutic intent.
- The benefits of manual therapy are considered to be those general and subjective changes that are beneficial to overall health and well-being, regardless of the form utilized.
- Physiologic effects of manual therapy are the specific, objective, and quantifiable tissue, organ, or system changes created by a particular form or technique. Effects can be categorized as structural, meaning that qualities of muscle, fascia, or other connective tissue have been changed, or systemic. Systemic effects are cellular, circulatory, and/or nervous system–mediated responses that occur regionally or throughout the body.
- See TABLE 1-1 to review the basic categories of manual therapy based on common terminology and therapeutic intent.

REVIEW QUESTIONS

Short Answer

1. A group of cells working together to perform a common function is called a(n)___________.
2. A system is a group of ___________ working together to perform a set of common functions.
3. Any change in the internal or external environment is called ___________.
4. Homeostasis is defined as___________.
5. ___________ feedback is when the response counteracts the original stimulus.

Matching

Match the system name with its primary function(s).

_____ 6. Integumentary	a. creates movement
_____ 7. Skeletal	b. regulates growth, puberty, and reproductive cycles
_____ 8. Muscular	c. fluid and protein return to blood
_____ 9. Nervous	d. protective covering of the body
_____10. Endocrine	e. break down food into basic nutrients
_____11. Cardiovascular	f. provides supportive framework for the body
_____12. Lymphatic	g. coordination and control of other body systems
_____13. Respiratory	h. helps balance fluids and blood pH
_____14. Digestive	i. transports nutrients and waste
_____15. Urinary	j. oxygen and carbon dioxide exchange

Multiple Choice

16. Which of the following is a primary component of the urinary system?
    a. pancreas
    b. kidney
    c. liver
    d. gall bladder
17. The spleen is a primary component of which body system?
    a. digestive
    b. urinary
    c. lymphatic
    d. endocrine

Continued on page 14

18. The pancreas is a key organ in which two systems?
    a. endocrine and digestive
    b. urinary and cardiovascular
    c. digestive and respiratory
    d. endocrine and lymphatic
19. Which of the following would be classified as a movement therapy?
    a. Aston Patterning®
    b. Wellness massage
    c. Feldenkrais®
    d. Active Release Technique™
20. Which category of manual therapy is defined as any technique focused on stretching, loosening, or broadening fascia and other connective tissues?
    a. Swedish/relaxation massage
    b. myofascial/deep tissue
    c. neuromuscular
    d. reflex/zone therapy
21. Which category of manual therapy applies light or deep pressure to stimulate defined energy zones, dermatomes, or points of the body?
    a. Swedish/relaxation massage
    b. myofascial/deep tissue
    c. neuromuscular
    d. reflex/zone therapy
22. Reduction of stress and improved mental focus are both examples of ___________ of manual therapy.
    a. structural effects
    b. systemic effects
    c. placebo effects
    d. benefits
23. Reduction of pain and improved range of motion are both examples of ___________ of manual therapy.
    a. benefits
    b. neuromuscular effects
    c. physiologic effects
    d. myofascial effects
24. Which of the following is an example of a structural effect of manual therapy?
    a. stretching muscles
    b. improved sleep patterns
    c. reduction of pain
    d. reduction of edema
25. Which category of manual therapy applies sliding and gliding techniques with a lubricant to create relaxation and reduce stress?
    a. energy
    b. Swedish
    c. myofascial
    d. zone/reflexive

References

1. Braverman DL, Schulman RA. Massage techniques in rehabilitation medicine. *Phys Med Rehabil Clin N Am.* 1999;10(3):631–648.
2. Prentice WE. *Therapeutic Modalities for Physical Therapists*. 2nd ed. New York: McGraw-Hill; 2002.
3. Field T. Massage therapy. In: Davis CM, ed. *Complementary Therapies in Rehabilitation*. Chapter 22. Thorofare, NJ: SLACK, Inc., 1997.
4. Field T, Henteleff T, Hernandez-Reif M, et al. Children with asthma have improved pulmonary functions after massage therapy. *J Pediatr*. 1998;132(5):854–858.
5. Field T, Morrow C, Valdeon C, et al. Massage reduces anxiety in child and adolescent psychiatric patients. *J Am Acad Child Adolesc Psychiatry*. 1992;31:1, 125–131
6. Juhan D. *Job's Body: A Handbook for Bodywork*. Expanded Edition. New York: Barrytown, Ltd.; 1992.
7. Verhoef MJ, Page SA. Physician's perspectives on massage therapy. *Can Fam Physician*. 1998;44:1018–1020, 1023–1024.
8. Zeitlin D, Keller S, Shiflett S, et al. Immunological effects of massage therapy during academic stress. *Psychosom Med.* 2000;62:1, 83–86.
9. Dryden T, Baskwill A, Preyde M. Massage therapy for the orthopaedic patient: a review. *Orthop Nurs.* 2004;3(5):327–332.
10. Pornratshanee W, Hume PA, Kolt GS. The mechanisms of massage and effects on performance, muscle recovery, and injury prevention. *Sports Med.* 2005;35(3):235–256.
11. Crosman LJ, Chateauvert SR, Weisberg J. The effects of massage to the hamstring muscle group on range of motion. *J Orth Sports Phys Ther.* 1984;168–172.
12. Lucas KR, Polus BI, Rich PA. Latent myofascial trigger points: their effects on muscle activation and movement efficiency. *J Bodyw Mov Ther.* 2004;8:160–166.
13. Simons DG, Travell JG, Simons LS. *Myofascial Pain and Dysfunction: The Trigger Point Manual*. 2nd ed. Vol. 1: Upper Half of the Body. Philadelphia, PA: Lippincott Williams & Wilkins; 1999.
14. Cyriax JH, Cyriax PJ. *Illustrated Manual of Orthopedic Medicine*. 2nd ed. Boston, MA: Butterworth & Heinemann; 1993.
15. Cook JL, Khan KM, Maffulli N, et al. Overuse tendinosis, not tendonitis. Part 2: Applying the new approach to patellar tendinopathy. *Phys Sports Med.* 2000;28:31.
16. Hernandez-Reif M, Ironson G, Field T, et al. Breast cancer patients have improved immune and neuroendocrine functions following massage therapy. *J Psychsom Res.* 2004;57:45–52.

17. Sunshine W, Field TM, Quintino O, et al. Fibromyalgia benefits from massage therapy and transcutaneous electrical stimulation. *J Clin Rheumatol.* 1996;2:18–22.
18. Casley-Smith JR, Casley-Smith JR. *Modern Treatment of Lymphoedema.* Adelaide, Australia: Henry Thomas Laboratory, Lymphoedema Association of Australia; 1994.
19. Foldi E, Foldi M. *Textbook of Foldi School.* Straussborg, Austria: self published, English translation by Heida Brenneke; 1999.
20. Forchuk C, Baruth P, Prendergast M, et al. Post-operative arm massage: a support for women with lymph node dissection. *Cancer Nurs.* 2004;27(1):25–33.
21. Cherkin DC, Sherman KJ, Deyo RA, et al. A review of the evidence for the effectiveness, safety, and cost of acupuncture, massage therapy, and spinal manipulation for back pain. *Ann Intern Med.* 2003;138(11): 898–906.

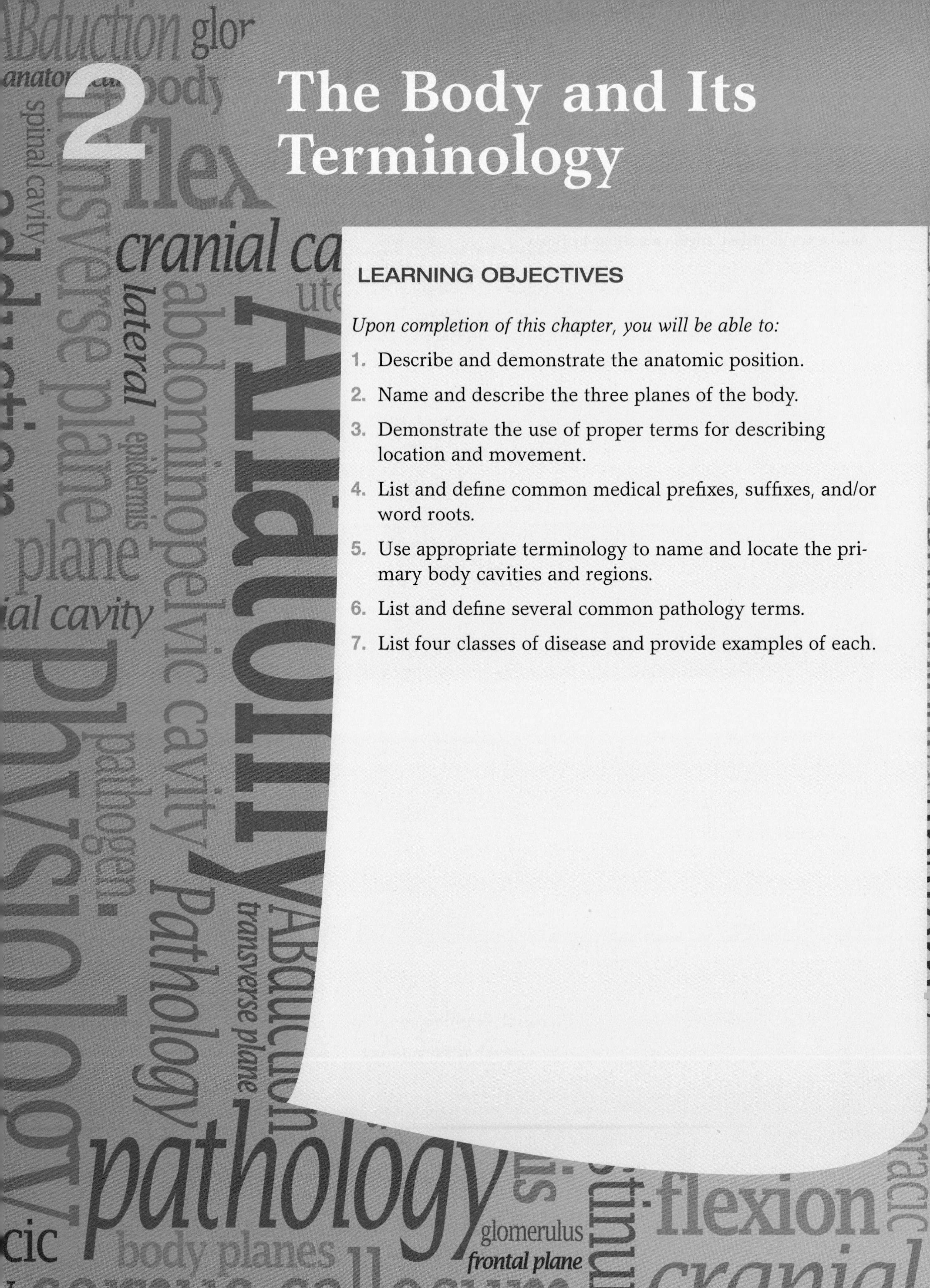

2 The Body and Its Terminology

LEARNING OBJECTIVES

Upon completion of this chapter, you will be able to:

1. Describe and demonstrate the anatomic position.
2. Name and describe the three planes of the body.
3. Demonstrate the use of proper terms for describing location and movement.
4. List and define common medical prefixes, suffixes, and/or word roots.
5. Use appropriate terminology to name and locate the primary body cavities and regions.
6. List and define several common pathology terms.
7. List four classes of disease and provide examples of each.

> "The difference between the right word and the almost-right word is the difference between lightning and the lightning bug."
>
> MARK TWAIN

KEY TERMS

abdominopelvic cavity (ab-dom-ih-no-PEL-vik KAV-ih-te)
abduction (ab-DUK-shun)
adduction (ad-DUK-shun)
anatomic position (an-ah-TOM-ik po-ZIH-shun)
body planes
circumduction (sur-kum-DUK-shun)
cranial cavity (KRA-ne-al KAV-ih-te)
extension (ek-STEN-shun)
flexion (FLEK-shun)
frontal (FRUN-tal) **plane**
mediastinum (me-de-AS-tin-um)
pathogen (PATH-o-jen)
pathology (path-OL-o-je)
rotation (ro-TA-shun)
sagittal (SAJ-ih-tal) **plane**
spinal cavity (SPI-nal KAV-ih-te)
thoracic cavity (thor-AS-ik KAV-ih-te)
transverse (TRANZ-vers) **plane**

Learning new words and their pronunciations is an integral part of becoming proficient in any field of study, including anatomy and physiology. The study of the human body requires learning scientific names for structures, locations, and the mechanisms that regulate body functions in both health and disease. For many learners, becoming familiar with this "foreign language" is one of the most challenging parts of their study. This chapter will introduce you to the scientific and medical terms used to map the body. Words that describe the location and relationship of structures and basic body movements are covered, as well as the body cavities and regions. Common word parts are provided to create a basic lexicon for your study of A&P, as are classifications of disease and basic pathology terms. For manual therapists, using the shared terminology of the medical professions allows us to communicate clearly with one another and other health care providers. It also provides the foundational language necessary for understanding pathology and scientific research. As manual therapies gain increased acceptance and use within the allied health care professions, the use of standardized language is an essential professional skill.

ORIENTATION TO THE BODY

The study of A&P, like all scientific endeavors, must start by establishing foundational concepts that will serve as the key points of reference. Just as you'd establish true north on a compass before starting a journey into new territory, you need to establish a constant point of reference to map the body.

Anatomic Position

Anatomic position is the standard body position or reference point used to navigate anatomical terminology. In this position, the body is standing upright with face, feet, and palms forward and arms at the side (▶ FIGURE 2.1). All anatomical terminology is based on the assumption that the body is in the anatomic position. When most people stand upright, the palms of their hands fall naturally into a position facing their sides or to the back. But in the anatomic position, the palms face *forward* in the same direction as the face and feet, so the palm side of the forearm is considered the *front* of the arm. Only by referring to our true north, anatomic position, do we achieve a consistent standard for applying descriptive terms to individual body structures.

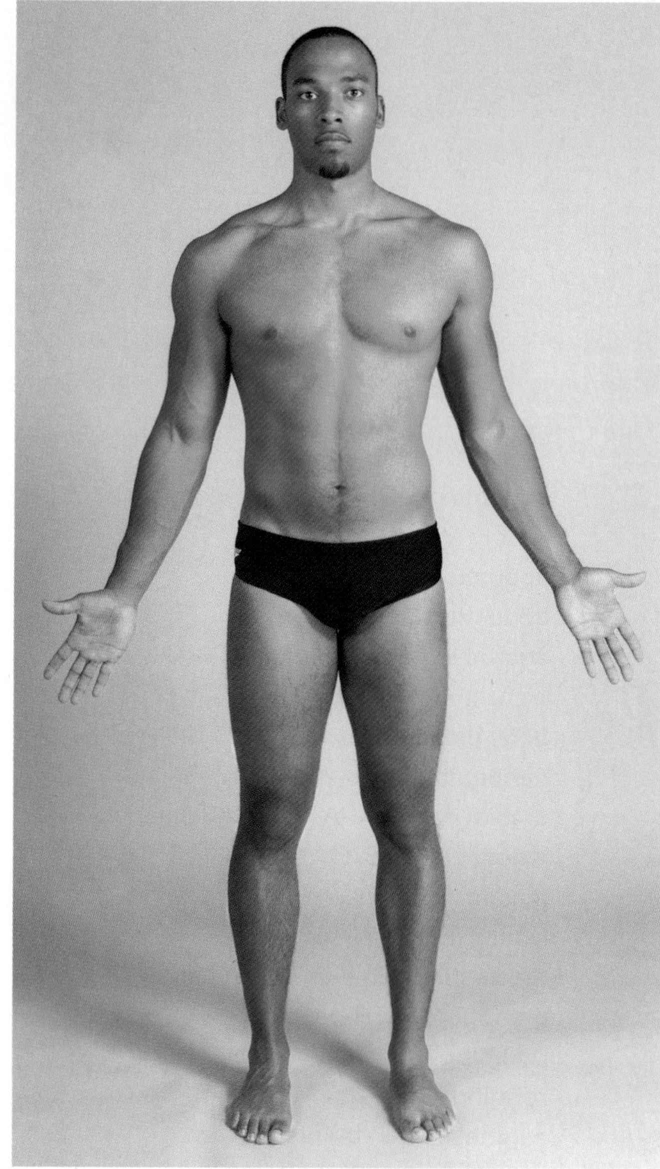

FIGURE 2.1 ▶ Anatomic position. This position with face, feet, and palms facing forward and arms at the sides is the standard position of reference for anatomical terminology.

Body Planes

The **body planes** serve as the remaining navigational points for the body's compass (▶ FIGURE 2.2). Just as lines of latitude and longitude establish east, west, north, and south on a map, the body planes establish front, back, top, bottom, right, and left on the body. The body is divided by three planes:

- **Sagittal plane**—a vertical plane that divides the body or a body part into right and left.
- **Frontal** (or **coronal**) **plane**—a vertical plane that divides the body or a body part into front and back.
- **Transverse** (or **horizontal**) **plane**—a horizontal plane that divides the body or a body part into top and bottom.

Notice that the planes shown in Figure 2.2 are considered ***cardinal planes***, meaning that each represents the absolute center of the body and divides it into equal halves. However, the terms sagittal, frontal, or transverse are also used when the planes divide the body into unequal

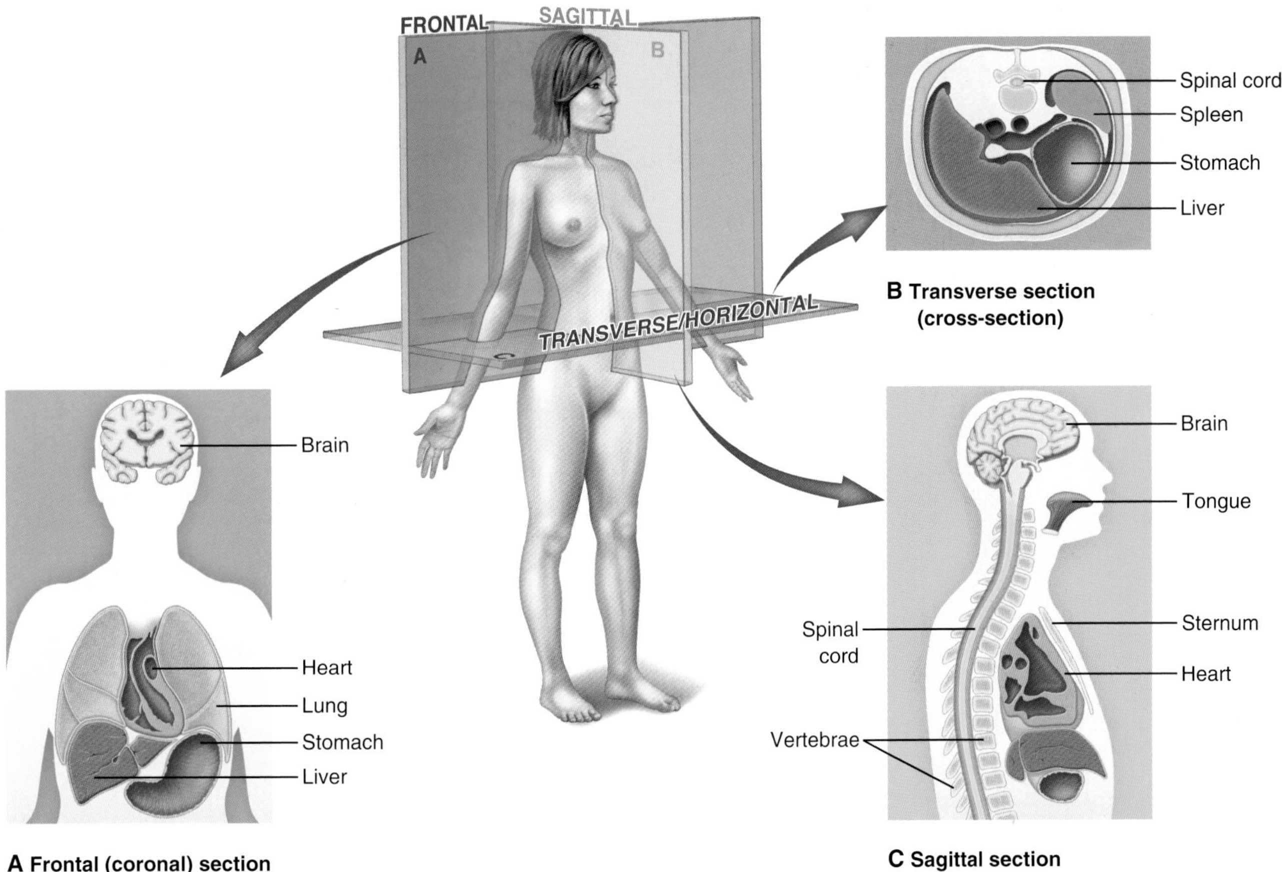

FIGURE 2.2 ▸ Body planes and sections. Three planes divide the body into specific sections. **A.** Frontal (coronal) section. **B.** Transverse section (cross section). **C.** Sagittal section.

portions. For instance, a sagittal plane still divides left from right whether it is on the midline (the midsagittal plane) or moved to the extreme left to separate the arm or the ear from the rest of the body. The same is true for the frontal and transverse planes when they are moved away from absolute center.

Planes help us describe the location and position of various structures and serve as the basis for standardized movement terminology. For example, the midsagittal plane is more commonly known as the midline of the body, and structures such as the nose, sternum (breastbone), and umbilicus (navel) are described as being on the midline. Additionally, body movements are described as occurring within a specific plane or toward or away from the midline. For instance, if you raise your arms overhead as if to do jumping jacks, you are moving in the frontal plane, and your arms move away from and back toward the midline.

Describing Location

Directional terms are words that clearly communicate the position of one body part or component in relation to another. For example, if you describe the location of a scar as "on the face," it would be unclear exactly where the scar is located. But if the scar is described as "just superior and lateral to the left nostril," its location would be clearly understood. ▸ Figure 2.3 shows how these terms are used to describe precise locations on the body or the relationship between structures.

Directional terms are easier to understand and remember if you learn pairs of contrasting terms together. Each term describes whether it is closer to or farther from a cardinal plane or other point of reference. ▸ Table 2-1 includes a list and definitions of common directional terminology used by manual therapists.

Describing Movement

While the unique motions of specific joints will be discussed in detail in Chapter 5, basic movement terminology is an important part of the foundational language of A&P. Movement terminology describes how the body's structures move in general and in relation to other parts. Like directional terminology, movement terms are easiest to learn in contrasting pairs, as shown in ▸ Table 2-2.

Terms describing movement are based on a relationship to the three body planes. Movements along the

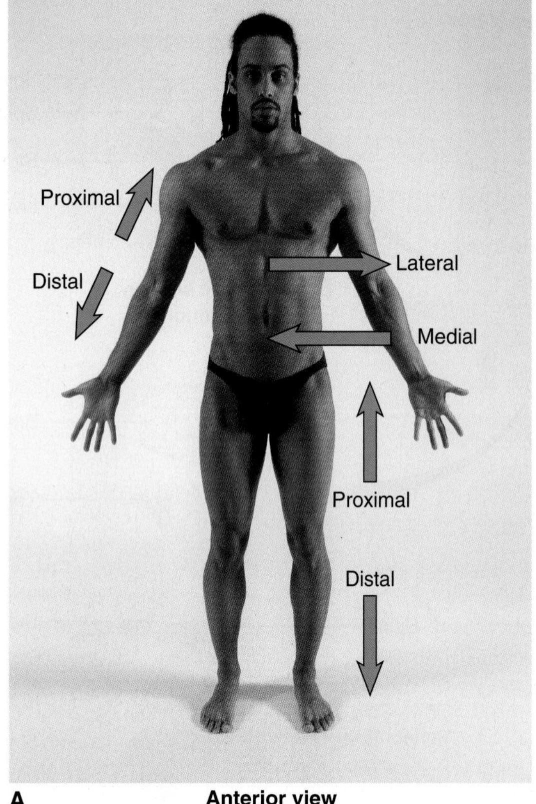

A Anterior view

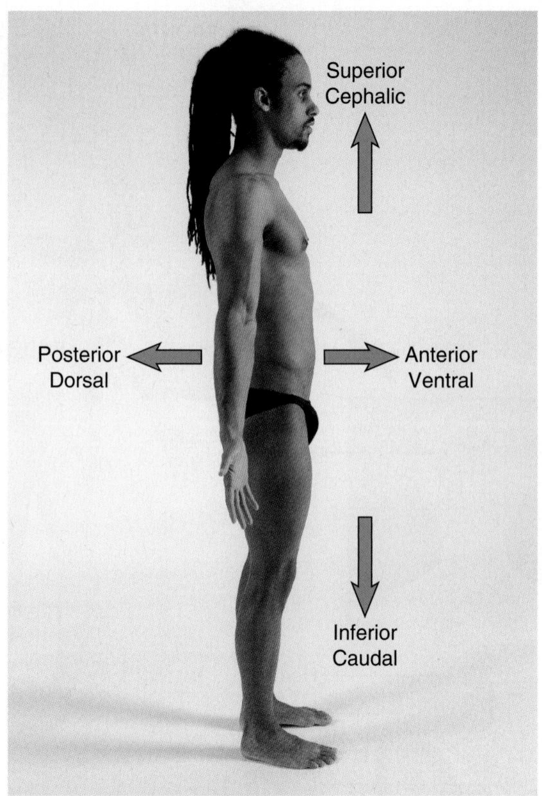

B Lateral view

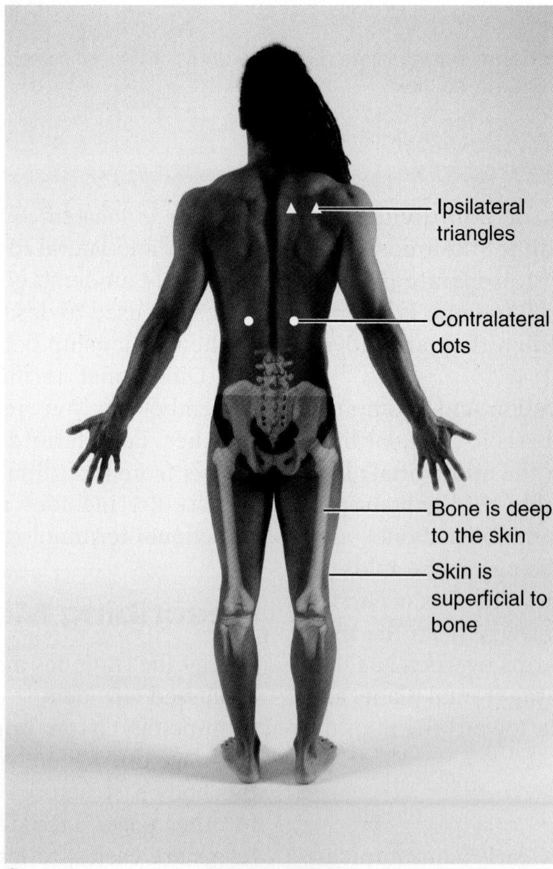

C Posterior view

FIGURE 2.3 ▸ Directional terms. These terms are used to describe precise locations or the positional relationship between structures. **A.** Anterior view. **B.** Lateral view. **C.** Posterior view.

Table 2-1 Directional Terms

Term	Plane or Point of Reference	Definition
anterior (ventral)	Frontal (coronal) plane	Front
posterior (dorsal)		Back
lateral	Midsagittal plane	Farther from midline
medial		Closer to midline
superior (cephalad)	Transverse (horizontal) plane	Above; closer to head
inferior (caudal)		Below; closer to feet
proximal	In extremities, point of attachment to torso (e.g., shoulder or hip)	Closer to attachment point
distal		Farther from attachment point
superficial	Body surface	Closer to the surface
deep		Farther from the surface
contralateral	Midsagittal plane	Opposite side of the median
ipsilateral		On the same side of the median

frontal plane move toward—**adduction**—or away from—**abduction**—the midline (▶ FIGURE 2.4). In the movement described in the previous example of doing jumping jacks, both arms and legs are first abducted and then adducted to complete one repetition of the exercise. If the plane of movement shifts from the frontal plane to the transverse plane as when you swing a tennis racket, movement away from or toward the midline would still be described as abduction or adduction. However, the term horizontal must be included to designate that the movement is occurring on the transverse rather than the frontal plane. When movement away from the midline occurs along the transverse plane (think backswing), it is called ***horizontal abduction***. Continuing this example, contact with the tennis ball and follow-through with the racket (movement toward the midline) would be called ***horizontal adduction*** (▶ FIGURE 2.5).

Along the sagittal plane, movements in an anterior direction (to the front) are termed **flexion,** while movements in a posterior direction (to the back) are called **extension** (▶ FIGURE 2.6). For example, nodding your head, bowing, and lifting a cup to your lips are all flexion movements of different body parts. With one exception, the flexion motion makes the anterior angle between two bones smaller, and extension increases that angle. The knee joint is the one exception: flexion and extension are defined according to the change in angle on the posterior side of the joint.

Rotation is a pivoting motion that may occur in any body plane but must occur on a single axis. Shaking your head "no" is a simple example of rotation. In comparison, **circumduction** is a multiaxial rotation that passes through all three planes around a single point.

By the Way

Because the words *abduction* and *adduction* look and sound very much the same, many health care professionals use pronunciations and writing styles that emphasize the difference. In saying *abduction*, it is common practice to give the first two letters their own syllable: *A-B*–duction. When writing the word, the first two letters may be capitalized: ABduction.

What Do You Think? 2.1

- Practice your understanding of the body planes by using the proper terminology to describe the orientation of pictures in a magazine or to describe your slicing plane when cutting fruit or vegetables.
- What everyday activities (such as twisting a doorknob or raising your hand to ask a question) can you think of to illustrate each of the basic movement terms?

Table 2-2 General Movement Terminology

Term	Plane of Movement	Definition
abduction	Frontal (coronal) plane	Away from the midline
adduction		Toward the midline
horizontal abduction	Transverse (horizontal) plane	Away from the midline
horizontal adduction		Toward the midline
flexion	Sagittal plane	Anterior motion; decreases joint angle
extension		Posterior motion; increases joint angle
rotation	Any plane, single axis	Pivot
circumduction	Multiple planes	Circular motion around a fixed point

Twiddling your thumbs and drawing a circle on a white board while holding your arm straight are examples of circumduction (▸ FIGURE 2.7).

COMMON ANATOMICAL TERMINOLOGY

Although learning a whole new set of terms for the study of anatomy and physiology may initially seem tedious and overwhelming, the value of your efforts will become apparent as you deepen your study of the body and its systems. For example, the meaning of common prefixes, suffixes, and word roots can help you decode unfamiliar scientific terms for body processes, pathologies, or anatomic structures. Additionally, anatomical terminology for body regions and cavities will help you describe the location and relative position of many structures with greater precision. Remember, anatomical terminology is a large portion of the universal medical language used by health care professionals. Manual therapists must

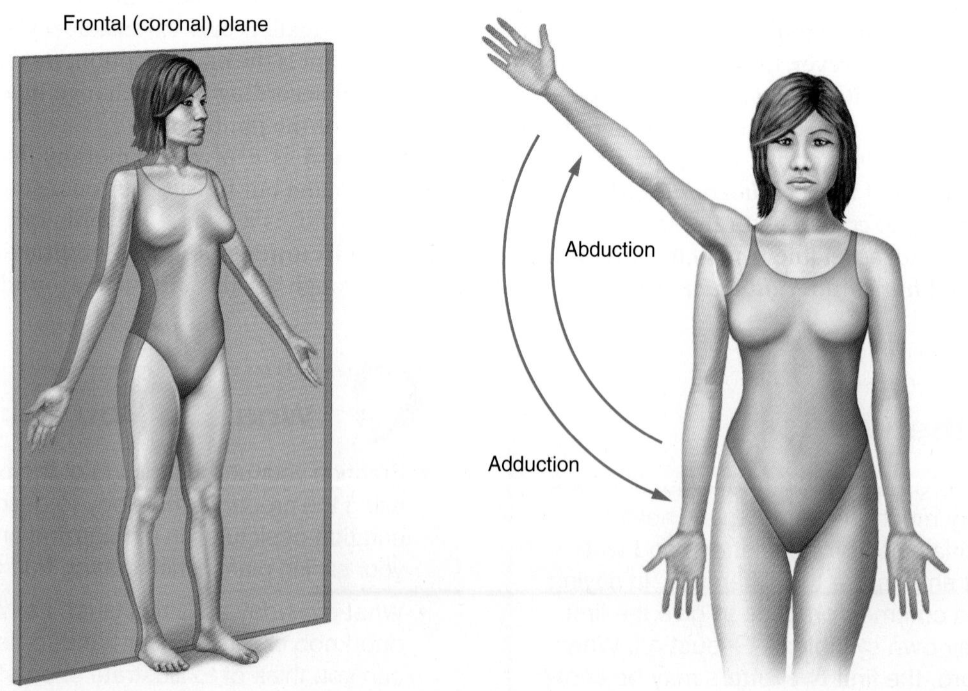

FIGURE 2.4 ▸ Abduction and adduction. Abduction and adduction occur within the frontal plane.

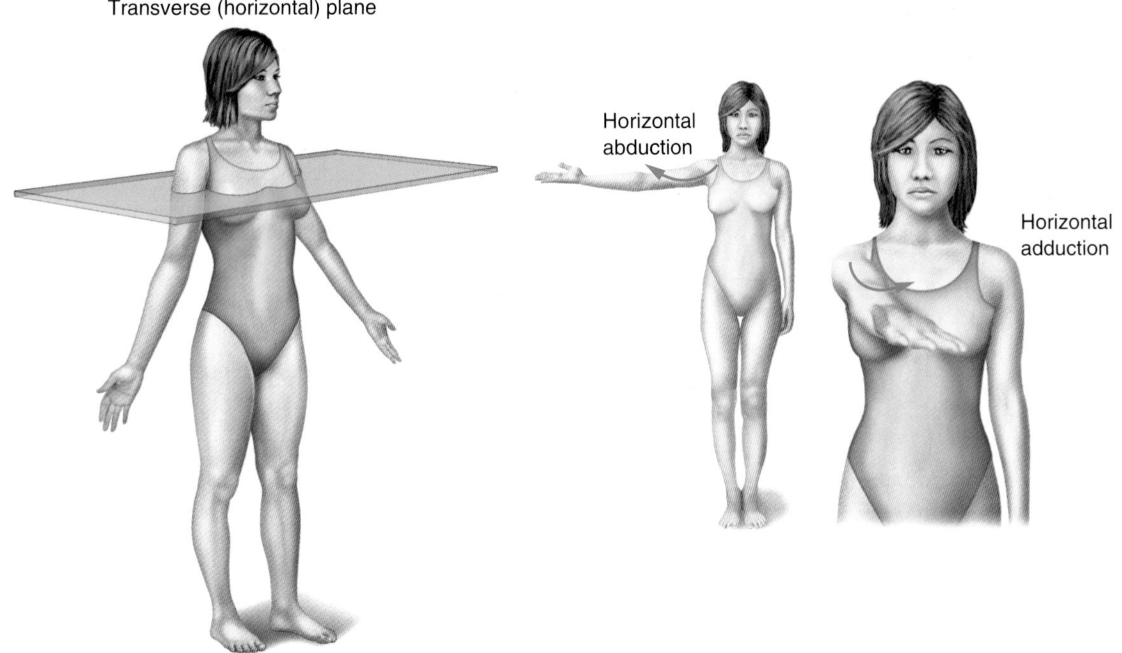

FIGURE 2.5 ▸ **Horizontal abduction and adduction.** Horizontal abduction and adduction occur within the transverse plane.

be fluent in this language to be both understood and respected within the wider health care community.

Prefixes, Suffixes, and Word Roots

Many anatomy and physiology terms are formed by combining word roots, prefixes, and suffixes that are of Greek or Latin origin. Familiarity with these common word parts makes it easier to understand the basic meaning of complex terms without needing to look up their precise definitions. For example, the prefix *arthro-* means joint, and the suffix *-itis* means inflammation. Therefore, you can determine that *arthritis* is an inflammatory condition of a joint. ▸ Table 2-3 provides a quick reference list of word parts frequently used in manual therapy practice.

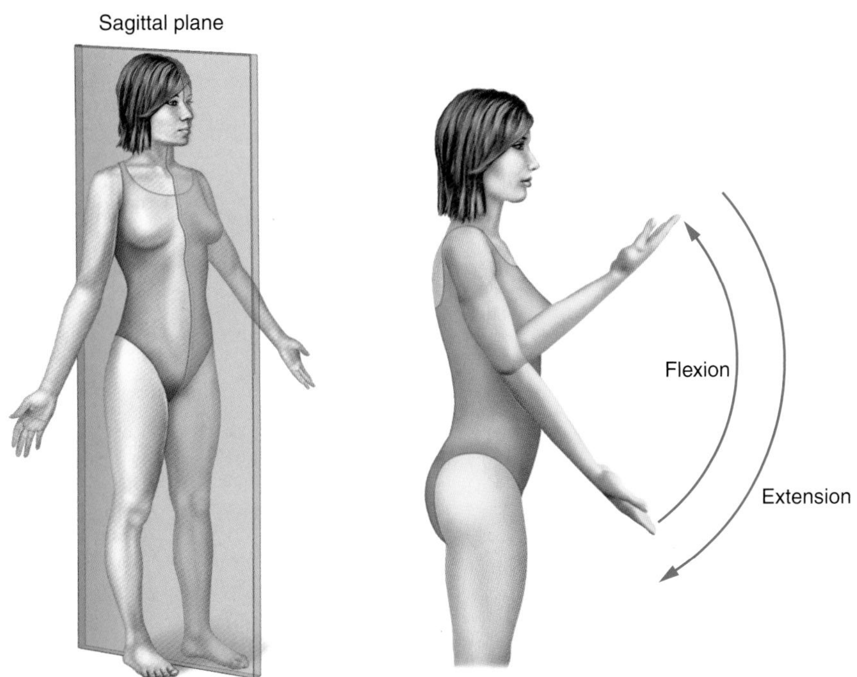

FIGURE 2.6 ▸ **Flexion and extension.** Flexion and extension occur within the sagittal plane.

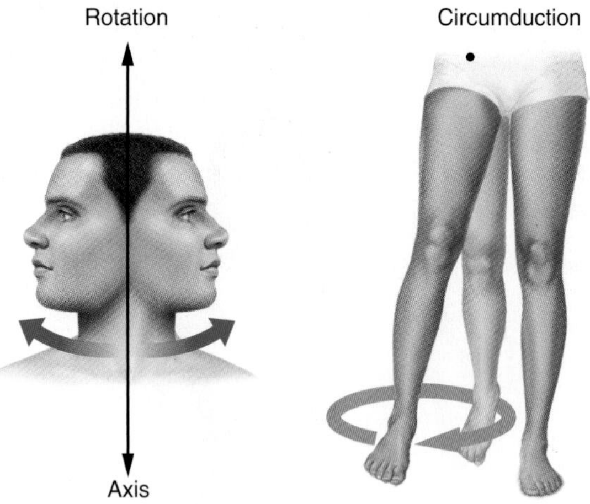

FIGURE 2.7 ▸ Rotation and circumduction. Rotation occurs on a single axis, while circumduction is movement in a complete circle that passes through all planes.

Body Regions

Regional terminology serves as the foundation for naming areas and divisions of the body and is more precise and descriptive than the common vernacular. For example, referring to the "brachial artery" is much more succinct and accurate than referring to "the big blood vessel on the inside of the arm that carries blood away from the heart." ▸ TABLE 2-4 lists commonly used regional terms and their definitions. ▸ FIGURES 2.8 through 2.10 show three views of the body and provide a visual reference for each term in ▸ TABLE 2-4.

By the Way

Although the origins of most A&P terms are Latin or Greek, there are a few common words taken from other languages. For example, the Spanish term *con* meaning *with* and the French term *sans* meaning *without* are often used in medical charting. *With* is abbreviated as w/ or $\bar{c}$. *Without* is abbreviated as w/o or $\bar{s}$.

Table 2-3 Common Word Parts

Word Part	Definition
a- or an-	Without; lack of; deficient
-algia	Painful condition
ante-	Before
anti-	Against
-asis or -asia or -osis	Condition or state of
arth- or arthro-	Joints
aut- or auto-	Self
bi-	Two; double
blast-	Germ or bud
cardio-	Heart
-cid or -cide	To cut, kill or destroy
chon- or chondro-	Cartilage
circum-	Around
clast-	Break or destroy
co-, com-, or con-	With or together
contra-	Against; opposite
cryo-	Cold
cyt- or -cyte	Cell
derma-	Skin
dys-	Painful; difficult
e- or ef-	Out from; out of
ecto- or exo-	Outside
end- or endo-	Within; inside
epi-	Upon; on; above
extra-	Outside; beyond; in addition to
gastr- or gastro-	Stomach; gastrointestinal tract
-genic	Producing
glyco-	Sugar
hemi-	Half
hist- or histo-	Tissue
homeo- or homo-	Unchanging; the same; steady
hep- or hepato-	Liver
hydro-	Water
hyper-	Over; above; excessive
hypo-	Under; beneath; deficient
infra-	Beneath

Table 2-3 Common Word Parts *(Continued)*

Word Part	Definition
inter-	Among; between
intra-	Within; inside
ipsi-	Same
-ism	Condition or state
iso-	Equal; like
-itis	Inflammation
-logy	Study or science of
-lysis	Dissolution; loosening; destruction
macro-	Large
mal-	Bad
micro-	Small
mono-	One
my- or myo-	Muscle
necro-	Corpse or dead
neo-	New
nephr- or nephro-	Kidney
neur- or neuro-	Nerve
-oma	Tumor
-osis	Condition; disease
os- or osteo-	Bone
para-	Near; beside; beyond
path- or patho-	Disease
-penia	Deficiency
peri-	Around
phleb-	Veins
pneumo-	Lung; air
poly-	Much; many
post-	After; beyond
pre- or pro-	Before; in front of
pseudo-	False
retro-	Backward; behind
scler- or sclero-	Hard; hardening
semi-	Half
soma- or somato-	Body
stasis or stat-	Stand still
sten-	Narrow
sub-	Under; beneath; below
super- or supra-	Above; over; beyond
syn- or sym-	With; together
therm-	Heat
thromb-	Clot
trans-	Across; through
vas- or vaso-	Vessel; duct

Table 2-4 Body Regions

Term	Pronunciation	Description/Region	Figure Reference
abdominal	*ab-DOM-i-nal*	Belly	Figure 2.8
acromial	*ah-KRO-me-al*	Point of shoulder	Figures 2.9 and 2.10
antebrachial	*an-teh-BRA-ke-al*	Forearm	Figure 2.8
antecubital	*an-teh-KU-beh-tal*	Anterior elbow	Figure 2.8
axillary	*AK-sil-air-e*	Arm pit	Figure 2.8
brachial	*BRA-ke-al*	Upper arm	Figure 2.8
calcaneal	*kal-KA-ne-al*	Heel	Figure 2.10
carpal	*KAR-pal*	Wrist	Figures 2.8 and 2.9
cephalic	*seh-FAL-ik*	Head	Figure 2.10

(Continued)

Table 2-4 Body Regions *(Continued)*

Term	Pronunciation	Description/Region	Figure Reference
cervical	*SER-vih-kal*	Neck	Figure 2.10
coxal	*KOKS-al*	Hip; lateral pelvis	Figure 2.9
cranial	*KRA-ne-al*	Head	Figure 2.10
crural	*KRUH-ral*	Lower leg	Figure 2.8
cubital	*KU-bih-tal*	Elbow	Figure 2.9
digital	*DIH-jih-tal*	Fingers or toes	Figure 2.8
femoral	*FEM-or-al*	Thigh	Figures 2.8 and 2.10
frontal	*FRONT-al*	Forehead	Figure 2.8
gluteal	*GLU-te-al*	Buttocks	Figure 2.10
inguinal	*ING-gwih-nal*	Groin	Figure 2.8
lumbar	*LUM-bar*	Low back	Figure 2.10
manual	*MAN-u-al*	Hand	Figure 2.9
nasal	*NA-sal*	Nose	Figure 2.8
occipital	*ok-SIP-ih-tal*	Posterior head	Figure 2.10
olecranal	*o-LEK-crah-nal*	Posterior, point of elbow	Figure 2.10
oral	*OR-al*	Mouth	Figure 2.8
orbital	*OR-bih-tal*	Eye	Figure 2.8
otic	*AH-tik*	Ear	Figure 2.9
palmar	*PALM-ar*	Palm of hand	Figure 2.8
patellar	*pa-TEH-lar*	Anterior knee	Figure 2.8
pectoral	*PEK-to-ral*	Chest	Figure 2.8
pedal	*PE-dal*	Foot	Figures 2.8 and 2.9
pelvic	*PEL-vik*	Hip girdle	Figure 2.8
peroneal	*peh-RO-ne-al*	Lateral lower leg	Figure 2.9
plantar	*PLAN-tar*	Sole of foot	Figure 2.10
popliteal	*pop-leh-TE-al*	Posterior knee	Figure 2.10
pubic	*PU-bik*	Genital	Figure 2.8
sacral	*SA-kral*	Area of tail bone	Figure 2.10
scapular	*SKAP-u-lar*	Area of shoulder blade	Figure 2.10
spinal	*SPI-nal*	Backbone or spinal cord	Figure 2.10
sternal	*STER-nal*	Breastbone	Figure 2.8
sural	*SUR-al*	Calf; posterior lower leg	Figure 2.10
tarsal	*TAR-sal*	Ankle	Figures 2.8 and 2.9
temporal	*TEMP-or-al*	Temples of head	Figure 2.9
thoracic	*thor-AS-ik*	Rib cage	Figure 2.8
umbilical	*um-BIL-ih-kal*	Navel	Figure 2.8

Table 2-4 Body Regions *(Continued)*

Term	Pronunciation	Description/Region	Figure Reference
vertebral	*ver-TE-bral*	Backbone	Figure 2.10
volar	*VO-lar*	Posterior hand	Figure 2.10

Body Cavities

Specialized terminology is used to designate the interior regions of the head and torso. These hollow spaces, or cavities, can be divided into two general areas. The chest and abdomen create the two cavities in the front of the torso called the ***ventral*** or ***anterior cavities***. The two hollow spaces inside the skull and spinal column are designated as the ***dorsal*** or ***posterior cavities*** (▶ FIGURE 2.11).

Ventral Cavities

On the ventral side, a large skeletal muscle called the diaphragm forms a floor at the bottom of the rib cage that separates the chest and abdomen into two distinct cavities. The cavity superior to the diaphragm is called the **thoracic cavity**. It contains the right and left lungs as well as the heart, which is located in a central canal called the **mediastinum**. The inferior region is called the **abdominopelvic cavity**, which extends from the diaphragm to the pelvic floor (▶ FIGURE 2.11). Because the abdominopelvic cavity contains both abdominal and pelvic organs, it is helpful to subdivide this cavity into several regions. These regions provide for more precise descriptions of a specific organ's location. The two major methods for subdividing the cavity are

- The ***quadrant method***—commonly used by manual therapists to locate a general region of pain, discoloration, or other problem (▶ FIGURE 2.12A).

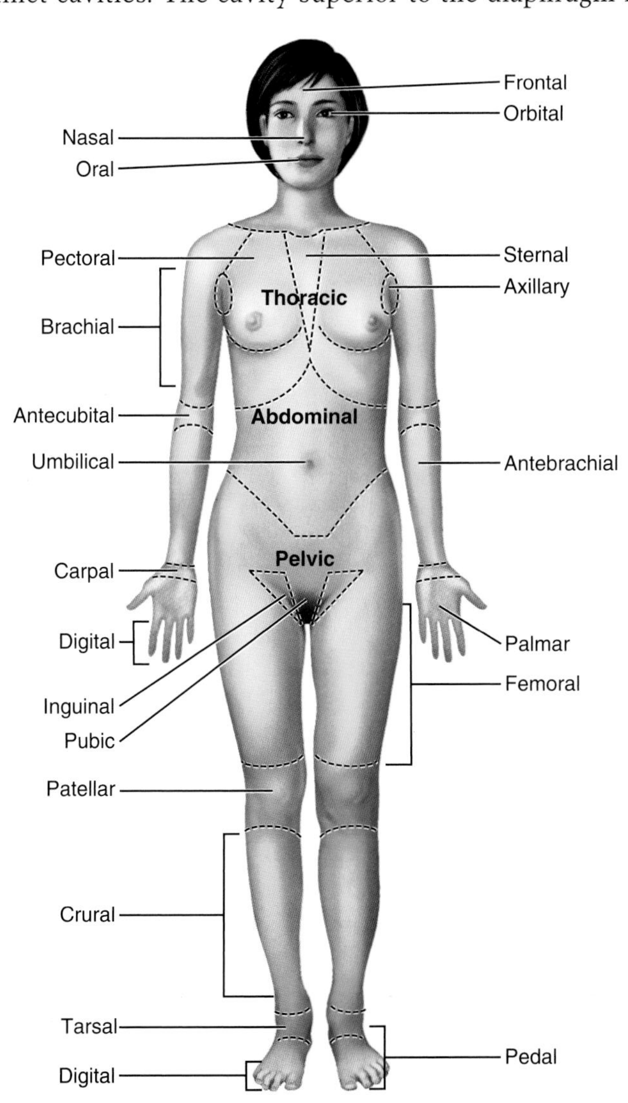

FIGURE 2.8 ▶ Body regions (anterior view).

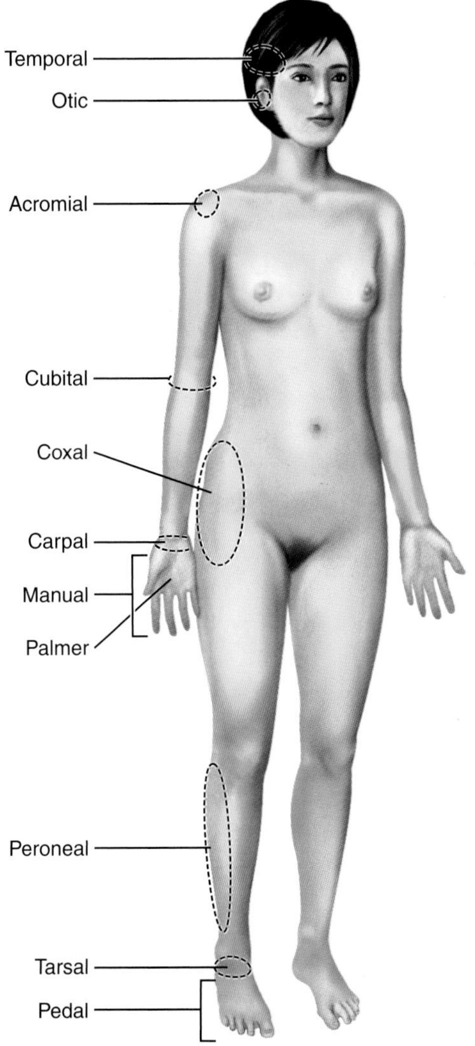

FIGURE 2.9 ▶ Body regions (lateral view).

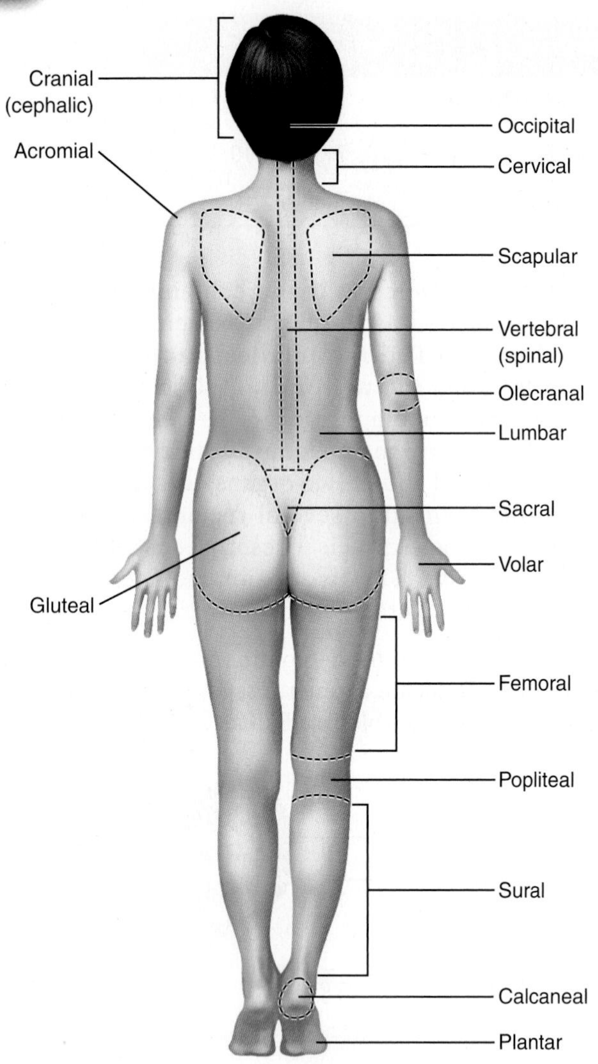

FIGURE 2.10 ▸ Body regions (posterior view).

- The ***nine-region method***—most often used for precise anatomical studies or surgeries (▸ FIGURE 2.12B)

Dorsal Cavity

The two dorsal cavities include the **cranial cavity**, which houses the brain, and the **spinal cavity**, which houses the spinal cord. The base of the skull divides these two cavities, with an opening that allows a connection between the brain and spinal cord (▸ FIGURE 2.11A).

What Do You Think? 2.2

- *Antitheft* device and *asymmetry* are examples of everyday words that use the prefixes, suffixes, or word roots in TABLE 2-3. What other examples can you think of?
- How do you think knowing the terms that describe body cavities and regions will help you in the remainder of this course and in your profession?

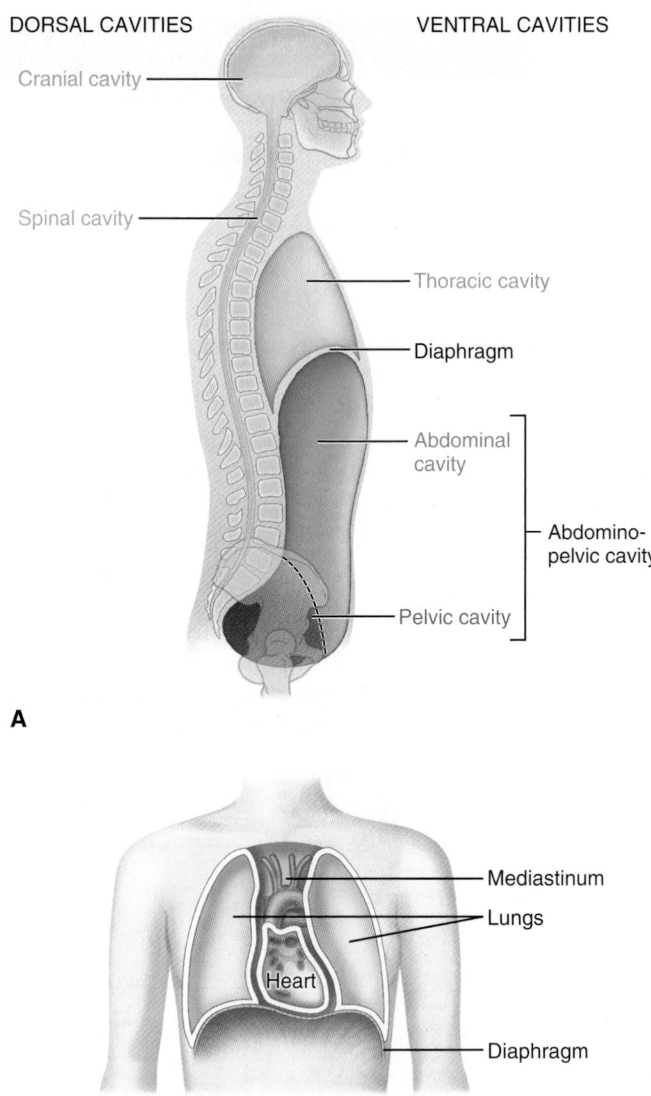

FIGURE 2.11 ▸ Body cavities. A. There are two ventral and two dorsal cavities. Note that the abdominopelvic cavity has two distinct regions. **B.** Thoracic cavity. The mediastinum is the central partition of the thoracic cavity that contains the heart. The lungs sit on either side of the mediastinum.

PATHOLOGY BASICS: CLASSIFICATIONS AND TERMINOLOGY

Disease can be defined as an impairment of, or disruption to, the normal structures or functions of the body. **Pathology** is the study of disease. Sometimes the structure or function of a particular body system is better understood by exploring how structural damage or disruption of function affects health. Although an A&P textbook cannot offer an exhaustive review of pathologies, commonly used classifications and terminology that is important for manual therapists are presented in the following sections.

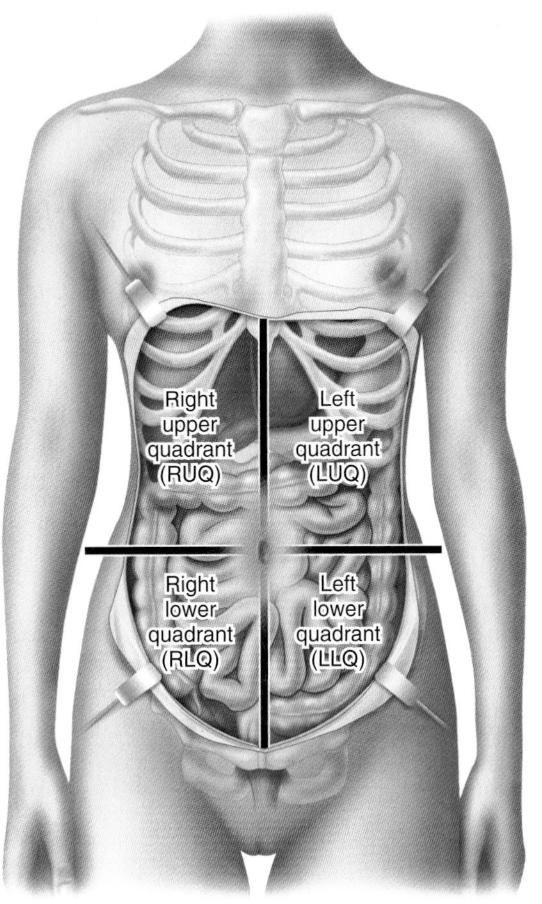

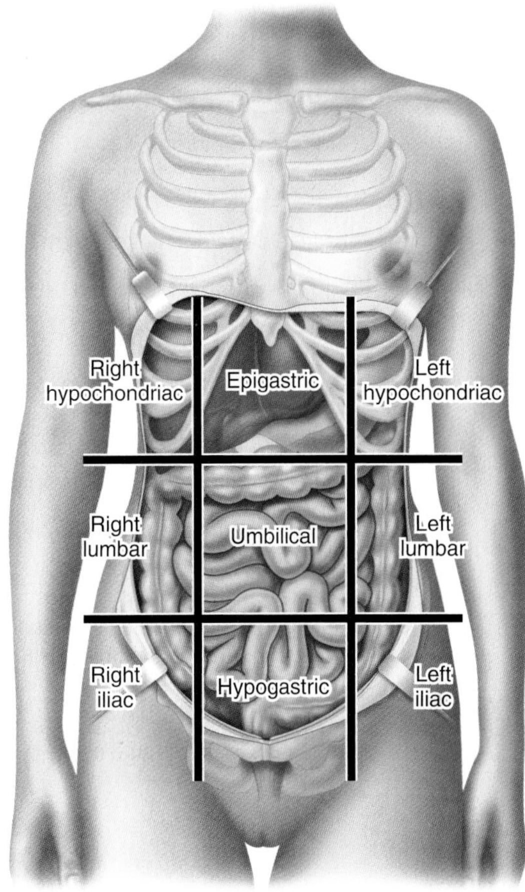

FIGURE 2.12 ▸ Regions of the abdominopelvic cavity. A. Abdominal quadrants. The umbilicus is the center point for dividing the cavity into four regions. **B.** Nine abdominal regions.

Classifications of Disease

Diseases can be broadly classified into one of four groups according to their general cause:

- ***Infectious***—An infectious disease is one in which specific disease-causing agents, called **pathogens,** create a physiologic disruption. Pathogens include bacteria such as those that cause tuberculosis, viruses such as those that cause the flu and the common cold, fungi such as those that cause yeast infections or athlete's foot, and parasites such as tapeworms or lice.
- ***Environmental***—Exposure to harmful substances found in the environment, such as coal dust, cigarette smoke, or pesticides, can also lead to a wide variety of illnesses. Diseases caused by these pathogens are classified as having environmental causes.
- ***Hereditary***—A hereditary disease is caused by a particular genetic trait that is passed from one generation to the next. This is one of the reasons that health care professionals spend time obtaining and reviewing a complete health history before treatment. Hemophilia and sickle cell anemia are examples of blood diseases that are hereditary.
- ***Nutritional and lifestyle***—Many diseases are related to poor nutrition and an unhealthy lifestyle. For example, dietary deficiencies in essential vitamins and minerals are related to pathologies such as scurvy, anemia, and osteoporosis. Lifestyle choices like smoking or eating large amounts of fast food, coupled with lack of exercise and a high-stress work environment, have been clearly linked to heart disease, cancer, and stroke.

General Terminology

Understanding pathology requires that manual therapists learn some new terms. ▸ Table 2-5 lists and defines some basic medical terminology related to pathology.

What Do You Think? 2.3

- Name at least two diseases or conditions for each category shown in Table 2-5.

Table 2-5 Basic Pathology Terms

Term	Definition
acute	Sudden or rapid onset
chronic	Gradual or long term onset
congenital	Disease or defect present from the time of birth
contraindication	Sign or symptom that could be made worse with a specific course of treatment; negative effect is likely
diagnosis	Determination of the nature of a disease, injury, or defect; naming the disease, disorder, or dysfunction
etiology	Cause and specific course of development for a disease
idiopathic	Disease of unknown origin
indication	Sign or symptom that would improve with a specific course of treatment; positive effect is likely
lesion	Any pathogenic change in normal tissue structure, such as a tear, tumor, scar, or wound
prognosis	Expected progression or outcome of a disease or dysfunction
sequela	Lasting result of a specific disease or dysfunction, such as a shortened leg after a healed fracture or low respiratory capacity after tuberculosis
sign	Objective indicator of disease that is usually measurable
symptom	Subjective indicator of disease that is not easily measured or quantified

SUMMARY OF KEY POINTS

- Anatomic position is the body standing with arms at the sides, feet shoulder width apart, and the face, feet, and palms facing forward; it is the basis for all anatomical terminology. All location, position, and movement terms are based on the assumption that the body is in this position.
- Three anatomic planes divide the body into different portions. The sagittal plane divides the body into left and right sides, the frontal (coronal) plane divides it into front and back, and the transverse (horizontal) plane divides it into top and bottom. Knowing the planes also helps to accurately describe movement and the location and relative position of different structures.
- See Table 2-1 for a review of directional terminology and Table 2-2 to review movement terminology. Table 2-3 provides a review of key prefixes, suffixes, and word roots. See Table 2-4 and Figures 2.8 through 2.10 for a review of the major body regions.
- The body has four major cavities: two dorsal and two ventral. The dorsal cavities are the cranial cavity that houses the brain and the spinal cavity that houses the spinal cord. The ventral cavities are the thoracic cavity containing the lungs and heart and the abdominopelvic cavity. The location of organs in the abdominopelvic is specified by either the quadrant method or 9-region method of dividing the cavity.
- Pathology is the study of disease. Diseases are categorized as being infectious, environmental, hereditary, or related to nutrition and lifestyle.
- Pathogens are the microscopic substances that cause disease. The most common types of pathogens are viruses, bacteria, fungi, parasites, chemicals, and particles.
- See Table 2-5 for a review of common pathology terminology.

REVIEW QUESTIONS

Short Answer

1. The plane that divides the body into front and back is called the frontal or ___________ plane.
2. The ___________ plane divides the body into right and left sides.
3. In anatomic position, arms are at the side with _________, _________, and _________ facing forward.
4. The two ventral cavities are the ___________ and the ___________.
5. The brain is contained within the _________ cavity.

Multiple Choice

6. Which terms are used to describe the location of a structure in relationship to the midline of the body?
   a. anterior, posterior
   b. proximal, distal
   c. dorsal, ventral
   d. medial, lateral
7. The directional term *superior* is used to describe a structure as being
   a. on the opposite side
   b. above or in an upward direction
   c. closer to the point of attachment
   d. closer to the body surface
8. Which of the following terms is used to describe the position of two points that are on opposite sides of the midline?
   a. contralateral
   b. ipsilateral
   c. medial
   d. distal
9. The elbow can be described as being ___________ to the wrist.
   a. superior
   b. posterior
   c. proximal
   d. inferior
10. What is the technical term for the region of the head?
    a. cranial
    b. cervical
    c. caudal
    d. occipital
11. The posterior aspect of the knee is called the ___________ region.
    a. sural
    b. femoral
    c. crural
    d. popliteal
12. What is the technical term for the armpit region?
    a. acromial
    b. axillary
    c. humeral
    d. olecranal
13. The inguinal body region is more commonly known as the ___________.
    a. groin
    b. buttocks
    c. hip
    d. chest
14. An anterior movement along the sagittal plane that decreases the angle between two bones is called ___________.
    a. extension
    b. abduction
    c. flexion
    d. adduction
15. Movement about a fixed point and on a single axis is known as ___________.
    a. circumduction
    b. opposition
    c. flexion
    d. rotation
16. Which word part means below or beneath?
    a. infra-
    b. supra-
    c. hyper-
    d. epi-
17. The word part *cyt-* or *-cyte* means
    a. cold
    b. small
    c. cell
    d. tissue

Continued on page 32

18. What word part could be used to identify the innermost layer of tissue in a particular structure?
    a. epi-
    b. peri-
    c. endo-
    d. exo-

19. The liver is an organ located in which quadrant of the abdominopelvic cavity?
    a. upper right
    b. upper left
    c. lower right
    d. lower left

20. The common term for the cubital region of the body is
    a. shoulder
    b. arm
    c. wrist
    d. elbow

21. Which term refers to any type of disease-causing agent?
    a. pathogen
    b. congenital
    c. virus
    d. bacteria

22. A lung disease caused by inhaling asbestos fibers would be classified as what type of disease?
    a. hereditary
    b. environmental
    c. psychological
    d. infectious

23. The cause and factors involved in the development of a disease are the
    a. prognosis
    b. diagnosis
    c. etiology
    d. sequela

24. A disease with a sudden or rapid onset is said to be a(n) ___________ disease.
    a. chronic
    b. congenital
    c. epidemic
    d. acute

25. Tapeworms or lice are examples of which type of pathogen?
    a. parasites
    b. fungi
    c. bacteria
    d. viruses

3 Chemistry, Cells, and Tissues

LEARNING OBJECTIVES

Upon completion of this chapter, you will be able to:

1. Discuss the importance of understanding basic chemistry, structures, and functions of cells and tissues as they relate to the practice of manual therapy.
2. Name and explain the key inorganic and organic components of cells and tissues and explain the primary role of each.
3. Name the three structural components common to all cells and describe the general function of each.
4. Name and explain the function of the organelles in a cell.
5. Compare and contrast passive and active transport.
6. List and describe the various types of active and passive transport mechanisms.
7. Define, compare, and contrast cellular metabolism, anabolism, catabolism, cell division, and cell differentiation.
8. List the four categories of tissue found in the body, describe their distinguishing characteristics, and explain the general function and location of each.
9. Explain the different classifications of epithelial cells and tissues.
10. Name three types of muscle tissue and explain the distinguishing characteristics of each.
11. Name the two types of nervous tissue and explain the function of each.
12. Name the common components of all types of connective tissue.
13. List the different types of connective tissue and describe where they are located.

...the biochemical mechanisms employed by cellular organelle systems are essentially the same mechanisms employed by our human organ systems. Even though humans are made up of trillions of cells... there is not one "new" function in our bodies that is not already expressed in the single cell."

Bruce H. Lipton, PhD
The Biology of Belief

KEY TERMS

active transport
atom
compound (KOM-pound)
cytokinesis (si-to-kin-E-sis)
cytology (si-TOL-o-je)
cytoplasm (SI-to-plaz-um)
diffusion (deh-FYU-zhun)
electrolyte (e-LEK-tro-lite)
element (EL-eh-ment)
filtration (fil-TRA-shun)
histology (his-TOL-o-je)
ion (I-on)
inorganic (in-or-GAN-ik) **compound**
matrix (MA-triks)
metabolism (meh-TAB-o-lizm)
mitosis (mi-TO-sis)
molecule (MAHL-eh-kule)
organelle (or-gah-NEL)
organic (or-GAN-ik) **compound**
osmosis (oz-MO-sis)
passive transport
pH scale

One of the amazing aspects of life is that each of us is at once an individual and, at the same time, a member of the larger human community. While we are each responsible for living our own lives, we are not isolated units. Everything we think and do, to some degree, affects other people. Similarly, each cell of the body is a single functional unit and also a member of a complex community, the human organism. Like people, cells influence their community by communicating, cooperating, defending, irritating, hurting, and destroying one another. To survive, cells must respond and adapt to a constantly changing environment, just as people must adapt to their surroundings and the society in which they live. The goal of this chapter is to explain the fundamental structures, functions, and behaviors of individual cells and tissues to better comprehend the whole cooperative cellular community that is the human body.

In this chapter, you will learn about the basic chemical components that make up the body's cells and tissues. The basic structure and components of cells and tissues are identified, and the key functions of each are explored. As manual therapists, it may be easy to lose sight of the purpose and application of this rather detailed information. However, when we speak with clients and other health care professionals about what is happening in the body and tissues, at some level we are speaking about what is happening in and around the cells. Developing a basic understanding of cells and tissues lays the necessary foundation for comprehending the body's more complex levels of organization.

CHEMICAL COMPONENTS OF CELLS

Most of the physiologic processes that take place in the body involve chemical reactions that occur in and around the cells. While it is not imperative for manual therapists to have in-depth knowledge of the body's complex chemistry, it is helpful be to able to identify the most common chemicals that make up cells and their main characteristics as a basis for understanding organ and system processes.

The periodic table of elements sometimes displayed in science classrooms is a graphic depiction of the **elements** that make up *all* matter. **Atoms** are microscopic particles that are the units of energy that make up these elements. When atoms bind together, they form **molecules**, a larger but still microscopic particle. The matter that we see and touch, such as a book, tree, grain of salt, or drop of water, is made up of molecules by the thousands, millions, and more.

When atoms combine to create molecules, they bond with either an identical atom to form a molecule of one of the basic elements on the periodic table (e.g., oxygen, hydrogen, or gold) or *different* types of atoms to create substances called **compounds**. The molecules that make up water and air are examples of compounds. In each element or compound, the molecule is represented by a simple formula with letters that designate the type(s) of atoms and subscript numbers that identify how many of that particular atom is in the molecule. For example, oxygen is a gas containing millions of molecules, each made of two oxygen atoms bound together. Therefore, the element oxygen is designated as O_2. The molecular formula for the compound known as water is H_2O, to designate that each molecule of water has two atoms of hydrogen and one atom of oxygen (▶ FIGURE 3.1). You will need to be familiar with a number of chemical symbols to fully understand the structural features and physiologic processes explained in later sections of this chapter and throughout the text. ▶ TABLE 3-1 lists the chemical symbols for elements commonly found in the human body.

The majority of chemicals in our bodies are bound together to form two broad groups of compounds: inorganic and organic. Both types of compounds play important roles in the composition and physiologic processes of cells and tissues.

By the Way

Of the 92 elements that exist in nature, only 26 are found in living organisms. Of these, only hydrogen, oxygen, carbon, and nitrogen are found in significant quantities in the human body. The combination of these four elements makes up approximately 96% of our body by weight.

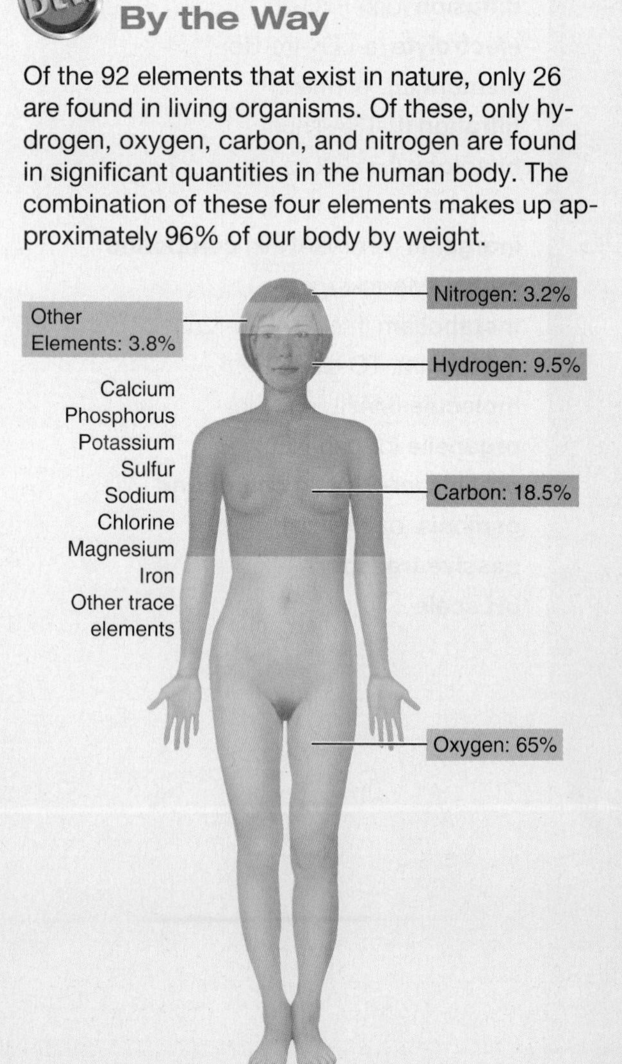

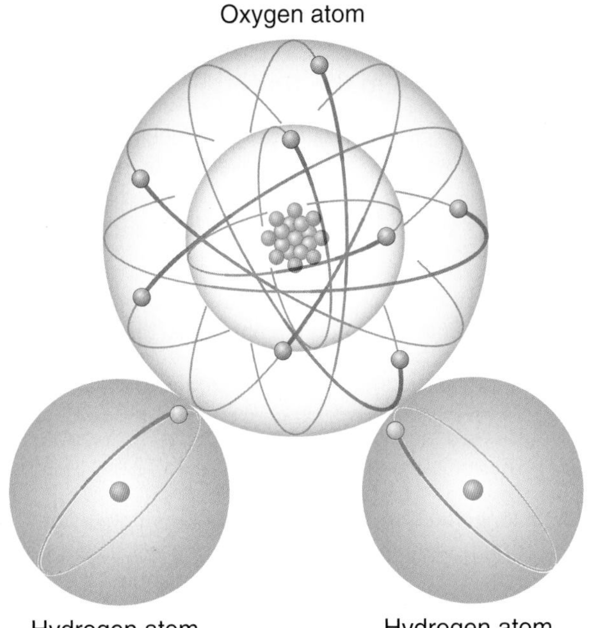

FIGURE 3.1 ▶ Atoms combine to form a molecule. Every molecule of water (H_2O) is made up of one atom of oxygen and two atoms of hydrogen.

Inorganic Compounds

Although there are a few exceptions, **inorganic compounds** are made of molecules that do not contain carbon atoms. The most common and important inorganic compounds found within the body's cells and tissues are water, salts, acids, and bases.

Water

Water (H_2O) represents about 60% of overall body weight and is vital to all physiologic processes. It is the medium for most chemical exchanges between cells and tissues and the primary component of all body fluids, including the fluid inside the cells. This makes water an essential compound for the transportation of nutrients, wastes, and other byproducts formed during the millions of chemical reactions that take place throughout the body. One of the key byproducts transported by water is the heat generated by muscle contraction and other cellular activities. Water has the unique ability to absorb and retain unusually high levels of heat, which makes blood an ideal fluid for transporting heat and helping to regulate body temperature. Water also serves as an important lubricant to reduce friction between various body surfaces.

Inorganic Salts

Inorganic salts are compounds that break apart in water to release either positively or negatively charged atoms or molecules called **ions**. Ions are designated by placing a plus sign (positive charge) or minus sign (negative charge) after their molecular symbol, as shown in the following examples: sodium (Na^+), calcium (Ca^{2+}), potassium (K^+), and bicarbonate (HCO_3^-). Ions in water conduct an electrical current, and therefore, these inorganic salts are often called **electrolytes**. Electrolytes are essential for a wide variety of cellular processes, including cellular transportation, nerve impulse conduction, and muscle contraction. Sports drinks such as Gatorade® or Propel® are advertised as electrolyte drinks because they contain ready sources of essential ions such as sodium, calcium, and potassium.

Acids and Bases

Electrolyte compounds that specifically release hydrogen ions (H^+) when dissolved in water are called **acids**. Compounds that release another type of ion known as the hydroxide ion (OH^-) in water are called **bases**. The balance between the number of hydrogen and hydroxide ions is measured using the **pH scale** that has values from 0 to 14 (▶ FIGURE 3.2). The midpoint of the pH scale (7) designates a neutral pH substance having equal numbers of H^+ and OH^-. Water is the primary example of a neutral substance. Any substance with a pH value below 7 is **acidic** (containing large numbers of hydrogen ions) while those with measurements above 7 are basic or **alkaline**. The normal chemical activities of cells and tissues can only take place within a small slightly alkaline pH range of 7.35 to 7.45. This homeostatic range is ensured via numerous mechanisms that constantly assess and make adjustments to maintain the pH balance within the body.

Organic Compounds

Organic compounds are those that always contain carbon molecules and generally contain hydrogen as well. These compounds are not only important to cellular processes but are also the true structural building blocks of cells. Carbohydrates, lipids, proteins, and nucleic acids are the organic substances that make up all cells and tissues in the body.

Carbohydrates

An organic compound is considered a **carbohydrate** when a mixture of hydrogen and oxygen atoms is joined with long or short chains of carbon atoms. A short chain of carbon atoms makes the carbohydrate a sugar. **Glucose** is the simplest sugar used and stored by the body, and it serves as the richest source of energy for cellular activities. The specific methods by which carbohydrates and glucose are utilized and stored in the body are discussed in greater detail in Chapters 6 and 14.

Lipids

Like carbohydrates, **lipids** are a group of organic compounds made of carbon, hydrogen, and oxygen atoms, but they occur in different ratios and arrangements. Lipids are used by the body to form important cellular structures as well as hormones produced by endocrine tissues and organs. Lipids can also be stored as fat

Table 3-1 Common Elements in the Human Body

	Chemical Name	Chemical Symbol	Importance in the Body
Primary Elements	Hydrogen	H	A major constituent of organic molecules and water (H_2O). As a free ion (H^+) makes fluids more acidic.
	Nitrogen	N	Primary component of organic compounds, proteins, and nucleic acids.
	Oxygen	O	A part of many organic molecules and water; used by cells in production of energy (ATP).
	Carbon	C	The primary component of all organic molecules.
Secondary Elements	Phosphorus	P	Important part of nucleic acids and ATP; needed for normal bone structure.
	Sodium	Na	In ionic form, Na^+ is the most abundant chemical component in extracellular fluid; essential for nerve impulse conduction and water balance.
	Calcium	Ca	As an ion, Ca^{2+} is necessary for muscle contraction and blood clotting; needed for normal bone structure.
	Potassium	K	In ionic form, K^+ is the most abundant chemical component of intracellular fluid; needed for nerve impulse conduction.
	Chlorine	Cl	As an ion, Cl^- is the most abundant negatively charged particle in extracellular fluid; needed for maintaining optimal water balance.
	Magnesium	Mg	As an ion, Mg^{2+} supports the actions of many enzymes in the body.
	Iron	Fe	In ionic form, Fe^{2+} and Fe^{3+} are part of the oxygen-carrying molecule (hemoglobin) found in red blood cells.
	Sulfur	S	Performs a number of metabolic functions; is important for healthy hair, nails, and skin and helps to maintain oxygen balance for optimal brain function and cellular respiration.

to insulate and protect the body or to serve as a future energy source. In fact, fat deposits are the body's largest source of stored energy.

By the Way

The amount of lipids stored in and around body tissues is measured as a percentage of overall body mass. This body fat percentage should be above 10% in women or 6% in men to support normal body processes and maintain health.

Proteins

Proteins are the most versatile of the organic molecules made of carbon, hydrogen, oxygen, and nitrogen. These compounds are large and complex chains of smaller molecules called **amino acids**. Proteins contain between 50 and 200 linked amino acids and serve as the building blocks for the structural framework of all cells and tissues.

Nucleic Acids

Nucleic acids are very large complex molecules that differ from proteins due to the presence of phosphorus in addition to the carbon, hydrogen, oxygen, and nitrogen atoms. There are two kinds of these molecules found

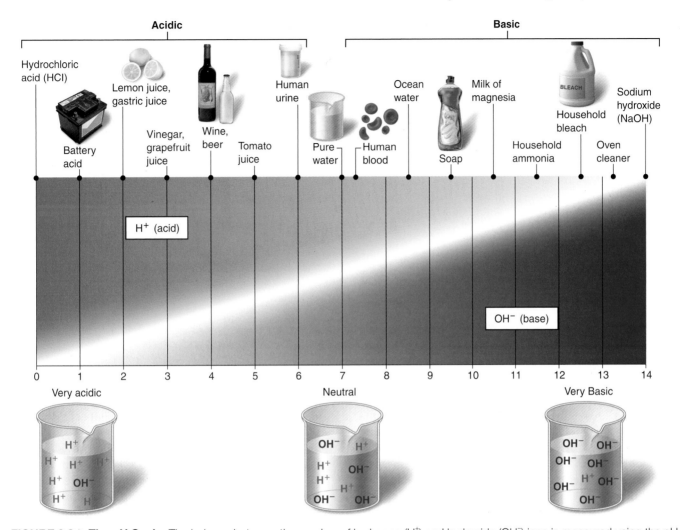

FIGURE 3.2 ▸ The pH Scale. The balance between the number of hydrogen (H^+) and hydroxide (OH^-) ions is measured using the pH scale. Acidic substances have more hydrogen ions, while basic, or alkaline, substances have larger numbers of hydroxide ions.

in the nucleus of cells: **deoxyribonucleic acid (DNA)** and **ribonucleic acid (RNA)**. The specific roles of these nucleic acids will be discussed later in this chapter.

What Do You Think? 3.1

- Do you think most of the substances you come in contact with every day are elements or compounds? Explain your answer.
- Is it accurate to say that all the vegetables available at your local supermarket are organic? Explain your answer.

STRUCTURE OF THE CELL

As previously discussed, cells are the simplest level of organization in the body. However, this basic building block of life is a miniature masterpiece; all of the different systems and physiologic processes that occur in the whole human organism appear in microscopic form within the cells. The study of cells, or **cytology**, begins by looking at the anatomy of a cell.

The trillions of cells in the body have a seemingly endless variety of shapes, sizes, and unique functional characteristics that distinguish one type from another. However, just as the immense variety of flowers share certain structural elements—stem, leaves, and petals—all types of cells share three primary structural features: the plasma membrane, cytoplasm, and nucleus.

Plasma Membrane

The outer covering that defines and encloses each cell is called the **plasma membrane**, also known as the **cell membrane**. This membrane plays a central role in most cellular activities, including the regulation of substances entering and leaving the cell. It is primarily made up of lipids, including ***cholesterol, glycolipids*** (lipids with attached carbohydrates), and lipid molecules attached to phosphorus called ***phospholipids***. Phospholipid molecules have two portions—one charged and the

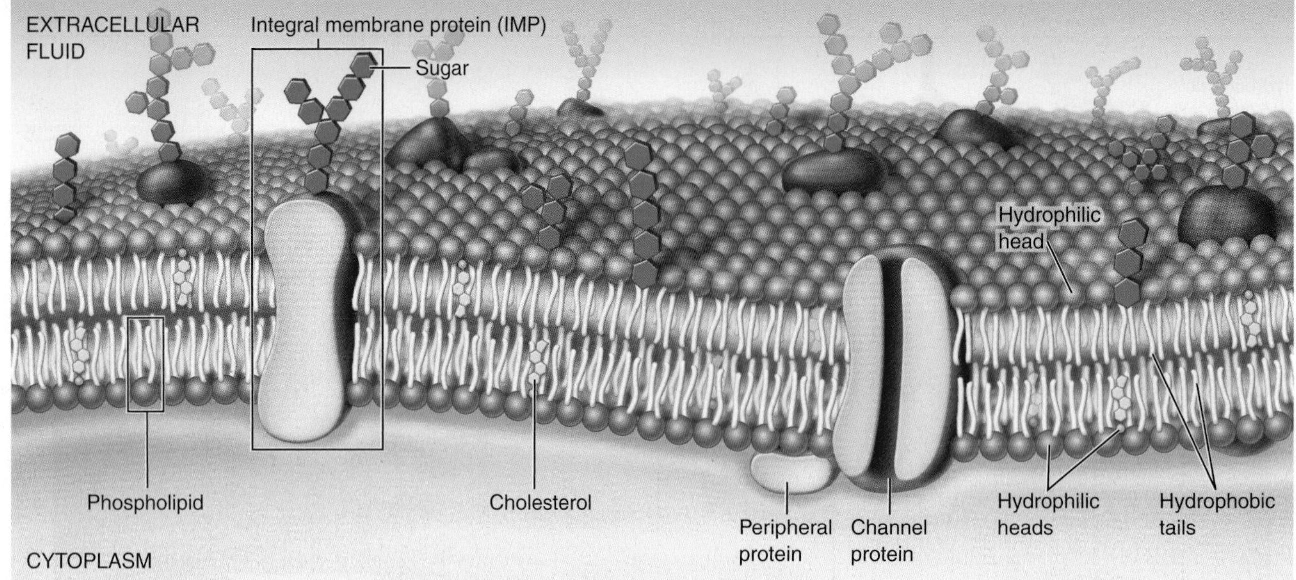

FIGURE 3.3 ▶ Plasma membrane. The plasma membrane is a thin, semipermeable membrane made up of proteins and lipids that encloses the cytoplasm of a cell. It contains a double layer of lipids and several types of proteins, including IMPs.

other uncharged—that are uniquely arranged into a double layer to create the basic framework of the plasma membrane (▶ FIGURE 3.3). The charged phosphorous end of the molecule is ***hydrophilic***, meaning "water-loving," and is oriented to face either the extracellular or intracellular environments, making the cell comfortable in its watery home. The lipid portion of these molecules that is not charged is classified as ***hydrophobic***, or "water-fearing." It lies in the center of the membrane and serves as an electrical insulator, keeping the cell from being overwhelmed by the many different ions within the surrounding fluid environment.

Some cells have microscopic extensions on the outer surface of the plasma membrane. **Microvilli** are hair-like projections that increase the surface area of the membrane in order to increase the cell's absorptive ability. Cells that line the small intestine of the digestive tract have microvilli. **Cilia** are the other type of plasma membrane extensions. These projections are longer and larger than microvilli and serve a different function; they wave in a coordinated fashion to move or brush substances from one area to another. Cells that line the trachea (windpipe) have cilia to help move mucus and trapped particles out of the respiratory tract.

Embedded within the phospholipid bilayer are a variety of proteins called **integral membrane proteins (IMPs)**. These unique proteins allow the cell to read and respond to its environment. IMPs can be divided into two general categories: receptor and effector proteins.

- ***Receptor proteins***—These specialized protein molecules monitor the internal and external environment of the cell to keep it informed about what is happening in and around it. Some receptor proteins function as cellular identity markers that help cells recognize one another during tissue formation and assist the body in identifying potentially dangerous foreign cells.
- ***Effector proteins***—These proteins direct the responses of the cell using information provided by the receptor proteins. These specialized protein molecules include ***channel proteins*** that create a pore to allow specific ions through the membrane, ***transport proteins*** that shuttle nutrients and wastes across the membrane, ***cytoskeletal*** or ***linker proteins*** that regulate the shape and movement of the cell, and ***enzymes***, which speed up the breakdown and synthesis of various molecules within the cell.

At one time, the cell membrane was thought to be a simple organic envelope with pores or channels through which nutrients and wastes could enter or leave the cell. In reality, it functions in a much more sophisticated and intelligent manner. The cell membrane is ***selectively permeable***, meaning it allows certain things in and out of the cell while restricting others. Additionally, it directs the cell's activities through the interactions between the receptor and effector IMPs. These molecules within the

By the Way

A good example of identity markers is the unique protein markers found on red blood cells, which are used to identify blood types. This is why blood transfused between recipient and donor must be of the same blood type or of type O, which doesn't have these identity markers. Otherwise, the transfused blood cells will be identified as foreign and destroyed by the recipient's immune system.

cell membrane allow cells to read, respond, and adapt to the changing conditions in and around it.

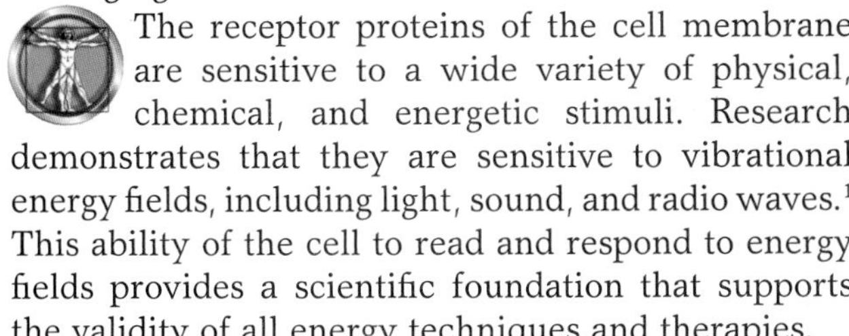

The receptor proteins of the cell membrane are sensitive to a wide variety of physical, chemical, and energetic stimuli. Research demonstrates that they are sensitive to vibrational energy fields, including light, sound, and radio waves.[1] This ability of the cell to read and respond to energy fields provides a scientific foundation that supports the validity of all energy techniques and therapies.

Bruce Lipton points out in his groundbreaking text, *The Biology of Belief*, that "for thousand of years ... Asians have honored energy as the principle factor contributing to health and well-being. In Eastern medicine, the body is defined by an elaborate array of energy pathways called meridians. In Chinese physiologic charts of the human body, these energy networks resemble electronic wiring diagrams. Using aids like acupuncture needles, Chinese physicians test their patient's energy circuits in exactly the same manner that electrical engineers troubleshoot a printed circuit board, searching for electrical pathologies."[2]

Cytoplasm

The **cytoplasm** refers to the cellular contents between the plasma membrane and the nucleus. It has two primary components: **cytosol**, a gel-like medium that is 75% to 90% water with variable amounts of organic compounds, plus a variety of small structures called **organelles** (▸ FIGURE 3.4). Organelles are like miniature organs that carry out the physiologic processes that sustain life for the cell, just as organ systems do for the entire body. The organelles and their functions are:

- **Mitochondria**—small sausage-shaped organelles that are the "powerhouses" of the cell. Each mitochondrion breaks down glucose to produce adenosine **triphosphate (ATP)** molecules, which are the energy source for all cellular work.
- **Endoplasmic reticulum (ER)**—a system of twisting canals within the cytoplasm that provides a pathway for substances to move throughout the cell. The ER is either *rough*, because it is studded with tiny round protein factories called **ribosomes**, or *smooth* if it is not. Smooth ER is involved in the production of lipids and steroids, as well as in the metabolism of many different compounds.
- **Golgi apparatus**—an organelle that looks like a stack of flattened sacs, generally located close to the nucleus. Its role is to modify and package proteins for export out of the cell.
- **Lysosomes**—tiny bags of digestive enzymes (literally translated as "breakdown bodies") that destroy foreign substances that enter the cell and digest worn-out and used-up cell structures.
- **Cytoskeleton**—a series of protein filaments and tubules that extend through the cytoplasm forming a type of internal scaffolding that provides shape, strength, and mobility to cells.
- **Centrosome**—made up of two bundles of microtubules called ***centrioles***, the centrosome plays an important role in cell division and in producing and organizing the cytoskeleton.

Nucleus

Most cells have a **nucleus**, a smaller structure with its own membrane. The nucleus contains DNA, which stores the genetic code, or blueprint, for each cell and for the body as a whole. DNA contains tens of thousands of coded segments known as **genes**. Genes determine the traits or characteristics that we inherit from our parents and carry instructions for the production of all the specific proteins needed to sustain life. A small spherical structure inside the nucleus called the ***nucleolus***, or "little nucleus," is formed by a large number of tightly packed DNA and RNA strands. This structure is the primary site for the partial assembly of ribosomes.

Once thought to be the brain or control center of the cell, the nucleus functions more like a library, housing information about our past and present, as well as possibilities for the future, to be referenced and copied when needed. The nucleus does not direct which genetic code should be copied or when that should occur. Instead, the process of making new protein is controlled and directed by the IMPs of the plasma membrane. Effector proteins responding to signals picked up by the receptor proteins control which portions of the DNA should be read and copied. This process enables the cell to adapt and respond to changes in its environment.

What Do You Think? 3.2

- Which organ systems do you associate with the functions of the organelles of a cell?
- How does your understanding of the plasma membrane's function as the "director" of cellular adaptation and change factor into your opinion on the age-old nature versus nurture debate?

CELLULAR PROCESSES

All the essential functions of the living organism—use of nutrients, excretion of wastes, movement, growth, response to changes in environment, and reproduction—are founded in several processes carried out by all types of cells. These common processes are

- Transporting nutrients and wastes across the plasma membrane
- Breaking down glucose to produce energy for cellular work

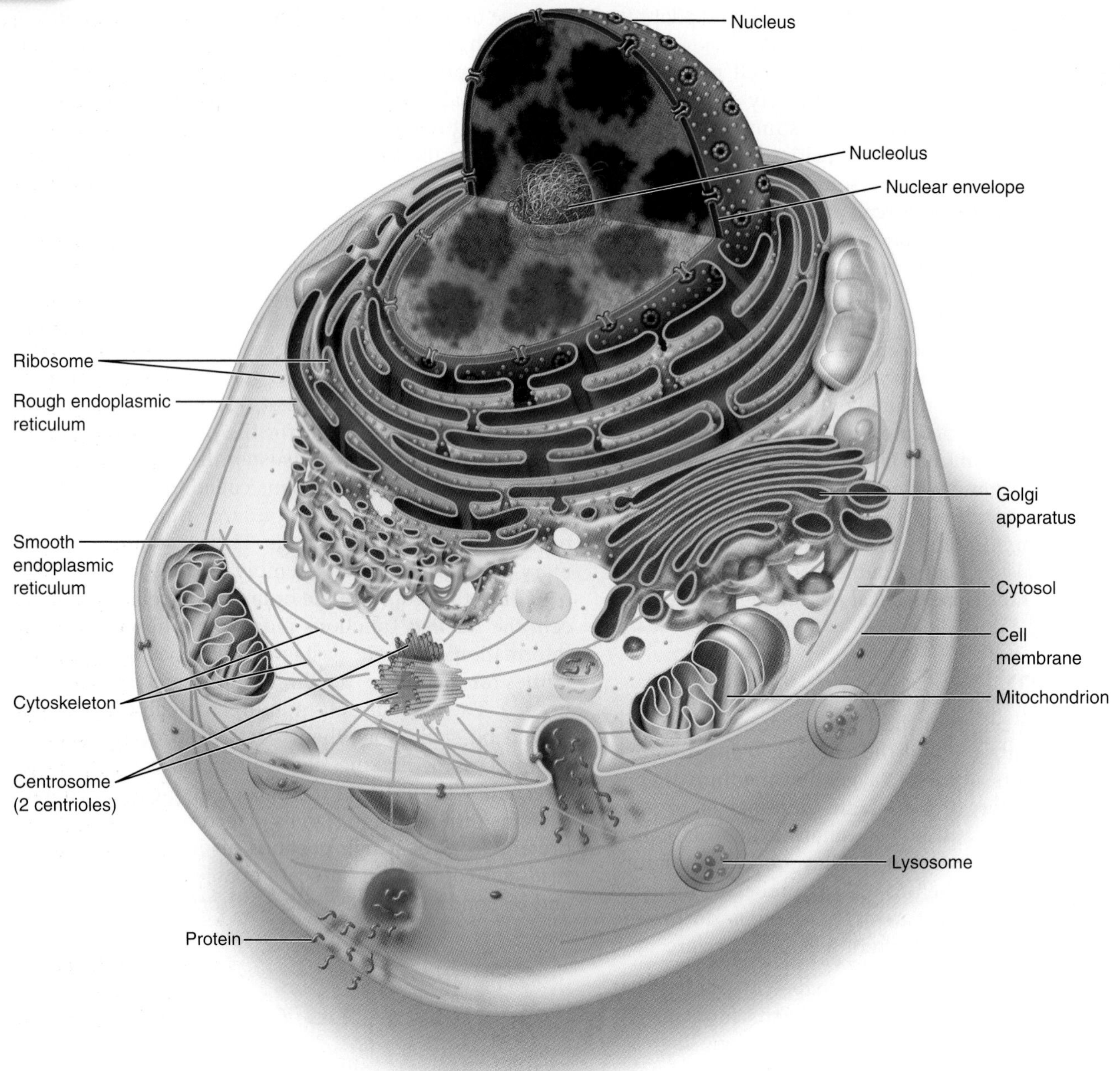

FIGURE 3.4 ▸ Key structures of the cell.

- Building essential proteins for growth and repair
- Adapting to changes in the environment
- Reproduction

Transport Mechanisms

Both the inside and outside of the cell membrane are fluid environments. Two-thirds of the total water in the human body can be found inside its cells; this fluid is known as **intracellular fluid**. The remaining one-third, **extracellular fluid**, makes up the liquid component of blood, lymph, and the fluid that fills the small spaces between cells, called **interstitial fluid**. The area between cells where the fluid is located is known as the **interstitium**. In all tissues, interstitial fluid constantly circulates around the cells providing a dynamic medium from which nutrients are extracted and wastes released to support cellular activities. Movement of substances across the cell membrane can be divided into two general categories: passive and active transport.

Passive Transport

If the cell does not have to expend energy (doesn't need to use ATP molecules) to move a substance through the cell membrane, this process is considered a **passive transport** mechanism. In normal cell activity, small

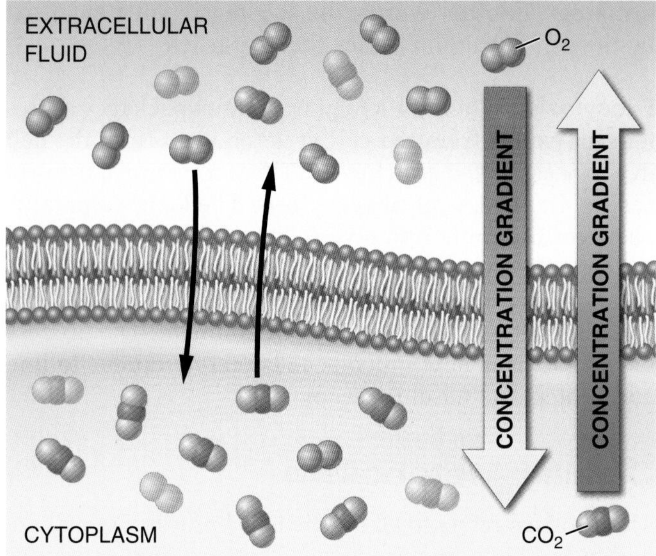

FIGURE 3.5 ▶ Diffusion. Diffusion is a passive transport mechanism in which atoms or molecules move across the plasma membrane from an area of high concentration to an area of lower concentration. Shown is the diffusion of oxygen (into the cell) and carbon dioxide (out of the cell) along their concentration gradients.

ions and molecules such as oxygen, carbon dioxide, water, and glucose are moved across the cell membrane via passive transport.

Diffusion. **Diffusion** is a passive transport mechanism that relies on differences in the level of *concentration* between substances. In diffusion, molecules move with or according to the concentration gradient, from an area of high concentration to an area of low concentration (▶ FIGURE 3.5). Imagine skating on a crowded ice rink. In an attempt to keep from bumping into others, you move to an area with fewer people, an area with a lower concentration of skaters. If the other skaters do the same, eventually all the skaters will be equally spread across the rink. This simple process is how molecules of oxygen, carbon dioxide, and electrolytes move across the cell membrane. Some particles, such as glucose, are too large to pass through the semipermeable plasma membrane and must be helped across by a special transport protein or carrier molecule. When a carrier molecule is involved in the diffusion process, it is called **facilitated diffusion**.

In some cases, cells must move water across the plasma membrane rather than dissolved particles, or ***solutes***, in an attempt to balance the concentration of molecules. When water moves across the membrane from an area of high to low concentration, the diffusion process is referred to as **osmosis** (▶ FIGURE 3.6). This transport mechanism is essential to maintaining fluid balance throughout the body, including the movement of fluid between the blood and interstitium and the process of urine formation by the kidneys.

Filtration. The process of **filtration** is driven by differences in *pressure*. When fluids press against a barrier, they create **hydrostatic pressure** (also known as fluid pressure) that pushes the fluid and smaller solutes through any opening from an area of high pressure to an area of lower pressure. A good analogy for filtration is to think about the audience at a rock concert. When the crowd at the back pushes forward to get a better view of the performers, the pressure pushes people at the front up against and sometimes through the openings of barriers in front of the stage. The people in front did not expend their own energy to move forward; they were pushed, or filtered, closer to the stage.

Active Transport

Active transport mechanisms require the cell to use energy (break down ATP molecules) to move substances across the plasma membrane. Energy is required when

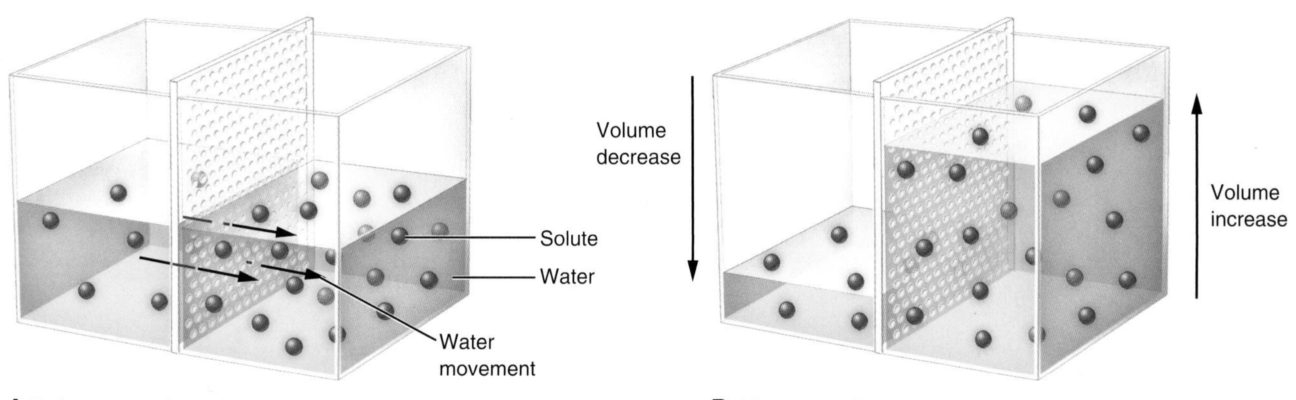

FIGURE 3.6 ▶ Osmosis. Each container has two compartments divided by a membrane through which only water can pass. **A.** Before osmosis, the volume of water in each compartment is the same, while the concentration of solutes is higher in the right compartment. **B.** Through osmosis, water has moved across the membrane from the left to the right compartment to balance the concentration of the fluids. While the solute concentration is now balanced, the fluid volume of the right compartment has increased.

the cell must transport particles *against* the concentration gradient: from an area of low concentration to one of high concentration. The most common examples of active transport mechanisms in the body are ion pumps, phagocytosis, and exocytosis.

Ion Pumps. All cells employ special transport proteins called **ion pumps** to move ions such as sodium (Na^+), potassium (K^+), and calcium (Ca^{2+}) across the cell membrane. While this sounds similar to the process of facilitated diffusion, in these situations, the ion being carried is moving *up* or *against* the concentration gradient and therefore requires the use of energy. An analogy for this process would be when a celebrity steps out of a limousine and is escorted into a packed nightclub by her bodyguard. In this case, the bodyguard is acting as a pump, pushing forcefully against the high concentration of people in the crowd to allow the star to enter the club. In the body, cells must actively pump K^+ into and Na^+ out of the intracellular environment to maintain the proper concentrations of these essential ions.

Phagocytosis. **Phagocytosis** literally means "cell eating." The process is carried out by a specialized group of cells, called **phagocytes**, generally found in blood, lymph, and connective tissue. The cell membranes of phagocytes engulf a particle, fold around it, and then pinch off that portion inside the cell. This particle-filled sac called a **vesicle** is then delivered to a lysosome for digestion and destruction (▶ FIGURE 3.7). **Pinocytosis** is a process similar to phagocytosis that means "cell drinking." In other words, the substance being engulfed by the cell is a liquid rather than a particle.

FIGURE 3.7 ▶ Phagocytosis. A. In this active transport mechanism, particles are engulfed and brought into the cell, forming a vesicle. A lysosome attaches to the vesicle and releases digestive enzymes to dissolve the particles within the vesicle. **B.** White blood cell engulfs a parasite.

Exocytosis. When the Golgi apparatus packages a protein for export from the cell, it is removed from the cell in a process called **exocytosis**. This active transport process is the reverse of phagocytosis. The Golgi apparatus packages the protein in a vesicle, which moves outward through the cytoplasm to the cell membrane, where it bonds. After bonding, the vesicle then discharges its contents into the extracellular environment. The process of exocytosis is how mucous cells secrete mucus to line and protect the nasal passages.

Cellular Metabolism

Metabolism refers to all the chemical processes that happen in the body to sustain life. There are two categories of metabolic processes: those that synthesize or build up substances and those that break down or decompose substances. Substances are either chemically broken down into smaller more usable particles by the cells or utilized as building blocks to make or repair cells and tissues. Building-up processes are called anabolism, while breaking-down processes are known as catabolism (▶ FIGURE 3.8).

Anabolism

Anabolism occurs when the body uses molecules as building blocks to repair and build new tissue, or to store nutrients for use at a later time. **Protein synthesis**, in which amino acids are chained together to produce proteins, is an example of anabolism. In this process, a portion of the genetic code stored in DNA is read and copied by RNA, the second type of nucleic acid. Next, the RNA attaches to a ribosome within the cytoplasm, which is able to translate the code in the RNA into a specific amino acid chain (protein). Depending on the specific

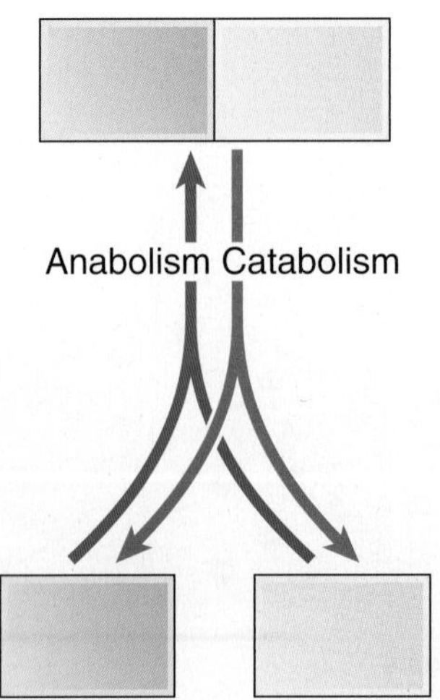

FIGURE 3.8 ▶ Metabolism. Metabolic processes are categorized as anabolism or catabolism. Anabolic processes (*blue arrow*) build up or combine substances, while catabolic processes (*red arrow*) are those that break substances apart.

protein created, this new molecule may be used to build cells, tissues, or hormones needed by the body.

A second example of anabolism involves storing a nutrient. When the body recognizes that it has more glucose than it can immediately use, it bonds and chemically alters the extra molecules to form another simple sugar, **glycogen**, which is then stored in the skeletal muscles and liver. When no glucose is being delivered from the digestive tract, stored glycogen can be broken down and used for energy.

Catabolism

Catabolism is any chemical process the body uses to break down nutrients or molecules. For example, the breakdown of glucose in the mitochondria to create ATP molecules is a ***catabolic*** process, as is the breakdown of ATP to release energy. These processes are explained in more detail in the Energy Production section of Chapter 6.

Cell Division and Differentiation

One of the most wondrous processes to consider is how every individual begins as a single cell and develops into a complete human organism. This amazing transformation involves two processes: cell division and cell differentiation. When an egg is fertilized, **cell division** immediately begins to produce more cells. All body cells except sperm and eggs divide through the linked processes of mitosis and cytokinesis. In **mitosis**, the nucleus duplicates its DNA and then divides to form two identical nuclei that carry the same genetic blueprint. Simultaneously, the cell divides its cytoplasm through a process called **cytokinesis**. The end result of these processes is two daughter cells with the same genetic code as the parent cell (▶ FIGURE 3.9). Sperm and egg cells are duplicated through a different form of cell division called **meiosis**, which is discussed in Chapter 16.

The cells of the embryo are a group of nonspecialized cells called **stem cells**. Stem cells are a unique type of cell that not only reproduces itself but can also produce different types of cells. This process of specialization is called **cell differentiation**. There are three general classes of stem cells based on their ability to produce more specialized cells (▶ FIGURE 3.10):

- ***Totipotent stem cells***—The first eight cells produced through division of a fertilized egg are totipotent stem cells. Each of these cells is capable of producing a complete human organism.
- ***Pluripotent stem cells***—While unable to divide and produce an entire human organism, these stem cells are able to reproduce themselves or differentiate into any type of body tissue.
- ***Multipotent stem cells***—These cells are the most limited in their differentiation capacity. Each is capable of producing only certain types of cells. For example, some multipotent stem cells produce blood and marrow cells, while others make fat, bone, ligament, and muscle cells.

As embryonic stem cells continue to divide and differentiate, they organize themselves into communities that function together. Initially, the cells form three layers of tissue: the ***ectoderm, mesoderm***, and ***endoderm*** (▶ FIGURE 3.11). Each of these layers develops into specific tissues or organs of the body. Cells in the ectoderm become the specialized cells, tissues, and organs of the skin and nervous system. Mesodermal cells give rise to the muscles and connective tissues, while the internal organs develop from the cells of the endoderm.

What Do You Think? 3.3

- What everyday analogies or examples can you think of that illustrate diffusion and filtration?
- Which word parts make up the term *cytokinesis*? What is the meaning of each word part?

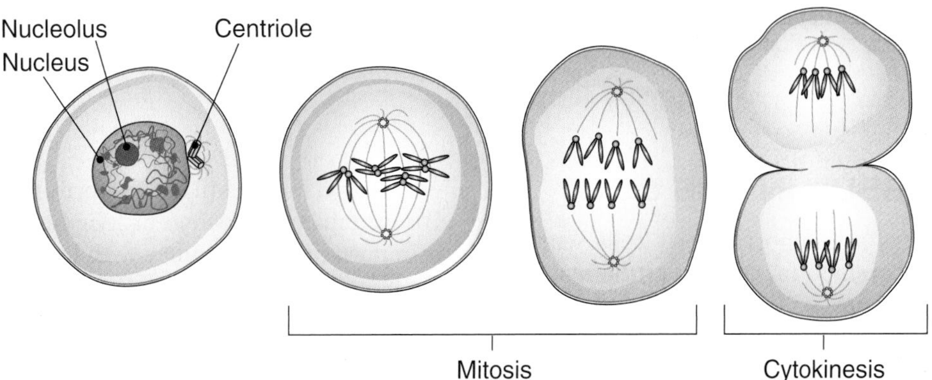

FIGURE 3.9 ▶ Cell division. When a cell divides to create two daughter cells, it copies and separates its genetic material through mitosis and divides its cytoplasm through cytokinesis.

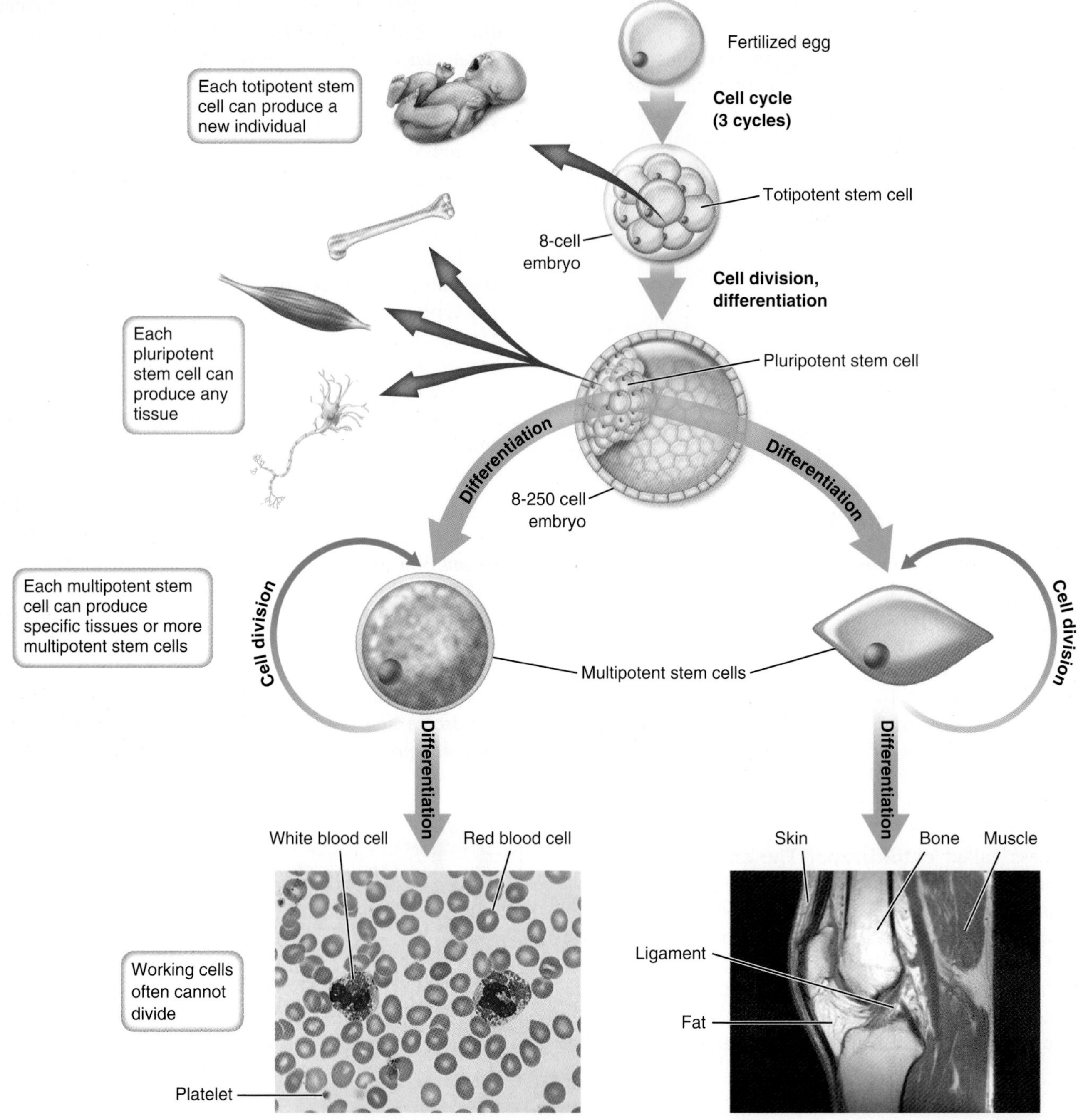

FIGURE 3.10 ▶ Cell differentiation. The entire human organism develops from a single fertilized egg. As this initial cell divides it produces stem cells, which further divide and differentiate to produce all the specialized cells of the body.

TYPES OF TISSUE IN THE BODY

Recall that a tissue is a group of similar cells working together to perform common functions. For manual therapists, the study of tissues, or **histology**, is of particular interest because the intention, techniques, and effects of many forms of manual therapy are directly linked to the characteristics of specific tissue types. Knowing the characteristics of a particular tissue type influences how and why a specific technique or form of therapy is applied. For example, to broaden or lengthen connective tissue, it is important to be familiar with the linear arrangement of the fibers within that specific tissue. It is helpful to become familiar with the location and characteristics of the four types of tissue in the body before studying the detailed anatomy of the body's organs and systems. ▶ Table 3-2 summarizes the general characteristics and functions of epithelial, muscle, nervous, and connective tissue.

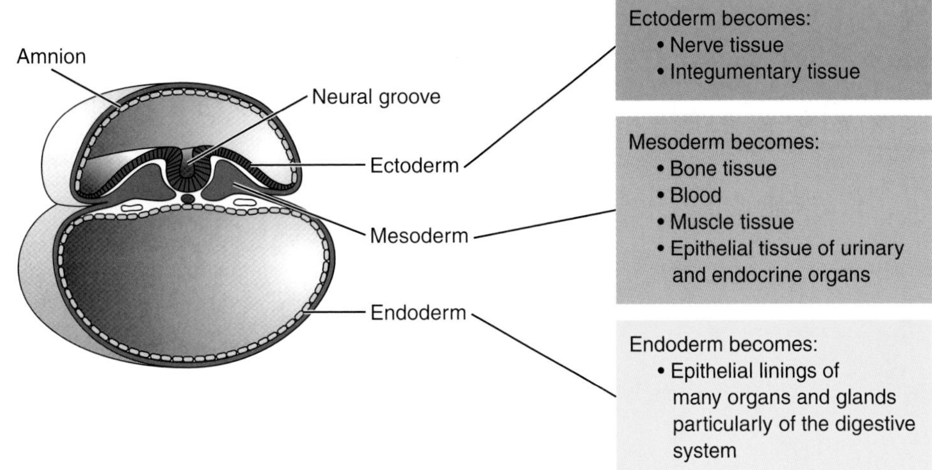

FIGURE 3.11 ▸ Embryonic layers. Stem cells of the developing embryo organize into three distinct layers, the ectoderm, mesoderm, and endoderm. Each embryonic layer differentiates into a specific set of cells and tissues.

Epithelial Tissue

Epithelial tissue, or **epithelium**, serves as the internal lining and external covering for the body. Most epithelial tissue functions both as a tough protective layer and as a selective barrier. Additionally, these tissues are characterized as **avascular**, meaning without blood vessels. Epithelial tissues are classified by the shape and arrangement of their cells. Shapes include flat cells that are referred to as ***squamous***, square cells called ***cuboidal***, or long and cylindrical cells called ***columnar***. A "hybrid" cell called ***transitional*** is a specialized type of cuboidal or columnar cell that changes its shape in response to tension. These transitional cells are found in structures such as the stomach and urinary bladder that need to stretch or distend as they fill, yet retain their smaller shape when empty. Epithelial cells are arranged to form either single layers called ***simple epithelium*** or multilayered tissues called ***stratified epithelium***. According to these criteria, a single layer of square cells would be classified as simple cuboidal epithelium, while an epithelial

Table 3-2 Characteristics and Functions of Tissues

Tissue Type	Characteristics	General Function	Locations
Epithelial	Avascular Described by shape and arrangement of cells: Flat (squamous) Square (cuboidal) Columnar Single layer (simple) or multilayered (stratified)	Lines Covers Protects Secretes	Skin Lining body cavities Covering organs
Muscle	Described by appearance and nervous system control: Striped (striated) or smooth (nonstriated) Voluntary or involuntary control	Contracts to create: Posture Movement of body parts Movement of fluid and substances through the body	Skeletal muscles Heart Walls of internal organs and vessels
Nervous	Two types of cells: Nonconductile (glial cells) Conductile (neurons)	Supports, protects and insulates (glial) Conducts electrical impulses (neurons)	Brain Spinal cord Nerves
Connective	Three common structural elements: cells, fibers, and ground substance	Supports and provides structure Connects Protects Transportation	Bone and cartilage Tendons and ligaments Blood and lymph Fat

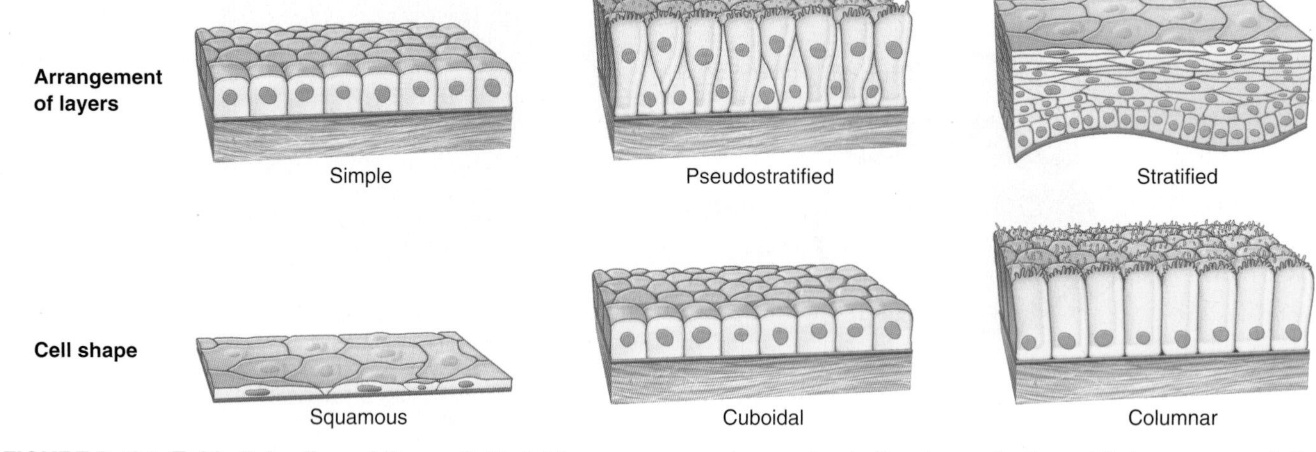

FIGURE 3.12 ▸ Epithelial cells and tissue. Epithelial tissues are named according to the shape of cells and their arrangement. Cell shapes can be squamous (flat), cuboidal, or columnar. Cells can be organized in simple, pseudostratified, or stratified layers.

tissue with multiple layers of flat cells would be called stratified squamous. Because of their shape, columnar epithelial cells are not easily arranged into multiple layers, and single layers sometimes look stratified because the nuclei of the cells are positioned at different levels. These tissues are called ***pseudostratified epithelium***, and they are found in the tubes and canals of the respiratory and reproductive systems (▸ FIGURE 3.12).

A highly specialized type of epithelial tissue, ***glandular epithelium***, is composed of specialized cells that produce and expel substances through the exocytotic process called **secretion**. Most glandular epithelium is embedded within the columnar and cuboidal epithelium of the integumentary, digestive, respiratory, and reproductive systems.

Muscle Tissue

The unique quality of all **muscle tissue** is its ability to contract or shorten to generate force for movement of body parts, fluids, and other substances. Because of their elongated shape and slightly tapered ends, the cells that make up these tissues are often called ***muscle fibers***. There are three different types of muscle tissue in the body: skeletal, cardiac, and visceral (▸ FIGURE 3.13). They are distinguished from one another by appearance and type of nervous system control.

By the Way

Small organs known as **glands** are aggregations of glandular epithelial cells. There are two types of glands in the body: ***exocrine*** glands, which have a specific canal or duct for the substances they secrete, and ***endocrine*** glands, which secrete directly into the tissue space or blood. Sweat and salivary glands are examples of exocrine glands, while the pituitary and adrenals are endocrine glands.

- **Skeletal muscle tissue**—This tissue makes up the muscular system that holds the skeleton upright and moves various body parts. It is made up of muscle fibers with long, thin fibrils in alternating light and dark shades, giving the cell a striped or ***striated*** appearance. These cells also have multiple nuclei situated just inside the cell membrane throughout their length. Because skeletal muscle contractions can be consciously controlled, it is described as ***voluntary*** muscle tissue.
- **Cardiac muscle tissue**—Found only in the heart, cardiac muscle cells are also striated. They are arranged in an end-to-end manner with a unique connecting junction called an ***intercalated disk*** between each fiber. These connections create a bending and branching arrangement of the cells, giving cardiac muscle a woven appearance. Cardiac muscle is characterized as ***involuntary*** because its contractions are not consciously controlled.
- **Visceral** or **smooth muscle tissue**—This tissue is found in the walls of hollow organs such as the airways within the lungs, the stomach, intestines, and in the blood vessels and has a ***nonstriated*** or ***smooth*** appearance. Like cardiac muscle, smooth muscle is classified as involuntary.

Nervous Tissue

Nervous tissue is found in the brain, spinal cord, and nerves. It is classified as either ***conductile*** or ***nonconductile*** tissue. Conductile nervous tissue is made up of ***neurons***, or nerve cells (▸ FIGURE 3.14). Neurons conduct the electrical impulses by which the nervous system carries out its communication functions.

Nonconductile nervous tissue includes a group of cells called ***neuroglia*** or **glial cells**. "Glial" is the Latin term for glue and reflects their function as specialized connecting cells. In addition to binding neurons together to form the nervous system organs, some glial cells also

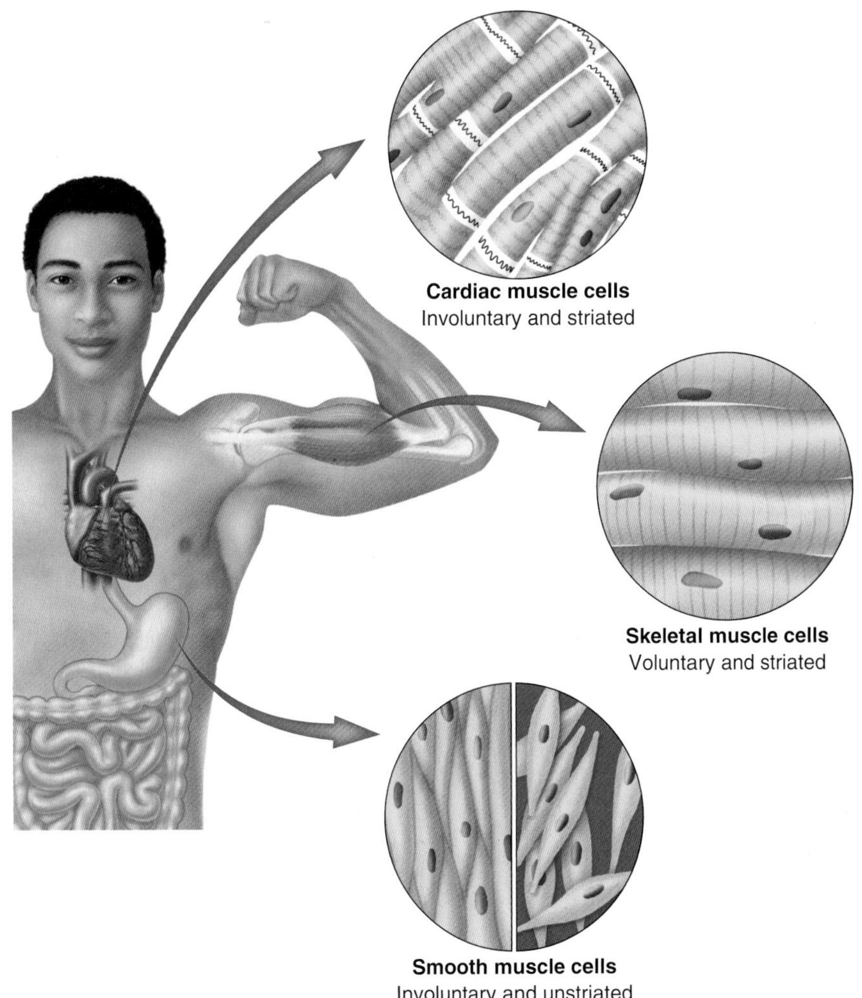

FIGURE 3.13 ▸ Muscle tissue. The three types of muscle tissue are cardiac, skeletal, and smooth.

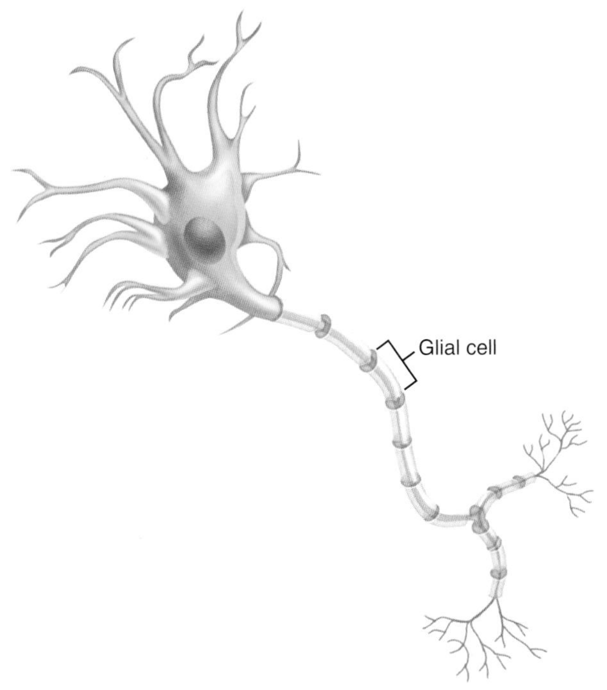

FIGURE 3.14 ▸ Nervous tissue. A single neuron with one type of glial cell that creates a fatty sheath to protect and insulate the nerve cell.

insulate and protect the neurons, while others connect neurons to blood vessels or carry out phagocytosis.

Connective Tissue

Connective tissue is the most abundant tissue in the body and includes a wide variety of types of tissue such as bone, cartilage, fat, and blood. As its name implies, this tissue connects, binds, supports, and protects all the other structures in the body. Connective tissue throughout the body can be organized into five basic types: liquid, loose, fibrous, cartilage, and bone (▸ FIGURE 3.15). ▸ TABLE 3-3 lists the major types of connective tissue, along with examples and common locations of each.

While each type of connective tissue makes its own unique contribution to the structure and function of the body, all connective tissues have three common structural features: cells, fibers, and ground substance.

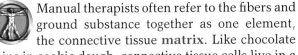

Manual therapists often refer to the fibers and ground substance together as one element, the connective tissue **matrix**. Like chocolate chips in cookie dough, connective tissue cells live in a "dough" called the matrix.

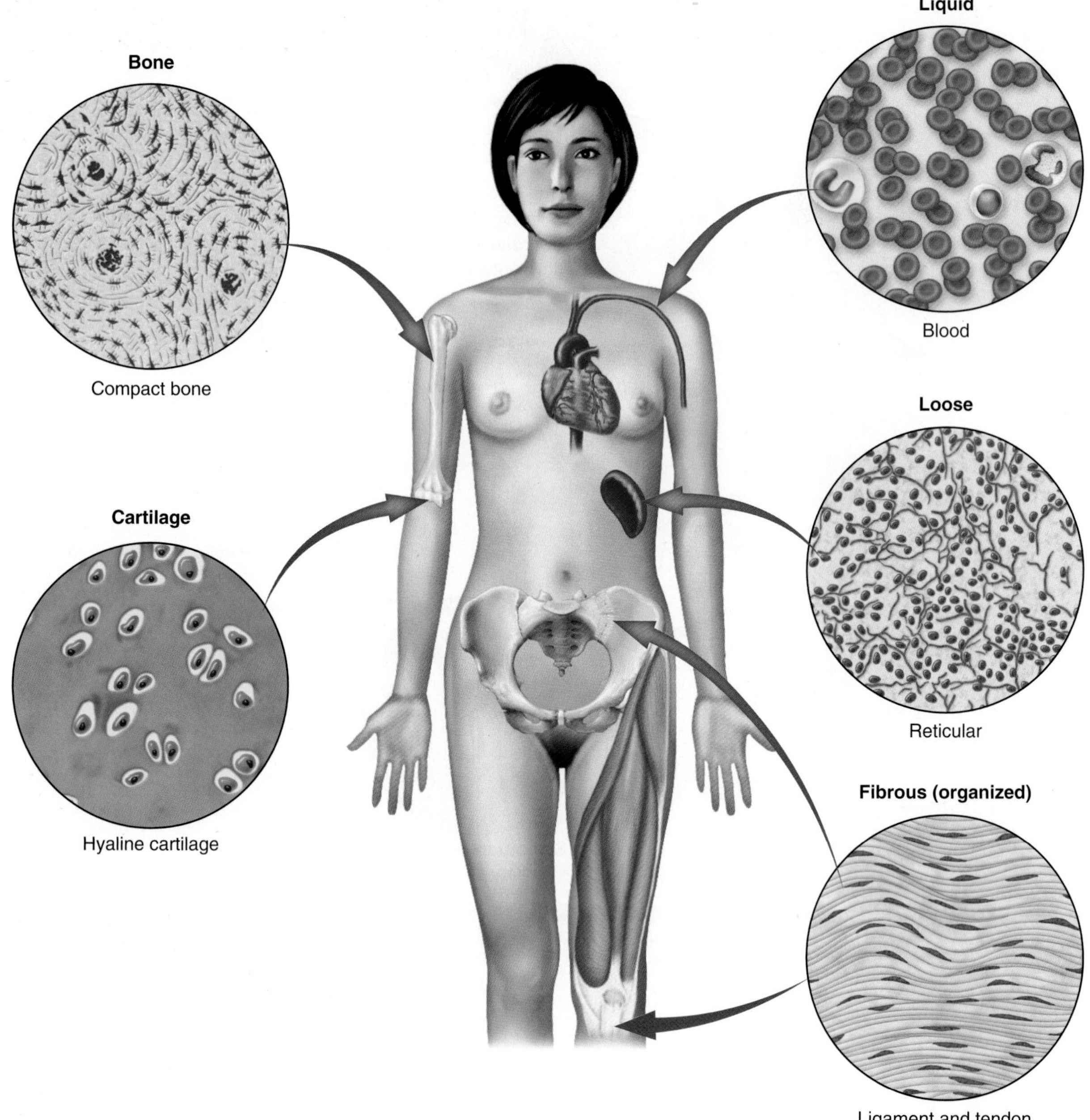

FIGURE 3.15 ▸ Types of connective tissue. There are five types of connective tissue found in various locations within the body.

Cells

Three kinds of cells are present in significant numbers in all connective tissue: fibroblasts, mast cells, and macrophages. **Fibroblasts** are the most numerous and commonly occurring connective tissue cells. These cells are responsible for producing the various types of fibers found in all connective tissues. Fibroblasts have specialized names specific to the type of tissue they are in. For example, fibroblasts in bone are called ***osteoblasts*** and those in cartilage are ***chondroblasts***. **Mast cells** are widely distributed throughout most types of connective tissue, but the largest numbers are found next to blood vessels. This is because mast cells secrete two important substances that affect blood and blood vessels: ***heparin***, which prevents blood clotting, and ***histamine***, which increases the permeability of surrounding tissues and causes blood vessels to open wider. The changes promoted by histamine and heparin are associated with the blood and the inflammatory and healing processes of the body (more detail on their importance is provided in Chapter 10). There are almost as many **macrophages** in connective tissue as there are fibroblasts, but they

Table 3-3 Types of Connective Tissue

Tissue Type		Examples	Locations
Liquid		Blood Lymph	Cardiovascular system Lymphatic system
Loose		Adipose Areolar Reticular	Subcutaneous layer of skin; around joints and organs Subcutaneous layer of skin; basement membranes Network inside organs
Fibrous	Organized	Tendons Ligaments	Attachments of skeletal muscles to bone Joints (attaches bone to bone)
	Disorganized	Fascia	Superficial and deep layers surrounding all structures of the body
Cartilage		Hyaline Fibrocartilage Elastic	Joints Fibrous disks in spinal column, knee, and between pubic bones Nose and ears
Bone		Dense (compact) Spongy (cancellous)	Bones of the skeletal system

serve a very different function. The literal translation of macrophage is "big eater," which clearly describes their role as highly active phagocytes.

Fibers

There are three varieties of connective tissue fibers made up of protein chains secreted by the fibroblasts and embedded in the ground substance of connective tissues. These fibers provide variable amounts of strength, support, and elasticity depending on the ratio and pattern of their arrangement. **Collagen fibers**, made up of the protein **collagen**, are the thickest, strongest, and most abundant fiber in the body. Because collagen is very tough and resistant to stretch, it is most abundant in structures that must withstand a lot of force and tension, such as tendons, ligaments, and bones. Detailed coverage of the structure and function of collagen is presented in Chapter 8.

Reticular fibers are made from a thinner and more delicate type of collagen known as **reticulin**. Like collagen, reticulin is slightly extensible and tolerates moderate to high levels of tensile stress without tearing. The connective tissue framework of organs has a high number of reticular fibers, and the connective tissues that surround and support nerves and smooth and skeletal muscles also contains some reticular fibers.

Elastic fibers are made up of a protein called **elastin**, which is much smaller and more flexible than either collagen or reticular fibers. When tensile stress is applied, elastic fibers have the unique ability to stretch up to 150% of their normal length without tearing, then quickly recoil to normal resting length when the tension is released. Therefore, connective tissues high in reticular and elastic fibers such as skin, lungs, and blood vessels are much more flexible and resilient than connective tissues such as tendons and ligaments that are high in collagen.

Ground Substance

Ground substance is the intercellular fluid component of all connective tissues. It is mostly water, but it also contains unique molecules called ***glycoaminoglycans (GAGs)*** that serve as water magnets. It is the presence of GAGs that keeps the ground substance fluid and allows it to fulfill its function as the spacer and lubricant in connective tissue. When connective tissue is dehydrated or injured, the fluid capacity of ground substance is diminished, causing fibers to stick to one another and form **adhesions**.

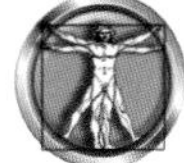

The ground substance within connective tissues gives these tissues a unique characteristic called **thixotropy**. Thixotropy refers to the ability of the ground substance to shift between a more solid or **gel** state and a more liquid or soluble state called **sol**. When connective tissue is cold, it is stiffer and less pliable because the ground substance is in a gel state. However, when heat or movement is applied, the ground substance in connective tissues shifts to a more sol state. This is why we often apply heat to relieve stiffness in our bodies and part of the reason why athletes and others who exercise warm up before vigorous activity. Kneading, compressing, rolling, or other techniques that stir the ground substance are used in many forms of manual therapy to achieve softer, looser, and more pliable tissues.

TISSUE REPAIR AND REGENERATION

In addition to their structure and functions, epithelial, muscle, nervous, and connective tissues also differ in their ability to repair and regenerate after damage has occurred. In general, epithelial tissue, smooth muscle, and bone are the only types of tissues capable of ***regeneration***, meaning they fully repair a damaged area with like cells. Epithelium regenerates quickly and easily, while the process is slower in the more complex smooth muscle and bone tissues. Although nerve tissue is capable of regeneration, the complexity of this tissue makes the process extremely slow, and severe nerve damage generally leads to a complete loss of function in the body part or organ controlled by the damaged nerve. In contrast, skeletal and cardiac muscle tissues are not easily regenerated; when damaged, fibrous connective tissue is used as repair tissue instead of the appropriate muscle fibers. This repair, or ***scar tissue***, often interferes with the normal function of the muscle tissue. More detail on tissue healing and repair is provided in Chapter 10.

As we age, we experience an overall degeneration of all body tissues. Bones become more brittle, cartilage becomes less resilient, epithelial tissues thin, and the general ability to repair and restore tissue is diminished. The effects of aging on specific tissues and organ functions are discussed within each body system chapter.

SUMMARY OF KEY POINTS

- Atoms are small units of energy that are the basic building block of all matter. When atoms combine, they form molecules. Elements contain molecules formed by a group of like atoms. Compounds contain molecules formed by different types of atoms.
- See Table 3-1 for a summary of the primary and secondary elements in the human body.
- The majority of chemicals in the body are bound together to form two broad groups of compounds: inorganic and organic. Inorganic compounds do not have carbon molecules, while organic compounds do. Both types of compounds play important roles in the composition and physiologic processes of body cells and tissues.
- Key inorganic compounds of cells and tissues include water, salts, acids, and bases. Key organic compounds are carbohydrates, lipids (fats), proteins, and nucleic acids.
- The ratio of acid to base of compounds is measured using the pH scale, which has values from 0 to 14. A pH value below 7 is acidic, and measurements above 7 are basic or alkaline. The homeostatic range for pH in the body is slightly alkaline at 7.35 to 7.45.
- Cells have many different sizes, shapes, and functions, but all share these common components:
  - Plasma membrane—the selectively permeable outer membrane (also known as the cell membrane) that defines cell boundaries, regulates substances entering or leaving the cell, and monitors internal and external environmental changes.
  - Cytoplasm—the cellular contents between the plasma membrane and the nucleus. It has two primary components: cytosol, a gel-like medium that is 75% to 90% water, plus a variety of small structures called organelles.
  - Nucleus—a smaller structure with its own membrane inside the cell. It contains DNA, which stores the genetic code or blueprint for the body.
- The key organelles of a cell and their functions are
  - Mitochondria—serves as the "powerhouse" of the cell; produces the majority of ATP, the predominant energy molecule.
  - Endoplasmic reticulum—serves as the transportation system for the cell; rough ER carries the ribosomes.
  - Ribosomes—synthesize proteins for the cell.
  - Lysosomes—tiny bags of digestive enzymes that destroy foreign substances and digest worn-out cell structures.
  - Cytoskeleton—a series of filaments and tubules in the cytoplasm that form internal scaffolding to provide shape, strength, and mobility to cells.
  - Centrosome—made of two bundles of centrioles, plays an important role in cell division, producing and organizing the cytoskeleton.
  - Golgi apparatus—located close to the nucleus, modifies and packages proteins for export out of the cell.
- The two mechanisms for moving substances in and out of the cells and tissues are passive and active transport. In passive transport, no energy

is expended; substances move according to a concentration gradient or a pressure gradient. Passive transport mechanisms include diffusion, filtration, and facilitated diffusion. When molecules need to be moved against the concentration or pressure gradient, active transport mechanisms that require energy—breaking down ATP—must be used. Phagocytosis, ion pumps, and exocytosis are common forms of active transport mechanisms.

- Metabolism is all chemical processes that happen in the cells to sustain life. It can be divided into two categories: anabolism, a building-up process used to replace, repair, or store cells and tissues, or catabolism, a breaking-down process that provides more usable forms of nutrients and creates or releases energy.
- The four types of tissues in the body are epithelial, muscle, nervous, and connective tissue. See TABLE 3-2 for a summary of their key characteristics and locations.
- Epithelial tissue is classified according to the shape of its cells and the number of layers.
- There are three types of muscle tissue: (1) Skeletal muscle is striated and voluntary, (2) Cardiac muscle is striated and involuntary, and (3) Visceral or smooth muscle is nonstriated and involuntary.
- There are two types of cells in nervous tissue: neurons that conduct impulses and neuroglia (glial cells) that support and protect the neurons.
- See TABLE 3-3 for a summary of the types of connective tissue and their locations.

REVIEW QUESTIONS

Short Answer

1. Name the three cellular components common to all cells, and explain the general function of each.
2. Explain the difference between organic and inorganic compounds and give two examples of each type.
3. List and explain the names of epithelial cells and tissues.
4. List the different types of connective tissue and give an example of the location of each.

Multiple Choice

5. When two or more like atoms combine to form a molecule, it is called a(n)
   a. compound
   b. element
   c. acid
   d. base
6. If a compound has a pH below 7, it is
   a. acidic
   b. alkaline
   c. neutral
   d. basic
7. Which organic compound is broken down into amino acids?
   a. carbohydrates
   b. proteins
   c. lipids
   d. starches
8. What is the role of the receptor proteins in the plasma membrane?
   a. shuttle substances across the membrane
   b. monitor the internal and external environment of the cell
   c. regulate the shape and movement of the cell
   d. break down and synthesize various molecules within the cell
9. Which cell organelle is responsible for intracellular transportation?
   a. mitochondria
   b. lysosomes
   c. endoplasmic reticulum
   d. cytoskeleton
10. What cell organelle is responsible for synthesizing the energy molecule ATP?
    a. lysosomes
    b. ribosomes
    c. mitochondria
    d. endoplasmic reticulum
11. Protein synthesis is the primary function of which organelle?
    a. ribosomes
    b. mitochondria
    c. lysosomes
    d. cytoskeleton

Continued on page 54

12. What is the term for the passive transport mechanism that moves a substance due to a pressure gradient?
    a. diffusion
    b. filtration
    c. osmosis
    d. phagocytosis
13. What is the name of the diffusion process in which only fluid moves across the semipermeable membrane?
    a. facilitated diffusion
    b. filtration
    c. osmosis
    d. dialysis
14. Which of the following is an example of an active transport mechanism?
    a. facilitated diffusion
    b. osmosis
    c. filtration
    d. sodium pump
15. Which of the following is another term for cell division?
    a. differentiation
    b. mitosis
    c. osmosis
    d. splitosis
16. What is the process in which cells begin to specialize into the various types of body cells and tissues?
    a. differentiation
    b. cell division
    c. meiosis
    d. cytolytic mitosis
17. What are the four basic types of tissue in the body?
    a. muscle, skeletal, epithelial, connective
    b. epithelial, muscle, adipose, connective
    c. connective, muscle, adipose, nervous
    d. muscle, epithelial, connective, nervous
18. What are the characteristics that describe visceral muscle tissue?
    a. smooth and involuntary
    b. smooth and voluntary
    c. striated and involuntary
    d. striated and voluntary
19. Which type of nervous tissue functions to connect, protect, and support neurons?
    a. neuroglial
    b. neuronal
    c. neuro-connective
    d. neuro-matrix
20. Which type of tissue is avascular?
    a. visceral muscle
    b. neuronal
    c. epithelial
    d. loose connective

References

1. Tsong TY. Deciphering the language of cells. *Trends Biochem Sci.* 1989;14:89–92.
2. Lipton BH. *The Biology of Belief.* Santa Rosa, CA: Mountain of Love/Elite Books; 2005:108.

4 Body Membranes and the Integumentary System

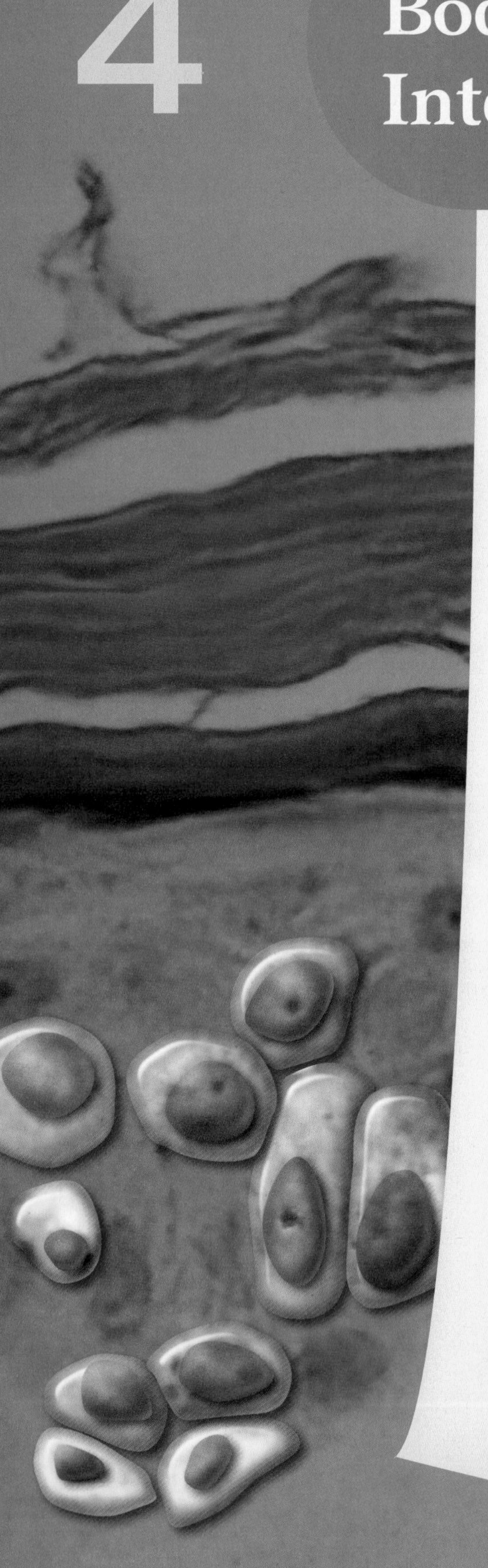

LEARNING OBJECTIVES

Upon completion of this chapter, you will be able to:

1. Discuss the importance of understanding the anatomy and physiology of the integumentary system as it relates to the practice of manual therapy.
2. Name the four types of membranes in the body and provide an example of each.
3. Describe the structure and list the general functions of each type of membrane.
4. Explain the functions of the integumentary system.
5. Describe the structure and function of each layer of the skin.
6. List the primary accessory organs of the skin and explain the function of each.
7. Explain the functions of hair, nails, and glands of the skin.
8. Explain the specific functions of cutaneous receptors and describe their general location.
9. Discuss the links between the integumentary and nervous systems and the importance of touch for neurological development and overall health.
10. List four types of contagious skin conditions and provide an example of each.
11. List and describe the general signs and symptoms of several common noncontagious skin disorders and the ABCD warning signs of skin cancer.
12. Describe common integumentary system changes associated with aging and their implications for the practice of manual therapy.

On one hand, the skin is a barrier, effectively containing within its envelope everything that is ourselves and sealing out everything that is not. On the other hand, it is a window, through which our primary impressions of the world around us enter into our consciousness and structure our experience. The nature of this envelope itself provides many ways for us to feel good in our own skins."

DEANE JUHAN

Job's Body: A Handbook for Bodyworkers

KEY TERMS

cutaneous (ku-TA-ne-us) **membrane**
dermatome (DERM-ah-tome)
keratin (KARE-ah-tin)
melanin (MEL-ah-nin)
membrane (MEM-brain)
mucous (MU-kus) **membrane**
parietal (pah-RI-ah-tal) **layer**
sebum (SE-bum)
serous (SEER-us) **membrane**
strata (STRAH-tah)
synovial (sin-O-ve-al) **membrane**
visceral (VIS-er-al) **layer**

Primary System Components

- **Cutaneous membrane**
  - Epidermis
  - Dermis
- **Hypodermis**
- **Accessory organs**
  - Sebaceous glands
  - Sudoriferous glands
  - Hair
  - Nails
  - Sensory receptors

Primary System Functions

- **Provides superficial covering and protective layer for the entire body**
- **Serves as the body's largest sensory organ**
- **Helps regulate body temperature via sweating**
- **Excretes trace levels of metabolic byproducts and water via sweating**
- **Absorbs substances through pores of skin**
- **Synthesizes vitamin D**

The skin is a window that often provides a view of our inner workings and general health. We flush with fever and excitement; pale with pain, loss of circulation, or fear; and erupt in rashes due to systemic dysfunction or emotional distress. With most forms of manual therapy, the skin serves as the primary interface between the client and the therapist. It allows for a two-way conversation in which the therapist reads and manipulates a client's skin and tissue, and the client reads and responds to the therapist's touch. In other words, no one can touch without being touched in turn. Therapists may feel differences in the temperature, tension, texture, and mobility of tissues or sense a shift in the client's energy or emotions. Simultaneously, the client feels the status of their tissue as well as the quality of the therapist's touch, which may be interpreted as confident, nurturing, and therapeutic or tentative, unfocused, or unsettling. Understanding the complexities and subtleties of the nature and function of skin is essential to explain and appreciate the profound nature and impact of touch.

In this chapter you will be introduced to four classifications of body membranes. Three of these classes include membranes that function as primary organs in various systems of the body. The fourth class, the cutaneous membrane, is the foundational organ of the skin. This membrane, combined with several accessory organs (nails, hair, general sensory receptors, and specific epithelial glands), makes up the skin, or **integumentary system.**

MEMBRANES

A **membrane** is a broad flat sheet of at least two layers of tissue. The human body has four primary types of membranes, classified according to the main tissue type in their layers and designated as either connective tissue or epithelial membranes. The synovial membrane is classified as a connective tissue membrane, while the other three are epithelial membranes: the mucous, serous, and cutaneous membranes. All four types share the same general functions of covering or lining, but each secretes a different type of lubricating fluid. An overview of the four membrane types and a summary of their functions and locations are given in TABLE 4-1.

Connective Tissue Membranes

Synovial membranes line the fibrous connective tissue capsules found in the joints of the skeletal system (FIGURE 4.1). These membranes have a thick fibrous connective tissue layer on the outside and a thin internal layer of simple epithelium. The fluid secreted by the epithelial layer, **synovial fluid,** serves to reduce friction and wear to the bone ends during joint movement. Additionally, synovial fluid serves as the medium for nutrient and waste exchange for the cartilage that covers the ends of bones.

Epithelial Tissue Membranes

Each epithelial membrane is comprised of a layer of epithelial tissue, either simple or stratified, attached to a connective tissue layer referred to as the **basement membrane.** There are three types of epithelial membranes: mucous, serous, and cutaneous membranes. Because they are composed of epithelial cells, these membranes are capable of rapid regeneration, making them ideal for protecting organs and lining body cavities and passages.

Mucous Membranes

Mucous membranes are made up of a simple epithelial layer attached to a thin basement membrane.

Table 4-1 Overview of Body Membranes

Membranes	General Function	Location	Examples
Mucous	• Protection • Production of mucus	Lining cavities open to external environment	• Respiratory tract • Gastrointestinal tract • Urinary tract • Female reproductive tract
Serous	• Protection • Production of serous fluid	Lining cavities closed to the external environment and covering organs	• Peritoneum • Pleura • Pericardium
Synovial	• Protection • Production of synovial fluid	Lining joint capsules of synovial joints	• Shoulder • Elbow • Hip • Knee • Ankle
Cutaneous	• Protection	Covering the body	• Skin

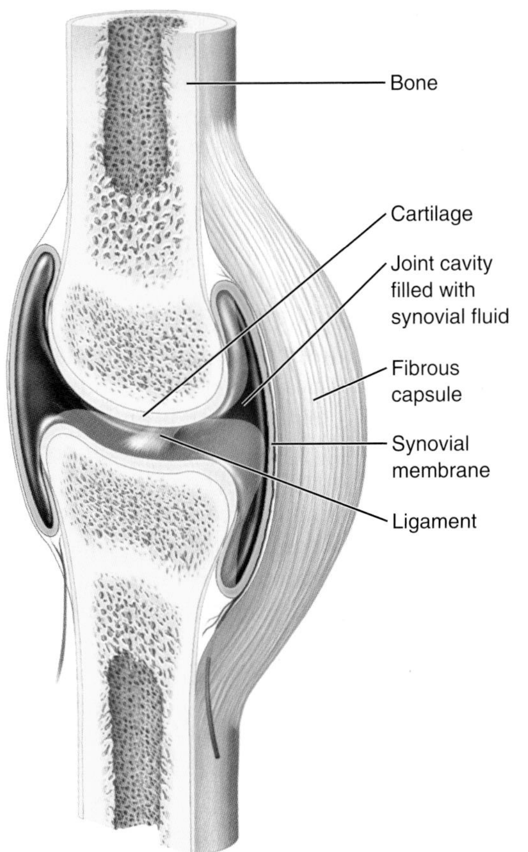

FIGURE 4.1 ▶ Synovial membrane. This connective tissue membrane lines the joint capsule of synovial joints and secretes the synovial fluid that lubricates the joint.

They line cavities open to the external environment such as the respiratory, digestive, urinary, and vaginal tracts. These membranes secrete a thick, usually clear substance called **mucus.** In the respiratory tract, mucus helps protect the body from invading pathogens by trapping them and facilitating their movement out of the body. Everyone has experienced the removal of these pathogens through increased coughing or nose blowing when they suffer a cold, flu, or allergies. Additionally, mucus helps warm and humidify the air drawn through the cavities on the way to the lungs for gas exchange. In the digestive, urinary, and vaginal tracts, mucous membranes line the passageways and protect them from wear or erosion. For example, the stomach contains high levels of hydrochloric acid, and the mucous membrane prevents the acid from eating through the stomach wall. If this membrane breaks down, the wall of the stomach will become damaged by the highly acidic environment, most likely leading to the formation of an **ulcer.**

Serous Membranes

Serous membranes are the epithelial membranes found in cavities that do *not* have openings to the external environment: the thoracic and abdominopelvic cavities. Like mucous membranes, the serous membranes are also made up of a single epithelial layer attached to a thin basement membrane. Serous membranes are continuous but folded into two distinct layers:

- The **parietal layer** that lines the cavity
- The **visceral layer** that covers the organs

To visualize the position and relationship of these layers, see ▶ Figure 4.2. This figure shows a partially inflated balloon inside a bowl. A fist is pushing into the balloon. The outer surface of the balloon pressing against the bowl represents the parietal layer of a serous membrane. The surface of the balloon enclosing the fist represents the visceral layer of the membrane as it covers an organ. The small

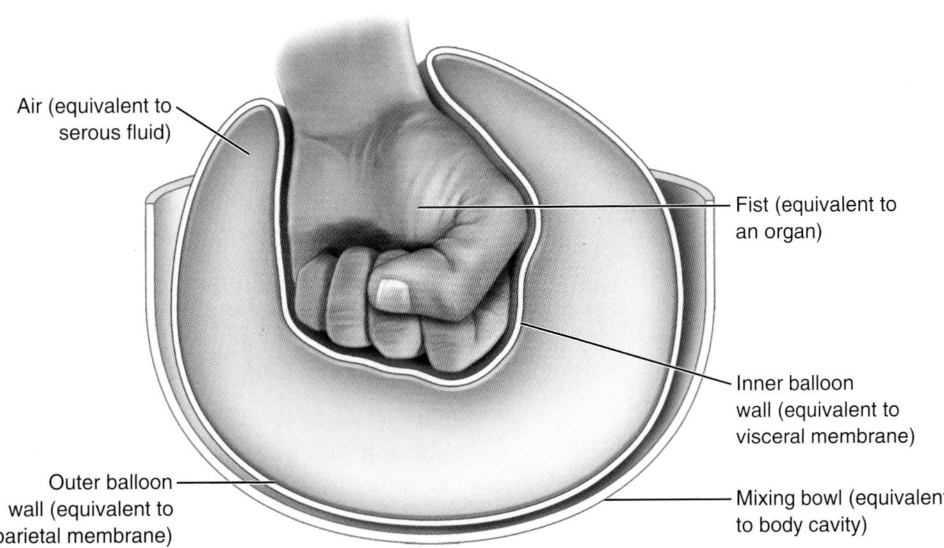

FIGURE 4.2 ▶ Layers of a serous membrane. Just as the balloon folds around the fist inside the bowl, serous membranes fold around organs within body cavities to form two different layers. The visceral layer covers the organs, and the parietal layer lines the cavity.

space between the parietal and visceral layers contains **serous fluid,** which is secreted by the epithelial layer of the membrane. This fluid provides lubrication to reduce friction between the two layers, which allows for smooth expansion, contraction, or movement of organs within the cavity. In the figure, serous fluid is represented by the air inside the balloon.

In the thoracic cavity, the serous membrane, also called the **pleura**, protects the lungs as they expand and contract during breathing (FIGURE 4.3). Serous fluid provides lubrication, allowing the membrane layers to slide easily across one another. Without this fluid, there would be friction between the visceral and parietal layers, which would make each breath inefficient and painful. In the abdominopelvic cavity, the serous membrane that lines the cavity and covers the organs is called the **peritoneum**.

Cutaneous Membrane

The **cutaneous membrane** is the body's outer covering and the primary organ of the integumentary system. It consists of a stratified layer of epithelial tissue attached to a thick connective tissue layer.

THE INTEGUMENTARY SYSTEM

The integumentary system or skin is comprised of the cutaneous membrane and accessory organs, including hair, nails, sudoriferous (sweat) glands, sebaceous (oil) glands, and general sensory receptors. The functions of skin are based on the combined actions of both the membrane *and* the accessory organs imbedded within it. For this reason, skin is considered to be synonymous with the entire integumentary system, meaning the cutaneous membrane *plus* the accessory organs. Functions of the skin include:

- ***Protection***—The skin serves as a physical barrier that protects the body from invading organisms and environmental contaminants.
- ***Temperature regulation***—Sudoriferous glands produce sweat that helps to cool the body via evaporation. Additionally, there is an insulating layer of fat under the skin that helps to retain body heat.
- ***Excretion and absorption***—The process of sweating helps the body eliminate trace amounts of metabolic byproducts through pores in the skin. These pores also allow for absorption of substances such as lotions, creams, oils, herbal extracts, or medications when applied to the skin's surface.
- ***General sensory organ***—A wide variety of general sensory organs are found in every square centimeter of the skin. Similar to the receptor integral membrane proteins (IMPs) of the cell membrane, the sensory receptors of the skin collect information about what is happening on the body's surface.
- ***Synthesis of vitamin D***—When ultraviolet (UV) rays of light are absorbed by the skin, they stimulate the synthesis of vitamin D. The active form of vitamin D (calcitriol) is necessary for the absorption of calcium and phosphorous from the digestive tract. These nutrients are essential elements for proper bone growth and nerve and muscle function.

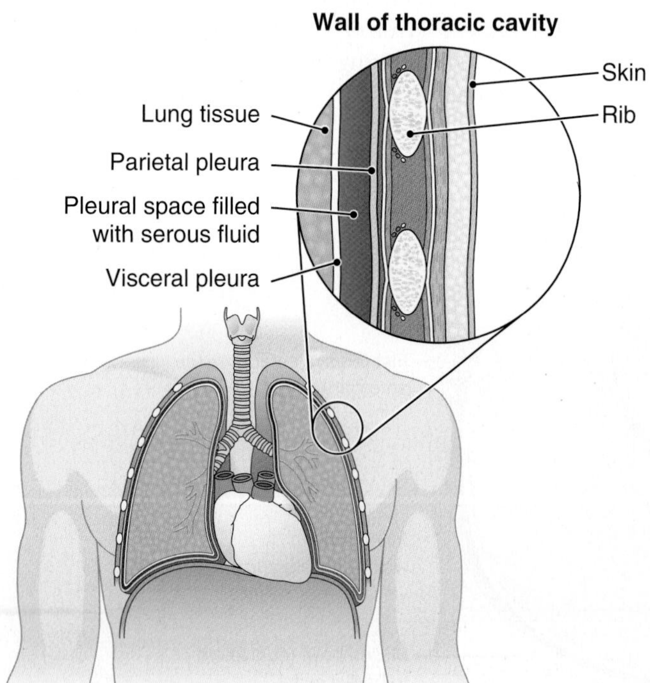

FIGURE 4.3 Pleura. The pleura is the serous membrane of the thoracic cavity.

What Do You Think? 4.1

- Why is it important that in addition to providing a protective covering for the body, the skin also functions as a sense organ?
- What everyday examples can you give of the skin's excretion and absorption functions?

LAYERS OF THE SKIN

Skin is made up of the two-layered cutaneous membrane plus a group of accessory organs. The external epithelial layer is the epidermis, and the deeper dermis is a thicker connective tissue layer. The dermal layer is attached to underlying muscles with a connective tissue layer made of adipose and areolar tissue referred to as the subcutaneous layer or hypodermis (FIGURE 4.4).

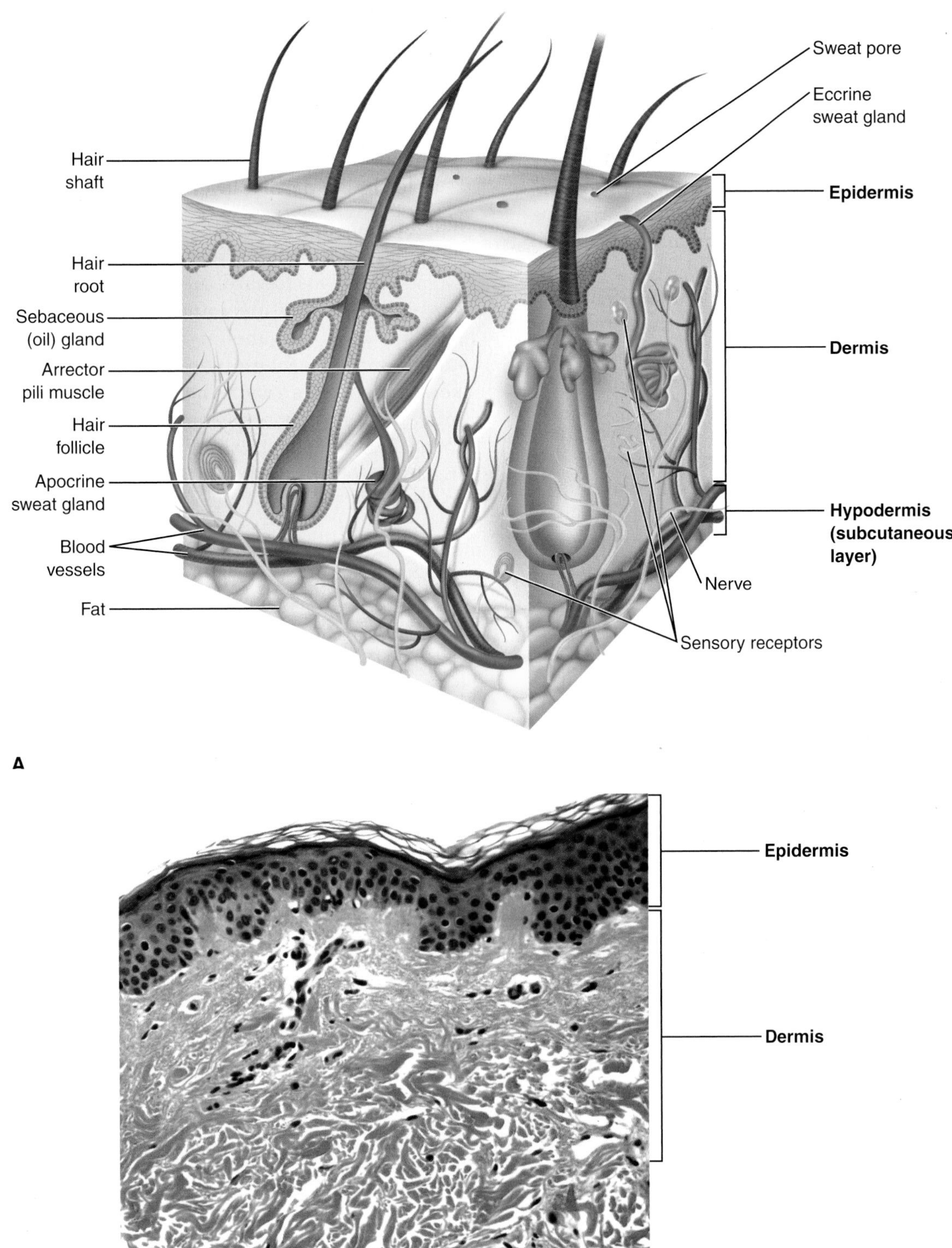

FIGURE 4.4 ▸ **Anatomy of the skin. A.** Layers of the skin and its many accessory organs. **B.** Micrograph of the skin clearly shows the cell distinction between the epidermis and dermis.

Epidermis

The **epidermis** is the outermost layer of skin composed of stratified squamous epithelium. Ninety percent of cells in this layer are **keratinocytes** that produce and contain a tough water-resistant substance known as **keratin**. The entire epidermis is quite thin, between 0.08 and 0.5 mm, which is about the thickness of one or two sheets of paper. Even so, the numbers, arrangement, and characteristics of the keratinocytes divide the epidermis into five distinct layers, or **strata** (▸ FIGURE 4.5).

- ***Horny layer*** **(stratum corneum)**—This outermost stratum consists of 20 to 30 layers of dead keratinized cells that form a protective barrier; surface cells are sloughed off as the skin renews itself.
- ***Clear*** or ***glassy layer*** **(stratum lucidum)**—This layer is present in areas of the body with the thickest skin, such as the palms, fingertips, and soles of the feet. The densely packed cells of this layer are dead, flat, and full of keratin, similar to the cells of the horny layer.
- ***Granular layer*** **(stratum granulosum)**—This layer is named for specialized granules in the keratinocytes that release a lipid secretion that gives them their water-resistant quality. Cells in this layer actively replace their cytoplasm with keratin.
- ***Spiny layer*** **(stratum spinosum)**—This layer has about 10 layers of cells. It contains specialized immune cells called **Langerhans cells** that help skin protect the body from organisms that penetrate the superficial epidermis. In living tissue, these cells have a plump, rounded appearance, but when they are prepared for viewing through a microscope, they dry out and appear to have spiny, thorn-like edges, hence the name.
- ***Germinating layer*** **(stratum germinativum** or **stratum basale)**—Germination is the process of sprouting or growing. The keratinocytes in this layer continually undergo mitosis to create new cells, which allows skin to heal rapidly. The germinating layer is the deepest layer of the epidermis and is also known as the basal layer. Its continuously dividing cells require a steady supply of oxygen and other nutrients that are provided via diffusion from capillaries in the dermal layer.

The epidermis continually renews and replaces itself as cells produced in the germinating layer gradually work their way toward the surface. As new cells rise through the strata, they undergo a developmental process in which organelles and nucleus are dissolved and keratin is accumulated. Eventually, these fully keratinized cells reach the outer layer of the epidermis, where they are sloughed off. This process of forming new cells, pushing through the strata, becoming fully keratinized, and being sloughed off takes 2 to 4 weeks.

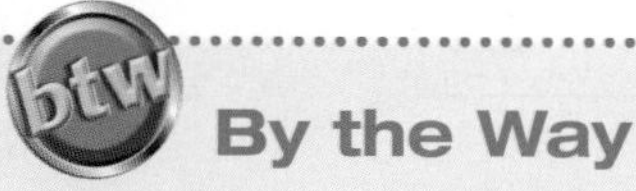

By the Way

Skin makes up about 16% of a person's total body weight, with an area of approximately 22 ft^2.

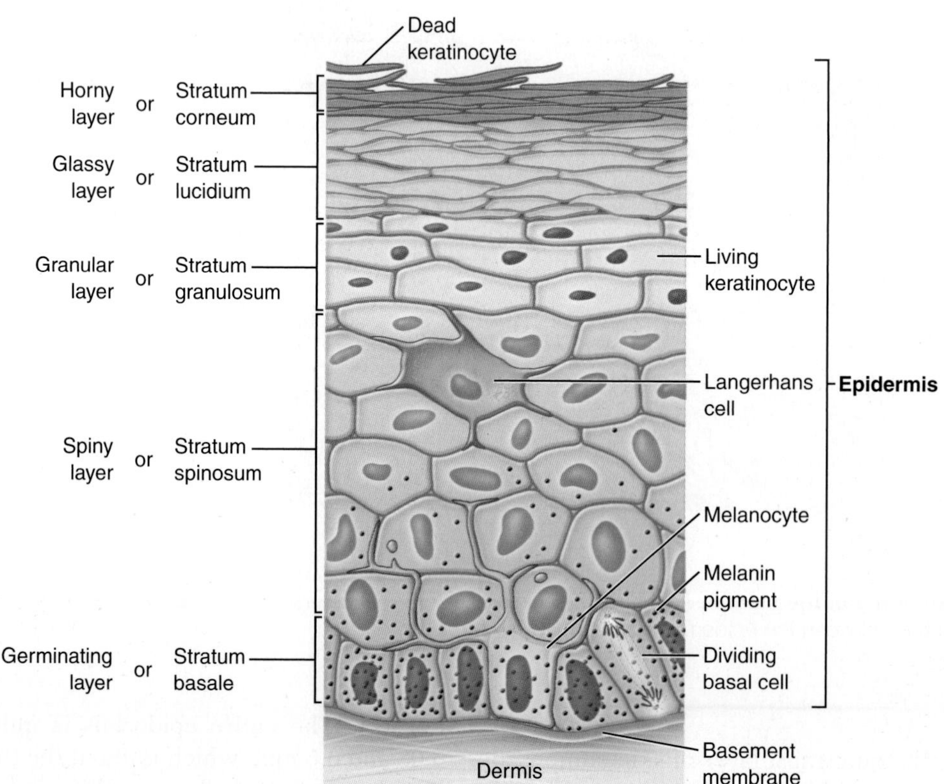

FIGURE 4.5 ▸ Layers of the epidermis. Shown is an area of skin where the epidermis is thick and includes the stratum lucidum. New cells are produced in the germinating layer, the stratum basale. They are keratinized as they rise to the outermost layer.

Approximately 8% of all epidermal cells are specialized cells called **melanocytes**. These cells produce **melanin**, a dark pigment that protects the skin from the damaging effects of UV radiation. While everyone has about the same number of melanocytes, the amount of melanin produced by these cells varies. This difference in melanin levels explains why there are so many different shades of skin color. Sometimes too much melanin is produced in one area of skin, causing pigment spots, or freckles. A mole results when melanocytes are overproduced in a concentrated area.

The epidermis contains thousands of microscopic openings, or **pores**, through which accessory glands secrete oil and sweat onto the surface of the skin. These pores also allow skin to absorb certain substances through the epidermis and into superficial circulation. It is now common practice for some drugs or hormones, such as nicotine and estrogen, to be administered via adhesive patches placed on the skin. However, the rate of absorption through skin is usually quite slow, and superficial irritation of the skin can occur.

Because skin is the outermost covering of the body, it serves as the defining barrier between a person's internal and external environment. Any disruptions in this barrier, whether in the skin of the client or therapist, may allow pathogens to enter, which is why universal precautions and standards of hygiene are of paramount importance in our practices. Even undamaged skin allows substances into the internal environment due to the

Pathology Alert Skin Cancer

The most common type of skin cancer is **basal cell carcinoma (BCC)**, a cancer with slow-growing superficial tumors. Tumors associated with BCC commonly appear as small, clear, or dark brown lumps with rounded edges and a sunken middle. They are usually not dangerous if identified and removed but can encroach upon and threaten surrounding tissues if left untreated. A skin lesion that won't heal may be an indication of BCC. These lesions tend to bleed, crust over, or even fall off but reappear in the same place.

Malignant melanoma is a more serious condition and the most frequent cause of death from skin cancer. Malignant melanoma often starts with a mole that changes color, thickens, or elevates as melanocytes mutate and begin to replicate rapidly. The key warning signs for melanoma, known as the ABCDs, are

- ***Asymmetry***—Moles are usually round or oval, while a melanoma has an irregularly shaped appearance.
- ***Border***—The borders of a melanoma may be indistinct and appear to blend into the surrounding skin.
- ***Color***—Moles are usually a consistent black or brown color. Melanomas are generally multicolored, including a mix of black, brown, and even purple.
- ***Diameter***—Melanomas are large, therefore any mole that is bigger than one-quarter inch (6 mm) in diameter should be checked by a physician.

Another warning sign that is not a consistent indicator of melanoma, but should also be considered, is elevation (sticking out from the skin). Any mole that exhibits these characteristics or changes should be examined by a physician. Because a manual therapist may be one of the few people to regularly view large segments of a client's skin, it is important to be familiar with these signs to help clients identify suspicious skin growths or conditions and seek diagnosis and treatment.

The ABCD's of Malignant Melanoma

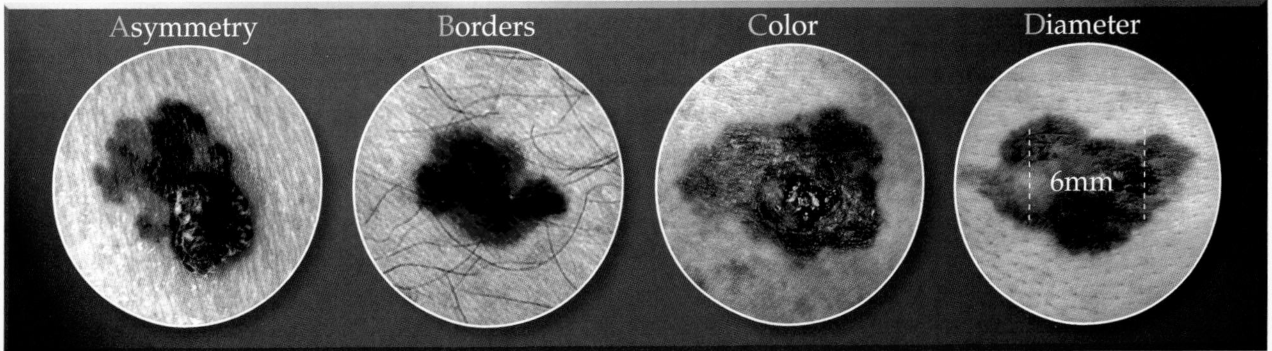

absorbent nature of the cutaneous membrane. Therefore, therapists must exhibit care in their choice of emollients to minimize the risk of allergic responses in clients and prevent the risk of long-term damage to their own skin.

The skin also provides an avenue for the excretion of substances. Therefore, the choice of emollient must also avoid blocking the pores or glands. Additionally, manual therapists who provide spa treatments such as exfoliations, or mineral, mud, or seaweed wraps must pay close attention to the ingredients and purity of the products applied during these treatments. Caution should also be used when hydrotherapy treatments such as sauna or steam are administered because they open the pores, enhancing the skin's absorbency and excretory functions.

Dermis

The **dermis** is the thicker connective tissue layer of skin beneath the epidermis. It houses blood and lymph vessels, nerves, and several accessory organs (▶ FIGURE 4.6). The composition of the dermal layer, rich with collagen and elastin fibers, makes the skin both **extensible** (able to stretch) and **elastic** (able to quickly return to original shape after stretch). The cells present in the dermis are mostly fibroblasts with a few macrophages and mast cells. This layer of skin is divided into a thin superficial zone called the **papillary region**, with a thicker **reticular region** underneath.

- ***Papillary region***—This areolar connective tissue zone made up of thin collagen and elastin fibers gets its name from finger-like projections called **dermal papillae** that extend into the germinating layer of the epidermis. These projections help to increase the surface area of the dermis and attach it firmly to the epidermis. They also contain blood and lymph vessels plus many of the sensory receptors of the skin.
- ***Reticular region***—This deeper, thicker region of the dermis attaches the cutaneous membrane to the fat tissue below. It is disorganized fibrous connective tissue that contains bundles of collagen and most of the dermal fibroblasts. Most of the skin's accessory organs are in this layer.

Subcutaneous Layer

The **subcutaneous layer**, or **hypodermis**, attaches skin to the underlying tissues and organs (FIGURE 4.4). This layer, also known as the **superficial fascia**, is

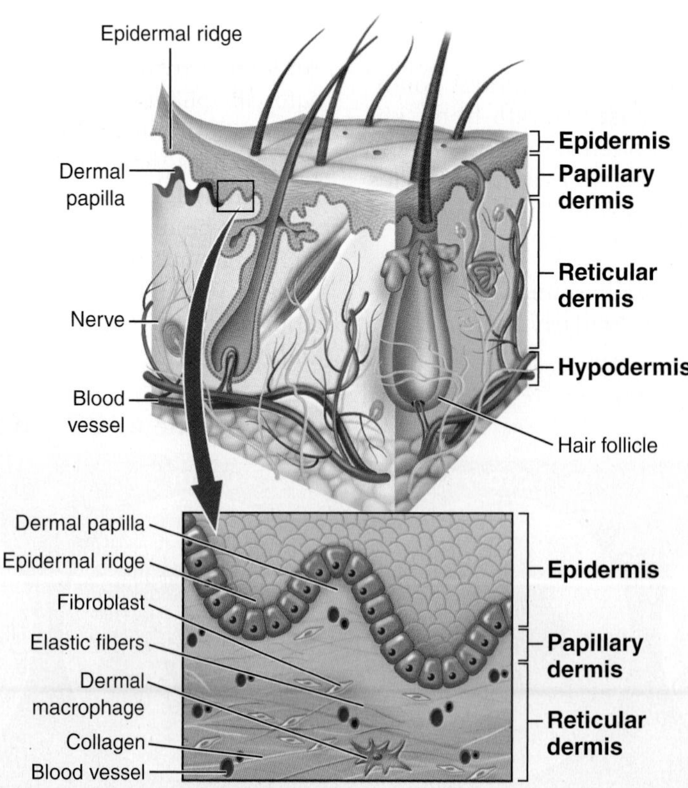

FIGURE 4.6 ▶ Layers of the dermis. The dermis has two layers: the thin outer papillary layer and thicker deep reticular layer. Notice that most of the accessory organs are in the reticular layer.

By the Way

The skin is a billboard of signs and symptoms that can indicate common pathologies. For example, one of the primary indicators of liver and gall bladder disease is a jaundiced (yellow) appearance to the skin. Poor circulation can be identified by skin pallor or a bluish tinge in the nail beds and lips, while imbalances in insulin, estrogen, or testosterone can lead to changes in the skin's healing capacity or the clarity of the complexion.

made of loose areolar connective tissue that contains fat deposits (adipose tissue) that insulate the body and store energy. The hypodermis is also rich with blood and lymph vessels, nerves, and several general sensory receptors.

What Do You Think? 4.2

- Why is it important that the epidermis has so many layers?
- Why do we bleed when we cut ourselves but not when we lightly scrape our knee or knuckles?
- Why do deep cuts or lacerations leave a scar, while superficial scrapes and cuts do not produce scarring?

ACCESSORY ORGANS OF THE SKIN

Most of the accessory organs of the skin are either wholly located within the dermis or arise from that layer and protrude through the epidermal layer. Each square inch of skin contains approximately 100 sebaceous (oil) glands, 700 sudoriferous (sweat) glands, and more than 20,000 sensory receptors.

Hair and Nails

Hair on the head and body and the nails on fingers and toes are accessory organs that serve a protective function. Hair is found on most skin surfaces except on the palms, lips, and the soles of the feet. The heaviest distribution of hair is on the scalp, eyebrows, and around the external genitalia of adults. Genetics, hormone production, and nutrition all influence the pattern of hair distribution over the body.

Each hair is a dead keratinized thread of cells extending from the dermis through the epidermis. The portion of each hair below the skin, the root, is surrounded by a tube called a **follicle** (Figure 4.4). Each follicle is actually a pocket of epithelial cells from the germination (basal) layer of the epidermis. As live germinating cells at the base of the follicle produce new cells, hair grows. Keratinocytes are pushed up through the follicle to create a rod-shaped **hair shaft** that projects above the surface of the skin. Each hair follicle has its own blood vessels that provide its cells with nutrients. As with epidermal cells, old hairs are shed and new hairs grow to replace them unless the follicle has been destroyed.

Hair color is due to the production of melanin from the melanocytes in the follicle. Gray hair indicates a decline in the synthesis of melanin, which is why hair grays with age. Hairs repel or keep substances off the surface of the skin,

Pathology Alert Burns

Burns occur when heat, radiation, chemicals, or electricity damage and destroy the skin or mucous membranes. The severity of a burn is determined by which body areas and how much surface area is involved. Burns are graded by how many layers of skin are damaged.

- ***First-degree burns***—These burns affect the superficial layers of the epidermis. They appear red, hot, mildly inflamed and can be uncomfortable and painful. Sunburn is a good example of this type of mild burn that generally heals within a few days.
- ***Second-degree burns***—All layers of the epidermis are damaged with second-degree burns. They appear red, swollen, and blistered and are quite painful.
- ***Third-degree burns***—In adults, these burns may appear leathery and often show white or black charred edges. In children, they often appear beefy red. Third-degree burns destroy the epidermis and dermis, including the accessory organs and glands. These burns tend to initially be less painful than second-degree burns because the sensory receptors in the dermis are destroyed. However, because the protective covering of the body is breached, there is a high risk of infection and excessive fluid loss.

Several forms of manual therapy may be a helpful adjunct to the rehabilitation of burn patients. Gentle manipulation and movement techniques can help to improve tissue mobility in areas of severe scarring.

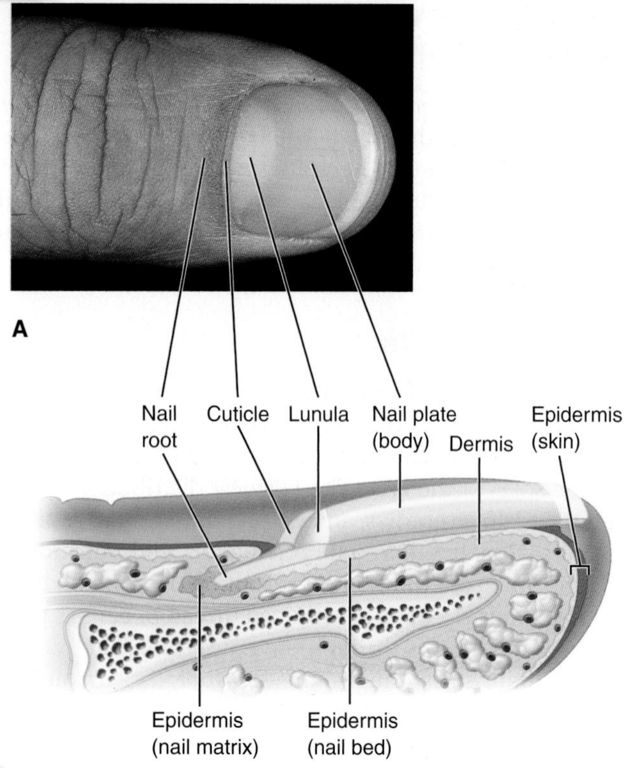

FIGURE 4.7 ▶ **Fingernail. A.** External view. **B.** Anatomy of a nail.

forming a protective barrier for the epidermis. Additionally, a small smooth muscle called the **arrector pili** attached to the side of each hair follicle contracts to raise the hair when we are cold or frightened (Figure 4.4). When hairs are raised by this muscle, it creates a superficial layer of "trapped air" for insulation and also creates goose bumps.

Nails are an extension of the epidermis composed of a densely packed sheet of dead keratinized cells. The ***nail body*** or ***nail plate*** is the portion of the nail that is visible, while the ***nail root*** is the portion of the nail hidden by the ***cuticle***, a specialized fold of the horny layer of epidermis (▶ Figure 4.7). The ***nail bed*** is a thickened region of the horny layer that secures the nail plate to the tips of the fingers and toes. The ***lunula***, a crescent-shaped region next to the cuticle, is lighter in color because the blood vessels of the nail bed below are obscured by the thicker layer of epidermal cells in this region. Functionally, nails protect the ends of fingers and toes and provide the ability to scratch and to grasp small objects. The shape and color of the nails can be a primary indicator of systemic disorders. Nails may turn yellow, become pitted or ridged, or more concave or convex as a result of blood, respiratory, or thyroid disorders.

Glands

Two specialized types of glands originate in the dermis and have structural extensions to carry their secretions to the surface of the epidermis. Formed by glandular epithelium, their secretions play important roles in the protective and excretory functions of the skin.

Sebaceous Glands

Sebaceous glands, or oil glands, secrete an oily or waxy fluid that keeps the skin soft and pliable. Most sebaceous glands are connected to hair follicles; therefore, there are no oil glands on the palms or soles of the feet. **Sebum**, the oily secretion from these glands, keeps hair and the epidermis soft and pliable and inhibits the growth of certain bacteria. Sebaceous gland activity increases during adolescence, which is why acne commonly occurs during the teen years.

Sudoriferous Glands

Sudoriferous glands secrete sweat through the pores of the skin. This process supports homeostasis by helping to regulate body temperature, water balance, and elimination of select metabolic byproducts. Because sweat is 99% water, it is easy to understand the role of sweat in temperature and water regulation. Evaporation of sweat cools the skin; in fact, as much as 1,600 ml (just under a half gallon) of water per hour may be lost during strenuous exercise. Skin is protected from certain types of bacterial growth due to the slightly acidic pH of sweat. This decreased pH is due to trace amounts of urea, uric acid, lactic acid, sodium chloride, and other organic waste products excreted through sweating.

There are two types of sudoriferous glands with differing distribution patterns and secretions. **Eccrine glands** are the primary sweat glands that give the skin its thermal and water regulatory ability. These glands are scattered over the entire surface of the body. The highest concentrations of eccrine glands are found in palms of the hands, soles of feet, forehead, and upper lip. This is why the accumulation of sweat is first felt in these areas. In contrast, **apocrine glands** are concentrated in only a few specific locations; the axilla (armpit), groin, areolar breast tissue, and the beard area on men. Apocrine gland secretions begin at puberty and are very different than sweat from eccrine glands. This secretion is a more viscous milky fluid, with a characteristic odor created by traces of lipids and proteins known as **pheromones**. Bacteria on the skin feed on these organic substances, magnifying the odor of sweat and giving each person a unique scent.

Sensory Receptors

The **sensory** or **cutaneous receptors** in skin include a wide variety of sensory organs that are sensitive to the **general senses**: touch, temperature, pain, vibration, and pressure. ▶ Table 4-2 provides an overview of the sensory receptors and categorizes them by their general sense

Table 4-2 Sensory Receptors of the Skin

General Sense	Specific Sensation	Associated Receptors
Tactile	Touch	Meissner corpuscles Hair root plexuses Merkel discs Ruffini corpuscles
	Pressure	Ruffini corpuscles Pacinian corpuscles
	Vibration	Meissner corpuscles Pacinian corpuscles
	Light touch (itch and tickle)	Free nerve endings
Temperature	Hot	Free nerve endings
	Cold	Free nerve endings
Pain	Acute pain	Nociceptors
	Chronic pain	Nociceptors

categories. These receptors are scattered throughout the papillary and reticular layers of the dermis (FIGURE 4.8).

Three types of receptors located in the papillary zone are sensitive to superficial stimuli. They are

- **Free nerve endings**—These are the most numerous and widespread of the cutaneous receptors. They are sensitive to light touch, light pressure, and temperature, with separate receptors being sensitive to heat or cold. **Nociceptors** or pain receptors are a specialized type of free nerve ending that are highly sensitive to the chemicals released by damaged cells.
- **Merkel discs**—These flat disc-shaped free nerve endings, also called ***tactile discs*** are highly concentrated in the fingertips, lips, and external genitalia. They are sensitive to light touch.

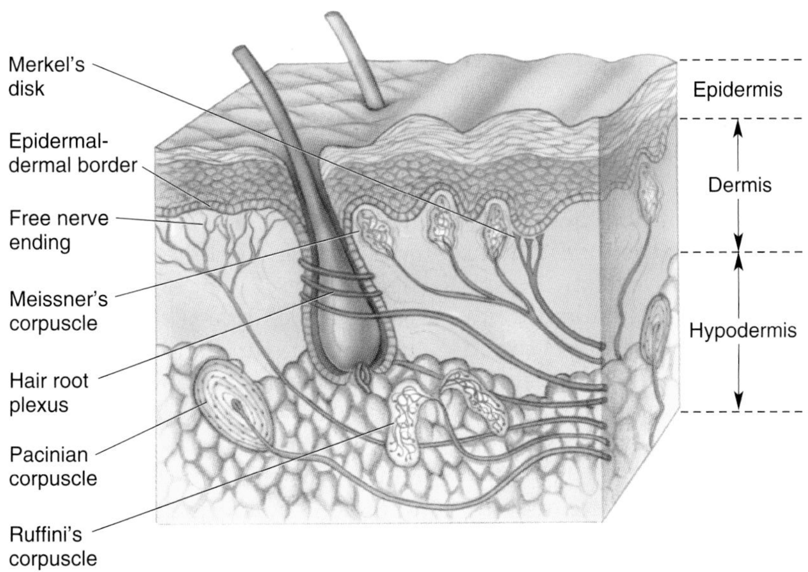

FIGURE 4.8 Cutaneous receptors. Light touch, temperature, and pain receptors are found in the upper dermis, while deep touch and pressure receptors are generally located deeper in the dermis or in the subcutaneous layer.

- **Meissner corpuscles**—Shaped like the end of a cotton swab, these receptors are sensitive to vibration and light touch. Like Merkel discs, they are most concentrated in the fingertips, lips, and external genitalia.

The remaining cutaneous receptors are located in the deeper reticular layer of the dermis and therefore are more sensitive to deep tactile stimulation. These receptors include

- **Pacinian corpuscles**—These egg-shaped receptors are sensitive to high-frequency vibrations and deep pressure. While many are found in the dermis, they are widely distributed throughout the body, most notably in the subcutaneous layer (the superficial fascia), as well as around joints, tendons, and muscles.
- **Ruffini corpuscles**—In the same way that the leaf canopy of a tree branches out from its trunk, the tips of this sensory receptor branch horizontally from a thin fibrous base. They are sensitive to deep touch, pressure, and tissue distortion.
- **Hair root plexus**—These sensory receptors wrap around the hair follicles and are sensitive to any movement of the exposed hair shaft.

LINKING THE INTEGUMENTARY AND NERVOUS SYSTEMS

The cutaneous receptors of the skin are connected to nerve cells or fibers that carry sensory information to the spinal cord and brain, where it is perceived and interpreted. These nerve fibers carry the sensory information from a specific body region to the spinal cord via a nerve. These nerves enter the spinal cord at specific levels. For example, the nerves with sensory fibers from the lower extremities enter the spinal cord at the sacral or lumbar levels, while those from the head and neck enter at the cranial or cervical levels. In this way, each region of skin can be correlated to specific nerves and spinal levels, creating a type of "sensory map" of the nervous system. ▸ Figure 4.9 shows the specific regions of skin, or **dermatomes**, numbered according to the level at which their nerve fibers enter the spinal cord. Because the dermatomes correlate skin sensations to a specific level of the spinal cord, they can be useful for evaluation and treatment. For example, numbness or a loss of sensation within a dermatome means there is an increased likelihood of damage to the associated segment of the spinal cord or to the nerve that carries sensory information to the cord.

A German massage technique developed by Elisabeth Dicke, called ***bindegewebsmassage***, uses knowledge of dermatomes to treat a wide variety of chronic pain and systemic conditions. Bindegewebsmassage strokes are applied in a specific pattern along the dermatomes to create systemic changes in the muscles, organs, or glands that are innervated by the spinal segment (level) for that dermatome. Although bindegewebsmassage is more likely to be known by its English translation, ***Connective Tissue Massage (CTM)***, it should not be considered a myofascial technique. The intention of bindegewebsmassage is to create systemic changes via dermatome stimulation, and the structural effect on connective tissue is of secondary importance.[1]

The intricate and profound association between the integumentary and nervous systems goes beyond the existence of dermatomes. Recall from Chapter 3 that both of these systems develop the ectoderm, the outermost cells of the embryo. In fact, the sensory receptors of the skin form *before* the spinal cord or brain. Through chemical coding that is yet to be well understood, tactile information gathered by the cutaneous receptors of the embryo and developing fetus appears to be the guiding factor for the organization of the nervous system, which creates a neural map of the body on the brain. ▸ Figure 4.10 shows the mapping that correlates areas of the brain with the density of sensory receptors in specific body regions. This type of representation that superimposes a distorted image of a human on the surface of the brain is called a ***homunculus***, which is Latin for "little man." Regardless of how it is illustrated, it is clear that the brain and skin are intricately linked and cannot be functionally separated.

The collective information provided to the brain by the general sensory receptors is essential for the body's healthy development and function. While humans can survive and thrive without the special senses (sight, hearing, taste, or smell), they cannot function fully without the touch and movement information provided to the brain from cutaneous receptors.

The term *failure to thrive* is applied to infants and children whose weight or rate of weight gain is significantly below that of others of the same age and gender. There are multiple medical and environmental causes of failure to thrive. Primary among environmental causes is neglect by parents or caregivers. This neglect usually includes the child not receiving adequate nourishment, which results in nutritional deficiencies and developmental problems. In some cases, the child is also deprived of adequate tactile or movement stimulation.

Research has shown that infants, children, and the elderly can develop extensive physical and psychological dysfunctions when deprived of sufficient tactile and movement stimulus. Studies conducted from the mid-19th century to the present day detail a variety of detrimental effects from long-term deprivation of tactile stimulus. Some of the most common are organ failure, severe depression, anxiety disorders, and high infant mortality rates.[2–6]

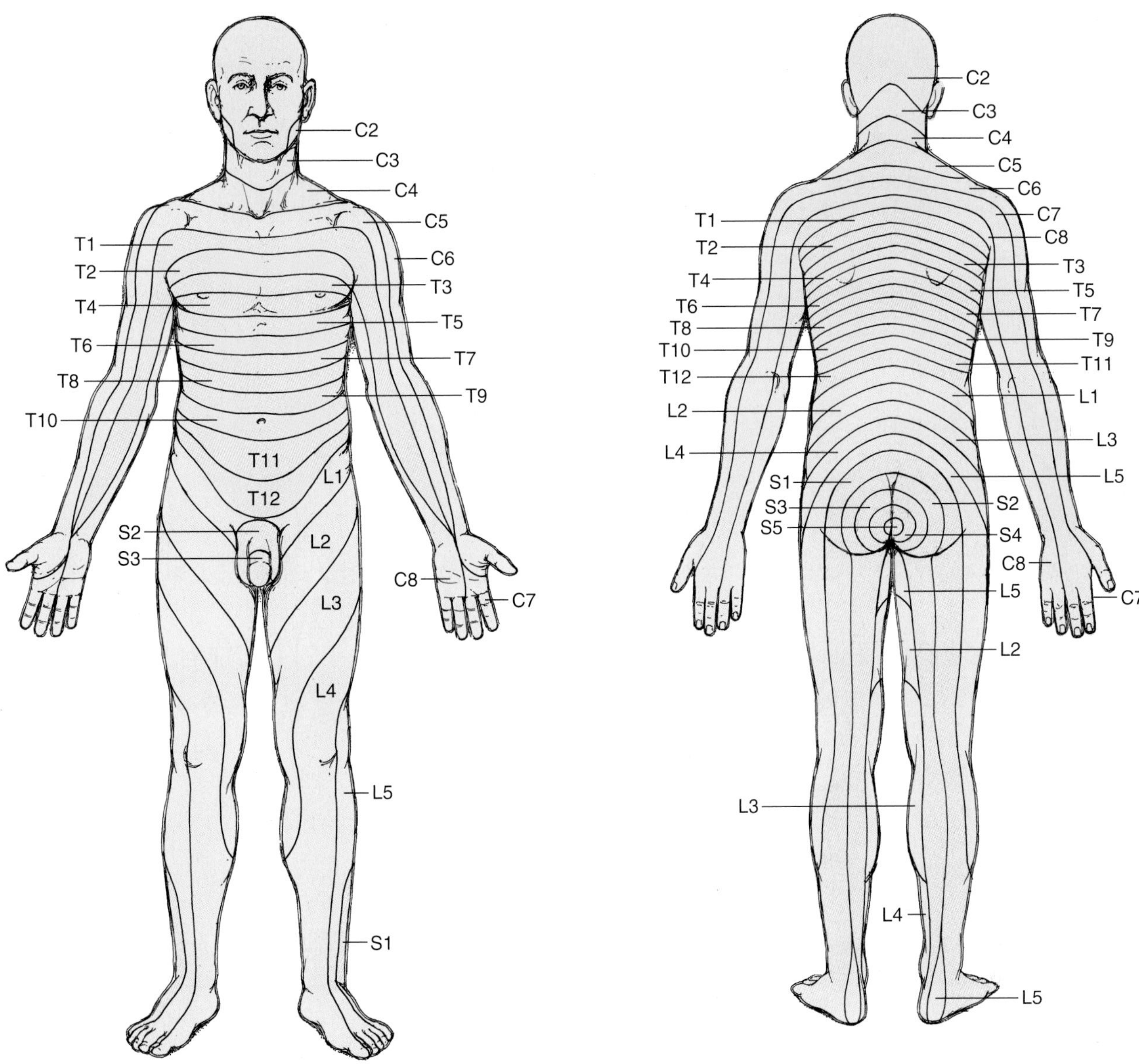

FIGURE 4.9 ▸ Dermatomes. Specific regions of skin are innervated by the branches of a specific spinal nerve. A "sensory map" as shown correlates each area of the skin with its related spinal nerve.

In a series of studies dating from 1986, the profound connection between tactile stimulus, emotional well-being, and physiologic function was demonstrated by Dr. Tiffany Field and colleagues. Their research showed that infants who received regular massage experienced fewer colds, less diarrhea, grew faster, gained more weight, and exhibited less agitation and excitability.[7-9] In a study of violence among adolescents, Dr. Field demonstrated that empathetic behavior increased and aggressive behavior decreased in adolescents receiving regular massage.[10] Although the sample size in this study was small, the suggestion that positive physical touch may play a significant role in moderating aggressive behavior is important to acknowledge.

We can observe examples of the connection between the integumentary and nervous systems every day. Emotions and thoughts become visible when we flush with embarrassment, pale with fear, or get goose bumps of excitement. Babies explore their bodies and their world through tactile stimulus; they touch and pick up everything within reach. Similar types of tactile stimulation can create different thoughts and emotional responses. For example, the pressure of a hand on a shoulder can be a comfort or a warning. It depends on the amount of pressure or grip used, the emotional status of the people involved, as well as the circumstances and environment surrounding them. In the end, whether or not we are consciously aware of the physiological explanations of why touch and movement are so profound, we intuitively acknowledge and experience its impact in our lives all the time. As manual therapists, understanding the anatomic and physiologic links between the integumentary and nervous systems is essential to create and maintain the therapeutic benefits of touch and movement.

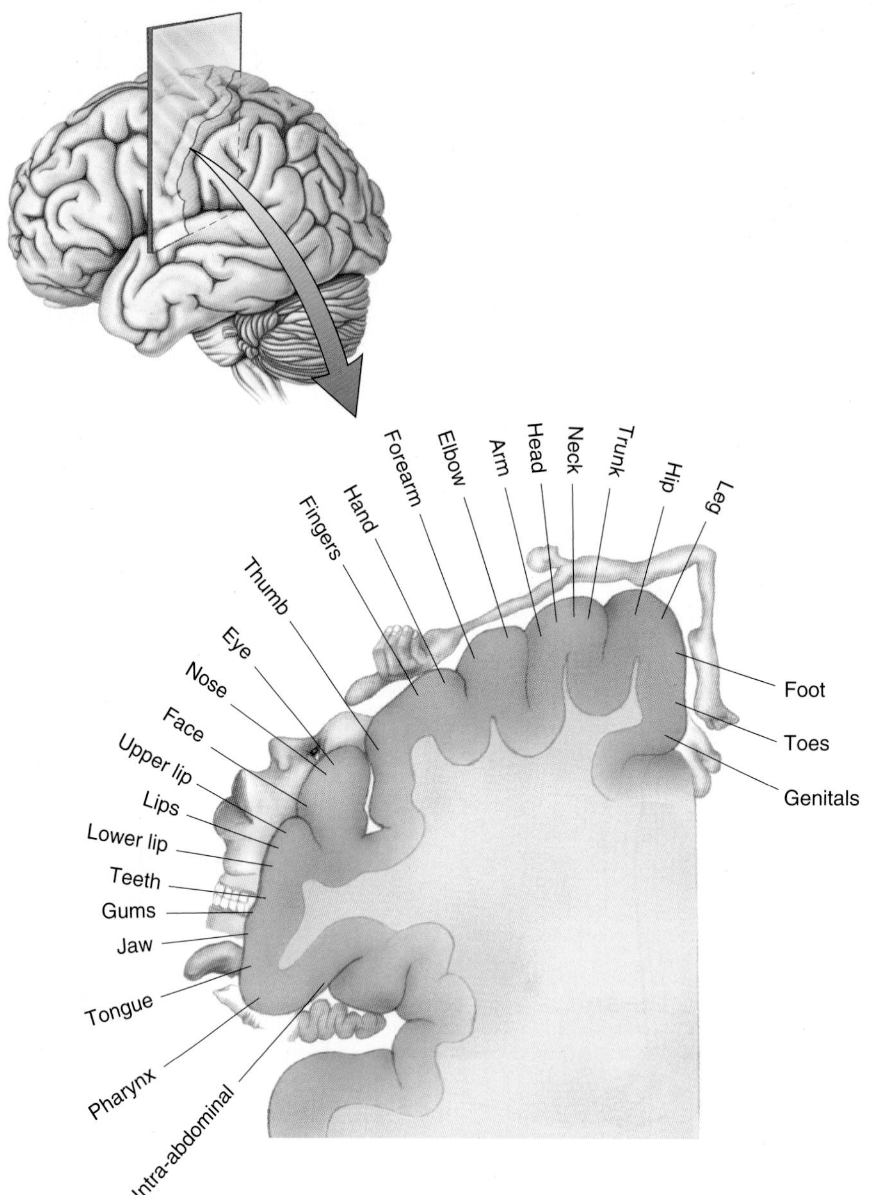

FIGURE 4.10 ▸ Homunculus. Regions of the brain correlate with sensory receptors in specific body areas. The parts of the human figure that appear largest in the illustration represent areas with the highest density of sensory receptors in the skin.

What Do You Think? 4.3

- How do you explain the phenomena of blind people "seeing" using their hands or deaf people "hearing" with their fingers, hands, or feet?
- Provide some everyday examples of how your thoughts and emotions are displayed or projected by your skin.
- What kind of touch, pressure, and temperature would you use to create feelings of fear, compassion, or joy?

SKIN CONDITIONS AND DISEASES

Since many manual therapy techniques involve direct contact between the skin of the client and that of the practitioner, it is important for therapists to be familiar with conditions and diseases of the skin. Therapists may view areas of skin that clients are not able to see on their own and can monitor changes in skin tone or appearance in clients seen on a regular basis. Practitioners must be able to readily identify a wide variety of skin conditions to advise clients and refer them for proper medical attention when needed. Ultimately, therapists should avoid direct contact with any break in the skin, and treat new irregularities or undiagnosed outbreaks with caution. The following sections list and describe examples of skin disorders commonly seen by manual therapists.

Common Contagious Skin Disorders

It can be challenging to differentiate common disorders of the skin. However, if a therapist suspects that a skin condition (a client's or their own) is contagious, it is important to reschedule the appointment after the condition has been diagnosed and properly treated. While proper hygiene, sanitary practices, and universal precautions should be practiced at all times, it is particularly important in cases of contagious skin conditions. Contagious skin disorders include those caused by fungi, bacteria, viruses, or parasites (FIGURE 4.11).

- ***Fungal infections***—These infections are generally referred to as tinea infections or ringworm (FIGURE 4.11A). Several of the most common fungal infections are athlete's foot (***tinea pedis***), jock itch (***tinea cruris***), and ringworm of the head (***tinea capitis***) or body (***tinea corporis***).
- ***Bacterial infections***—The most common bacterial skin infection is due to a staphylococcal (staph) organism. These organisms adapt and mutate rapidly and are related to a wide variety of infections. Those that are frequently seen by manual therapists include boils (FIGURE 4.11B), carbuncles (a group of boils), cellulitis, impetigo, and an antibiotic-resistant

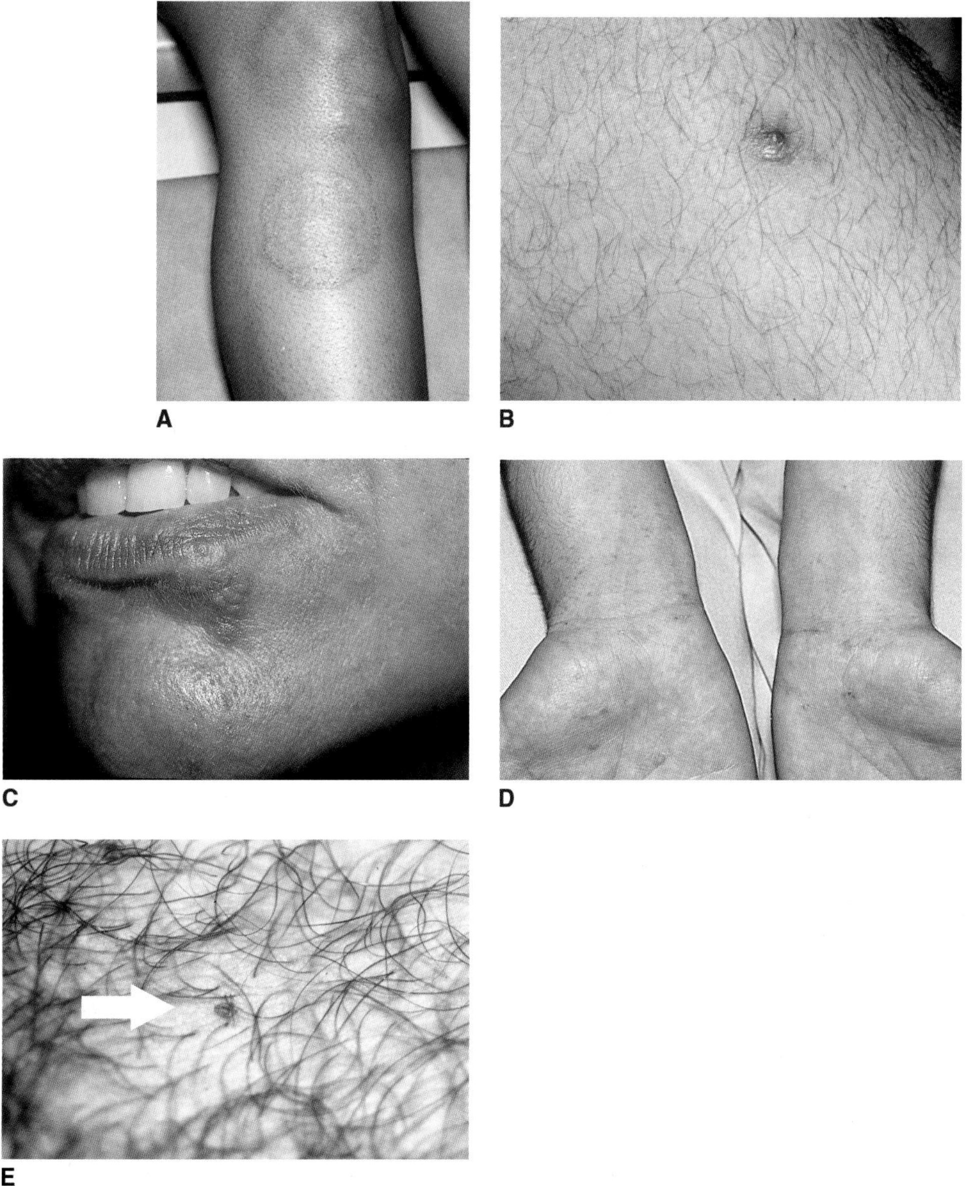

FIGURE 4.11 Contagious skin conditions. A. Ringworm is one of the most common fungal infections of the skin and can occur anywhere on the skin's surface. **B.** A *Staphylococcus aureus* infection can show up as a single boil or a group of boils called a carbuncle. **C.** Cold sores are caused by the herpes simplex virus. **D.** Scabies is a parasitic infection. **E.** Lice are a common parasitic infection found in hair.

super-bug known as MRSA (methicillin-resistant *Staphylococcus aureus*).

- ***Viral infections***—These include the infections that cause warts, mouth sores, and genital lesions. Warts are caused by the human papillomavirus. Oral sores (▶ FIGURE 4.11C) are usually caused by the herpes simplex 1 virus but can be caused by herpes simplex 2. Shingles are caused by herpes zoster, and genital sores are most commonly caused by herpes simplex 2, but can also be caused by herpes simplex 1.
- ***Parasitic infections***—The two most common parasitic infections are scabies (▶ FIGURE 4.11D) and lice (▶ FIGURE 4.11E). Both are extremely contagious, and while improper hygiene can be part of the etiology, they are easily spread by simple contact or sharing of clothes, linens, hair brushes, or combs.

Common Noncontagious Skin Disorders

Many rashes and skin eruptions are not contagious. While there is no concern of spreading the disease, direct contact may still be contraindicated to avoid irritating the condition or introducing a secondary infection through an open lesion. Noncontagious skin conditions include

- ***Acne***—This inflammatory disorder occurs when pores become clogged with oil, dead skin cells, and bacteria. Acne is common during adolescence when hormonal activity is high and is characterized by plugged pores (whiteheads and blackheads, usually on the face) and red papules and pustules on the face, chest, shoulders, and/or upper back.
- ***Eczema***—This disorder can be the result of allergies, hypersensitivities, or immune dysfunction and falls under the umbrella term of ***dermatitis***. It can appear as simple patches of dry flaky skin or more severe yellow oily patches of weeping blisters (▶ FIGURE 4.12A).
- ***Hives***—Also called ***uticaria***, hives are usually a result of stress, anxiety, or an allergic response. They appear as raised red spots that are generally itchy and often hot to touch (▶ FIGURE 4.12B).
- ***Psoriasis***—Characterized by pink or red patches of skin that sometimes have silvery scales on top, psoriasis is caused by an overproduction of new skin cells. It occurs most often on the elbows and knees (▶ FIGURE 4.12C).

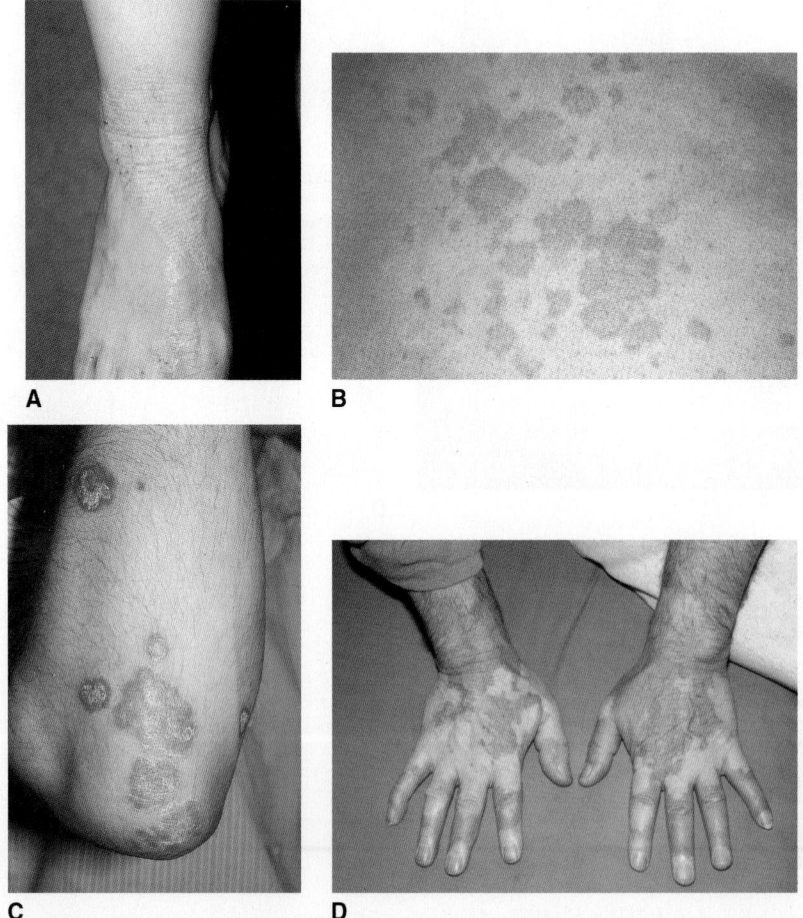

FIGURE 4.12 ▶ Noncontagious skin conditions. A. Eczema is a common form of dermatitis. **B.** An allergic reaction can produce hives. **C.** Even though psoriasis is not contagious, severe outbreaks can cause open lesions that may be painful and susceptible to secondary infection. **D.** Vitiligo is an autoimmune disorder that damages the melanocytes, resulting in smooth white patches of skin.

- *Vitiligo*—This is most often an autoimmune condition in which the body produces antibodies that destroy the melanocytes. It can also be caused by severe burns or wounds that damage the melanocytes. The result is widespread patches of skin without pigment (FIGURE 4.12D).

AGING AND THE INTEGUMENTARY SYSTEM

The integumentary system is one of the systems in which the signs of aging are most easily recognized. As we age, there is a general dehydration and breakdown of the collagen and elastin fibers in the body. As a result, the dermis begins to stiffen, wrinkle, and thin. Decreases in melanin production cause graying hair and changes in skin pigmentation. Some melanocytes increase in size and production, creating brownish age spots. The thinning of the dermal and subcutaneous layers often gives skin a translucent and fragile appearance. With aging, migration of new cells from the epidermis slows, and the skin is less able to regenerate and repair itself. Both men and women tend to experience thinning hair as they age. More advanced baldness is related to aging, heredity, and the amount of testosterone produced by the body.

What Do You Think? 4.4

- What do you think you should do if yesterday's client calls and informs you they have been diagnosed with ringworm?
- What precautions would you take in a manual therapy session with a client who has severe acne on their face, chest, and upper back?
- Based on your understanding of age-related changes in skin, what adjustments might you consider in a manual therapy session for a geriatric client?

SUMMARY OF KEY POINTS

- A membrane is a broad flat sheet of epithelial and connective tissue that serves as a lining or covering. See TABLE 4-1 for a review of the four categories of body membranes, their function, and locations.
- The integumentary system consists of the cutaneous membrane (skin) and accessory organs.
- The functions of skin are protection, temperature regulation, excretion and absorption, acting as a general sensory organ, and synthesis of vitamin D.
- The cutaneous membrane has two primary layers of tissue, the epidermis and dermis.
- The epidermis is the most superficial layer of the skin and is made up of stratified epithelium. This layer continually renews and replaces itself by producing new epithelial cells in its deepest layer (germinating or basal layer) that gradually rise to the outermost surface of the skin (horny layer) and are sloughed off. Specialized cells located within the epidermis include keratinocytes (water resistance and protection), and melanocytes (skin tone/color), and Langerhans cells (immune cells).
- The dermis is the deeper connective tissue layer of skin that contains the blood and lymph vessels, nerves, and accessory organs.
- The hypodermis or subcutaneous layer is a layer of loose areolar and adipose connective tissue that anchors the skin to underlying structures. The hypodermis is also known as the superficial layer of fascia.
- Sebaceous (oil) glands also originate in the dermal layer of skin; they produce an oily substance called sebum that keeps skin and hair soft and pliable.
- There are two types of sudoriferous (sweat) glands in the dermal layer. Eccrine glands that secrete a colorless and odorless fluid are the most numerous and widespread; apocrine glands that secrete a milky colored fluid with a slight musky odor are found only in the axilla (armpit), groin, and areolar breast tissue.
- Skin serves as the body's largest sensory organ because it contains abundant cutaneous sensory receptors. These receptors are sensitive to stimuli such as temperature, light and deep touch, vibration, pressure, and pain, which are classified as general senses.
- The integumentary system has integral links to the nervous system. Both systems develop from the same embryonic layer (ectoderm). Dermatomes are regions of the skin that are innervated by the branches of a particular spinal nerve. Tactile and movement input from the cutaneous receptors is essential for the development of the nervous system and the overall health of the body.
- Manual therapists must be able to recognize a wide variety of skin pathologies, both contagious and noncontagious. Any break or opening in the skin should be considered a precaution to direct touch. Because manual therapists regularly view areas of skin that a client may not be able to see, it is also essential to know the ABCD warning signs of skin cancer.

REVIEW QUESTIONS

Short Answer

1. Define the term *membrane.*
2. Name the four types of body membranes and provide an example of each.
3. List the primary functions of the integumentary system.
4. Name and describe the general location of the two layers of a serous membrane.
5. Name four of the six types of general sensory receptors found in the skin.
6. Describe the embryologic connection between the integumentary and nervous systems.

Multiple Choice

7. The ___________ membrane, plus its associated accessory organs, makes up the integumentary system.
   a. synovial
   b. serous
   c. cutaneous
   d. mucous
8. What is the name of the most superficial layer of skin?
   a. dermis
   b. epidermis
   c. integument
   d. superficial fascia
9. The deepest layer of the epidermis is called the __________ layer because that is where the epithelial cells are produced.
   a. germinating
   b. horny
   c. granular
   d. spiny
10. What is the function of keratin?
    a. lubricate and soften the skin
    b. eliminate heat and metabolic byproducts
    c. toughen and waterproof the skin
    d. protect the skin from UV radiation
11. Which substance in skin determines its color?
    a. keratin
    b. sebum
    c. corneum
    d. melanin
12. Which of the following is a defining characteristic of the epidermis?
    a. contains many accessory organs
    b. composed of numerous layers of epithelial cells
    c. has a rich blood supply
    d. provides energy storage
13. Which layer of skin is primarily connective tissue and contains nerves and blood and lymph vessels?
    a. epidermis
    b. dermis
    c. hypodermis
    d. superficial fascia
14. What is the name of the most superficial layer/region of the dermis?
    a. hyperdermis
    b. stratum corneum
    c. exodermis
    d. papillary
15. What is the role and function of the hypodermis?
    a. insulation and energy storage
    b. protection and sebum production
    c. secretion of metabolic byproducts
    d. synthesis of vitamin D
16. Which of the following glands produce oil to keep skin soft?
    a. sudoriferous
    b. sebaceous
    c. apocrine
    d. eccrine
17. What types of glands are found only in certain areas of the body such as the axilla and groin?
    a. sebaceous
    b. eccrine
    c. apocrine
    d. nociceptors

18. Which of the accessory organs is most responsible for creating the temperature and water regulation function of our skin?
    a. apocrine glands
    b. pacinian corpuscles
    c. eccrine glands
    d. sebaceous glands
19. Which of these specific sensations is properly categorized as a touch or tactile sense?
    a. hot
    b. acute pain
    c. vibration
    d. olfaction
20. What is the term for an area of skin that is innervated by the branches of a specific spinal nerve?
    a. dermatome
    b. myotome
    c. enterotome
    d. neuraltome

Matching

Place a **C** on the line next to the skin conditions that are contagious and an **N** next to those that are noncontagious.

_____21. Eczema

_____22. Psoriasis

_____23. Tinea pedis

_____24. Ringworm

_____25. Acne

_____26. Vitiligo

_____27. Impetigo

_____28. Herpes zoster

_____29. Scabies

_____30. Uticaria

REFERENCES

1. Dicke E, Schliack H, Wolff A. *A manual of Reflexive Therapy of the Connective Tissue: Bindegewebsmassage*. Scarsdale, NY: S.S. Simon; 1978.
2. Holden C. Small refugees suffer the effect of early neglect. *Science*. 1996;274:1076–1077.
3. Hockenberry M, Wilson D, Winkelstein M. *Wong's Essentials of Pediatric Nursing*. St. Louis, MO: Elsevier Mosby; 2005.
4. Carvell GE, Simons DJ. Abnormal tactile experience early in life disrupts active touch. *J Neurosci.* 1996;16(8): 2750–2757.
5. Lin SH, et al. The relationship between length of institutionalization and sensory integration in children adopted from Eastern Europe. *Am J Occup Ther.* 2005;59(2):139–147.
6. Cosgray RE, Hanna V. Physiological causes of depression in the elderly. *Perspect Psychiatr Care*. 1993;29(1): 26–28.
7. Field T. *Touch*. Cambridge, MA: MIT Press; 2001.
8. Field T, et al. Massage therapy by parents improves early growth and development. *Infant Behav Dev.* 2004;27(4):435–442.
9. Field T, Schanberg S, Scafidi F. Tactile/kinesthetic stimulation on pre-term neonates. *Pediatrician*. 1986;77: 654–658.
10. Field T. Violence and touch deprivation in adolescents. *Adolescence*. 2002;37(148):735–749.

5 The Skeletal System

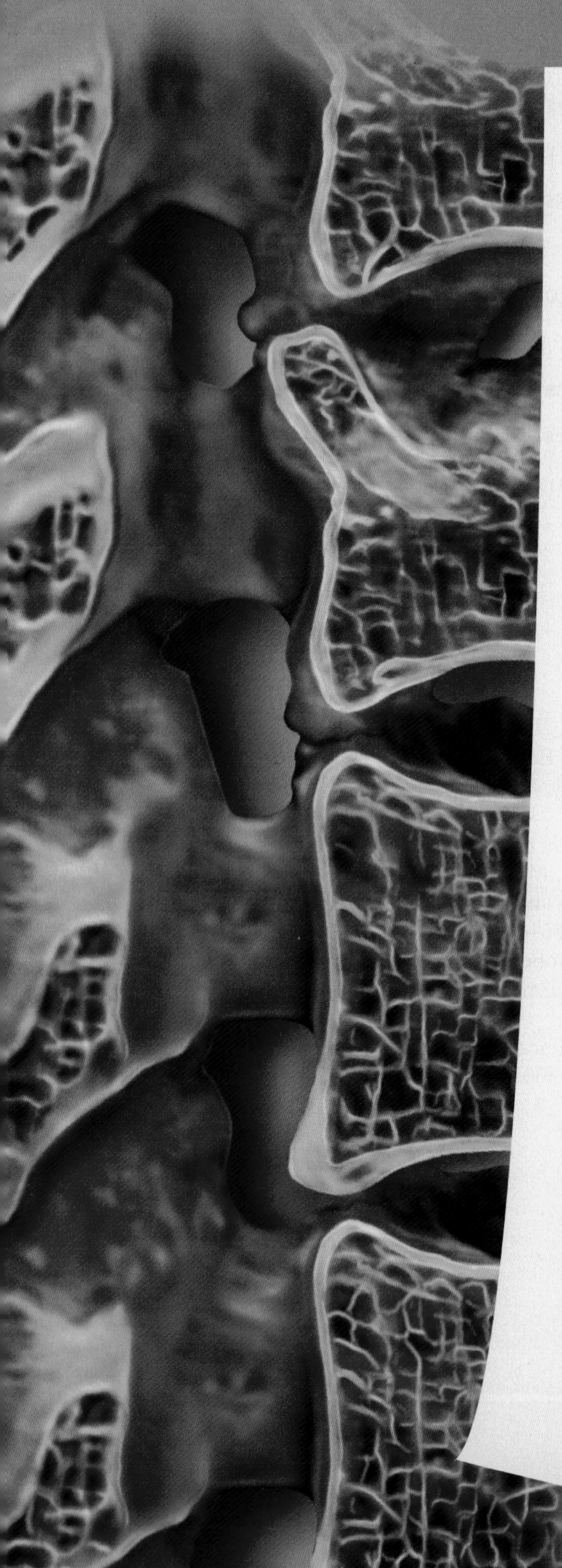

LEARNING OBJECTIVES

Upon completion of this chapter, you will be able to:

1. Provide several examples of how knowledge of bone and joint structures is applied in manual therapy practices.
2. List and explain the primary functions of the skeletal system.
3. Name the primary bones of the skeleton and identify them as axial or appendicular.
4. Name the two types of bone tissue and describe the key features of each.
5. Name and describe each of the four bone classifications according to their shape.
6. Identify the key parts of a long bone and describe the composition and function of each.
7. Name the key bone landmarks of the body and locate them on a diagram.
8. Name and define the three structural and functional categories of joints and provide examples for each.
9. Name the five common structures of all synovial joints and explain the general function of each.
10. List the six types of synovial joints and provide examples for each.
11. List and demonstrate the movement capabilities of each type of synovial joint.
12. Describe common skeletal system changes associated with aging and their implications for the practice of manual therapy.

It is our bones, our structure, which frees us to dance, to make love. Without our structure we would be an imprisoned, amorphous blob of flesh, incapable of response. The amoeba has a minimum of structure, but I doubt if it has much fun."

MADELEINE L'ENGLE
A Circle of Quiet

KEY TERMS

amphiarthrosis (am-fe-ar-THRO-sis)
appendicular (ap-pen-DIK-u-lar)
articular (ar-TIK-u-lar) **cartilage**
axial (AKS-e-al)
cancellous (KAN-sel-us)
condyle (KON-dile)
cortical (KOR-teh-kal)
crest
diaphysis (di-AF-eh-sis)
diarthrosis (di-ar-THRO-sis)
epicondyle (ep-eh-KON-dile)
epiphyseal (eh-pif-eh-SE-al) **plate**
epiphysis (eh-PIF-eh-sis)
facet (FAS-et)
foramen (for-A-men)
fossa (FOS-sah)
fovea (FO-ve-ah)
haversian (hah-VER-zhen) **system**
marrow (MARE-o)
meatus (me-A-tus)
medullary (MED-u-lar-e) **cavity**
osteon (OS-te-on)
periosteum (pair-e-OS-te-um)
process (PRAH-ses)
skeleton (SKEL-eh-ton)
sinus (SI-nus)
synarthrosis (SIN-ar-THRO-sis)
synovial (sin-O-ve-al)
trochanter (TRO-kan-ter)
tubercle (TU-ber-kul)
tuberosity (tu-ber-OS-eh-te)

Primary System Components

- **Bones**
  - Axial bones
  - Appendicular bones
- **Joints**
  - Hyaline cartilage
  - Fibrocartilage
  - Joint capsule
  - Ligament
  - Bursa
  - Synovial membrane

Primary System Functions

- **Acts as a supportive frame for the body**
- **Protects vital organs**
- **Provides the levers (bones) and fulcrum points (joints) that allow movement when muscles contract**
- **Stores calcium and other minerals in bone tissue matrix**
- **Produces blood cells in the bone marrow**

The term structure can be challenging. It can conjure up images of a rigid collection of rules that seem designed to restrict individuality and confound freedom of expression. On the other hand, structure can remind us of what is secure, comforting, and reassuring by defining roles in relationships and outlining procedural guidelines and agreements to provide a sense of pattern and order in our lives. For a structure to be effective, it cannot represent either of these extremes. It cannot be so rigid that it is unyielding or brittle, nor can it be so pliable that it is amorphous or discretionary. Instead, like the human skeleton, it must be strong and firm yet resilient and somewhat flexible, to provide a definitive frame that moves and responds to the challenges of life.

For many manual therapists, postural and movement assessments are a significant part of the objective information gathering that informs their therapeutic choices. Without understanding normal structural relationships between body parts, recognizing structural imbalances would be impossible. Bone landmarks provide important reference points for these evaluations. Additionally, the soft tissues cannot be understood without being familiar with the structural frame to which they attach. In this chapter, you will learn the names, locations, and composition of the bones and joints that make up the human skeleton. You will explore how these structures work together to elegantly support, protect, and allow freedom of movement. Additionally, you will learn how bones serve as dynamic repositories for minerals such as calcium and phosphorous and as production centers for the blood cells.

FUNCTIONS OF THE SKELETAL SYSTEM

While bones are often thought of as inert rigid beams, they are also living organs, a combination of different types of tissue working together as the primary organs of the skeletal system. Bones and cartilage, a type of connective tissue, collectively form the skeleton, the internal scaffolding that supports the body. While support and protection are easily recognized as functions of this system, there are other important functions that occur in specialized tissue within the bones that are not so apparent. The functions of the skeletal system are

- ***Framework and support***—The skeleton provides an internal framework that anchors the soft tissues and organs to provide both form and structural support for the body. The shape and individual features of each person's face and head are primarily due to bone structure. This is why forensic pathologists can accurately reconstruct an individual's face from the skull. The weight of the torso is supported through the alignment of the weight-bearing bones in the lower extremities and spine.
- ***Protection***—This rigid framework provides protection for the internal organs in both the ventral and dorsal cavities. Additionally, bones provide protective channels and openings for the eyes, ears, nose, blood vessels, and nerves.
- ***Levers and fulcrums***—The bones and joints of the skeleton serve as the levers and fulcrum points needed for human locomotion. It is the muscular system that generates the power for movement; however, without both systems working together, no movement would be possible.
- ***Mineral storage***—Bones serve as the primary storage site for calcium, phosphorus, and a few other minerals. This calcium storage in the bone matrix plays a major role in giving bones their rigidity and strength.
- ***Blood cell production***—The process of blood cell formation, or **hematopoiesis,** is carried out by a soft connective tissue mass inside the bones called red **marrow**. ***Red bone marrow*** produces all three types of blood cells; the process is discussed in detail in Chapter 10.

THE SKELETON

The 206 bones of the **skeleton** are organized into two divisions based on their location (FIGURE 5.1). Since the head and torso form the core or ***axis*** of the body, this region makes up the **axial** skeleton, while the bones in the arms and legs, or ***appendages***, are called the **appendicular** skeleton. As seen in FIGURE 5.1, there are a few bones located in the torso region of the body—scapula, clavicle, and pelvic bones—that actually *attach* or *append* the arms and legs to the torso, and as such, they are classified as appendicular. TABLE 5-1 lists the bones of the axial and appendicular skeleton with their pronunciations.

Axial Skeleton

The axial skeleton includes the bones in the skull, spine, and rib cage, including the cartilage that attaches the ribs to the breast bone. The axial bones are

- ***Cranial and facial***—Bones in the skull are divided into two regions: the 8 cranial bones that make up the round superior portion of the skull and the 14 facial bones.
- ***Hyoid***—This small U-shaped bone in the anterior neck serves as support for the tongue and is the attachment point for several throat muscles essential for swallowing.
- ***Ossicles***—These three tiny bones inside each ear play a vital role in hearing by transmitting the vibration of sound waves.
- ***Vertebrae***—The spinal column consists of 33 individual ***vertebra*** with fibrocartilage pads between each. There are 7 cervical, 12 thoracic, 5 lumbar, 5 sacral, and 3 or 4 ***coccygeal*** vertebrae.

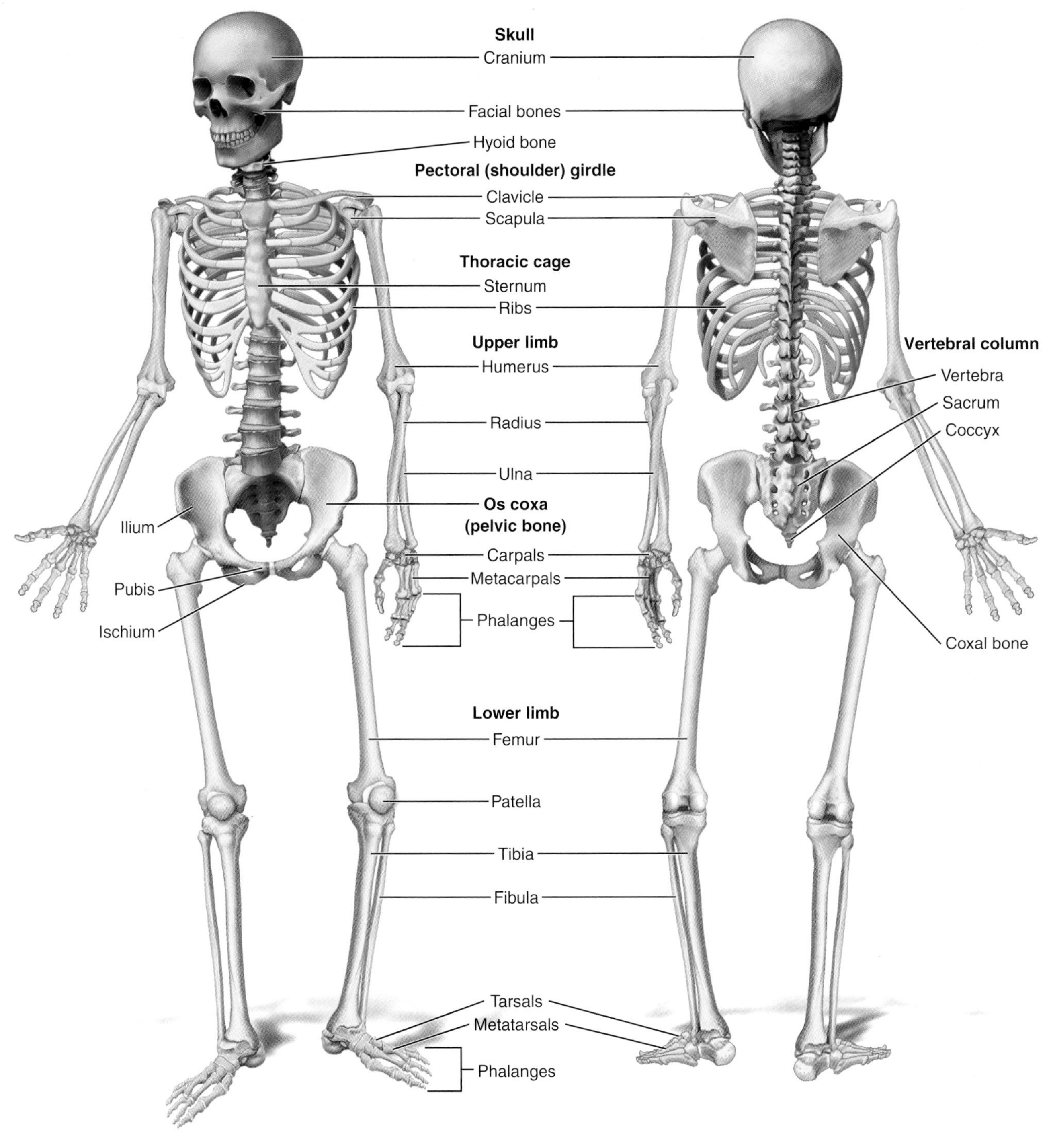

FIGURE 5.1 ▸ Axial and appendicular skeleton. Axial bones are shown in green, and the appendicular bones are shown in light tan. **A.** Anterior view. **B.** Posterior view.

- ***Rib cage***—The ***sternum***, or breastbone, plus 12 pairs of ***costal*** (rib) bones attached posteriorly to the vertebrae, make up the rib cage. Most but not all of the ribs also attach via fibrocartilage to the sternum to complete the protective enclosure for the lungs and heart.

Appendicular Skeleton

The remaining bones of the skeleton are considered to be appendicular, even though those in the pectoral and pelvic girdle may appear to be part of the torso. The appendicular bones are

Table 5-1 Bones of the Axial and Appendicular Skeleton

Division	Body Region	Bone Name and Number	Pronunciation
Axial	**Skull: cranial**	Frontal (1) Parietal (2) Occipital (1) Temporal (2) Ethmoid (1) Sphenoid (1)	FRON-tal pah-RI-eh-tal ok-SIP-eh-tal TEM-por-al ETH-moid SFE-noid
	Skull: facial	Nasal Zygomatic Maxilla Mandible Lacrimal Vomer Palatine	NA-zal zi-go-MAH-tik MAKS-il-ah MAN-dib-el LAH-kreh-mal VO-mer PAL-ah-tine
	Anterior neck	Hyoid	HI-oyd
	Spine	Vertebra (33)	VER-teh-brah
	Thoracic/rib cage	Sternum Rib (12 pairs)	STER-num
Appendicular	**Pectoral girdle**	Clavicle Scapula	KLAV-ah-kel SKAP-u-lah
	Upper extremity	Humerus Ulna Radius Carpal (8 per wrist) Metacarpal (5 per hand) Phalanges (14 per hand)	HU-mer-us UL-nah RA-de-us KAR-pel meh-tah-KAR-pel fah-LAN-geez
	Pelvic girdle	Ilium (2) Ischium (2) Pubis (2)	IL-e-um ISH-e-um PU-bis
	Lower extremity	Femur Tibia Fibula Patella Tarsal (7 per foot) Metatarsal (5 per foot) Phalanges (14 per foot)	FE-mer TIB-e-ah FIB-u-lah pah-TEL-ah TAR-sel meh-tah-TAR-sel fah-LAN-geez

- ***Pectoral girdle***—The ***clavicle***, or collar bone, and ***scapula***, or shoulder blade, form the bone arrangement that attaches the upper extremity to the axial skeleton. The clavicle is the anterior bone of the pectoral girdle. Its medial tip attaches to the sternum, and its lateral end attaches to the scapula, the posterior component of the pectoral girdle.
- ***Upper extremity***—The most proximal bone of the upper extremity is the ***humerus***. There are two bones in the forearm: the ***radius*** on the lateral side and the ***ulna*** on the medial. There are eight wrist bones, called ***carpals***, while the bones in the palms are ***metacarpals***, and finger bones are ***phalanges***.
- ***Pelvic girdle***—There are two ***coxal*** (hip) bones called ***os coxae*** that attach to each other anteriorly and with the sacrum posteriorly to form the pelvic girdle. Each os coxa is made up of three different bones: ilium, ischium, and pubis. The ***ilium*** is the wide, wing-shaped bone just below the waistline; the ***ischium*** is the looped bone in the buttocks on which we sit, and the ***pubic*** bones are the anterior connecting point for the two sides of the pelvic girdle.

- ***Lower extremity***—The largest and most proximal bone in the lower extremity is the ***femur***, the long bone of the thigh. The medial bone of the leg is the ***tibia***, and the thinner non-weight-bearing bone on the lateral side of the leg is the ***fibula***. The ***patella*** is a smaller rounded bone in front of the joint between the femur and tibia. The ankle bones are the ***tarsals***, the foot bones are ***metatarsals***, and like the fingers, the toes are ***phalanges***.

By the Way

Remember your word parts? *Osteo-* or *os-* refers to bone, while the suffix *-blast* means germ or bud. The suffix *-cyte* means cell, *-clast* means to break or destroy, and *-genic* means to produce.

BONE TISSUE AND STRUCTURE

Although the bones may seem to be simple structural beams, they are actually quite dynamic organs. Bones are capable of growth and repair, and consistently undergo a process of self-destruction and reformation. This natural cycle, called **remodeling,** has two parts: resorption and deposition. Bone **resorption** is the breakdown of bone tissue, and **deposition** is the building up of new bone tissue. Both resorption and deposition are on-going processes; dynamic adaptations to stress and strain put on the bones by growth and development, exercise, diet, injury, and aging.

Bone remodeling serves as an important homeostatic mechanism to balance the levels of calcium and phosphorus ions in the blood. When blood levels of these ions are low, bone tissue is resorbed, and when their blood levels are high, bone deposition occurs. When bones are fractured, resorption is inhibited and bone deposition occurs at the site of injury. However, a balance between bone resorption and deposition is essential. If too much bone is resorbed, bones can become weak and brittle as in osteoporosis. When too much bone is deposited in an area, bone spurs can occur. The homeostatic balance between resorption and deposition is controlled and regulated by hormones from the endocrine system (discussed in detail in Chapter 9).

Bone Tissue

Like all connective tissues, bone or **osseous** tissue is made of cells, fibers, and ground substance. The composition of the matrix (fibers plus ground substance) makes bone hard as well as resistant to stretching and tearing, making it the perfect tissue from which to fashion the structural timber of the skeleton. About 50% of the matrix is equally composed of ground substance and collagen fibers arranged to form a structural framework for the bone tissue. Calcium and phosphorous mineral salts deposit, crystallize, and harden, or **calcify**, within the collagen frame to complete the other 50% of the tissue. These mineral deposits make bone hard, while the collagen fibers make bone resistant to tension, torque, and shearing forces.

There are four types of cells found in bone tissue:

- **Osteoblasts**—These cells, sometimes called "bone builders," make and secrete collagen fibers and ground substance to create the framework for all bone tissue. As they secrete more and more matrix, they become trapped in it and mature into osteocytes. Osteoblasts are the cells responsible for bone deposition.
- **Osteocytes**—These are mature bone cells responsible for the continual exchange of nutrients and wastes that maintains bone as living tissue.
- **Osteoclasts**—These "bone breakers" are large cells that make and secrete acids and other strong lysosomal enzymes to break down bone. Their secretions play a primary role in bone matrix resorption to maintain a normal growth cycle.
- **Osteogenic cells**—Unlike other bone cells, osteogenic cells are the only bone cells capable of mitosis. The new cells they create develop into osteoblasts. A thin layer of these cells resides in the outer covering of each bone.

While bone tissue may seem to be completely solid, there are actually many spaces between the cells and matrix. Differences in the size, arrangement, and distribution of these spaces give rise to two different types of bone tissue: compact (dense) bone and spongy bone (▸ FIGURE 5.2).

Compact Bone

Compact bone is dense, with a hard matrix making it very resistant to the stresses of body weight and movement. Also referred to as **cortical** bone, it forms the outer layer of all bones and makes up most of the shaft of long bones such as the femur, tibia, and humerus.

Small units called **osteons** or **haversian systems** are repeated throughout compact bone like logs strapped together. Each osteon has concentric rings or **lamellae** arranged around the central or **haversian canal**, a channel through which blood vessels, lymphatic vessels, and nerves travel. Osteocytes are located in small spaces around the lamellae called **lacunae.** Like the city of Venice, known for its extensive network of canals, these lacunae are connected to one another and to the central canal through an intricate network of tiny channels, or **canaliculi,** which allow for the passage of nutrients and wastes to and from the osteocytes through the mineralized bone matrix. Transverse canals called **perforating** or **Volkmann's canals** push through from the surface of the bone and connect vessels and nerves

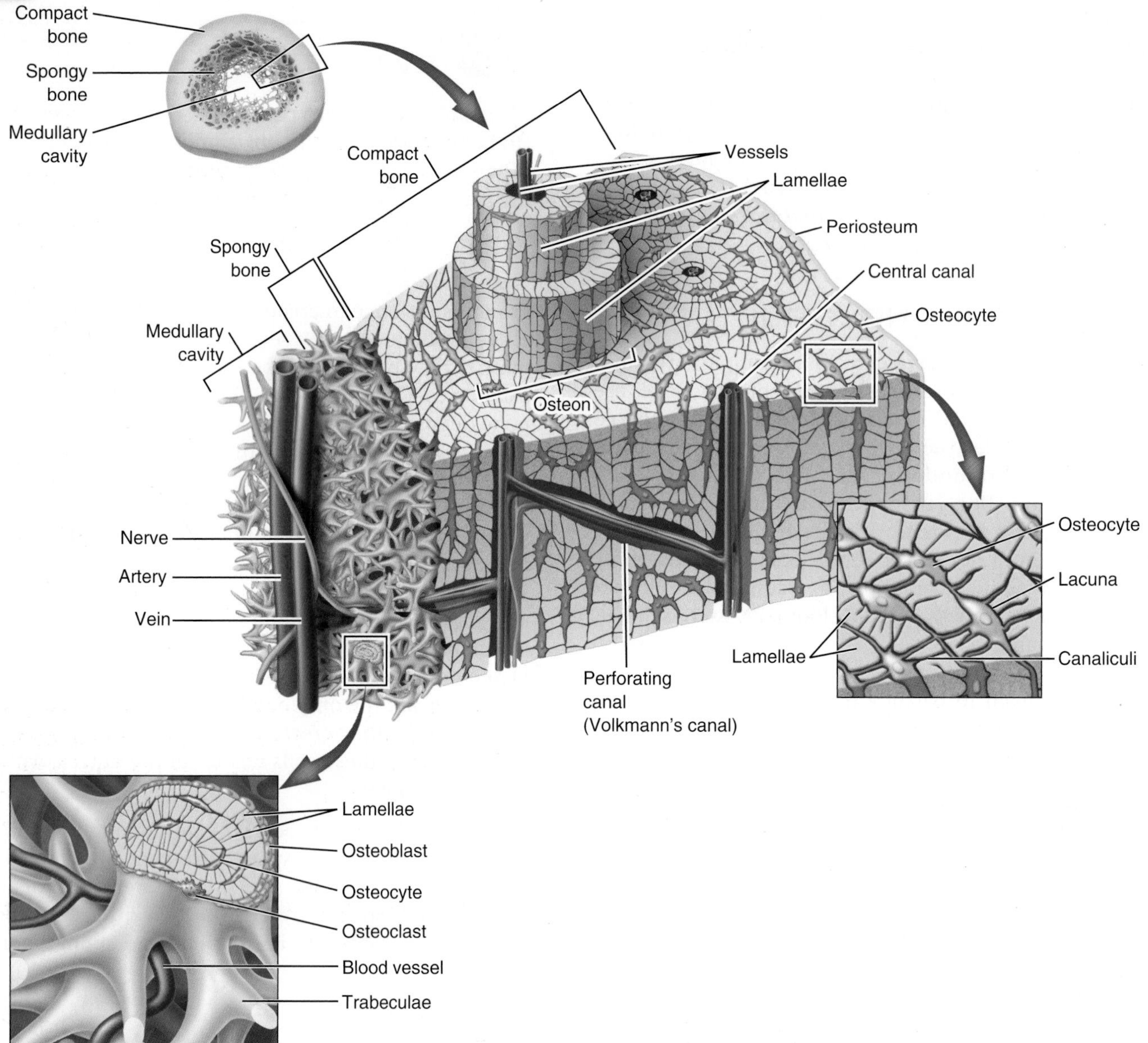

FIGURE 5.2 Bone tissue. Compact bone is organized into units called haversian systems or osteons. Spongy bone is characterized by a lattice-like framework of bone structures called trabeculae.

from outside the bone to the vessels and nerves within the central canals.

Spongy Bone

Spongy bone tissue actually looks like a sponge. It is also called **cancellous bone**, referring to its lattice-like appearance. Spongy bone reduces the weight of individual bones and the skeleton as a whole. This type of bone tissue is found at the ends of long bones and throughout the ribs, sternum, vertebral column, and hip bones. Spongy bone is made up of small structural beams called **trabeculae**. Osteocytes contained in lacunae reside within the trabeculae and, similar to cortical bone, the lacunae are connected by canaliculi. Red bone marrow fills the spaces of spongy bone tissue.

Bone Classifications by Shape

Bones come in many different sizes and shapes. However, most bones fall into one of four basic types according to their shape (FIGURE 5.3):

- ***Long bones***—These bones are found in the appendicular skeleton and are longer than they are wide. They have distinct ends and a long central shaft. The femur, tibia, fibula, humerus, ulna, radius, metatarsals, metacarpals, and phalanges of the feet and hands are long bones.
- ***Short bones***—Sometimes called ***cuboid*** bones, these bones are squat and cube shaped. The carpals of the wrist and tarsals of the ankle are short bones.

Pathology Alert Osteoporosis

Osteoporosis, literally "porous bones," occurs when bone resorption outpaces bone deposition. This causes a decrease in the mineral component of the bone matrix, leading to a decrease in overall bone density. Because the lack of mineralization affects spongy bone more than cortical bone, the incidence of fractures to the head of the femur and vertebral bodies greatly increase in people with osteoporosis. Leading causes of osteoporosis include endocrine imbalances and challenges with calcium metabolism or absorption. Although it is most common in the elderly, osteoporosis can occur anytime in adulthood. Manual therapists must use extreme caution when working with clients with osteoporosis. Clients may require frequent adjustments in position to maintain comfort, and any lifting or moving of body parts must be done with great care. Additionally, deep pressure or compressive techniques, especially in the back and chest, are contraindicated.

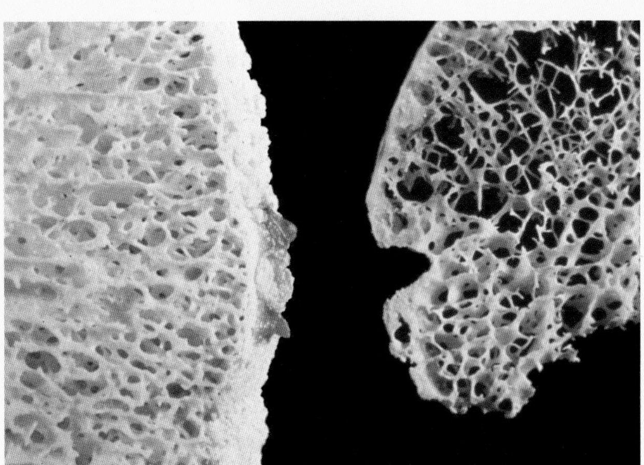

- ***Flat bones***—These are thin bones that protect the vital organs and provide large surface areas for muscle attachments. The bones of the skull, pelvic girdle, sternum, ribs, and scapula are flat bones.
- ***Irregular bones***—Bones like the vertebrae and some of the facial bones don't fit any of the other categories, and are simply called irregular. Included in this category are ***sesamoid*** bones such as the patella, which are embedded within tendons.

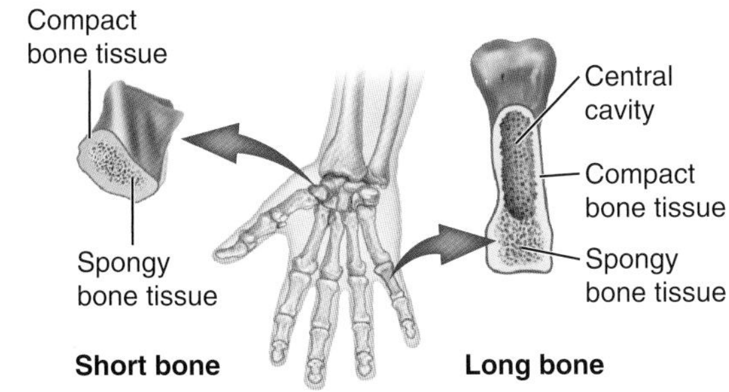

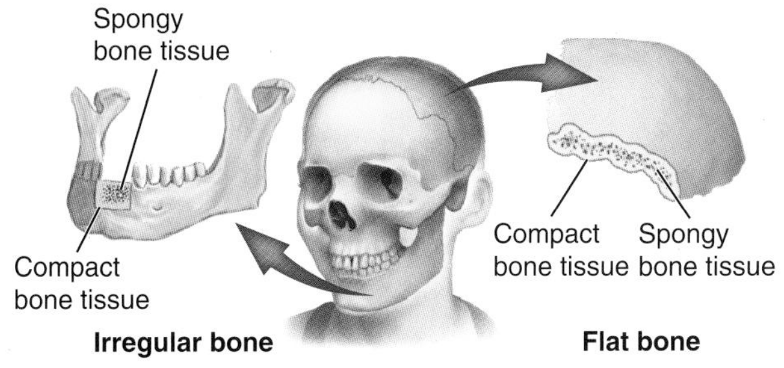

FIGURE 5.3 Bone shapes. Bones are classified into four categories based on their shape.

Anatomy of a Bone

Bones are composed of several types of connective tissue: bone, cartilage, and disorganized fibrous connective tissue. It is helpful to examine a typical long bone to identify the location and explain the unique contribution of each of these tissues. The parts of a long bone (as shown in FIGURE 5.4) are

- **Diaphysis**—The long shaft or main body of the long bone that is composed of compact bone tissue.
- **Medullary cavity**—The hollow, cylindrical space within the diaphysis, which in adults is filled with ***yellow bone marrow***. Yellow marrow is fatty connective tissue, and unlike red marrow, it plays no role in blood cell production.
- **Epiphysis**—Each long bone has two ends, or epiphyses, made primarily of spongy bone.
- **Metaphysis—**This flared area where the diaphysis joins the epiphysis is the location of the epiphyseal plate during the years of bone growth.
- **Epiphyseal plate**—Also called the growth plate, this cartilaginous region between the epiphyses and diaphysis is where bone growth occurs in children and adolescents. Bone growth occurs when cartilage cells, or **chondrocytes**, on the epiphyseal side of this plate divide and create a new layer of cells (FIGURE 5.5). Older chondrocytes are pushed toward the diaphyseal side of the plate, where they are replaced by bone, thus lengthening the diaphysis. As we reach adulthood, hormones signal the cells in this region to stop producing new cartilage cells and matrix, and the area eventually completely calcifies and growth stops. In mature bone, this calcified area is called the ***epiphyseal line***.

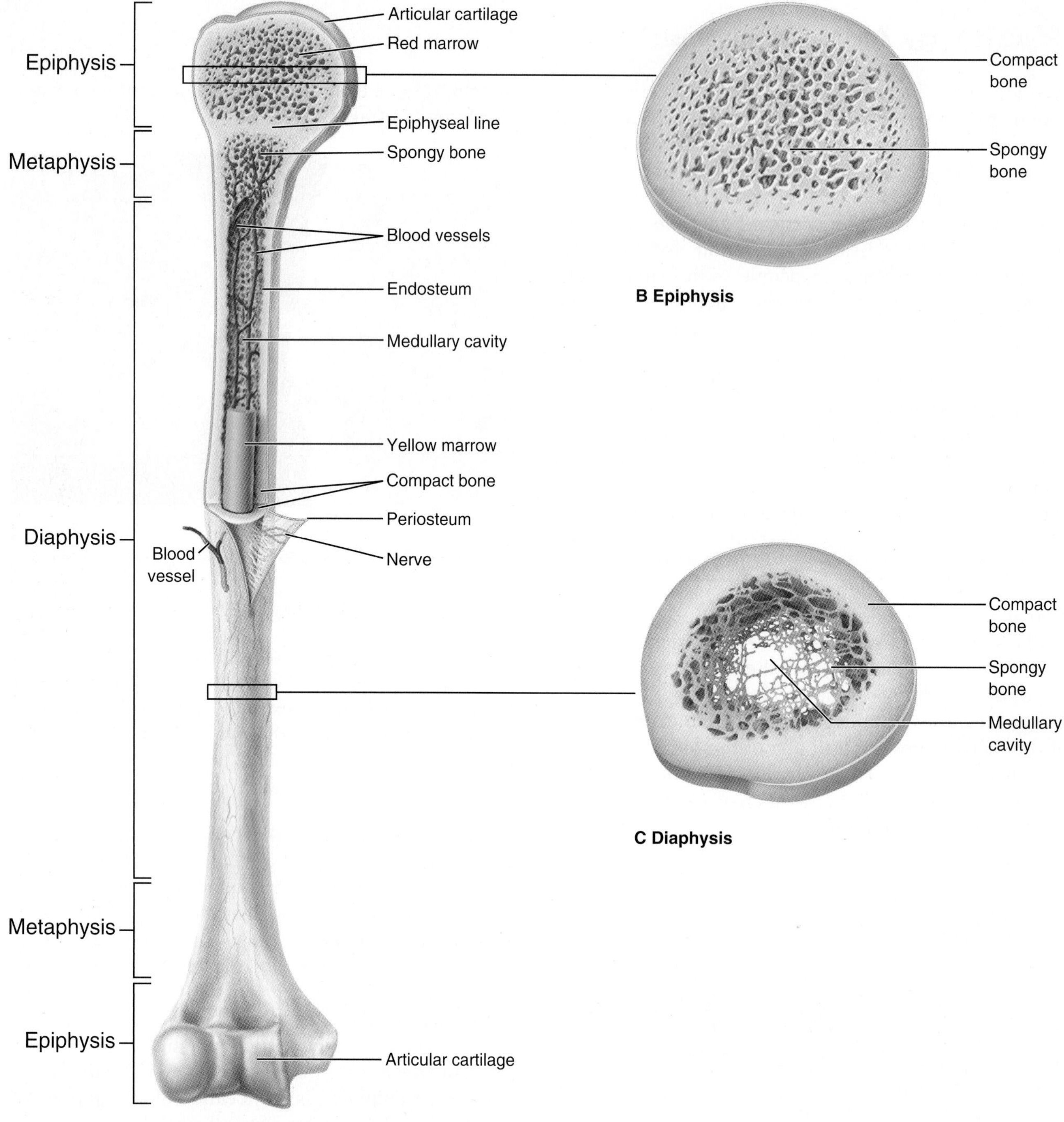

FIGURE 5.4 **Anatomy of a long bone.** The parts of a long bone are shown, along with the types of tissue located in its various regions.

- **Articular cartilage**—Each bone end is covered with a thin layer of **hyaline cartilage**. This cartilage decreases wear and tear to the bone caused by bearing weight and the friction from movement.
- **Periosteum**—This disorganized fibrous connective tissue sheet covers the outer surfaces of the bone that are not covered with articular cartilage. The periosteum serves as the attachment point for ligaments and tendons. It has a rich blood, lymph, and nerve supply. The inner lining of periosteum contains osteogenic cells that produce osteoblasts, which create new bone tissue to thicken the shaft during bone growth.
- **Endosteum**—This is a thin membrane lining the medullary cavity. It contains a single layer of osteoclasts.

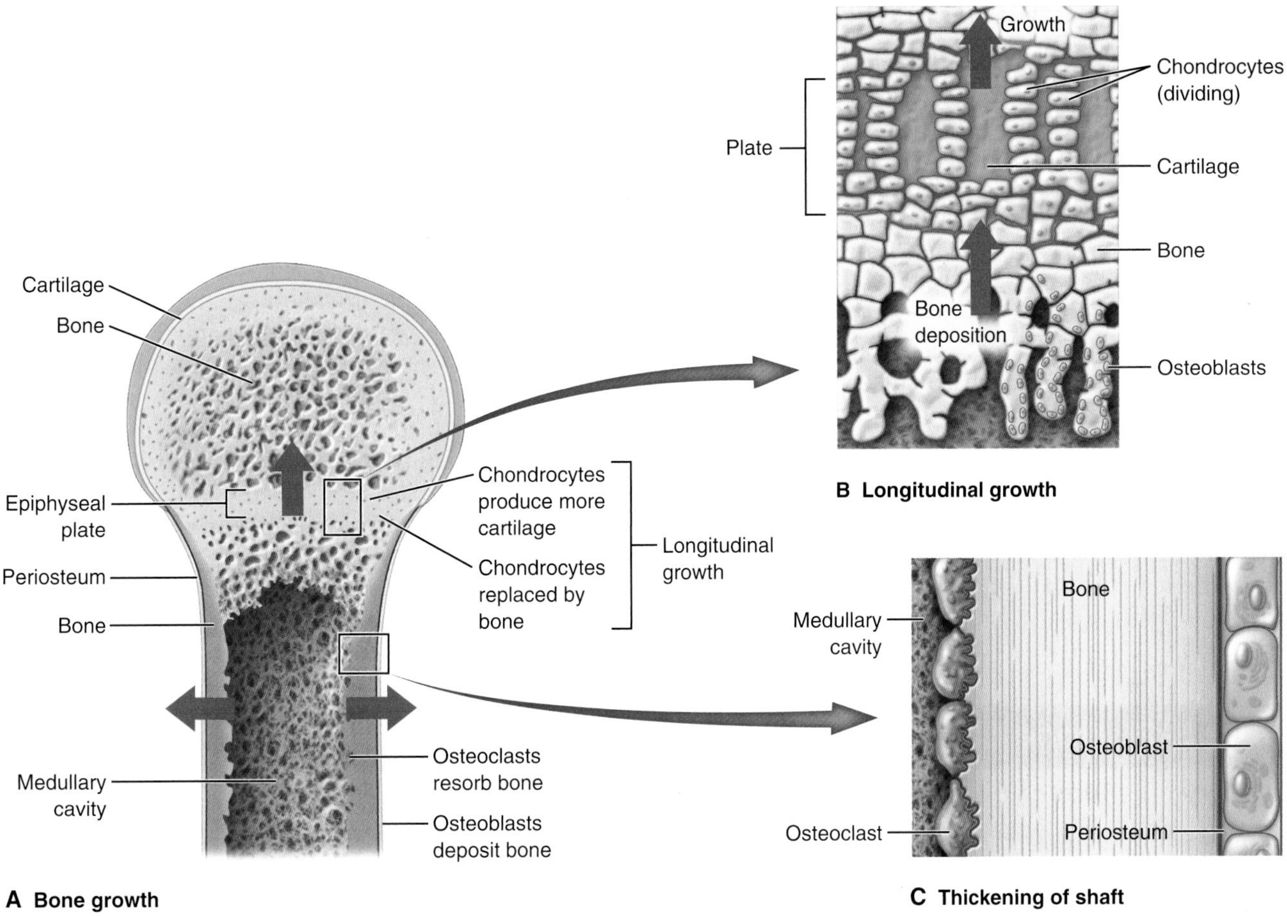

A Bone growth

B Longitudinal growth

C Thickening of shaft

FIGURE 5.5 ▸ Bone growth. Bones grow in length at the epiphyseal plate. The bone shaft grows and thickens as osteoblasts add new layers.

What Do You Think? 5.1

- What overall structural advantage does the medullary cavity provide for the skeleton?
- Knowing the dynamic activity that occurs at the epiphyseal plate, how much truth do you think there might be in the concept of "growing pains"?

BONES AND THEIR LANDMARKS

While bones can be categorized by shape, there are distinctive lines, grooves, bumps, and projections called **bone landmarks** that make each unique. These landmarks generally serve as attachment points for muscles or passageways for blood vessels and nerves.

Some landmarks can be felt when palpating the superficial tissue. These are of particular importance to manual therapists because they help to precisely locate or approximate the location of internal organs, nerves, and blood vessels. For example, the channel for the sciatic nerve is approximated as being halfway between the ischial tuberosity and the tip of the sacrum. Additionally, palpable landmarks provide cues that help therapists identify structural imbalances in the body. For instance, by placing the hands on both iliac crests to see if they are level, a practitioner can identify common problems related to chronic low back pain.

Many forms of manual therapy rely on static and dynamic postural assessments prior to developing appropriate treatment sessions for clients. Accurate assessment is based on the knowledge of proper alignment between specific bone landmarks. For example, when viewing a client's profile, an imaginary plumb line can be drawn from the mastoid process just behind the ear through several other landmarks to the lateral malleolus of the fibula. Unless a therapist is familiar with these specific landmarks, it is impossible to identify subtle postural misalignments. Likewise, when abnormalities are found in the soft tissue, we must rely on our knowledge of bone landmarks to accurately identify what we feel and document that on a chart. Unfortunately, it is not unheard of for an inexperienced practitioner to palpate and document a lump or bump in the neck as a knot in the muscle or a fibrotic adhesion, when in fact it is the bony transverse process of a cervical vertebrae or the edge of the first rib.

General Landmark Terminology

While each bone has its own unique set of landmarks, standard terminology is used to describe the general shape or type of marking, regardless of which bone it is on.

Bone landmarks are named according to shape or type of marking and can be classified into three broad categories: projections, depressions, and holes. ▶ TABLE 5-2 lists and defines some common terms and provides examples of specific landmarks for each. Becoming familiar with this general terminology will help you as you learn about specific bones and their unique landmarks.

Head and Face

The bones of the skull are designated as either cranial or facial bones (▶ FIGURE 5.6). The cranial bones of the skull are

- ***Frontal***—This single bone forms the front of the skull and the upper portion of the orbital cavities (eye sockets). The two small cavities inside the bone are the *frontal sinuses*.
- ***Parietal***—Located at the top of the skull, there is one parietal bone on each side, situated just posterior to the frontal bone.
- ***Temporal***—Situated just inferior to the parietal bones, there is one temporal bone on each side of the skull. The *mastoid process* is a short round projection situated on the inferior edge of the bone just behind the ear. The *auditory meatus* is the opening for the ear, and immediately inferior to the meatus there is a needle-like projection called the *temporal styloid process*. Just superior to the auditory meatus there is a short anterior projection called the *zygomatic process* that joins with the zygomatic bone of the face. A small depression between the auditory meatus and zygomatic process called the *mandibular fossa* is the point at which the mandible (jaw bone) attaches to the skull.

Table 5-2 Bone Landmarks

Category	Term	Description	Examples
Projections	Tubercle	Small bump	Deltoid tubercle on the humerus; gluteal tubercle on the femur
	Tuberosity	Large bump	Tibial tuberosity; ischial tuberosity
	Head	Knob-like, rounded end of a long bone connected to the shaft via a slender region called the neck	Head of the humerus, femur, and metacarpals
	Condyle	Rounded projection at the end of a bone; usually articulates with another bone	Condyles of humerus and femur
	Epicondyle	Small rise or bump just superior to a condyle	Epicondyles of humerus and femur
	Process	Prominent projection; may be given a secondary descriptor for shape, such as *styloid* for needle-like	Styloid process on temporal bone; acromion process on the scapula
	Crest	Sharp ridge-like border of a bone	Crest of ilium; tibial crest
Depressions	Fossa	Saucer-like depression	Gluteal fossa of ilium; infraspinous fossa of scapula
	Fovea	Small pit or tiny depression	Fovea capitis in the head of the femur
	Facet	Small flat articular surface	Rib facets; superior and inferior facets on vertebrae
Holes	Foramen	Hole or opening in a bone for nerves or blood vessels	Foramen magnum of skull; obturator foramen of pelvic girdle
	Meatus	Short tube or passageway within a bone	Auditory and nasal meatus in skull
	Sinus	Small cavity in certain cranial or facial bones	Sinuses in the frontal, ethmoid, sphenoid, and maxillary bones

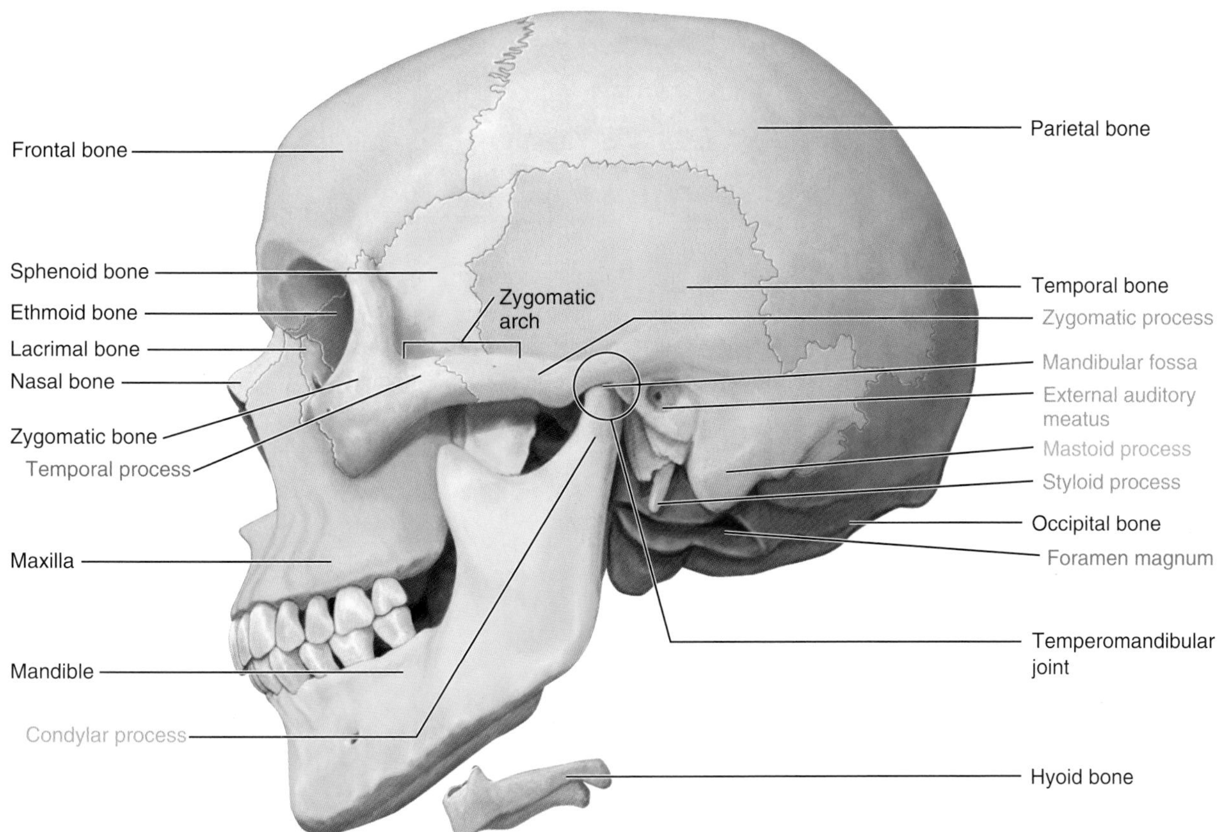

FIGURE 5.6 ▸ Lateral view of skull and hyoid. Cranial and facial bones and some of their major landmarks are shown. Note the position of the hyoid.

- ***Occipital***—This one bone forms the posterior portion of the skull (▸ FIGURE 5.7). There are two roughened ridges across the inferior aspect of the bone; the *superior nuchal line* and the deeper, more difficult to palpate *inferior nuchal line*. There is a small protrusion in the middle of the superior nuchal line known as the *external occipital protuberance*, or *inion*. At the base of the occipital bone, there is an opening called the *foramen magnum* through which the spinal cord passes to connect to the brain. On either side of the foramen magnum are two round protrusions, the *occipital condyles*, which are the points of attachment between the skull and the first cervical vertebra.
- ***Sphenoid***—The sphenoid is a butterfly-shaped bone in the floor of the cranium that can only be seen by looking inside the cranial cavity (▸ FIGURE 5.8). In the central portion of the sphenoid, there is a unique saddle-shaped depression called the *sella turcica*, which is the location of the pituitary gland. This bone also contains the *sphenoid sinus*.
- ***Ethmoid***—This small bone is also inside the cranial cavity, situated just anterior to the sphenoid. It serves as the anterior portion of the floor of the cranium and the roof of the nasal cavity. There are two sinuses in the ethmoid bone (▸ FIGURE 5.9).

It is not uncommon for professionals in any of the health care fields to use less-scientific terminology for several common bone landmarks. For example, manual therapists often refer to the superior nuchal line as the *occipital ridge.* This term is more descriptive of what therapists actually palpate on clients, a pronounced ridge on the back of the occiput (FIGURE 5.7). This region is a primary attachment point for many spinal muscles that are frequently involved in tension headaches and neck pain. When clients complain of a headache that seems to be localized over their ears, therapists will often investigate the muscles below the occipital ridge for tension or possible trigger points.

An anterior view of the face and skull shows both the cranial and facial bones (▸ FIGURE 5.10). The facial bones include

- ***Lacrimal***—These two bones are situated posterior to the nasal bones and form the medial wall of the eye socket. They are the smallest bones in the face. Lacrimal bones get their name from a small groove where they meet the nasal bones; this groove houses a small sac that gathers tears and passes them into the nasal cavity.

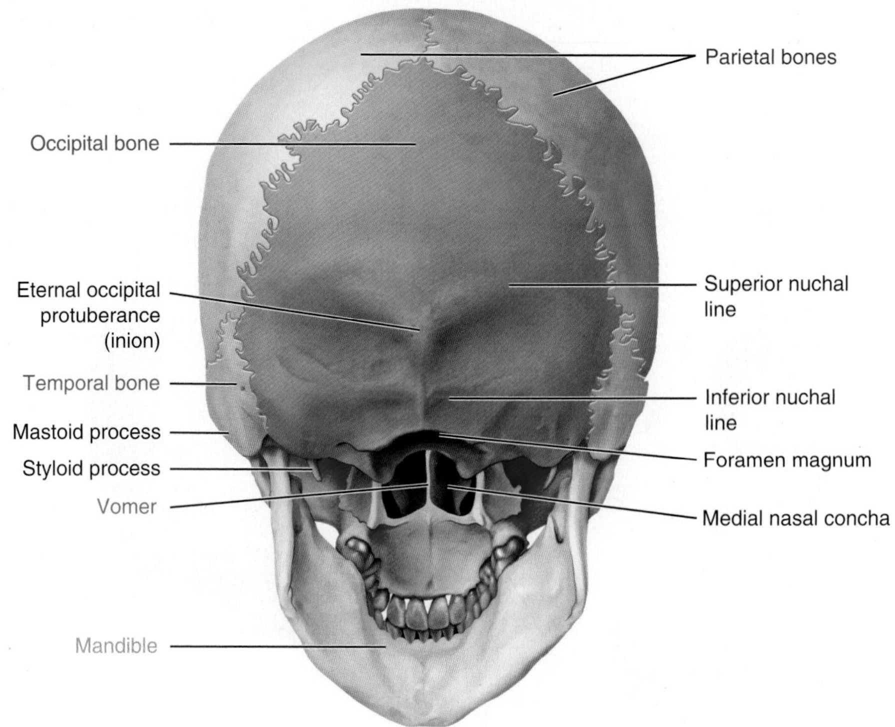

FIGURE 5.7 ▸ Posterior view of the skull. This view clearly shows the occipital bone landmarks.

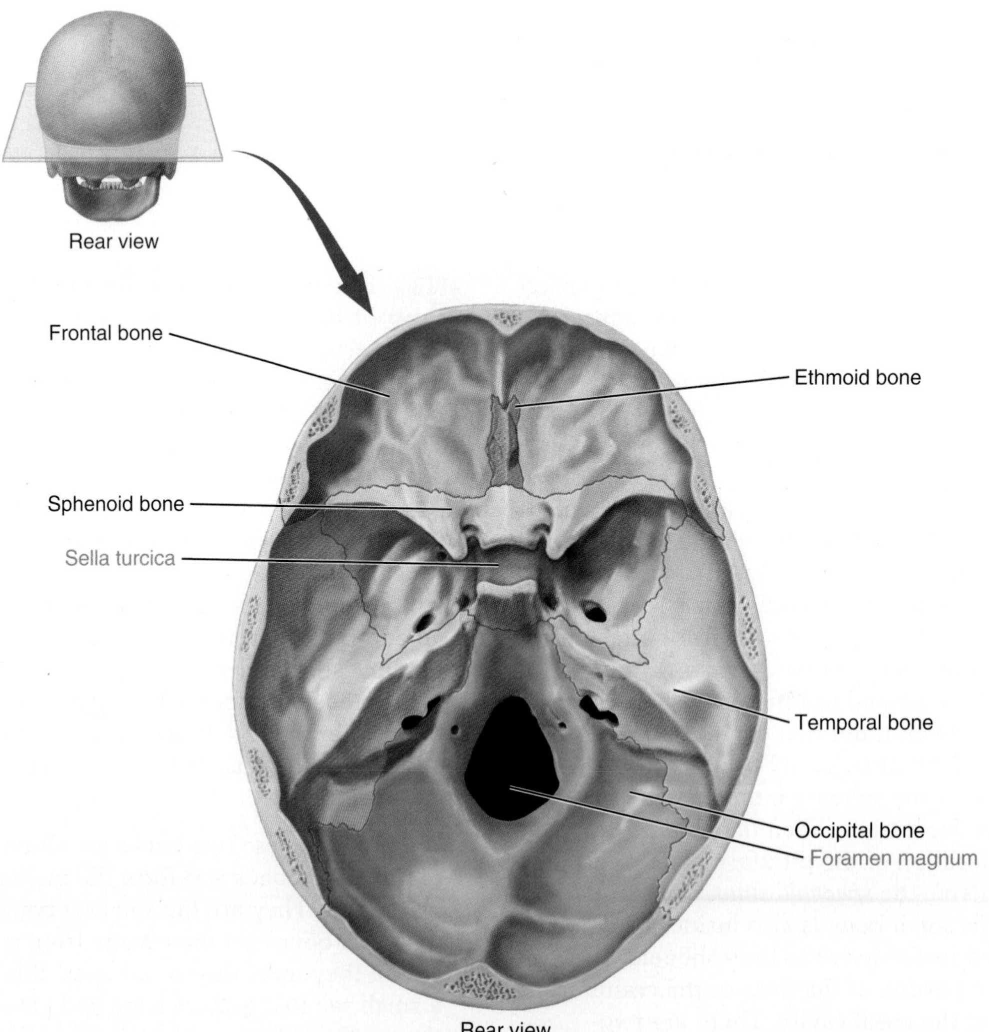

FIGURE 5.8 ▸ Superior view of the cranial cavity. This view of the cranial floor, clearly shows the position of the ethmoid and sphenoid.

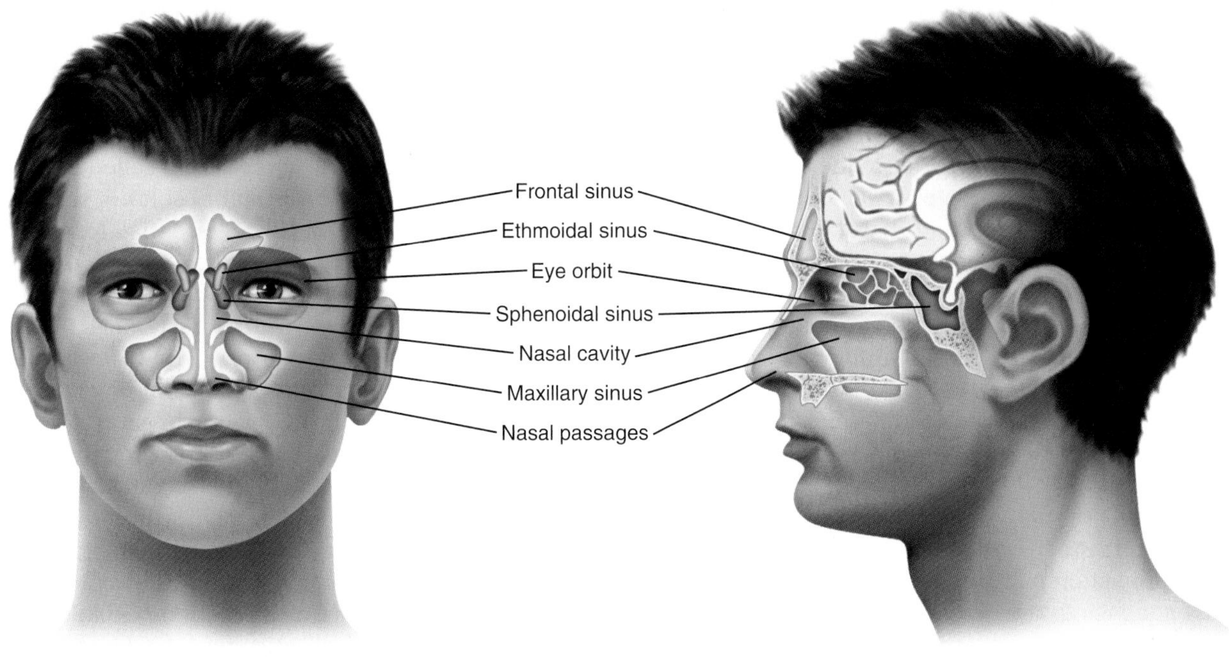

FIGURE 5.9 ▸ Sinuses of the skull. Sinuses are air-filled cavities within certain cranial and facial bones. **A.** Frontal view. **B.** Lateral view.

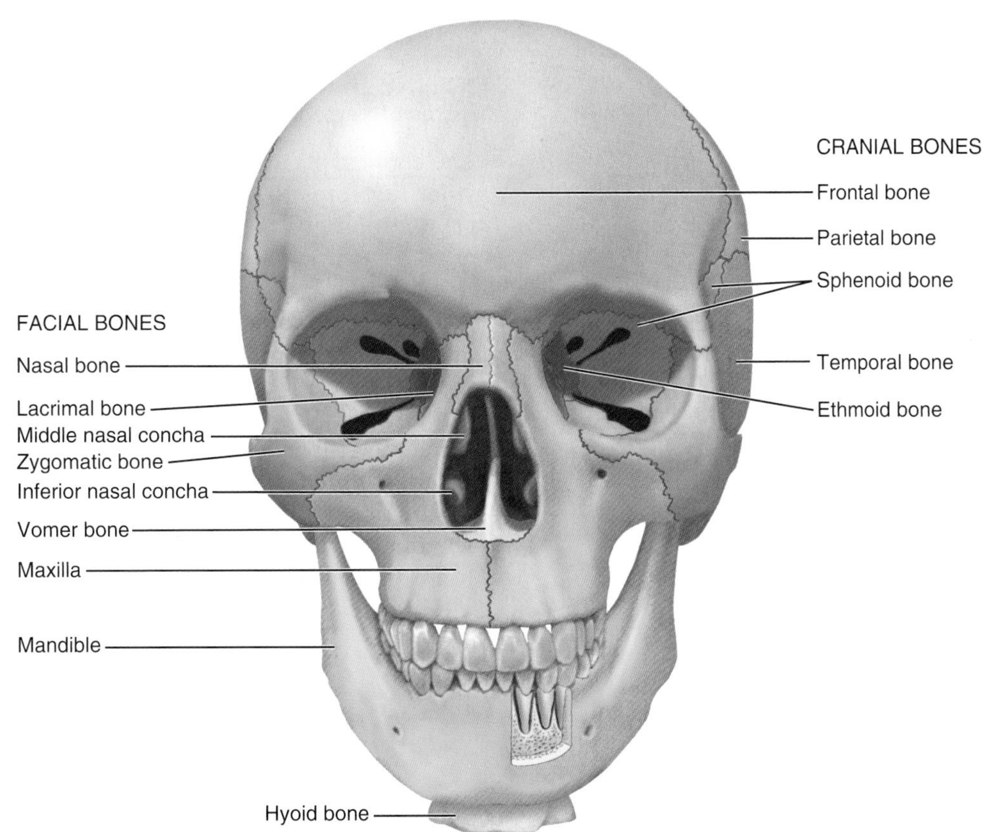

FIGURE 5.10 ▸ Anterior view of the skull. This view shows both the cranial and facial bones. Note the position of the hyoid.

- ***Nasal***—These paired bones join at the midline, forming the bridge of the nose. The anterior portion of the nose is soft cartilage.
- ***Nasal concha (turbinate)***—Each nostril contains a superior, middle, and inferior turbinate. These are separate bones, each situated inferior to the ethmoid bone to form part of the inferior lateral wall of the nasal cavity. The nasal conchae direct air flow and assist in heating, humidifying, and filtering air in the nasal cavity.
- ***Vomer***—This triangular bone sits vertically inside the nasal cavity, connecting with the ethmoid, palatine, and maxillary bones to form the inferior portion of the nasal septum (▶ FIGURE 5.11).
- ***Palatine***—These bones form the hard palate at the roof of the mouth as well as a small portion of the inferior eye socket.
- ***Zygomatic***—Commonly known as cheek bones, these two bones form the ridge at the top of the cheeks via a lateral projection called the *temporal process*. This process attaches to the zygomatic process of the temporal bone to form the *zygomatic arch*.
- ***Maxilla***—These two bones join in the middle of the face and form the region between the nose and lips. The maxillae attach to the zygomatic bones to form the cheeks. In fact, this bone attaches to every bone in the face except the mandible. There are two paranasal sinus cavities within the maxillae (FIGURE 5.9).
- ***Mandible***—Completing the facial bones is the mandible (jaw bone), the only moving bone in the skull. There are two *condylar processes* on the posterior edge of the mandible that fit into the mandibular fossa of each temporal bone. This junction forms the ***temporomandibular joint (TMJ)***, shown in FIGURE 5.6.

Spinal Column

The 33 ***vertebrae*** of the spinal column are divided into five regions (▶ FIGURE 5.12A):

- Cervical region—7 vertebrae designated as C_1 through C_7
- Thoracic region—12 vertebrae designated as T_1 through T_{12}
- Lumbar region—5 vertebrae designated as L_1 through L_5
- Sacral—5 vertibrae designated as S_1 through S_5
- Coccygeal—3 or 4 vertibrae designated as Co_1 through Co_4

Between the ages of 16 and 30, the five sacral vertebrae fuse into one triangular-shaped bone called the sacrum (FIGURE 5.12B). Similarly, the coccygeal vertebrae partially fuse to form the coccyx (tail bone). While the sacrum and coccyx are part of the axial skeleton, the sacrum is a functional part of the pelvic girdle and forms an essential foundation for the torso.

From a lateral view, the bone arrangement of the vertebrae creates curves in the adult spinal column. The cervical and lumbar regions of the spine have a concave appearance or anterior curve called a **lordotic** curve. In contrast, the thoracic and sacral regions curve posteriorly to give them a convex appearance called a **kyphotic** curve. These curves give the spinal column the strength and shock absorption that are essential for maintaining balance and an upright position during movement. Interestingly, the lordotic curves of the cervical and lumbar regions are not present at birth (FIGURE 5.12C) but develop over time as infants progress from holding up their head, to crawling, and then to standing upright and walking. For this reason, the thoracic and sacral curves in the spine are referred to as primary curves, and the cervical and lumbar are said to be secondary curves.

Cervical, thoracic, and lumbar vertebrae have the same general structure: a round drum-like anterior portion called the *body*, one posterior projection called a *spinous process*, two lateral projections called *transverse processes*, a *pedicle* that attaches the transverse processes to the body, and a flat region of bone between the transverse and spinous processes called the *lamina* (▶ FIGURE 5.13). The pedicle, lamina, transverse and spinous processes collectively form a bone arch that attaches to the vertebral body, creating the *vertebral foramen* that houses the spinal cord. Vertebrae also have several flat surfaces called *facets*, or *articular processes*, where they connect to one another to form the spinal column. The shape and density of each individual vertebra vary from region to region. Some distinguishing features of vertebrae include

- The cervical vertebrae are flatter and wider than other vertebrae to support the head and provide greater mobility to the neck. They have a two-pronged spinous process and a small hole in the transverse process called the *transverse foramen,* which provides a pathway for blood vessels.
- In thoracic vertebrae the spinous processes have a sharp downward angle that helps lock them together and provide a firm anchor for the rib cage. These vertebrae also have an additional costal facet on their transverse processes that is the point of attachment for the ribs.
- The lumbar vertebrae are the primary weight-bearing bones in the spine; therefore, they are the thickest and broadest vertebrae.

▶ FIGURE 5.13 shows the difference in shapes and density of the cervical, thoracic, and lumbar vertebrae, as well as the important bone landmarks. It also highlights the first two cervical vertebrae, the ***atlas*** (C_1) and ***axis*** (C_2), because they are the most uniquely shaped vertebrae of the entire spinal column. Notice the atlas does not have a body or a spinous process; it is simply an osseous ring. However, it has two prominent *articular processes* that are the point of connection between the occipital

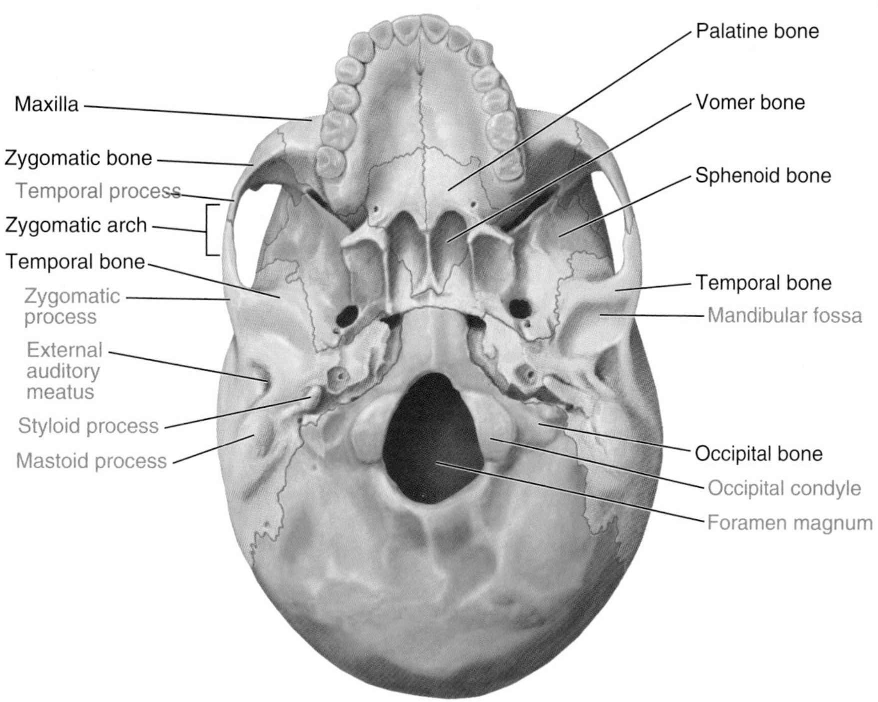

FIGURE 5.11 ▸ Inferior view of the skull. Note the locations of inferior bone landmarks of the cranial bones.

B Posterior view of sacrum and coccyx

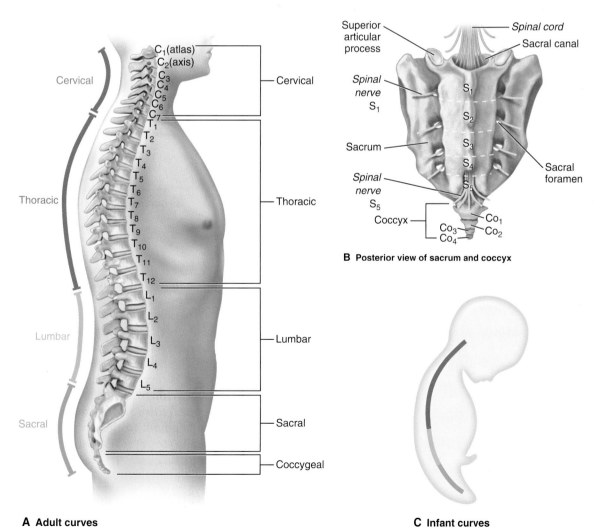

A Adult curves

C Infant curves

FIGURE 5.12 ▸ Spinal column. A. Lateral view shows the number of vertebrae in each spinal region and their curvatures. **B.** Posterior view of the sacrum and coccyx shows the spinal levels and foramen of these fused vertebrae. **C.** Infants have a C-shaped spine; its curvature changes as the baby develops.

Pathology Alert Spinal Deviations

The spinal column can develop exaggerated lordotic or kyphotic curves as functional adaptations or in conjunction with diseases such as osteoporosis. A hyperlordotic curve in the lumbar spine is commonly called *sway back*, while the term *humpback* is sometimes used to describe hyperkyphosis of the thoracic region. An abnormal lateral curve called **scoliosis** can also develop. In many cases, scoliosis involves more than a simple S-curve. There is an actual twist to the spinal column, referred to as rotoscoliosis, that increases the complexity of the condition. Any deviation from normal spinal structure can be congenital or caused by long-term mechanical stress on the vertebrae due to muscular strain, imbalance, or disease. Spinal deviations decrease the overall strength, resilience, and flexibility of the spinal column and can cause nerve irritation, restricted breathing, or lead to severe dysfunction in the organs of the ventral cavities. Several forms of manual therapy have been shown to be effective ancillary modalities for reducing pain, muscle spasm, and movement limitations associated with these conditions. In some cases of severe spinal deviation, therapists may need to adjust the client's position and use supportive cushions or bolsters to stabilize and assure client comfort.

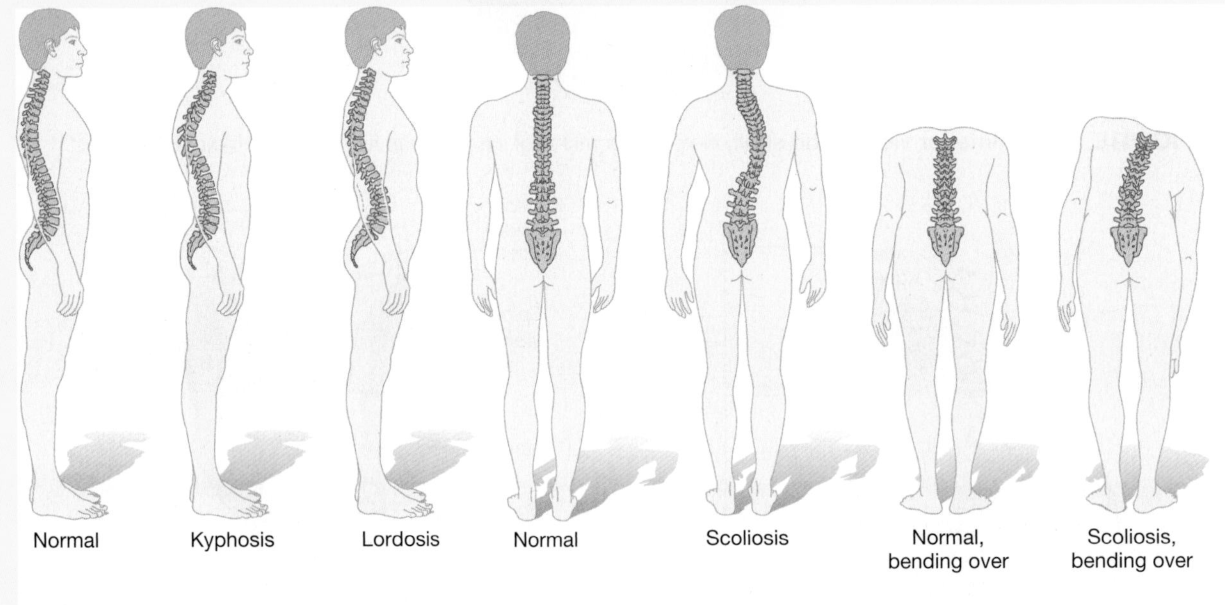

condyles of the skull and the spinal column. The axis has a large projection on the superior aspect of its body called the *dens*, or *odontoid process*. The dens projects upward into the atlas and is held in place by a strong ligament. This design allows the atlas to rotate easily on the axis.

Thorax

The thorax or chest is the region between the neck and abdomen encased by the ribs and containing the heart and lungs. The rib cage (▶ FIGURE 5.14) consists of the *sternum* and 12 pairs of *costals* (ribs). In childhood, the sternum is actually made up of three different segments held together with fibrous connective tissue until they eventually fuse into one bone. The most superior octagon-shaped region of the sternum is the *manubrium*, the long flat portion is the *body*, and the sharp, dagger-like projection off the inferior tip of the bone is called the *xiphoid process*.

The costals are flat bones that make up the protective cage of the thorax. All the ribs attach to thoracic vertebrae at the costal facets, and most of them also attach to the sternum. The first seven pairs are called ***true ribs*** because they are directly attached to the sternum by individual ***costocartilage*** bridges. The remaining five pairs are considered ***false ribs*** because they either share one cartilage attachment to the sternum (pairs 8–10) or do not attach to the sternum at all, making them ***floating ribs*** (pairs 11 and 12).

Pectoral Girdle

Two bones form the pectoral girdle, the clavicle and scapula (▶ FIGURE 5.15). The ***clavicle*** lies horizontally across the top of the rib cage connecting the sternum to the scapula;

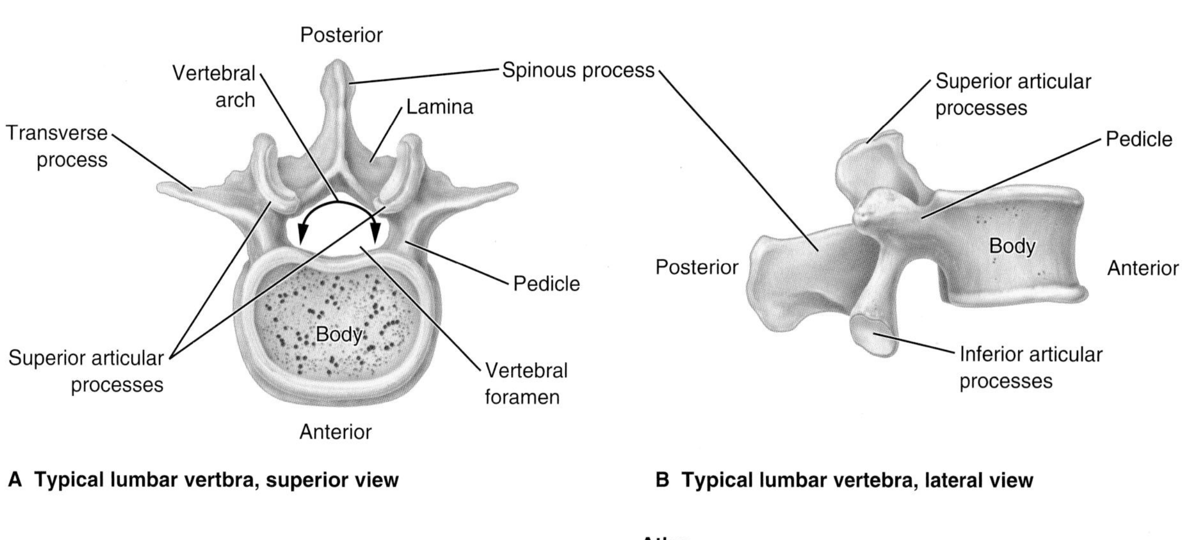

A Typical lumbar vertbra, superior view

B Typical lumbar vertebra, lateral view

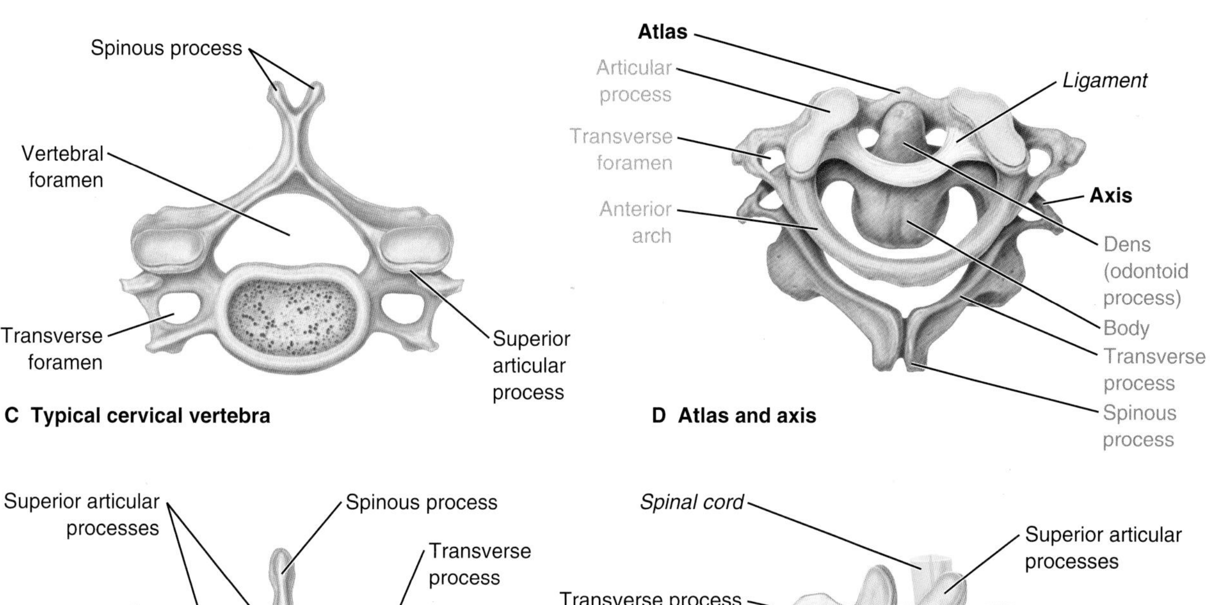

C Typical cervical vertebra

D Atlas and axis

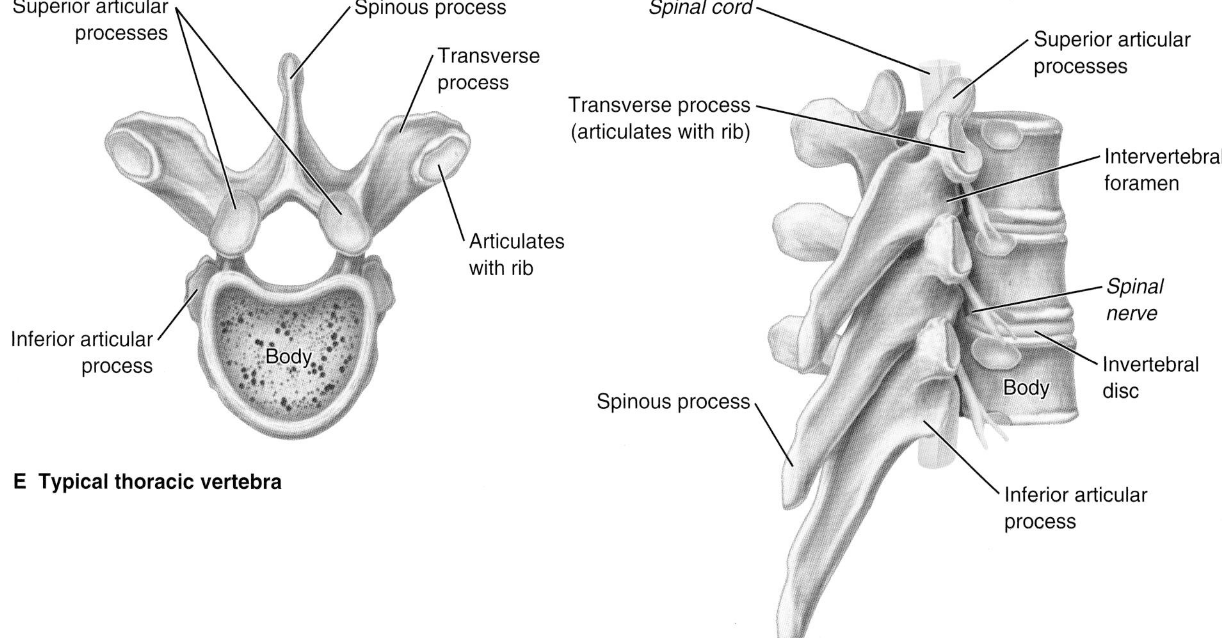

E Typical thoracic vertebra

F Articulated thoracic vertebrae

FIGURE 5.13 ▸ Vertebral structure. The shape and size of vertebrae differ according to their location. **A.** Lumbar vertebra, superior view. **B.** Lumbar vertebra, lateral view. **C.** Cervical vertebra, superior view. **D.** Atlas and axis. **E.** Thoracic vertebra, superior view. **F.** Vertebrae stack to form the spinal column. Note the intervertebral discs and the intervertebral foramen, the site where a spinal nerve exits. Visit http://the point lww.com/Archer-Nelson for a video showing features of a typical vertebra.

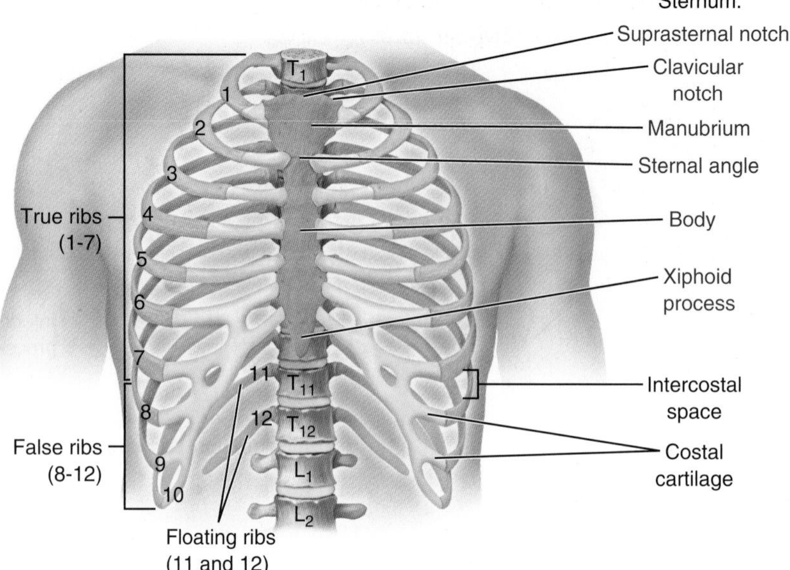

FIGURE 5.14 ▸ Rib cage. Bones of the thorax.

it functions as a suspension strut for muscles that move the humerus. It has a unique S-shape and also changes contour from one end to the other. The medial portion of the bone is round and curves outward from the rib cage and then bows inward and flattens as it projects laterally. The rounded medial end of the clavicle connects to the manubrium at the *clavicular notch* to form the ***sternoclavicular (SC) joint***. The flatter lateral end of the bone connects to the scapula at the ***acromioclavicular (AC) joint***.

The ***scapula*** is the triangular bone that forms the posterior portion of the pectoral girdle (FIGURE 5.15). The sharp corner at the top of the bone is called the *superior angle*, and the sharp point at the bottom is the *inferior angle*. The lateral edge of the scapula is called the *axillary border*, while the medial edge is the *vertebral border*. The anterior surface of the scapula, the *subscapular fossa*, lies against the muscles of the posterior rib cage. The posterior surface is divided into two unequal regions called the *supraspinous* and *infraspinous fossae* by the *spine*, a ridge-like projection. The lateral edge of the spine forms a flat shelf called the *acromion process* where the clavicle attaches forming the *AC joint*. The *coracoid process* is a finger-like projection off the anterior aspect of the bone. There is a shallow depression in the lateral superior portion of the scapula called the *glenoid fossa* that forms the socket of the shoulder joint. On the upper and lower edges of the glenoid fossa, there are two small bumps, the *supraglenoid* and *infraglenoid tubercle*, that serve as attachment points for muscles of the brachium.

By the Way

The xiphoid process is an important bone landmark to be aware of when performing CPR or the Heimlich maneuver. Those administering first aid are cautioned to avoid pressure at the distal end of the sternum to keep from forcing the sharp xiphoid process into the internal organs.

Upper Limb and Forearm

The largest and most proximal bone in the upper extremity, the ***humerus***, has several very important bone markings (▸ FIGURE 5.16). The *head* of the humerus is the proximal knob-like portion that joins with the glenoid fossa of the scapula to form the ***glenohumeral (shoulder) joint***. The *neck* of the humerus is the region that connects the head to the shaft. Just inferior to the neck are two pronounced bumps: the *greater tubercle* on the lateral side and the *lesser tubercle* on the medial side. The deep groove between the tubercles is called the *intertubercular groove*, or more commonly, the *bicipital groove*, because the tendon of the long head of the biceps brachii muscle passes through it before attaching to the scapula. The *deltoid tuberosity*, the attachment point for the deltoid muscle, is on the lateral side of the humerus about mid-shaft. The *medial* and

By the Way

Because the clavicle changes both shape and contour in the middle, it tends to break easily and is one of the most commonly fractured bones in the body.

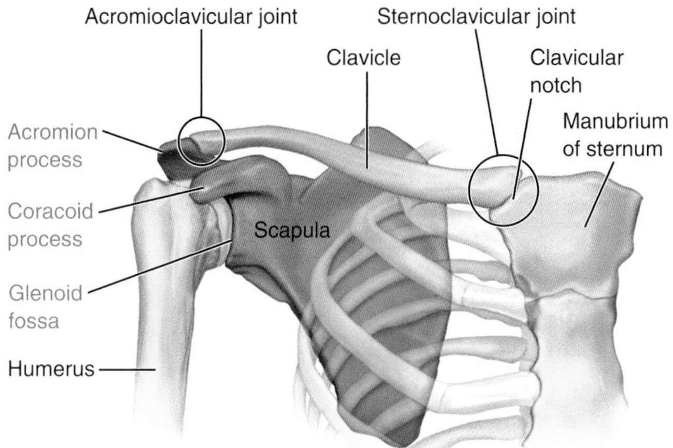

A Pectoral girdle, anterior view

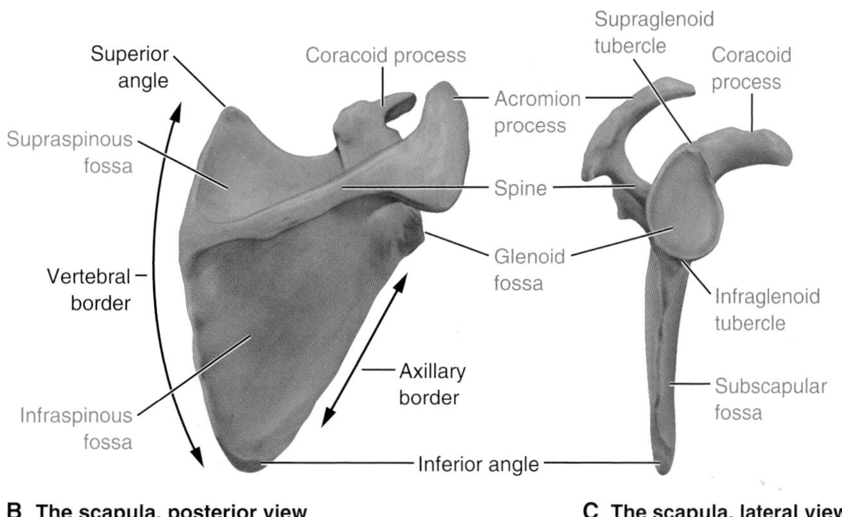

B The scapula, posterior view

C The scapula, lateral view

FIGURE 5.15 ▶ Pectoral girdle. A: Anterior view. The clavicle and scapula attach the upper extremity to the axial skeleton. **B.** Scapula, posterior view. **C.** Scapula, lateral view.

lateral epicondyles are located at the distal end of the bone, and just superior to each are the *medial* and *lateral supracondylar ridges*. The condyles of the humerus have names related to their shape and action. The medial condyle is the *trochlea,* based on the Greek term for a system of pulleys, and the lateral condyle is the *capitulum,* a rounded protuberance. The trochlea and capitulum attach to the ulna and radius at the elbow joint. There are also three important fossae in the distal humerus: the *olecranon fossa* on the posterior side, and the *coronoid* and *radial fossae* on the anterior. These depressions in the humerus are named for the ulnar and radial projections that move into them during elbow flexion and extension.

In anatomic position, the ***radius*** is the lateral bone and the ***ulna*** is the larger bone on the medial side of the forearm (▶ Figure 5.17). The *head* of the radius is the round proximal tip that joins with the capitulum of the humerus. Just inferior to the head on the medial edge of the bone closest to the ulna is the *radial tuberosity*. The broader distal end of the radius has a needle-like projection off the lateral tip called the *radial styloid process*.

In contrast to the radius, the *head* of the ulna is at the distal end of the bone. The proximal end of the ulna has several important landmarks beginning with the *olecranon process,* a posterior knob on the elbow informally known as the "funny bone." The deep hook-like depression on the anterior side of the olecranon is the *trochlear notch* that connects to the trochlea. The sharp projection at the inferior lip of the trochlear notch is the *coronoid process*. Another small projection, the *ulnar tubercle,* is just lateral and inferior to the coronoid process. The small depression just lateral to the trochlear notch is the *radial notch,* so named because this is where the radial head meets the ulna. Figure 5.17 also shows the tough fibrous connective tissue that fills the space between the radius and ulna, the *interosseous membrane*. Functionally this membrane serves both as an attachment point for muscles and as a stabilizing structure between the two bones.

Greater tubercle
Head
Bicipital or intertubercular groove
Neck
Lesser tubercle
Deltoid tuberosity
Lateral supracondylar ridge
Medial supracondylar ridge
Radial fossa
Coronoid fossa
Lateral epicondyle
Trochlea
Capitulum
Medial epicondyle
Condyle
A Anterior view
Greater tubercle
Lateral supracondylar ridge
Olecranon fossa
Lateral epicondyle
B Posterior view

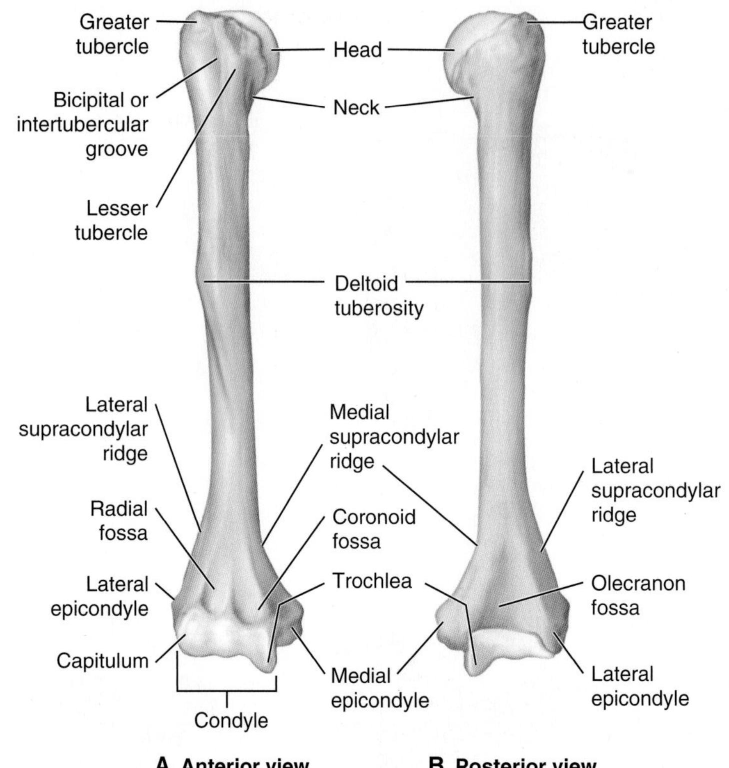

FIGURE 5.16 ▶ **Humerus. A.** Anterior view. **B.** Posterior view.

Olecranon process
Trochlear notch
Radial notch
Coronoid process
Head of radius
Radial tuberosity
Ulnar tubercle
LATERAL
MEDIAL
Interosseous membrane
Ulnar notch
Head of ulna
Radial styloid process
Ulnar styloid process
A Anterior View
Olecranon process
Head of radius
Medial surface
Posterior surface
Posterior surface
MEDIAL
LATERAL
Interosseous membrane
Head of ulna
Ulnar styloid process
Radial styloid process
B Posterior View

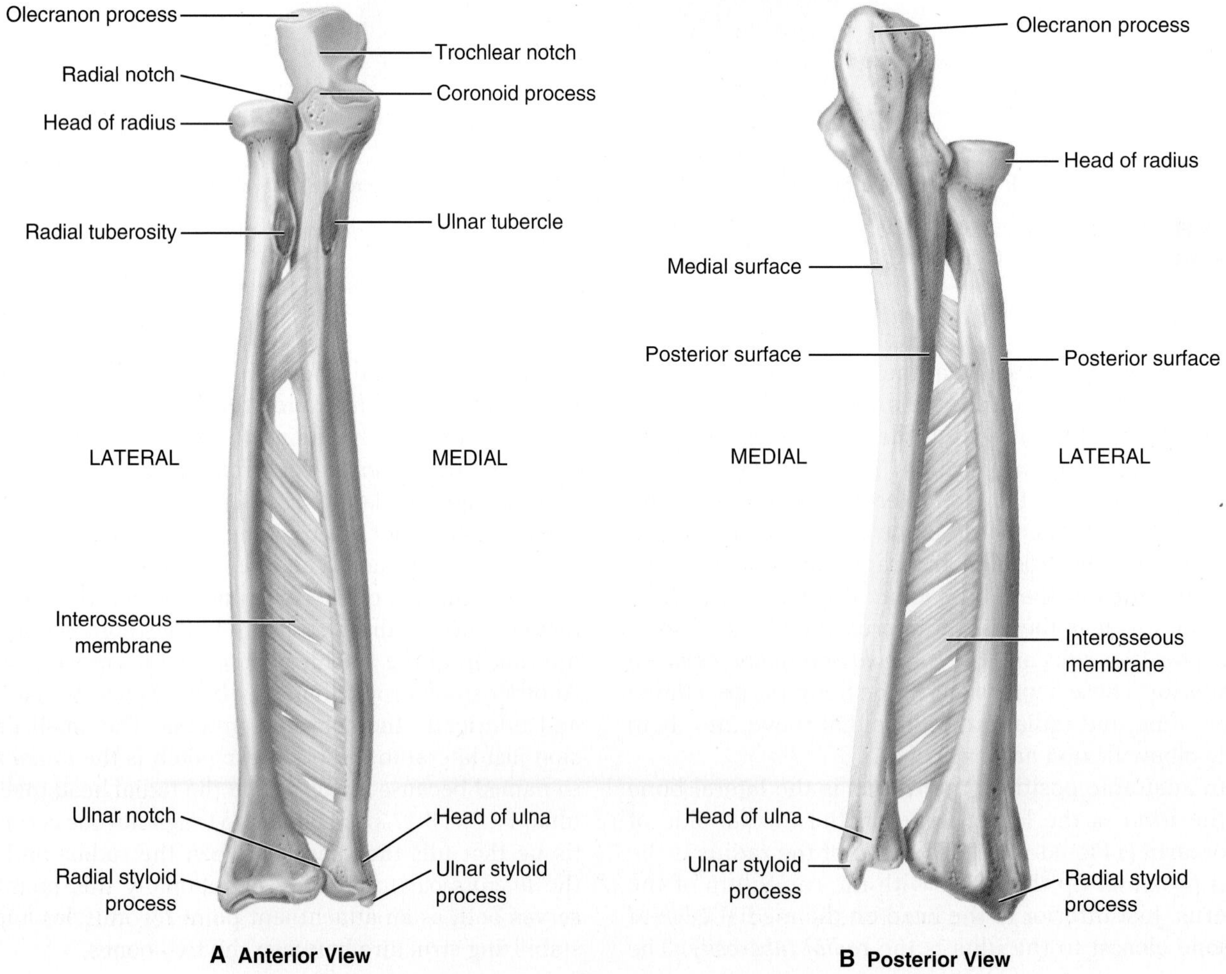

FIGURE 5.17 ▶ **Radius and ulna. A.** Anterior view. **B.** Posterior view.

Wrist and Hand

The wrist has eight ***carpal*** bones that are loosely arranged into a proximal and distal row. The proximal row of carpals includes the ***pisiform***, ***triquetral***, ***lunate***, and ***scaphoid*** bones, and the ***hamate***, ***capitate***, ***trapezoid***, and ***trapezium*** form the distal row. The long bones in the palm of the hands, the ***metacarpals***, are numbered 1 to 5 from thumb to little finger. The proximal end of each metacarpal is considered the *base* of the bone, and the round distal end is the *head*. Each hand has a total of 14 ***phalanges*** in the fingers. As shown in ▸ Figure 5.18, each finger has three individual bones: one proximal, one middle, and one distal ***phalanx***. The thumb has only two bones, a proximal and distal phalanx. Knuckles are the joints between the metacarpals and phalanges. The first row of knuckles consists of the ***metacarpophalangeal (MP) joints***. The knuckles in the fingers are named according to their position as ***proximal*** or ***distal interphalangeal*** joints, most often shortened to ***PIP*** and ***DIP joints***.

Pelvic Girdle

The pelvic girdle is formed by the sacrum and the os coxae. Each os coxa is formed when three bones, the ***ilium, ischium***, and ***pubis***, fuse into one at approximately age 17 (▸ Figure 5.19). A lateral view of the pelvic girdle shows a large fossa called the *acetabulum* at the junction between the three bones. This fossa holds the head of the femur to form the hip joint. An anterior view of the girdle reveals three large openings: two *obturator foramen* between the ischium and pubic bones and the *pelvic inlet*, a large opening defined by the *pelvic brim* of the os coxae. The ilium has a broad ridge across the top edge, the *iliac crest*, which ends anteriorly with a sharp projection called the *anterior superior iliac spine* and posteriorly with the *posterior superior iliac spine*. The broad depression on

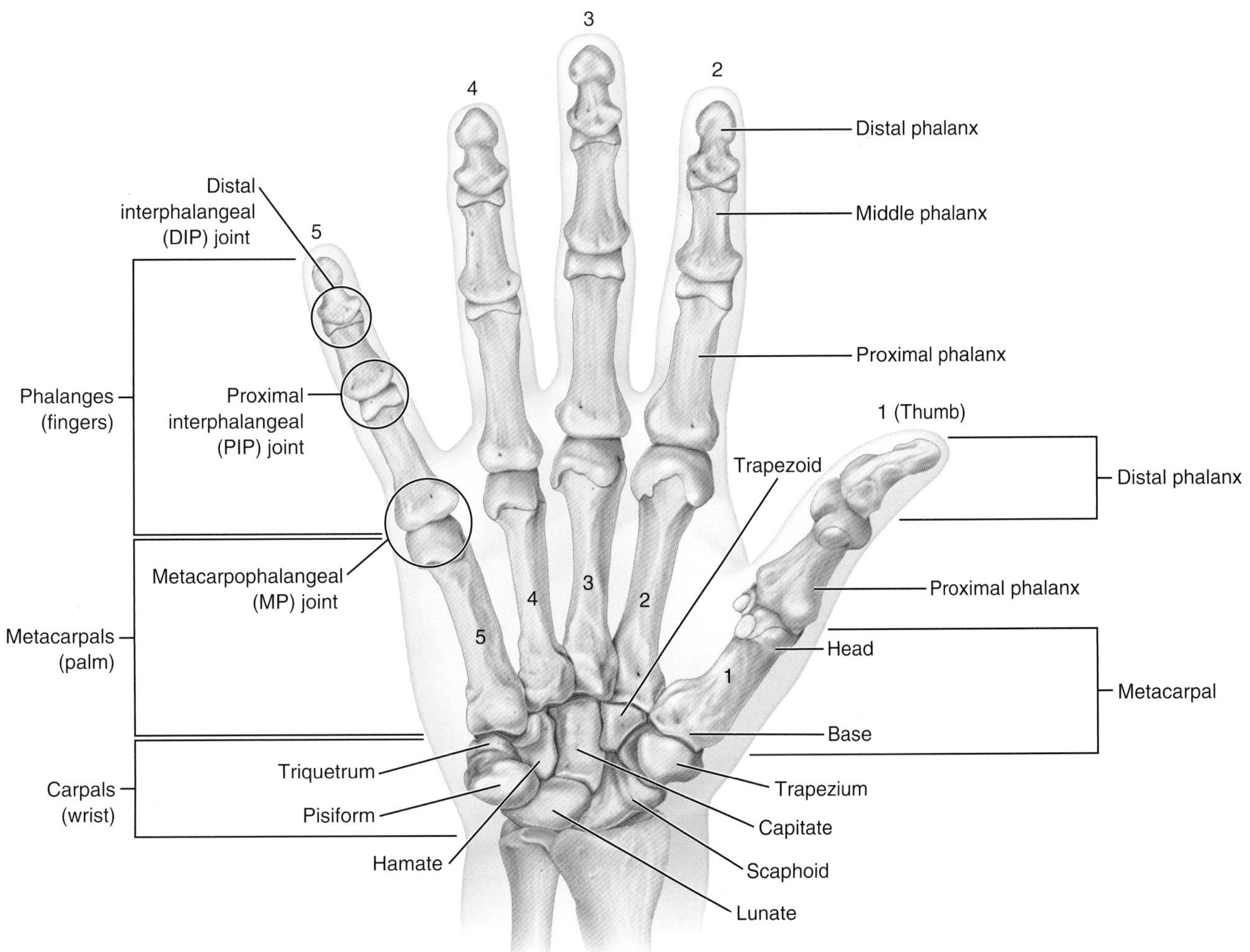

FIGURE 5.18 ▸ Bones and joints of the wrist and hand. Palmar view of the right hand shows the 8 carpals, 5 metacarpals, and 14 phalanges.

the anterior or inner side of the ilium is the *iliac fossa*. A major landmark on the ischium is the inferior projection that we sit on, the *ischial tuberosity*. There are several other important landmarks to be aware of in the coxal bones, including the *greater and lesser sciatic notch,* and *anterior* and *posterior inferior iliac spines.*

Thigh and Leg

The ***femur*** (thigh bone) is the longest, heaviest, and strongest bone in the body (▶ FIGURE 5.20). It curves anteriorly from hip to knee, giving it additional strength. Similar to the humerus, the proximal end has a knob-like *head* that

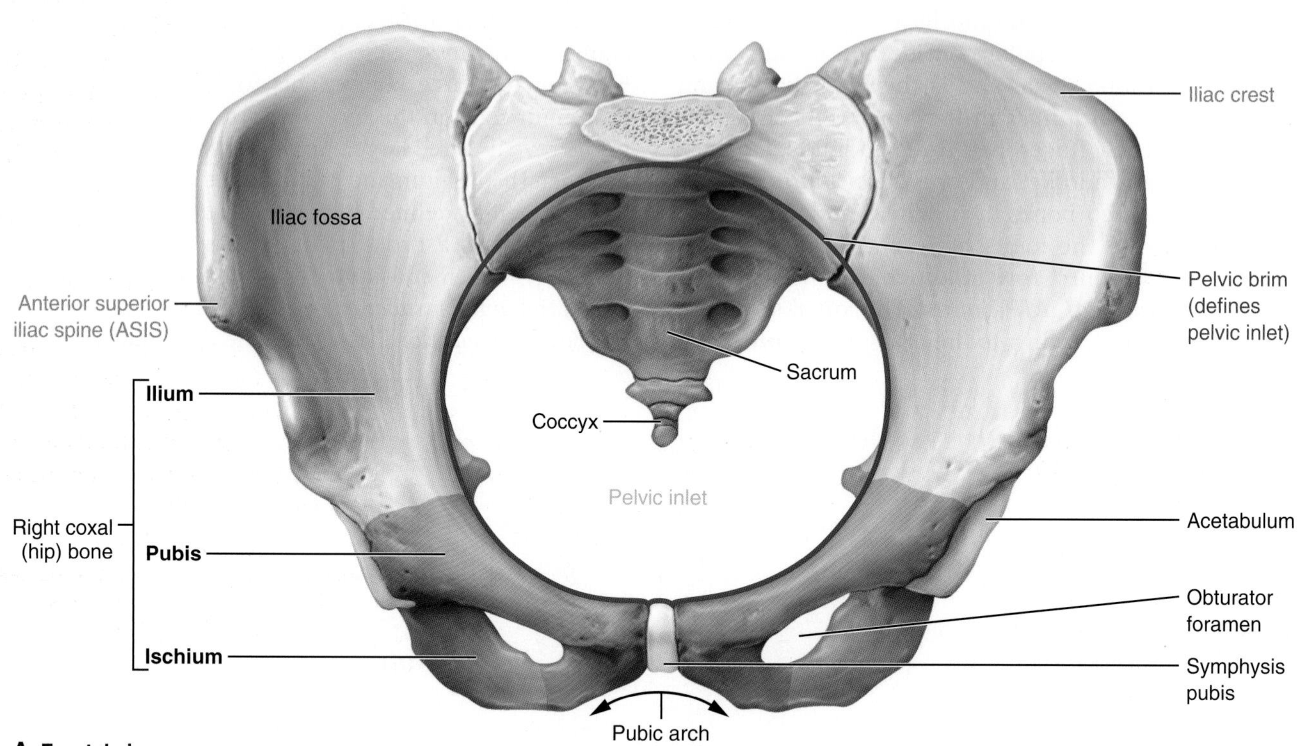

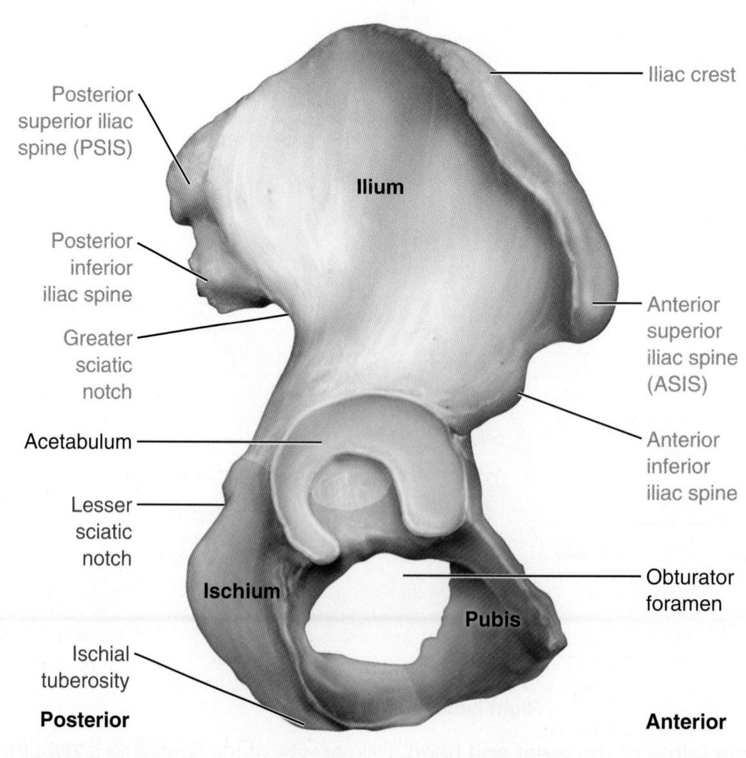

FIGURE 5.19 ▶ Pelvic girdle. A. Frontal view. **B.** Right lateral view.

By the Way

The female pelvic girdle is flatter, broader, and more shallow than that of a male. This shape makes the pelvic inlet wider to accommodate a baby's head during childbirth.

fits into a socket, the acetabulum of the hip. However, the neck of the femur is more pronounced than that of the humerus. There is a tiny depression on the superior aspect of the head called the *fovea capitis*. A special ligament emanating from the acetabulum attaches to the fovea capitis. In addition to stabilizing the joint, this ligament contains a small artery that feeds the head of the femur. At the proximal end of the bone, there are two projections: the *greater trochanter* is the large lateral projection and the *lesser trochanter* is the smaller projection on the medial-posterior shaft. On the posterior femur, there is a rough, broad line called the *linea aspera* that runs almost the full length of the shaft and serves as an attachment point for several major muscles of the thigh. The distal end of the femur has *medial* and *lateral condyles* and *epicondyles* just superior. The *adductor tubercle* is a small bump on the medial epicondyle where the largest of the adductor muscles attach. There are depressions between the condyles on both the anterior and posterior femur. The shallow anterior depression called the *patellar groove* or *patellar surface* is where the patella rests. The deeper posterior groove protects major blood vessels and nerves and is referred to as the *intercondylar fossa*.

The large weight-bearing bone on the medial side of the leg is the ***tibia*** (FIGURE 5.21). The proximal end is thick and broad with rounded sides called the *medial* and *lateral condyles*. The tibial condyles are unique because they are the only instance of condyles that are *not* an articular surface. The flat, slightly concave surface that the femur sits on is the *tibial plateau*. A sharp thorn-like projection in the middle of the plateau, the *intercondylar eminence*, is an important attachment point for ligaments in the knee joint. The prominent anterior projection just inferior to the tibial condyles is the *tibial tuberosity*, which serves as the attachment point for the quadricep muscles. The *tibial crest* is a long sharp line on the anterior shaft, and the distal projection of the tibia, sometimes misrepresented as an "ankle bone," is called the *medial* or *tibial malleolus*.

The thin non–weight-bearing bone on the lateral side of the leg is the ***fibula*** (FIGURE 5.21). The proximal head of the fibula is attached to the tibia just inferior to the lateral tibial condyle. The distal end extends a little further than the end of the tibia and ends with a projection called the *lateral* or *fibular malleolus*.

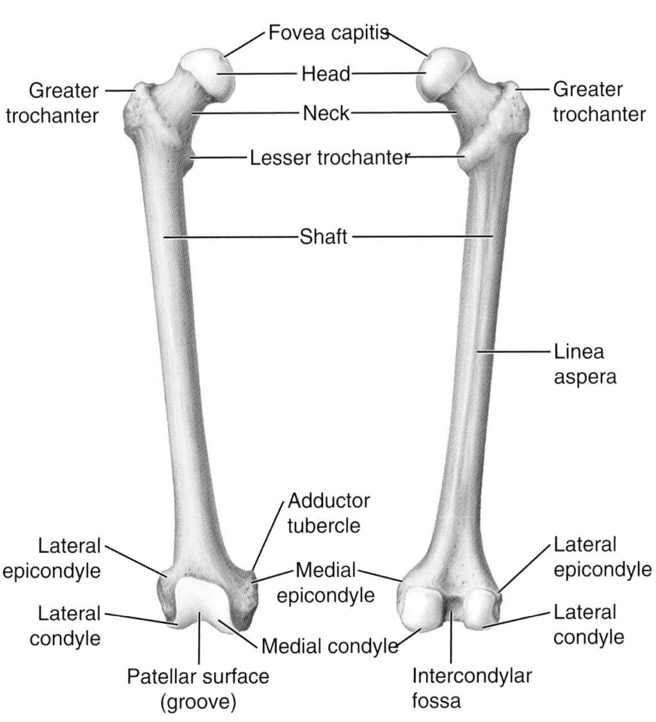

FIGURE 5.20 **Femur. A.** Anterior view. **B.** Posterior view.

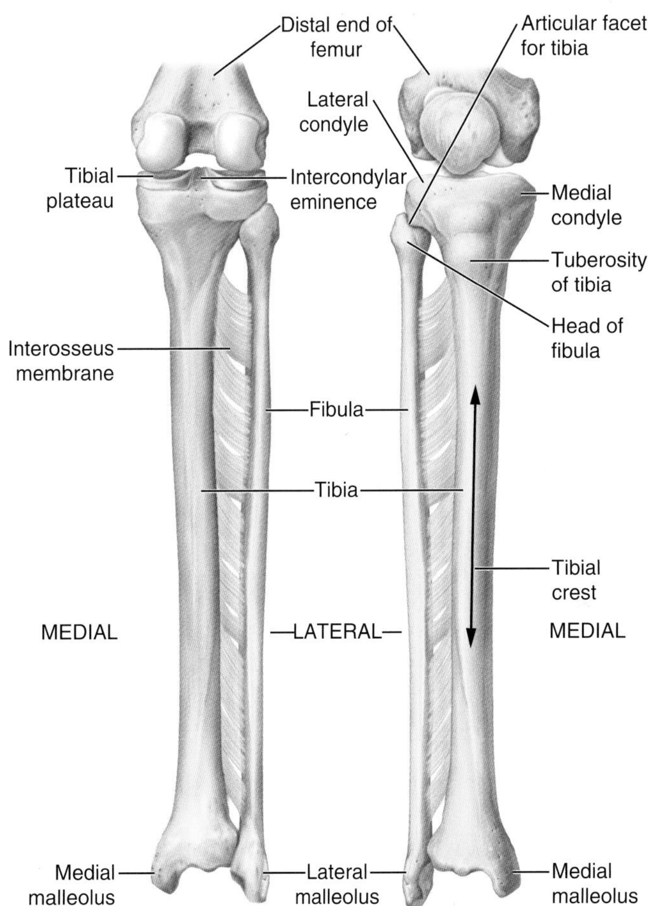

FIGURE 5.21 **Tibia and fibula. A.** Posterior view. **B.** Anterior view.

Box 5.1 SURFACE ANATOMY FOR MANUAL THERAPISTS

Manual therapists often need to visualize the relationship of key skeletal structures and landmarks to effectively assess and treat their clients. The following surface anatomy illustrations align key bone landmarks with the external contours of the body and are designed to help you improve your visualization and palpation skills.

Upper Body

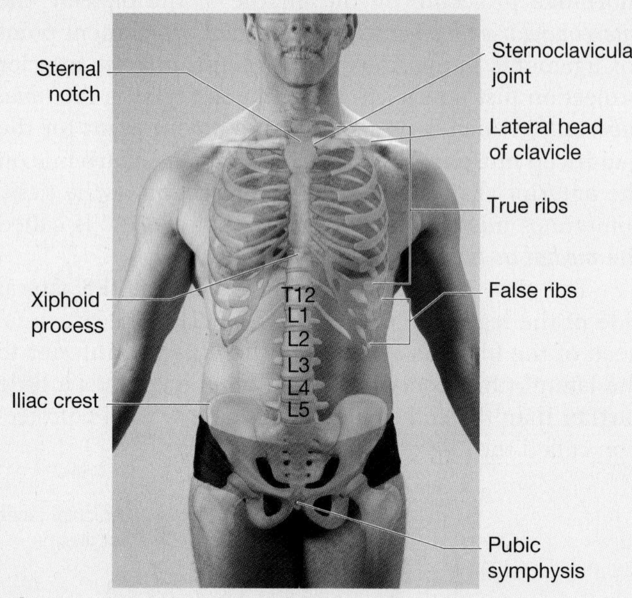

A

Occipital ridge
Spine of scapula
Acromion process
Superior angle of scapula
Inferior angle of scapula
T12
Posterior superior iliac spine (PSIS)
Iliac crest
Sacrum
Coccyx
Ischial tuberosity

A = Horizontal line between the superior angles approximates T2.
B = Horizontal line between the inferior angles approximates T7.
C = Horizontal line between iliac crests approximates L4-5 space.

Lower Body

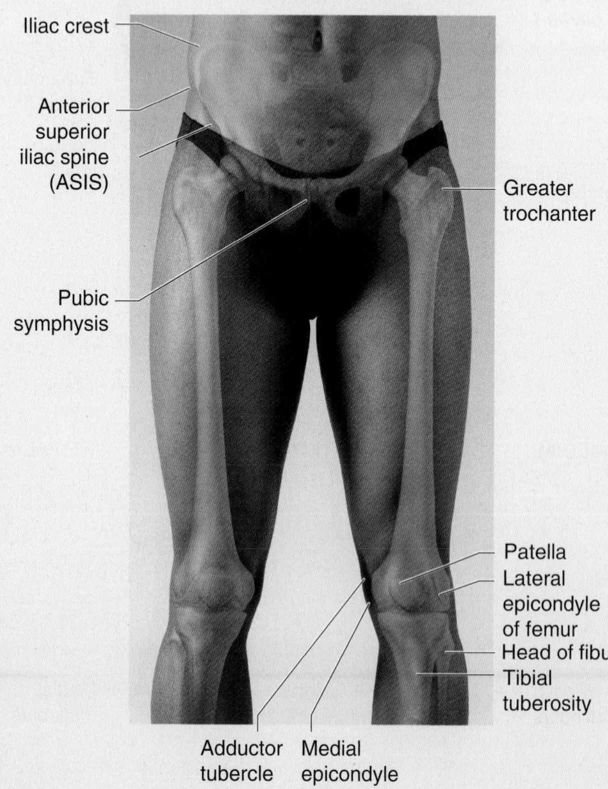

A

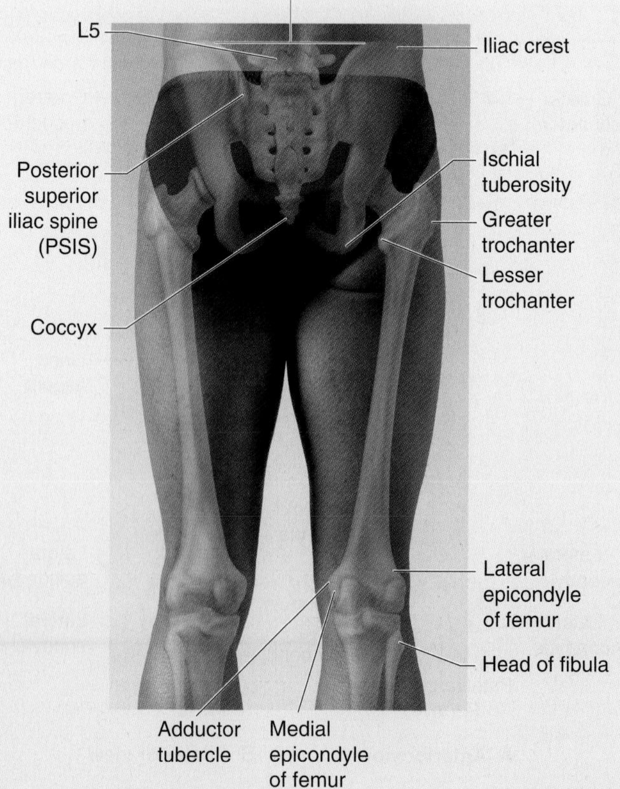

B

Ankle and Foot

The bones of the ankle and foot are shown in FIGURE 5.22. There are seven ***tarsal*** (ankle) bones. The easiest to identify is the ***calcaneus*** (heel bone). On top of the calcaneus is the ***talus***, which has the most direct relationship to the tibia. Directly anterior to these two bones there are two more large bones, the ***navicular*** on the medial side and the ***cuboid*** on the lateral. The final three tarsals are the smaller ***cuneiforms***. They are anterior to the navicular and are designated as ***medial, intermediate***, or ***lateral***. The long bones in the feet are ***metatarsals*** and like metacarpals, the proximal end is referred to as the *base*, and the round distal end is the *head*. The toe bones are ***phalanges*** that have the same proximal, middle, and distal arrangement as the fingers.

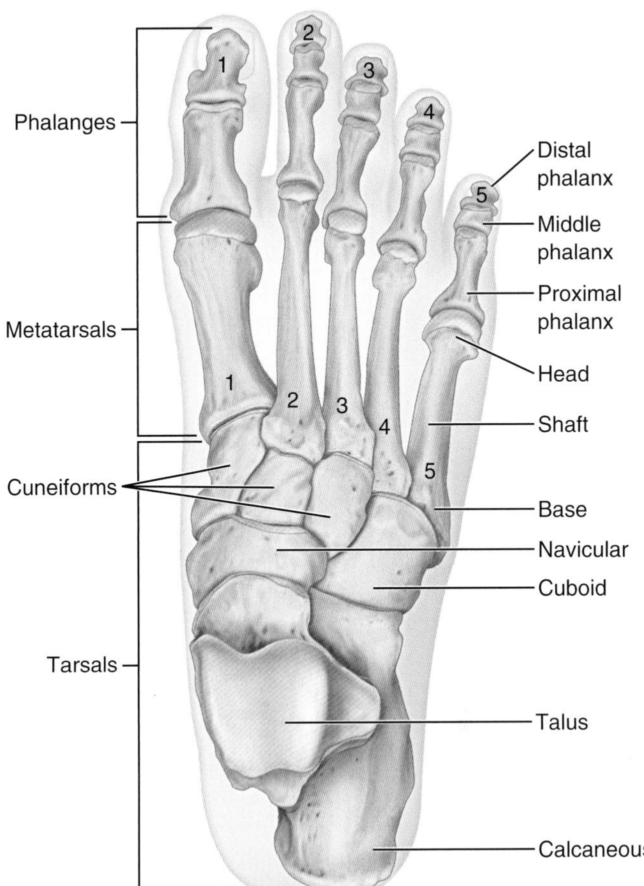

FIGURE 5.22 **Bones of the ankle and foot.** Superior view shows the 7 tarsals, 5 metatarsals, and 14 phalanges.

What Do You Think? 5.2

- A shoulder dislocation and shoulder separation are different injuries that occur in two different joints. Between the glenohumeral, AC, and SC joints, which joint do you think is the site for each of these and why?
- Since the fibula is a non–weight-bearing bone, what other functional purposes do you think it might serve?

JOINTS

Any junction in the skeleton where two bones meet is called a **joint** or **articulation**. The bones of the skeleton provide rigid levers needed for movement, and the joints are the fulcrums or points where movement occurs. This creates a more flexible frame for the body and allows humans to move in meaningful and graceful ways.

By the Way

Some of the joints in the body are not easily classified by structure or movement capability. For example, the sacroiliac (SI) joint has structural characteristics of both fibrous and synovial joints. While historically it was classified as either synarthrotic or amphiarthrotic, more current thinking recognizes that enough gliding motion occurs at the joint to more correctly categorize it as a **diarthrosis.**

Joints can be classified either by their structure or by their function (FIGURE 5.23). The primary consideration when classifying a joint structurally is the type of connective tissues found in the joint. When the bone ends are joined by fibrous connective tissue, a joint is classified as a **fibrous joint. Cartilaginous joints** have a fibrocartilage pad between the two bone ends, and a **synovial joint** is one in which the bone ends are connected by a joint capsule lined with a synovial membrane. The functional classification of joints uses the amount of movement allowed by the joint to create three groupings. **Synarthroses** are immovable joints, **amphiarthroses** are partially movable, while **diarthroses** are freely movable joints. With few exceptions, there is a direct correlation between the structural and functional classifications: fibrous joints are immovable, cartilaginous joints are slightly movable, and synovial joints are freely movable (TABLE 5-3).

Fibrous Joints

The best example of a fibrous joint is found in the skull where the cranial bones are held together by dense fibrous connective tissue. The small seams or **sutures**

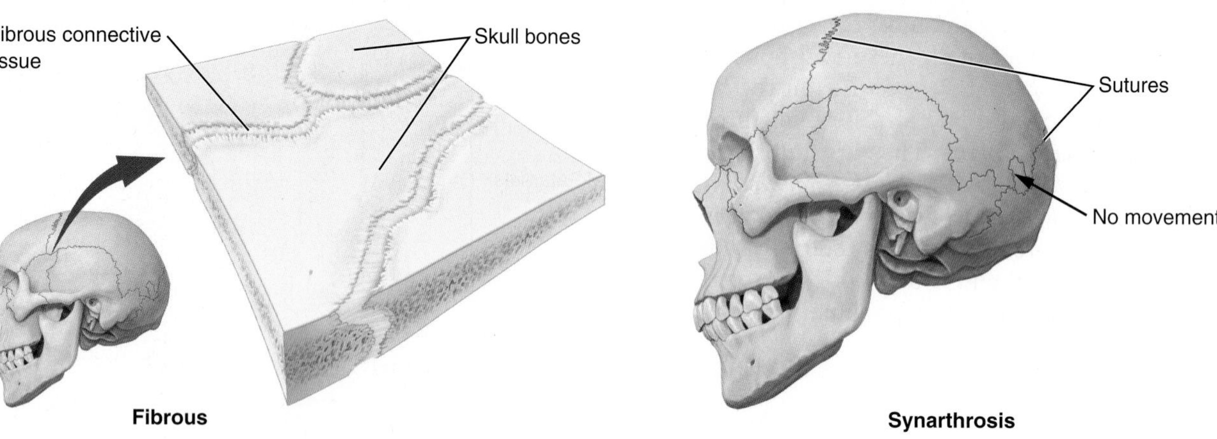

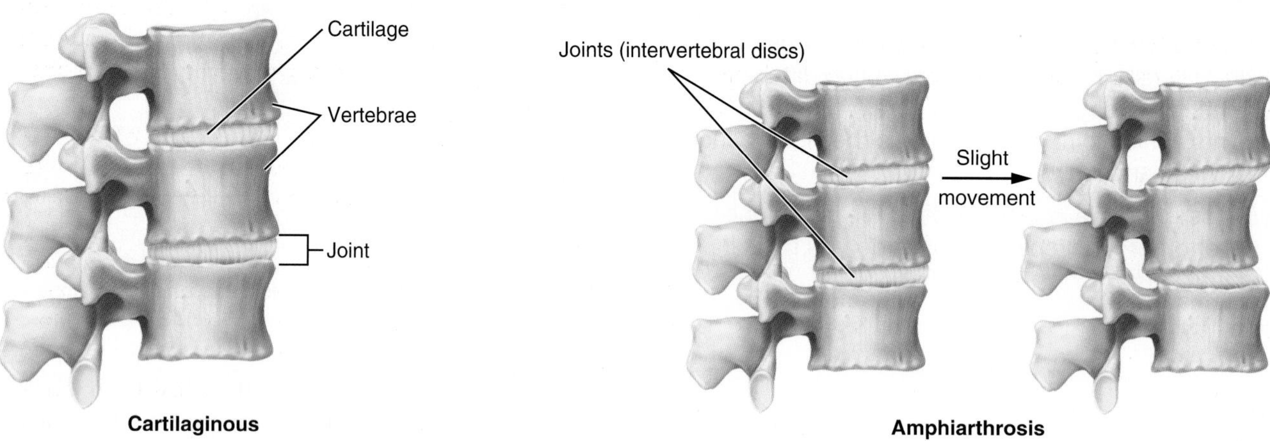

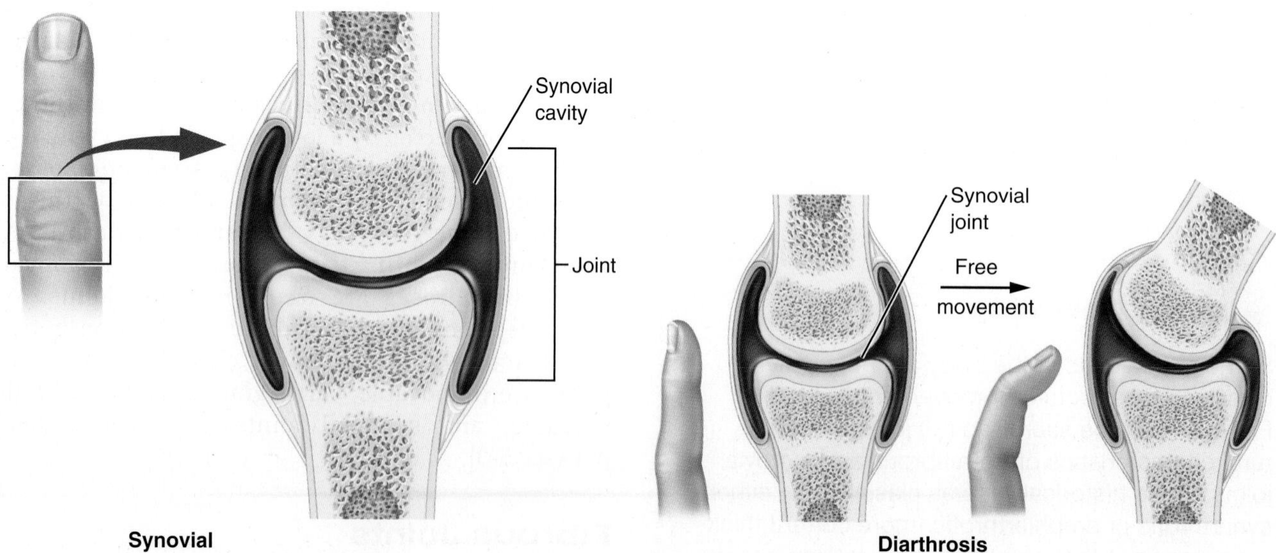

FIGURE 5.23 ▸ Joint classifications. Joints can be classified either by their structural or functional characteristics.

Table 5-3 Joint Classifications

Structural	Functional	Examples
Fibrous	Synarthrotic (immovable)	Sutures of the skull Tibiofibular articulations
Cartilaginous	Amphiarthrotic (slightly movable)	Pubic symphysis Intervertebral joints
Synovial	Diarthrotic (freely movable)	Elbow (humeroulnar joint) Knee (tibiofemoral joint) Hip (iliofemoral joint)

between the flat skull bones have jagged and irregular ends that interlock like puzzle pieces. The thin layer of dense fibrous tissue along the sutures creates a very firm junction between the cranial bones. While some shifting can occur at the sutures, these joints are basically immovable and thus considered synarthrotic. The proximal and distal tibiofibular articulations provide additional examples of fibrous joints. However, as long bones, the tibia and fibula do not have the same type of firm interlocking arrangement found in the cranial sutures. This allows the tibiofibular joints to shift and function more like an **amphiarthrosis** even though they are fibrous joints.

Cartilaginous Joints

Cartilaginous joints have a fibrocartilage pad between the bone surfaces that allow partial movement of the joint. Therefore, all cartilaginous joints in the body are also amphiarthrotic. These include the joint between the two pubic bones of the os coxae called the ***pubic symphysis*** and the intervertebral joints between the bodies of the vertebrae where ***intervertebral discs*** lie.

Synovial Joints

The synovial joint is named for the synovial membrane that lines the joint capsule. As shown in FIGURE 5.24, synovial joints are characterized by the presence of a number of common structural elements:

- **Articular cartilage**—The articular surface of the bone ends are covered with a layer of hyaline cartilage. This cartilage provides a smooth surface that eases movement and protects the bone ends from the wear and tear of friction and compression due to movement and weight-bearing activities.
- **Joint capsule**—This fibrous connective tissue sleeve surrounds and encloses the entire joint. It weaves into the periosteum of the connecting bones to firmly stabilize the joint.
- **Synovial membrane**—The fibrous joint capsule is lined with a synovial membrane that secretes

By the Way

Technically, the epiphyseal plates in the bones of children are cartilaginous joints. However, unlike other cartilaginous joints, the epiphyseal plate is not an articulation that allows movement; it is simply the site of bone growth and therefore, not really a joint at all.

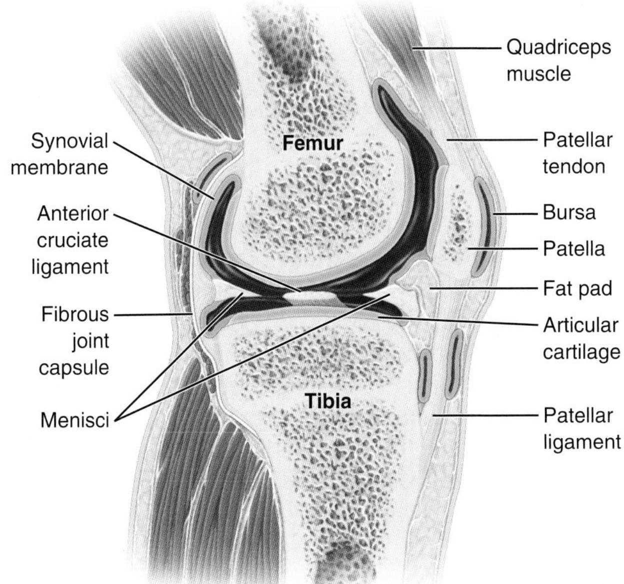

FIGURE 5.24 Synovial joint. All synovial joints share the same basic structural components as shown in this sagittal section of a knee joint.

Pathology Alert Spinal Disc Disorders

Intervertebral discs serve as shock absorbers and stabilizing spacers between the vertebrae. These discs have a tough outer ring, the ***annulus fibrosus***, and a central gelatinous core called the ***nucleus pulposus***. When discs are repetitively compressed or stressed beyond normal limits, they can distort or bulge into the vertebral canal or intervertebral foramen. Sometimes, the nucleus pulposus pushes through the annulus like the center of a jelly-filled doughnut squishing through the pastry. This condition, called a ***herniated disc***, can cause severe neck or back pain depending on the location of the damaged disc. Both bulging and herniated discs can cause neurological symptoms if the spinal cord or nerve roots are impinged by the portion of the disc invading the vertebral canal or intervertebral foramen. Common neurologic symptoms include numbness, tingling, or burning pain that shoots down the arm(s) or leg(s). If a client's low back or neck pain is accompanied by any of these neurological symptoms, therapists need to refer the client for a definitive diagnosis and may need to alter their session to ensure client comfort and safety.

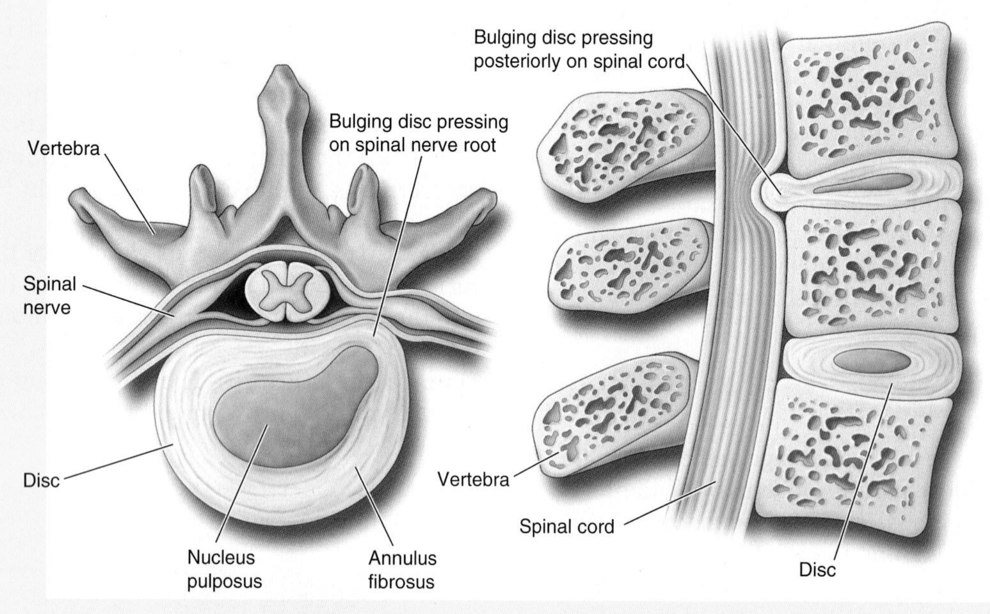

synovial fluid into the joint cavity. This fluid lubricates the joint, cushions bone ends, and supplies the articular cartilage with nutrition.

- **Ligaments**—These bands of organized fibrous connective tissue connect bones to one another. Ligaments may be located inside or outside the joint capsule and together with the capsule, they support and stabilize the joint. When ligaments are stretched or torn, the injury is called a **sprain**, and the joint is mildly or severely destabilized.

There are a few other structures that are common, but not present in all synovial joints. Small sacs of synovial-like fluid called **bursae** are situated between a bone projection or edge and tendons, ligaments, muscles, or skin. Their purpose is to reduce friction and cushion these structures during movement. When a bursa is inflamed due to irritation, injury, or infection, it is called **bursitis**. The most common sites for bursitis to occur are the calcaneus, knee, olecranon process of the elbow, greater trochanter of the femur, ischial tuberosity, and under the acromion process at the shoulder. Another important structure found in the knee and temporal-mandibular joint is a cartilage disc called a **meniscus** that cushions and provides additional support to the joint.

Types of Synovial Joints

The fibrous and cartilaginous joints contribute to the stabilizing and supportive capacity of the skeleton because their structures allow little or no movement. In contrast, the synovial joints are diarthrotic articulations that give the skeleton its role in locomotion. Synovial joints are classified according to the shape of their articulating bone surfaces. Like a machine-tooled piece of metal, it

Pathology Alert Arthritis

The literal translation of arthritis is "joint inflammation." Affected joints become hot, red, swollen, and painful. There are three types of arthritis that differ in etiology and the joint structures involved.

- ***Osteoarthritis***, also called degenerative joint disease (DJD), involves the articular cartilage of synovial joints. It is caused by wear and tear due to injury or normal life activities or the thinning of cartilage due to age. It generally occurs later in life and is most common in the weight-bearing joints such as knees, hips, and lower back.
- ***Rheumatoid arthritis (RA)*** is an autoimmune condition in which the body attacks its own synovial membranes. It particularly affects the joints of the hands and feet but can also affect blood vessels, serous membranes, the heart, lungs, and liver. RA can occur in children as well as adults and is characterized by intermittent episodes of acute pain and disability followed by periods of remission.
- ***Gout*** is caused by the collection of uric acid, a byproduct of protein metabolism, in and around a joint. When the body is unable to metabolize it effectively, excess uric acid crystallizes into sharp shards that accumulate in the joint capsule and synovial membrane, irritating the joint structures. Gout most often affects the joints of the feet, particularly the metatarsophalangeal joint of the big toe.

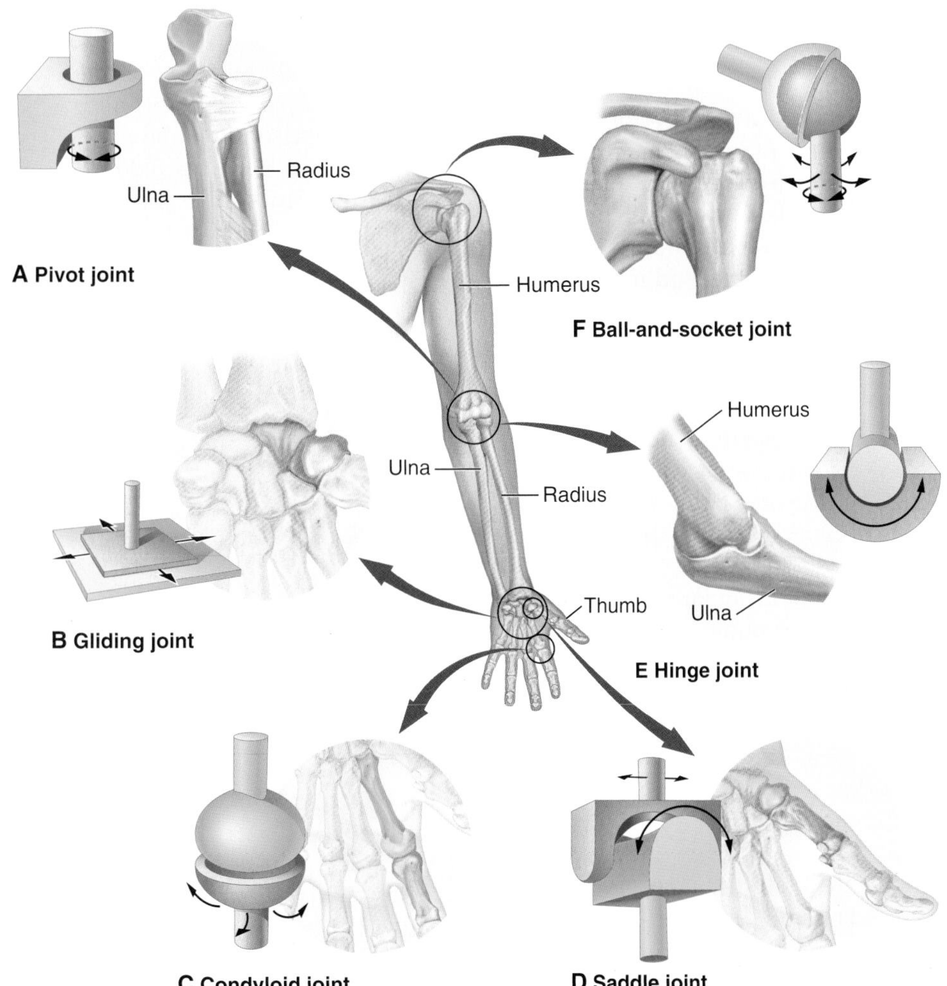

FIGURE 5.25 ▸ Types of synovial joints. Synovial joints are classified according the shape of the bone ends, how they articulate, and the movements they allow.

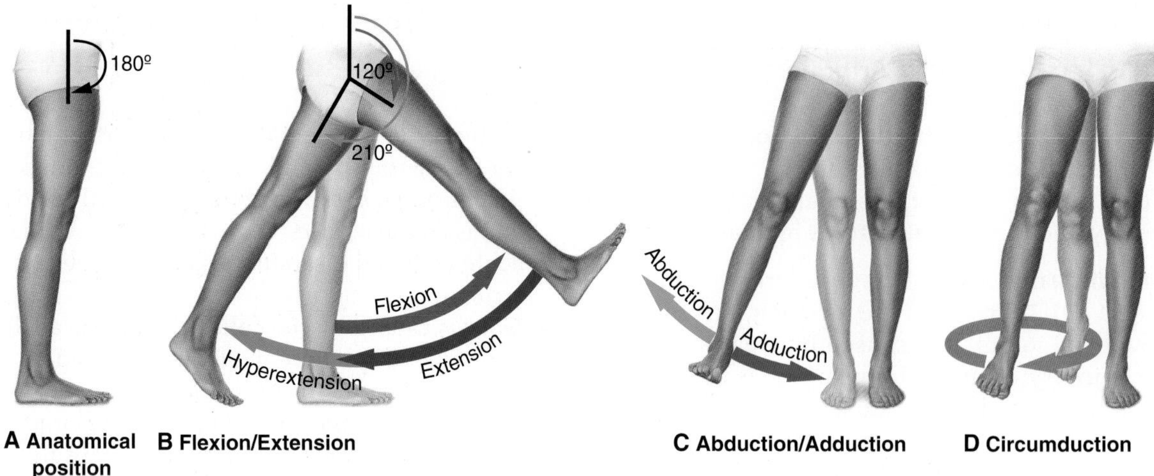

FIGURE 5.26 ▸ Basic movements. As a ball-and-socket joint, the hip allows for movement in all planes. **A.** Hip in anatomic position. **B.** Hip flexion and extension. **C.** Hip abduction and adduction. **D.** Hip circumduction.

is the shape and positioning of the bone ends, as well as the contour and angle of the articular surfaces that dictate the direction and variety of movements allowed at each diarthrotic joint. The six types of synovial joints (▸ FIGURE 5.25) are

- **Hinge joints**—Like a hinge on a door, the rounded end of one bone articulates with a groove or trough in another to create this type of joint. This hinged arrangement limits movement to only one plane, the sagittal plane. The movements allowed in a hinge joint are **flexion** (bending) and **extension** (straightening). The jaw, elbow, knee, ankle, and interphalangeal joints are all examples of hinge joints.
- **Condyloid joints**—These joints are characterized by two oval-shaped articular surfaces, one convex and one concave, that fit into one another. This allows for movement in two planes, the sagittal and coronal. Condyloid joints can flex and extend within the sagittal plane and **abduct** or **adduct** within the frontal plane. The wrist, MP, and metatarsophalangeal joints are all examples of condyloid joints.
- **Pivot joints**—When a round articular surface of one bone fits into the bone ring or notch of another, it allows the bones to **rotate** around a single axis like a wheel on an axle. Examples of pivot joints include the articulation between the atlas (C_1) and the axis (C_2) at the atlantoaxial joint. The proximal radioulnar joint is another good example.
- **Saddle joints**—Both bone surfaces in this joint are shaped like a saddle. They are oriented perpendicular to each other to create a joint that allows basically the same motions as a condyloid joint: flexion, extension, abduction, and adduction. The best example of a saddle joint is the carpometacarpal joint in each thumb.
- **Ball-and-socket joints**—Just as the name implies, these joints are shaped like a rounded ball fitted inside a shallow bowl or socket. This arrangement allows movement through all planes, giving these synovial joints the greatest movement capability. The two ball-and-socket joints are the shoulder and hip. These joints can flex and extend, abduct and adduct, horizontally abduct and adduct, rotate, and **circumduct** (▸ FIGURE 5.26).
- **Gliding joints**—The bones in these joints have smooth flat articular surfaces called **facets** that slide or glide across one another. These movements can occur on any axis or plane, but the amount of motion is generally quite limited. Gliding joints most often occur where short bones are clustered, as in the wrist and ankle, or are stacked in a series, as in the spinal column. The accumulated short-glide and shift between all the bones in the group

By the Way

The SC joint is easily recognized as a synovial joint, but experts disagree about its shape and movement capabilities. Most classify it as a gliding joint because the medial head of the clavicle seems to slide across the articulating surface of the manubrium. Others describe it as a saddle joint because the articular surfaces of the clavicle and manubrium can be described as shallow saddles. Still others argue that it is a pivot joint because the head of the clavicle appears to rotate when the arm is raised; however, ligaments around the joint keep it from fully rotating like other pivot joints.

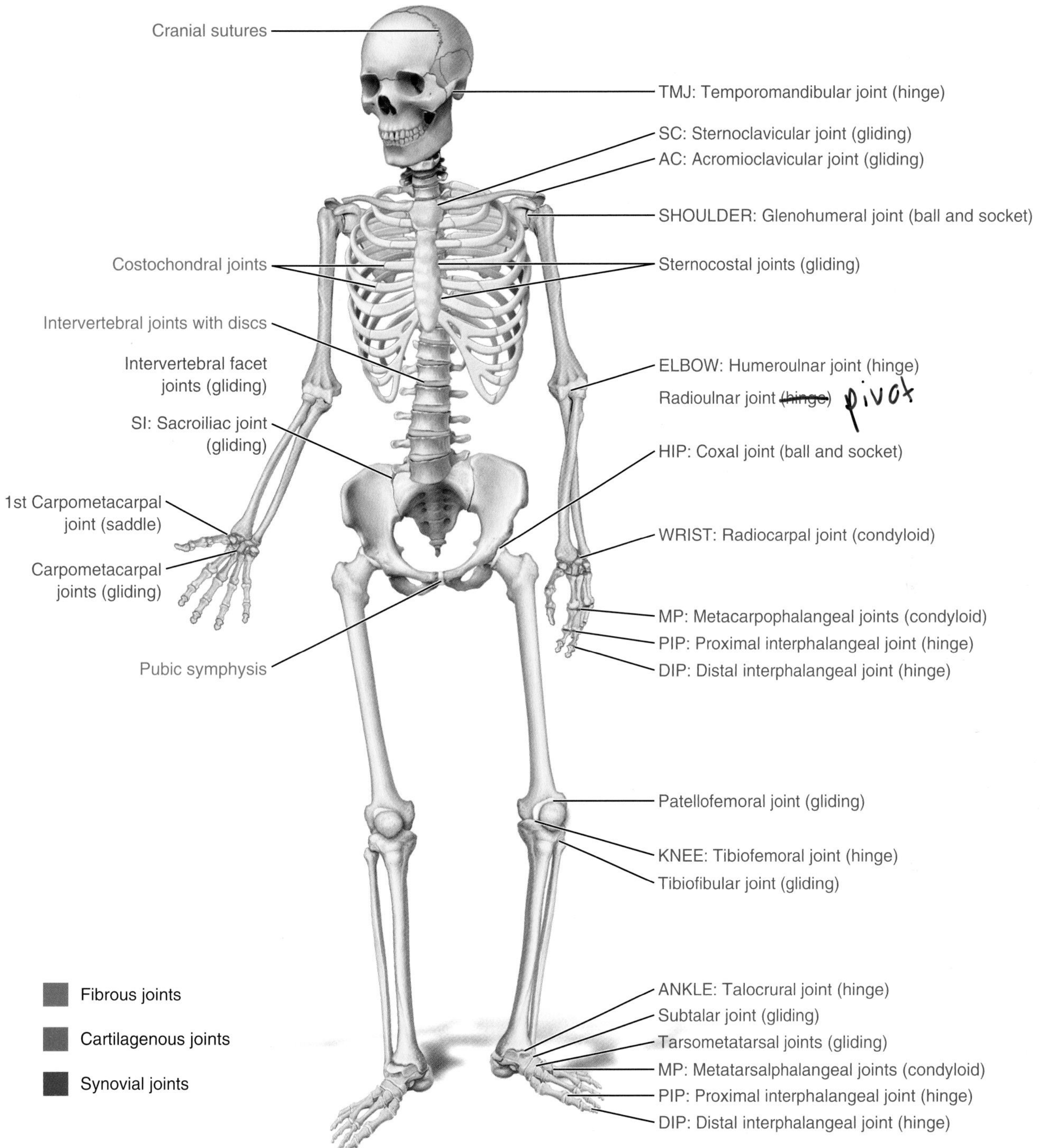

FIGURE 5.27 ▸ Joints classified by structure. For each synovial joint, the common name, anatomic name, and specific diarthrotic classification are shown.

allows for complex movements and adjustments in these areas. For example, the gliding action between the tarsals in the foot allows it to adapt to uneven surfaces during walking. The sliding that occurs between intervertebral facets allows the spinal column to bend, twist, and turn in multiple directions.

FIGURE 5.27 provides a summary of the joints classified by structure. For each synovial joint, the figure also provides the common name, anatomic name, and specific diarthrotic classification.

Osteokinematics is the movement of bones around a joint axis, consisting of the basic joint movements: flexion, extension, abduction, adduction, rotation, and circumduction. **Arthrokinematics** is the normal and necessary joint play or the shift, slide, and rotational movements that occur between the articular surfaces of the bones. Normal joint movements are greatly restricted when conditions such as arthritis or cartilage degeneration impede the arthrokinematics of a joint. However, addressing deficiencies in arthrokinematics may be outside the scope of practice for many manual therapists. Most therapeutic interventions are directed toward improving osteokinematics by decreasing muscle tension and fibrotic adhesions in the joint capsule, ligaments, and tendons. Therapeutic movement techniques may also enhance osteokinematics by improving the emotional, mental, or proprioceptive sense of ease with movement.

Special Movements

The general movement terminology introduced in Chapter 2 is all that is needed to describe the basic motions in most synovial joints. Yet there are several joints for which the position of the bones or the unique direction of movement must be described more specifically. For example, the ankle is a hinge joint, so the movements it allows are flexion and extension. However, the ankle hinge is already at 90° making it difficult to decide which direction is bending and which straightening, so a special descriptor is needed. Pulling the foot and toes upward (toward the nose) is called **dorsiflexion** because the movement is in the direction of the dorsum of the foot. Pointing the toes like a ballerina is called **plantar flexion** because the foot is moving in the direction of the plantar surface. TABLE 5-4 lists special movement terms, their definitions, and shows an example of each.

AGING AND THE SKELETAL SYSTEM

Over time, significant changes occur within skeletal tissue. While the natural cycle of bone remodeling continues throughout our lives, the balance between bone deposition and resorption shifts. From birth through adolescence, bones make more bone tissue than they break down to meet the demands of a growing body. As we reach adulthood, bone production and resorption rates equalize. However, when we enter middle age, the rate of bone loss begins to outpace the rate of bone production. Over time, this leads to a gradual demineralization of bone tissue, decreasing overall bone density. Bone loss tends to be more significant for women because demineralization begins around age 30, as compared to age 60 in men. Additionally, women generally have smaller bones, so any loss of tissue is more significant.

Along with the demineralization of the bone matrix, there is also a decrease in the production of new collagen fibers and ground substance. Together, these decreases change the bone matrix, and the bones become thinner and more brittle. This often leads to a decrease in height or increase in thoracic or lumbar curves as spinal vertebrae and discs thin and weaken. These changes also make us more susceptible to tooth loss and bone fractures. Additional changes occur in the synovial joints. Articular cartilage thins, the joint capsule and ligaments stiffen, and less synovial fluid is produced. These changes can cause joint pain, decreased joint flexibility, decreased range of motion, and challenges with locomotion and balance.

Manual therapists may need to take extra precautions to ensure the safety and comfort of older clients. For example, clients with extreme kyphosis may require special bolstering, or they may not be able to lie prone at all. Some geriatric clients may require assistance getting on or off a treatment table, and caution should always be used with any deep tissue techniques, range of motion exercises, or stretching.

Table 5-4 Special Movement Terms

Term	Definition	Example
Inversion	Turn inward	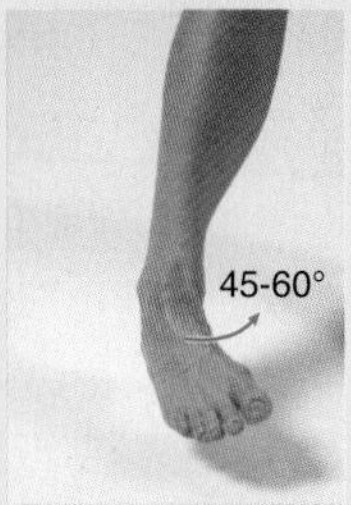 45-60°
Eversion	Turn outward	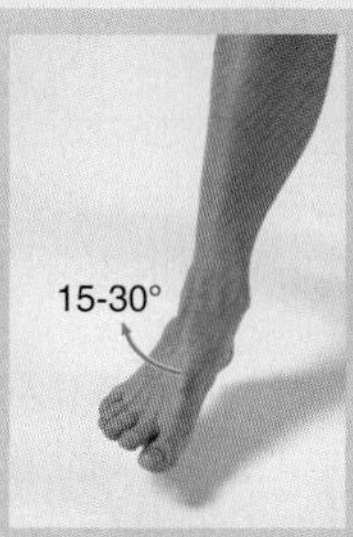 15-30°
Dorsiflexion	Move the ankle in the direction of the dorsum of the foot	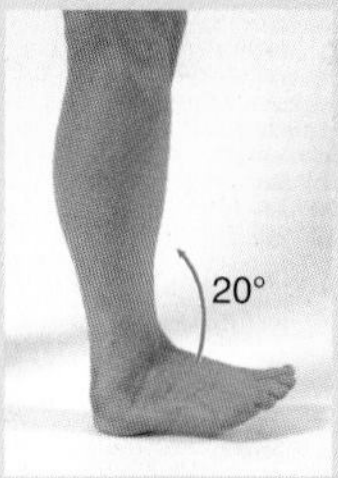 20°
Plantar flexion	Move the ankle in the direction of the plantar surface of the foot	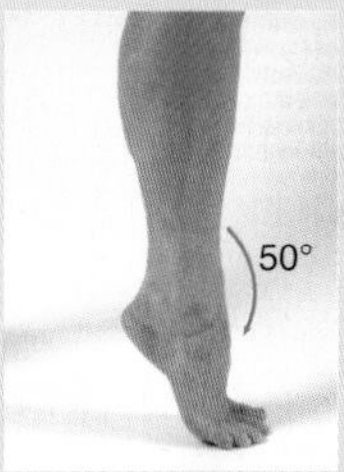 50°
Lateral flexion	Bend to the side, from the trunk or neck	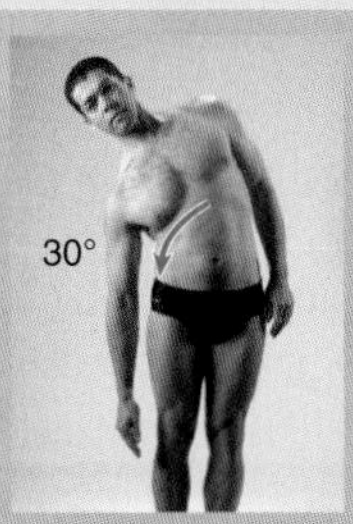 30°
Elevation	Move upward	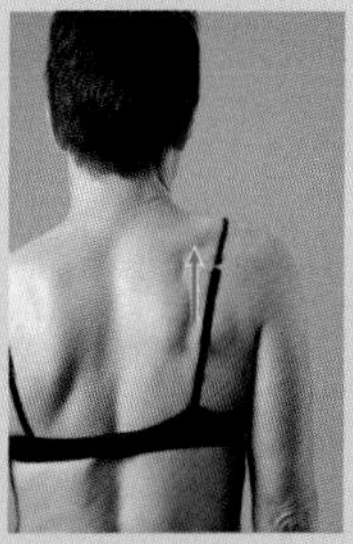

(Continued)

Table 5-4 Special Movement Terms *(Continued)*

Term	Definition	Example
Depression	Press downward	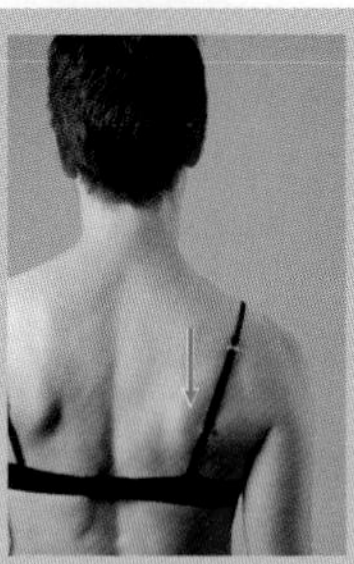
Protraction	Move a body part forward	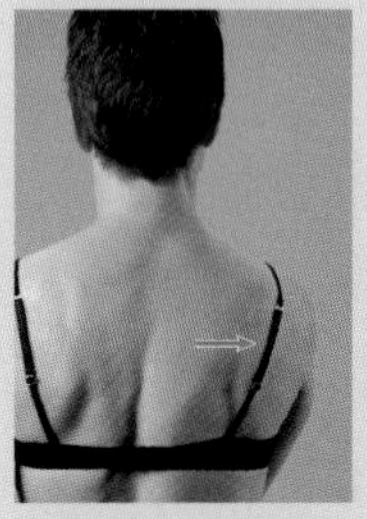
Retraction	Draw a body part back	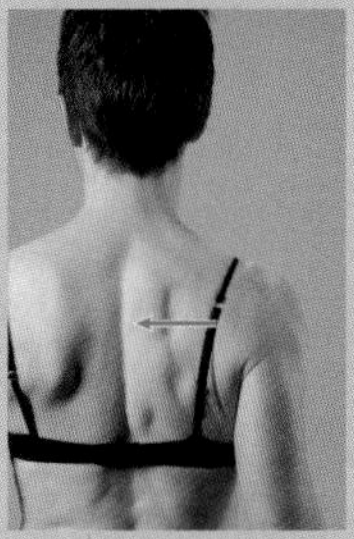
Upward rotation (scapula)	Glenoid fossa and acromion rotate upward	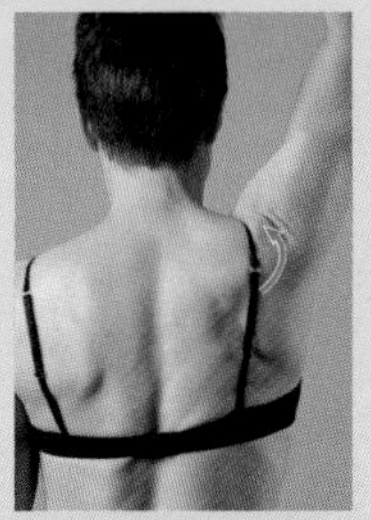
Downward rotation (scapula)	Glenoid fossa and acromion rotate downward	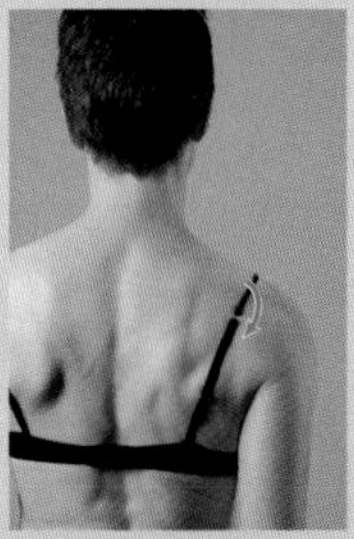
Medial or internal rotation	Rotate in toward the midline	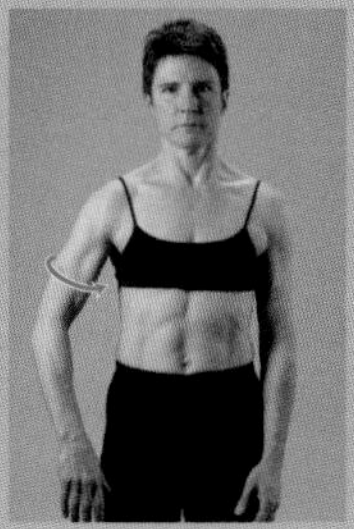

Table 5-4 Special Movement Terms *(Continued)*

Term	Definition	Example
Lateral or external rotation	Rotate out away from the midline	
Supination	Turn the forearm so the palm faces forward	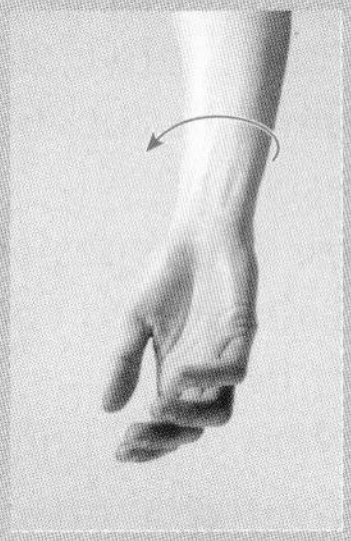
Pronation	Turn the forearm so the palm faces backward	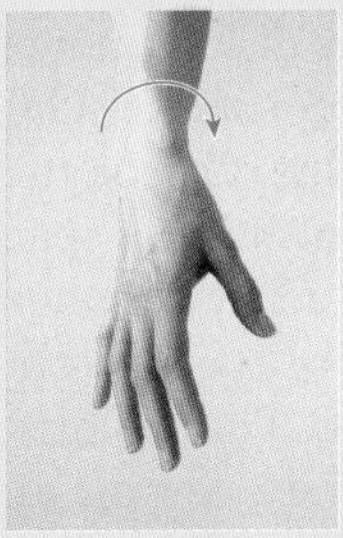
Radial deviation	Abduction of the wrist (anatomic position)	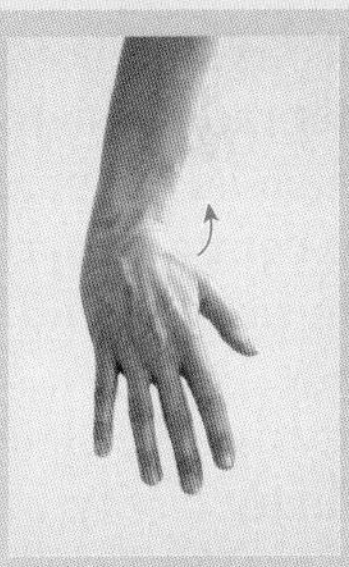
Ulnar deviation	Adduction of the wrist (anatomic position)	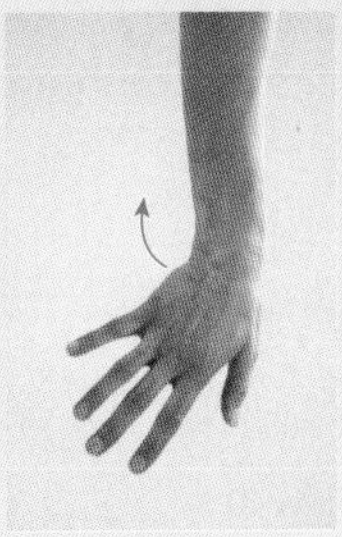

SUMMARY OF KEY POINTS

- The skeletal system provides framework, protection, levers for movement, calcium and other mineral storage, and is the site for blood cell production in the body.
- The bones of the skeleton are organized into two major divisions: axial and appendicular. See TABLE 5-1 to review the axial and appendicular bones.
- There are four types of bone cells: osteoblasts that build up bone, osteoclasts that break down bone, osteocytes that carry out nutrient-waste exchange to keep bones alive, and osteogenic cells that are the only bone cells capable of mitosis (developing into osteoblasts).
- There are two types of bone tissue: compact (dense), also called cortical, and spongy (cancellous). Compact bone forms the shaft of long bones and the outer layers of all bones. It is made up of osteons, or haversian systems, a group of concentric rings of bone tissue called lamellae. Spongy bone is found in the ends of long bones and throughout the flat bones. It is composed of a lattice work of bone tissue called trabeculae. Osteocytes that keep bone tissue alive reside in the tiny spaces of both types of bone tissue, connected by canaliculi, a network of canals.
- Bones are classified by shape as flat, long, short, or irregular. The majority of appendicular bones are long bones; carpals and tarsals are short or cuboid; the sternum, scapula, and bones in the skull and pelvic girdle are flat bones; and irregular bones include the vertebrae, some facial bones, and the sesamoid bones such as the patella.
- The five common parts of long bones are
  - Epiphysis—the end of the bone; contains red marrow
  - Diaphysis—the shaft
  - Medullary cavity—cavity in the middle of the shaft; contains yellow marrow
  - Articular or hyaline cartilage—covers the articular surfaces of the bone
  - Periosteum—outer connective tissue covering
  - Endosteum—lining of the medullary cavity
- Each bone has a variety of depressions, bumps, grooves, and projections called bone landmarks that serve as attachments for other bones, muscles, and ligaments, or as passageways for nerves and blood vessels. See TABLE 5-2 to review these landmarks.
- Joints (articulations) can be classified by structure and function. The structural classification is based mostly on the connective tissues of the joint and the manner in which they stabilize the bone ends. The structural classification of joints is as fibrous, cartilaginous, or synovial. The functional method of classification is based on the movements allowed at the joint. Fibrous joints are called synarthroses and are immovable, and cartilaginous joints are amphiarthrotic; allow partial movement. All synovial joints are diarthrotic (freely movable) joints.
- All synovial joints have four common structural features:
  - A stabilizing fibrous capsule
  - Stabilizing ligaments; can be inside or outside the joint capsule
  - A synovial membrane that secretes synovial fluid into the joint cavity
  - Articular or hyaline cartilage to protect the bone ends and stabilize the joint
- Synovial joints are classified according to shape and their movement capability as hinge, condyloid, pivot, ball-and-socket, gliding, or saddle joints.
- Common changes in the skeletal system related to aging include wearing away of the articular cartilage, decreased range of motion, decreased density of bone tissue matrix, and decreased production of blood cells.
- Manual therapists use their knowledge of the names and locations of bones, bone landmarks, types of joints and their movement capabilities to assess and treat clients.

REVIEW QUESTIONS

Short Answer

1. Name and explain the five functions of the skeletal system.
2. List the four types of bones.
3. Name and define the three structural classifications of joints and describe the movement capability of each.
4. Name and explain the movement capability of the six types of synovial joints.

Classification

Place an **A** next to the names of the appendicular bones and an **X** next to the names of the axial bones.

5. ________ sacrum
6. ________ radius
7. ________ calcaneus
8. ________ sternum
9. ________ fibula
10. ________ ilium
11. ________ clavicle
12. ________ scapula
13. ________ humerus
14. ________ rib

Place a **C** next to the names of the cranial bones and an **F** next to the names of the facial bones.

15. ________ occipital
16. ________ frontal
17. ________ mandible
18. ________ ethmoid
19. ________ sphenoid
20. ________ zygomatic
21. ________ lacrimal
22. ________ maxilla
23. ________ parietal

Multiple Choice

24. What is the anatomic term for the collar bone?
    a. humerus
    b. zygomatic
    c. clavicle
    d. scapula
25. What is the anatomic name for the bones of the fingers and toes?
    a. metacarpals
    b. phalanges
    c. fontanels
    d. tarsals
26. What is the anatomic name for the end of a long bone?
    a. epiphysis
    b. diaphysis
    c. epiphyseal plate
    d. medullary end
27. Which type of bone tissue gives the shaft of a long bone its strength?
    a. compact
    b. spongy
    c. fibrous
    d. cancellous
28. The hollow cavity in the shaft of a long bone is called the ____________ cavity.
    a. diaphysis
    b. yellow marrow
    c. epiphyseal
    d. medullary
29. Blood cell production occurs in which part of the long bone?
    a. medullary cavity
    b. epiphysis
    c. diaphysis
    d. periosteum
30. Which type of tissue covers the articular surface of the bone ends?
    a. fibrocartilage
    b. hyaline cartilage
    c. periosteum
    d. synovial membrane
31. Which bone in the lower extremity is not a weight-bearing bone?
    a. femur
    b. tibia
    c. fibula
    d. talus
32. What is the name of the only movable bone in the skull?
    a. maxilla
    b. mandible
    c. zygomatic
    d. nasal

Continued on page 114

33. How many vertebrae make up the cervical region of the spinal column?
    a. 10
    b. 5
    c. 7
    d. 9
34. How many pairs of "true ribs" are in the rib cage?
    a. 7
    b. 10
    c. 12
    d. 5
35. What is the term for the most proximal portion of the sternum that articulates with the clavicle?
    a. body
    b. xyphoid process
    c. coracoid process
    d. manubrium
36. Which of the following joints is amphiarthrotic?
    a. acromioclavicular
    b. atlantoaxial
    c. pubis symphysis
    d. intervertebral facet joints
37. Which of the following is the best example of a condyloid joint?
    a. metacarpophalangeal
    b. proximal radioulnar
    c. distal interphalangeal
    d. iliofemoral
38. What is the name of the flat shelf-like process projecting from the lateral end of the spine of the scapula?
    a. coracoid
    b. coranoid
    c. acromion
    d. deltoid
39. What is another term for compact (dense) bone tissue?
    a. cancellous
    b. trabeculae
    c. lacunae
    d. cortical
40. The system of concentric rings that make up compact bone are osteons or the ___________ system.
    a. haversian
    b. trabecular
    c. osseous
    d. cancellous

Matching

Match the bones in the right column with their associated bone landmarks in the left column. (Bones in the right column may be used more than once or not at all.)

	Landmark	Bone
_______	41. mastoid process	a. scapula
_______	42. coracoid process	b. temporal bone
_______	43. linea aspera	c. humerus
_______	44. axillary border	d. radius
_______	45. medial malleolus	e. ulna
_______	46. bicipital groove	f. femur
_______	47. greater trochanter	g. tibia
_______	48. trochlear notch	h. fibula
_______	49. deltoid tubercle	i. clavicle
_______	50. sternal notch	j. sternum
_______	51. obturator foramen	k. os coxae
_______	52. trochlea	l. patella
_______	53. olecranon process	m. occipital
_______	54. glenoid fossa	n. parietal
_______	55. acetabulum	
_______	56. lesser tubercle	
_______	57. coranoid process	
_______	58. infraspinous fossa	
_______	59. patellar groove	
_______	60. radial fossa	
_______	61. foramen magnum	
_______	62. greater sciatic notch	

6 The Skeletal Muscle System

LEARNING OBJECTIVES

Upon completion of this chapter, you will be able to:

1. Give several examples of how knowledge of the muscular system is applied in manual therapy practice.
2. List and describe the key characteristics and functions of skeletal muscle.
3. Name and locate the major parts of a skeletal muscle.
4. Name and describe the distinguishing characteristics between fascia and tendons.
5. Describe the microscopic arrangement of a skeletal muscle fiber.
6. Name the parts of a motor unit and explain the role of each in muscle contraction.
7. Explain the key physiologic principles that govern the function of skeletal muscle, including sliding-filament mechanism, all-or-none response, threshold stimulus, and motor unit recruitment.
8. List and explain the three major types of muscle contraction.
9. Explain the three primary mechanisms of producing energy for muscle contraction.
10. Explain muscle fatigue and oxygen debt.
11. Name and describe the different types of fiber arrangements found in muscles of the body.
12. Name and describe the four major roles muscles play in creating and controlling movement.
13. Explain how muscles get their names.
14. Describe the general location of the major muscles and the prime function(s) of each.
15. Describe common skeletal muscle system changes associated with exercise and aging and explain their implications for manual therapy practices.

> "Holding on to anger, resentment, and hurt only gives you tense muscles, a headache, and a sore jaw from clenching your teeth."
>
> JOAN LUNDEN AND LAURA MORTON
>
> *Joan Lunden's Healthy Living*

KEY TERMS

actin (AK-tin)
aerobic (ah-RO-bik) **metabolism**
agonist (AG-on-ist)
all-or-none response
anaerobic (AN-ah-RO-bik) **metabolism**
antagonist (an-TAG-ah-nist)
aponeurosis (AP-o-nu-RO-sis)
concentric (kon-SEN-trik)
contractile (kon-TRAK-tile)
direct phosphorylation (fos-for-ah-LA-shun)
eccentric (e-SEN-trik)
end feel
fulcrum (FUL-krum)
graded response
hypertrophy (hi-PER-tro-fe)
inert (in-ERT) **tissue**
insertion (in-SIR-shun)
isometric (i-so-MET-rik)
isotonic (i-so-TON-ik)
motor tone
muscle tone
myosin (MI-o-sin)
origin (OR-eh-jin)
oxygen (OKS-eh-jin) **debt**
prime function
reciprocal inhibition (re-SIP-ro-cal in-hi-BISH-un)
sliding filament mechanism (FIL-ah-ment MEK-ah-niz-em)
stabilizer (STA-bil-i-zer)
synergist (SIN-er-jist)
threshold (THRESH-hold) **stimulus**
tonic (TAH-nik)
tropocollagen (tro-po-KAHL-ah-jin)

Primary System Components

- **Skeletal muscle**
  - Muscle belly
  - Fascicle
  - Muscle fiber
  - Myofibrils
  - Myofilaments
  - Sarcomere
- **Fascia**
  - Epimysium
  - Perimysium
  - Endomysium
- **Tendons**
  - Aponeurosis
  - Tenoperiosteal junction
  - Musculotendinous junction
- **Motor units**
- **Neuromuscular junction**

Primary System Functions

- Moves body parts via coordinated muscle contractions
- Maintains posture
- Helps stabilize joints
- Generates heat to maintain body temperature

While there are three types of muscle tissue in the body (skeletal, smooth, and cardiac), only skeletal muscle tissue is organized into a single coordinated system. More than 600 skeletal muscles that make it possible for us to crawl, walk, skip, or run, as well as to actively demonstrate what we think and feel. Through the coordinated contraction of teams of muscles we put pen to paper, share ideas with colleagues, text friends, smile at strangers, embrace loved ones, and sometimes dance with abandon. These same muscles also express our feelings through subtle movements, facial expressions, and posture. Together with the skin and connective tissue, skeletal muscles make up the soft tissue medium most directly affected by manual therapies. Indeed, most clients who schedule appointments with manual therapists are seeking to have sore, stiff, tense, or tired muscles loosened, relaxed, and stretched.

In this chapter, you will learn the anatomy of skeletal muscles and the microscopic structure of their individual cells. The mechanism of contraction and its physiologic principles are explained, and different types of contractions, the mechanics of movement, and roles that muscles play during movement are explored. However, the cooperative function between the muscular and nervous systems that fully coordinates movement, including proprioception and several specific muscle reflexes, will not be fully discussed until Chapters 7 and 8. In this chapter, the location and actions of the major skeletal muscles are provided, along with a special boxed feature on surface anatomy for manual therapists. Cardiac and smooth muscle will be discussed in detail in Chapter 10.

CHARACTERISTICS AND FUNCTIONS OF SKELETAL MUSCLE

Skeletal muscles are strong force generators that are both flexible and resilient due to the unique properties of muscle tissue. Muscle tissue fibers are particularly **excitable**, designed to respond quickly to stimulation from the nervous system. Their inner organization makes them extremely **contractile**, allowing each fiber to forcefully shorten when stimulated. They are also **extensible** and **elastic**, giving them the ability to lengthen and rebound back to their original shape and length. These characteristics enable the skeletal muscle system to efficiently carry out four primary functions:

- Movement of body parts—Muscles are attached to bones on either side of one or more joints. When a muscle contracts it generates tension, a pulling force that moves the bones.
- Posture—The skeleton provides a rigid frame for the body. However, it cannot sustain an upright position against gravity without the tension created by muscle contraction.
- Stabilization of joints—Skeletal muscles work with ligaments and joint capsules to maintain the structural stability of each individual articulation.
- Production of heat—Muscle contraction requires the breakdown of adenosine triphosphate (ATP) to produce energy, and one of the primary byproducts of this chemical breakdown is heat. This is why we shiver when we're cold; the brain signals the muscles to contract to create heat that will help maintain body temperature.

STRUCTURE OF SKELETAL MUSCLE

Skeletal muscle tissue and fibrous connective tissue are the two primary tissues that make up the organs known as skeletal muscles. Each muscle is covered by an outer connective tissue layer called the **epimysium**, which forms an outer envelope that anchors the hypodermis of the skin to the muscle (▶ FIGURE 6.1). A second layer of connective tissue, the **perimysium**, divides each muscle into several large sections known as **fascicles**. These sections are large bundles of muscle fibers covered by another thin layer of connective tissue called the **endomysium**. A good way to envision this kind of fascial layering is to peel an orange. The rind of the fruit represents the epimysium surrounding the entire muscle. Like the perimysium divides a muscle into fascicles, a thin membrane divides the orange into multiple sections. When you pull the sections apart, you see that the individual "cells" of the fruit are covered by a thin membrane, just as the endomysium covers each individual muscle fiber.

The central portion of every muscle is the **muscle belly**. As the muscle belly tapers toward its attachment points, the epimysium, perimysium, and endomysium converge to form a single band of fibrous connective tissue called a **tendon**. Similar to the way in which ligaments attach one bone to another, tendons attach muscles to bones and transmit the pulling force of muscle contraction to the skeleton. To understand the mechanism of contraction, it is important to take a deeper look into the microscopic structure of skeletal muscle fibers, as well as the specific cellular composition of the fibrous connective tissues integrated into each muscle. This information also provides helpful insights into the intention and specific techniques of several forms of manual therapy.

Skeletal Muscle Fibers

Skeletal muscle fibers have the same structural components as other cells, but also have two very unique features; they have more than one nucleus (multinucleated) and two to three times more mitochondria than other cells. The multiple nuclei are formed during embryonic development when several immature cells

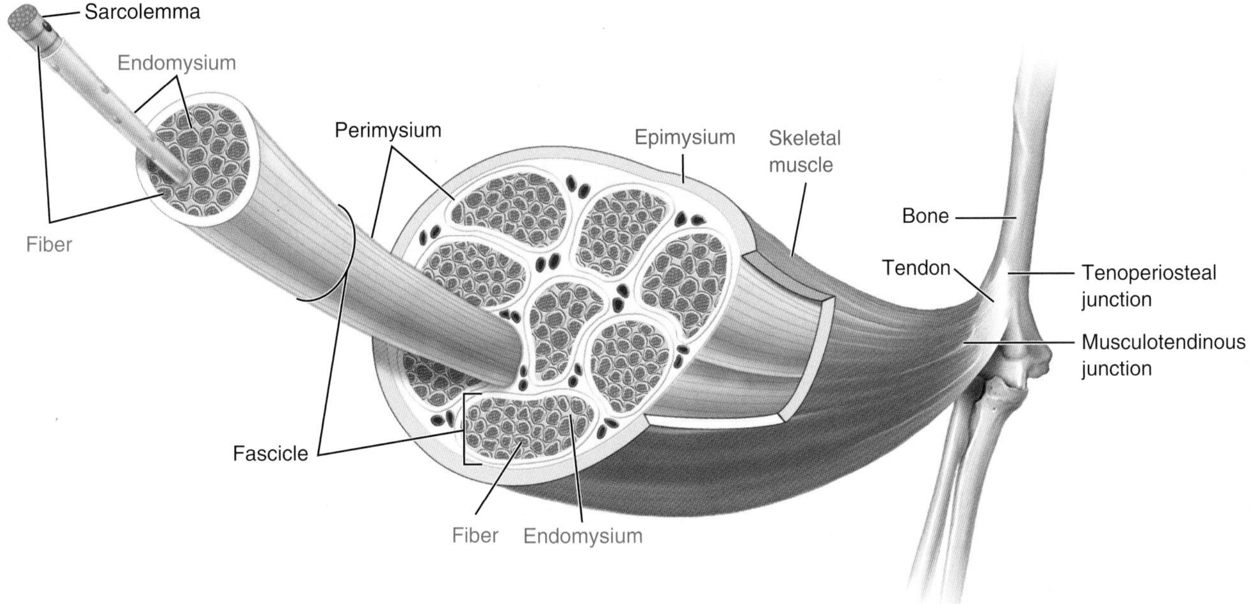

FIGURE 6.1 ▸ Structure of a skeletal muscle. Skeletal muscles are composed of muscle fibers organized by fibrous connective tissue layers. The endomysium surrounds each muscle fiber. Multiple fibers are bundled together and covered by the perimysium, forming fascicles, which are held together by the epimysium.

with individual nuclei merge to form one multinucleated muscle fiber. To meet the higher energy demand of muscle contraction, fibers also develop a high number of mitochondria. Like other specialized cells in the body, several of the organelles in a muscle fiber are given specific names that designate them as part of a muscle cell. Because muscles make up the fleshy bulk of the body, the prefix *sarco-* (meaning flesh) is used for these terms. The plasma membrane is referred to as the **sarcolemma**; the cytoplasm is called the **sarcoplasm**; and the endoplasmic reticulum is the **sarcoplasmic reticulum**.

Inside each muscle fiber, there are bundles of small cylindrical organelles that extend the entire length of the cell in parallel alignment called **myofibrils** (▸ FIGURE 6.2). In turn, the myofibrils are composed of two even smaller **myofilaments**, one thick and the other thin. The thick filaments are made of the protein **myosin** and the thin filaments are made of another protein, **actin**. Myosin filaments have a central shaft with multiple extensions. In contrast, actin filaments look like thin threads spiraled around each other. Neither of the myofilaments runs the entire length of the myofibril. Instead, they are arranged in an overlapping pattern and divided into short segments called **sarcomeres** that attach end-to-end along the length of the myofibril. The number of sarcomeres in any myofibril will vary according to the length and diameter of the fiber and can be altered by regular use or disuse.

Under the microscope, sarcomeres have dark and light areas that give skeletal muscle tissue a striated, or striped, appearance. The dark areas, called ***A bands***, are created by the thick myosin filaments. Light areas, ***I bands***, appear where there are only thin actin filaments. I bands extend between two sarcomeres and include a zigzagged "fence line" called a ***Z line***, which separates one sarcomere from another. Functionally, the Z line serves as the anchor point for the actin filaments.

By the Way

Muscle tissue makes up about two-fifths or 40% of overall body mass and consumes the most energy and oxygen of all body tissues.

Connective Tissue Components

Skeletal muscles include both disorganized and organized fibrous connective tissue components. The epimysium, perimysium, and endomysium are considered **fascia**, and classified as disorganized fibrous connective tissue. In contrast, the tendons and aponeuroses that attach muscles to bone are categorized as organized. Recall from Chapter 3 that these fibrous connective tissues differ in the density and arrangement of their fibers, as well as in the amount of ground substance in the tissue matrix.

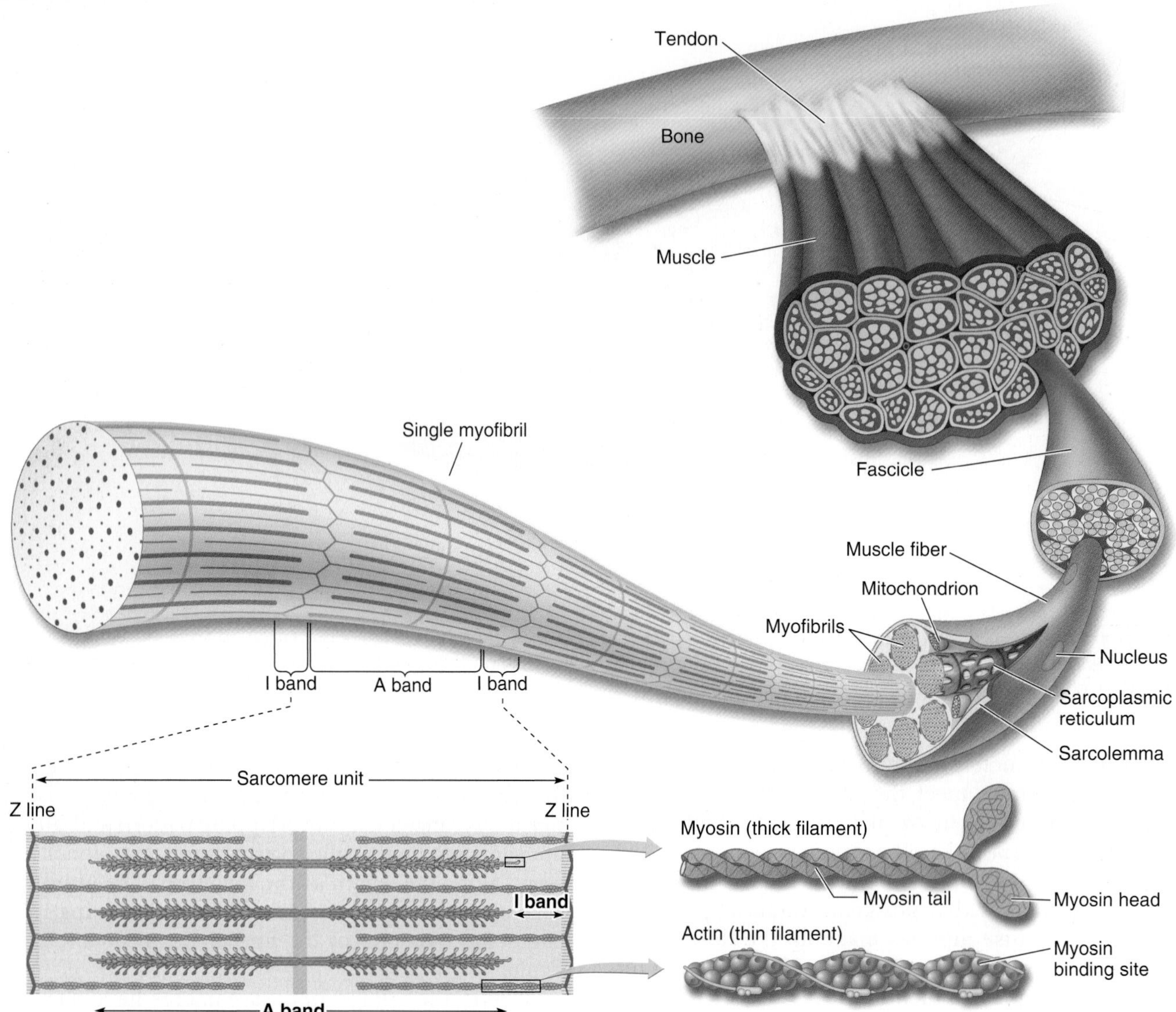

FIGURE 6.2 ▸ Structure of a skeletal muscle fiber. Each muscle fiber consists of myofibrils organized into functional units called sarcomeres. The sarcomeres are created by a highly organized arrangement of actin and myosin myofilaments.

All fibrous connective tissues have a high number of collagen fibers. These tough fibers are composed of bundles of smaller fibrils made of a protein called **tropocollagen**, which gives collagen its stretch-resistant quality. When tension is applied, the triple-helix (corkscrew) shape of tropocollagen causes collagen fibers to slowly extend or unwind, rather than stretch. The ground substance in these connective tissues is a viscous gel that prevents cross-links or adhesions from forming between collagen fibers. It also lubricates the muscle and connective tissue fibers to reduce friction during movement and provides a cushion that protects the muscle from compressive forces. The characteristics of the collagen and ground substance in fibrous connective tissues provide important physiologic rationale for several of the structural effects of manual therapy. More detailed explanations of these characteristics and their impact on manual therapy techniques are provided in the myofascial section of Chapter 8.

Fascia

Fascia is the tough yet pliable sheet of connective tissue that covers, divides, and supports all the structures of the body. It is classified as disorganized fibrous, irregular, or loose connective tissue because the fibers in it are neither strictly arranged nor tightly packed. The fibers in fascia are widely spaced by ground substance and are rarely in parallel alignment. Additionally, fascia contains higher numbers of reticular and elastic fibers than organized fibrous or dense connective tissues

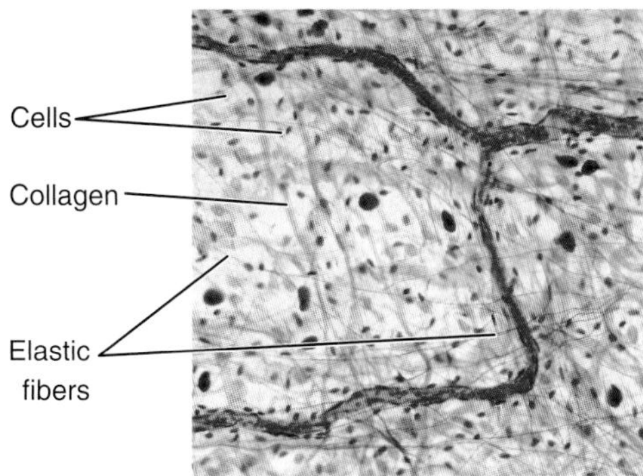

FIGURE 6.3 ▶ Fascia. This micrograph shows the disorganized fibrous connective tissue of fascia.

(▶ FIGURE 6.3). As a result of its fibers running in different directions and the presence of so many "stretchy fibers," fascia is able to withstand repeated bouts of multidirectional stresses without damage. As you can imagine, this characteristic is of particular importance for the fascial layers in skeletal muscle.

Although named as if they are distinctly different layers, the epimysium, perimysium, and endomysium are actually one contiguous layer of fascia. From a structural standpoint, these fascial layers attach the skin to muscles, compartmentalize the muscle into fascicles, and hold all the muscle fibers together. Functionally, these fascial layers play a key role in focusing and transferring the force of contraction to the bones. It has been demonstrated that muscle loses up to 15% of its strength if there is a slit in the fascial covering.[1] Another function of the fascial layers in muscle is to lubricate the fibers and fascicles to reduce friction during contraction and lengthening. Therefore, if the fascia is dehydrated, the fibers adhere to one another and limit the muscle's functional capacity.

Tendon and Aponeurosis

Muscles attach to bone via a cordlike tendon or a broad flat sheet called **aponeurosis**. These structures are classified as organized fibrous, regular, or dense connective tissue because the collagen fibers are arranged linearly and tightly packed. These tissues contain thicker and more densely packed collagen fibers, as well as less ground substance than fascia (▶ FIGURE 6.4). Additionally, the collagen fibers are highly organized in a parallel arrangement, making these connective tissues strong enough to withstand the internal stress of muscle contraction; they are also highly resistant to external tensile stresses. Ligaments contain slightly higher numbers of elastic fibers than do tendons because light stretch is required for the joints to achieve full range of motion. Other fascial structures that play an important role in the function of the muscular system are thickened superficial bands called ***retinaculi***. These bands are not structural components of muscles; rather, they serve as supportive straps that hold tendons in place. Retinaculi are found in the wrist, ankle, and knee.

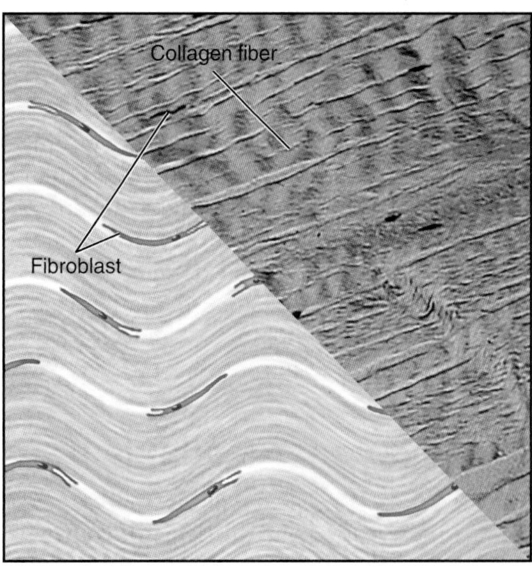

FIGURE 6.4 ▶ Tendon. This combined micrograph and illustration shows the organized fibrous connective tissue that forms a tendon.

There are two important transitional zones between muscles, tendons, and bones. The **tenoperiosteal junction** is the connective tissue zone where the tendon weaves with the periosteum of the bone. The transitional zone between the fascia that surrounds the muscle and the tendon is called the **musculotendinous junction** (see FIGURE 6.1). Because there is a change in tissue flexibility and extensibility as the fascia transitions to tendon and the tendon weaves into the periosteum, these tissue junctions are common sites of injury. When a muscle or tendon is stretched or torn, the injury is called a **strain**. Because muscle tension adds external stability to the joints, it is common for there to be some associated muscle or tendon strain when the ligaments of a joint have been sprained. However, this does not mean that a muscle strain always destabilizes a joint. It may simply result in weakened or painful movement.

What Do You Think? 6.1

- How does the highly organized nature of skeletal muscle affect its ability to generate force and tension?
- How would you apply your knowledge of the components and characteristics of fascia and tendons in your manual therapy practice?

PHYSIOLOGY OF MUSCLE CONTRACTION

Contraction of skeletal muscle requires a cooperative effort between the nervous and muscular systems. While skeletal muscles have all the structural components necessary to contract, they are unable to do so unless signaled by nerves. In this chapter, the basics of this relationship are explained. A more detailed explanation of neuromuscular coordination and integration is presented in Chapter 8. For an animation of muscle contraction, visit http://thePoint.lww.com/Archer-Nelson.

Sliding Filament Mechanism

The physiology of muscle contraction is explained through a model called the **sliding filament mechanism**. This model lays out the key physiologic events that create the shortening of the sarcomeres during muscle contraction. A contraction occurs when a chemical bond is formed between the actin and myosin myofilaments that pulls them together and increases the overlap between them (▶ FIGURE 6.5). This action only occurs when muscle cells are signaled to do so by the nervous system. When stimulated, calcium stored in the sarcoplasmic reticulum is released into the sarcoplasm. This calcium-rich environment exposes bonding sites on the actin filaments. The myosin heads then attach to these sites, forming bridges between the myofilaments. In the same way that an oar pulls a boat through the water, these chemical bridges cause the filaments to slide over one another. Notice that the length of the actin and myosin does not change. Rather, the overlap between the myofilaments is increased to shorten the distance between Z lines, which shortens the sarcomere. This microscopic contraction in the sarcomeres translates to a shortening of myofibrils and the muscle fiber as a whole.

The cascade of events that occur when a muscle is stimulated to contract take place in the following order:

1. Calcium ions (Ca^{2+}) stored in the sarcoplasmic reticulum are released into the sarcoplasm.
2. The presence of Ca^{2+} causes myosin binding sites on the actin molecules to be exposed.
3. The myosin heads bind to the exposed sites, forming cross-bridges that pull the two myofilaments across one another.
4. ATP is used to detach the myosin heads so that they can flip forward to the next binding site. The net result is continued sliding of the filaments across one another.
5. When the stimulus is removed, more energy is expended to pump calcium back into the sarcoplasmic reticulum, and chemical bridges can no longer form.

When a muscle is in a longer or shorter than normal position, it affects the strength of contraction because the number of binding sites available between actin and myosin is altered. In a muscle that is too long, the overlap between actin and myosin is decreased, so that fewer myosin heads are able to bond and initiate the sliding filament mechanism. In contrast, a contraction in a muscle that is already shortened will also be weak because the sarcomere is already shortened and the overlap is maximized. This relationship between the length of the muscle and the strength of contraction is called the **length-strength ratio**. You can experience the effects of this relationship by doing a simple exercise:

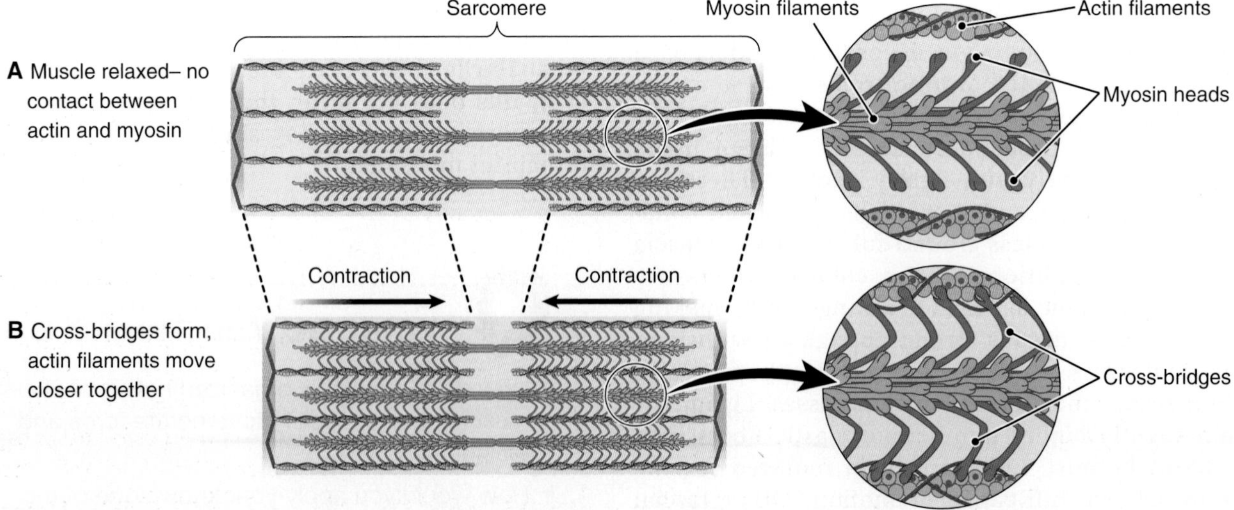

FIGURE 6.5 ▶ Sliding filament mechanism. Skeletal muscle contraction is produced by the bonding and sliding of actin and myosin filaments. Cross-bridges between the myofilaments are formed and the sarcomere shortens as the filaments slide over one another. This only occurs when calcium is released into the sarcomere.

First, stand up straight. Lift your heels and rise up onto the balls of your feet. Notice the strong contraction you feel in your calf muscles as you do this. Now, try the exercise again, but first bend your knees and squat down slightly. Maintain the squat position as you again rise up onto the balls of your feet. With the calf muscles pre-shortened by the squat you will notice that the strength of contraction in your calf muscle is dramatically decreased.

The Motor Unit: Stimulating Muscle Contraction

The coordination between the nervous and muscular systems occurs through the functional structure called a **motor unit** (FIGURE 6.6A), which is made up of a single nerve cell, a **motor neuron**, and the multiple muscle fibers it innervates. The actual number of fibers in a motor unit depends on the size and function of the muscle. Fine control muscles such as those in the face and hands are particularly suited for precision movement, because they have as few as 10 muscle fibers per motor neuron. In contrast, larger gross movement muscles that must generate more force than precision (such as those in the thighs) average between 100 and 150 muscle fibers per motor unit. This means that any single muscle can be controlled by hundreds or thousands of motor units.

The structural interface between the motor neuron and fibers of each motor unit is called the **neuromuscular junction** (FIGURE 6.6B). This junction is actually a microscopic space between the knobby ends of the neuron and a highly sensitive area of the muscle fiber called the **motor end plate**. Located within the knobs at the end of the neuron are vesicles that hold specialized "communicating" chemicals called **neurotransmitters**. A muscle contraction is initiated when neurotransmitters are released from the vesicles into the neuromuscular junction to stimulate the motor end plate.

A contraction only occurs when enough stimulation is delivered from the neuron to the motor end plate to initiate the required chemical changes within the muscle cell. This minimum amount of stimulus is called a **threshold stimulus**. When threshold stimulus is reached, all the muscle fibers in the motor unit will fully contract. If less than threshold stimulus is transmitted, none of the fibers will contract. This physiological principle is known as the **all-or-none response**. The all-or-none response applies to each individual motor unit, not the muscle as a whole. This allows the nervous system to regulate the force of each individual contraction by controlling the number of motor units stimulated within any given muscle. For example, only one motor unit might be stimulated to lift a pencil, while hundreds of motor units in the same muscle would be engaged to lift a 20-lb barbell. The regulation of a muscle's effort by increasing or decreasing the number of motor units stimulated is called **graded response**, or **motor unit recruitment**. This pattern of motor unit recruitment is regulated by the central nervous system and is necessary for creating smooth coordinated movement and reducing muscle fatigue.

What Do You Think? 6.2

- Choose one of the following analogies and explain why it is helpful for understanding the sliding filament mechanism: (a) pushing two hair combs together; (b) an eight-man crew rowing through the water; or (c) a tug-of-war between two teams.
- How do you think temperature might impact threshold stimulus?
- What would life be like if all the fibers in a muscle were controlled by a single motor unit?

ENERGY FOR CONTRACTION

Muscle contraction, like other body functions, requires energy in the form of ATP molecules. When one phosphate is broken off the ATP molecule, it releases the energy required for cellular work; in this case, muscle contraction. The amount of energy actually required by muscles varies greatly throughout the day according to the level of activity. There is a small amount of ATP stored in muscle cells, but only enough to supply energy for a few seconds. If muscle activity continues longer, there are three ways to produce ATP to meet the energy demands of contraction: creatine phosphate, anaerobic cellular metabolism, and aerobic cellular metabolism.

Creatine Phosphate

When a muscle is at rest, the amount of ATP produced by its fibers is more than needed. While some ATP is simply stored within the cell, other ATP is converted into a substance called **creatine phosphate**, a compound made of an amino acid–like molecule called **creatine** with an attached phosphate ion. When muscle activity increases, phosphate ions are broken off ATP molecules to release energy, leaving only two phosphate ions behind, which creates **adenosine diphosphate (ADP)**. As the level of ADP within working fibers rises, an enzyme known as creatine kinase breaks the phosphate group off stored creatine phosphate and attaches it to ADP, converting it back to ATP. This ATP can then be broken apart to release energy for contraction. While creatine phosphate is readily available for this quick exchange process called **direct phosphorylation** (FIGURE 6.7), it provides less than 15 seconds of energy;

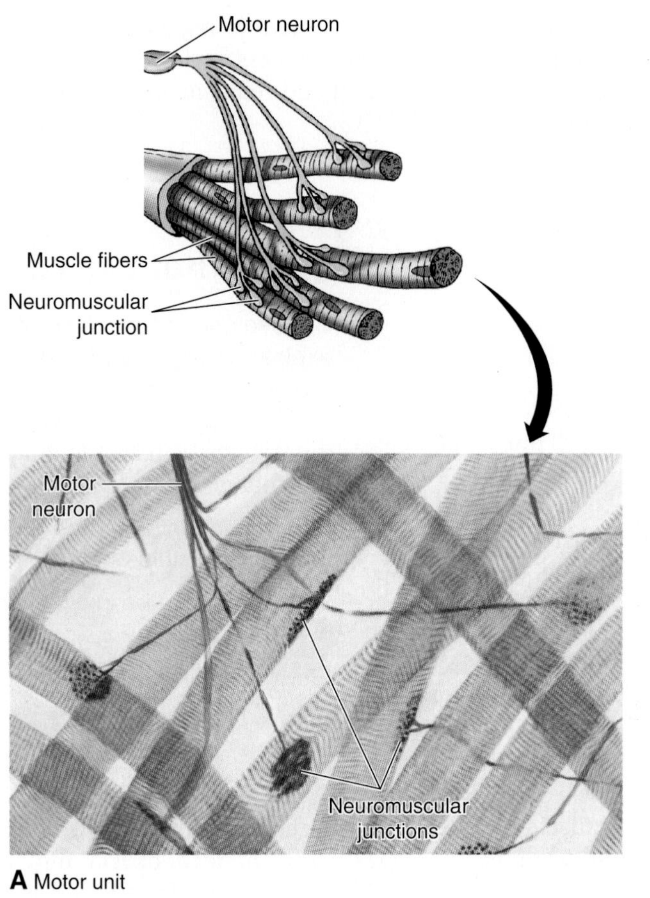

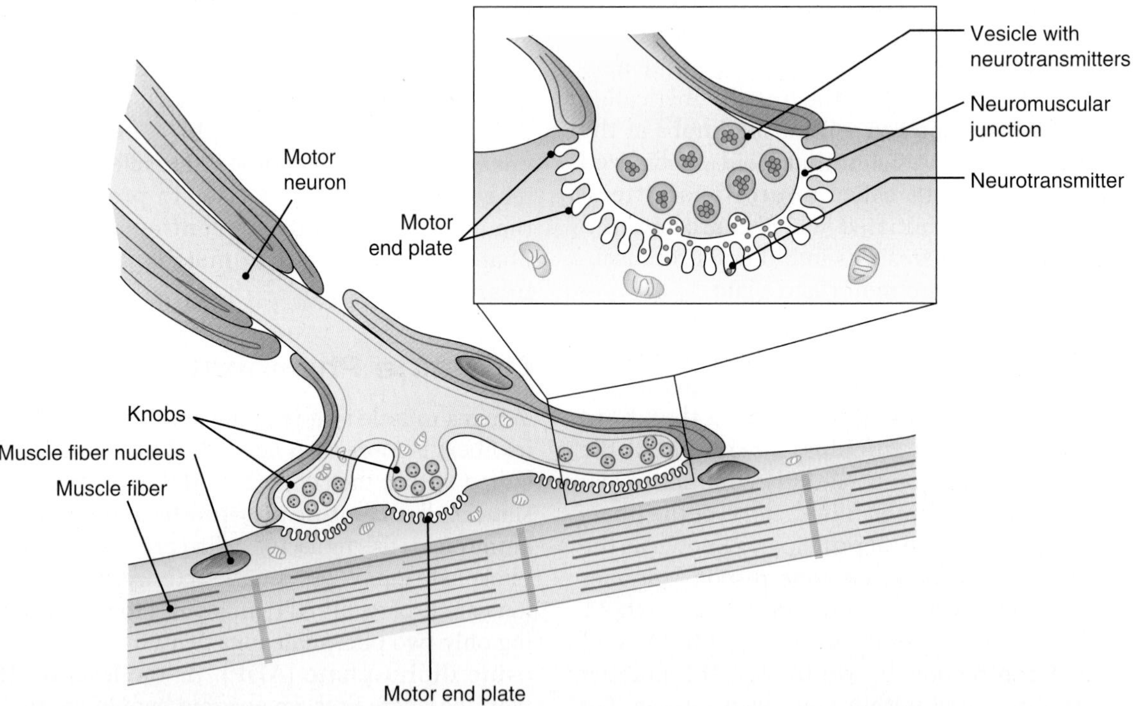

FIGURE 6.6 ▶ Motor unit. A. A motor unit is made up of a single motor neuron and the multiple muscle fibers it innervates. **B.** At the neuromuscular junction, neurotransmitters are released from knobs at the end of the motor neuron to stimulate the motor end plate of the muscle fiber.

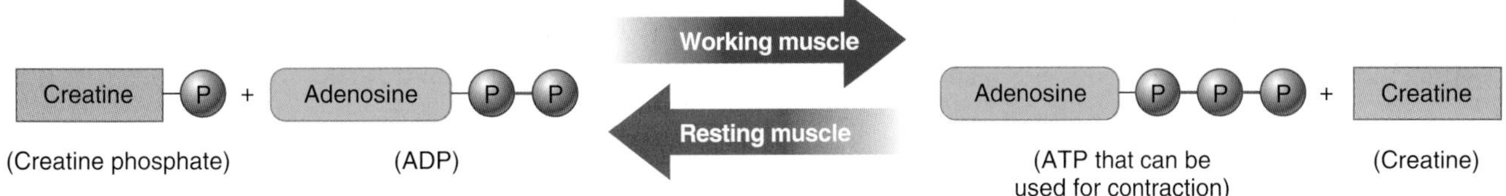

FIGURE 6.7 ▸ Direct phosphorylation. This process provides a quick but very limited source of ATP for muscle contraction. It is used for activities that require short bursts of energy.

just enough for a short burst of activity such as dashing up stairs or sprinting down the street.

Anaerobic Cellular Metabolism

After a muscle uses up its creatine phosphate and ATP stores, it must begin to utilize glucose for energy through a process called **glycolysis** (FIGURE 6.8). Because this process does not require the presence of oxygen, it is also known as **anaerobic cellular metabolism**. The glucose used for this process is obtained either from the blood or from the breakdown of glycogen stored within the muscle. Remember, glycogen is the storage form of glucose, and the liver and skeletal muscles are the primary storage sites. While glycolysis is another quick source of energy for muscle contraction, it only provides enough energy for 30 to 40 seconds of activity. For every molecule of glucose, glycolysis produces two molecules of ATP. Another byproduct of glycolysis is **pyruvic acid**. Without the presence of oxygen, pyruvic acid is converted to **lactic acid**, a more difficult byproduct to metabolize. While most lactic acid diffuses out of the cell to be excreted via the urinary system or metabolized in the liver, it can remain in the muscle cells for short periods of time and interfere with efficient muscle contraction.

Aerobic Cellular Metabolism

If sufficient levels of oxygen are present in the muscles, the pyruvic acid produced during glycolysis can be converted into more ATP by the mitochondria via a process known as the **Krebs cycle**, or **citric acid cycle** (▸ FIGURE 6.8). Because oxygen is required, this cycle of chemical reactions is referred to as **aerobic cellular metabolism**. Oxygen is made available to muscles from oxygen-binding proteins either in the blood or stored in muscle fibers. **Hemoglobin** is the oxygen-binding protein found in red blood cells that gives them their red color and transports oxygen from the lungs to the tissues. **Myoglobin** is the protein present in the sarcoplasm of muscles that binds to oxygen, providing an immediate source of O_2 to the cell when needed. Although aerobic cellular metabolism is slower than glycolysis, it supplies much more ATP (30–32 ATP molecules for each molecule of pyruvic acid) and can continue to provide energy as

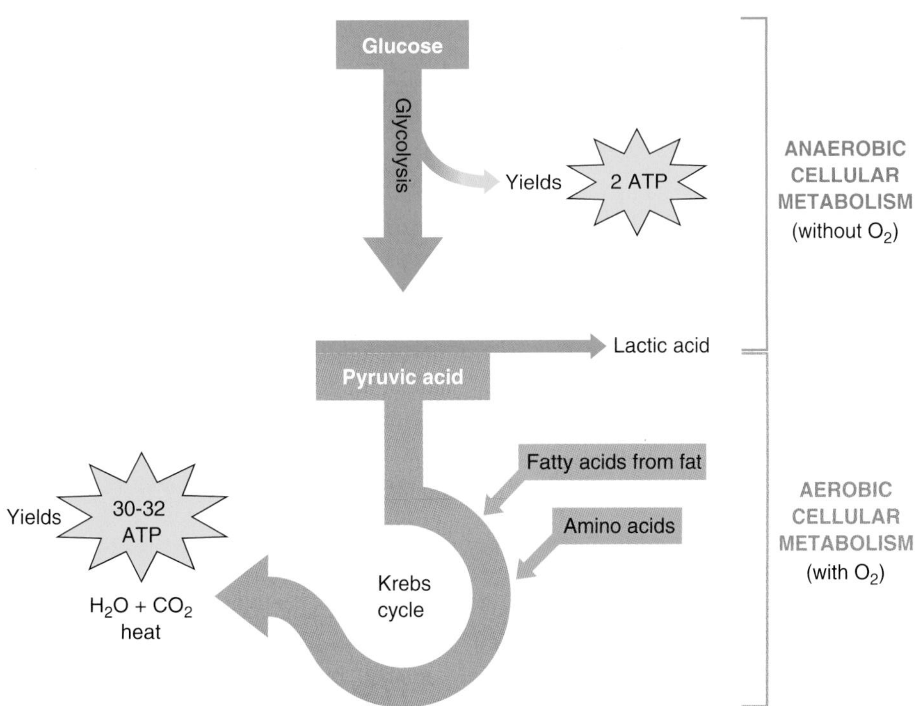

FIGURE 6.8 ▸ Glycolysis and aerobic cellular metabolism. Most energy for muscle contraction is provided through these two methods of ATP production.

By the Way

Multiple research studies have concluded that lactic acid does not accumulate in muscle tissue to cause delayed onset muscle soreness (DOMS). Instead, it appears that soreness is related to the inflammatory response initiated by microtrauma to the muscle fibers and epimysium during exercise.

long as enough nutrients and oxygen are available. In fact, in muscle activity lasting more than 10 minutes, more than 90% of ATP is produced via the Krebs cycle. Besides being a rich source of energy, another advantage to this process is that its only byproducts are heat, water, and CO_2, which are easily recycled for use or removed from the body.

Although the methods of energy production are described as three distinct processes, the muscles switch efficiently from one to the other according to the intensity and duration of exercise and the oxygen supply. Once activity is sustained beyond 15 seconds and the ATP and creatine phosphate stores are depleted, the body begins to use glucose to produce energy. During moderate exercise, there is generally an adequate supply of oxygen. Therefore, the mitochondria are able to take the pyruvic acid formed during glycolysis and move into the Krebs cycle. With aerobic metabolism, the body is also able to utilize fats as a fuel source. This is why moderate aerobic exercise such as walking, jogging, or swimming is recommended for weight loss. During strenuous exercise, the demand for energy outpaces the body's ability to supply the mitochondria with oxygen. Therefore, the body must fall back on the anaerobic process of glycolysis to meet the energy demand, and lactic acid begins to build up.

Muscle Fatigue and Oxygen Debt

Muscle fatigue is the inability of muscle to contract forcefully after prolonged activity, even when stimulated to do so. Fatigue can be caused by a variety of factors, such as a lack of oxygen, a decrease in the calcium supply needed for myofilament bonding, or the depletion of glycogen and other fuels used for contraction. Other possible causes of muscle fatigue can be a build-up of lactic acid or ADP or insufficient release of neurotransmitters from the motor neurons.

After activity, breathing and heart rates remain elevated for a period of time in an effort to provide the oxygen needed to restore muscles to their resting metabolic condition. The body needs this added oxygen to "pay back" the **oxygen debt** that occurs during exercise. Oxygen is required to metabolize lactic acid and to replenish glycogen, creatine phosphate, and ATP supplies.

By the Way

Muscle fibers contract at different speeds, allowing them to be categorized as either slow-twitch (type I) or fast-twitch (type II) fibers. They also vary in the mechanisms they use to create ATP, which determines the rate at which the fibers fatigue. Slow-twitch fibers have a high capacity for generating ATP through aerobic cellular metabolism and as such, are slow to fatigue. Postural muscles have high numbers of slow-twitch fibers. In contrast, fast-twitch fibers, such as those in the muscles of the limbs, have a lower capacity for generating ATP and fatigue more quickly.

Additionally, the oxygen removed from myoglobin during exercise must be replaced.

TYPES OF MUSCLE CONTRACTION

Muscular contractions are classified according to the mechanical changes of a muscle as a whole, rather than the cellular shortening that is constant in all types of contractions. Muscle contraction is a physiologic process in which the stimulated muscle fibers shorten and create tension. However, the mechanical results of contraction on the muscle as a whole vary. While the contraction creates two constant mechanical results: (a) tension or pulling force at a muscle's attachments, and (b) bunching or broadening at the middle of the muscle as a result of overlapping myofilaments, contraction may or may not generate movement. The three basic types of muscle contraction are tonic, isometric, and isotonic.

Tonic Contractions

Tonic contractions create a constant low-grade tension and firmness in the muscles but do not create movement. In a tonic contraction, a few motor units in a muscle are stimulated to contract, creating only enough tension to stabilize joints and to maintain posture and a state of readiness for voluntary contractions. The postural muscles in the neck, torso, and legs have a higher level of tonic contraction than other muscles. They work to resist the pull of gravity to maintain the body's upright position. The erector spinae and psoas muscles of the torso are good examples of postural muscles. The exact number and location of motor units involved in tonic contractions is controlled and coordinated by a specific part of the brain, the cerebellum, which is explained in detail in Chapter 7.

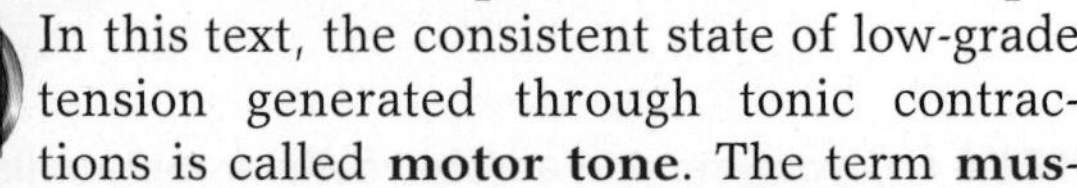

In this text, the consistent state of low-grade tension generated through tonic contractions is called **motor tone**. The term **muscle tone** is used to describe the natural firmness of muscle tissue created by the combination of its fluid and connective tissue elements. While many authors use the term *muscle tone* to describe tonic contractions, Lederman, Chaitow, and Walker-Delaney have suggested that the term *motor tone* more accurately describes this neuromuscular activity[2–4]. It is important for manual therapists to make a distinction between motor tone and muscle tone because a muscle may feel tight and stiff either because of problems with its fluid and connective tissue elements (increased muscle tone) or because it is contracted (increased motor tone). These two different conditions require differing approaches and treatment techniques. For example, a cramped or spasmed muscle has increased motor tone and may be effectively addressed using neuromuscular techniques. In contrast, conditions such as anterior compartment syndrome and adhesive capsulitis display tight and stiff muscles because of connective tissue changes. These muscle tone dysfunctions will more likely respond to treatment modalities such as myofascial or lymphatic techniques that address the connective tissue and fluid components of the region.

Isometric Contractions

Isometric contractions, like tonic contractions, do not produce movement; however, the tension generated by the contraction is much higher than in tonic contractions (▶ FIGURE 6.9). Imagine trying to push down a cement wall. Your contracting muscles bulge with the effort, but the joints are locked into one position and the muscles do not change in length. It is easy to understand why no movement occurs in this example; the weight of the objects creates resistance that clearly exceeds the amount of force that can be generated by the muscles. Bodybuilders use isometric contractions when they pose, contracting not for movement but to make the muscle bulge.

Isotonic Contractions

Isotonic contractions are those that create movement. There are two subcategories of isotonic contractions: concentric and eccentric. In a **concentric contraction**, the muscle as a whole *shortens* as the attachment points move closer together. In contrast, in an **eccentric contraction**, the muscle *lengthens* as the attachments move farther away from each other (▶ FIGURE 6.9). A good way to demonstrate the difference between the two types of isotonic contractions is to lift and then slowly put down any object of medium weight. Focus your attention on the major muscle on the front side of the upper arm (biceps) by placing your other hand lightly on top of it. When you pick up the object and lift it, you can feel the muscle "tighten" with the contraction and see the attachment points on the ulna and humerus move closer

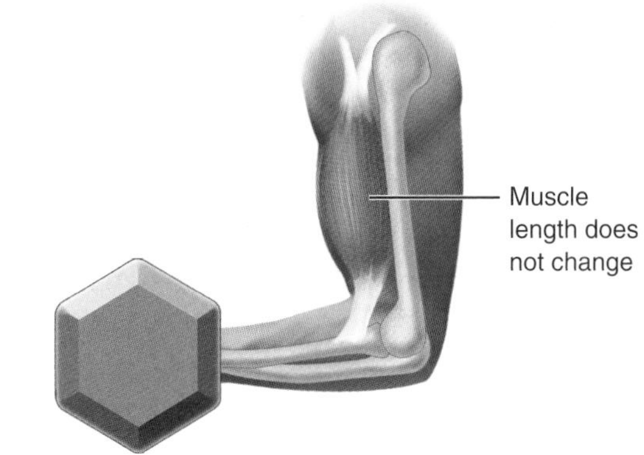

A Isometric contraction

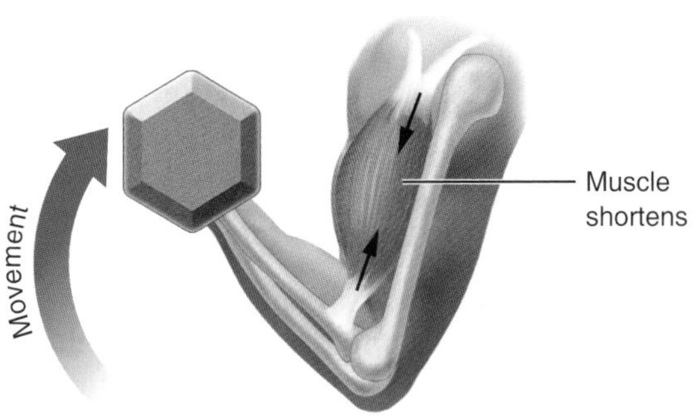

B Concentric isotonic contraction

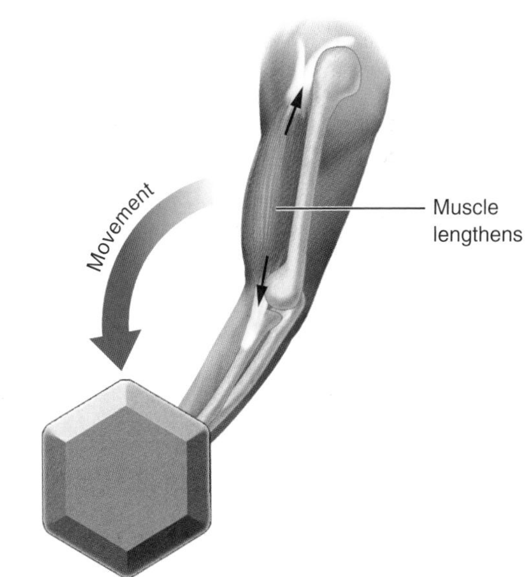

C Eccentric isotonic contraction

FIGURE 6.9 ▶ Isometric and isotonic muscle contractions. **A.** Isometric contractions generate force but no movement. **B.** During a concentric isotonic contraction, the muscle shortens as it generates force. **C.** In an eccentric isotonic contraction, the muscle generates force as it lengthens.

By the Way

A quick review of word parts can help us remember the distinctions between isometric and isotonic contractions. Isometric means "same or equal measure," indicating that the length of the muscle does not change. Isotonic can be translated as "equal tension or force." In isotonic contractions, the length of the muscle changes but tension is maintained as movement occurs.

together, indicating a concentric isotonic contraction. When you slowly lower the object, you can still feel the muscle tension that confirms contraction, but the muscle lengthens as the biceps performs an eccentric isotonic contraction.

Twitch and Tetanic Contraction

While tonic, isometric, and isotonic contractions explain how skeletal muscles accomplish normal movement and posture, there are other types of contractions that can be considered nonproductive, since they do not contribute to the system's functions. A muscle **twitch** is a sudden small involuntary contraction that can occur due to a subconscious command, such as during sleep, or when a fatigued muscle doesn't have the energy to pump calcium out of the sarcoplasm.

In trigger point therapy, a common response to palpation of an active trigger point is a visible twitch called ***fasciculation*** in which all fibers in a particular motor unit involuntarily contract.

A **tetanic contraction** is another type of nonproductive contraction in which the muscle is bombarded with constant stimuli. This causes a sustained contraction of multiple fibers that effectively "locks" the muscle. Tetanic contractions are only produced in lab experiments or occur as the result of bacterial infections such as tetanus or botulism; they should not be confused with muscle cramps or spasms.

What Do You Think? 6.3

- Based on what you know about energy for muscle contraction, explain why a track coach has different training regimes for sprinters and long-distance runners?
- Why is it important for manual therapists to understand that muscle contraction both increases tension at the attachment points and bunches or broadens the belly of the muscle?
- Do you think isometric contraction is an effective means of increasing one's overall strength? Why or why not?
- Strength coaches and fitness trainers often recommend negative lifting (lowering a weight slowly) as an important part of building strength. Which type of muscle contraction is involved in this exercise? Why is this type of training effective at building strength?

MOVEMENT AND MUSCLE ASSIGNMENTS

While the bones and joints of the skeletal system provide the levers and fulcrums that *allow* movement, it is the contraction of muscles attached to these levers that *create* movement. Additionally, the nervous system controls and coordinates muscular contractions, making movement a highly integrated and interdependent effort

Pathology Alert Muscle Cramps and Spasms

Muscle cramps are acute involuntary muscle contractions that generally last for several minutes. These involuntary contractions form a "knot" in a large section of the muscle that can be can be seen or felt by the individual and therapist. Although cramps can be stimulated by physiological stress, the most common causes are muscle fatigue and metabolic imbalances. Cramps are generally of short duration and do not present any long-term problems.

Muscle spasms are also involuntary contractions, but they are sustained over hours, days, weeks, or months. Muscle spasms can lead to long-term challenges such as postural adaptations, limited or painful movement, and poor circulation. Many clients seek the help of manual therapists to relieve tight or painful muscles. To make effective treatment choices, therapists need to determine whether the client's complaint is caused by muscle spasms (increased motor tone) or is due to connective tissue and fluid changes (increased muscle tone).

between all three systems. Whether a muscle attaches via a tendon or aponeurosis, the points of attachment are given different names based on which bone moves when the muscle contracts. A muscle's **origin** is the fixed (non-moving) attachment, while the **insertion** is attached to the bone that is moved by the force of the contraction. This section describes how the skeletal and muscular systems create different types of lever systems and how a muscle's shape and fiber direction determine the specific movement role or assignment during different motions. This section also defines the categories of range of motion and describes how manual therapists use this information for assessment and therapeutic applications.

Lever Systems

In any leverage system, there are three key components: the **fulcrum**, the ***resistance*** (also called the ***load***), and the ***force*** (also called the ***effort***). In the body, joints serve as fulcrums, the resistance (load) is the weight of the body part or object being moved, and the attachment point of a muscle provides the force. For example, when we lift a glass to our lips to take a drink of water, the elbow joint is the fulcrum, the resistance is the weight of the arm, hand, and glass, and the force is created by the biceps and other muscles in the arm that flex the elbow.

Levers fall into one of three categories according to the position of the fulcrum in relationship to the point of force and the resistance:

- ***First-class levers***—Scissors and see-saws/teeter-totters are examples of this category. In first-class levers, the fulcrum is *between* the force and the resistance. There are very few first-class levers in the body. However, one example occurs with extension of the head. In this case, the fulcrum is the articulation between the skull and the atlas (occipitoatlanto joint), the force is created by the posterior cervical muscles, and the resistance is the weight of the skull (FIGURE 6.10A).
- ***Second-class levers***—A wheelbarrow is the simplest example of a second-class lever. In this case, the resistance lies between the fulcrum and the point of force. Again, there are only a few examples of second-class levers in the body. Exercises to strengthen the calf muscles provide the best example. When we do heel raises, the balls of our feet serve as fulcrums, and the force is created by contraction of the calf muscles to lift the heel off the ground. Because we are standing, the weight of the body is the resistance, which is centered in the foot between the fulcrum and force (FIGURE 6.10B).
- ***Third-class levers***—Most levers in the body fall into this category in which the force is applied between the fulcrum and the resistance. Tweezers are an everyday example of a third-class lever. Arm curls (elbow flexion) offer a simple example of a third-class lever within the body. The force is provided by the anterior brachial muscles that cross the elbow joint (fulcrum) to flex the arm, moving the weight. (FIGURE 6.10C).

In the lever systems described previously, movement is easy to understand because the levers are straight, the fulcrums are unidirectional, and force is applied to only one point on the lever. However, in the body, the bones are rarely straight levers, the joints are often multidirectional, and several muscles attach to the same bone to apply pulling force at multiple points and angles. This creates more complex and interesting movements, and it can be challenging to recognize the basic mechanical principles at work. However, we can simplify the mechanical concepts of the body's lever systems into a few basic rules of movement:

1. A muscle must originate and insert on different bones and will create movement in the joint(s) it crosses.
2. The force of muscle contraction always pulls on the insertion point.
3. The location of the muscle and the angle at which it crosses the joint determine the specific movement created by the contraction.

These basic rules, combined with an understanding of the influences of a muscle's specific architecture, help us understand each muscle's movement assignment, capabilities, and limits.

Muscle Architecture

Whether considering a building, a bridge, or skeletal muscle, the term *architecture* refers to the design of a structure that dictates its functional capacity. A muscle's architecture is defined by the direction of its fibers and the manner in which fascia divides, aligns, and arranges the fascicles in relation to the muscle's tendon. This determines what role the muscle plays in creating specific body movements, the amount of power it can generate, and the specific direction, quantity, and quality of movement that occurs with contraction.

As seen in FIGURES 6.11 and 6.12, muscles have different shapes and sizes and the fibers in adjacent muscles often run in different directions. There are two basic categories of fiber arrangements: parallel and pennate. Recall that skeletal muscle fibers are bundled together in parallel alignment to form a fascicle. Therefore, the parallel and pennate patterns of different muscles are based on the arrangement of their fascicles and tendons.

Parallel Muscles

In **parallel muscles**, the fibers are all the same length and in parallel arrangement, allowing them to shorten equally and pull in the same direction. For this reason, contractions in parallel muscles tend to produce movement over

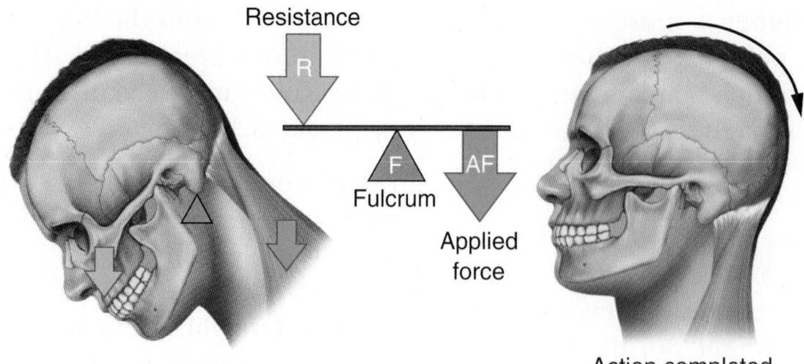

A **First-class lever**

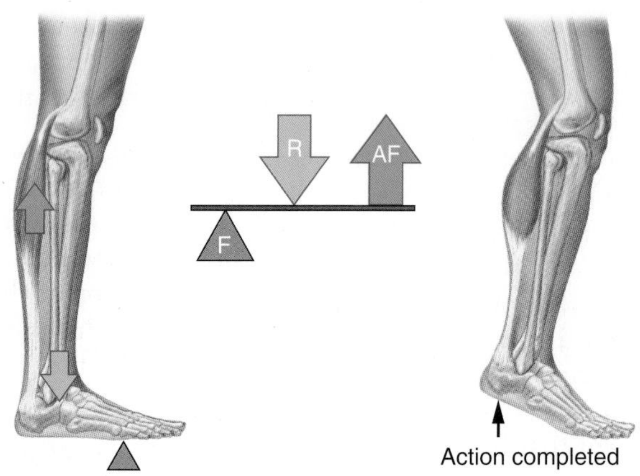

B **Second-class lever**

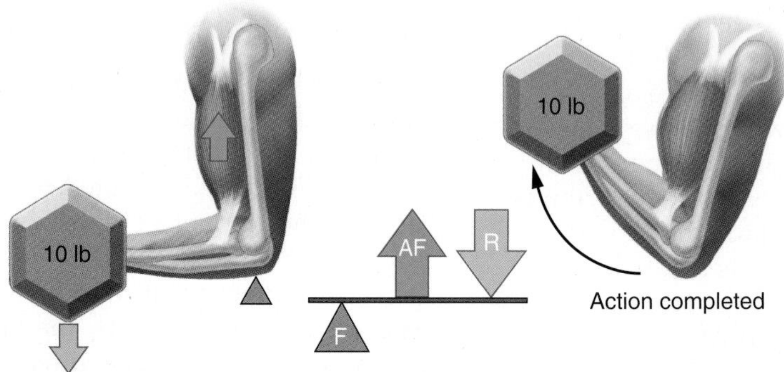

C **Third-class lever**

FIGURE 6.10 ▶ Lever systems. A. First-class lever. **B.** Second-class lever. **C.** Third-class lever.

a wide range of motion. Although their tendons can be either flat or more rounded and cordlike, they are always at the ends of the muscle rather than within the muscle belly. Different arrangements of fascicles result in three subclasses of parallel muscles (▶ FIGURE 6.13A):

- ***Fusiform***—The arrangement of fascicles in fusiform muscles give it a thicker or broader belly that tapers at the ends to form cord-like tendons. This type of tendinous attachment focuses the force of a contraction to a specific point or landmark on a bone. As shown in Figure 6.12, the gastrocnemius is a good example of a fusiform muscle.
- ***Circular***—The fascicles of circular muscles are arranged in a circular pattern. Contraction and relaxation of these muscles closes or opens an orifice. An example of a circular muscle is the orbicularis oris that surrounds the mouth (FIGURE 6.11).
- ***Triangular***—In triangular muscles, one end (usually the origin) is spread out like a fan and attached over a broad area of bone. On the opposite end, the fascicles converge and form one thick tendon of insertion. This gives the muscle multiple lines of pull that can produce several different and sometimes opposing movements. The pectoralis major (FIGURE 6.11) is a triangular muscle.

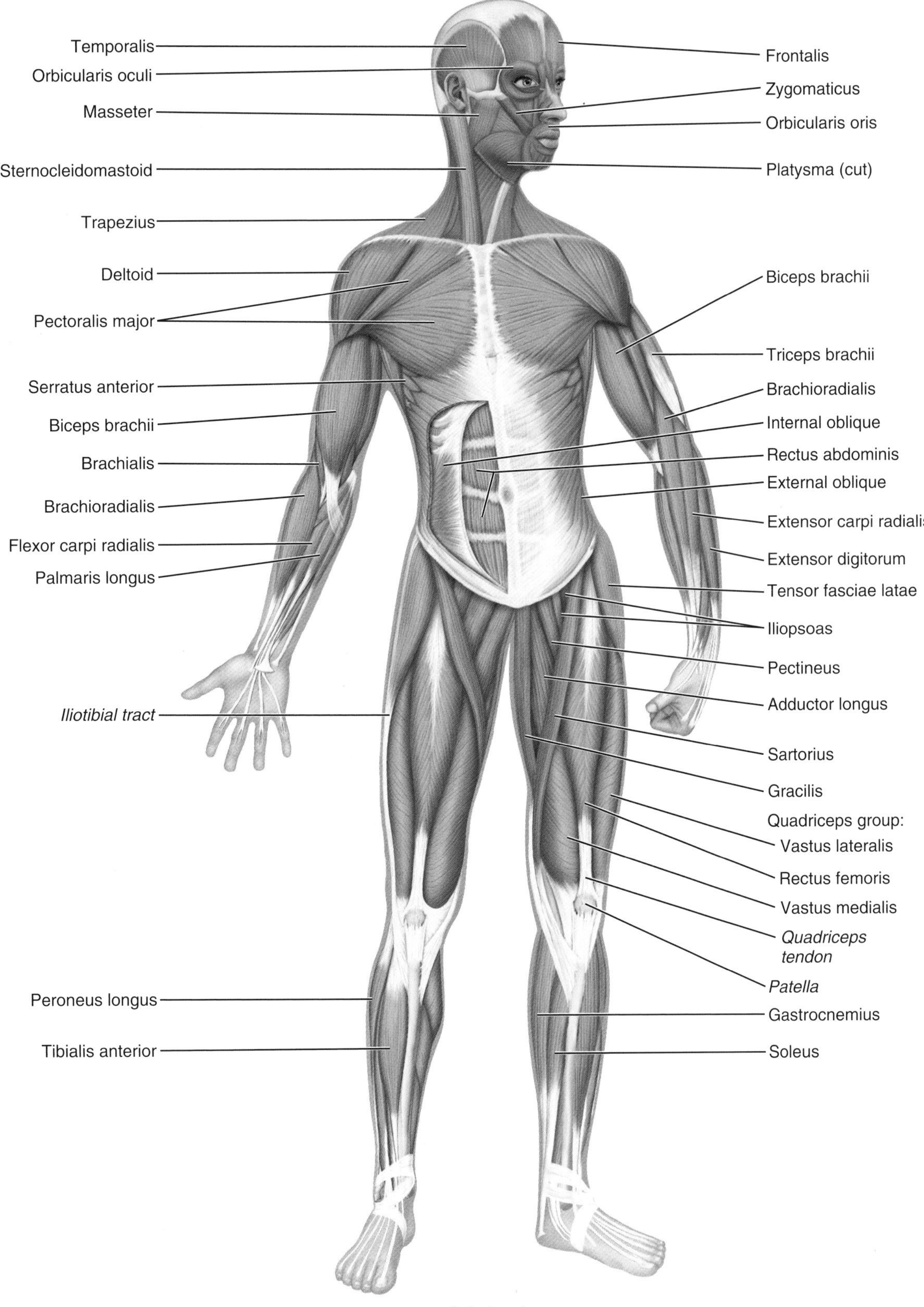
Temporalis
Orbicularis oculi
Masseter
Sternocleidomastoid
Trapezius
Deltoid
Pectoralis major
Serratus anterior
Biceps brachii
Brachialis
Brachioradialis
Flexor carpi radialis
Palmaris longus
Iliotibial tract
Peroneus longus
Tibialis anterior
Frontalis
Zygomaticus
Orbicularis oris
Platysma (cut)
Biceps brachii
Triceps brachii
Brachioradialis
Internal oblique
Rectus abdominis
External oblique
Extensor carpi radialis
Extensor digitorum
Tensor fasciae latae
Iliopsoas
Pectineus
Adductor longus
Sartorius
Gracilis
Quadriceps group:
Vastus lateralis
Rectus femoris
Vastus medialis
Quadriceps tendon
Patella
Gastrocnemius
Soleus
Anterior view

FIGURE 6.11 ▸ Superficial muscles of the body, anterior view.

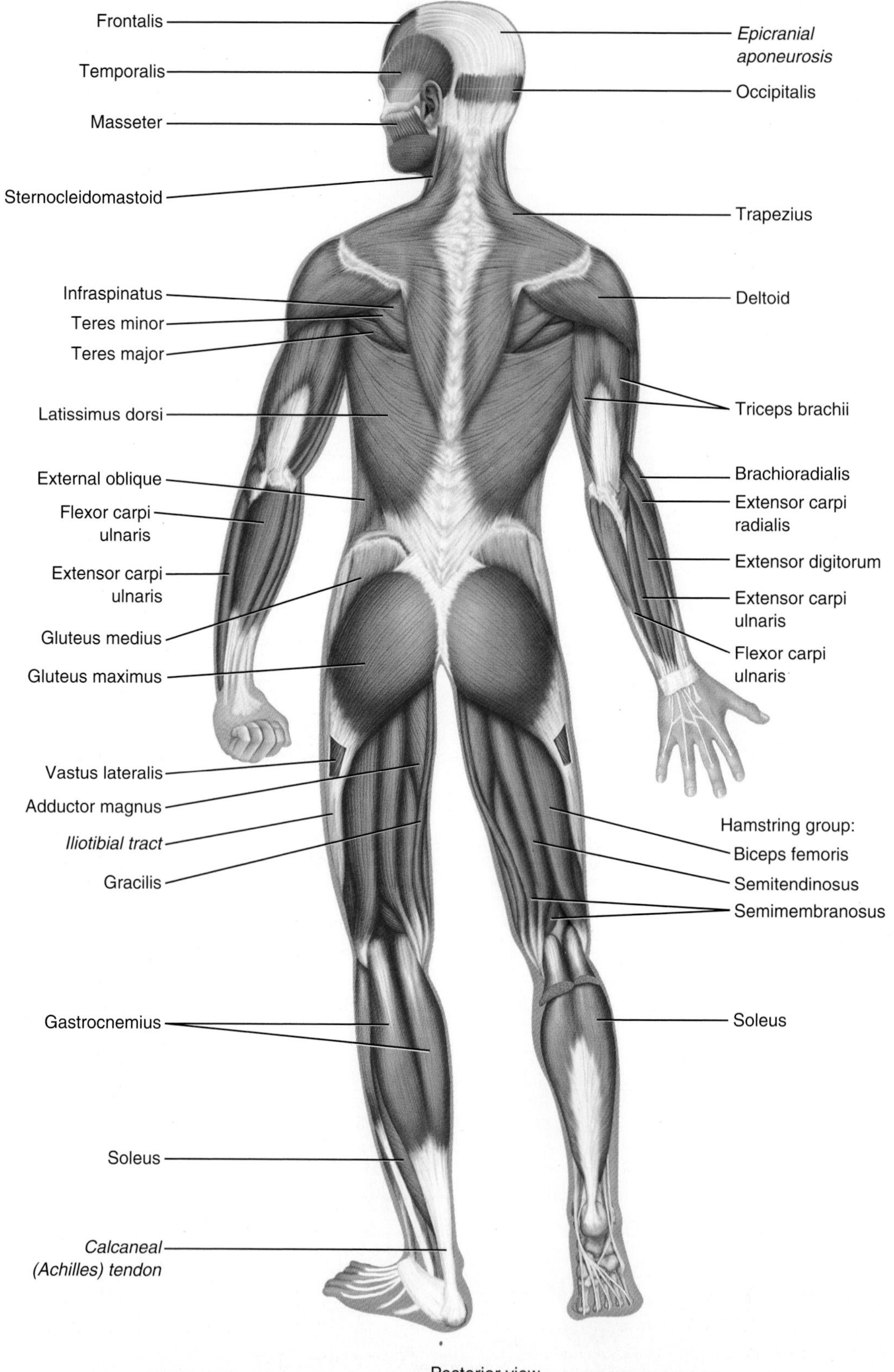
Frontalis
Temporalis
Masseter
Sternocleidomastoid
Epicranial aponeurosis
Occipitalis
Trapezius
Infraspinatus
Teres minor
Teres major
Deltoid
Latissimus dorsi
Triceps brachii
External oblique
Flexor carpi ulnaris
Extensor carpi ulnaris
Brachioradialis
Extensor carpi radialis
Extensor digitorum
Extensor carpi ulnaris
Flexor carpi ulnaris
Gluteus medius
Gluteus maximus
Vastus lateralis
Adductor magnus
Iliotibial tract
Gracilis
Hamstring group:
Biceps femoris
Semitendinosus
Semimembranosus
Gastrocnemius
Soleus
Soleus
Calcaneal (Achilles) tendon
Posterior view

FIGURE 6.12 ▶ Superficial muscles of the body, posterior view.

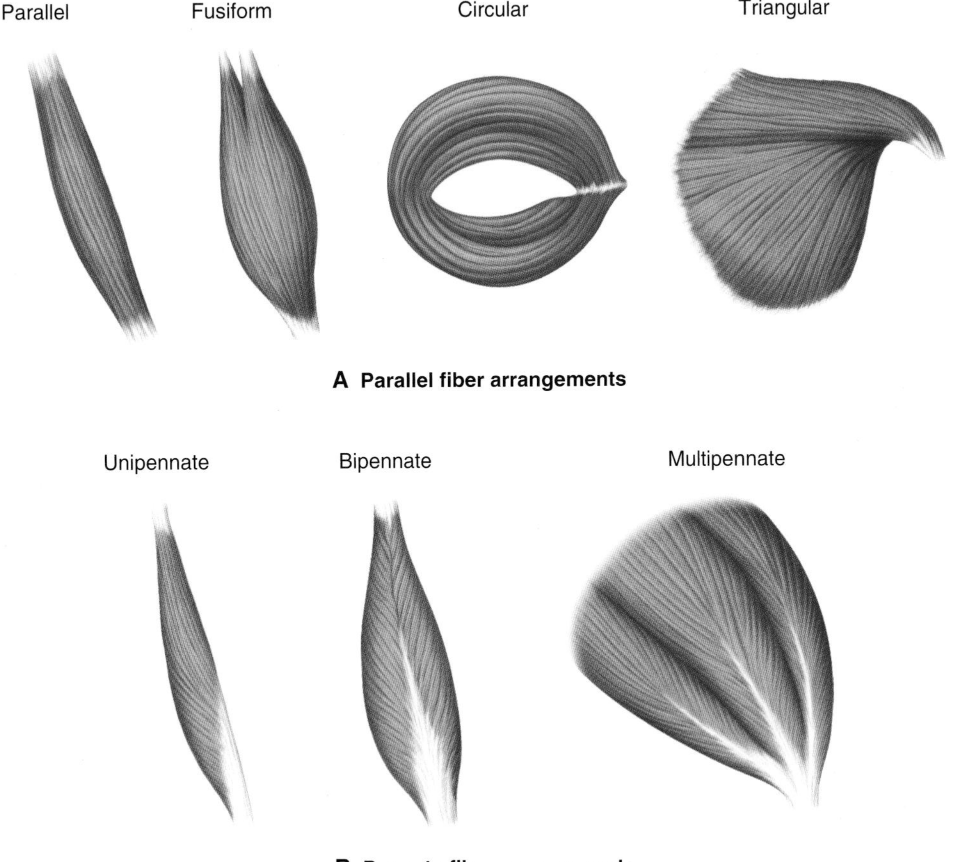

FIGURE 6.13 ▸ Muscle shapes and fiber arrangements. A. Parallel fiber arrangements: fusiform, circular, and triangular. **B.** Pennate fiber arrangements: unipennate, bipennate, and multipennate.

Pennate Muscles

Pennate muscles have shorter fibers within each fascicle that run in an oblique line to attach to a central tendon. This high concentration of muscle fibers along the tendon tends to produce a powerful contraction, but only over a small range of motion. There are three different types of pennate muscles (▸ FIGURE 6.13B):

- ***Unipennate***—Like a flag on the main mast of a sail boat, the fascicles of unipennate muscles converge into the central tendon from one side only. This means that the force of muscle contraction can only produce movement in one direction. An example of a unipennate muscle is the extensor digitorum longus that extends the fingers of the hand (see FIGURE 6.23B).
- ***Bipennate***—In bipennate muscles, the fascicles run in an oblique line from the sides toward the central tendon, giving them a feather-like appearance. The force of contraction pulls on the tendon from both sides, and the tendon directs that accumulated force to the point of insertion. An example of a bipennate muscle is the rectus femoris (FIGURE 6.11).
- ***Multipennate***—In multipennate muscles, the central tendon divides into two or more branches. Each branch then has its own bipennate arrangement, with the fascicles attaching obliquely from both sides to a central tendon. An example of a multipennate muscle is the deltoid (FIGURE 6.12).

Understanding the shape, fiber direction, and fascial divisions within a muscle has several therapeutic implications. For example, knowing the general shape of a muscle helps in identifying the fiber direction. This information guides the therapist in making conscious choices either to stroke across the fibers to broaden the muscle or to use parallel strokes to stretch the fibers. Visualizing the fascial divisions and junctions in the muscle also helps therapists quickly locate and treat the musculotendinous and tenoperiosteal junctions, the most common stress points and probable sites of restriction or injury.

Muscle Assignments during Movement

Coordinated movement is accomplished through the cooperative effort of many muscles. Each individual movement, such as lifting a finger or bowing the head, requires muscles in a specific area to contract and release in a synchronized and complimentary manner. One or two muscles generate most of the power for motion, while others stabilize adjacent bones and maintain the body's position in relation to gravity. A muscle's location and architecture determines the amount of force it contributes and its specific movement assignment, that is, the role it plays in creating a particular movement. A muscle's movement assignment may be agonist, antagonist, synergist, or stabilizer. A single muscle can be cast in any one of these roles depending on the specific movement involved.

- **Agonist**—A muscle in this role is also called a **prime mover**. Because more than one muscle is involved in creating a single movement, the agonist is generally the muscle that is the largest, strongest, or has the best angle of pull across the joint based on its attachment point. It is not uncommon for more than one muscle to be considered a prime mover in any one motion. For example, in hip extension, the gluteus maximus and all three of the hamstrings could be considered agonists because all are large powerful muscles with efficient angles of pull. When in the role of agonist, muscles are generally engaged in concentric contraction.
- **Antagonist**—This term describes a muscle that works in opposition to the agonist. When a muscle is acting as antagonist, it needs to relax slightly and yield to the action of the agonist to allow movement. For example, during an arm or biceps curl, the biceps brachii on the anterior humerus is the agonist, and the triceps brachii on the posterior humerus is the antagonist. In this example, the triceps must relax and lengthen as the biceps contract to bring the elbow into flexion. If the triceps did not yield, there would be no movement. In other cases, it is important for the antagonist to maintain some opposing tension, to act as a brake that controls movement. For example, when we bend over to tie our shoes, the braking action of the trunk extensors, acting as antagonists, keeps us from falling on our face. This balancing act between agonist and antagonist is mediated by the nervous system via a process called **reciprocal inhibition.** Simply stated, reciprocal inhibition means that when the agonist is signaled to contract, the antagonist receives a simultaneous signal that inhibits its contraction.
- **Synergist**—Recall from Chapter 2 that *syn-* means "with or together," so a synergist is a muscle that works with or assists the prime mover during a particular movement. Very few body movements occur as the result of a single muscle contracting. Synergists are signaled to contract with the agonist, creating a smooth and coordinated motion. A synergist may be smaller and less powerful than the agonist or may cross the joint in a manner that gives it a poor pulling angle on the bone. For example, in shoulder abduction, the smaller and more sharply angled supraspinatus muscle acts as a synergist to the deltoid.
- **Stabilizer**—Also called a **fixator**, this role describes muscles that stabilize the proximal or origin end of the prime mover to make the movement more efficient. For example, movement of the humerus at the shoulder joint requires that both the scapula and clavicle are stabilized against the rib cage because the origins of several large shoulder movers are found on these two bones. In this case, the muscles that hold the scapula and clavicle in place are playing the role of stabilizers.

Range of Motion

Range of motion (ROM) refers to the amount or degrees of movement used to measure and describe the osteokinematics of a synovial joint. This range is dependent on a variety of factors, including the type of joint, the overall muscle structure and tension, as well as the pliability of the connective tissue elements in the muscle and joint. Manual therapists use ROM to assess tissue quality, flexibility, and to identify injured tissues. ROM can also be applied as a treatment technique to ease muscle tension, increase body awareness, and create general relaxation. As an assessment tool, ROM is commonly divided into three categories named according to the client's involvement in the motion:

- ***Passive range of motion (PROM)***—During PROM, the client is completely relaxed as the therapist moves a joint through its range of motion. PROM demonstrates how much motion is actually *available* at the joint.
- ***Active range of motion (AROM)***—When a client actively contracts the muscles to create movement, it is considered AROM. During AROM, the client is demonstrating the range of motion they are currently able to use: their *usable* ROM.
- ***Resistive range of motion (RROM)***—When a therapist applies manual resistance to a client's active movement, it is considered RROM. The resistance can be applied throughout the range of motion, or as an isometric resistive, meaning the therapist matches the client's effort with their resistance to stop or hold the movement at a particular point.

As assessment tools, each category of ROM provides manual therapists with useful information about the inert connective tissues that support the joint and the contractile tissues that create movement. By observing and feeling the quantity and quality of motion, therapists get a better idea of which body regions and specific tissues might be contributing to pain and limitation. For example, PROM is used to assess problems in **inert** (noncontractile) tissues such as joint capsules, ligaments, and cartilage since the client is not actively engaging the muscles to create movement. In contrast, RROM engages the **contractile** tissues (e.g., muscle, tendon, and fascia). However, an isometric resistive must be used to minimize movement of the inert tissues, so that pain or loss of strength implicates injury to the contractile tissues. During AROM, both inert and contractile tissues are engaged, so pain or limitation during AROM could originate from either type of tissue. For this reason, an AROM assessment simply helps to identify the client's willingness to move, which movements are problematic, and the general area of tension or pain.

At the end of a normal range of motion, the tissues surrounding the joint create resistance to stop movement and protect the structural integrity of the joint. The quality of this resistance, called the **end feel**, can be assessed during PROM and is generally described as being soft, firm, or hard.

- ***Soft end feel*** is created when a specific joint motion is limited by soft tissue running into soft tissue. For example, during elbow flexion, the motion is limited when the soft tissue of the forearm bumps into the bulk of the biceps.
- ***Firm end feel*** is the most common type of end feel. When movement is limited by the tension of soft tissue being stretched and pulled taut, it creates a firm end feel. For example, dorsiflexion is limited by the extensibility of the Achilles tendon and the amount of tension in the calf muscles.
- ***Hard end feel*** is created when movement is stopped by one bone end hitting another. This only occurs naturally in two motions: elbow extension, when the olecranon process runs into the olecranon fossa, and when the teeth collide in closing the mouth.

Many manual therapists also use ROM to ease muscle tension, increase body awareness, improve overall mobility, or create general muscle relaxation. For example, therapists can improve a client's limited AROM by passively moving the client's joint through its full available ROM. This increases the client's awareness of any conscious or unconscious tension that may be limiting their movement. Other examples of using ROM for therapeutic benefit include the PROM used during pin-and-stretch techniques or the use of AROM for active release methods. In both cases, the amount of soft tissue release is enhanced by adding movement to the manipulation.

What Do You Think? 6.4

- What are some everyday objects that would serve as analogies for the different types of parallel and pennate muscle arrangements?
- Where would you find the prime movers for knee flexion versus knee extension? Explain your reasoning.
- Where would you find the antagonists for elbow flexion? Explain your reasoning.
- What kind of changes would you expect to find in PROM, AROM, and RROM in a sprain versus a strain injury?

MAJOR MUSCLES OF THE BODY

The following sections provide the name, location, and primary movement functions for the major skeletal muscles of the body. The superficial muscles that can be directly manipulated, as well as the deeper or more intrinsic muscles associated with postural distortion and movement dysfunctions that a manual therapist is likely to see are also included. Although muscle information can be organized in many different and useful ways, we have chosen to organize the muscles by body regions rather than by joint or movement classifications. For each body region, a summary table lists the origin, insertion, and action(s) of each major muscle. In general, the actions listed are the muscle's prime function(s) only. A **prime function** is the strongest movement created by concentric contraction of the muscle; however, a muscle may be playing the role of either agonist or synergist when performing this function. To reference the specific nerves and spinal segments that provide innervation to each muscle, see Appendix B.

Muscle Names

Muscle names provide helpful clues about the anatomic characteristics, location, or functions of a muscle. In other words, knowing the name of a muscle may provide information about where it is, its shape and fiber direction, or what it does. Typical themes among muscle names include:

- ***Size***—The terms *maximus* or *major* refer to a large muscle, while *minimus* or *minor* are used to describe smaller muscles. For example, the large muscle in the buttocks is called the gluteus maximus. The terms *longus* (long) or *brevis* (short) describe the length of a muscle. The peroneus longus and peroneus brevis muscles in the leg are good examples.
- ***Shape***—Several muscles such as the deltoid muscle in the shoulder are given names that describe their shape. The deltoid is triangular shaped like a river delta or the Greek letter Delta (Δ). Muscles with the term *quadratus*, such as the quadratus lumborum (QL), are square shaped, and the rhomboid muscle has the geometric shape of a rhombus or diamond.
- ***Function***—Several muscles have names that clearly describe their function. It is not difficult to guess that

the prime function of the supinator is supination or that the flexor carpi radialis flexes the wrist.

- ***Fiber direction***—The fiber orientation of a muscle in relation to the midline is also used in naming muscles. The term *rectus* describes a muscle with fibers that run parallel to the midline, while *oblique* and *transverse* describe a diagonal or perpendicular orientation of muscle fibers. The best examples of this naming method are in the abdominal muscles: rectus abdominus, transverse abdominus, and the internal and external obliques.
- ***General location***—Names of body regions and bones are commonly used in a muscle's name to indicate its general location. For example, the tibialis anterior is located on the front of the leg, and the brachialis is found in the upper arm. Knowing the general location of a muscle also helps in identifying its most likely function.
- ***Origin or insertion***—The name of a muscle can also give precise information about its attachment points. The best example of this is the sternocleidomastoid (SCM), which originates from the sternum (*sterno-*) and clavicle (*cleido-*) and inserts on the mastoid process. The coracobrachialis, originating on the coracoid process of the scapula and inserting on the shaft of the humerus (*brachium*), is another example of how attachment points are used in muscle names.
- ***Number of origins***—Muscles whose names include the term *biceps* such as the biceps femoris have two origins, while the triceps brachii muscle includes the term *triceps* to indicate its three heads.

Muscles of the Head and Face

Muscles on the head and face are responsible for moving the head and neck and for **mastication** (chewing), speech, and creating facial expressions. It is helpful to know the location and general function of these muscles, because they are commonly involved in stress-related problems that manual therapists frequently address, such as headaches and temporomandibular joint dysfunction (TMJ). There are two broad flat muscles anchoring the scalp to the skull, the ***occipitalis*** on the posterior side and the ***frontalis*** on the forehead. Several smaller muscles, called ***suboccipitals***, run from the first two cervical vertebrae to the occipital ridge. Collectively, the suboccipitals extend, rotate, and laterally flex the head. The ***temporalis*** is located on the cranium, but it inserts on the mandibular condyle, making it an important muscle for mastication (FIGURE 6.14).

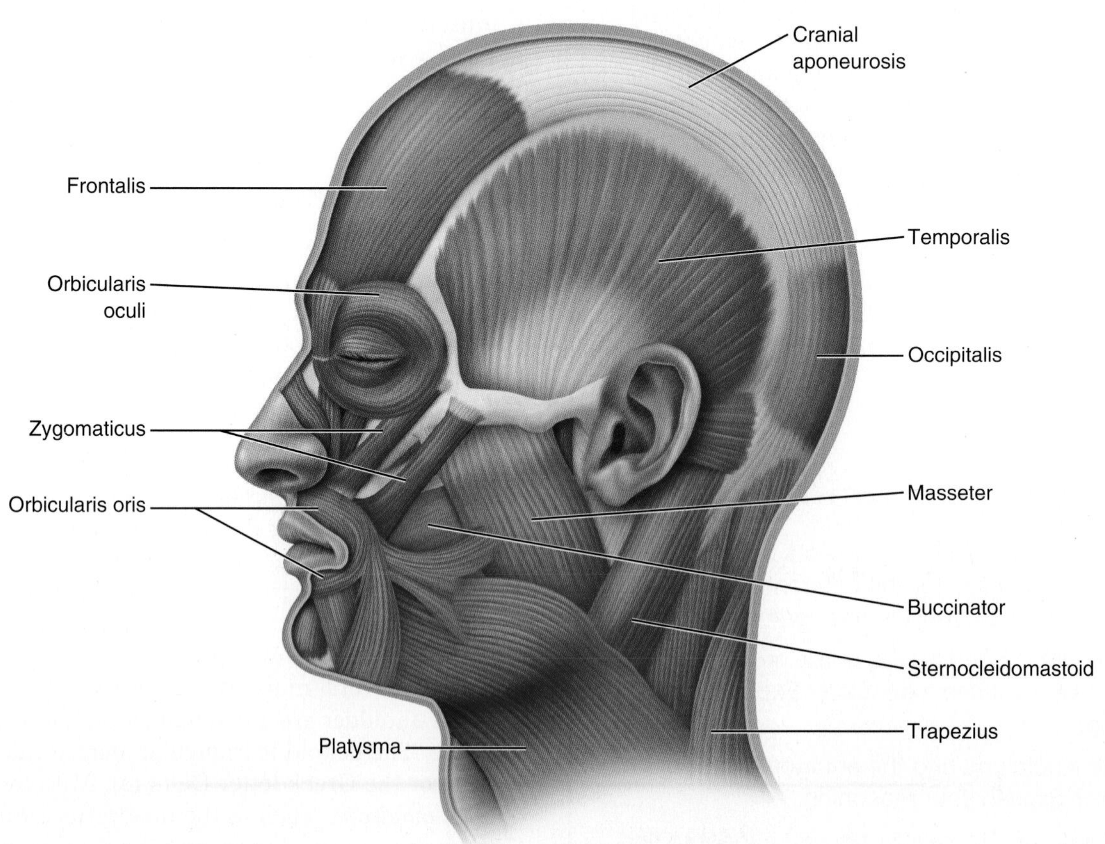

FIGURE 6.14 Muscles of the head and face. A. Lateral view.

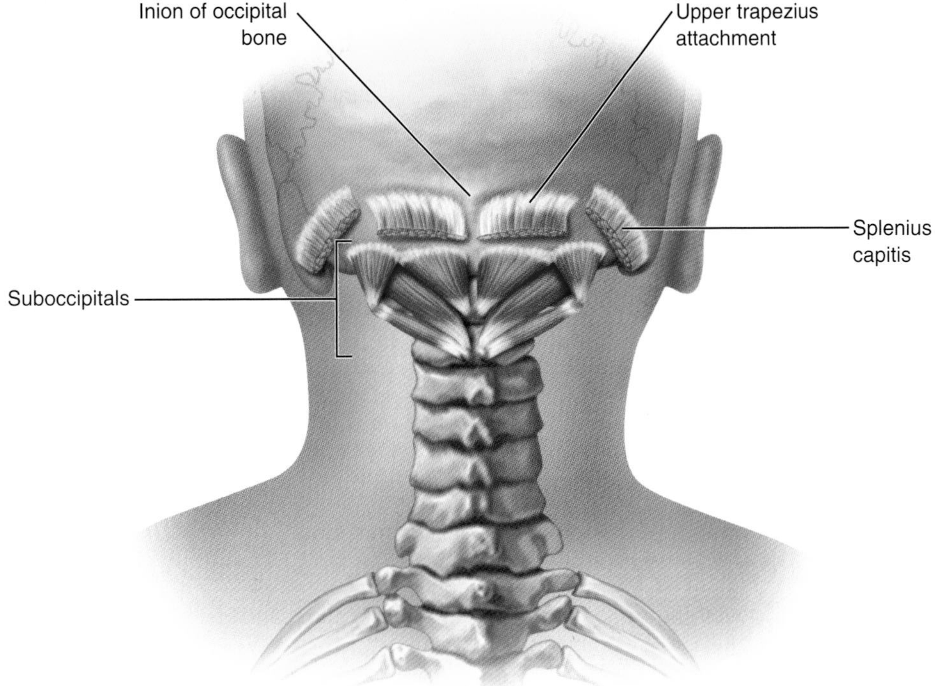

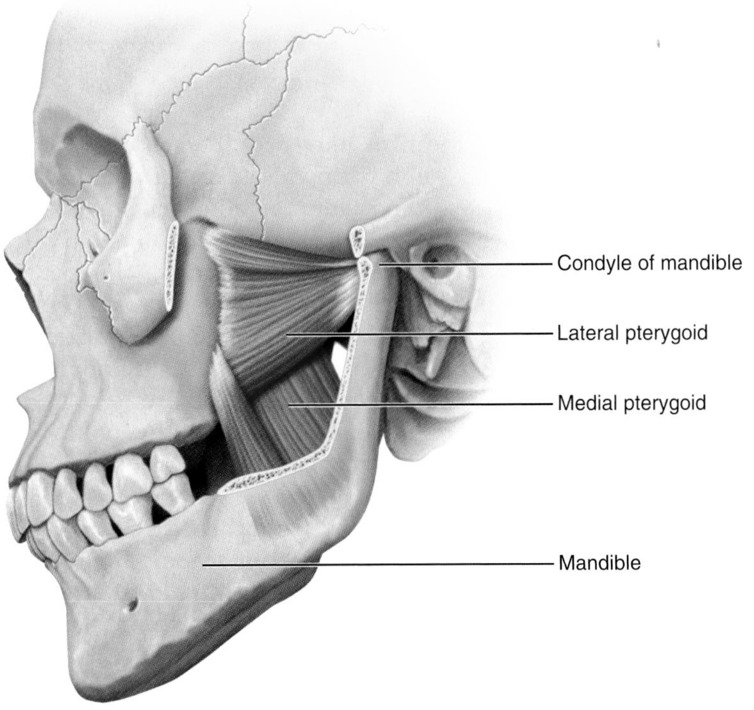

FIGURE 6.14 ▸ *(Continued)* **Muscles of the head and face. B.** Suboccipitals. **C.** Pterygoids.

Table 6-1 Muscles of the Head and Face

Group	Muscle	Action(s)	Origin	Insertion
Skull	Occipitalis	Anchors and draws scalp posteriorly	Occipital ridge	Base of scalp
	Suboccipitals	Extension, rotation, and lateral flexion of the head	Spinous and transverse processes of C1 & C2	Lower occipital bone and transverse processes of C1
	Frontalis	Anchors and draws scalp anteriorly; raises eyebrows	Scalp	Skin at supraorbital rim
	Temporalis	Mastication	Temporal fossa	Coronoid process of mandible
Face	Masseter	Mastication	Zygomatic arch	Angle of mandible
	Medial pterygoid	Mastication and lateral deviation of mandible	Sphenoid and maxilla	Medial angle of mandible
	Lateral pterygoid	Protrusion, depression, and lateral deviation of mandible	Sphenoid	Mandible and articular disk of TMJ
	Buccinator	Moves cheeks for whistling, blowing, and sucking	Maxilla and mandible	Orbicularis oris
	Orbicularis oris	Closes and shapes lips during speech	Muscle fibers around the mouth	Skin around the lips
	Orbicularis oculi	Closes the eyes	Medial orbital socket	Skin around the eye
	Zygomaticus • Major • Minor	Raises upper lip and corner of the mouth to expose the teeth and smile	Muscle fibers around the mouth	Skin of upper lip and corner of the mouth

The other major muscles of mastication are facial muscles: the masseter and pterygoids. The ***masseter*** originates on the zygomatic arch and inserts on the mandible, allowing it to close and retract the jaw. Two other important muscles used for chewing are the ***medial*** and ***lateral pterygoids***, which lie deep to the masseter. In Figure 6.14A, you can see that the eyes and mouth are surrounded by circular muscles. The ***orbicularis oris*** encircles the mouth and is nicknamed the "kissing muscle" because it draws the lips together as in a kiss. The muscle that surrounds and closes the eye is the ***orbicularis oculi***. The ***buccinator*** is the large muscle that attaches to the orbicularis oris; it draws in the cheeks when we whistle or drink from a straw. The ***zygomaticus major*** and ***minor*** muscles that extend from the zygomatic bone to the corner of the mouth are the "smiling muscles" (TABLE 6-1).

Muscles of the Neck

The most superficial of the anterior neck muscles is the ***platysma*** (FIGURE 6.14A). This thin, sheet-like muscle extends from the fascia of the chest muscles over the clavicle to the mandible. While it is not considered a prime mover for the head, face, or neck, it pulls the lips into a frown, assists in opening the jaw, and plays an important protective and stabilizing role for the anterior neck. The ***sternocleidomastoid (SCM)***, has all points of attachment in its name (FIGURE 6.15). Originating on the sternum and clavicle and inserting on the mastoid process, the SCM creates flexion, lateral flexion, and rotation of the head to the opposite side. Deeper in the anterior neck, there are two groups of small muscles that tense the floor of the mouth and stabilize the hyoid bone during swallowing. The ***suprahyoids*** run from the mandible to the hyoid, and the ***infrahyoids*** come up to the hyoid from their origins on the manubrium, clavicle, and superior scapula (FIGURE 6.15B).

The ***anterior, middle, and posterior scalenes*** are the group of muscles situated lateral to the sternocleidomastoid (FIGURE 6.15A). All three muscles arise from the transverse processes of the cervical vertebrae and attach to either the first or second ribs. A bilateral contraction of this group flexes the head and neck, while unilateral contraction creates slight lateral flexion and rotation to the same side. The scalenes also elevate the first and second ribs for inhalation, making them important muscles for breathing.

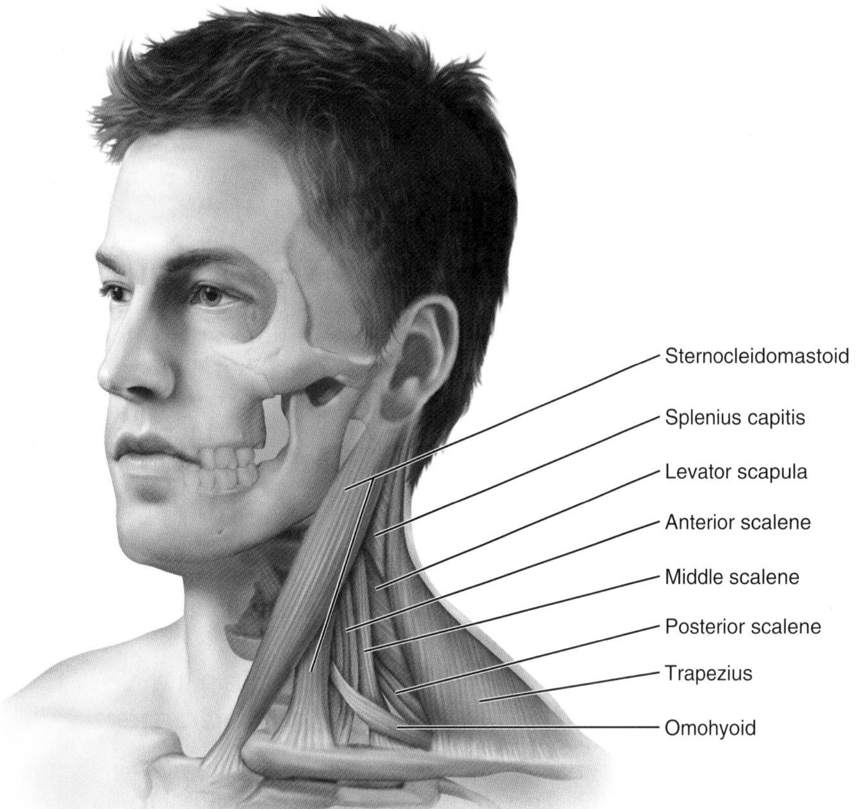

A

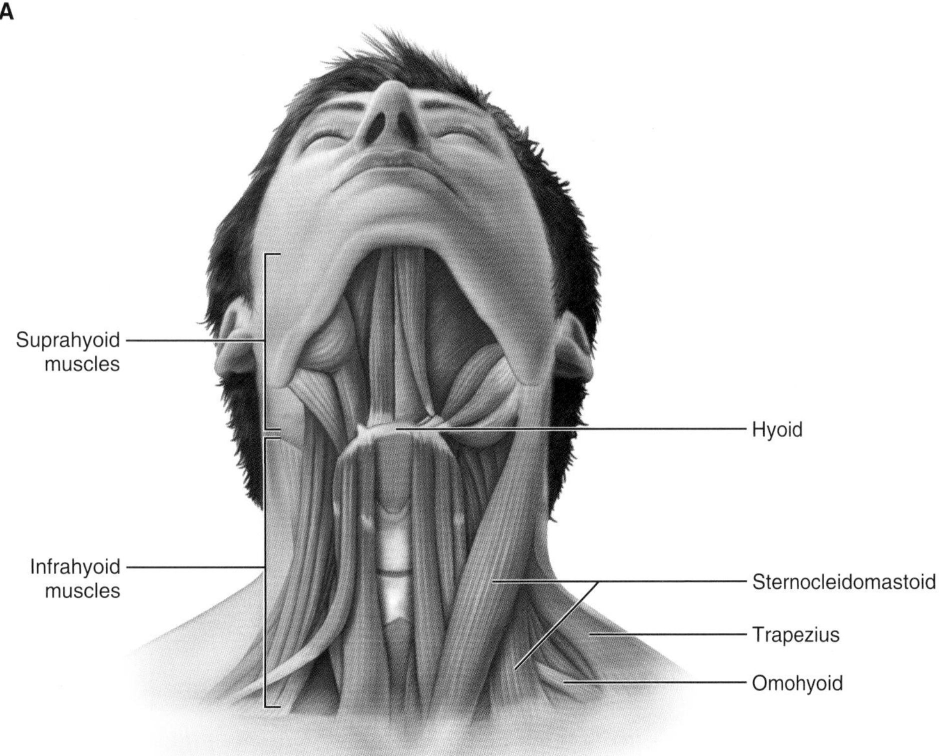

B

FIGURE 6.15 ▸ **Muscles of the anterior neck. A.** Lateral view. **B.** Frontal view.

Pathology Alert **Temporomandibular Joint Dysfunction**

While most hinge joints only allow flexion and extension, the shape and structure of the temporomandibular joint also allow the mandible to shift forward, backward, and side-to-side within the joint. Similar to the knee joint, the TMJ also has a fibrocartilage disk that helps to stabilize the joint and cushion the bones. These unique structural and functional characteristics, coupled with its repetitive use in chewing and speaking, make this joint a frequent site of pain and dysfunction. While TMJ disorders can be due to trauma such as motor vehicle accidents or falls, more common causes are muscle tension due to physical or emotional stress, dysfunctional bite patterns, or teeth grinding. The pain and muscle tension associated with TMJ dysfunction is both a symptom and a cause of the problem. Several different types of manual therapy have been shown to be effective interventions in relieving the pain and other symptoms associated with this condition. In many states, therapists are required to have a special intraoral endorsement or certificate in order to use techniques that approach this joint from inside the mouth or ear.

The extensors for the head and neck are found on the posterior side. The ***splenius capitis*** and ***splenius cervicis*** both originate along the cervical and thoracic vertebrae and work together to extend, laterally flex, and rotate the head (capitis) and neck (cervicis). The ***levator scapula*** is unique among the muscles of the posterior neck; it originates on the cervical vertebrae but inserts on the superior angle of the scapula (▶ FIGURE 6.16). Therefore, its primary function is to elevate and downwardly rotate the scapula, and it is only a weak neck extensor (▶ TABLE 6-2).

Muscles of the Chest and Abdomen

When looking at the chest or pectoral region, the large superficial muscle running from the clavicle, sternum, and

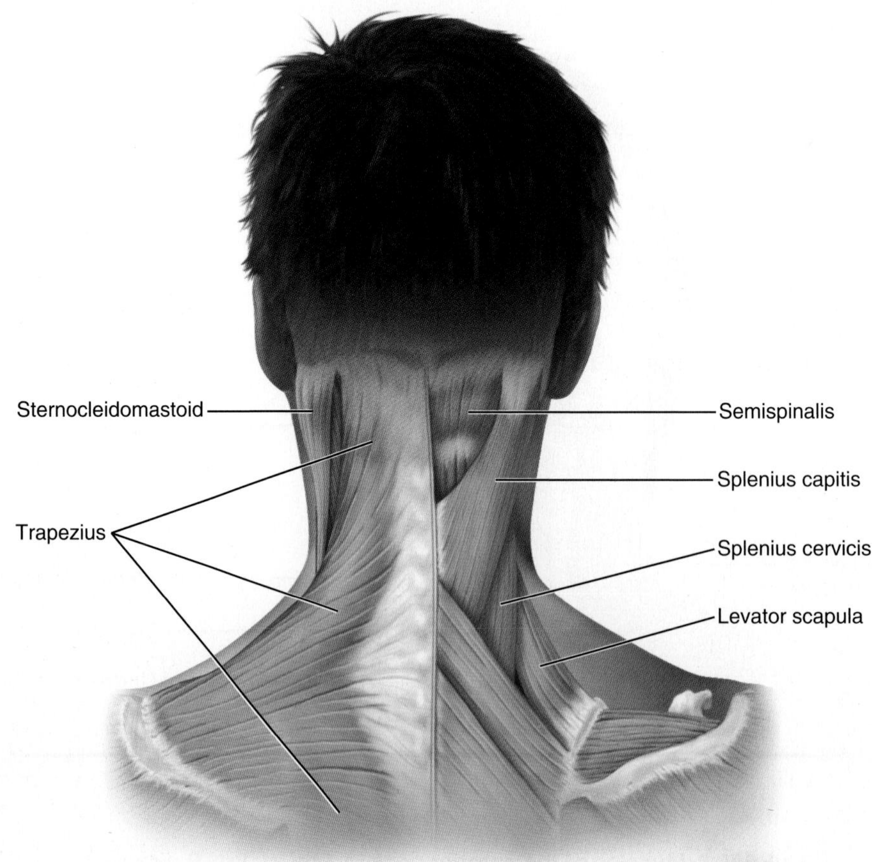

FIGURE 6.16 ▶ Muscles of the posterior neck.

Table 6-2 Muscles of the Neck

Group	Muscle	Action(s)	Origin	Insertion
Anterior neck	Platysma	Pulls mouth into frown and assists depression of the mandible	Fascia of chest muscles	Lower mandible and skin at corners of mouth
	Sternocleidomastoid	Flexion, lateral flexion, and rotation of the head and neck to opposite side	Manubrium of sternum and medial clavicle	Mastoid process
	Suprahyoids	Elevation of the hyoid and tenses floor of mouth when swallowing	Mandible, styloid process and mastoid process	Hyoid
	Infrahyoids	Depression and stabilization of the hyoid and thyroid cartilage	Manubrium, medial clavicle and superior border of scapula	Hyoid and thyroid cartilage
	Scalenes • Anterior • Middle • Posterior	Elevation of ribs 1 & 2; flexion, lateral flexion and rotation of neck	Transverse processes of C2–C7	Ribs 1 & 2
Posterior neck	Splenius capitis	Extension, lateral flexion, and rotation of the head	Nuchal ligament of C3–C6 and spinous processes of C7–T3	Mastoid process and occipital ridge
	Splenius cervicis	Extension, lateral flexion, and rotation of the neck	Spinous processes of T3–T6	Transverse processes of C1–C3
	Levator scapula	Elevation and downward rotation of the scapula	Transverse processes of C1–C4	Superior angle to the root of the scapular spine

ribs to the humerus is the ***pectoralis major*** (FIGURE 6.17). Its multiple points of origin create multiple lines of pull on the humerus, making it a prime mover in several shoulder motions: flexion, adduction, horizontal adduction, medial rotation, and extension of the shoulder from a flexed position. The ***pectoralis minor*** is much smaller and lies deep to the pectoralis major. It originates on the anterior aspect of ribs 3 through 5 and inserts on the coracoid process, making its prime function protraction of the scapula.

A small muscle underneath the clavicle, the ***subclavius***, originates at the costocartilage of the first rib and inserts on the inferior surface of the mid-clavicle. It is primarily a stabilizer, holding the clavicle firmly in place when the humerus and scapula are moved. On the anterior-lateral aspect of the ribcage, we see the ***serratus anterior***, so named because the multiple origin points along the first eight ribs give it a jagged or serrated appearance. Considering its location, the serratus anterior has a surprising prime function. It protracts the scapula because it wraps around the rib cage and passes under the scapula to insert on its vertebral border. Found at the base of the rib cage, the ***diaphragm*** is the large dome-shaped muscle that divides the ventral cavities into thoracic and abdominopelvic regions (FIGURE 6.18). It is the primary muscle used in breathing because it expands the thoracic cavity when it contracts. Two other important respiratory muscles in the chest, the ***external*** and ***internal intercostals***, fill the space between each pair of ribs and work together to help elevate and depress the ribs during breathing.

In the abdominal region, the most superficial and central muscle is the ***rectus abdominus***, which runs vertically from the pubic bone to the xiphoid process and lower ribs (see FIGURE 6.17). Often-admired "six-pack abs" are the result of a well-developed rectus abdominus. There are three other major abdominal muscles: from superficial to deep they are the ***external obliques, internal obliques***, and ***transverse abdominus***. Their names describe the general fiber direction of each muscle as they attach from various points on the pelvic girdle and the central cord of the abdominal fascia, the ***linea alba***. The rectus abdominus and obliques are all strong trunk flexors, while the obliques also rotate the trunk. Because of its fiber direction, the transverse abdominus does not contribute to any major movements of the trunk.

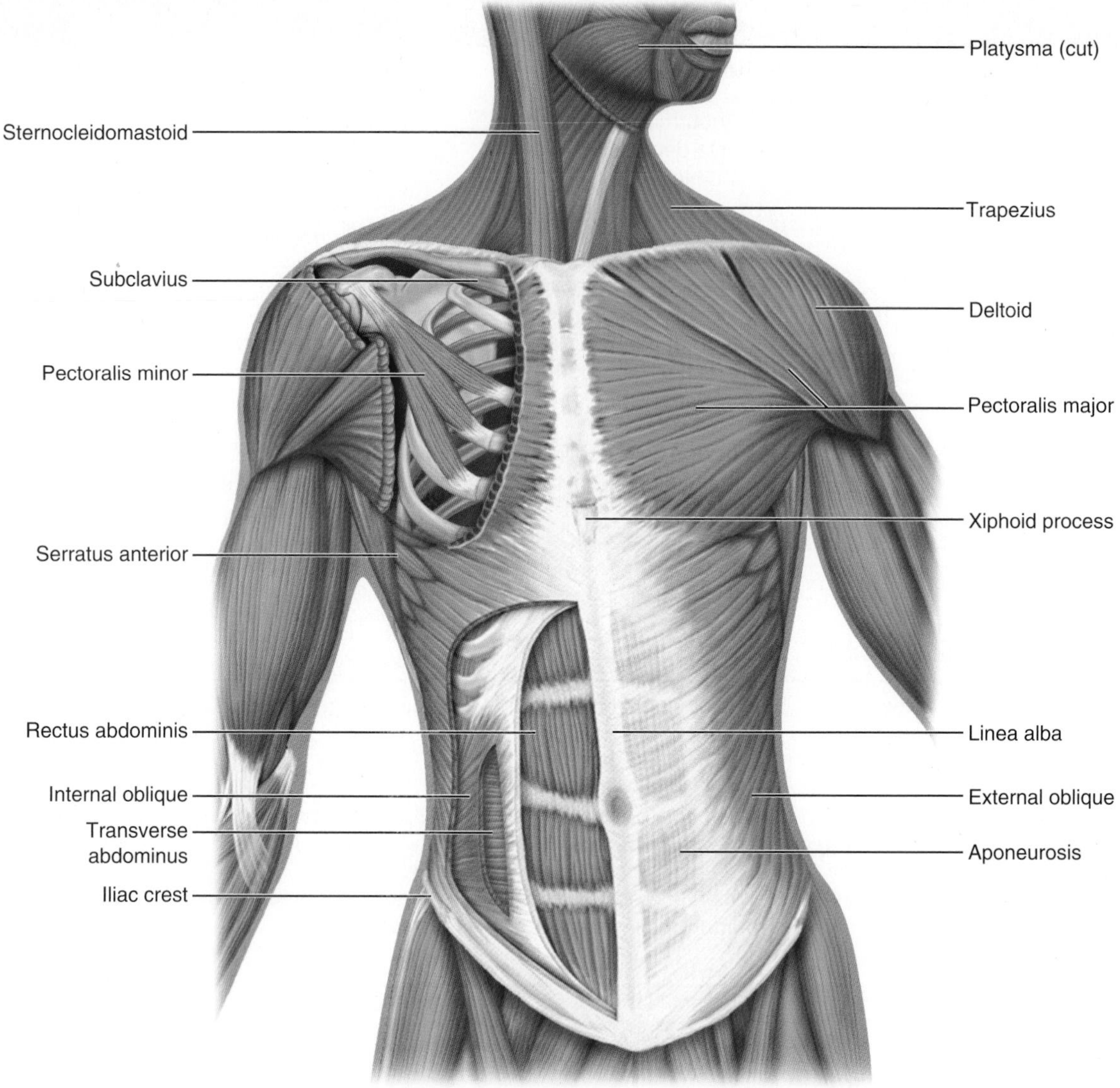

FIGURE 6.17 ▸ Chest and abdominal muscles.

Instead, its primary function is to compress and stabilize the contents of the abdominal cavity. Although they are not key muscles of respiration, all four abdominals can be recruited to assist with breathing.

The deepest muscle in the abdominal cavity is the ***psoas***. A strong trunk flexor, it originates from the anterior aspect of the lumbar vertebrae and attaches to the iliac fossa, where it entwines with the ***iliacus*** muscle of the pelvic girdle. Because these muscles are so deeply invested with each other, many refer to these muscles as one combined functional unit called the ***iliopsoas*** (▸ FIGURE 6.19). Depending on which attachment point is held stable, the iliopsoas either flexes the trunk or flexes the hip (▸ TABLE 6.3).

The Paraspinal Muscles

The two groups of muscles running the length of the spine are collectively referred to as the ***paraspinals.*** The most superficial paraspinal is the ***erector spinae*** group, which has three vertical divisions that all extend the spine. From lateral to medial, these divisions are the ***iliocostalis***, the ***longissimus***, and the ***spinalis*** (▸ FIGURE 6.20).

By the Way

Like the iliopsoas, there are several other muscles in the body where the designation of origin and insertion is ambiguous. With muscles such as the rectus abdominus, the obliques, scalenes, and quadratus lumborum (QL), either attachment can be fixed or moved. Therefore, it may be more useful to think of their origins and insertions as attachments rather than rigidly adhere to the technical definitions of those terms.

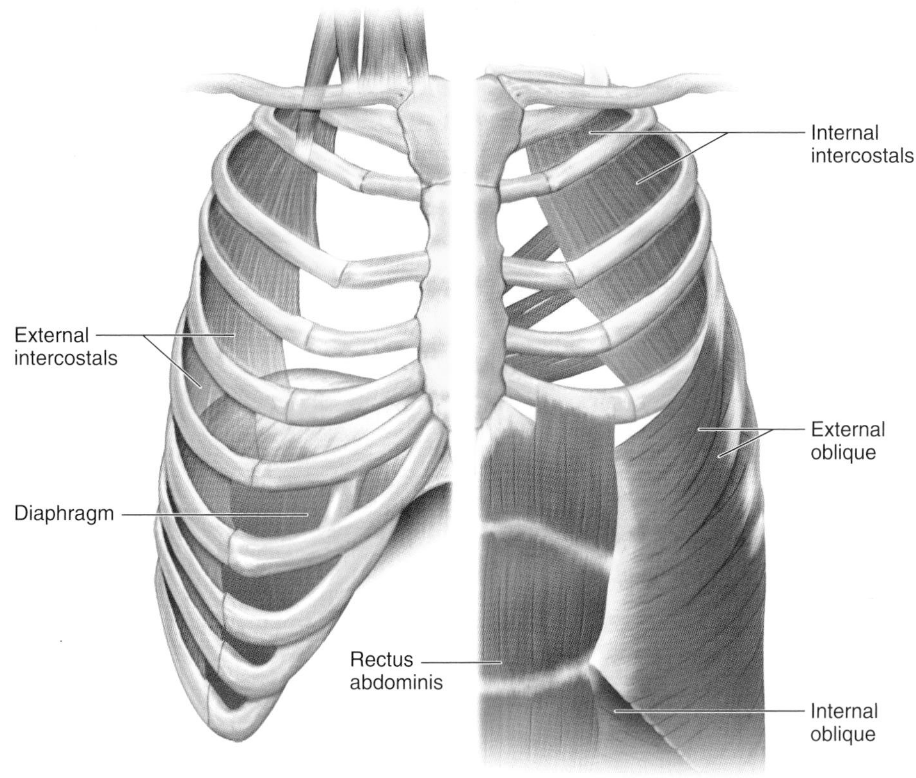

FIGURE 6.18 ▶ Diaphragm and intercostals. These are the primary muscles used in breathing.

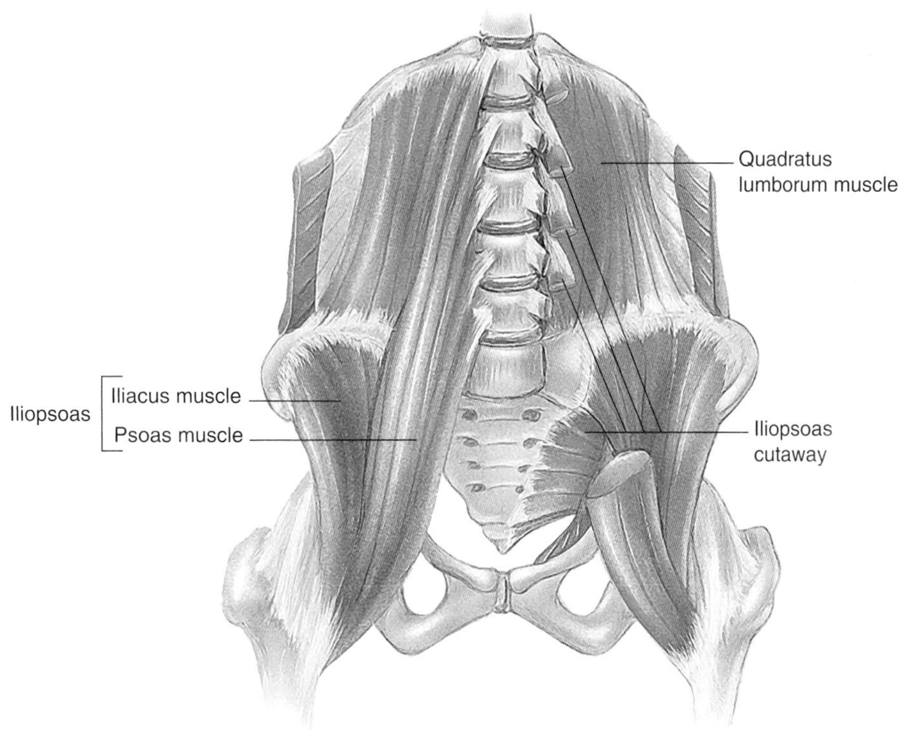

FIGURE 6.19 ▶ Iliopsoas and QL. These muscles form the posterior abdominal wall. The QL originates from the 12th rib and the first four lumbar vertebrae and inserts along the posterior crest of the ilium.

Table 6-3 Muscles of the Chest and Abdomen

Group	Muscle	Action(s)	Origin	Insertion
Chest	Pectoralis major	Flexion, adduction, horizontal adduction, and medial rotation of the shoulder; extends shoulder from flexed position	Sternum and costal cartilages of ribs 1–6 and medial half of clavicle	Lateral lip of bicipital groove
	Pectoralis minor	Protraction and downward rotation of the scapula	Anterior surface of ribs 3–5	Coracoid process
	Subclavius	Depression of the clavicle and stabilization of the sternoclavicular joint	First rib	Inferior edge of the clavicle
	Serratus anterior	Protraction and upward rotation of the scapula; stabilization of scapula during shoulder movement	Anteriolateral ribs 1–8	Vertebral border of scapula
	Intercostals • Internal • External	Movement of ribs during breathing	Between all ribs	
	Diaphragm	Major muscle of breathing; expansion of thoracic cavity	Xiphoid process, inferior edges of ribs 6–12, and bodies of upper lumbar vertebrae	Central tendon of the diaphragm
Abdominal	Rectus abdominis	Flexion of the trunk and spine	Pubic symphysis and crest	Xiphoid process and costocartilage of ribs 5–7
	External obliques	Flexion, lateral flexion, and rotation of trunk to opposite side	Lateral surface of ribs 5–12	Iliac crest, inguinal ligament, and linea alba
	Internal obliques	Flexion, lateral flexion, and rotation of trunk to same side	Iliac crest, inguinal ligament, and lumbar fascia	Linea alba and costocartilage of ribs 7–10
	Transverse abdominis	Compression of abdomen; forces exhalation	Lumbar fascia, inguinal ligament, iliac crest, and costocartilage of ribs 5–10	Linea alba from xiphoid process to pubis
	Psoas • Major • Minor	Flexion of trunk and hip	Bodies and transverse processes of T12–L5	Anterior iliac fossa

The name for each vertical division comes from the insertion points—the iliocostalis attaches to ilium, ribs, and lower cervical vertebrae; the longissimus is the longest muscle, running from the lumbosacral aponeurosis to the mastoid process and occipital ridge; and the spinalis is attached along the spinous processes of all the vertebrae.

The ***transversospinalis*** muscles are the deeper paraspinals that consist of several short muscles between small vertebral segments (▶ Figure 6.20). The most superficial muscle in this group is the ***semispinalis***. This muscle can be segmented into cranial, cervical, and thoracic regions, but all serve the same function in their particular region: extension and rotation to the opposite side. Deep to the semispinalis are the ***multifidi*** and ***rotatores*** muscles that extend and rotate the spine to the opposite side. The ***interspinales*** and ***intertransversarii*** form the deepest layers of the transversospinales group, and they only exist in the cervical and lumbar regions of the spine. The interspinales are short muscles that lie between spinous processes and extend the spine, while the intertransversarii attach to

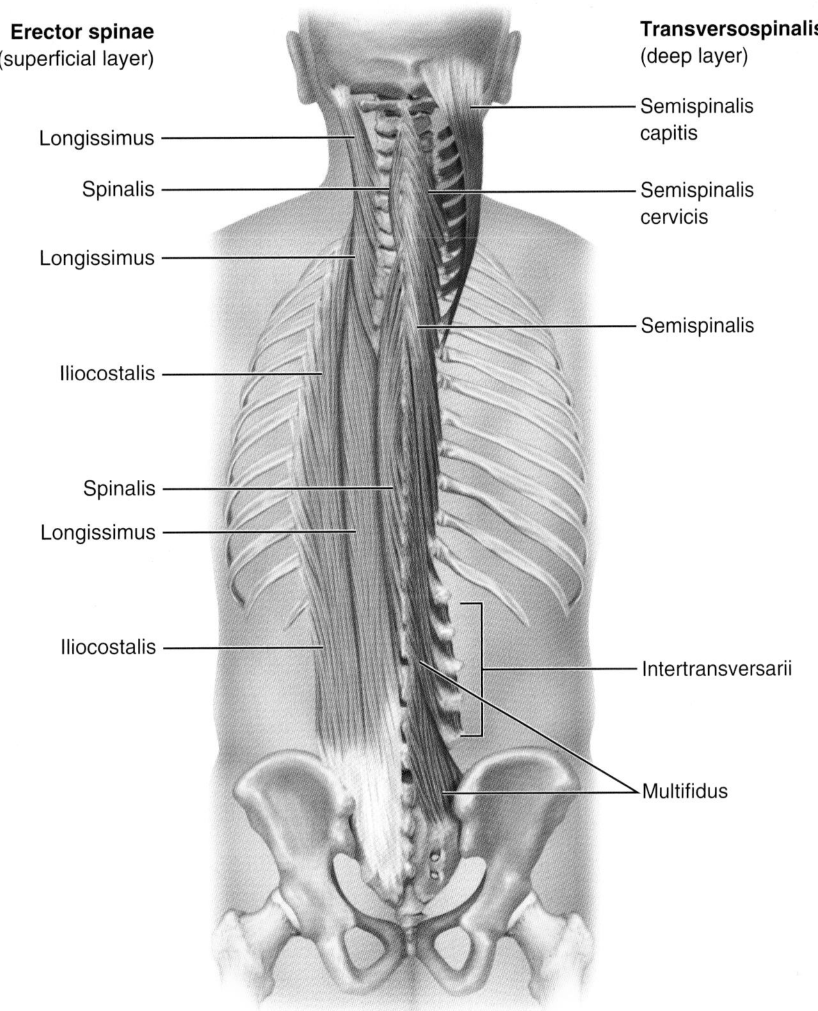

FIGURE 6.20 ▶ Paraspinals. The paraspinals consist of two groups of muscles: the superficial erector spinae and the deeper transversospinalis muscles. For a video of the paravertebral muscles of the spine, go online to http://thePoint.lww.com/Archer-Nelson.

the transverse processes to create extension and lateral flexion (▶ TABLE 6-4).

Muscles of the Back

The posterior torso is dominated by two large superficial muscles: the trapezius and latissimus dorsi. Named for its kite-like or trapezoid shape, the ***trapezius*** (nicknamed the "traps") originates along the spine, running from the occipital ridge to T-12 then attaches to the scapula and lateral clavicle (▶ FIGURE 6.21). Like the pectoralis major, the multiple attachment points of the trapezius give it multiple functions; it retracts, elevates, and upwardly rotates the scapula.

The ***latissimus dorsi*** is the large superficial muscle seen in the lower back. It originates along the spine, running from T-6 to the lumbar fascia and out along the posterior iliac crest (▶ FIGURE 6.21). The body of the muscle fans laterally around the rib cage and through the axilla, merging into a common tendon that inserts on the anterior humerus at the bicipital groove. Contraction of the "lats" creates extension, adduction, and medial rotation of the humerus. A small muscle originating on the inferior angle of the scapula, the ***teres major***, is nicknamed "lats little helper" because it inserts at the same point on the humerus and assists the latissimus dorsi in all its actions. Deep to the trapezius in the interscapular region of the back, the ***rhomboids*** originate from the spinous processes of C-7 through T-5 and attach to the medial border of the scapula (▶ FIGURE 6.21). The primary functions of the rhomboids are to retract and downwardly rotate the scapula, as occurs when hanging from a bar and doing pull-ups.

The ***quadratus lumborum*** (***QL***) lies deep to the latissimus dorsi and gets its name from its square shape and location. It originates from the 12th rib and the first

Table 6-4 Paraspinal Muscles

Group	Muscle	Action(s)	Origin	Insertion
Erector spinae	Iliocostalis	Extension and lateral flexion of the spine	Lumbar fascia, iliac crest, and sacrum	Posterior surface of all ribs and transverse processes of lower cervical vertebrae
	Longissimus	Extension and lateral flexion of the spine	Lumbar fascia, iliac crest and sacrum	Ribs 9 and 10, transverse processes of cervical and thoracic vertebrae, and mastoid process
	Spinalis	Extension of spine	Spinous processes of upper lumbar and lower thoracic vertebrae	Spinous processes of all cervical and upper thoracic vertebrae
Transversospinales	Semispinalis • Capitis • Cervicis • Thoracis	Extension of the head and spine; rotation of the spine to opposite side	Transverse processes of thoracic and lower cervical vertebrae	Occipital bone and spinous processes of the five vertebrae superior to their origin
	Multifidi	Extension and rotation of the spine to opposite side	Sacrum and transverse processes of lumbar, thoracic and lower cervical vertebrae	Spinous processes of the lumbar through second cervical vertebrae, spanning two to three vertebrae above
	Rotatores	Extension and rotation of the spine to opposite side	Transverse processes of lumbar through cervical vertebrae	Lamina of vertebrae, spanning one or two vertebrae above
	Interspinales • Cervical • Lumbar	Extension of the spine	Found between the spinous processes of cervical and lumbar vertebrae; C2–T3 and T12–L5	
	Intertransversarii • Cervical • Lumbar	Extension and lateral flexion of the spine	Found between the transverse processes of cervical and lumbar vertebrae; C2–T7 and L1–L5	

four lumbar vertebrae and inserts along the posterior crest of the ilium (see Figure 6.19). Like the iliopsoas, contraction of the QL can lead to movement of the trunk or the pelvic girdle depending on which attachment is stabilized during contraction. When the trunk is stabilized, unilateral contraction of the QL elevates the hip, giving the muscle its nickname as the "hip hiker." When the hip is stabilized, the QL laterally flexes the trunk.

A group of smaller muscles collectively referred to as the "rotator cuff" originate on the scapula and insert at various points around the head of the humerus. These four muscles merge with the joint capsule of the shoulder to stabilize that joint and rotate the humerus in the glenoid fossa. The ***supraspinatus*** fills the supraspinous fossa and inserts on the superior greater tubercle. The prime function of the supraspinatus is to initiate abduction of the humerus. The ***infraspinatus*** originates from the infraspinous fossa, and the ***teres minor*** from the axillary border of the scapula (see Figure 6.21). Both insert on the posterior greater tubercle and work together to laterally rotate the humerus. The final muscle of the rotator cuff, the ***subscapularis***, originates on the anterior aspect of the scapula and inserts on the lesser tubercle of the humerus (Figure 6.22). This makes the subscapularis the only medial rotator of the group (Table 6-5).

Muscles of the Brachium

As the most proximal muscle of the brachium, the ***deltoid*** sits on top of the humerus like a shoulder pad. It originates on the spine of the scapula, acromion process, and lateral clavicle, and inserts on the proximal lateral shaft of the humerus. As seen in Figures 6.21 and 6.22, the deltoid has a triangular shape and wraps around the anterior, lateral, and posterior aspects of

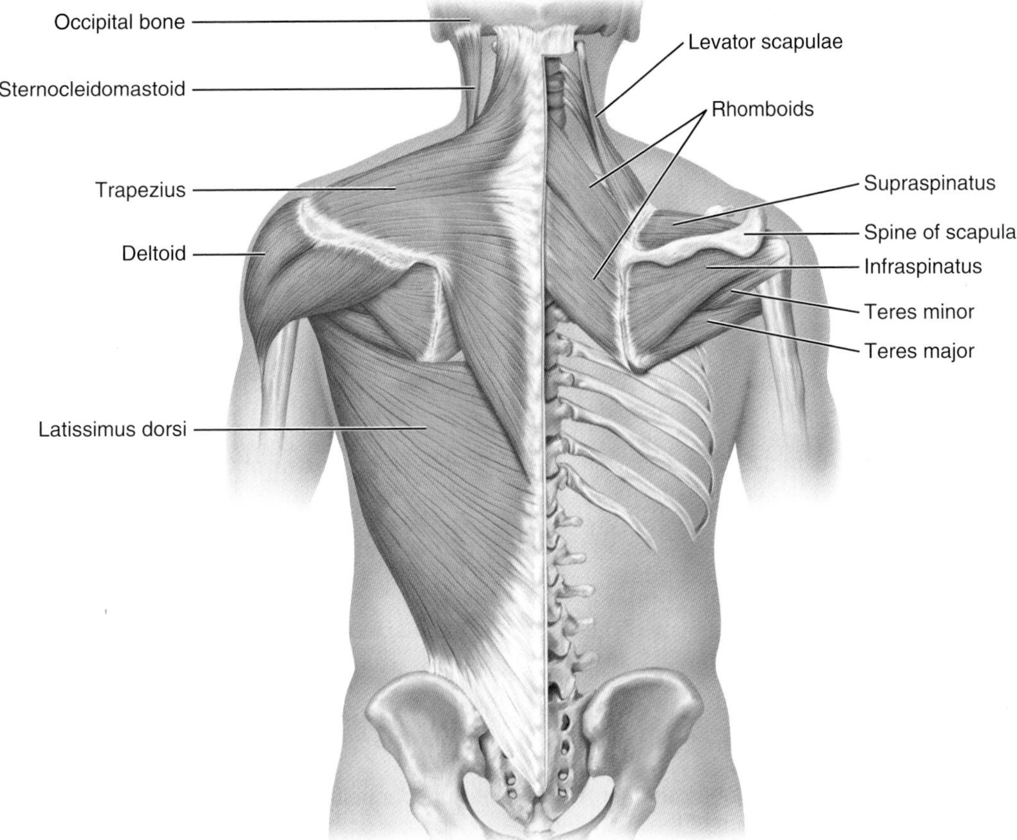

FIGURE 6.21 ▸ **Muscles of the back.**

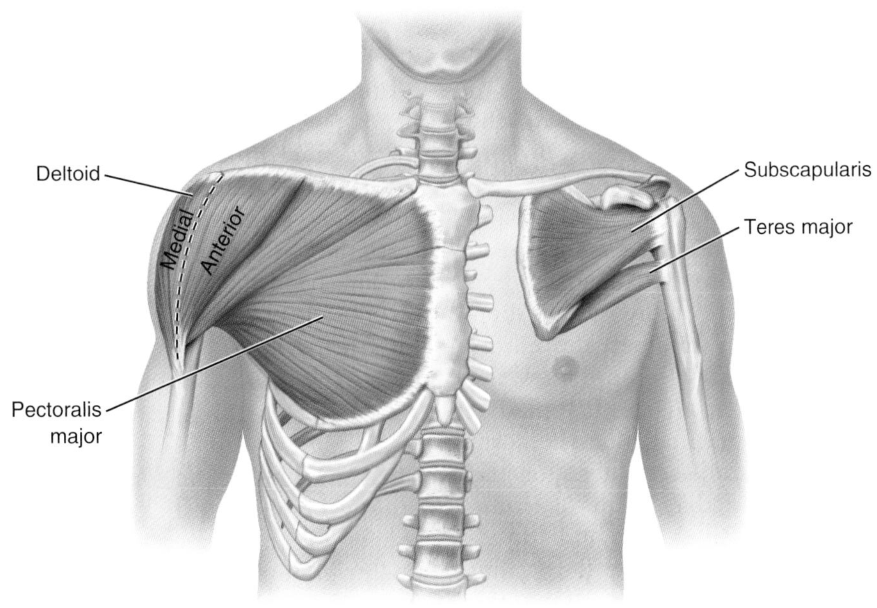

FIGURE 6.22 ▸ **Subscapularis.** This view shows the subscapularis as if the chest muscles and rib cage have been removed. The subscapularis originates on the anterior surface of the scapula (subscapular fossa). Together with the supraspinatus, infraspinatus, and teres minor that originate on the posterior scapula, it forms the rotator cuff.

Table 6-5 Muscles of the Back

Group	Muscle	Action(s)	Origin	Insertion
Back	Trapezius	Elevation, retraction, depression, and upward rotation of the scapula	Occiput, nuchal ligament and spinous processes of C7–T12	Lateral 1/3 of clavicle, acromion process and full spine of scapula
	Latissimus dorsi	Extension, adduction and medial rotation of the shoulder	Spinous processes of T6–L5, posterior aspect of ribs 10–12, lumbar fascia and posterior iliac crest	Medial lip of bicipital groove
	Teres major	Extension, adduction and medial rotation of the shoulder	Inferior angle axillary border of the scapula	Medial lip of bicipital groove
	Rhomboids	Retraction and downward rotation of the scapula	Spinous processes of C7–T5	Medial border of scapula
	Quadratus lumborum	Elevation of the ilium and lateral flexion of the trunk	12th rib and transverse processes of L1–L4	Posterior iliac crest
	Supraspinatus	Initiates abduction of the shoulder	Supraspinous fossa	Superior greater tubercle of the humerus
	Infraspinatus	Lateral rotation and horizontal abduction of the shoulder	Infraspinous fossa	Posterior greater tubercle of the humerus
	Teres minor	Lateral rotation and horizontal abduction of the shoulder	Superior half of the axillary border of the scapula	Posterior greater tubercle of the humerus
	Subscapularis	Medial rotation of the shoulder	Subscapular fossa	Lesser tubercle of the humerus

the glenohumeral joint, making it a prime mover in all shoulder movements *except* adduction.

The largest and most superficial muscle on the anterior side of the brachium is the ***biceps brachii*** (▶ FIGURE 6.23A). The name biceps refers to its two points of origin or heads, with the long head arising from the supraglenoid tubercle and the short head from the coracoid process. These two heads merge into a common tendon that inserts on the radial tubercle. Because it crosses both the shoulder and elbow joints, the biceps flexes both and also supinates the radioulnar joint. The ***coracobrachialis*** lies just medial to the biceps brachii and also originates on the coracoid process. In contrast to the short head of the biceps, it inserts on the mid-medial humerus, making its primary functions shoulder flexion and adduction. The ***brachialis*** is a broad flat elbow flexor that lies deep to the biceps brachii. It originates on the mid-anterior shaft of the humerus and inserts on the coronoid process and tuberosity of the ulna. The brachialis is sometimes called the "true flexor" because its location makes it the most powerful elbow flexor from the anatomic position.

The ***triceps brachii*** is the major muscle of the posterior brachium, and as the name implies, it has three heads (▶ FIGURE 6.23B). The long head originates at the infraglenoid tubercle and is the only one that crosses the shoulder joint, making the triceps a synergist in shoulder extension. The lateral and short heads arise from the posterior shaft of the humerus and merge with the long head to form a common tendon that inserts on the olecranon process, making the primary function of the triceps elbow extension (▶ TABLE 6-6).

Muscles of the Forearm

Most muscles of the forearm create movement in the wrists, hands, and fingers. As seen in Figure 6.23B, the wrist and finger extensors are grouped on the posteriolateral aspect of the forearm and originate from the lateral epicondyle of the humerus. These muscles include the ***extensor carpi radialis longus, extensor carpi radialis brevis, extensor digitorum,*** and the ***extensor carpi ulnaris***. The ***brachioradialis*** originates from the lateral supracondylar ridge just superior to the extensor group. However, because it

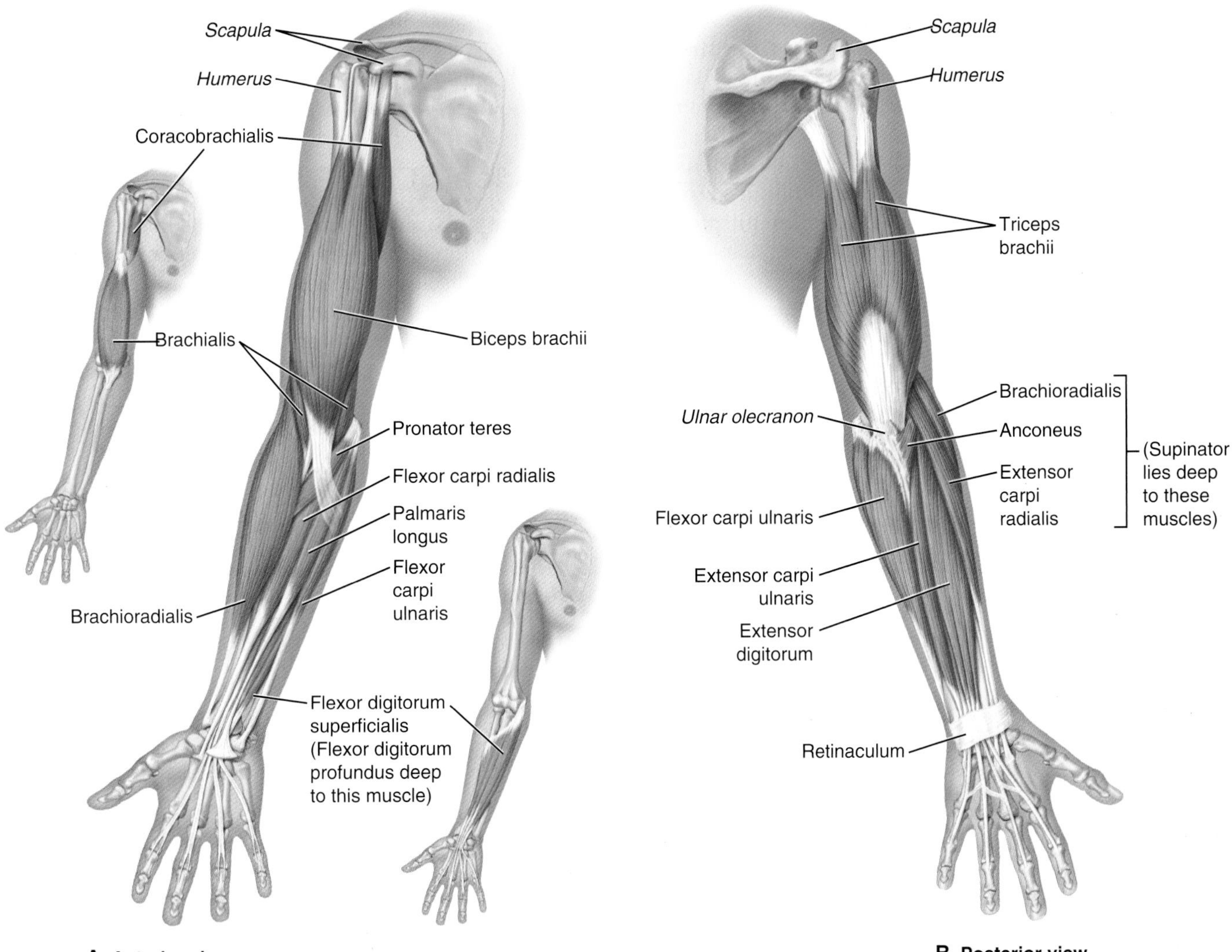

FIGURE 6.23 ▶ **Muscles of the arm. A.** Anterior view. **B.** Posterior view. See online video of the muscles passing from the scapula to the humerus.

Table 6-6 Muscles of the Brachium

Group	Muscle	Action(s)	Origin	Insertion
	Deltoid	Abduction, horizontal abduction, flexion, extension, medial and lateral rotation of the shoulder	Lateral 1/3 of clavicle, acromion process and spine of scapula	Deltoid tuberosity of the humerus
Anterior brachium	Biceps brachii	Flexion of the elbow and shoulder; supination	Coracoid process and supraglenoid tubercle	Radial tuberosity
	Coracobrachialis	Flexion and adduction of the shoulder	Coracoid process	Mid medial shaft of the humerus
	Brachialis	Flexion of the elbow	Anterior distal half of humerus	Coronoid process and ulnar tuberosity
Posterior brachium	Triceps brachii	Extension of the elbow and shoulder	Infraglenoid tubercle, proximal posterior and distal shaft of humerus	Olecranon process
	Supinator	Supination	Lateral epicondyle of humerus and posterior proximal ulna	Proximal lateral shaft of radius

inserts on the distal radius and does not cross the wrist joint, its prime function is elbow flexion. The ***supinator*** is another muscle that arises from the lateral epicondyle; however, a larger portion of its origin is from the posterior ulna. The supinator wraps laterally around the elbow joint before inserting on the proximal lateral shaft of the radius and as its name specifies, it supinates the radioulnar joint. It lies deep to the extensors and another small muscle, the ***anconeus***, which works with the triceps to extend the elbow.

The wrist and finger flexors originate from the medial humeral epicondyle to form the anteriomedial group of forearm muscles (FIGURE 6.23A). These muscles include the ***flexor carpi radialis, palmaris longus, flexor carpi ulnaris,*** and ***flexor digitorum superficialis.*** Another strong finger flexor, the ***flexor digitorum profundus***, originates from the ulna deep to the other flexors. Although it is not a wrist or finger flexor, the ***pronator teres*** is the most superior muscle in the anteriomedial group. As its name implies, this muscle pronates the radioulnar joint (TABLE 6-7).

Muscles of the Pelvic Girdle

There are two major muscle groups on the posterior ilium: the gluteals and the external rotators of the hip. The three muscles of the gluteal group form the bulk of the buttocks, arising from the posterior ilium and inserting on the greater trochanter. The largest and most superficial muscle is the ***gluteus maximus***, which extends and laterally rotates the hip (FIGURE 6.24). The ***gluteus medius*** and deeper ***gluteus minimus*** work together to create abduction and medial rotation of the hip.

Table 6-7 Muscles of the Forearm

Group	Muscle	Action(s)	Origin	Insertion
Posterior Lateral Forearm	Brachioradialis	Flexion of the elbow	Lateral supracondylar ridge of the humerus	Styloid process of the radius
	Extensor carpi radialis longus	Extension and radial deviation of the wrist	Lateral supracondylar ridge	Dorsal base of Second metacarpal
	Extensor carpi radialis brevis	Extension of the wrist	Lateral epicondyle of the humerus	Dorsal base of third metacarpal
	Extensor digitorum	Extension of the fingers	Lateral epicondyle of the humerus	Dorsum of middle and distal phalanges 2–5
	Extensor carpi ulnaris	Extension and ulnar deviation of the wrist	Lateral epicondyle of the humerus	Dorsal base of the fifth metacarpal
Anterior Medial Forearm	Pronator teres	Pronation	Medial epicondyle of the humerus	Mid lateral shaft of the radius
	Flexor carpi radialis	Flexion and radial deviation of the wrist	Medial epicondyle of the humerus	Palmar base of second and third metacarpals
	Palmaris longus	Flexion of the wrist	Medial epicondyle of the humerus	Palmar aponeurosis
	Flexor carpi ulnaris	Flexion and ulnar deviation of the wrist	Medial epicondyle of the humerus and proximal posterior ulna	Pisiform, hamate and palmar base of fifth metacarpal
	Flexor digitorum superficialis	Flexion of the fingers at the PIP and MP joints	Medial epicondyle of the humerus, coronoid process of the ulna and radial shaft	Lateral aspects of middle phalanges
	Flexor digitorum profundus	Flexion of the fingers at the DIP, PIP and MP joints	Proximal medial ulna	Palmar aspect of distal phalanges

A group of six small muscles deep to the gluteals have closely associated attachments along the sacrum, ischium, and greater trochanter. Because the ***piriformis, gemellus superior, obturator internus, gemellus inferior, obturator externus***, and ***quadratus femoris*** also function together to create lateral rotation of the hip, these muscles are sometimes collectively called "the deep six" (▶ FIGURE 6.24B). An easy way to remember the superior to inferior order of these muscles is the mnemonic "Please GO-GO-Q."

The ***tensor fasciae latae (TFL)*** and ***iliacus*** are the major muscles on the anterior side of the pelvic girdle (▶ FIGURE 6.25). The TFL originates on the anterior iliac crest and has a long, broad aponeurosis called the ***iliotibial tract*** or ***band (ITB)*** as its tendon. The gluteus maximus also attaches to the ITB, which runs down the lateral thigh and attaches to the lateral tibial condyle. The primary functions of the TFL are abduction and flexion of the hip. The iliacus is another hip flexor that originates on the anterior ilium and inserts on the lesser trochanter. As discussed earlier, the iliacus muscle forms the hip flexor portion of the functional unit referred to as the iliopsoas (▶ TABLE 6-8).

Muscles of the Thigh

The fascial layers around the thigh divide it into three major muscle groups: the four ***quadriceps*** muscles on the anterior side, a medial group of five hip ***adductors***, and three ***hamstring*** muscles on the posterior aspect. The four muscles in the quadriceps are the ***rectus femoris, vastus medialis, vastus intermedius***, and ***vastus lateralis***. The "quads" blend together distally to form the patellar tendon that attaches them to the patella and

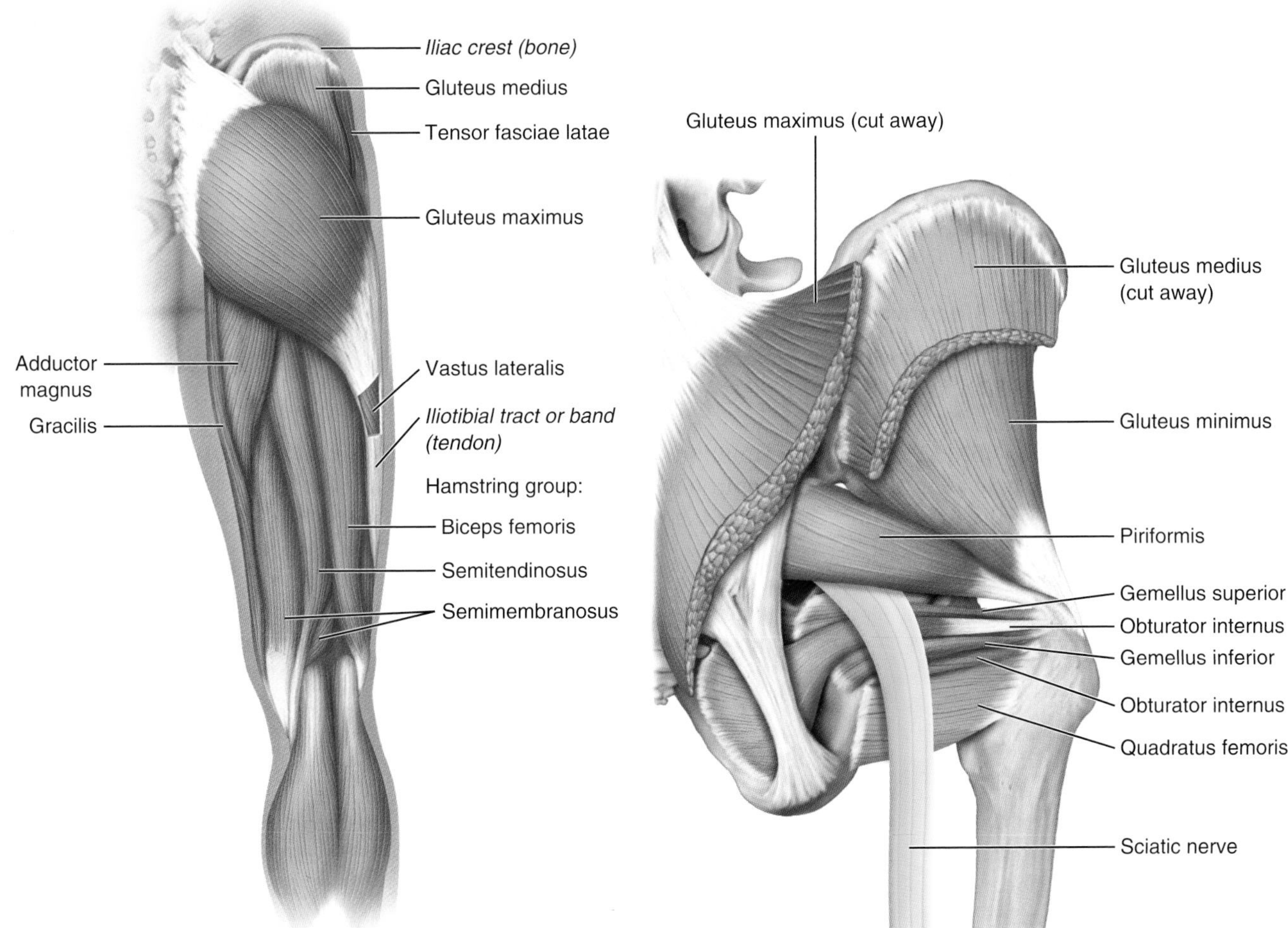

FIGURE 6.24 ▶ **Posterior muscles of the hip and thigh. A.** Superficial muscles. **B.** Deep muscles.

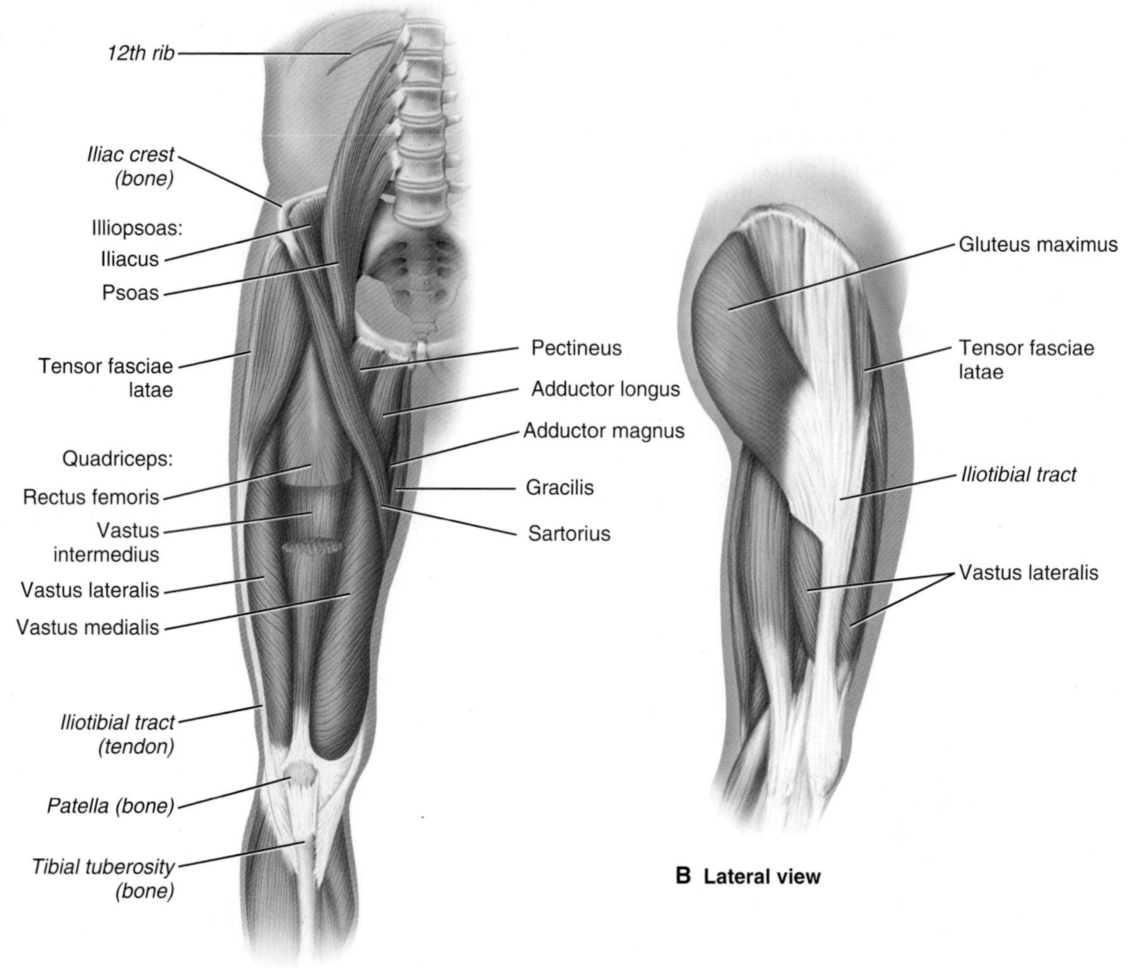

FIGURE 6.25 ▸ Muscles of the anterior and lateral hip and thigh. A. Anterior view. **B.** Notice how the gluteus maximus and TFL both merge into the iliotibial tract, which is superficial to the vastus lateralis.

tibial tuberosity (▸ FIGURE 6.25A). The common function of the quadriceps is knee extension. Due to its origin on the pelvic girdle instead of the femur, the rectus femoris is also a synergist in hip flexion.

Effectively dividing the quadriceps from the adductor group, the ***sartorius*** is a long thin muscle running diagonally across the anterior thigh from the ASIS to the medial anterior proximal shaft of the tibia. It is nicknamed the "tailor's muscle" because its primary functions are to externally rotate the hip and flex both hip and knee, which is the crossed-leg position that a tailor traditionally used while sewing. In an anterior view of the thigh (▸ FIGURE 6.25A), only three of the five adductor muscles can be clearly

Pathology Alert Tendinopathies

Tendinopathy refers to a group of dysfunctions in which tendons are strained or degenerated. Traditionally, the term ***tendonitis*** was used to describe any injury to a tendon. However, more recent research has revealed that inflammation of the tendon, as the suffix *-itis* implies, is actually quite rare. Therefore, the term tendonitis is reserved for acute tendon injuries with associated inflammation, edema, and pain. When overuse, chronic strain, and microtearing occur, the term ***tendinosis*** more accurately describes the chronic degenerative process of the collagen in tendons. ***Tenosynovitis*** is a term applied to describe irritation and inflammation of tendons that have a synovial sheath (e.g., wrist, ankle, and bicipital tendon). Common sites for tendinopathy include the Achilles, patellar, rotator cuff, peroneal, and the extensor pollicis tendons.

Table 6-8 Muscles of the Pelvic Girdle

Group	Muscle	Action(s)	Origin	Insertion
Posterior	Gluteus maximus	Extension and lateral rotation of the hip	Sacrum and posterior iliac crest	Posterior greater trochanter and iliotibial band
	Gluteus medius	Abduction, medial and lateral rotation of the hip	Inferior aspect of the posterior and lateral iliac crest	Greater trochanter
	Gluteus minimus	Abduction and medial rotation of the hip	Inferior aspect of lateral ilium	Anterior greater trochanter
	Piriformis	Lateral rotation of the hip	Anterior sacrum	Posterior aspect of the greater trochanter
	GO-GO-Qs • Gemellus superior • Obturator internus • Gemellus inferior • Obturator externus • Quadratus femoris	Lateral rotation of the hip	Ischium for gemelli and quadratus; obturator foramen for the obturators	Posterior aspect of the greater trochanter
Anterior	TFL	Flexion, abduction, and medial rotation of the hip	Anterior iliac crest	Iliotibial band to the lateral condyle of the tibia
	Iliacus	Flexion and external rotation of the hip	Anterior iliac fossa	Lesser trochanter

seen: the ***pectineus, adductor longus***, and ***gracilis***. The ***adductor brevis*** is deep to the pectineus and longus. All four of these muscles originate on the anterior pubic bone, with the gracilis being the most medial. The fifth adductor, the ***adductor magnus***, originates on the inferior pubic ramus and ischial tuberosity, so it is best seen in posterior views of the thigh (FIGURE 6.24A). All five muscles adduct the hip, and with the exception of the gracilis, they also flex the hip. Because the origin of the ***gracilis*** is so medial on the pelvis (pubic symphysis), it does *not* contribute to hip flexion. Instead, it is the only knee flexor in the group because it inserts alongside the sartorius and semitendinosus at the medial anterior proximal shaft of the tibia.

> **btw By the Way**
>
> The three muscles that insert at the medial anterior proximal shaft of the tibia, or ***MAPS*** point, sartorius, gracilis, and semitendinosus, are collectively referred to as the pes anserine muscles. *Pes anserine* is Latin for "goose foot," which is exactly what the common insertion looks like from a medial view—three tendons or toes with connective tissue webbing between them.

The three muscles of the hamstring group are the ***biceps femoris, semimembranosus,*** and ***semitendinosus*** (FIGURE 6.24A). While all three originate from the ischial tuberosity, the biceps, as its name implies, also has another point of origin: the mid-linea aspera. The long and short heads of the biceps femoris blend together and insert on the head of the fibula. Both of the *semi-* muscles pass to the medial side of the knee, where the deeper *-membranosus* inserts on the posterior medial tibial condyle, while the *-tendinosus* inserts on the medial anterior proximal shaft of the tibia. All three of these muscles are strong hip extensors and knee flexors (TABLE 6-9).

Muscles of the Leg

Similar to the thigh, the fascial layers of the leg divide the muscles into anterior, lateral, and posterior compartments. The dorsiflexors and invertors of the foot and ankle, along with the toe extensors, are the anterior compartment muscles: the ***tibialis anterior, extensor hallucis longus***, and ***extensor digitorum longus*** (FIGURE 6.26). While all three muscles contribute to

Table 6-9 Muscles of the Thigh

Group	Muscle	Action(s)	Origin	Insertion
Anterior	Rectus femoris	Flexion of the hip and extension of the knee	AIIS and acetabulum	Patella and tibial tuberosity via patellar tendon
	Vastus medialis	Extension of the knee	Medial linea aspera	Patella and tibial tuberosity via patellar tendon
	Vastus intermedius	Extension of the knee	Anterior shaft of femur	Patella and tibial tuberosity via patellar tendon
	Vastus lateralis	Extension of the knee	Lateral linea aspera	Patella and tibial tuberosity via patellar tendon
	Sartorius	Flexion and external rotation of the hip and flexion of the knee	ASIS	Medial anterior proximal shaft of the tibia
Medial	Pectineus	Adduction and flexion of the hip	Anterior pubic ramus	Proximal posterior femur
	Adductor brevis	Adduction and flexion of the hip	Anterior inferior pubic ramus	Proximal linea aspera
	Adductor longus	Adduction and flexion of the hip	Anterior inferior pubis	Mid linea aspera
	Adductor magnus	Adduction and flexion of the hip	Inferior pubic ramus and ischial tuberosity	Full linea aspera and adductor tubercle
	Gracilis	Adduction of the hip and flexion of the knee	Pubic symphysis	Medial anterior proximal shaft of the tibia
Posterior	Biceps femoris	Flexion of the knee and extension of the hip	Ischial tuberosity and mid linea aspera	Head of the fibula
	Semimembranosus	Flexion of the knee and extension of the hip	Ischial tuberosity	Posterior medial tibial condyle
	Semitendinosus	Flexion of the knee and extension of the hip	Ischial tuberosity	Medial anterior proximal shaft of the tibia

dorsiflexion of the ankle, it is the primary function of the tibialis anterior that also inverts the foot due to its insertion point at the base of the first metatarsal. The names extensor digitorum and extensor hallucis clearly describe the prime function and imply the insertion points of each.

The lateral compartment of the leg contains two peroneal muscles that share the primary functions of eversion and plantarflexion. The ***peroneus longus*** (also known as the ***fibularis longus***) is the larger of the two, running from the anterior proximal shaft of the fibula, behind the lateral malleolus and under the foot to finally insert on the first metatarsal and medial cuneiform (FIGURE 6.27). Because the tibialis anterior and fibularis (peroneus) longus insert on opposite sides of the first metatarsal and medial cuneiform, the pair are sometimes referred to as the "stirrup muscles." The shorter ***peroneus brevis*** (also known as the ***fibularis brevis***) originates from the distal two-thirds of the fibula and inserts at the base of the fifth metatarsal.

The largest and most superficial muscles in the posterior compartment of the leg are the gastrocnemius and soleus (FIGURE 6.28). They are the prime movers for plantarflexion, while the other posterior compartment muscles are considered synergists in this action. The ***gastrocnemius*** has two heads that originate from the posterior aspect of the medial and lateral femoral condyles. These two heads merge to form the Achilles tendon before inserting on the calcaneus. The ***soleus***

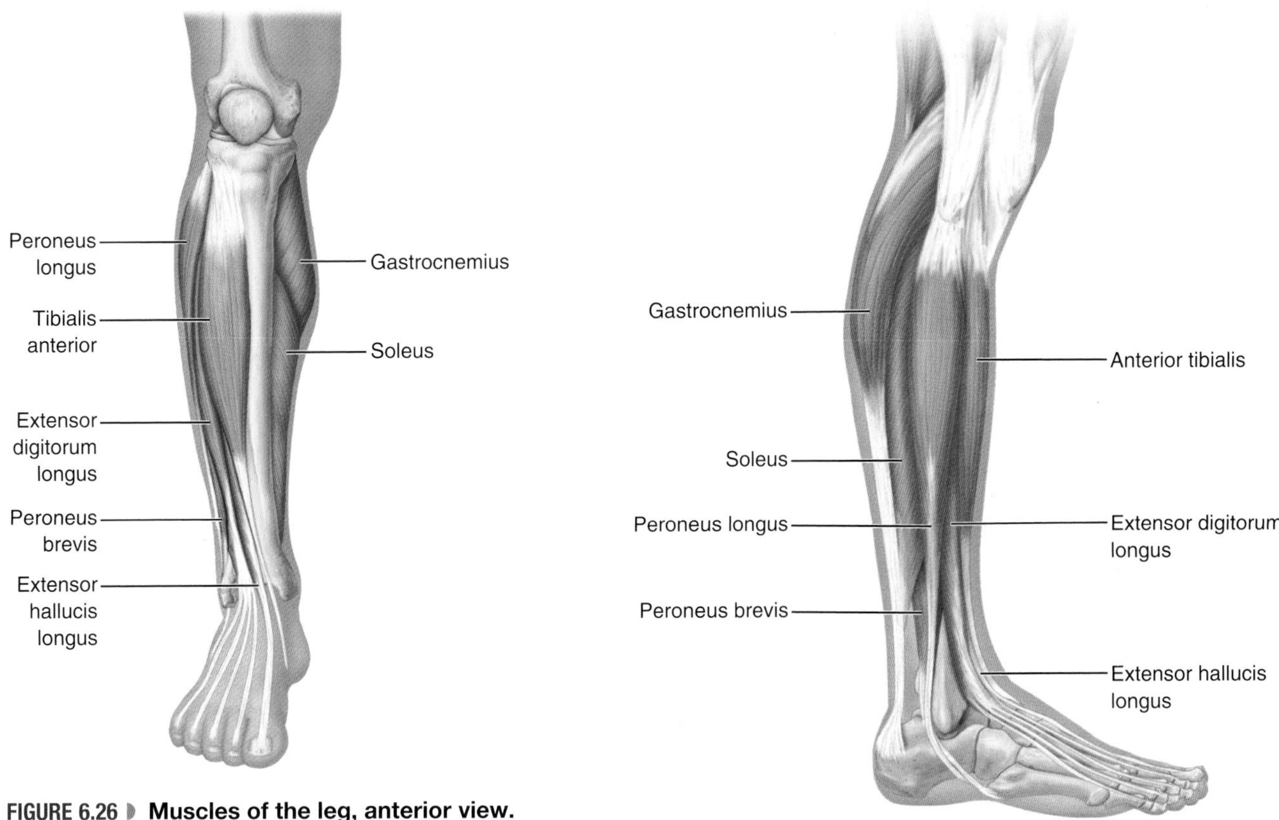

FIGURE 6.26 ▸ Muscles of the leg, anterior view.

FIGURE 6.27 ▸ Muscles of the leg, lateral view.

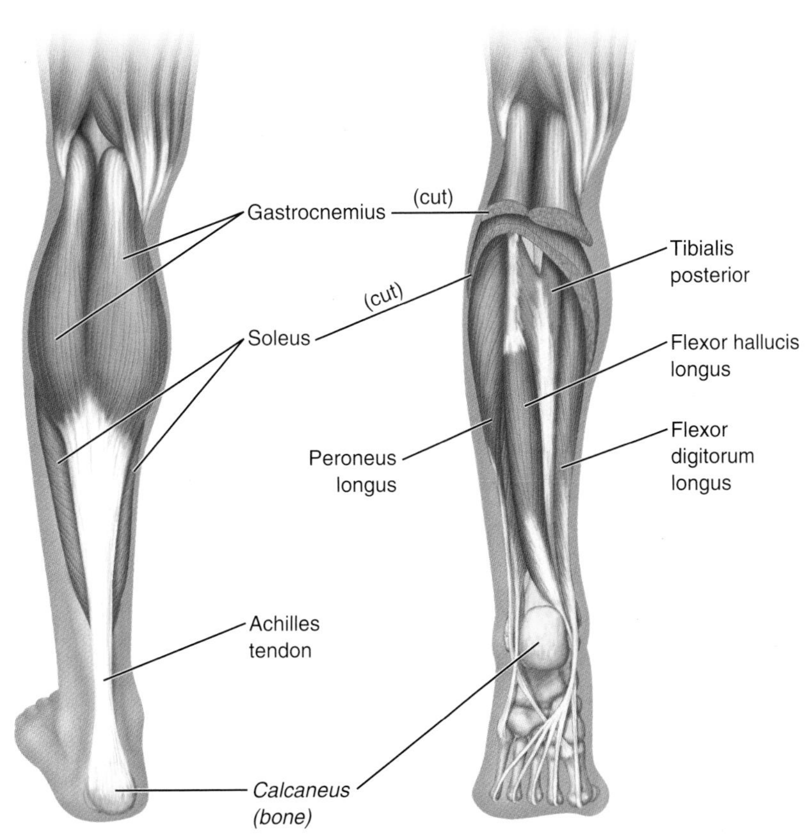

FIGURE 6.28 ▸ Superficial and deep muscles of the leg, posterior view.

Table 6-10 Muscles of the Leg

Group	Muscle	Action(s)	Origin	Insertion
Anterior	Tibialis anterior	Dorsiflexion and inversion of the foot and ankle	Lateral condyle and anterior shaft of tibia	Medial aspect of first cuneiform and base of first metatarsal
	Extensor hallucis longus	Extension of the big toe (hallux); assists dorsiflexion	Anterior mid fibular shaft	Dorsal base of distal phalanx of hallux
	Extensor digitorum longus	Extension of toes 2–5; assists dorsiflexion	Lateral condyle of tibia and proximal two-thirds of fibula	Middle and distal phalanges of toes 2–5
Lateral	Peroneus longus	Eversion of the foot and ankle; assists in plantarflexion	Head and proximal shaft of the fibula	Medial aspect of first cuneiform and base of first metatarsal
	Peroneus brevis	Eversion of the foot and ankle; assists in plantarflexion	Distal shaft of fibula	Base of the fifth metatarsal
Posterior	Gastrocnemius	Plantarflexion and flexion of the knee	Posterior medial and lateral condyles of the femur	Calcaneus via the Achilles tendon
	Soleus	Plantarflexion	Posterior tibia and fibula	Calcaneus via the Achilles tendon
	Tibialis posterior	Inversion of the foot and ankle, assists plantarflexion	Posterior tibia, fibula and interosseous membrane	Plantar aspect of the navicular, cuneiforms, cuboid and bases of second to fourth metatarsals
	Flexor hallucis longus	Flexion of the hallux	Mid posterior fibula	Plantar distal phalanx of the hallux
	Flexor digitorum longus	Flexion of toes 2–5	Mid posterior tibia	Plantar distal phalanges of toes 2–5

is a broad and flat muscle deep to the "gastrocs" that originates from both the tibia and fibula. It also inserts on the calcaneous via the Achilles tendon. The other three muscles in the posterior compartment, the ***tibialis posterior, flexor hallucis longus***, and ***flexor digitorum longus***, lie deep to the gastrocnemius and soleus. The tibialis posterior works with the tibialis anterior to create inversion, while the flexor hallucis and flexor digitorum flex the toes (TABLE 6-10).

EXERCISE, AGING, AND THE MUSCULAR SYSTEM

The exact type and number of muscle fibers in skeletal muscles are predetermined by individual genetics and, for the most part, does not change over time. However, the characteristics of those fibers, including their length, diameter, and number of sarcomeres, *can* change in response to the amount and type of regular exercise. **Hypertrophy** is the term used to describe a general increase in the size and bulk of muscle. When subjected to regular resistance exercise, the muscle responds by increasing the number of sarcomeres and myofilaments. This increases both the length and diameter of the myofibrils within the muscle cells but does not change the actual number of fibers within the muscle. Aerobic exercises such as walking, swimming, or cycling do not have as dramatic an effect on muscle size; instead, this type of activity increases the vascularity of the tissue and its overall aerobic capacity.

Extreme muscle hypertrophy does not always mean the muscle is stronger and more powerful. If adequate attention is not given to maintaining flexibility, strength may be inhibited because the natural broadening action

Box 6.1 SURFACE ANATOMY FOR MANUAL THERAPISTS

Each person has their own unique shape. However, each body has the same bony framework and soft tissue contours with visual and/or palpable landmarks. Using their eyes and hands, therapists must be able to recognize the standard anatomic divisions between major muscle groups as well as the specific boundaries and edges of muscles and tendons within a group. Without this knowledge, it can be difficult to be specific with our work or to properly position clients for the application of deep techniques. The following illustrations are designed to help develop your visualization and palpation skills.

Anterior Torso

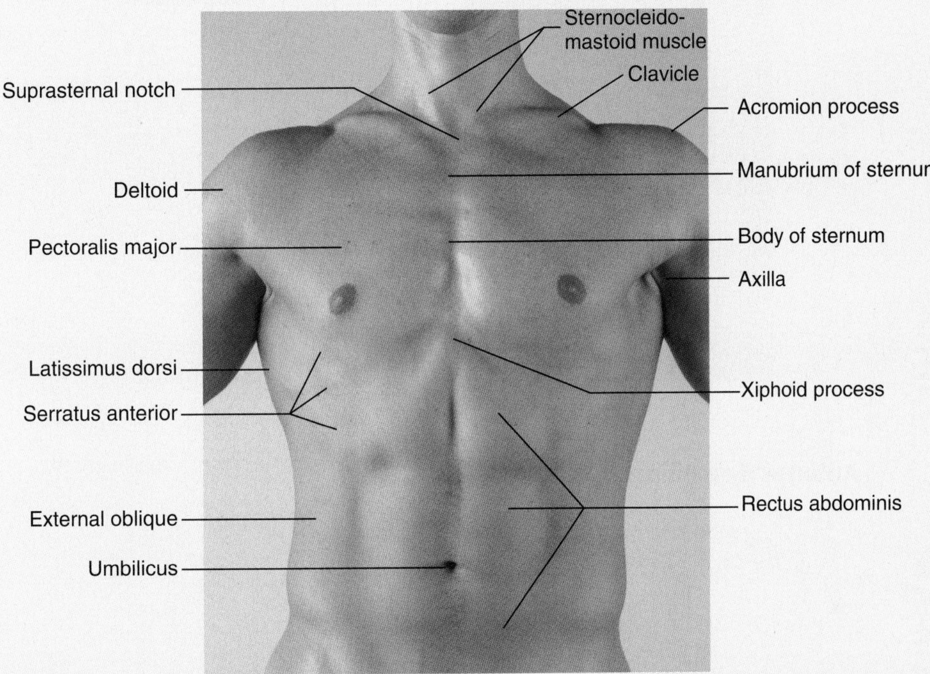

Posterior Torso

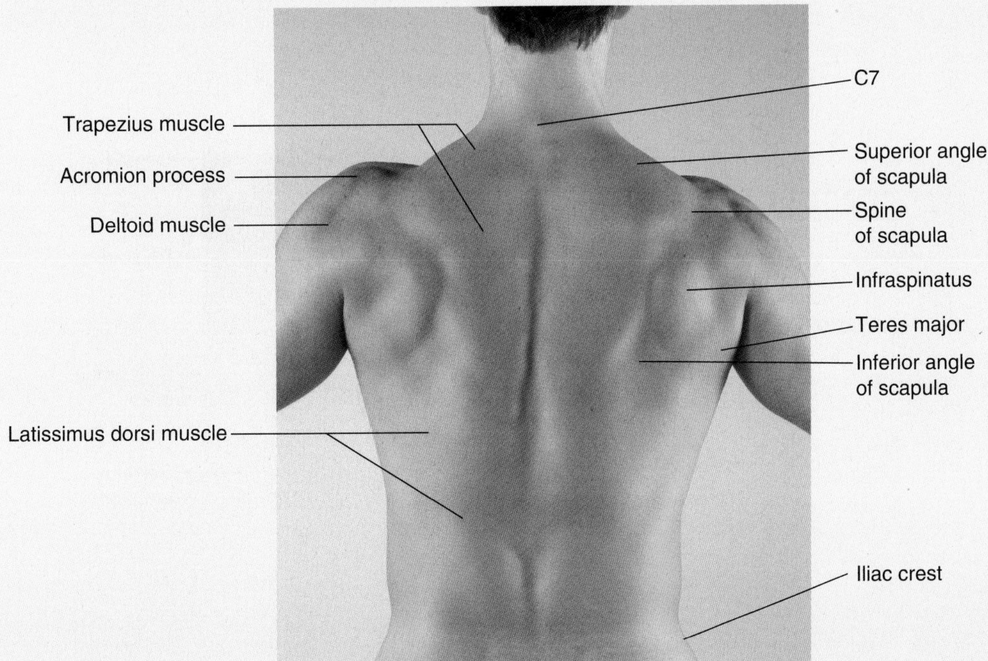

(Continued)

Box 6.1 SURFACE ANATOMY FOR MANUAL THERAPISTS *(Continued)*

Posterior Brachium

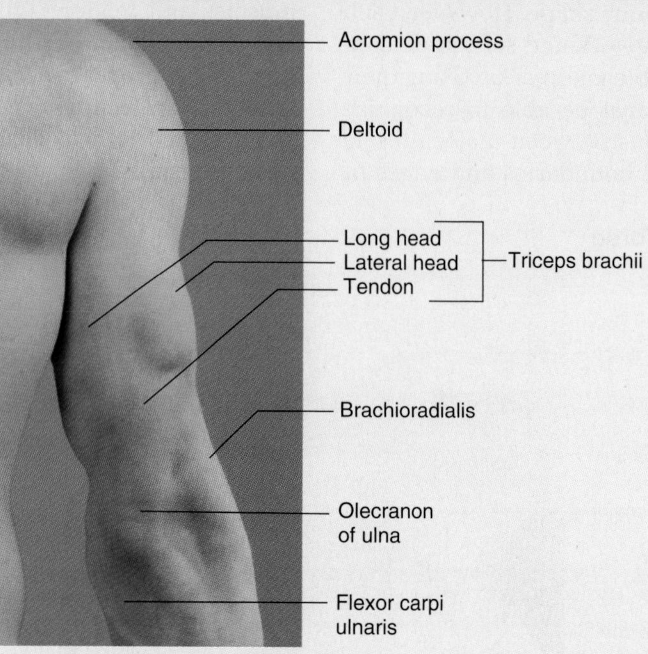

Anterior Brachium

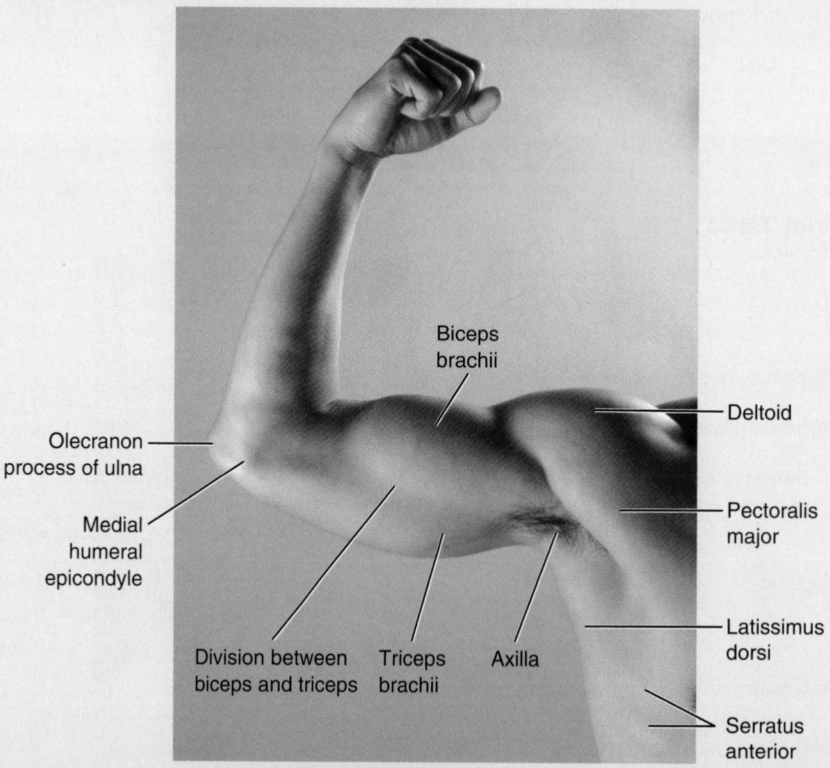

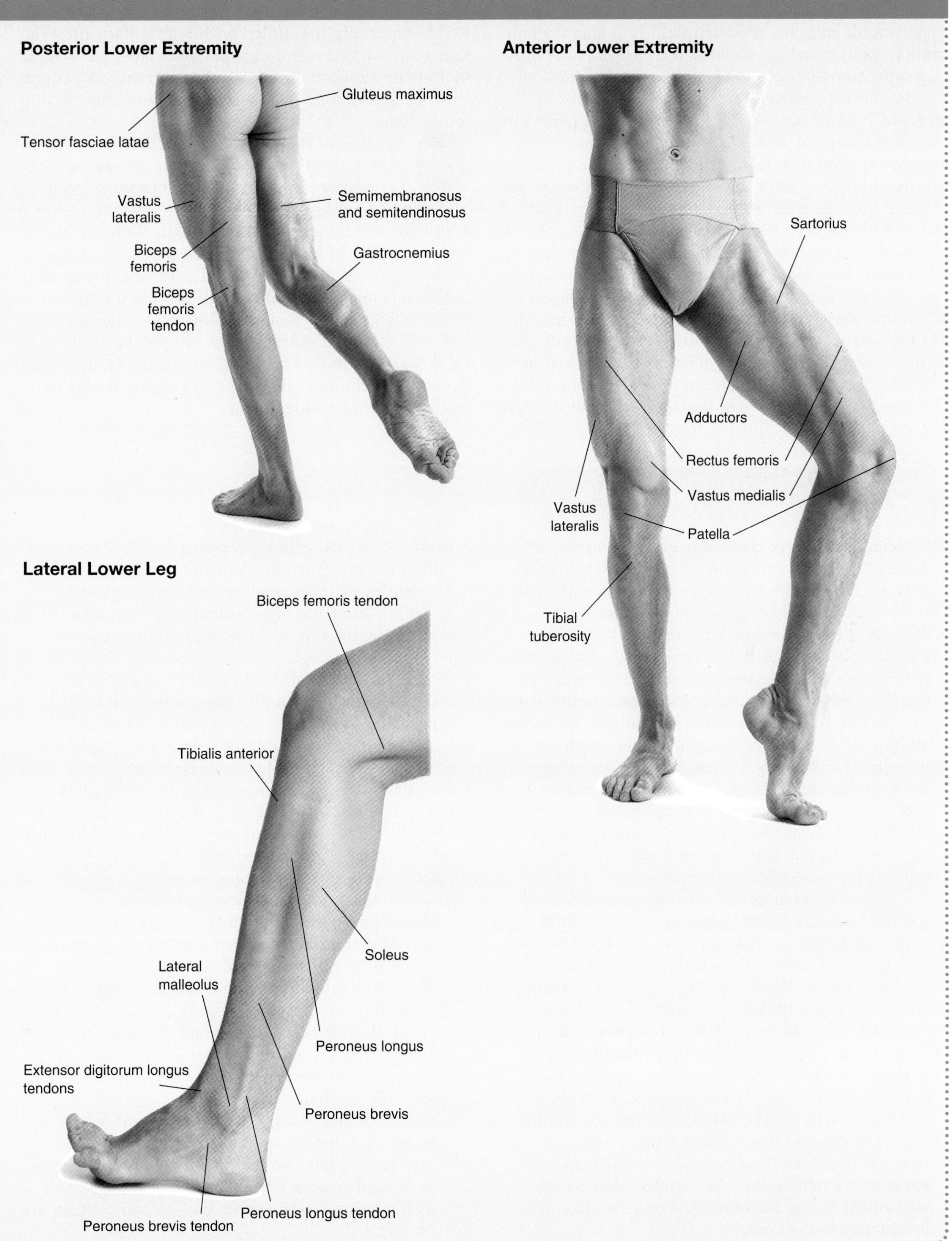
Posterior Lower Extremity
Gluteus maximus
Tensor fasciae latae
Vastus lateralis
Semimembranosus and semitendinosus
Biceps femoris
Gastrocnemius
Biceps femoris tendon
Anterior Lower Extremity
Sartorius
Adductors
Rectus femoris
Vastus medialis
Vastus lateralis
Patella
Tibial tuberosity
Lateral Lower Leg
Biceps femoris tendon
Tibialis anterior
Soleus
Lateral malleolus
Peroneus longus
Extensor digitorum longus tendons
Peroneus brevis
Peroneus longus tendon
Peroneus brevis tendon

associated with muscle contraction is restricted by tight connective tissue within the muscles. Additionally, hypertrophic muscles are often held in a shorter-than-normal position, which weakens the force that the muscles can generate due to the length–strength ratio.

In contrast, muscle **atrophy** is a decrease in the size and bulk of muscle. Lack of exercise, a period of immobilization, or loss of stimulus from the nervous system can lead to a decrease in the size of muscle fibers. In this case, actin and myosin myofilaments get thinner, and the number of sarcomeres at the ends of the myofibrils may be decreased. Muscle atrophy is common in older individuals primarily due to a decline in regular exercise, particularly in resistance activities. Therefore, as we age, there is a progressive loss of muscle mass and tone as muscle fibers shrink and are replaced by fibrous connective tissue and adipose. Additionally, as we age, we tend to be less hydrated overall, changing the consistency of the ground substance of all the connective tissues, including the fascial components of our muscles. This causes stiffness, general soreness, decreased flexibility, and decreased ROM, especially if coupled with a sedentary lifestyle. Other muscular system changes associated with aging include diminished strength and endurance, a general slowing of reflexes, and changes in both muscle and motor tone.

Because of the muscle tissue changes associated with aging, manual therapists may need to modify some of their therapeutic techniques when working with older clients. Muscles generally become thinner and more fragile with age and may feel more fibrous or wiry. The depth and speed of soft tissue manipulation overall must be adjusted to the client's tolerance, and more time is often required to warm and soften tissues. Additionally, due to the slowing of muscle reflexes, stretching often needs to be slower and more gradual. Some of the facilitated stretching techniques discussed in Chapter 8 may be less effective with older clients.

SUMMARY OF KEY POINTS

- Skeletal muscles are characterized as being excitable, extensible, elastic, and contractile (voluntary). These muscles have four functions: posture, movement, stabilization of joints, and generation of heat.
- Muscles are organized by three distinct fascial layers: the epimysium surrounds the muscle as a whole; perimysium wraps several muscle fibers together into fascicles; and endomysium envelops each individual muscle cell. All the fascial layers blend together at the ends of the muscle to form tendons that attach the muscle to the bone by weaving into the periosteum.
- Each muscle cell or fiber is composed of smaller myofibrils, which are in turn composed of even smaller myofilaments: actin and myosin. The actin and myosin are arranged to create the contractile units of a muscle called sarcomeres.
- The fascial layers in muscle are disorganized fibrous connective tissue characterized by a large amount of ground substance that keeps the cells and fibers widely spaced and by fibers that are not organized in parallel patterns. Fascia contains elastic and reticular fibers that give it the ability to stretch in several directions and return to its original shape without damage.
- The tendons and aponeuroses that attach muscle to bone are classified as organized fibrous connective tissue. All of their fibers are densely packed and highly organized in parallel alignment with the muscle fibers, and they contain fewer elastic and reticular fibers than fascia. Tendons and aponeuroses tend to extend rather than stretch, making them well-suited for the job of transferring the force of contraction to the bone.
- Muscle contraction is a physiologic event that occurs within the sarcomeres of a myofibril. The sliding filament mechanism explains how myosin bonds to actin, sliding the myofilaments over one another and shortening the sarcomeres to produce muscle contraction.
- The motor unit of skeletal muscle is made up of a single motor neuron plus the multiple fibers it innervates. When threshold stimulus is applied, all the fibers in a motor unit must contract completely; this is known as the all-or-none response.
- The amount of force generated by muscle contraction is moderated by changing the number of motor units stimulated; this process is known as motor unit recruitment, or graded response.
- Energy for muscle contraction requires the breakdown of ATP. ATP is produced as follows:
  - Direct phosphorylation uses creatine phosphate and ADP to produce ATP, but only enough to sustain a short burst of activity, 15 seconds or less.
  - Anaerobic cellular metabolism (glycolysis) involves breaking down glucose to pyruvic acid and occurs when effort is sustained beyond the capacity of the CP-cycle and when there is an insufficient supply of oxygen. Anaerobic metabolism produces only a few ATP molecules.
  - Aerobic metabolism is used when effort is sustained and there is a sufficient oxygen supply. This is the preferred method of energy production because it creates 30 to 32 molecules of ATP and its byproducts (water, heat, and carbon dioxide) are harmless.

- Skeletal muscle contraction generates increased tension within the muscle and a "pulling force" at the attachment points. Contraction also creates a slight bunching or broadening at the muscle belly.
- Skeletal muscle contractions are categorized according to the mechanical changes that occur in the muscle as a whole and whether the contraction does or does not generate movement. The three basic types of muscle contraction are tonic, isometric, and isotonic (concentric or eccentric).
- A muscle's architecture is defined by the direction of its fibers and the manner in which fascia divides, aligns, and arranges the fascicles in relation to the muscle's tendon. This determines what role the muscle plays in creating specific body movements (muscle assignment), the amount of power it can generate, and the specific direction and range of motion.
- The different shapes of muscles are classified according to their fiber arrangement and the location of the tendons.
    - Parallel muscles have tendons at the ends of the muscle and parallel fiber alignment. There are three types of parallel muscle: fusiform, circular, and triangular.
    - Pennate muscles have a central tendon in the middle of the muscle and shorter muscle fibers that run in a diagonal line toward the tendon. There are three types of pennate muscles: unipennate, bipennate, and multipennate.
- During movement, muscles are described as playing the role of agonist (prime mover), antagonist, synergist, or stabilizer/fixator. While muscles are generally categorized according to their prime function (the movement they are best at creating), each muscle is capable of playing any of these roles.
- In order for movement to occur, the antagonist muscle must release or be inhibited by the nervous system when the agonist is signaled to contract. This muscle reflex is called reciprocal inhibition.
- Muscle and fascial responses to exercise include increased efficiency in energy production, hypertrophy, and increased extensibility in response to regular stretching.
- Muscle and fascial responses to aging include atrophy, loss of strength, and decreased extensibility and flexibility, which leads to decreased ROM.

REVIEW QUESTIONS

Short Answer

1. List the functions of the skeletal muscle system.
2. What are the key characteristics that allow skeletal muscles to carry out these functions?
3. Define the term *aponeurosis* and name three examples of aponeuroses in the muscular system.
4. Explain the all-or-none response of muscle contraction.
5. Give two examples of muscles named according to:

Size:	a.
	b.
Shape:	c.
	d.
Function:	e.
	f.
Location:	g.
	h.

6. Name and explain the key physiologic events of a muscle contraction.

Multiple Choice

7. What is the smallest contractile unit of a muscle?
    a. fascicle
    b. motor unit
    c. muscle fiber
    d. sarcomere
8. What is the name for the large grouping of muscle fibers surrounded by the perimysium?
    a. belly
    b. muscle compartment
    c. fascicle
    d. myofibrils
9. Which of these tissue characteristics is true of fascia?
    a. densely packed fibers
    b. low ratio of elastic and reticular fibers
    c. highly organized parallel arrangement of fibers
    d. large amount of ground substance that keeps fibers widely spaced

Continued on page 162

10. A motor unit consists of one motor neuron and
    a. a single muscle fiber
    b. multiple muscle fibers
    c. two fascicles
    d. several sarcomeres
11. The graded response theory explains how the force of muscle contraction is changed by
    a. increasing or decreasing the number of motor units stimulated
    b. increasing the amount of calcium released into the sarcomere
    c. altering the amount of stimulus applied to the muscle
    d. shifting the role of the muscle from agonist to synergist
12. Which term is used to describe the minimum amount of stimulus required for muscle contraction?
    a. minimum stimulus
    b. contractile stimulus
    c. threshold stimulus
    d. tetanic stimulus
13. Which method of energy production for muscle contraction provides the quick energy needed for short bursts of activity?
    a. aerobic glycolysis
    b. anaerobic cellular metabolism
    c. aerobic cellular metabolism
    d. direct phosphorylation (ATP-CP)
14. Which method of energy production for muscle contraction creates lactic acid and the largest oxygen debt?
    a. direct phosphorylation
    b. anaerobic cellular metabolism
    c. aerobic cellular metabolism
    d. Krebs cycle
15. What type of muscle contraction maintains posture and each muscle's state of readiness for contraction?
    a. isometric
    b. tonic
    c. concentric
    d. tetanic
16. A muscle contraction that radically increases the tension in the muscle but does *not* result in movement of a body part is classified as
    a. tonic
    b. isotonic
    c. eccentric
    d. isometric
17. The pectoralis major and biceps brachii are both examples of __________ fiber arrangements.
    a. fusiform
    b. parallel
    c. triangular
    d. bipennate
18. Which movement role is assigned to an antagonist muscle?
    a. oppose the primary movement
    b. create the primary movement
    c. assist in creating the primary movement
    d. stabilize an adjacent bone during movement
19. What is the name for the muscle attachment that is generally fixed or stabilized during movement?
    a. insertion
    b. origin
    c. fixed attachment
    d. primary attachment
20. Which term is used to describe the natural state of firmness of a muscle due to the status of the tissue and fluid elements?
    a. motor tone
    b. muscle tonicity
    c. muscle tone
    d. muscle spasm
21. Which type of contraction occurs when a muscle is functioning as the prime mover?
    a. isometric
    b. tonic
    c. concentric isotonic
    d. eccentric isotonic
22. Which category of ROM is best for assessing the available range of movement and the status of the inert tissues around a joint?
    a. active
    b. assistive
    c. passive
    d. resistive

23. What is the most widely found class of lever in the body?
    a. first
    b. second
    c. third
    d. fourth
24. What is the primary function of the quadriceps muscle group?
    a. knee extension
    b. hip extension
    c. knee flexion
    d. hip extension
25. The muscles of mastication (chewing) include the temporalis, pterygoids, and
    a. frontalis
    b. obicularis oris
    c. hyoids
    d. masseter
26. The three muscles that are prime movers in elbow flexion are the biceps brachii, brachioradialis, and
    a. coracobrachialis
    b. brachialis
    c. triceps brachii
    d. flexor carpi radialis
27. The "rotator cuff" muscles that move the humerus include the supraspinatus, infraspinatus, subscapularis, and
    a. teres major
    b. deltoid
    c. rhomboids
    d. teres minor
28. Which muscle in the hip adductor group also flexes the knee?
    a. pectineus
    b. adductor longus
    c. gracilis
    d. adductor magnus
29. Which two muscles attach to the lateral tibial condyle via the iliotibial band?
    a. gluteus maximus and TFL
    b. semimembranosus and semitendinosus
    c. gracilis and sartorius
    d. gluteus medius and minimus
30. Which of the following correctly describes hypertrophy?
    a. the muscle fibers increase in number
    b. the muscle fascicles increase in number
    c. the muscle fibers decrease in size and strength
    d. the muscle fibers increase in size and thickness

References

1. Barnes JF. The elasto-collagenous complex. *Phys Ther Forum*. 1988.
2. Chaitow L, DeLany JW. *Clinical Application of Neuromuscular Techniques*. Vol. 1: The Upper Body. Edinburgh, New York: Churchill Livingstone; 2000.
3. Chaitow L, DeLany JW. *Clinical Application of Neuromuscular Techniques*. Vol. 2: The Lower Body. Edinburgh, New York: Churchill Livingstone; 2000.
4. Lederman E. *Fundamentals of Manual Therapy*. Edinburgh: Churchill Livingstone; 1998

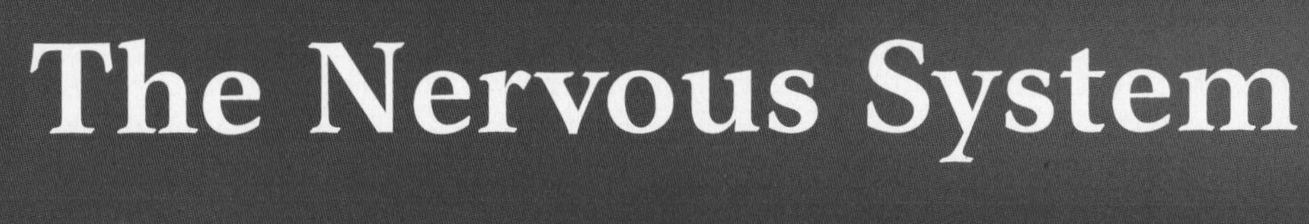

7 The Nervous System

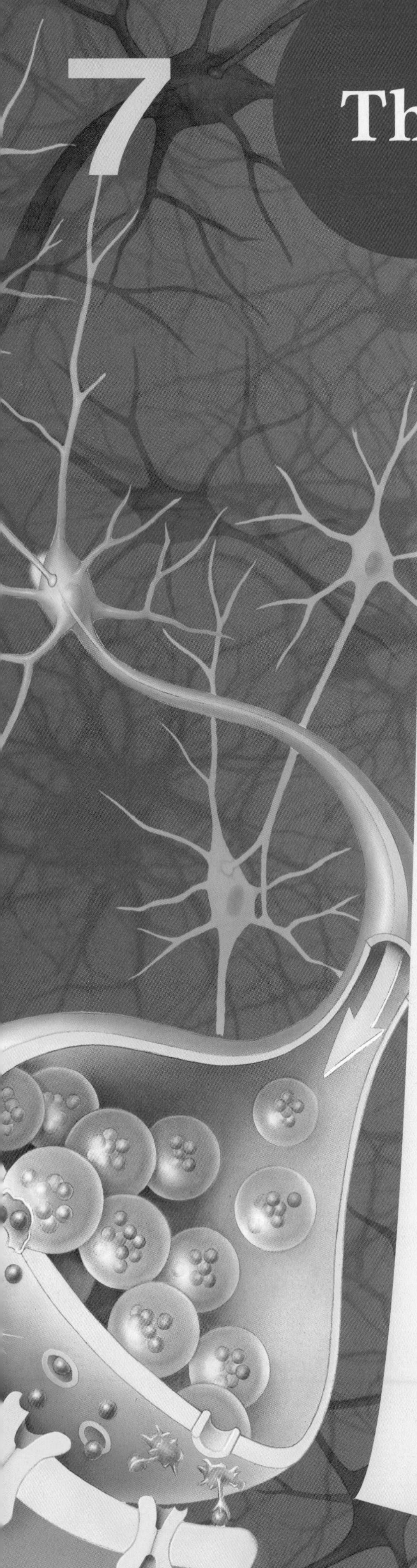

LEARNING OBJECTIVES

Upon completion of this chapter, you will be able to:

1. Discuss the importance of understanding the anatomy and physiology of the nervous system as it relates to the practice of manual therapy.
2. List and explain the primary functions of the nervous system, its two major divisions, and the key structural components of each.
3. Name and describe the key structural components of a neuron and the different types of neurons based on structure and function.
4. Name the different types of neuroglia plus the location and function of each.
5. Describe the general structure of a nerve and explain the difference between cranial and spinal nerves.
6. List the four cranial nerves that manual therapists need to know and explain why.
7. Name the four major nerve plexuses and the body regions they innervate.
8. List and explain the key events of nerve impulse conduction and synaptic transmission.
9. Explain the structure and function of a reflex arc.
10. List six categories of sensory receptors, explain the sensitivity of each, and give an example of their location.
11. Explain the location, structure, and functions of the meninges and cerebrospinal fluid.
12. Name the key structural features and regions of the spinal cord and explain the general functions of each.
13. Name the key structural features and regions of the brain and explain the general functions of each.
14. Compare and contrast the key structural features of somatic and autonomic motor pathways of the peripheral nervous system.
15. Name, compare, and contrast the structural features and functions of the two motor divisions of the autonomic nervous system.
16. Discuss the effects of aging on the nervous system.

Sometimes a thing in front of you is so big you don't know whether to comprehend it by first getting a dim sense of the whole and then fitting in the pieces or by adding up the pieces until something calls out what it is."

NORMAN MACLEAN

A River Runs Through it

KEY TERMS

action potential (AK-shun po-TEN-shul)
afferent (A-fer-ent)
autonomic (ah-to-NAH-mik)
axon (AKS-on)
cognition (kog-NIH-shun)
dendrite (DEN-drite)
depolarization (DE-po-lah-ri-ZA-shun)
effector (e-FEK-tor)
efferent (E-fer-ent)
ganglion (GANG-le-on)
impulse propagation (IM-puls prop-ah-GA-shun)
innervate (IN-er-vate)
integration (in-teh-GRA-shun)
myelin (MI-eh-lin)
neuronal pathway (nu-RON-al PATH-way)
neuronal pool (nu-RON-al pool)
neurotransmitter (nur-o-TRANS-mit-ter)
perception (per-SEP-shun)
plexus (PLEKS-us)
reflex arc (RE-fleks ARK)
repolarization (RE-po-lah-ri-ZA-shun)
saltatory conduction (SAL-teh-tor-e con-DUK-shun)
sensation (sen-SA-shun)
somatic (so-MAT-ik)
spinal segment (SPI-nal SEG-ment)
synapse (SIN-aps)
tract

Primary System Components

- **Neurons**
- **Glial cells**
- **Nerves (12 cranial; 31 spinal)**
- **Sensory receptors**
- **Spinal cord**
- **Brain**
  - Cerebrum
  - Diencephalon
  - Cerebellum
  - Brain stem
- **Meninges**
- **Cerebrospinal fluid**

Primary System Functions

- **Communication, coordination, and control of virtually all body processes by**
  - Sensing changes in the internal and external environment
  - Interpreting and integrating sensory information to determine motor responses
  - Stimulating motor responses of muscles, glands, and organs

The nervous system is the major regulatory system of the body. Along with the endocrine system, it directs and coordinates the functions of every other system. The nervous system incorporates amazing and sometimes contradictory qualities and characteristics, which together make an integrated whole. From birth, the brain is automatically programmed to coordinate and control breathing, heart rate, and millions of unconscious responses. Yet the mind also has the capacity to learn from new experiences, developing new neural connections, behaviors, and patterns of activity unique to each individual. Accounting for about 2% of the average human's total body weight, the brain is at once powerful and vulnerable. It has the power to influence the function of nearly every cell, yet requires multiple layers of protection and a reliable blood supply to keep its delicate neurons safe from damage and functioning optimally. Throughout history, humans have struggled to unravel and understand the fascinating inner workings of the nervous system. Even though we now understand how electrochemical signals allow neurons to communicate, it seems nothing short of magical that these energy impulses are translated into a vast array of thoughts, memories, actions, and emotions that can be expressed in poetry, movement, complex calculations, or meaningful touch.

The sheer number of nervous system components and the intricate anatomy of each may, at first, seem daunting. Additionally, to understand how the system works, we must simultaneously grasp key facts about its separate parts while fully appreciating the collaborative function of the integrated whole. To prevent becoming overwhelmed, it is important to keep a broader view of how the system functions. For manual therapy practitioners, it is essential to develop an understanding of the impact of touch, pressure, and movement on the nervous system as a whole. For example, therapists need to know how to translate reports of pain, decreased sensation, or reduced strength into specific indications or contraindications for manual therapy. Many therapists also need to know the location and functions of major nerves to avoid irritating them with deep tissue manipulation. While it is helpful to understand the basic system structures and how neurons communicate, as therapists, it is important to appreciate the mystery of the mind and how our intentions, intensity, and pace of work influence the emotional and physical well-being of our clients.

This chapter begins by organizing the nervous system according to its general functions. This functional framework will help you organize and connect the individual system components, learn the basic functions and role of each, and progress to a clear understanding of their interdependent relationships. From individual nerve cells to the nerves, spinal cord, and brain, you will explore the anatomy and physiology that allows these components to communicate, coordinate, and contribute to the function of all other organs of the body.

FUNCTIONAL ORGANIZATION OF THE NERVOUS SYSTEM

The brain, spinal cord, and nerves form the sophisticated and intricate communicating and coordinating network known as the nervous system. Whether we are smelling a bouquet of flowers, considering what to eat for lunch, being startled by a loud noise, or running to catch a bus, this system senses, interprets, and responds to a wide array of internal and external stimuli. Nervous system activity can be grouped into three basic functions:

- ***Sensory function*** refers to the system's ability to detect a broad spectrum of stimuli. From changes in blood pressure or stomach acidity to the brush of a hand across a cheek, sensory receptors react to changes in the internal and external environments and transport this information via nerves to the brain and spinal cord.
- ***Integrative functions*** encompass the system's capacity to process sensory information and direct single or multiple body responses. This includes the entire range of complex processes such as perceiving, feeling, analyzing, responding, and remembering.
- ***Motor function*** includes the transportation of commands from the brain via nerves to activate the muscles, glands, and organs of the body.

Together, the tasks of the nervous system do not seem to be that complex; the system senses, analyzes, and responds. Like a computer, it appears to be a simple equation of "information in and information out." Although certain simple reflexive pathways in the nervous system function in this manner, more often, information travels more complex routes. Intricate webs of nerve cells provide multiple pieces of information to the brain that must be sorted, prioritized, and interpreted to produce an appropriate coordinated response.

The nervous system is considered to have two major divisions. The **central nervous system (CNS)** is the integrative center that receives and interprets all sensory information and directs motor responses. The CNS has two components: the brain and spinal cord. The other division, the **peripheral nervous system (PNS)**, carries out both sensory and motor functions. It is composed of sensory receptors and a network of nerves that relay sensory information to, and motor information from, the CNS (FIGURE 7.1).

NERVOUS TISSUE

As described in Chapter 3, nervous tissue contains two types of cells. The cells that conduct electrical impulses are **neurons**. Cells that support, insulate, and protect the neurons are the **neuroglia**, also known as **glial cells**.

Neurons

All neurons have the same three basic parts: a cell body, dendrites, and an axon (FIGURE 7.2). Like all

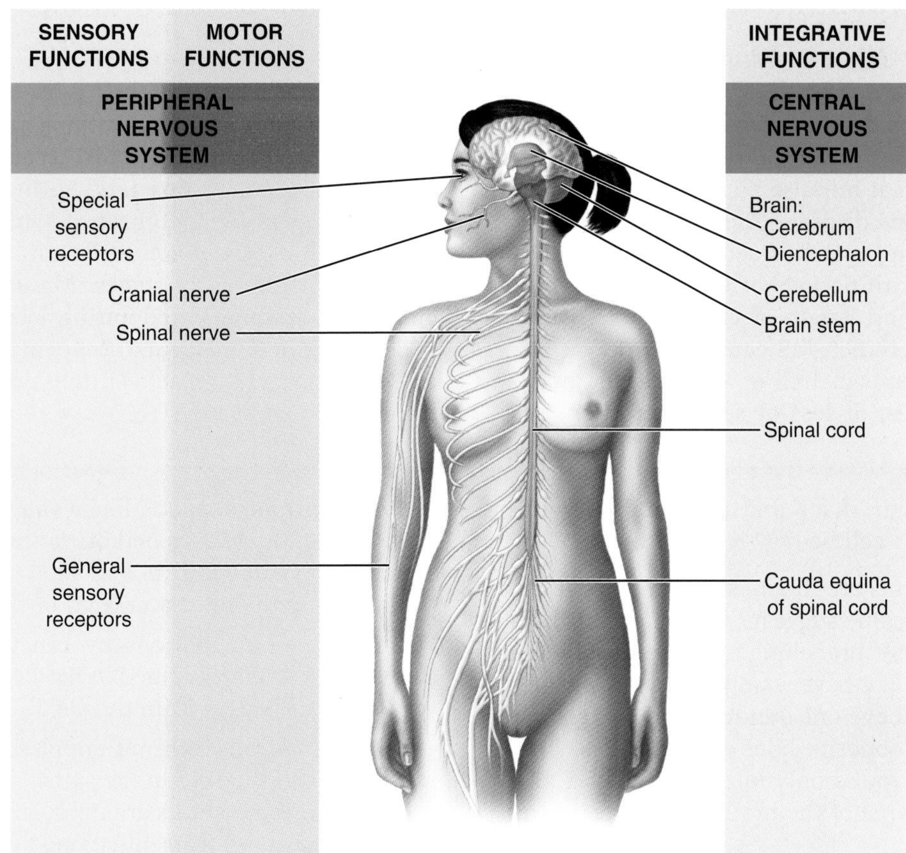

FIGURE 7.1 ▸ Functional organization of the nervous system. The CNS consists of the brain and spinal cord and serves as an integration center that interprets sensory input and determines motor responses. The PNS includes the sensory receptors and nerves that detect stimuli and bring sensory information *to* the CNS and carry motor commands *away from* the CNS to the effectors.

cells, the ***cell body*** of a neuron contains the nucleus and cytoplasm with its cellular organelles such as mitochondria, lysosomes, Golgi apparatus, ribosomes, and endoplasmic reticulum. The neuronal cytoskeleton is made of neurofibrils and microtubules that provide shape and support to the cell and give rise to the **nerve fibers**—dendrites and axons—that extend from the cell body.

Dendrites are generally short and highly branched. These fibers are the sensitive portion of a neuron that receive a stimulus and transmit it *to* the cell body. In contrast, an **axon** is a single long fiber that carries an impulse *away* from the cell body to another neuron, muscle, or gland. This structural arrangement dictates that impulse conduction along a neuron is always one-way: from dendrite to cell body to axon. Many axons are covered with an insulating lipid layer called **myelin**. Functionally myelin insulates, protects, and speeds up nerve impulse conduction across the axon. The distal portion of an axon splits into smaller multiple extensions called **axon terminals** that end in a small knob called a **synaptic bulb**. The bulbs contain specialized pouches, or vesicles, that store and release the chemicals known as **neurotransmitters**.

Neurons can be classified by their structure or function. Structurally, there are three types of neurons,

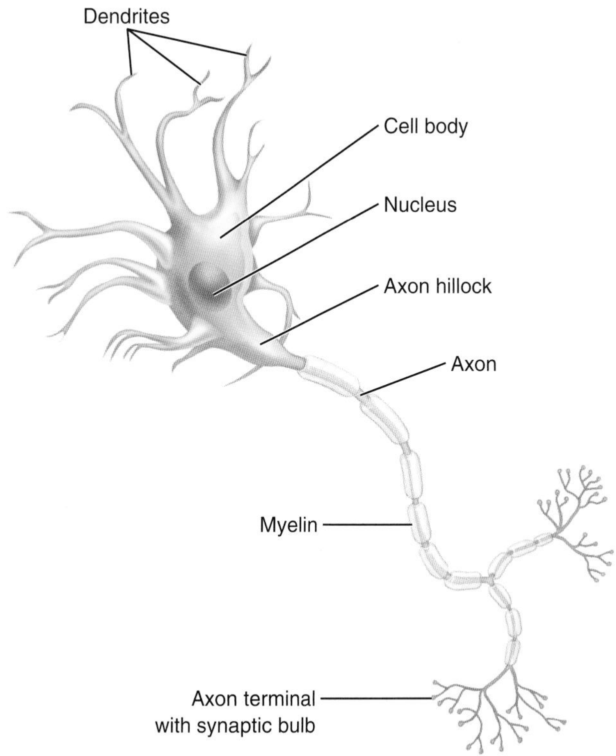

FIGURE 7.2 ▸ Neuron. A neuron is made up of a cell body, dendrites, and a single axon.

Pathology Alert Multiple Sclerosis

Multiple sclerosis (MS) is a degenerative condition in which myelin is destroyed and replaced by scar tissue, impairing neuronal impulse conduction. Believed to be an autoimmune disorder, the disease is characterized by alternating cycles of acute inflammation and demyelination with periods of remission and repair. The frequency and length of each cycle is variable and difficult to predict. MS causes a wide variety of signs and symptoms including weakness, paresthesia (pins and needles) or loss of sensation (numbness), extreme fatigue, difficulty walking and loss of coordination, as well as digestive disturbances. While most autoimmune conditions tend to have a few recognizable triggers for the onset of symptoms, MS does not. This makes the condition difficult to diagnose and treat in the early stages. Manual therapies are generally contraindicated during active inflammatory cycles but can be helpful in relieving symptoms such as depression and muscle stiffness during periods of remission.

distinguished by their shape and the number of fibers that extend from the cell body (▶ FIGURE 7.3):

- ***Multipolar neurons*** are the most common type of neuron. As their name suggests, these neurons have numerous dendrites branching off the cell body and like all neurons, they have a single axon.
- ***Bipolar neurons*** have one dendrite and one axon that extend from opposite sides of the cell body. This type of neuron is found only in the inner ear, the olfactory (smell) area of the brain, and in the retina of the eye.
- ***Unipolar neurons*** look like a single continuous fiber with a cell body attached off to the side. The fiber serves as both dendrite and axon, with the axon making up the largest portion of the fiber.

When classifying neurons by function, there are also three types: two are located in the PNS, and one is found almost exclusively within the CNS:

- ***Sensory neurons*** transmit impulses from the sensory receptors in the PNS to the spinal cord and brain. Most of these neurons are unipolar, with the exception of the bipolar neurons mentioned previously.

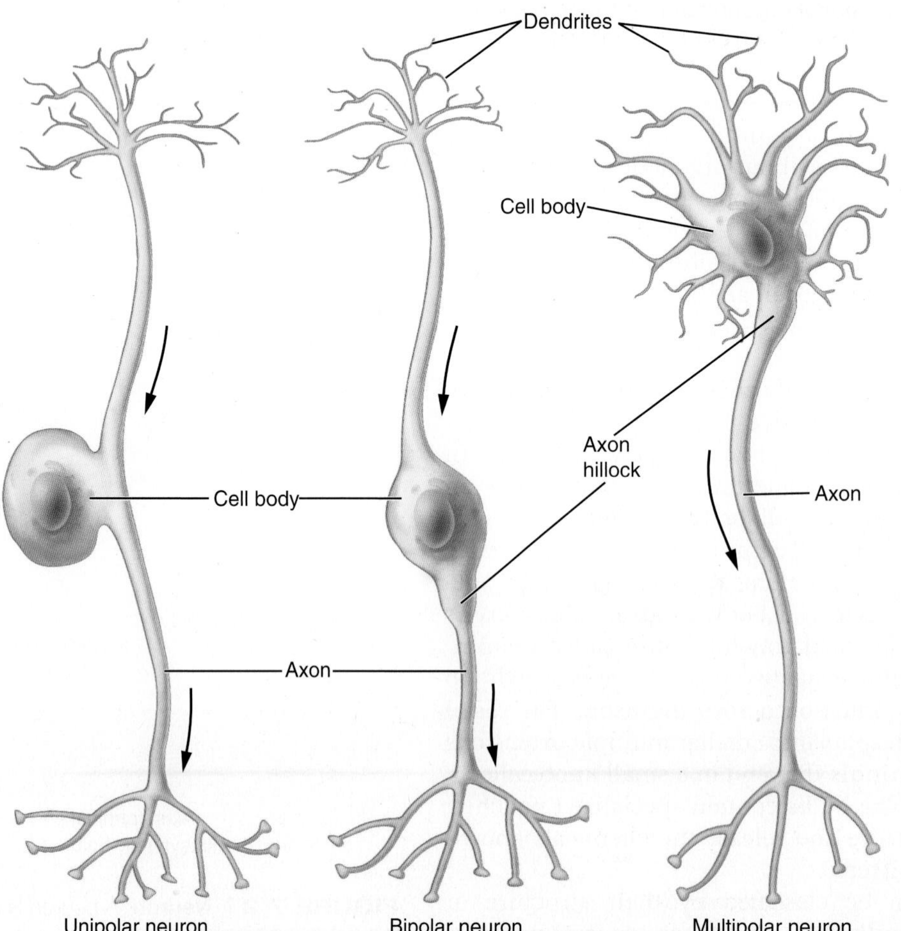

FIGURE 7.3 ▶ Classification of neurons by shape.

- ***Motor neurons*** carry motor commands away from the CNS to the muscles, glands, and organs, collectively known as **effectors**, throughout the body. Most motor neurons are multipolar.
- ***Interneurons*** or ***associative neurons*** are positioned between sensory and motor neurons and, therefore, are predominantly found within the CNS. These associative neurons are functionally unique because not only do they connect sensory and motor neurons, they also process and interpret sensory input to direct motor responses. Similar to motor neurons, most interneurons are multipolar in structure.

Neuroglia

The majority of nervous tissue is comprised of neuroglia (glial cells) rather than neurons (FIGURE 7.4). The Greek term *glia* means "glue," so the literal translation of

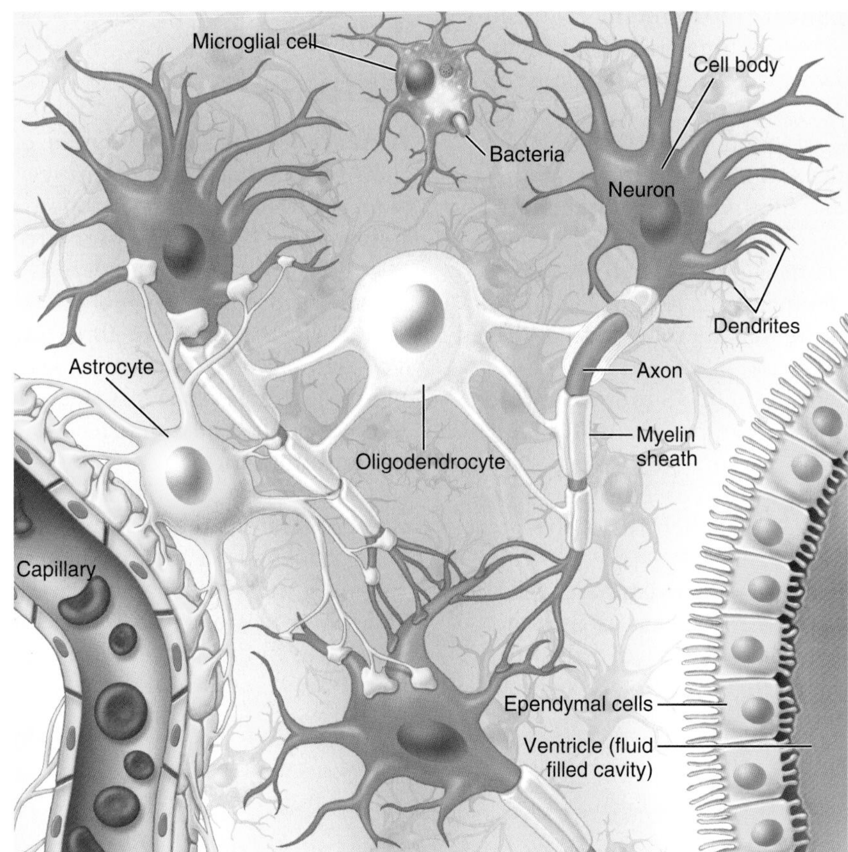

A Neuroglia of the CNS

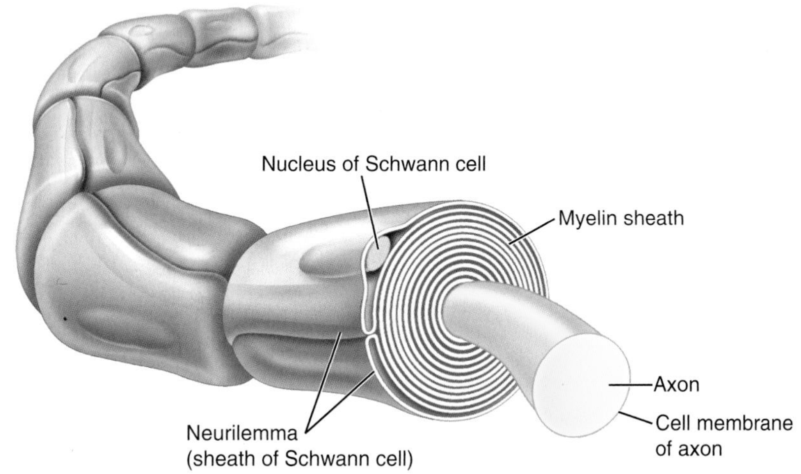

B Schwann cell

FIGURE 7.4 Neuroglia. A. There are four varieties of glial cells in the CNS: astrocytes, oligodendrocytes, microglia, and ependymal cells. **Neuroglia. B.** Schwann cells produce the myelin sheath that wraps around the axons of neurons in the PNS.

neuroglia is "nerve glue." The name was assigned to these cells by early histologists but is actually a misnomer because these cells do much more than simply hold nerve tissue together; they also support, nourish, and protect the neurons. There are six types of glial cells:

- ***Astrocytes***—These star-shaped cells are the largest and most numerous of all glial cells. Found only in the CNS, they serve several important functions, including structurally supporting neurons, maintaining the chemical environment required for the conduction of nerve impulses, and forming scar tissue following an injury to the nervous system.
- ***Oligodendrocytes***—Also found only in the CNS, oligodendrocytes make and maintain the insulating myelin that surrounds axons within the brain and spinal cord.
- ***Microglia***—As their name suggests, microglia are very small cells. They function as phagocytes, removing and eating cellular debris, microbes, and damaged nerve tissue from the CNS.
- ***Ependymal cells***—Within the CNS there are spaces filled with a clear fluid known as cerebrospinal fluid (CSF). This fluid is produced by a single layer of ependymal cells, which form an epithelial-like membrane that lines certain structures of the brain.
- ***Schwann cells***—Schwann cells form and maintain the myelin that insulates and protects neuronal axons in the PNS. These cells have an additional function that oligodendrocytes lack—the ability to form a protective sheath around the myelin, known as **neurilemma** (FIGURE 7.4B). The presence of this sheath in neurons of the PNS makes axon regeneration after injury much easier and faster than in neurons of the brain or spinal cord.
- ***Satellite cells***—There are several places within the PNS where dendrites and cell bodies of multiple neurons are bound together by satellite cells to form structures called **ganglia**. By surrounding the cell bodies and dendrites, the ganglia provide structural support and help regulate exchanges between the cell bodies and interstitial fluid.

NERVES

A **nerve** consists of a bundle of nerve fibers (axons), their connective tissue coverings, and blood vessels outside the CNS. Nerves function as highways, carrying information between sensory receptors, the CNS, and effectors. Structurally, a nerve is very similar to skeletal muscle, with three layers of connective tissue that organize neuronal fibers into three groupings (FIGURE 7.5). The outer layer is the **epineurium**, the **perineurium** bundles several axons into nerve fascicles, and the innermost layer that surrounds each individual axon is called the **endoneurium**.

Nerves are functionally classified according to the types of neurons they contain. A few nerves contain only one type of neuron, making them ***sensory*** or ***motor nerves***. However, most of the body's nerves contain both sensory and motor neurons and are referred to as ***mixed nerves***. All nerves in the PNS are categorized as either cranial or spinal nerves. **Cranial nerves** originate from the brain,

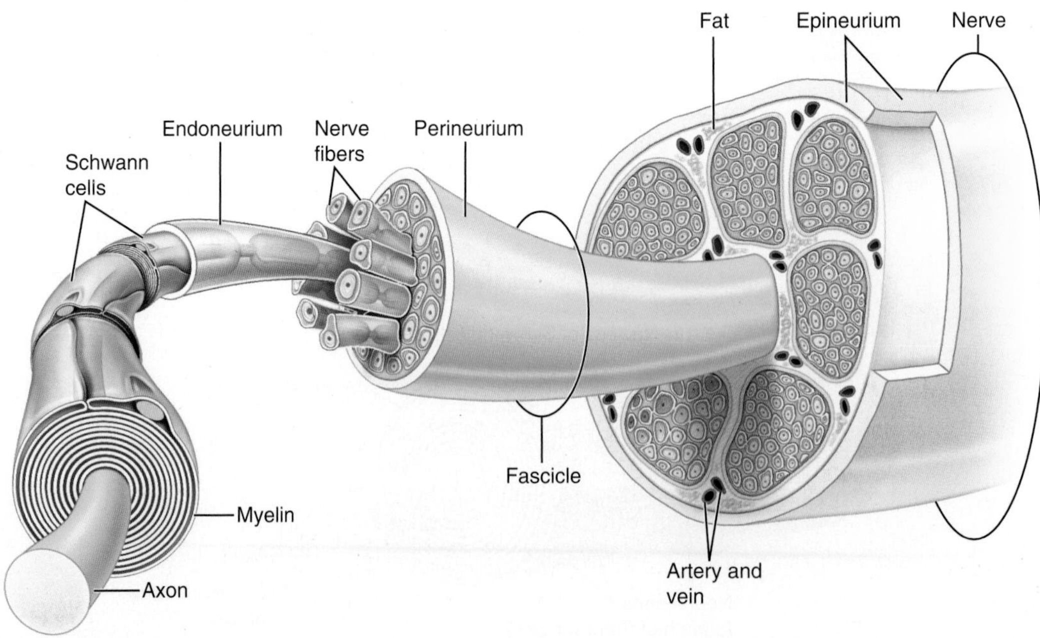

FIGURE 7.5 ▶ Cross section of a nerve. Axons covered with endoneurium are bundled together by the perineurium to form fascicles. Multiple fascicles, blood vessels, and fat are bundled together by the epineurium to form the nerve.

while **spinal nerves** arise from the spinal cord. Several of the cranial nerves are classified as sensory or motor only. However, all the spinal nerves are mixed nerves.

Cranial Nerves

There are 12 pairs of cranial nerves, and most originate from the brain stem. The cranial nerves deliver all sensory information from the special sense organs, abdominal viscera, and some areas of tissue in the neck, chest, and face. In addition, some provide motor control for muscles in the eyes, viscera, neck, and chest (▶ FIGURE 7.6). Cranial nerves are designated by Roman numerals that indicate the descending order of their appearance along the vertical axis of the brain stem. In addition, each cranial nerve is named according to its function, location, or appearance. The names, order, and general function of the cranial nerves are detailed in ▶ TABLE 7-1.

By the Way

Typically anatomy texts identify 12 pairs of cranial nerves, However, anatomists have identified a 13th nerve located just in front of cranial nerve I (olfactory). This nerve, christened "nerve zero" or the "terminal nerve," is sensitive to pheromones and plays a significant role in sexual attraction.

Cranial nerves V (trigeminal), VII (facial), X (vagus), and XI (accessory) are of primary interest to manual therapists. All four of these nerves exit the brain stem and become superficial in the musculature around the head and neck; therefore, direct tissue manipulation or movements of the head and neck may affect them. For example, both

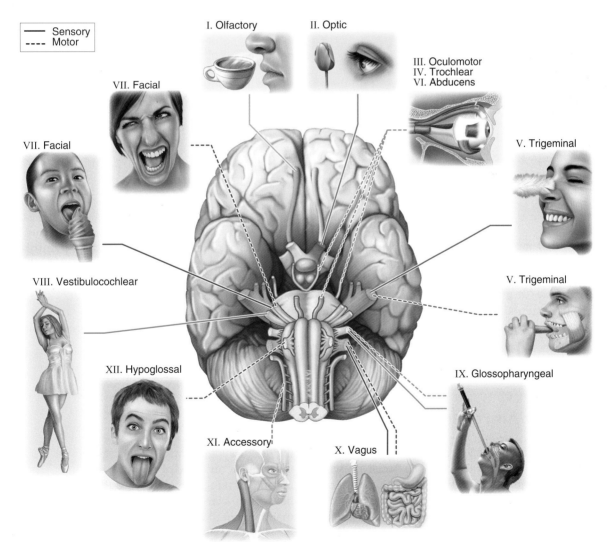

FIGURE 7.6 ▶ The cranial nerves. Most of the cranial nerves exit as pairs from underneath the brain. In this illustration, solid lines show sensory functions and dotted lines designate motor functions. Some nerves transmit only sensory signals, while others transmit only motor signals. Other nerves are mixed, having both sensory and motor neurons.

Table 7-1 Cranial Nerves

Name and Number	Nerve Type	Functions
I. Olfactory	Sensory only	Carries sensory impulses from nose to olfactory centers in the brain
II. Optic	Sensory only	Carries sensory impulses from the eyes to the vision centers in the brain
III. Oculomotor	Motor only	Conducts motor commands to muscles that move the eyes and adjust the lens and iris of the eyes
IV. Trochlear	Motor only	Conducts motor commands to one of the muscles that move the eyes
V. Trigeminal	Mixed	Carries sensory impulses from the skin of face and from the mucosa of the nose and mouth to the brain Carries motor commands to chewing muscles of the face
VI. Abducens	Motor only	Conducts motor commands to the external muscle that rolls the eye laterally
VII. Facial	Mixed	Conducts motor commands to the muscles of facial expression Carries sensory impulses from the anterior taste buds of tongue to the gustatory centers in the brain
VIII. Vestibulocochlear	Sensory only	Carries sensory impulses from cochlea to the auditory centers of the brain and sensory impulses from the semicircular canal and vestibule to the cerebellum
IX. Glossopharyngeal	Mixed	Carries sensory impulses from the posterior taste buds of tongue and pressure impulses from baroreceptors in carotid artery to the brain
X. Vagus	Mixed	Carries sensory impulses from mouth, throat and thoracic and abdominal viscera to the brain Conducts motor commands to digestive tract and heart
XI. Accessory	Motor only	Conducts motor commands to SCM and trapezius muscles
XII. Hypoglossal	Mixed	Conducts motor commands for movement of tongue

the trigeminal nerve (▶ FIGURE 7.7) and facial nerve (▶ FIGURE 7.8) are located in and around the temporomandibular joint and can be irritated with TMJ dysfunctions. Care must be taken with manual therapy treatments for this condition to avoid further irritation of the nerves. Damage or irritation of the trigeminal nerve can lead to a painful condition called trigeminal neuralgia or tic douloureux. The condition is characterized by sharp stabbing or zinging pains in the teeth or jaw. Due to the intense pain and hypersensitivity associated with trigeminal neuralgia, manual therapy is generally contraindicated.

Pathology Alert Bell Palsy

Bell palsy, named for a pioneer in neurological research, Sir Charles Bell, is a paralysis of the facial muscles caused by inflammation or damage to cranial nerve VII (facial). Most cases of Bell palsy are caused by viral infection from the herpes simplex virus that also causes cold sores and chicken pox. In rare cases, the condition can be caused by structural damage to the facial nerve from a broken or dislocated jaw. Signs and symptoms include a sudden onset of weakness and drooping of muscles on one side of the face, difficulty in swallowing and blinking, earache, hypersensitivity to sound, and headaches. Manual therapy is generally indicated to help maintain circulation and tissue flexibility in the affected facial muscles. However, therapists must remember that neuromuscular reflexes are impaired and avoid deep or overly vigorous manipulation of the tissue.

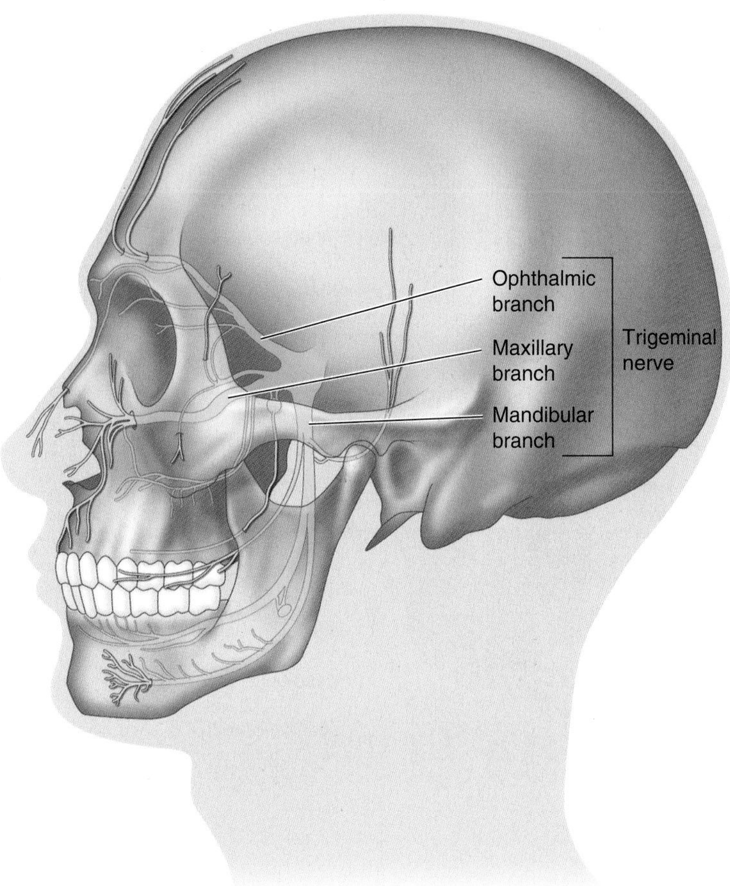

FIGURE 7.7 ▸ The trigeminal nerve. Three primary branches innervate the face, teeth and jaw.

Specific inflammation or impairment of the facial nerve leads to another condition called Bell palsy (see "Pathology Alert"). Additionally, the vagus and accessory nerves lie deep to cervical muscles, making them vulnerable to compression or irritation from trigger points or muscle spasms (guarding) associated with whiplash, headaches, and cervical manipulations (FIGURE 7.6).

Spinal Nerves

The spinal nerves arise from the spinal cord, exiting the CNS through the intervertebral foramen of the spinal column (▸ FIGURE 7.9). Named and numbered according to their location along the spinal column, there are 31 pairs of spinal nerves: 8 cervical, 12 thoracic, 5 lumbar, 5 sacral, and 1 coccygeal (▸ FIGURE 7.10). Spinal nerves are generally given the number of the vertebra just superior to the nerve; for example, the T1 nerve is found between the first and second thoracic vertebrae, the L2 nerve between the second and third lumbar vertebrae, and so on. However, the numbering system is different for cervical nerves because only the skull is superior to the first spinal nerve. Therefore, C1 through C7 are numbered for the vertebra *below* them, and C8 lies between the last cervical and first thoracic vertebrae.

Within a few centimeters of the intervertebral foramen, spinal nerves divide into trunks and branches. In specific regions, several branches from adjacent spinal nerves converge to form networks called **plexuses**. Each plexus is a functional unit that innervates muscles and organs in the same body region. Individual nerves that emerge from each plexus are named for the specific structures or body region they innervate. Even though there is some disagreement among anatomists over which specific spinal nerves are included in a particular plexus, the major plexuses are generally recognized as

- ***Cervical***—This plexus includes C1 through C5 spinal nerves and innervates the skin and several muscles in the posterior head, neck, and upper shoulder. The ***phrenic nerve*** that innervates the diaphragm is a major branch off this plexus (FIGURE 7.10).

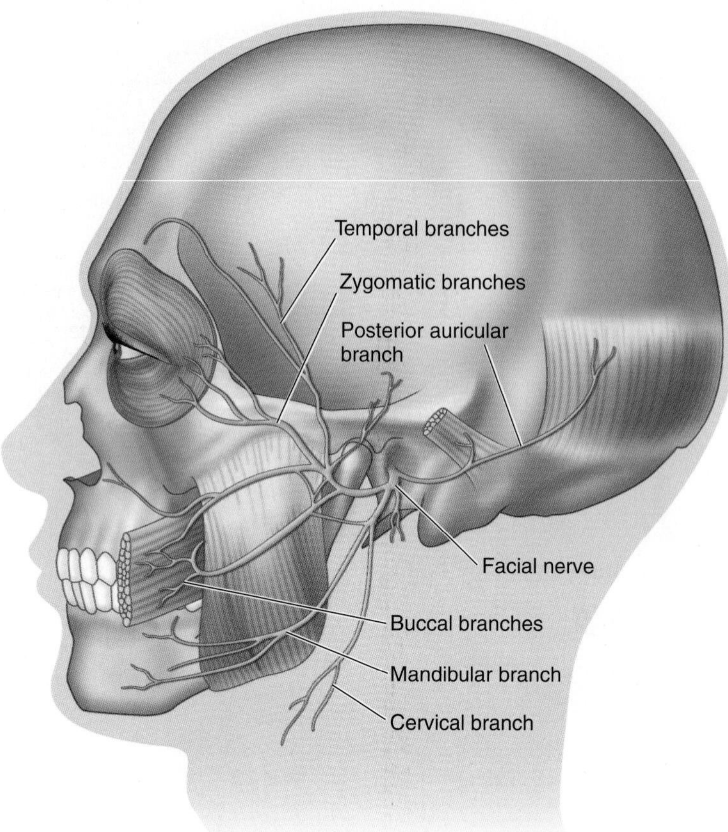

FIGURE 7.8 ▶ The facial nerve. Six primary branches innervate muscles in the face, head, and neck.

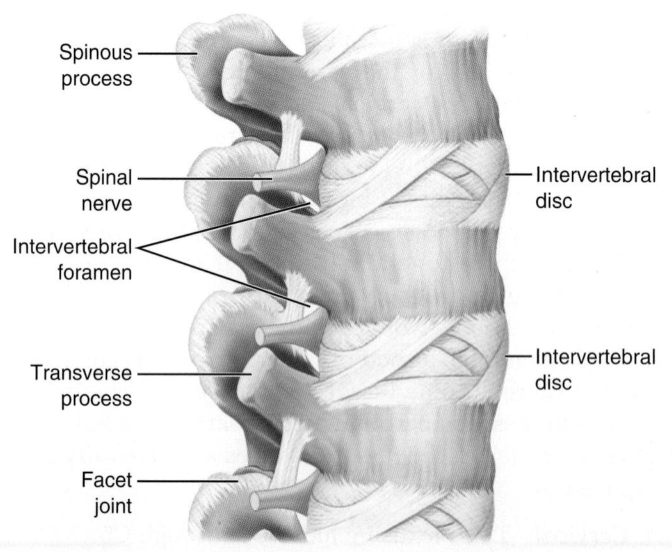

FIGURE 7.9 ▶ Spinal nerves. Each pair of spinal nerves exits the spinal column through intervertebral foramen.

- ***Brachial***—Spinal nerves C5 through C8 and T1 form this plexus that innervates the pectoral girdle and entire upper extremity. The major branches of the brachial plexus include the ***axillary, musculocutaneous, radial, ulnar,*** and ***median nerves*** (▶ FIGURES 7.10 and 7.11A).
- ***Lumbar***—The lumbar plexus consists of the L1 through L4 spinal nerves that innervate many effectors in the lumbar, abdominal, and thigh regions. Major branches of the lumbar plexus include ***ilioinguinal, obturator, femoral***, and ***lateral femoral cutaneous nerves*** (▶ FIGURES 7.10 and 7.11B).
- ***Sacral***—The sacral plexus is formed by the L4 through S4 spinal nerves, which innervate the pelvic region and full lower extremity. The major branches of this plexus are the ***gluteal, sciatic, posterior femoral, tibial***, and ***fibular nerves*** (▶ FIGURES 7.10 and 7.11C).

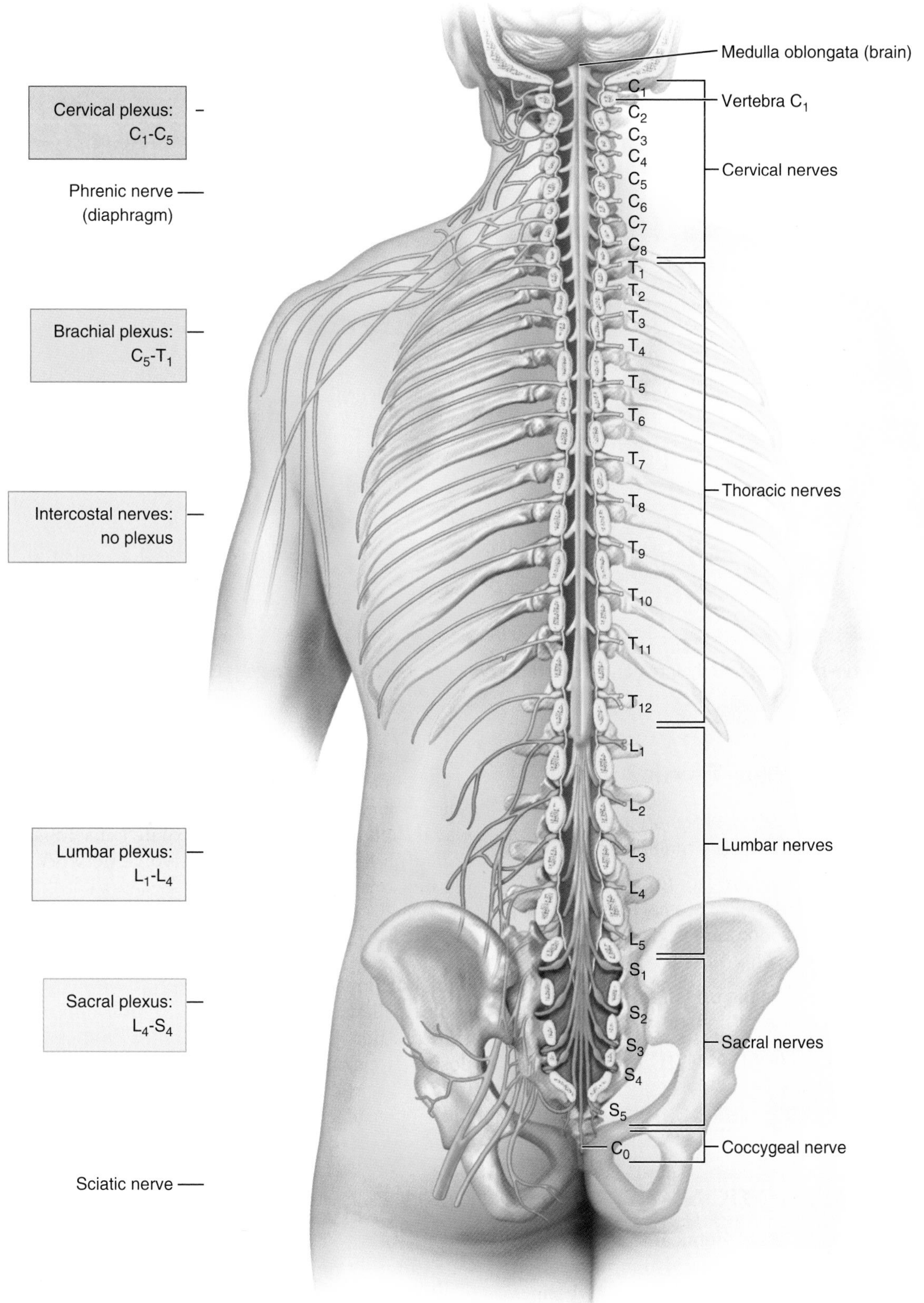

FIGURE 7.10 ▸ Spinal nerves and nerve plexuses, posterior view. The 31 pairs of spinal nerves and their regions are labeled on the right, and the four primary plexuses formed as these nerves branch, merge, and rebranch are labeled on the left. To view a video on the nerves of the shoulder region, visit http://thePoint.lww.com/Archer-Nelson.

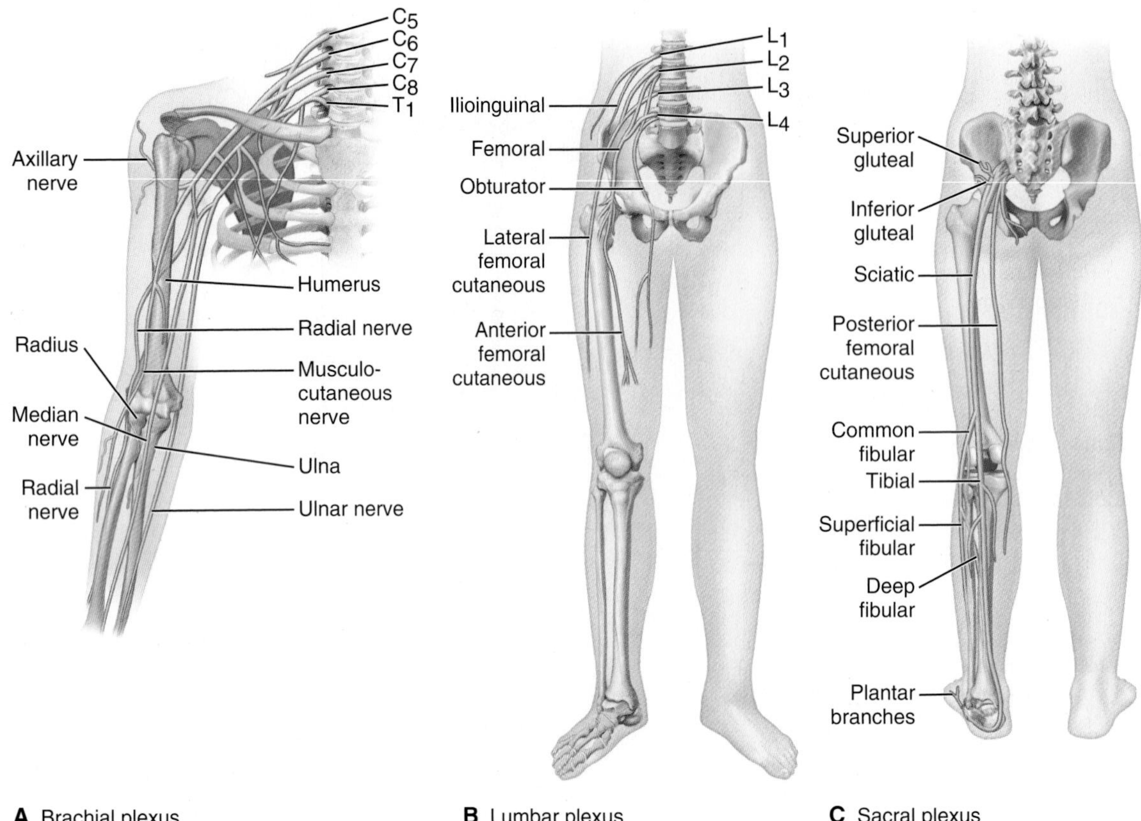

FIGURE 7.11 ▸ Nerve plexuses and their major peripheral nerve branches. A. Brachial plexus. **B.** Lumbar plexus. **C.** Sacral plexus.

What Do You Think? 7.1

- Why do you think sensory neurons are generally unipolar and motor neurons are multipolar?
- There is a common mnemonic for remembering the names and order of the 12 cranial nerves. Rather than relying on someone else's mnemonic, make up your own.
- What name would you make up for a nerve that innervates the lungs, the spleen, or the heart?
- In considering the definition of plexus, why is there no thoracic nerve plexus?

IMPULSE CONDUCTION

Nerve impulses are the electrochemical signals carried by neurons for both rapid and organized communication throughout the body. Impulses are created by changes in the environment that stimulate dendrites, initiating a sequence of electrical events that travel along a neuron's plasma membrane all the way down the axon. When the impulse reaches the axon terminals, it causes the release of neurotransmitters that build a chemical bridge between one neuron and another neuron or effector cell. Impulses traveling from one region of the nervous system to the next follow specific routes or pathways. Some of these pathways are very simple and direct, involving one sensory neuron and one motor neuron. Other pathways, such as the ones that mediate conscious activities, are much more complex and involve networks of multiple neurons working together.

Action Potentials

Recall from Chapter 3 that electrically charged molecules called ions are present in both intracellular and extracellular fluids. At rest, cells have an excess of positive ions outside the plasma membrane while inside, they have an excess of negative ions. This difference in the electrical charge across the membrane is referred

By the Way

Potentials are measured in volts. For example, a typical alkaline AA battery carries a 1.5-V charge. The membrane potential of body cells is measured in thousandths of a volt or millivolts (mV). Because there are more negative ions on the inside of a cell than on the outside, the resting membrane potential is negative, approximately −70 mV.

Pathology Alert Neural Compression-Tension Syndromes

Neural compression-tension syndromes (NCTS) include conditions such as sciatica, thoracic outlet syndrome (TOS), and carpal tunnel syndrome (CTS) in which tight fascia and/or muscle spasms cause peripheral nerves to be compressed or pulled taut. In ***sciatica***, the piriformis is most often tight, compressing the sciatic nerve that passes directly underneath it (Figure A). In ***thoracic outlet syndrome (TOS)***, the brachial nerve plexus and/or the subclavian artery can be irritated as they pass between the middle and anterior scalene muscles, between the clavicle and ribs (subclavius muscle tension), or underneath the pectoralis minor (Figure B). Tension and spasm in the finger flexors, and inflammation or restrictions in the flexor retinaculum and transverse carpal ligament are the likely cause of ***carpal tunnel syndrome (CTS)***.

Common signs and symptoms of NCTS include numbness or tingling, muscle weakness, decreased coordination, chronic aching, and shooting pain down the affected limb. Manual therapy can be helpful in relieving the pain and dysfunction associated with NCTS. However, therapists must recognize which structures are creating the compression-tension of the peripheral nerve and be sure to rule out more severe neurologic causes such as spinal stenosis or disc damage.

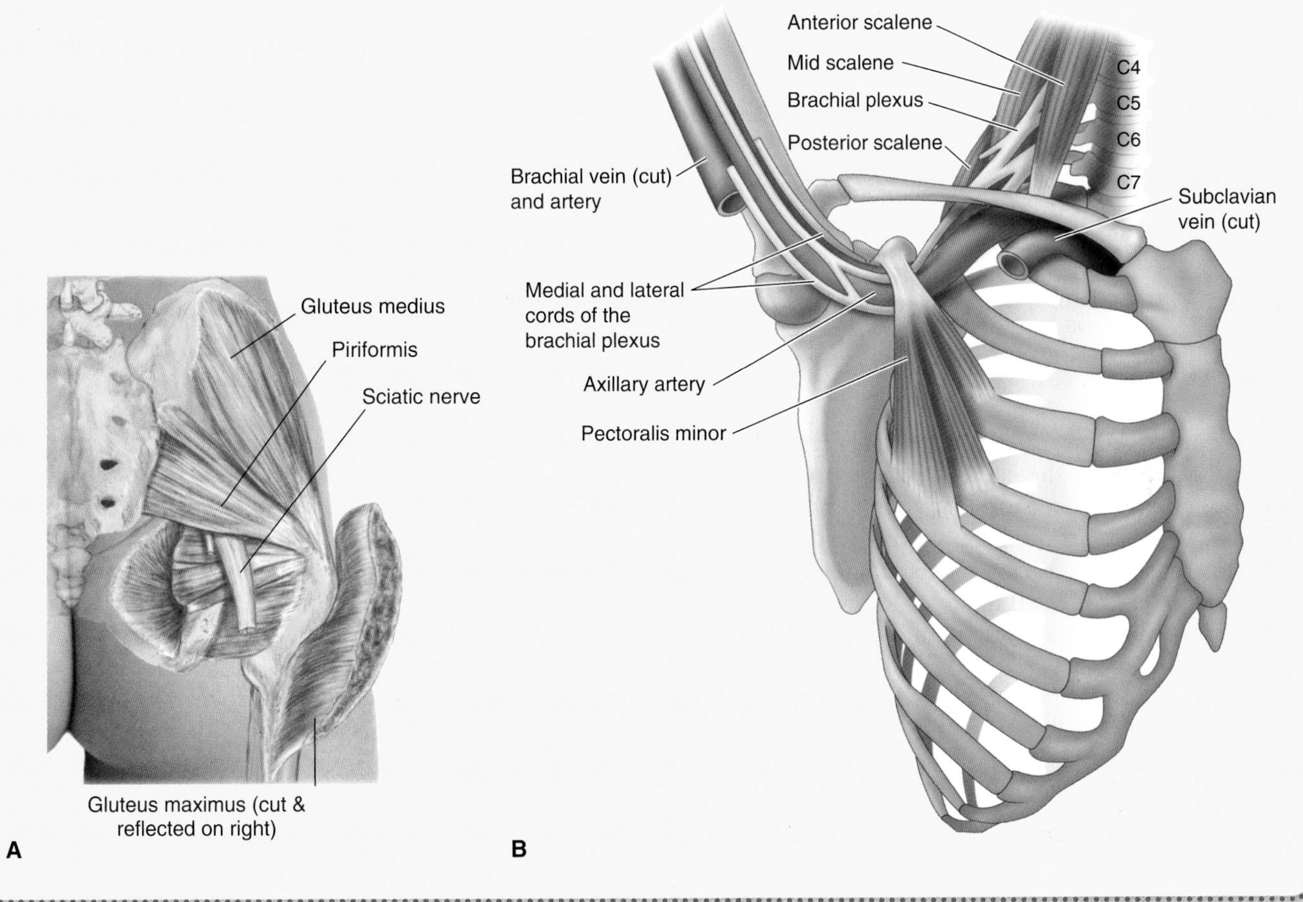

to as the cell's **resting membrane potential**. When cells sense a change in the environment (stimulus), they respond by opening ion channels in their membranes at the site of stimulation. These open channels allow ions to pass through the membrane, causing a change in the electrical charge across the plasma membrane. In neurons, when the stimulus causes a large enough shift in the ionic charge of the plasma membrane, it is called a **threshold stimulus**. Once this point is reached, the neuron generates an impulse, or **action potential**, at the **axon hillock**, a bump on the neuron where the axon connects with the cell body.

When an action potential is generated, the membrane of the neuron goes through two phases as shown in ▶ FIGURE 7.12:

1. **Depolarization** occurs when the neuron opens gated channels in response to the stimulus. These channels allow positive sodium (Na^+) ions to diffuse into the cell from the outside. The influx of positive

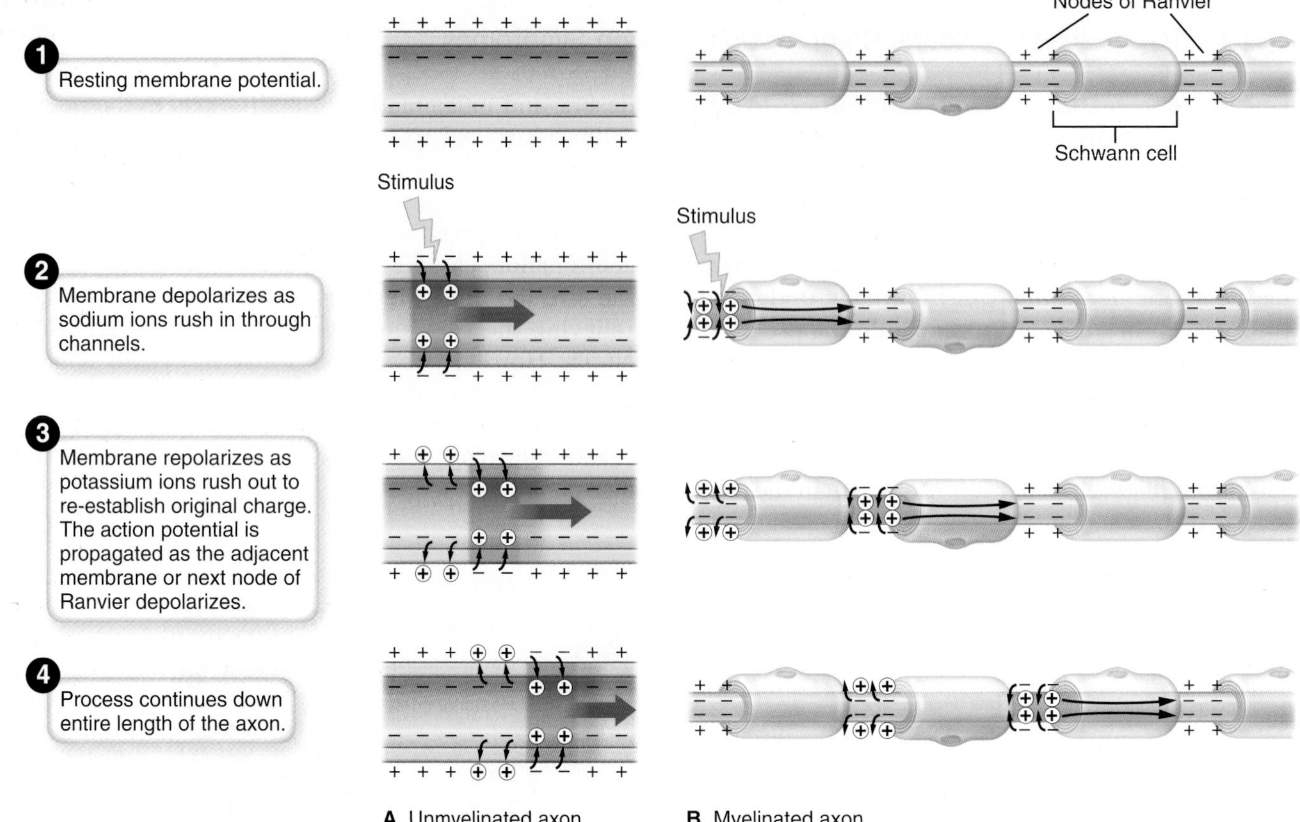

1
Resting membrane potential.
2
Membrane depolarizes as sodium ions rush in through channels.
3
Membrane repolarizes as potassium ions rush out to re-establish original charge. The action potential is propagated as the adjacent membrane or next node of Ranvier depolarizes.
4
Process continues down entire length of the axon.
Stimulus
Stimulus
Nodes of Ranvier
Schwann cell
A Unmyelinated axon
B Myelinated axon

FIGURE 7.12 ▶ Phases of an action potential. A. Depolarization and repolarization phases occur during conduction of an action potential (nerve impulse). **B.** Saltatory conduction along a myelinated axon.

ions causes an increase in the membrane potential; the charge inside the cell becomes more positive.

2. **Repolarization** occurs quickly after the first phase. During repolarization, potassium (K^+) ions flow out of the neuron to return the plasma membrane to its resting potential. Until the neuron has repolarized, it cannot respond to another stimulus.

The brief period of time before the neuron has fully repolarized is called a **refractory period**. This lull can be visualized by thinking of the motion-sensitive paper towel dispensers found in public restrooms. To dispense a length of paper towel, you wave your hand in front of a sensor. If you need more paper than was dispensed, you must wave your hand in front of the sensor again. Often, the dispenser does not respond immediately to this second request. This brief pause between your movement and the dispenser releasing another paper towel is analogous to the refractory period of a neuron. If you wait a moment and try again, the dispenser responds appropriately, just as a neuron can respond once it has repolarized.

The opening of Na^+ channels and influx of ions in a section of the plasma membrane during depolarization causes the adjacent portion of membrane to reach threshold and open more Na^+ channels. This creates a domino effect that continues along the length of the axon until the action potential reaches the axon terminals. This process is called **propagation** or **conduction** of a nerve impulse. In myelinated axons, action potentials are able to propagate along the axon faster because the opening Na^+ and K^+ channels are located primarily at the gaps between the Schwann cells. Because the exchange of ions occurs only at these gaps, known as the **nodes of Ranvier**, the action potential appears to leap from node to node, speeding up the transmission in a process referred to as **saltatory conduction** (▶ FIGURE 7.12B).

Synaptic Transmission

A **synapse** is the junction point at which an electrical signal passes from one cell to another. The **synaptic cleft** is a microscopic space between two communicating neurons or a neuron and its effector. When an impulse travels along a presynaptic neuron to the axon terminal, it causes the vesicles in the synaptic bulbs to release their neurotransmitters into the synaptic cleft (▶ FIGURE 7.13). The neurotransmitters cross the synaptic cleft and stimulate the postsynaptic (receiving) neuron or effector on the other side.

Neurotransmitters include chemicals such as ***acetylcholine, norepinephrine, dopamine, serotonin***, and

endorphins. Some of these substances have an *excitatory effect* on postsynaptic neurons, meaning that their presence creates a threshold stimulus and depolarization in those neurons. Other neurotransmitters have an *inhibitory effect*. Their presence prevents the receiving neuron from reaching threshold and generating an action potential.

In order for a postsynaptic neuron to return to its resting state, the neurotransmitters must be removed from the synaptic cleft. Neurotransmitters may simply diffuse away but more often, they are destroyed by special enzymes. Some are actively transported back into the presynaptic neuron through a process known as ***reuptake*** or transported into an adjacent glial cell through ***uptake***.

Neuronal Pathways

Nerve impulses (action potentials) follow specific one-way routes or **neuronal pathways** as they travel from one region of the nervous system to the next. Pathways that transmit sensory information *to* the spinal cord and brain are known as **afferent** pathways, while **efferent** pathways carry motor impulses *away from* the CNS. The complexity of a neural pathway is directly related to the number of interneurons (associative neurons) involved and the amount of processing that occurs in the CNS. For instance, the automatic involuntary response called a **reflex** requires little interpretation by the integrating center. Therefore, the pathways are simple, involving a sensory neuron, few or no interneurons, and a motor neuron. In contrast, conscious and voluntary responses require a great deal of interpretive processing by the brain. As a result, the pathway is much more complex and includes multiple associative neurons.

Reflex Arcs

Reflexes involve simple neuronal pathways called **reflex arcs**, which provide a predictable motor outcome for a specific sensory stimulus. For example, the deep tendon reflex in which an abrupt tap causes a reflexive contraction of a muscle involves only two neurons: one afferent and one efferent. The spinal cord serves as the integrating center with a single synapse as the junction between the sensory and motor components. The withdrawal reflex that causes us to quickly pull a hand away when a hot object is touched involves three neurons: a

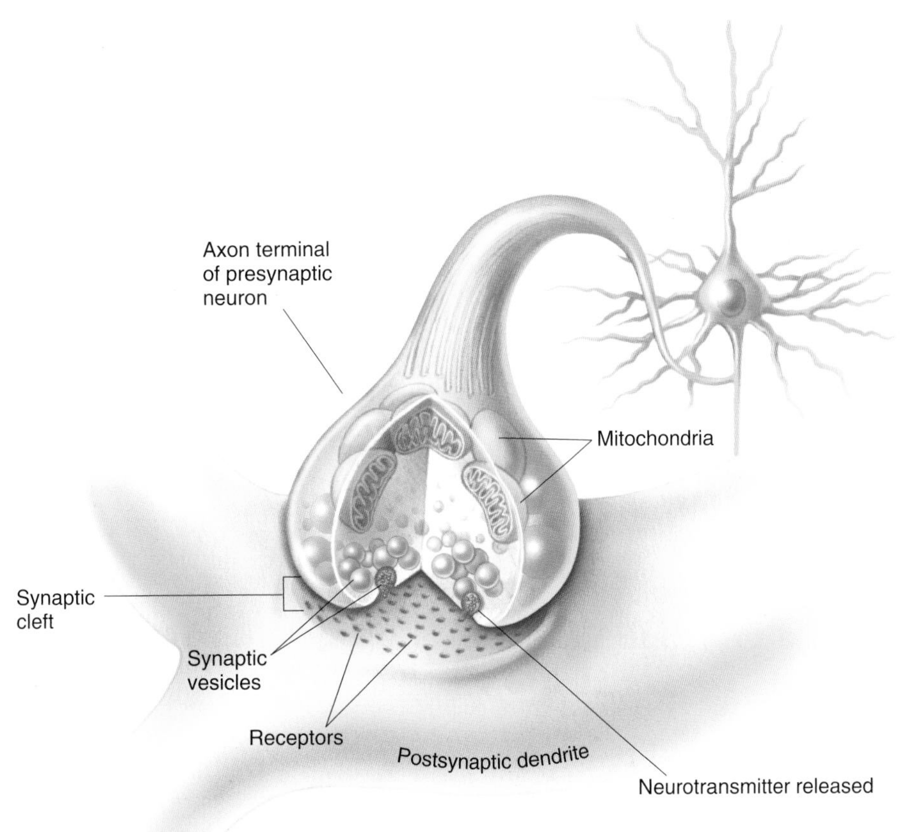

FIGURE 7.13 ▶ Synaptic transmission. Vesicles in the end bulb of the presynaptic neuron release neurotransmitters that cross the synaptic cleft to stimulate the postsynaptic neuron. This process is the same between a motor neuron and an effector.

sensory neuron, one interneuron, and a motor neuron (▶ FIGURE 7.14).

The term *reflex arc* should not be confused with the manual therapy term *reflex effect*. A reflex arc is a hard-wired pathway in which a stimulus produces a predictable motor response. In contrast, a reflex effect refers to any unconscious or automatic response to manual stimulus. These responses are common and probable, but not predictable. Additionally, reflex effects occur both in the tissue being worked and in far-removed areas. For example, a simple neck rub will relax the neck muscles but may also lower blood pressure, decrease heart rate, and also relax face, back, and arm muscles. Reflex effects require a lot of interpretive processing in the brain and numerous associative neurons in the pathway. Therefore, using the term *systemic effect* instead of reflex effect is less confusing and more descriptive of the widespread involuntary responses elicited through manual therapy.

Neuronal Pools and Circuits

In contrast to the simple pathway of a reflex arc, the pathways that control activities such as walking, talking, or thinking require complex networks of sensory, associative, and motor neurons. In the CNS, billions of interneurons are organized into functional groups called **neuronal pools**. These pools are like departments within a large corporation; each handles specific functions based on limited sources of input. Individual neuronal pools may have both excitatory and inhibitory neurons and may affect the activity of other neuronal pools, motor neurons, or effectors.

Neural networks involved in sensing and controlling complex responses communicate in specific patterns called **circuits**. While there are various types of circuits, the two simplest are convergence and divergence. **Convergence** describes a circuit in which impulses from several neurons converge onto a single neuron (▶ FIGURE 7.15). Converging circuits allow for processes that can be both voluntarily and involuntarily controlled. For example, breathing is controlled without your having to consciously think about it. Respiratory centers in the brain stem activate the diaphragm and other muscles of ventilation to contract. However, you can also consciously contract those muscles when you want to take a deep breath and hold it. The involuntary process and voluntary process originate from different regions of the brain, but both converge on the same motor neurons to cause contraction of the respiratory muscles.

Divergence describes a circuit in which impulses from one neuron spread out to several other neurons. This allows information to be spread and utilized by many different integrative areas of the CNS. For example,

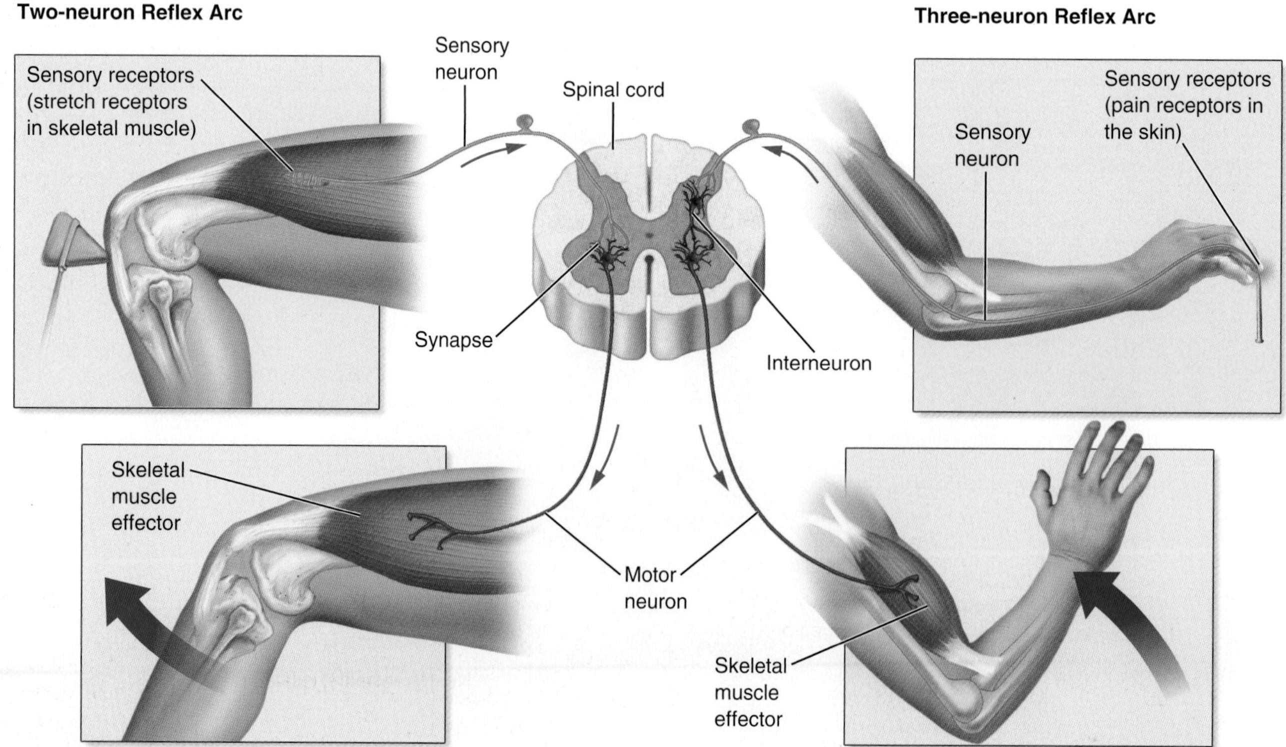

FIGURE 7.14 ▶ Reflex arcs. The patellar tendon stretch reflex (left) is an example of a two-neuron reflex arc, while the withdrawal reflex (right) is a three-neuron reflex arc.

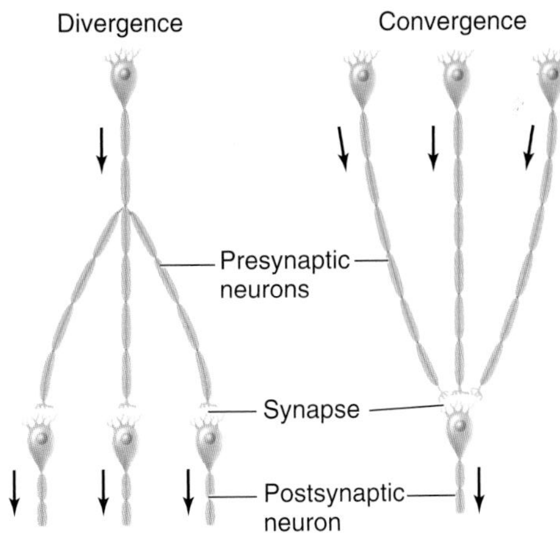

FIGURE 7.15 ▸ Neural circuits. Divergence and convergence are examples of simple neural circuits.

when you withdraw your hand from a hot object, you feel the sensation of pain, draw back from the heat source, and you may yell "Ouch!" (or some stronger exclamation) while racing to put your hand in cold water. While the initial withdrawal of your hand is mediated through a reflex arc, the other responses involve more complex pathways. This occurs because the impulse from the sensory neuron diverges out to reach multiple neurons in the CNS, allowing for multiple responses to the initial sensory information.

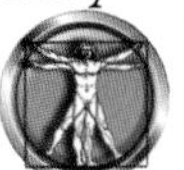

Divergence helps explain in part how widespread and varied systemic effects occur in response to manual therapy. The sensory information created by touch or movement in one area of the body travels to the CNS, where it diverges to communicate with multiple neuronal pools. Because several areas of the brain are stimulated, multiple responses can occur. Systemic effects can be either relaxing or invigorating depending on the intent, contact, pressure, and rhythm applied by the manual therapist and the experience and perception of the client.

What Do You Think? 7.2

- How does the phrase "a self-propagating wave of negativity" apply to nerve impulse conduction?
- Do you prefer to use the term *systemic effects* or *reflex effects* in relation to massage? Explain the reason for your choice.
- What examples can you give of systemic effects that occur locally in the tissue being worked versus elsewhere in the body?

THE PERIPHERAL NERVOUS SYSTEM: SENSORY STRUCTURES AND FUNCTIONS

The sensory components of the nervous system include the sensory receptors and neurons that detect and carry sensory information to the CNS. Millions of receptors, each sensitive to a particular type of stimulus, monitor the body for internal or external change. When sufficiently stimulated, an impulse is created and transmitted to the CNS along a sensory neuron. When stimulus is sustained, sensory receptors undergo a process called **adaptation**. In adaptation, the receptors are desensitized to stimuli and the rate of impulses sent to the CNS is decreased. Some sensory receptors adapt quickly while others are very slow or do not adapt at all. Adaptation can continue to the point of failure, when no impulse is sent from the receptor. Once a receptor has fully adapted, only an increase in the intensity of the stimulus restarts impulse conduction.

The impulses carried by sensory neurons are like Morse code: simple dots and dashes of energy that have no meaning until they are translated. In other words, the actual **sensation** or impression of heat, cold, pain, or movement is based on the *interpretation* of the impulse by the brain. When a specific region of the brain receives sensory impulses, it projects that impulse back to the apparent source of the input. This process, called **projection**, explains why we tend to attribute a sensation to the location of the receptors; we describe our eyes as seeing, ears as hearing, or locate the pain in our left foot. These and other integrative functions of the brain are discussed in more detail in other sections of this chapter.

Sensory receptors can be divided into two main categories: **general receptors** are scattered throughout the skin, muscles, fascia, and joints, and **special receptors** are those that are concentrated into a few complex sense organs: the eyes, nose, ears, and tongue (▸ TABLE 7-2). Regardless of its status as a general or special receptor, each is sensitive to only one type of stimulus and can be designated as one of six types:

- **Photoreceptors**—sensitive to light
- **Chemoreceptors**—sensitive to changes in chemical concentrations
- **Thermoreceptors**—stimulated by change in temperature
- **Nociceptors**—stimulated by tissue damage; also called pain receptors
- **Mechanoreceptors**—stimulated by changes in pressure and movement
- **Proprioceptors**—specialized mechanoreceptors in skeletal muscle and joints. Note that while proprioceptors are merely specialized types of mechanoreceptors, their importance to manual therapists is significant enough to classify and discuss them as a separate category.

Table 7-2 Sensory Receptors

	Type of Receptor	Sensation	Receptor Cells and Organs
Special senses	Photoreceptors	Vision	Cones and rods in retina of the eye
	Mechanoreceptors	Hearing	Cochlea of the inner ear
	Mechanoreceptors	Equilibrium	Semicircular tubes and vestibular receptors of the inner ear
	Chemoreceptors	Smell (olfaction)	Olfactory cells in nasal membrane of the nose
	Chemoreceptors	Taste (gustatory)	Gustatory cells in taste buds on the tongue
General senses	Proprioceptors	Muscle length and rate of lengthening	Muscle spindles (intrafusal fibers) in skeletal muscle
	Proprioceptors	Muscle and tendon tension	GTO concentrated in musculotendinous junctions of skeletal muscles
	Nociceptors	Pain	Free nerve endings within skin, viscera, bones, fascia, etc.
	Baroreceptors	Pressure	Hollow organs in respiratory, digestive, urinary tract, plus large arteries
	Tactile receptors	Touch, pressure, or movement	Free nerve endings, Pacinian corpuscles, Merkle discs, Meissner corpuscles, Ruffini corpuscles, hair root plexus in skin and fascia
	Thermoreceptors	Temperature	Free nerve endings in the skin
	Chemoreceptors	pH, CO_2, O_2 levels in blood and body fluids	Blood vessels of the body

With the exception of one type—photoreceptors—there are normally several kinds of receptors involved in both the general senses (touch, temperature, pressure, pain) and special senses (vision, hearing, smell, taste, equilibrium).

Photoreceptors

The light-sensitive photoreceptors are found only in the **retina**, the innermost layer of the wall of the eyeball. The retina functions as a sensory membrane and contains two types of photoreceptor cells named rods and cones because of their shape (FIGURE 7.16). **Rods** are sensitive to dim light and are most densely distributed at the periphery of the retina. This makes the rods essential for night vision as well as for peripheral vision. **Cones** are more sensitive to bright light and specific color wavelengths to provide greater visual acuity and the ability to distinguish colors. There are three types of cones, each sensitive to specific wavelengths of visible light: blue, red, and green. Both rods and cones are stimulated when light waves pass through the eye and reach the retina. They pass the impulse to sensory neurons situated in the deepest layer of the retina. These neurons travel out the back of the eyeball via the optic nerve (cranial nerve II).

Vision and color discrimination result from the brain's interpretation of sensory data and are not functions of the retina. Visual centers in the brain interpret sensory impulses that arrive via the optic nerve, enabling us to see images and colors. When particular cones are stimulated by light waves, the brain perceives a specific color; impulses generated by the "blue" cones are translated as blue, those from the "red" cones as red, and from the "green" cones as green. Colors other than blue, red, and green are perceived when the brain receives simultaneous impulses from more than one set of cones. We see purple when the red and blue cones are stimulated and teal when impulses from blue and green cones are combined. If impulses are received simultaneously from all three types of cones, we see white.

Chemoreceptors

The receptor cells for the special senses of smell and taste are chemoreceptors embedded in the epithelial and mucosal tissues of the nose and mouth. The

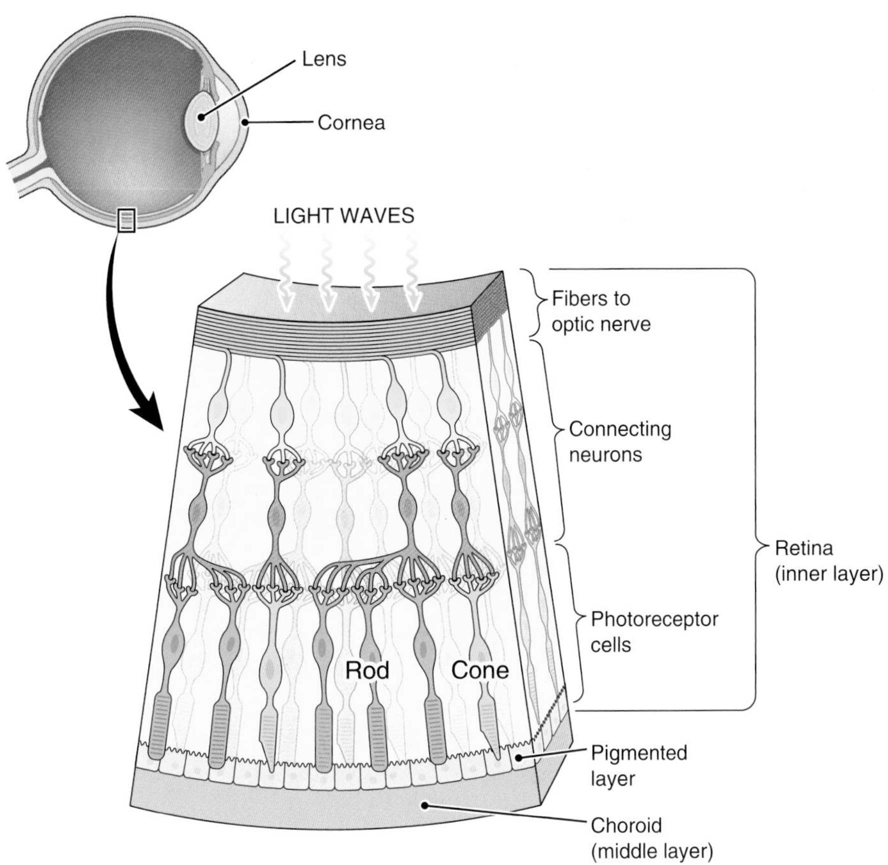

FIGURE 7.16 ▸ Photoreceptors. Light waves enter the eye through the cornea and are focused by the lens onto the retina. The light stimulates the rods and cones, which synapse with sensory neurons to send an impulse to the brain via the optic nerve.

sensory receptor cells for **olfaction** (smell) are located on the dendrites of specialized afferent neurons, which are embedded in the nasal membrane in the upper region of the nasal cavity. The distal ends of these dendrites have knobs with multiple **cilia**, or hair-like extensions, that project into the nasal cavity (▸ FIGURE 7.17A). When gaseous molecules of inhaled air dissolve in nasal fluids, the chemicals present are detected by the cilia, stimulating impulses that travel directly to the olfactory center of the brain via the olfactory nerves (cranial nerve I).

By the Way

Color blindness is a condition in which a person cannot distinguish colors normally; it is caused by one or more of the cone cell color-coding structures being absent or not functioning properly. The most common deficiency is in the perception of red and green tones. In this case, a person with red-green deficiency sees both colors as similar and cannot distinguish between them. Several visual tests are used to identify specific problems in color perception.

Gustatory cells are the chemoreceptors for the sense of taste. They are found in the **taste buds**, most of which are located in the ***papillae*** (tiny bumps) on the tongue's surface. Similar to olfactory cells, the gustatory cells have cilia that are sensitive to chemicals; in this case, to the chemicals dissolved in saliva. Taste buds in different regions of the tongue have higher sensitivities to different kinds of chemicals. There are four basic taste sensitivities: salty, sweet, sour, and bitter (▸ FIGURE 7.17B).

Our sense of taste and smell are closely linked because the proximity of the olfactory and gustatory receptor cells in the nose and mouth cause them to be stimulated at the same time. In fact, the smell of food either enhances or diminishes our sense of taste. This is why food can seem bland and tasteless when we have a head cold and why our mouth waters in anticipation when we smell a favorite dish cooking. Both olfactory and gustatory receptor cells undergo sensory adaptation easily and quickly, decreasing their sensitivity as any stimulus is sustained. In fact, the olfactory bulbs have been shown to decrease their reports by 50% within the first second of stimulus and ultimately adapt completely.

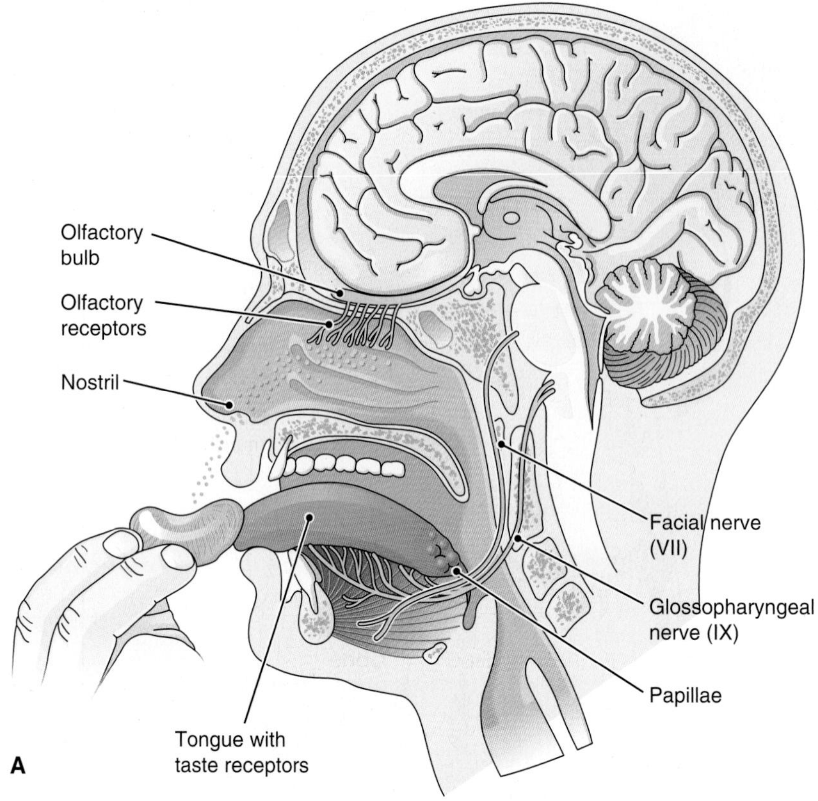

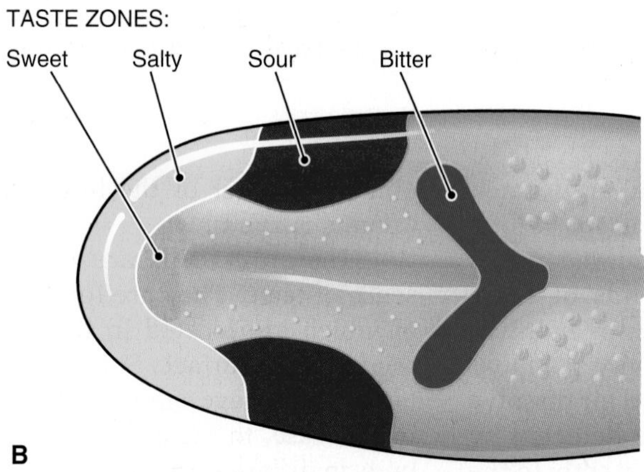

FIGURE 7.17 ▸ Chemoreceptors of the nose and tongue. A. Olfactory receptors are located in the upper nasal cavity, and taste receptors are found primarily in the papillae of the tongue. **B.** Taste zones of the tongue.

Chemoreceptors that are considered general sensory receptors include those that respond to water- and lipid-soluble substances dissolved in blood and cerebrospinal fluid. These chemoreceptors monitor pH, carbon dioxide, and oxygen levels, and stimulate some of the respiratory and cardiovascular reflexes that adjust breathing and cardiac rates. The mechanisms of these processes are discussed in detail in Chapters 10 and 13.

By the Way

While the brain senses and perceives pain, the brain itself has no nociceptors. Therefore, touching brain tissue does not cause pain. This explains why many brain surgeries can be performed with the patient conscious and fully aware of what is going on during the procedure.

Nociceptors

As described in Chapter 4, **nociceptors**, or pain receptors, are specialized free nerve endings distributed throughout the skin and other tissues. Joint capsules, fascia, periosteum, and the walls of blood vessels all have fairly high concentrations of nociceptors. There are three types of pain receptors, each sensitive to different stimuli:

- Mechanical damage to the tissue
- Extremes of temperature
- Chemicals, including those released by damaged cells

Because pain is an important warning signal of tissue damage, nociceptors do not adapt quickly and continue to send impulses to the brain for quite some time. Regardless of the frequency or duration of nociceptor stimulation, the actual perception of pain occurs in the brain and is modified by a wide variety of conscious and unconscious factors. A more detailed discussion of pain follows later in this chapter.

Thermoreceptors

There are two types of **thermoreceptors**, *warm* and *cold*, widely distributed throughout the skin. Both types are free nerve endings, with no structural distinctions between them. There are three to four times more cold receptors than warm. Cold receptors are located in the epidermal layer and are sensitive to a temperature range of 50°F to 105°F (10°C to 40°C). Warm receptors are located in the dermis and are stimulated by temperatures of 90°F to 118°F (32°C to 48°C). Temperatures below 50°F or above 118°F will cause tissue damage that stimulates nociceptors, producing pain. Both types of thermoreceptors are very active in response to initial changes in temperature but adapt quickly. For example, the initial shock of heat when entering a sauna is followed shortly after by a sense of tolerable warmth once the receptors have adapted.

Mechanoreceptors

As a broad category, **mechanoreceptors** are sensitive to touch, pressure, and movement stimuli such as stretch, compression, and torsion. Because of the large number and wide variety of mechanoreceptors, it is helpful to think of them grouped into three functional categories: tactile (touch) receptors, baroreceptors, and the mechanoreceptors in the ear.

Tactile Receptors

Tactile receptors were introduced in Chapter 4 as examples of cutaneous receptors found in the skin. As a quick review, these general sensory receptors include the free nerve endings, Merkel discs, and Meissner corpuscles found in the upper dermis that are sensitive to light touch, pressure, and vibration. They also include the Pacinian corpuscles, Ruffini corpuscles, and hair root plexi, which require stronger stimulus because they are located deeper in the dermis. As a group, the tactile receptors adapt easily, but not as rapidly as the olfactory and gustatory receptors.

Baroreceptors

Baroreceptors are specialized free nerve endings that act as pressure receptors for certain hollow organs. They are most abundant in the elastic tissues that form blood vessels and portions of the respiratory, urinary, and digestive tracts. Baroreceptors are sensitive to pressure changes inside the blood vessels, lungs, bladder, and digestive tract as they fill and empty. In this way, baroreceptors provide essential information for regulation of several vital functions. For example, the respiratory centers of the brain rely on information sent from the baroreceptors in the lungs to regulate the pace and depth of breathing. The same holds true for regulation of blood pressure; impulses from baroreceptors in the aortic artery send signals to the brain, so that appropriate adjustments can be made in cardiovascular structures.

Mechanoreceptors in the Ear

The mechanoreceptors for hearing and equilibrium are located in the bony labyrinth of the inner ear, a series of cavities filled with a fluid called ***perilymph***. The special receptors responsible for hearing are located in the **cochlea**, the snail-shaped portion of the bony labyrinth (▸ FIGURE 7.18). These mechanoreceptors are tiny hair-like extensions that are sensitive to vibrations caused by sound waves. As vibrations pass through the cochlea, the hairs bend, creating an impulse that is transmitted to the sensory neurons at the base of the mechanoreceptors. The impulse travels to the auditory region of the brain via the cochlear or auditory branch of the vestibulocochlear nerve (cranial nerve VIII).

The mechanoreceptors for equilibrium or balance are located within the semicircular canals and vestibule (▸ FIGURE 7.18). The mechanoreceptors of the **semicircular canals** are sensitive to changes in head or body position, especially rotational motions. When stimulated, they generate impulses that monitor and maintain balance during movement (***dynamic equilibrium***). At the base of the semicircular canals is the **vestibule**. Mechanoreceptors in the vestibule provide information used to maintain posture and stability when standing or sitting still (***static equilibrium***). Vestibular mechanoreceptors are sensitive to gravity and linear acceleration and deceleration, like the sensation caused by changes in speed when riding in a car. When static or dynamic equilibrium mechanoreceptors

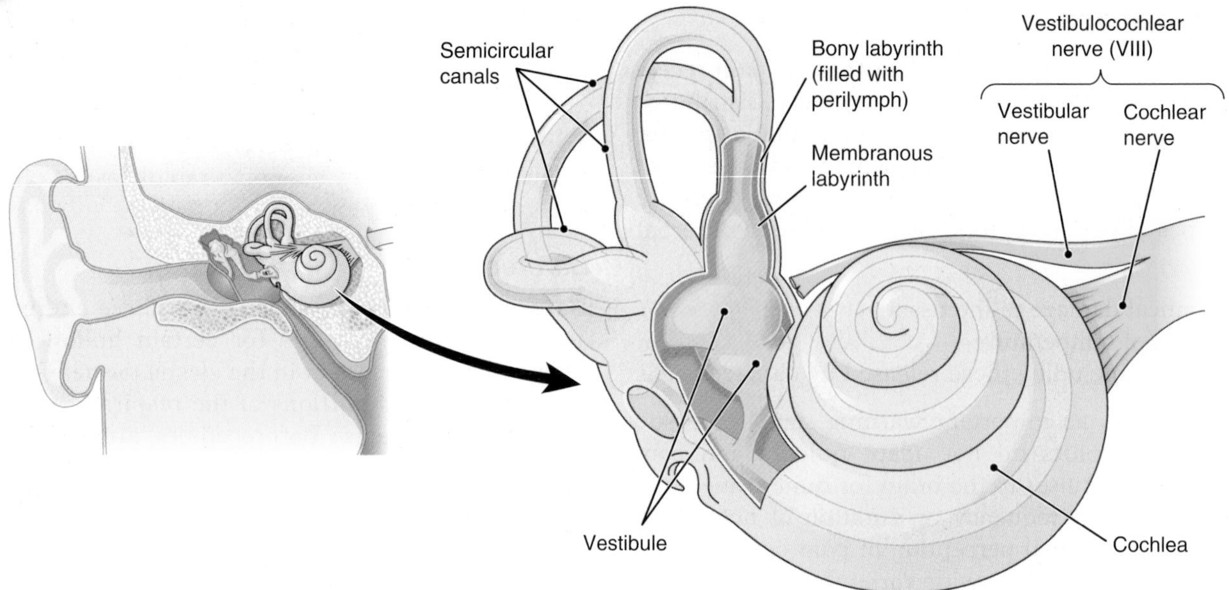

FIGURE 7.18 ▸ Mechanoreceptors of the ear. Mechanoreceptors for hearing and equilibrium are located in the bony labyrinth of the ear. Hearing receptors are found in the cochlea. Mechanoreceptors for dynamic equilibrium are found in the semicircular canals and those for static equilibrium are located in the vestibule.

detect movement or positional changes, they stimulate sensory neurons inside the bony labyrinth. These impulses are transmitted to the equilibrium centers of the brain via the vestibular branch of the vestibulocochlear nerve.

Proprioceptors

Proprioceptors are of particular interest to manual therapists because several of the benefits and effects of therapeutic techniques are derived from stimulation of this group of mechanoreceptors. Additionally, several pathologies commonly seen by therapists are related to proprioceptive dysfunction or stimulation. As general sense receptors, proprioceptors are scattered throughout skeletal muscles, tendons, joints, and fascia. They monitor movement; tension in muscles, tendons, and ligaments; and the length or state of contraction in skeletal muscles. The major types of proprioceptors and their primary locations are

- **Joint receptors** are embedded in the joint capsule and ligaments. They include Pacinian corpuscles and free nerve endings similar to those found in the dermal layer of skin. These receptors monitor the pressure, tension, and movement of a joint.
- **Muscle spindles** are bundles of modified muscle fibers called **intrafusal fibers** that are wrapped in their own connective tissue sheath and arranged in parallel alignment with skeletal muscle cells (**extrafusal fibers**) (▸ FIGURE 7.19). Sensory neurons associated with the muscle spindles continually monitor changes in muscle length. When skeletal muscles are stretched or lengthened rapidly, the muscle spindle signals a reflexive contraction of the muscle to protect it from tearing. This protective muscle reflex is called the **stretch reflex.**

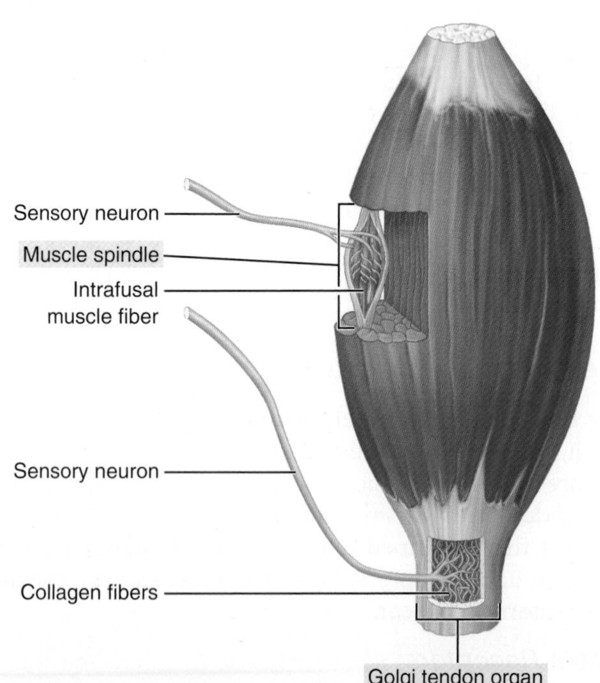

FIGURE 7.19 ▸ Proprioceptors. Muscle spindles are bundles of intrafusal fibers within the muscle belly that lie parallel to the extrafusal fibers and monitor muscle length. GTOs are concentrated in the tendon and musculotendinous junction and monitor tension of the muscles and tendons.

- **Golgi tendon organs**—GTOs are scattered throughout muscle but are most highly concentrated within tendons and at the musculotendinous junctions (▶ FIGURE 7.19). These proprioceptors monitor the tension level of tendons during muscle contraction and are stimulated when they sense excess tension. GTOs work closely with muscle spindles to protect muscles and tendons from tearing by *inhibiting* muscle contraction when stimulated. This muscle reflex is sometimes referred to as the **inverse stretch reflex** because it signals an action opposite to the stretch reflex.

Information from these three proprioceptors, along with information from mechanoreceptors in the inner ear, is processed subconsciously to create our ***kinesthetic sense***, which is the awareness of movement, body position, and the spatial relationship between parts. Proprioceptors assist us in moving safely through our environment by continuously sending information to the brain; they do not adapt to constant stimulation.

What Do You Think? 7.3

- Why is there a limitation to the range of temperatures that thermoreceptors are able to sense?
- Why do some sensory receptors fully adapt while others don't adapt at all?
- Based on your understanding of the action of muscle spindles and GTOs, why is slow movement and static stretching advised in a therapeutic session rather than fast or bouncy movements?

THE CENTRAL NERVOUS SYSTEM: INTEGRATIVE STRUCTURES AND FUNCTIONS

As previously stated, the CNS is the contact point between the sensory and motor portions of the PNS, performing all integrative functions. The CNS includes the brain and spinal cord, which are protected by the bones of the skull and vertebral column. Further protection is provided by the meninges, a connective tissue covering that completely encloses the brain and spinal cord, and the cerebrospinal fluid (CSF). The spinal cord carries sensory information *up to* and motor information *down from* the brain and serves as a center for reflex arcs. The various regions and structures in the brain serve as centers for higher-level processes such as **cognition**, memory, coordination of motor activity, and emotion.

Meninges and Cerebrospinal Fluid

Meninges is the collective name for three connective tissue membranes. Each membrane provides a slightly different form of protection for the brain and spinal cord (▶ FIGURE 7.20). The three membranes are

- **Dura mater**—This outermost meningeal layer, literally translated from Latin as "hard" or "tough mother," is made of two layers of strong disorganized fibrous connective tissue. Through most of the skull, these layers lie together. However, in several locations, they gap to form ***dural venous sinuses***. These sinuses drain blood and CSF from the brain and provide room for several large blood vessels.
- **Arachnoid mater**—This middle layer of the meninges has two portions. The outer portion is epithelial tissue; the inner consists of a web-like network of

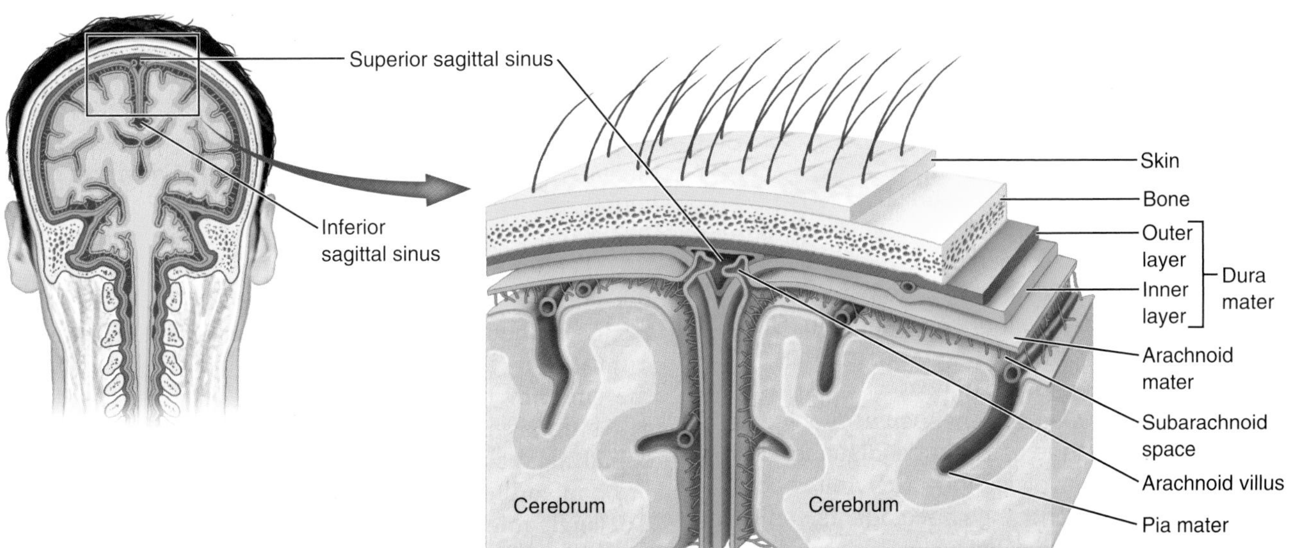

FIGURE 7.20 ▶ Layers of the meninges. Outer dura mater, middle arachnoid mater, and inner pia mater.

Pathology Alert Meningitis

As the word root and suffix imply, ***meningitis*** is an inflammation of the meninges caused by a bacterial, viral, fungal, or parasitic infection. It can also be caused by noninfectious processes such as cancer, autoimmune conditions, or certain drugs. Signs and symptoms include a very high fever, severe headache, a red or purplish rash, stiff neck, and sometimes nausea, vomiting, and delirium. These signs and symptoms can manifest suddenly or over a few to 24 hours. Because headache and stiff neck are common conditions seen by manual therapists, it is important to rule out the possibility of meningitis when clients seek an immediate appointment for relief of these conditions. It is wise to ask about onset, associated fever, and whether active neck flexion leads to extreme pain. The possibility of meningitis is increased when there is no traumatic episode related to the onset of the headache or stiff neck, and there is a positive pain response with neck flexion, regardless of whether other signs and symptoms can be confirmed. If therapists cannot rule out the possibility of meningitis, it is safest to immediately refer the client to a physician for diagnosis and treatment.

collagen and elastic fibers, hence the name based on the Greek word for spider. The small spaces created by this fibrous web are filled with CSF and are collectively referred to as the ***subarachnoid space***.
- **Pia mater**—Meaning literally "tender mother," the pia mater is the thin innermost membrane that adheres to the surface of the brain and spinal cord. The pia mater supports large blood vessels traveling along the surface of the brain and spinal cord, and it joins with the ependymal cells that produce CSF.

Cerebrospinal fluid is a colorless liquid that circulates through the subarachnoid space of the meninges, the central canal in the spinal cord, and through a series of chambers inside the brain called **ventricles** (FIGURE 7.21). Inside the ventricles, there is a network of specialized capillaries known as the **choroid plexus** covered with ependymal glial cells. CSF is produced from blood plasma by the ependymal cells and secreted into the ventricles to be circulated throughout the CNS. Eventually, CSF is reabsorbed back into the blood at the superior sagittal sinus of the dura mater via arachnoid granulations, or ***villi***, small projections of the arachnoid mater (FIGURE 7.21C).

By the Way

The lumbar puncture (also called a spinal tap) is a medical procedure in which a hollow needle is inserted into the subarachnoid space between L3 and L4 or L4 and L5 to withdraw a sample of CSF or to inject a dye for diagnostic purposes. The procedure is often used to diagnose inflammatory diseases of the nervous system such as meningitis and can also be used to assist in the diagnosis of stroke and cancers involving the CNS. Antibiotics, cancer medications, and anesthetic agents can also be delivered in this manner.

Functionally, CSF protects and supports the brain and spinal cord in three ways:

- Shock absorption—CSF in the subarachnoid space helps to cushion the tissues of the brain and spinal cord, protecting them from being jolted against the skull and vertebral column.
- Nutrition—CSF is the medium for nutrient and waste exchange between the blood and the nervous tissue of the CNS. Additionally, the specific ion composition of CSF is important for optimal neuronal activity.
- Physical barrier to blood-borne pathogens—The capillaries of the choroid plexus are the least permeable of any in the body. They are bound together tightly and create an important protective barrier, the ***blood–brain barrier*** that protects the brain from pathogens that might be present in blood. Only certain substances important to the composition of CSF can pass through the walls of the choroid plexus. All other substances are retained in the blood.

Spinal Cord

The spinal cord is approximately three-quarters of an inch (2 cm) in diameter and is housed within the vertebral foramen of the spinal column. At 16 in. to 18 in. (40–46 cm) long, the spinal cord is shorter than the vertebral column. It has a small channel in the center called the **central canal** that contains CSF. The spinal cord begins at the brain and extends from the foramen magnum (the large opening at the base of the cranium) to the superior border of the L2 vertebra. It is made up of 31 transverse segments that give rise to 31 pairs of spinal nerves. Because the cord ends at about the level of L2 in adults, the elongated nerve roots for L3 through Co1 angle inferiorly through the vertebral cavity and fan out like strands of hair to form the **cauda equina**, or "horse's tail" (see FIGURES 7.1 and 7.10).

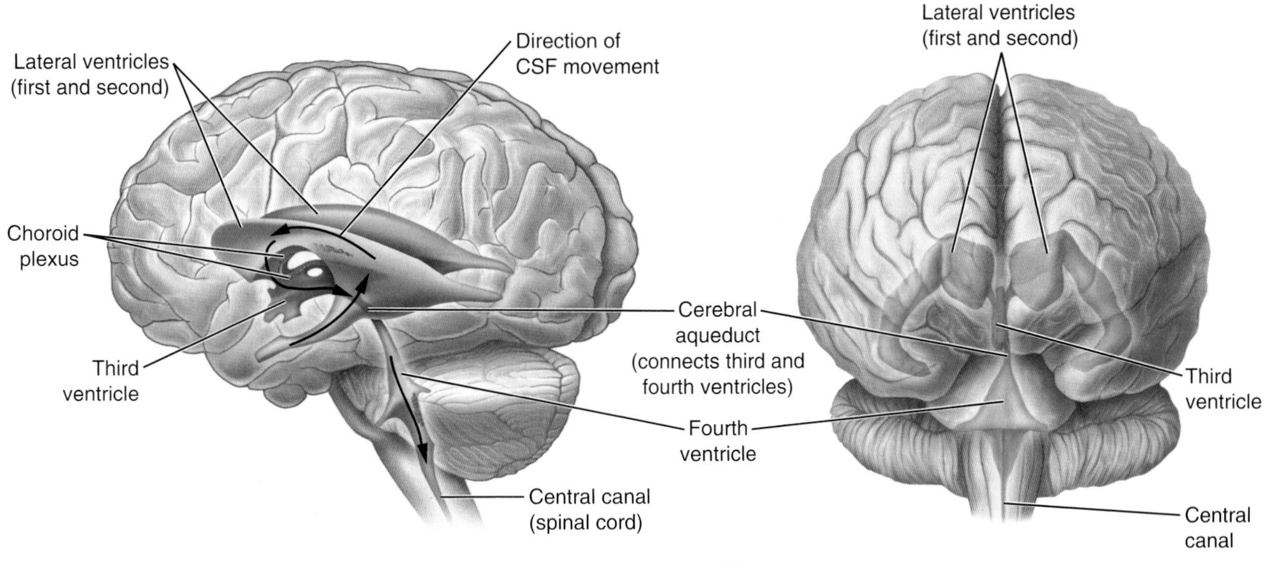

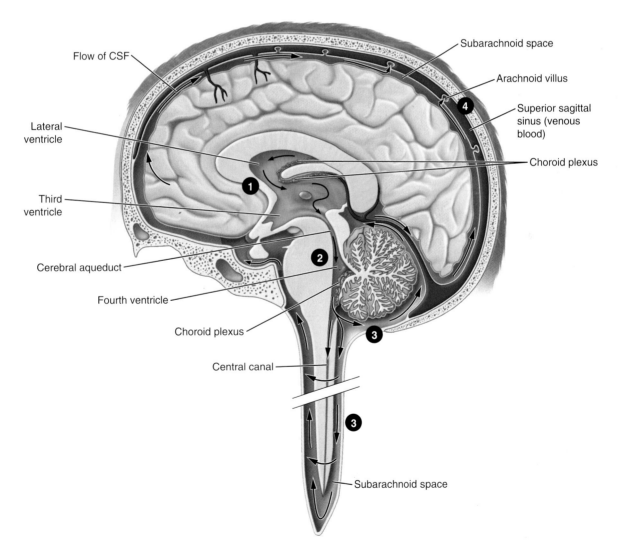

Flow of CSF
Subarachnoid space
Arachnoid villus
4
Superior sagittal sinus (venous blood)
Lateral ventricle
Choroid plexus
1
Third ventricle
Cerebral aqueduct
2
Fourth ventricle
Choroid plexus
3
Central canal
3
Subarachnoid space

C Cerebrospinal fluid production and circulation

FIGURE 7.21 ▸ Cerebrospinal fluid circulation. A. Left lateral view of the ventricles that contain the specialized capillaries of a choroid plexus that produces CSF. **B.** Anterior view. **C.** CSF circulates through the ventricles, the central canal of the spinal cord, and the subarachnoid space. It is reabsorbed from subarachnoid villi into the venous blood of the superior sagittal sinus of the dura mater.

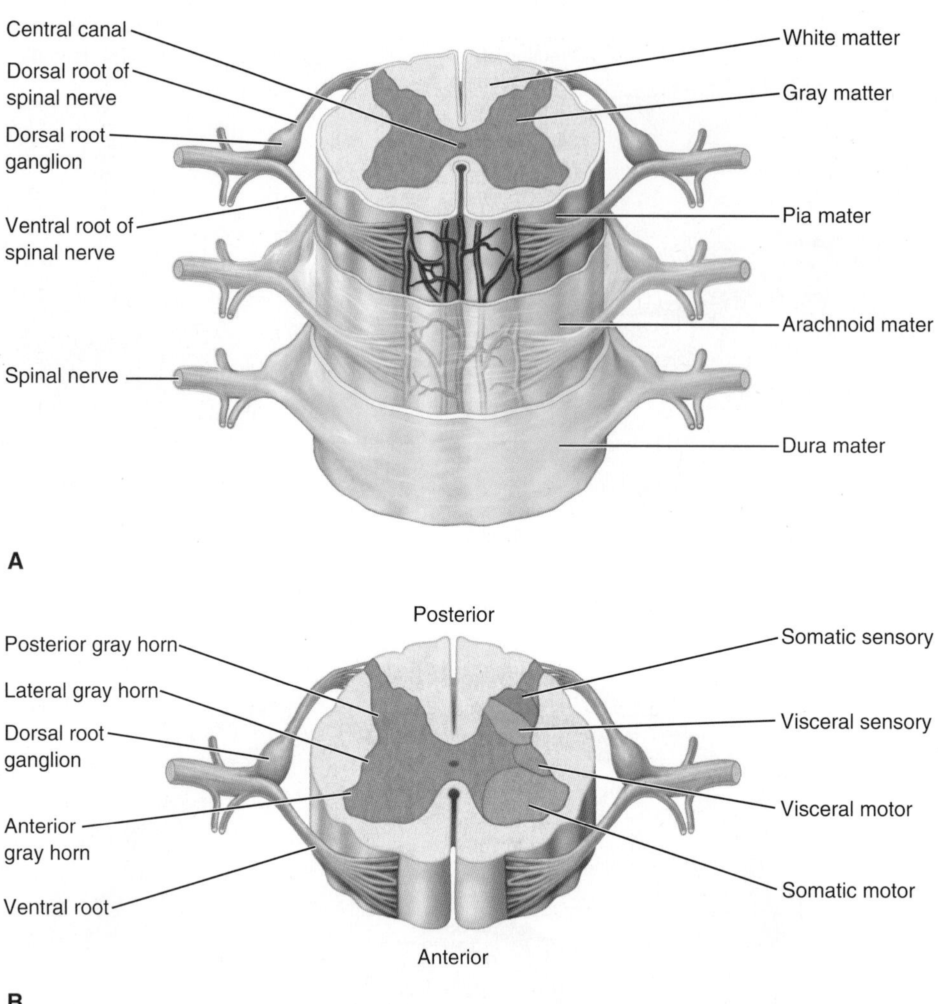

FIGURE 7.22 ▸ Sectional anatomy of the spinal cord. A. The meninges and the arrangement of gray and white matter are shown. The nerve roots that form the spinal nerves are also visible. **B.** Gray matter of the spinal cord is divided into three major regions, or horns. The posterior horn receives sensory information, while the lateral and anterior horns contain the cell bodies and dendrites of motor neurons. Visit http://thePoint.lww.com/Archer-Nelson for two videos of the spinal cord.

White and Gray Matter

In viewing a transverse section of the spinal cord, an outer mass of white matter surrounding a butterfly-shaped core of gray matter can be seen (▸ FIGURE 7.22). **White matter** is composed of myelinated axons, giving it a white appearance. In contrast, **gray matter** gets its coloration from the unmyelinated axons, cell bodies, dendrites, and glial cells it contains. The composition of white and gray matter divides and defines the major functions of the spinal cord. The myelinated axons of white matter are organized into large bundles called **tracts**, which act as impulse freeways carrying impulses either up to or down from the brain. Since sensory information must travel *up* the spinal cord to the brain for integration, tracts carrying sensory impulses can be called either ***ascending*** or ***sensory tracts***; these are predominantly located in the dorsal half of the cord. Similarly, ***descending tracts*** can be called ***motor tracts*** since they carry motor impulses from the brain *down* the spinal cord. These tracts are predominantly located in the ventral portions of the cord. The gray matter at the center of the cord contains interneurons for integrative functions and the cell bodies and dendrites of motor neurons. The "wings" of the gray matter are organized into dorsal (posterior), ventral (anterior), and lateral regions called **horns** (FIGURE 7.22). The ***posterior gray horns*** contain sensory neurons, while the ***anterior*** and ***lateral gray horns*** contain the cell bodies of motor neurons.

Dorsal and Ventral Nerve Roots

Nerve roots are extensions that connect the nerve fibers of each spinal nerve to their respective cord segment. Each root contains the axons of only sensory or motor neurons. The ***dorsal (posterior) root*** contains the axons of sensory neurons, which is why it can also be called the ***sensory root***. This dorsal root includes a "bubble" of nerve tissue called the ***dorsal root ganglion***, which contains the cell bodies of the sensory neurons. In contrast, the ***ventral (anterior) root*** or ***motor root*** contains the axons of motor neurons. Both dorsal and ventral roots join to form the spinal nerve that exits the CNS through the intervertebral foramen.

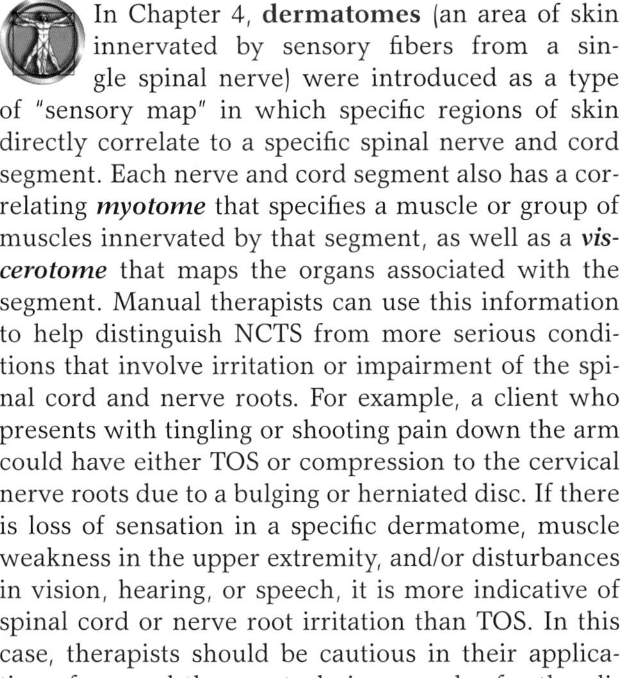

In Chapter 4, **dermatomes** (an area of skin innervated by sensory fibers from a single spinal nerve) were introduced as a type of "sensory map" in which specific regions of skin directly correlate to a specific spinal nerve and cord segment. Each nerve and cord segment also has a correlating ***myotome*** that specifies a muscle or group of muscles innervated by that segment, as well as a ***viscerotome*** that maps the organs associated with the segment. Manual therapists can use this information to help distinguish NCTS from more serious conditions that involve irritation or impairment of the spinal cord and nerve roots. For example, a client who presents with tingling or shooting pain down the arm could have either TOS or compression to the cervical nerve roots due to a bulging or herniated disc. If there is loss of sensation in a specific dermatome, muscle weakness in the upper extremity, and/or disturbances in vision, hearing, or speech, it is more indicative of spinal cord or nerve root irritation than TOS. In this case, therapists should be cautious in their application of manual therapy techniques and refer the client to a physician for definitive diagnosis.

What Do You Think? 7.4

- How would you explain the brain getting bruised on both sides in a whiplash accident?
- Do you think the cervical, thoracic, lumbar, or sacral segments of the spinal cord would innervate the kidneys and bladder? Why?

Brain

It has been estimated that the brain contains between 50 and 500 billion neurons and that a single gram of brain tissue may contain as many as 400 billion synaptic junctions. When we consider these numbers, the commonly heard statement that we use only 10% of our brain seems probable. However, with the advent of imaging techniques that allow us to observe the brain performing complex functions, it is clear that all regions of the brain are fully utilized. Still, when compared to the trillions of neuronal connections possible, the number of connections we actually form and use on a regular basis is relatively small.

The brain is the body's control center. Its primary function is to maintain homeostasis throughout the body. By registering, processing, and responding to all sensory input, it coordinates and controls the mechanisms that keep all systems functioning within normal parameters. While the stream of incoming information is endless, the brain efficiently prioritizes the constant flow of signals, allowing only a fraction of them to be registered at a conscious level. Information that confirms normal functioning does not demand any real attention and is handled at an unconscious or reflexive level. In contrast, we become consciously aware of thoughts, feelings, or experiences when new or important information is amplified and sent to specific regions of the brain for more complex processing.

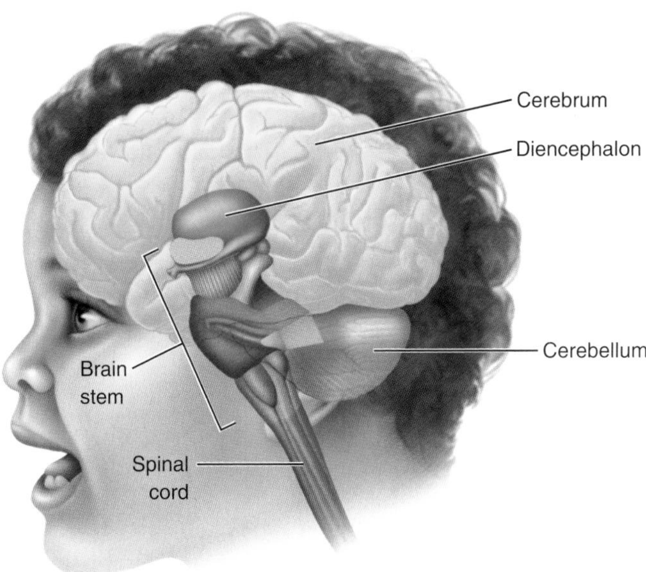

FIGURE 7.23 Four major regions of the brain.

The billions of interneurons within the brain are organized into several structural regions. The brain consists of four major parts: the brain stem, diencephalon, cerebrum, and cerebellum (FIGURE 7.23).

Brain Stem

The **brain stem** serves as a bridge or connector between the spinal cord and brain and as a reflex center for vital functions such as heart rate, blood pressure, and breathing. Most of the cranial nerves emanate from this region. The brain stem is continuous with the spinal cord and consists of three parts: the medulla oblongata, pons, and midbrain.

The **medulla oblongata** is the most inferior portion of the brain stem and is composed of both white and gray matter (FIGURE 7.24). The white matter includes ascending and descending tracts that connect the spinal cord with other areas of the brain. Some of these tracts form bulges called ***pyramids*** on the anterior side of the medulla. Just superior to the spinal cord, 90% of the tracts from the right pyramid cross over to the left, and 90% of the tracts from the left pyramid cross over to the right. Called the ***decussation of pyramids***, this crossing over explains why the movements of one side of the body are controlled by motor regions on the opposite side of the brain. The gray matter of the medulla oblongata consists of groups of synapsing neurons called ***nuclei*** that

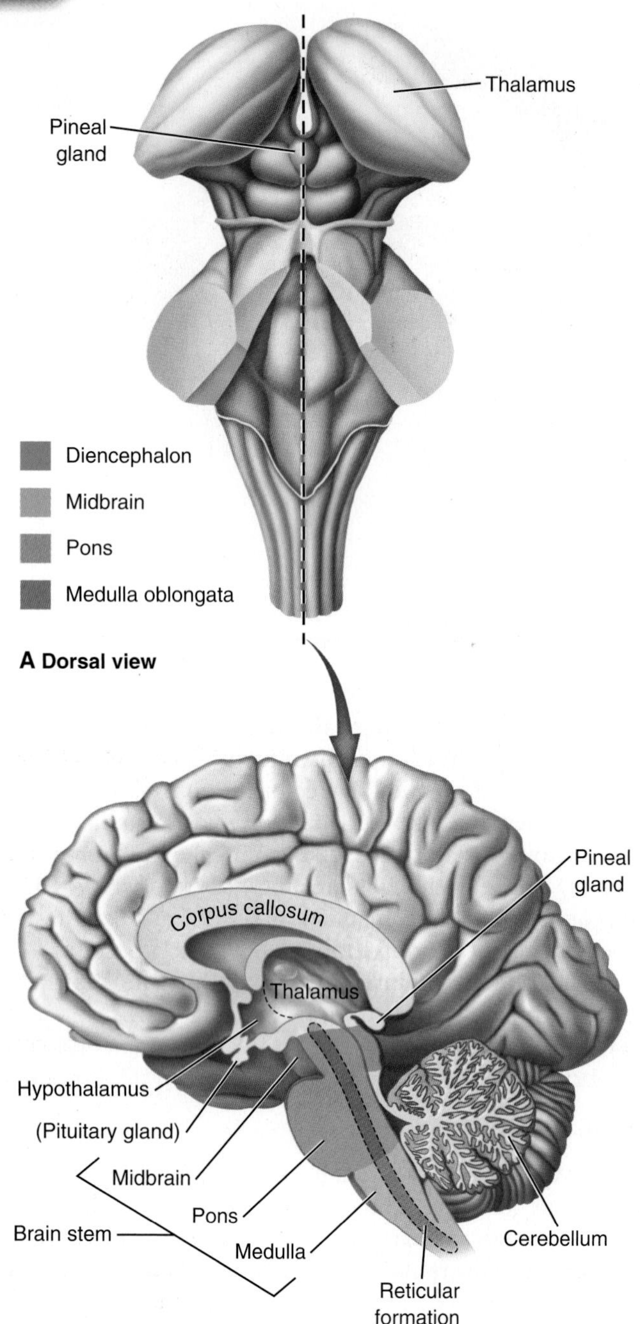

FIGURE 7.24 ▶ Structures of the brain stem, diencephalon, and cerebellum. A. Dorsal view of the brain stem and diencephalon (with cerebrum and cerebellum removed). **B.** Midsagittal section.

are associated with the sensations of touch and vibration. They control reflexes such as coughing, hiccuping, sneezing, and swallowing. These nuclei also include the ***medullary rhythmicity center***, which regulates breathing rate, and the ***cardiovascular center***, which monitors and adjusts heart rate and the diameter of blood vessels.

Just superior to the medulla oblongata is the region of the brain stem called the pons. The **pons**, the Latin word for "bridge," contains multiple tracts that serve as vital

By the Way

The ***substantia nigra***, which literally means "black substance," are dark nuclei in the midbrain. The neurons of these nuclei produce and release the neurotransmitter dopamine that stimulates the ***basal ganglia***, which are specific motor regions of the cerebrum. In individuals with Parkinson disease, deterioration of the substantia nigra cause decreased levels of dopamine, which leads to the muscular stiffness commonly experienced with this condition.

connecting links between the spinal cord, cerebrum, and cerebellum. It also contains nuclei that help control and regulate certain involuntary actions, including breathing.

The **midbrain** lies superior to the pons and connects the brain stem to the diencephalon. It consists of a large pair of tracts and several important nuclei that help coordinate muscle contractions and control movements of the eyes, head, and neck in response to visual stimuli. The midbrain also relays auditory impulses to the thalamus.

Running vertically through the brain stem is a network of mixed gray and white matter called the **reticular formation**. It contains ascending and descending pathways that manage several basic functions. The ascending fibers make up the ***reticular activating system*** (RAS) that transfers incoming sensory information to the brain to maintain an alert state in the cerebral cortex. The descending fibers help regulate resting muscle tone, digestion, salivation, urination, and sexual arousal.

Diencephalon

The **diencephalon** is located deep within the brain, just superior to the brain stem. It includes three small structures that are vital to human health and survival: the thalamus, hypothalamus, and pineal gland (▶ FIGURE 7.24).

The **thalamus** consists of two oval-shaped masses directly superior to the midbrain. It is sometimes called the "sensory clearinghouse" because its job is to sort and prioritize almost all incoming sensory input and relay it to appropriate centers of the cerebrum. As the thalamus sorts and prioritizes, it makes an initial interpretation of the sensory input. In addition to its sensory function, the thalamus also relays motor information from the cerebellum and other regions to the motor areas of the cerebrum.

The **hypothalamus** lies anterior and inferior to the thalamus. This small area of the brain serves as the primary connection between the nervous and endocrine systems, because it is both structurally and functionally connected to the pituitary gland. As the control center for the autonomic nervous system (ANS), it regulates circadian rhythms, state of consciousness, body temperature, hunger and thirst, as well as emotions and sensations

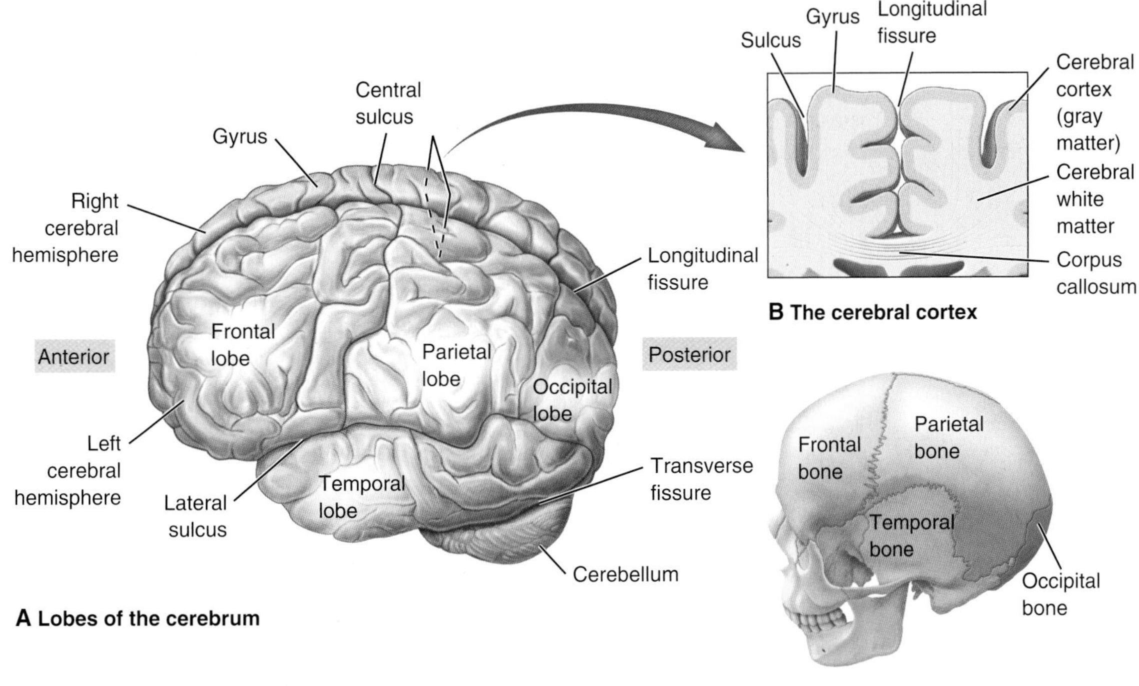

FIGURE 7.25 ▸ Lobes of the cerebrum. A. Sulci and deeper fissures divide the cerebrum into four lobes. **B.** The cerebral cortex is characterized by its many gyri (folds). **C.** Lobes are named for the overlying skull bones that protect them.

including anger, aggression, pain, and pleasure. Posterior and inferior to the thalamus is a pea-sized structure called the **pineal gland**. This small gland secretes the hormone ***melatonin*** that promotes sleepiness. Its location in the diencephalon creates another important link between the nervous and endocrine systems.

Cerebrum

The **cerebrum**, which fills most of the cranial cavity, is the largest part of the brain. The cerebrum consists of a thin outer rim of gray matter called the **cerebral cortex.** This layer surrounds a core of white matter with deep nuclei (islands of gray matter) called ***basal ganglia*** that help control subconscious muscle activities like walking or chewing. Characterized by numerous folds of tissue, or **gyri**, separated by shallow grooves called **sulci**, the cerebrum is what most people visualize when thinking of the brain (▸ Figure 7.25). There are several deeper grooves within the cerebrum called **fissures**; one of these fissures separates the cerebrum into left and right ***hemispheres***. The cerebral hemispheres are connected by a white-matter bridge of myelinated axons, the **corpus callosum**. Each hemisphere has four **lobes** (regions) named according to the cranial bones that protect them: frontal, parietal, temporal, and occipital.

By the Way

If all the folds of the cerebral cortex were flattened out, it would cover an area of about 2.5 ft^2.

The cerebrum is often called the "mastermind" of the brain. It is involved with complex processes, including conscious thinking and problem solving, learning and memory, emotions, sensory perception, and motor planning and movement. All of these activities are controlled by neuronal pools located in specific regions of the cortex. Each region is responsible for managing a specific sensory, motor, or associative function (▸ Figure 7.26). For example, precise muscle control, speech, and muscle coordination are handled by the motor areas of the frontal lobes, while an anterior association area is responsible for personality and conscious thought. General sensory information, including touch, pain, temperature, taste, and proprioception, is processed within the parietal lobes. Olfactory sensations and auditory stimuli such as speech and music are localized in the sensory and association areas of the temporal lobes, as is some memory storage. The occipital lobes are involved with visual perception, association, and memory.

One of the most amazing processes of the cerebrum is **perception**, the mental process of becoming aware of and recognizing an object or idea. The cerebrum's ability to form perceptions is based on an integrative process that links multiple sensations or impressions to create a singular complex concept. For example, how do we know that an object we are holding is a lemon? Photoreceptors in the eyes provide information that the object is yellow and somewhat round to the visual sensory region of the cortex. Touch receptors in the hands supply the information

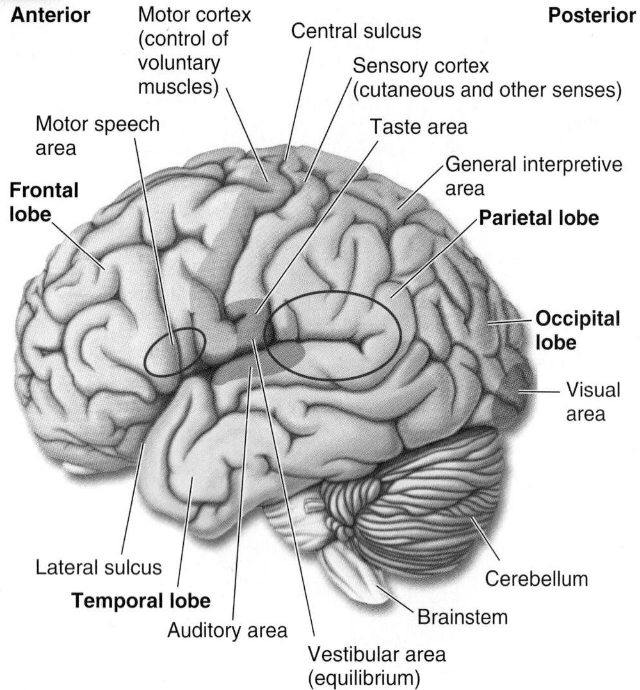

FIGURE 7.26 Regions of the cerebrum. The cerebral cortex has several different sensory, motor, and associative areas.

that it is three-dimensional and smooth and waxy in texture. The nose supplies the sensory input that registers in the olfactory centers as a distinct aroma. None of these sensations alone can determine that the object is a lemon. It is only via integrated processing in the cerebrum that all the sensory input is connected and compared with stored memory to identify the object as a lemon.

Remember that the brain is physically isolated within the cranial cavity. All the sensations that contribute to perception are detected by the body *not* the brain. The brain simply interprets the coded messages it receives into sensations and then connects them together to create our perception of a tree, a chiming bell, the smell of baking cookies, or a friend named Bob. Dr. George Sheehan, an American physician, writer, and philosopher put it this way, "The mind's first step to self-awareness must be through the body." The existence of this perceptual process provides profound meaning to the manual therapy arts. When we touch or move another person's body (or energy), we literally touch their mind. When we offer patterned and purposeful sensory data, we offer information that clients can use to support or change their lives.

Pathology Alert Headaches

A headache is one of the most common pathologies individuals can suffer, and manual therapy is often cited as one of the most effective treatments. There are three general categories of headaches:

- ***Stress or tension*** headaches are characterized by a constant aching or sometimes light throbbing pain. Symptoms are caused by muscle tension in the neck, shoulders, and/or jaw that stimulates nociceptors. The area of pain is often described as in the forehead, behind the eyes, or all around the head. Manual therapy is indicated to reduce stress and decrease muscle tension in the head, neck, and shoulders.
- ***Cluster*** headaches are a rare type of headache characterized by repeated short-term attacks of severe pain on one side of the head and around an eye. These headaches occur in clusters of four or five attacks per day with alternating pain-free periods. The period of active attacks can last several weeks and then go into remission for months or years. The cause of cluster headaches is unknown, but recent research shows that abnormal neuronal activity in the hypothalamus may be involved. Manual therapy may or may not be helpful in relieving pain associated with cluster headaches, and sufferers are not likely to seek help from a manual therapist during active periods because the attacks are so disabling.
- ***Migraine*** headaches involve intense throbbing pain, usually at the front or on one side of the head. Sensory disturbances and nausea are other common symptoms. There is evidence suggesting that migraines are caused by a surge of neural activity that stimulates the pain centers of the upper brain. While the exact cause of this surge is still unknown, fatigue, stress, dehydration, hormonal changes, and irregular eating patterns are common triggers. Manual therapy is contraindicated during migraine headaches, and a session in progress should be terminated if the client recognizes the onset of early-stage symptoms such as mood changes, difficulty concentrating, fatigue, or excessive energy.

Headaches can also be a warning sign of a more serious underlying pathology. Acute headaches such as ***traction headaches*** caused by eye strain that stretch or pull on the pain-sensitive areas of the brain or ***inflammatory headaches*** can indicate conditions such as high blood pressure, strokes, aneurysms, tumors, meningitis, or serious infections of the sinuses, ears, and teeth.

Cerebellum

The **cerebellum** is situated inferior to the cerebrum and posterior to the medulla and pons (see FIGURES 7.23 and 7.24B). This portion of the brain controls balance and posture, coordinates voluntary muscle activity, and together with the RAS, regulates resting muscle tone. While motor centers in the cerebrum plan voluntary muscular activities, the cerebellum compares and coordinates these plans with information from the muscles, proprioceptors, and visual receptors to create smooth and skilled motor movements.

Limbic System

The **limbic system**, sometimes called the "emotional brain," is a collection of connected structures in the brain that processes memory and emotion and controls unconscious aspects of behavior. While there is not complete agreement as to which structures make up this system, scientists generally include the cingulate gyri, hippocampus, amygdala, and mammillary bodies (FIGURE 7.27).

- The ***cingulate gyri*** are portions of the cerebral hemispheres that border the corpus callosum. These regions add emotional content to sensory input and are especially involved with the perception of pain.
- The ***hippocampus*** is actively involved in the process of creating new memories.
- The ***amygdala*** manages and mediates the overall emotional response by comparing sensory data with memories. It also plays an important role in impulse control.
- The ***mammillary bodies*** receive olfactory information and associate this input with memory.

As emotions are generated, they are experienced both unconsciously in the body and consciously by the mind. For example, when we are frightened, our fear is initially registered by the amygdala. The amygdala communicates with the hypothalamus to set off a series of physiologic responses that prepare the body for action. All these responses, such as dilation of the pupils, increased blood pressure, and increased breathing and heart rates occur *before* the cerebral cortex is even consciously aware that we are afraid. While the limbic system is able to quickly assess sensory data and cause us to *physically* "feel" the emotion, the cerebrum consciously perceives and analyzes the same sensory information to *mentally* "feel" the emotion and determine a conscious response.

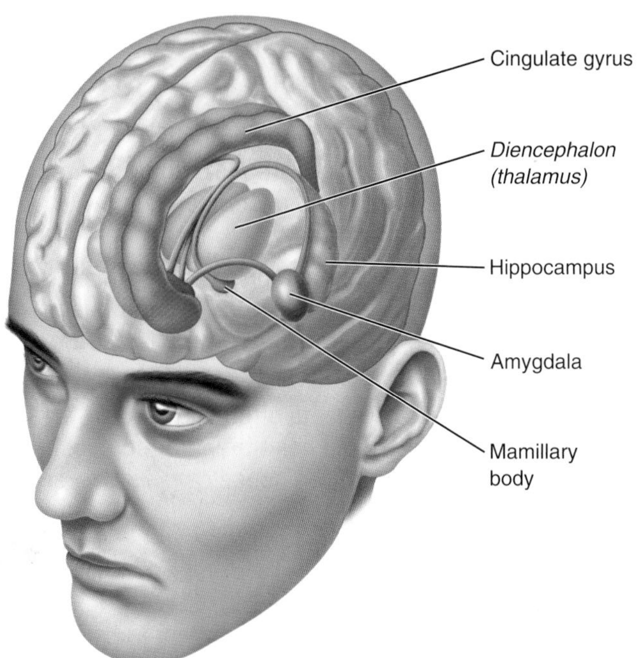

A Structures of the limbic system

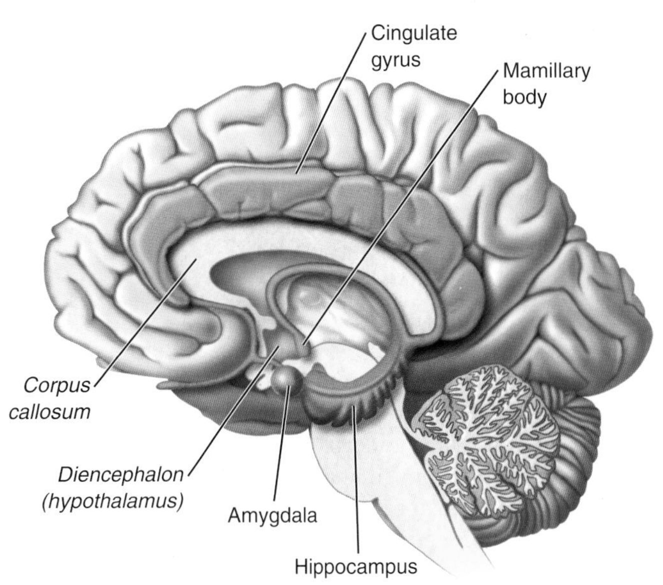

B Sagittal view

FIGURE 7.27 The limbic system. A. Limbic system brain structures involved in processing memory and emotions. **B.** Sagittal section.

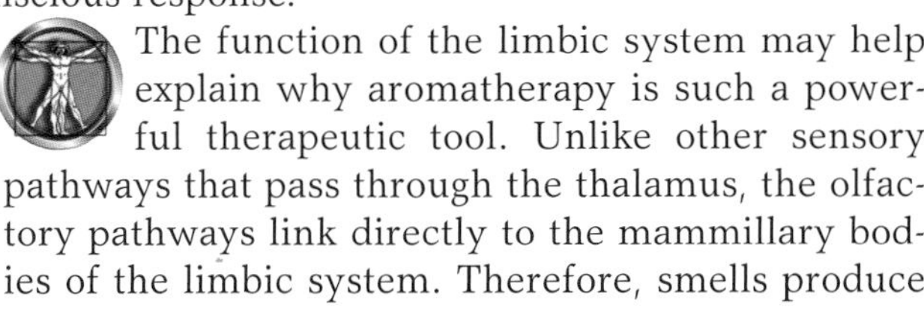

The function of the limbic system may help explain why aromatherapy is such a powerful therapeutic tool. Unlike other sensory pathways that pass through the thalamus, the olfactory pathways link directly to the mammillary bodies of the limbic system. Therefore, smells produce instantaneous and powerful emotional responses. Additionally, because the hippocampus is central to forming and recalling memories, aromas can be a powerful tool for evoking memories. While scent can be used intentionally as a therapeutic tool, manual therapists must also remember to exercise caution in their choice and use of perfumes, colognes, lotions, detergents, tobacco, incense, or candles.

What Do You Think? 7.5

- What structures and/or areas of the brain are likely to be involved when prioritization or interpretation of sensory input is confused, suppressed, or compromised?
- How does your understanding of perception help to explain sayings such as "Beauty is in the eye of the beholder," or "One man's trash is another man's treasure"?
- Why do real estate agents often suggest to sellers that they should bake bread or cookies before a showing or open house?

THE PERIPHERAL NERVOUS SYSTEM: MOTOR STRUCTURES AND FUNCTIONS

The PNS can be subdivided into two divisions: somatic and autonomic (FIGURE 7.28). While each division includes sensory receptors and sensory and motor neurons, it is the function and make-up of its motor pathways and the effectors they innervate that truly differentiate each division. The somatic division includes the motor pathways that innervate skeletal muscle tissue, while the autonomic division innervates smooth muscle, cardiac muscle, and glands.

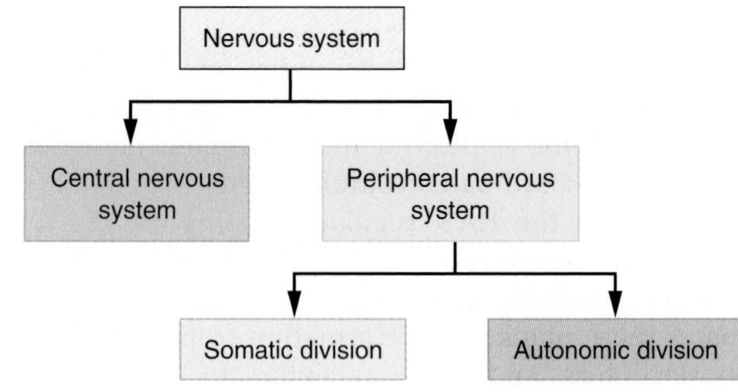

A Nervous system divisions

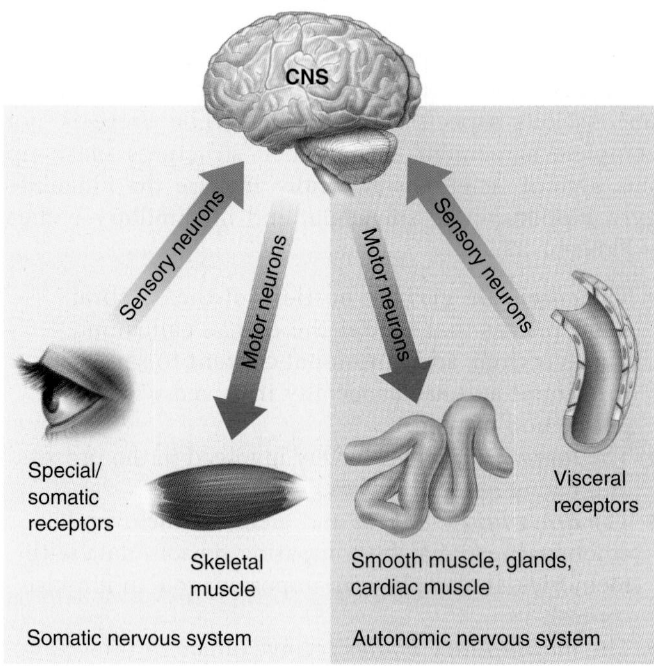

B Sensory and motor nerves

FIGURE 7.28 Somatic and autonomic divisions of the PNS.

Somatic Division

The **somatic** nervous system includes sensory neurons that carry information from the special sense receptors, tactile receptors, and general receptors of the joints and muscles. Its motor neurons are responsible for signaling the skeletal muscle contractions that produce movement and maintain balance. Since the motor responses of skeletal muscle can be consciously controlled, the somatic division is described as *voluntary*.

Somatic motor pathways contain only *one* motor neuron between the spinal cord and a skeletal muscle effector. As seen in FIGURE 7.29A, the cell body and dendrites of these motor neurons are located in the ventral horn of the spinal cord gray matter, and the axon carries the motor command to a specific skeletal muscle. Recall from Chapter 6 that each motor neuron stimulates a specific group of muscle fibers within a skeletal muscle: a **motor unit**. Therefore, nerves that innervate skeletal muscles contain hundreds to thousands of somatic motor neurons that control many different motor units within the muscle. Appendix B provides a quick reference of the major muscles of the body and the spinal segments and nerves that innervate them.

Autonomic Division

The **autonomic** nervous system (ANS) includes sensory neurons from the internal organs and fascia and motor neurons that stimulate smooth muscle, cardiac muscle, and glandular effectors.

The autonomic division is characterized as *involuntary,* since the actions of its effectors can occur automatically without conscious awareness. The motor activities stimulated by the ANS are subconscious responses regulated by the hypothalamus and brain stem. The autonomic division's motor pathways are more complex than those in the somatic division (FIGURE 7.29B). They differ in a few key ways:

- Autonomic pathways include *two* motor neurons.
- The two motor neurons connect with each other at a sort of "junction box" called an **autonomic ganglion**.

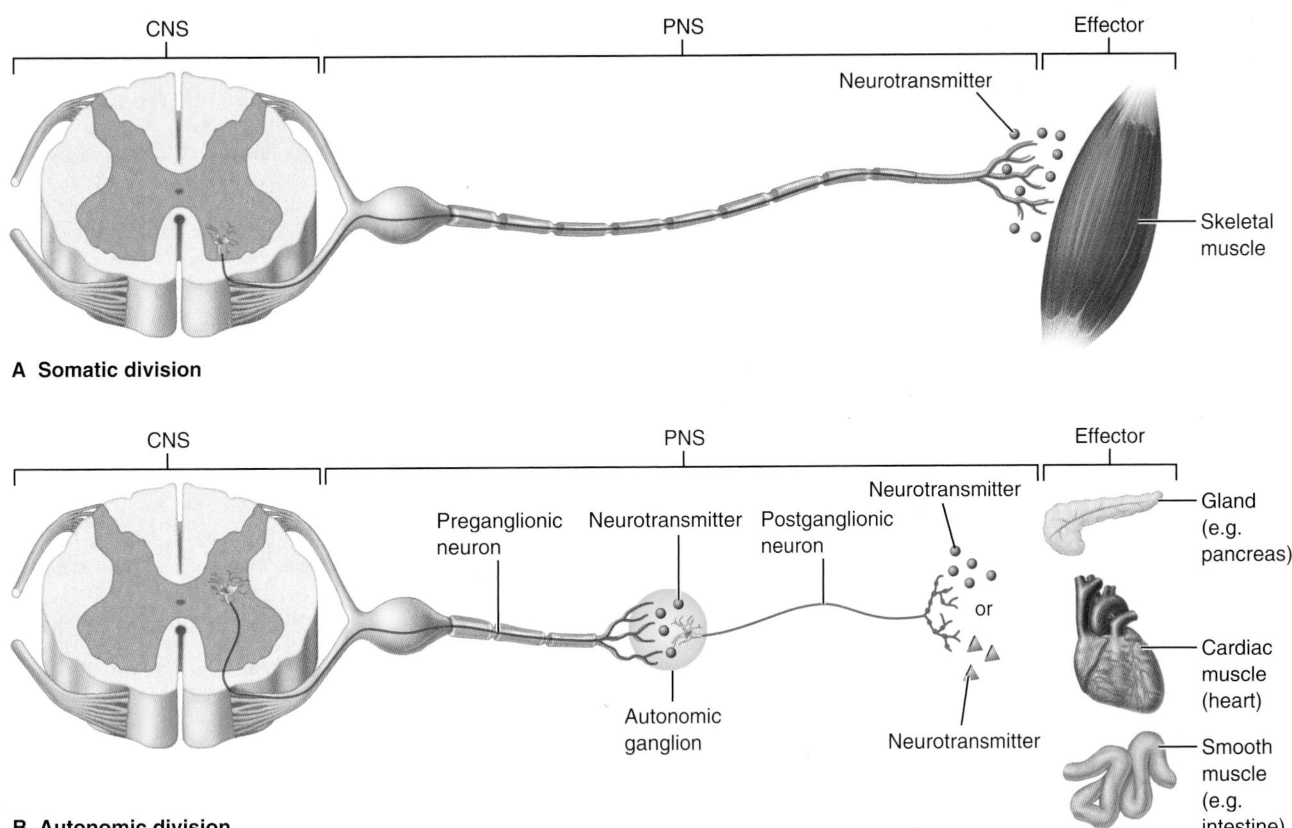

FIGURE 7.29 ▸ Somatic and autonomic motor pathways. A. Somatic motor pathways include one motor neuron that transmits signals from the spinal cord to a skeletal muscle effector. **B.** Autonomic pathways include two motor neurons between the spinal cord and visceral effector: a preganglionic neuron that coveys the motor impulse from the spinal cord to an autonomic ganglion and a postganglionic neuron that transmits the signal from the ganglion to a visceral effector.

- The first motor neuron of an autonomic pathway, the ***preganglionic neuron***, begins in the lateral horn of the spinal cord gray matter instead of in the ventral horn like somatic motor neurons. The ***postganglionic neuron*** runs from the ganglion to the effector.

The motor pathways of the ANS can be subdivided into two divisions—sympathetic and parasympathetic—that create different responses in the **visceral effectors**. **Sympathetic** pathways create an emergency response system that signals the body's *fight-or-flight* mechanisms such as increased heart rate, more rapid breathing, increased circulation in large skeletal muscles, and inhibition of digestive processes. These changes are meant to be short-term survival responses that serve as a protective mechanism when faced with physical or emotional stress. In contrast, **parasympathetic** pathways are responsible for maintaining homeostatic processes and returning organ function to normal levels once an emergency has passed. The parasympathetic system is nicknamed the "rest and digest" or "feed-and-breed" system, referring to its role in maintaining normal functions.

Most glands and organs have **dual innervation**, meaning they are innervated by both sympathetic and parasympathetic pathways. This dual innervation provides the means to respond to dangerous or stressful stimuli and then return to homeostasis once the stimulus is removed. For example, because the heart and lungs have dual innervation, stressful stimulus (such as having to sprint three blocks to catch the last bus home) signals an increase in heart and respiratory rates via sympathetic pathways. Once the stimulus is removed, parasympathetic stimulation returns the heart rate and breathing to normal levels. This balancing action between sympathetic and parasympathetic divisions is an essential feature of autonomic function. There are three major exceptions to dual innervation: the adrenal glands, sweat glands, and the smooth muscles inside blood vessels (▸ TABLE 7-3). These effectors only have sympathetic innervation because their stimulation is an important part of the body's stress response.

The function of the ANS provides another physiological explanation for the systemic effects that occur in response to manual therapy. Since these effects include changes in visceral function such as decreased blood pressure, heart rate, and respiration, stimulation through ANS pathways must be involved. However, the exact physiologic explanation of how the ANS mediates systemic effects is still debatable. Similar to maintaining skeletal muscle tone, the nervous system

Table 7-3 Effects of Dual Innervation on Visceral Effectors

Sympathetic Effects	Organ	Parasympathetic Effects
Sharpens close and centered vision; opens nasal septum	Eyes Nasal septum	Full peripheral and centered vision; normal opening of nasal septum
Increases heart rate	Heart	Maintains/returns resting heart rate
Increases breathing rate; dilation of bronchioles	Lungs	Maintains/returns resting respiratory rate; constricts bronchioles
Inhibits digestive processes; Constricts sphincters and decreases peristalsis, absorption, and elimination	Stomach Pancreas Small intestine Large intestine	Maintains returns normal digestion by increasing peristalsis, absorption, and elimination; relaxes sphincters
Increases glucose released from liver	Liver	No known effect
Constricts sphincters; prevents voiding	Urinary bladder	Allows voiding via relaxed sphincters
Increases metabolic rate and stimulates fat breakdown for energy	Cellular metabolism	Maintains normal homeostatic level of metabolism
Stimulates release of adrenalin Increases sweat production Constricts blood vessels in visceral and small muscles; Dilates blood vessels in large skeletal muscles; Increases blood pressure	Adrenal glands Sweat glands Smooth muscle in blood vessels (Vasomotor response)	None None None

provides a level of low-grade stimulation to visceral effectors. This creates a constant state of readiness in the body called ***sympathetic tone***. When we are subjected to long-term stress, sympathetic tone is increased. Some manual therapists attribute the systemic relaxation responses of manual therapy to increased parasympathetic stimulus. However, it seems more likely that these effects are created by decreasing sympathetic tone rather than increasing parasympathetic stimulation.

Regardless of the exact physiological explanation, it is important to remember that the client's individual experience and perception holds the key to the degree and type of systemic effects created. Therapists increase the probability of seeing positive effects by maintaining clear and appropriate intention, contact, pressure, and rhythm. The client's perception of the therapy is also influenced by previous experiences and mental and emotional status. Therefore, before a session begins, it is wise to ask the client how they are feeling and clarify expectations for the session. As the session progresses, a few verbal check-ins will help to ensure the therapeutic nature of the session.

Sympathetic Pathways

The preganglionic neurons of the sympathetic pathways exit the spinal cord via thoracic and upper lumbar (T1 through L2) spinal nerves. Therefore, the sympathetic branch of the ANS is sometimes identified as the ***thoracolumbar*** branch. Preganglionic neurons travel to the sympathetic ganglia, which are connected in a chain located anterior and lateral to the spine, called the **sympathetic chain** or **paravertebral chain** (▶ FIGURE 7.30). In the sympathetic chain, they can synapse with several different postganglionic neurons or travel through the chain to a ***collateral ganglion***. In this way, sympathetic stimuli travel up and down the paravertebral chain to create rapid and widespread motor responses in all the visceral effectors. This body-wide sympathetic response can be described as a "shotgun effect" in which one shot or stimulus produces a wide "spray of responses" over the entire body.

In ▶ FIGURE 7.31, notice that the preganglionic neurons all originate in the lateral horn of the spinal cord, and their axons exit via the ventral root of the spinal nerve. After a short distance, the preganglionic axon enters the paravertebral chain, where it splits to connect with several postganglionic neurons up and down the chain. The postganglionic neurons carry the motor signals out of the paravertebral chain or collateral ganglion to multiple visceral effectors.

Parasympathetic Pathways

In contrast to sympathetic pathways, the preganglionic neurons of the parasympathetic pathways exit the CNS via only a few cranial and sacral spinal nerves

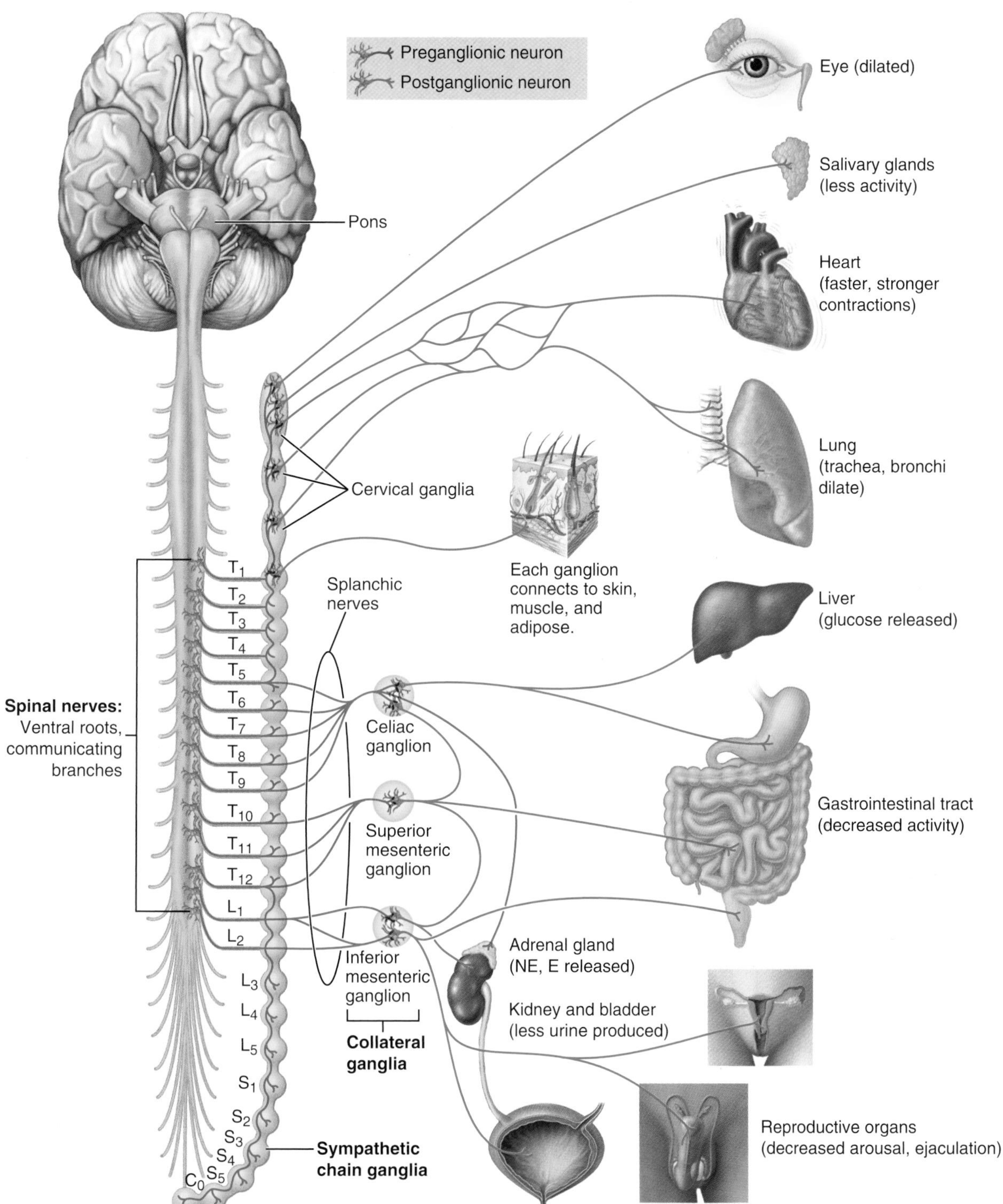

FIGURE 7.30 ▸ Sympathetic motor division of the ANS. The sympathetic nervous division has a paravertebral chain of ganglia and three collateral ganglia. Its preganglionic neurons originate in spinal cord segments T1 through L2.

(▸ FIGURE 7.32). Therefore, the parasympathetic branch of the ANS can also be identified as the ***craniosacral*** branch. Parasympathetic ganglia are located in or near the visceral effectors and are not connected to one another. Because of this, the motor signals from the postganglionic neurons are directed to a specific effector, giving the parasympathetic branch of the ANS a more targeted effect. In this way, the parasympathetic branch is able to make individual adjustments to visceral effectors to maintain homeostasis.

In ▸ FIGURE 7.33, the preganglionic neuron originates in the lateral horn of the spinal cord, and the axon exits at the ventral root. The axon travels to a terminal ganglion located in or near its autonomic effector, where it connects with a postganglionic neuron. The postganglionic neuron then carries the motor impulse to the visceral effector.

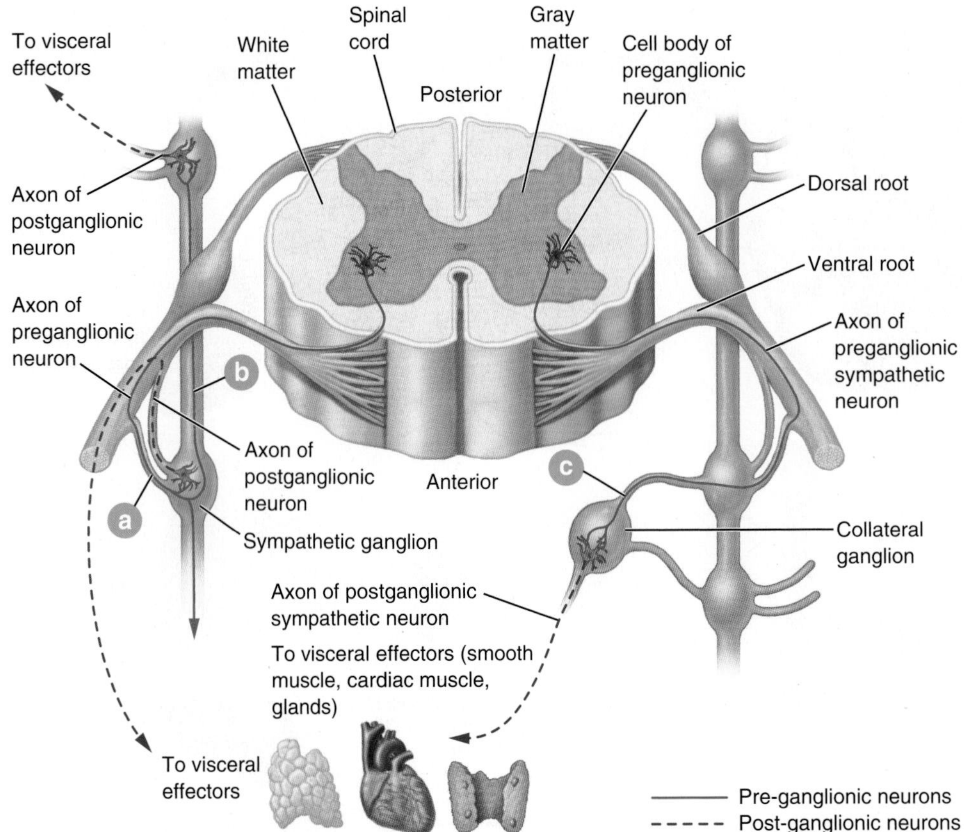

FIGURE 7.31 ▸ Sympathetic pathways. The preganglionic neurons in a sympathetic pathway begin in the lateral horn of the spinal cord. Their axons travel into the sympathetic chain where they synapse with postganglionic neurons (Path a and Path b). In some cases, the preganglionic neuron passes through the ganglionic chain to synapse in a collateral ganglion (Path c).

What Do You Think? 7.6

- What would you say are the longest and shortest motor axons in the peripheral system?
- During a session, how would you adjust your contact, pressure, and rhythm to create invigorating, toning responses rather than relaxation or stress-reduction responses?
- What is the possible intention of ending a back massage by placing one of your hands at the base of the client's skull and the other at their sacrum?

THE NERVOUS SYSTEM AT WORK

Each component of the nervous system serves an essential functional role within the integrated whole. The sensory data provided through receptors and afferent neurons inform us of the ever-changing conditions in our internal and external environments. The spinal cord and brain receive this constant stream of sensory information, organize it into sensations and perceptions, and formulate meaningful responses. Finally, efferent neurons deliver the motor commands of the CNS to a wide array of effectors.

While the names, locations, and functions of the system's components are known and the fact that communication occurs through electrochemical signals is understood, there is still much about the nervous system that cannot be explained. How exactly does the brain translate energy impulses into thoughts, ideas, and emotions? Why are some types of stimuli interpreted as painful and others as pleasurable? While perceptions require conscious awareness of sensations, how is it that they become subconscious through repetition? And how is it that the body functions so effectively without being

By the Way

The neurotransmitter released by both the motor neurons in a parasympathetic pathway is acetylcholine. This is also the neurotransmitter released by a preganglionic neurons in a sympathetic pathway. However, sympathetic postganglionic neurons release norepinephrine, also known as noradrenaline.

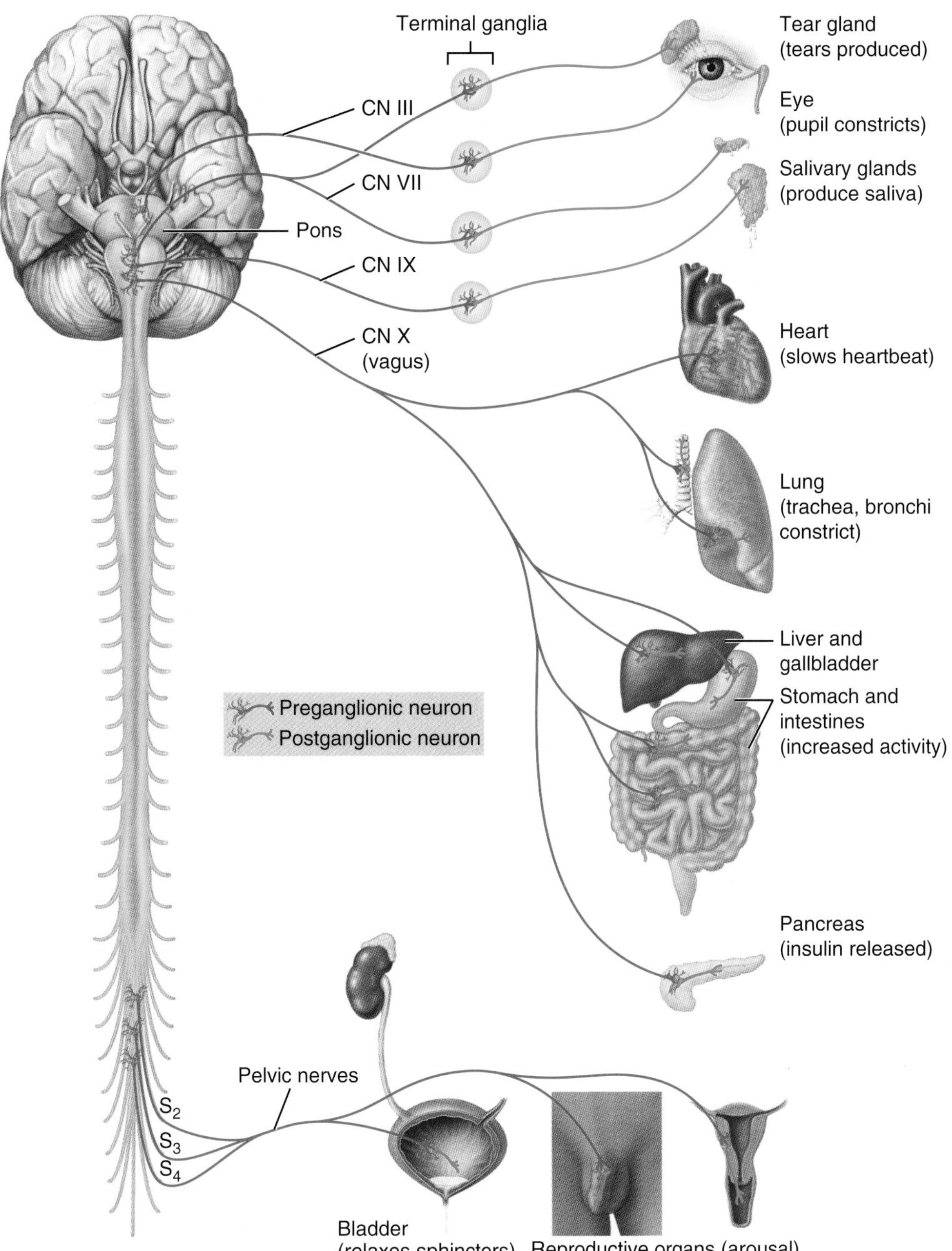

FIGURE 7.32 ▸ Parasympathetic motor division of the ANS. Preganglionic neurons of the parasympathetic division originate in the brain stem (cranial nerves III, VII, IX, and X) and spinal cord segments S2 through S4. Parasympathetic ganglia are located in or near the visceral effectors.

conscious of the vast majority of the brain's activity? Ultimately, understanding the nervous system must also include appreciating the wonder and awe of this incredibly integrated system.

Learning and Memory

Memory is a broad term that refers to the ability to recall a previous sensation, skill, knowledge, or event. Memory and learning are essentially the same physiologic process of developing new synaptic connections between neurons. While humans are born with nearly all the neurons needed for a lifetime, the pathways and networks that represent skills, knowledge, and memories are as yet undeveloped. By experiencing something new, we learn and form memories by making new cognitive connections between neurons. Through repetition, neuronal connections are strengthened and the pathways function more efficiently and subconsciously. Think of a path worn across a field. The path becomes more defined and more familiar each time you walk across, so that you can travel it faster and more confidently, even in fading sunlight.

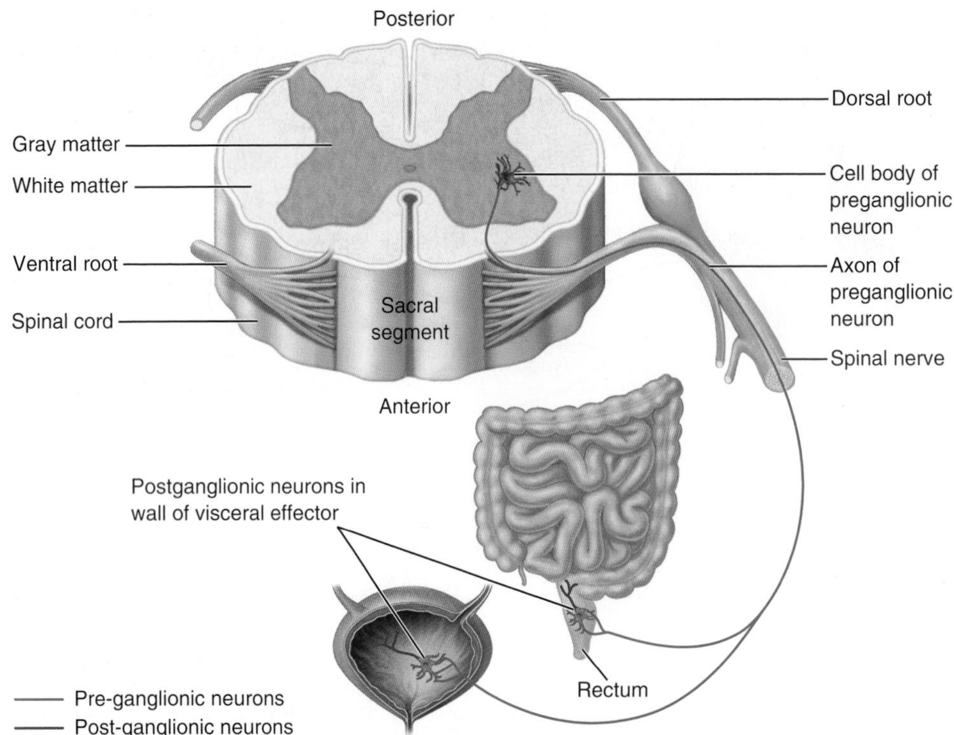

FIGURE 7.33 ▶ Parasympathetic pathways. The preganglionic neurons of parasympathetic pathways arise from the lateral horn of the spinal cord. Their axons travel to a ganglion that is in or near the visceral effector, where they synapse with the postganglionic neurons.

Pain

Pain is a common human experience yet it remains difficult to define and describe. The International Association for the Study of Pain defines pain as an "unpleasant sensory and emotional experience associated with actual or potential tissue damage or described in terms of such damage."[1] While this definition is a bit vague, it does attest to what is known about pain. Pain is generally considered an unpleasant experience that serves as an alarm that something is not right. It is a warning sign that calls us to protect ourselves first by withdrawing from the damaging stimulus and then by guarding the injured body as it heals. Every pain experience is also a learning opportunity. When the cause of pain is consciously considered, the pain can be treated more effectively, and behavior can be modified to avoid or mediate its cause in the future. If we fail to do this and simply silence the pain, it is like "disconnecting a ringing fire alarm to avoid receiving bad news."[2]

Theories of Pain

Throughout history there have been many theories of how pain happens. Proposed in 1965 by Melzack and Wall,[3] the **gate-control theory** was a giant step forward in the understanding of how pain occurs. While their theory was wrong in some of the structural and physiologic specifics, it does provide a useful overall concept of how pain is experienced. Basically, the pain experience can be broken down into three stages:

1. ***Signal***—This initial stage involves the stimulation of nociceptors. Remember that these receptors are sensitive to tissue damage caused by thermal, mechanical, or chemical agents.
2. ***Message***—Once the signal enters the dorsal horn of the spinal cord, its signal is directed along several different pathways. Besides the simple reflex arc for withdrawal, other pathways carry the message to the RAS and hypothalamus, causing arousal and ANS responses. More importantly, some pathways head to the thalamus, where pain signals and all other general sensory information are sorted and prioritized. The thalamus dispatches sensory messages to the limbic system and appropriate regions of the cerebral cortex.
3. ***Perception***—This final stage of the pain experience takes place in the neural network of the cerebral cortex. Here, sensory and associative regions locate the source of the painful stimulus, make links with stored memories, and communicate with the limbic system, producing an emotional response. It is not until this final stage of perception that pain actually exists.

Like all other sensory signals, pain signals enter the CNS in the electrochemical language of nerve impulses. These signals are not marked "Urgent," and they do not gain priority simply because they carry what one might consider important information. Since the spinal cord and lower brain have a limited capacity to deal with multiple sources of sensory stimuli simultaneously, this presents a unique opportunity. If sufficient levels of somatic stimulus such as touch, temperature, pressure, or movement are provided to the body, the thalamus will prioritize and send on those signals first. Because somatic afferent neurons have a larger diameter than pain neurons, they can transmit impulses faster, reaching the CNS first and effectively blocking the pain message to the brain. This is why animals lick their wounds and why people instinctively hold or rub an injured area as an initial response to pain. This gate-control phenomenon also explains how manual therapy can produce beneficial effects in managing pain.

In 1968, Melzack and Casey[4] presented a new conceptual model of pain that pushed neurophysiology and psychobiology to work together in the study of pain. They described pain as a result of the interactions between three dimensions of the nervous system:

- Sensory (sensation)
- Cognitive (thinking)
- Affective (emotional/feeling)

The theory helped to explain how placebos and emotional excitement modulate the pain experience. It also presented a more holistic perspective regarding the causes of pain and challenged physicians to explore less-invasive ways of managing pain by engaging the emotional and cognitive dimensions. This supports the approach of Complementary and Integrative Medicine (CIM) practices, which have always acknowledged the importance of treating the whole person: body, mind, and spirit.

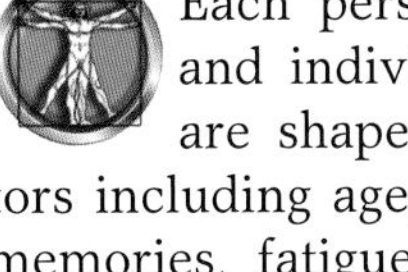

Each person experiences pain in a unique and individual way, since pain experiences are shaped and molded by a range of factors including age, gender, personality, culture, pain memories, fatigue, and overall stress levels. These factors can lead to differences in **pain threshold** (the amount of stimulus required to produce a pain signal) or **pain tolerance** (the amount of pain a person can withstand). This is why the level of pain a client expresses is actually a poor indicator of the severity of a musculoskeletal injury such as a sprain or strain. The severity (mild, moderate, severe) of an injury rates the degree of tissue damage *not* the degree of pain. However, a manual therapist must always respect and trust a client's report of pain or discomfort. For example, if a client experiences discomfort during a session, we need to make adjustments in pressure, pace, direction, or other aspects based on his or her feedback.

Types of Pain

Many terms and classifications have been devised to describe the sensation, cause, location, duration, or even intensity of pain (TABLE 7-4). In general, there are two broad categories based on their underlying etiology (cause). Pain classified as **somatogenic** arises from the physical body, while any pain that arises solely from the mind is considered **psychogenic**.

Somatogenic Pain. Somatogenic pain includes ***nociceptive pain*** caused by thermal, mechanical, or chemical stimulation of pain receptors. Since nociceptors are

Table 7-4 Pain Terminology

Term	Definition
Acute	Pain with sudden onset and/or for a duration of <6 months
Analgesia	Decreased sensation of pain
Analgesic	Substance that decreases pain
Anesthesia	A complete lack of sensation, including pain
Chronic	Pain with gradual onset and/or for a duration of >6 months or longer than expected for healing
Hyperesthesia	Increased sensitivity to stimulation
Hypoesthesia	Decreased sensitivity to stimulation
Idiopathic	Pain of unknown cause or etiology
Neuralgia	Pain along a specific nerve distribution (e.g., sciatica or trigeminal)
Neuritis	Inflammation of a nerve or nerves
Paresthesia	An abnormal prickling or pins-and-needles sensation
Phantom pain	Pain felt as though it arises from an amputated limb
Placebo effect	Decreased pain response due to the belief that the treatment will be effective
Psychogenic	Pain arising from the mind; no known physical cause
Radicular	Radiating pain that shoots out or down an extremity; "electric pain" generally associated with nerve root or cord irritation
Referred	Pain occurring in a predictable but separate region from the affected organ; generally associated with visceral pain
Somatogenic	Pain arising from the body

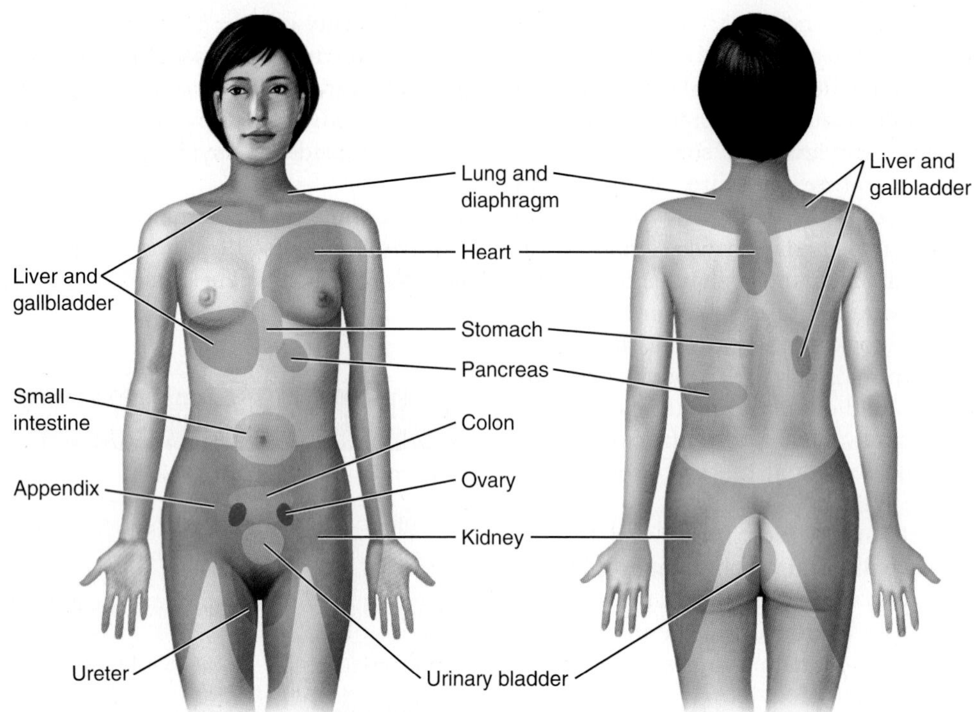

FIGURE 7.34 ▸ Common areas of referred pain. Visceral pain often manifests as pain felt in a body area that is distant from the affected organ.

found in all types of tissue, somatogenic pain can be further divided into three categories:

- ***Superficial*** pains are sharp and localized, caused by stimulation of the nociceptors in the skin and superficial fascia. They include pain experienced from bruising, lacerations, abrasions, or superficial burns.
- ***Deep somatic*** pains occur when the nociceptors in ligaments, tendons, muscles, fascia, bones, or blood vessels are stimulated. Caused by strains, sprains, fractures, or myofascial dysfunction, deep somatic pain tends to be dull, achy, and more diffuse than superficial pain.
- ***Visceral*** pains originate in the organs and are generally described as achy or cramping and are often accompanied by sweating, nausea, or other sympathetic responses. Some visceral pain is localized, such as the back pain that accompanies a kidney infection. However, visceral pain can also be felt in body areas that are distant from the location of the affected organ, a phenomenon called **referred pain**. The internal organs are well protected and have few nociceptors. Referred pain occurs because nerve signals from several areas of the body can travel along the same neural pathways to the spinal cord and brain. This explains why a heart attack can cause an aching or constricting pain in the neck, jaw, or left arm in addition to, or instead of, pain in the chest. ▸ FIGURE 7.34 presents a simplified view of some common locations of referred pain.

By the Way

The brain produces neuropeptides known as ***endorphins***, ***enkephalins***, and ***dynorphins***. These natural opiates can be up to 200 times stronger than morphine. It is thought that acupuncture may help to reduce pain by stimulating their release in the brain.

The somatogenic pain classification also includes ***neuropathic pain*** caused by irritation, compression, or damage to nerve tissue. This can be a burning, tingling, and shooting pain down an extremity or an aching, pins-and-needles sensation (parasthesia). Neuropathic pain can be caused by damage or irritation to the spinal cord and nerve roots, referred to as ***true*** or ***central neuropathic pain***. As discussed earlier in this chapter, ***peripheral neuropathic pain*** is caused by compression or tension on a nerve in the periphery of the body.

Psychogenic Pain. Psychogenic pain is physical pain that does not seem to have an underlying structural cause. Instead, emotional, mental, or behavioral factors cause physical symptoms such as headaches, stomach aches, or back and neck pain. Psychogenic pain is often judged as "less than real," especially by the Western medical community. However, there are common experiences that support its existence: the stomachache a child experiences on the first day of school, or an ache in the chest ("heartache") felt by someone grieving the loss of a spouse. How often can neck tension and soreness be attributed to a person or situation being "a pain in the neck"? These pains are no less real than the pain caused by a kick in the shin. They are simply a warning of a mental or emotional trauma instead of a signal from physical trauma and need to be managed with the same level of respect, attention, and care.

Pain Management

The treatment and management of pain is as varied as the types and descriptions of pain. Drug therapy is widely used and includes over-the-counter pain relievers as well as prescription medication such as opiates, muscle relaxants, sleep aides, antidepressants, and anti-anxiety drugs. Other methods of pain management such as surgery, acupuncture, hydrotherapy, biofeedback, hypnosis, art, music, meditation, visualization, prayer, deep breathing, and laughter have also been used effectively in various situations. So, what are the keys to effective pain management? First, it is important to take a thorough health history and spend time accumulating and documenting key signs and symptoms. Data such as onset, frequency, intensity, duration, and description, as well as factors that increase and decrease the pain are all important to identifying possible underlying pathology. It is also important to remember that pain is rarely **idiopathic**, without an underlying cause, but the cause may be difficult to identify.

If pain has a sudden onset or has been present 6 months or less, it is classified as **acute**. Acute pain is generally nociceptive in nature and is caused by trauma. The mechanism of injury and description of pain are usually obvious and can be precisely described by the client. Measurable objective signs, including swelling, heat, redness, and a loss of mobility and/or strength, are often present. Acute pain is also associated with rapidly developing infections such as appendicitis, tonsillitis, and meningitis.

Chronic pain is pain that has continued for more than 6 months or that has lasted longer than the expected healing period after an injury or trauma. Again, a thorough health history, interview, and physical assessment are essential, although physical findings are often absent or difficult to identify. Because chronic pain can be either somatogenic or psychogenic, curiosity and careful evaluation are key to determining the pain's etiology and identifying treatment options. Chronic pain syndromes can be divided into two general categories:

- ***Systemic conditions*** such as fibromyalgia, irritable bowel syndrome, chronic fatigue syndrome, and rheumatoid arthritis.
- ***Structural conditions*** such as plantar fasciitis, NCTS, and repetitive sprains/strains leading to chronic low back or neck pain.

Whether pain is acute, chronic, physical, mental, or emotional, manual therapy has proven to be an effective treatment for its reduction. Although the exact mechanisms of action have not been unequivocally determined, there are manual therapy techniques that can be used to intervene at each stage of the pain experience. For example, lymphatic techniques may decrease nociceptive stimulation in the signal stage by removing the irritating chemicals released by damaged cells and reducing swelling. As discussed previously, massage and movement therapies can override the message stage of pain by flooding the spinal cord and lower brain with other sensory input. Manual therapies may have their most powerful effect in the perception stage of pain by creating an environment and experience that help a client reconnect mind, body, and spirit to change the perception. The take-home lesson may be that even if the only measurable effect of manual therapy is that it helps a client "feel good," it is still a profound therapeutic modality for pain management.

What Do You Think? 7.7

- What common terms do you use to describe different kinds of pain?
- Why do you think some therapists use heat and/or cold therapies when treating clients for pain?
- Understanding the differences in pain perception, how do you think you should respond to a client's request for deeper pressure when you sense that the tissue being treated is at its limit or already resisting your pressure?

AGING AND THE NERVOUS SYSTEM

While most of the body's neurons remain healthy as we age, there is a slight decaying of the myelin sheaths and impulse conduction slows. Between the ages of 20 and 90, the overall mass of the brain decreases by 5% to 10% as the number of synaptic connections and pathways decrease. These common age-related changes lead to a general slowdown in thought processes, as well as declines in memory, balance, and coordinated movement. However, the rate at which these changes occur and the extent of their functional impact can be influenced by lifestyle choices. For example, regular aerobic exercise, adequate sleep, good nutrition, and engaging in enjoyable and mentally challenging activities have all been shown to be important for healthy aging of the brain and nervous system.

Pathology Alert Dementia

Dementia is caused by degeneration of neural tissue in the frontal and temporal lobes. It is characterized by a progressive loss of intellectual ability with impairments in judgment, cognitive processing, and memory. It may also be accompanied by personality changes that may lead to embarrassing social behaviors. The signs and symptoms of dementia are associated with several pathologies, including Alzheimer disease. It is *not* a normal part of age-related changes to the nervous system. Postmortem brain scans of patients with dementia show two identifiable changes: large deposits of ***plaque*** made of a naturally occurring cellular protein that is particularly adhesive to brain tissue, and a high number of neuronal ***tangles*** created by the degeneration and collapse of long neuronal fibers in the CNS. These two changes in brain tissue mean that dementia sufferers have a reduced number of normal-functioning neurons.

Manual therapy is neither indicated nor contraindicated in cases of dementia. The stress-reduction and nurturing effects of several manual therapies may be beneficial to some clients and perhaps improve their quality of life by decreasing the disruptive nature of their symptoms. However, unfamiliar surroundings or disruption of daily routines may disorient or threaten clients with dementia and negate the benefits of manual therapy.

SUMMARY OF KEY POINTS

- The sensory, integrative, and motor functions of the nervous system coordinate and control the functions of all body systems.
- The nervous system has two major divisions: central and peripheral. The brain and spinal cord are the primary structural components of the CNS, while sensory receptors and nerves are the major PNS components. Nervous tissue in the CNS and PNS is made up of impulse-conducting neurons, held together and protected by glial cells.
- Neurons have three structural elements: dendrites that receive stimulus, the cell body that houses the nucleus and organelles, and the axon, which carries nerve impulses away from the cell body. This means that impulse conduction is always one-way: dendrite to cell body to axon.
- Neuroglia support and protect the neurons. CNS glial cells include astrocytes, oligodendrocytes, microglia, and ependymal cells. PNS glial cells include the Schwann cells that make myelin and satellite cells that help form ganglia.
- A nerve is a bundle of neurons and their blood supply. Nerves in the PNS are divided into two major categories: cranial—12 pairs of nerves that emanate from the brain, and spinal—31 pairs of nerves emanating from the spinal cord.
- Nerves can have only sensory or motor neurons or can carry both types to form a mixed nerve. All spinal nerves are mixed. Cranial nerves include those that function as sensory only, motor only, or mixed.
- The cranial nerves of primary interest to manual therapists are cranial nerve V (trigeminal), VII (facial), X (vagus), and XI (accessory) because of their location and association with a few common pathologies.
- Spinal nerves in the cervical, upper thoracic, lumbar and sacral regions converge into common networks called plexuses that innervate the same region of the body. The four major plexuses are:
    - Cervical—innervates the neck, head, and face
    - Brachial—innervates the upper extremities
    - Lumbar—innervates the lumbar region of the back and gluteals region
    - Sacral—innervates the hips and lower extremities
- Nerve impulse conduction is all-or-none and a one-way transmission. The process begins when threshold stimulus is applied to the neuron. The conduction of an impulse, or action potential, is a complex process marked by two key events: depolarization and repolarization.
- Neuronal pathways are formed by impulses traveling along a specific route to a particular location. They range from simple pathways called reflex arcs, in which sensory stimulus immediately initiates a motor response, to more complex pathways involving neuronal pools and converging or diverging circuits.
- Sensory receptors are classified by the type of stimulus they are sensitive to, forming six categories: photoreceptors, chemoreceptors, nociceptors, thermoreceptors, mechanoreceptors, and proprioceptors. Proprioceptors are actually a highly specialized group of mechanoreceptors.
- General sense receptors are those involved in sensing touch, temperature, pressure, and pain. Special sense receptors are involved in vision, hearing, smell, taste, equilibrium.

- The spinal cord relays information between the brain and the body, and serves as the body's primary reflex center. The posterior (dorsal) portion of the spinal cord houses the sensory/ascending tracts, while the motor/descending tracts are located in the anterior (ventral) portion of the cord.
- The four regions of the brain and the primary functions of each are
  - Brain stem—Relays information between different regions of the brain and between the spinal cord and brain. Also serves as the junction point for the cerebrum and cerebellum and the point of emanation for most cranial nerves.
  - Diencephalon—Contains the thalamus, the sensory clearinghouse, and the hypothalamus, the control center for autonomic functions.
  - Cerebrum—Serves as the center for cognition, consciousness, and motor activity.
  - Cerebellum—Coordinates voluntary muscle activity and maintains muscle tone, posture, and equilibrium.
- The limbic system is a region of the brain that includes the cingulate gyri, hippocampus, amygdala, and mammillary bodies. It is called the "emotional brain" because it is responsible for processing memories and emotions and controlling the unconscious responses related to survival.
- The PNS has two divisions distinguished by differences in their motor pathways and effectors:
  - Somatic—Have one motor neuron between the spinal cord and effector, and only one type of effector; skeletal muscle.
  - Autonomic—Have two motor neurons and a ganglion at the synapse between the two that holds the dendrites and cell bodies of the postganglionic neurons. The glands, organs, and smooth muscles in the body are all autonomic effectors.
- The autonomic motor division has two branches:
  - Sympathetic—This branch is known as the fight-or-flight branch because it is the branch used to meet the demands of stress. The sympathetic ganglia are located in a connected chain situated just lateral and anterior to the spine. Because the ganglia are connected, the preganglionic impulse diverges to stimulate many different postganglionic neurons, producing a body-wide response.
  - Parasympathetic—This branch is in charge of maintaining and returning the body to homeostasis after stress; it is nicknamed the "rest and digest" or "feed-and-breed" branch. The parasympathetic ganglia are located in or near their visceral effector, which produces a more targeted response in the effectors.
- Most visceral effectors have dual innervation, that is, both sympathetic and parasympathetic branch innervation. The exceptions to dual innervation—adrenal glands, sweat glands, and vasomotor response (smooth muscles in blood vessels)—have only sympathetic innervation.
- There are three stages of pain: signal, message, and perception. The perception of pain varies widely and can be based on several factors, including the type and intensity of the signal, the emotional and psychological status of the individual, and memories of prior pain.
- Many terms and classifications are used to describe the types, origins, and sensation of pain. The two major types of pain are somatogenic pain (arising from the body) and psychogenic pain (arising solely from the mind).
- Somatogenic pain includes superficial, deep somatic, and visceral pain.
- Referred pain is visceral pain felt in one or more body areas separate from the affected organ.
- There are many pain management theories that help therapists understand and explain how manual therapy reduces pain, including the gate-control theory, decreasing sympathetic tone, and increasing the level of natural opiates or pleasure hormones.

REVIEW QUESTIONS

Short Answer

1. List and give a brief explanation of the three functions of the nervous system.
2. List the three types of neurons by function and the location of each.
3. Name four types of glial cells and describe the basic function of each.
4. Name the four regions of the brain.
5. What are the three key distinctions between cranial and spinal nerves?
6. List and briefly explain the key steps of nerve impulse conduction.

Continued on page 208

Multiple Choice

7. What is the name of cranial nerve V?
 a. facial
 b. trigeminal
 c. vagus
 d. optic
8. The C-8 nerve is positioned between which vertebrae?
 a. C-6 and C-7
 b. T-1 and T-2
 c. C-7 and T-1
 d. C-8 and T-1
9. Which spinal nerve plexus innervates the upper extremity?
 a. brachial
 b. cervical
 c. thoracic
 d. lumbar
10. Major nerve branches off the lumbar plexus include the ilioinguinal, obturator, and
 a. peroneal and tibial
 b. sciatic and femoral
 c. femoral and lateral cutaneous
 d. greater and lesser saphenous
11. Which region of the brain is responsible for coordinating voluntary muscles and maintaining posture?
 a. cerebrum
 b. pons
 c. medulla oblongata
 d. cerebellum
12. What is another term for a nerve impulse?
 a. polarization
 b. action potential
 c. threshold stimulus
 d. depolarization
13. What forms the chemical bridge that acts as a stimulus for an effector or postsynaptic neuron?
 a. neurotransmitters
 b. sodium ions
 c. neurilemma
 d. repolarized ions
14. When one neuron stimulates an entire group of postsynaptic neurons, it is an example of which type of neuronal circuit?
 a. postsynaptic pool
 b. convergence
 c. divergence
 d. neural spray
15. Which of the following is an example of a somatic effector?
 a. heart
 b. stomach
 c. blood vessels
 d. biceps femoris
16. Which of the following are the receptor cells for vision?
 a. cochlear villa
 b. olfactory cells
 c. rods and cones
 d. gustatory cells
17. Which of the special sense receptors are examples of chemoreceptors?
 a. taste and smell
 b. pressure and auditory
 c. vision and equilibrium
 d. taste and vision
18. Which type of neurons form the sympathetic and parasympathetic pathways?
 a. one sensory and one motor
 b. one each of sensory, motor, and associative
 c. one associative and one motor
 d. two motor neurons
19. Where are the specialized mechanoreceptors responsible for equilibrium located?
 a. muscles and tendons
 b. joint space and capsule
 c. vestibule and semicircular canals of inner ear
 d. cochlea of the inner ear
20. To what type of stimulus are muscle spindles most sensitive?
 a. rapid lengthening
 b. sudden shortening
 c. slow lengthening
 d. alternating short and long

21. Which type of tissue makes up the white matter of the spinal cord?
    a. fibrous connective tissue
    b. dendrites and cell bodies of motor neurons
    c. axons of sensory and motor neurons
    d. interneurons and microglia

22. Which of the following terms are synonymous to sensory tracts?
    a. descending and ventral tracts
    b. ventral and anterior tracts
    c. lateral and visceral tracts
    d. ascending and dorsal tracts

23. In which layer of meninges does the cerebrospinal fluid circulate?
    a. dura mater
    b. arachnoid
    c. pia mater
    d. subdural space

24. The functions of cerebrospinal fluid are to act as a shock absorber, create a physical barrier between blood and brain, and
    a. act as the medium for nutrient waste exchange
    b. carry out all immune processes in CNS
    c. provide lubrication to the brain and cranial nerves
    d. serve as the primary neurotransmitter for brain and spinal cord

25. What kind of dysfunction is most likely to occur when the anterior root of a spinal nerve is damaged?
    a. full loss of sensation in the associated dermatome
    b. partial loss of sensation in a few dermatomes
    c. loss of movement and/or muscle strength
    d. paresthesia and decreased smooth muscle function

26. Which part of the brain stem serves as the center for the cardiac, respiratory, and vascular motor reflexes?
    a. midbrain
    b. hypothalamus
    c. pons
    d. medulla oblongata

27. Virtually all sensory information is sorted, prioritized, and routed through which structure in the brain?
    a. hypothalamus
    b. thalamus
    c. medulla oblongata
    d. pineal gland

28. What is the function of the corpus callosum?
    a. produce cerebrospinal fluid
    b. relay information between cerebellum and cerebrum
    c. junction between the two cerebral hemispheres
    d. center for emotions such as anger, fear, and anxiety

29. Which region of the brain is known as the "emotional brain"?
    a. cerebral cortex
    b. corpus callosum
    c. hypothalamus
    d. limbic system

30. The control center for autonomic functions of the nervous system is the
    a. cerebellum
    b. hypothalamus
    c. thalamus
    d. medulla oblongata

References

1. International Association for the Study of Pain. http://www.iasp-pain.org.
2. Yancey P, Brand P. *The Gift of Pain*. Grand Rapids, MI: Zondervan Publishing House; 1997:188.
3. Melzack R, Wall PD. Pain mechanism: a new theory. *Science* 1965;150:971–979.
4. Melzack R, Casey KL. Sensory, motivational and central control determinants of chronic pain: a new conceptual model. In: Kenshalo D, ed. *The Skin Senses*. Springfield, IL: Chas C. Thomas;1968:432.

8 Neuromuscular and Myofascial Connections

LEARNING OBJECTIVES

Upon completion of this chapter, you will be able to:

1. Discuss the importance of understanding key neuromuscular and myofascial connections in the practice of manual therapy.
2. Describe the two neuronal loops utilized by muscle spindles to moderate muscle tension.
3. Explain how manual therapists can use their knowledge of reciprocal inhibition, stretch reflex, and gamma gain to reduce muscle tension.
4. Describe the different physiologic mechanisms involved in the development of tender points and trigger points and what implications they have on manual therapy choices.
5. Explain the concept of tensegrity as it applies to the human body.
6. Explain the general location and function of the layers, bands, and planes of the fascial system.
7. Explain the functional importance of myofascial chains and how knowledge of these chains might affect therapeutic choices.
8. Explain the key mechanical properties of fascia and how knowledge of these properties might affect therapeutic choices.
9. Name the four types of mechanoreceptors found in fascia.
10. Discuss how knowledge of the mechanoreceptors and smooth muscle cells in fascia might affect therapeutic choices.
11. Describe the neuromuscular and neurofascial mechanisms related to maintaining posture, coordinating movement, and regulating muscle and motor tone.
12. Explain the difference between motor unit and muscle recruitment and between myofascial and kinetic chains.

The body as a whole, the organs, tissues, cells, organelles, including the nucleus, and the strands of genetic material, DNA, can be viewed as a continuous and unbroken fabric: a matrix within a matrix within a matrix."

James L. and Nora H. Oschman
Somatic Recall

KEY TERMS

alpha (AL-fah) **loop**
deep fascia (FAH-shah)
fascial (FAH-shul) **band**
fascial (FAH-shul) **plane**
fascial plasticity (FAH-shul plas-TIS-eh-te)
fascial (FAH-shul) **tone**
gamma (GAM-a) **gain**
gamma (GAM-a) **loop**
hypertonicity (hi-pur-to-NIS-eh-te)
interstitial myofascial (in-tur-STIH-shul mi-o-FAH-shul) **tissue receptors**
myofascial (mi-o-FAH-shul) **chain**
neurofascial (Nu-ro-FAH-shul)
neuromuscular (nu-ro-MUSK-u-lar) **reflex**
phasic (FA-zik) **muscle**
piezoelectricity (PE-zo-e-lek-TRIS-eh-te)
postural muscle
superficial fascia
tender point
tensegrity (ten-SEG-rih-te)
thixotropy (THIKS-ah-tro-pe)
trigger point
viscoelasticity (VIS-ko-e-las-TIS-ih-te)

In the same way that an auto mechanic takes apart an engine to learn the structure and function of each part, subdividing the body into systems allows us to identify and study the cells, tissues, and organs that make up the human organism. Initially, this compartmentalized approach supports the learning of basic anatomy and physiology concepts. Yet this detail-oriented analysis can lead us to forget that everything in the body is connected. While it is necessary to consider the structure and functions of each individual system, it is equally important to broaden our understanding and comprehend the body as an integrated whole.

This chapter is designed as a holistic checkpoint, an opportunity to look at how the systems discussed thus far are structurally connected and how they work together. With manual therapy applications as a lens, this chapter outlines important neuromuscular and myofascial relationships and explores physiologic explanations for the methodology and effectiveness of established therapies. Regardless of whether the therapeutic approach is massage, other forms of bodywork, or movement or somatic education, therapists move and/or manually manipulate the skin, muscle, and fascia of clients, stimulating sensory receptors that constantly present new information to the brain. Therefore, a client's experience of relaxation and a greater sense of ease may be due to several factors: resetting the motor tone of tight muscles by accessing neuromuscular loops; loosening, stretching, and broadening connective tissues; or decreasing sympathetic tone through gentle movement and caring touch to produce a general state of relaxation. To understand the physiologic mechanisms of different forms of therapy, the connections between the integumentary, musculoskeletal, and nervous systems must be explored. In holistic practices, connections between body, mind, and spirit must also be considered to understand and respect the therapeutic power of touch and/or movement. This chapter is designed to be an exploration point, creating the opportunity to pause and ponder the connections between the integumentary, musculoskeletal, and nervous systems and how they help explain some of the benefits and effects of specific manual therapy techniques.

NEUROMUSCULAR REFLEXES

Neuromuscular reflexes that control and coordinate skeletal muscle contraction include simple reflex arcs as well as more complex neuronal pathways involving multiple interneurons in the spinal cord and brain. These reflex pathways are interesting because muscles serve as both the sense receptor and the effector that responds to the motor command. Muscles are involved in monitoring and moderating their own activity, since the brain is receiving signals from and sending signals to the muscles. In Chapters 6 and 7, the process of muscle contraction and the structures of a reflex arc were discussed. This chapter covers the neuromuscular connections that moderate skeletal muscle activity and how these principles can be applied in manual therapy.

Reciprocal Inhibition

The reflex mechanism that coordinates the effort between agonist and antagonist muscles is called **reciprocal inhibition**. For an agonist to create a desired movement, the antagonist must be sufficiently inhibited to allow smooth and coordinated movement. Because agonist and antagonist have reciprocal innervation when a motor command is given, the muscles receive simultaneous signals; the agonist is stimulated to contract, while contraction of the antagonist is inhibited. The antagonist cannot be fully inhibited (fully relaxed) because some tension is required to provide a braking action to control the speed and force of the movement created by the agonist. Recall the example from Chapter 6 in which the trunk extensors apply enough resistance to prevent a person from falling over when bending forward to tie his or her shoes.

Reciprocal inhibition can be utilized by therapists to relieve excessive muscle tension, a state called **hypertonicity**. In the case of acute muscle cramps, the use of reciprocal inhibition is safer than simply stretching the muscle because it allows the cramp to relax a little before fibers are pulled apart with a stretch. By engaging the antagonist of the cramping muscle with an isometric contraction, the target muscle (cramping) receives a relaxation signal (FIGURE 8.1). Reciprocal inhibition is also an effective tool for reducing the chronic hypertonicity of muscle spasms. For example, a common complaint is tension and pain in the back of the neck. Reciprocal inhibition is a very practical way to alleviate some of the tension before any soft tissue manipulation is done. Engaging the neck flexors (or the opposite side lateral flexors) in a light isometric contraction will inhibit the tension in the tight posterior muscles.

Stretch Reflex and Gamma Gain

Recall that muscle spindles are proprioceptors that sense lengthening or stretching of muscle fibers and signal a reflexive contraction of the muscle called the stretch reflex. Muscle spindles sense lengthening due to two key structural features: their parallel alignment with the extrafusal fibers and connective tissue attachments between the capsule of the spindle and endomysium and perimysium of the muscle.

The muscle spindle has two distinct portions linked to different neuronal pathways, one innervates extrafusal fibers and another innervates the intrafusal fibers (FIGURE 8.2). The **alpha loop** is made up of the ***alpha sensory neuron*** looped around the central region of the muscle spindle, and the ***alpha motor neuron*** that innervates the extrafusal fibers. The alpha loop is the reflex

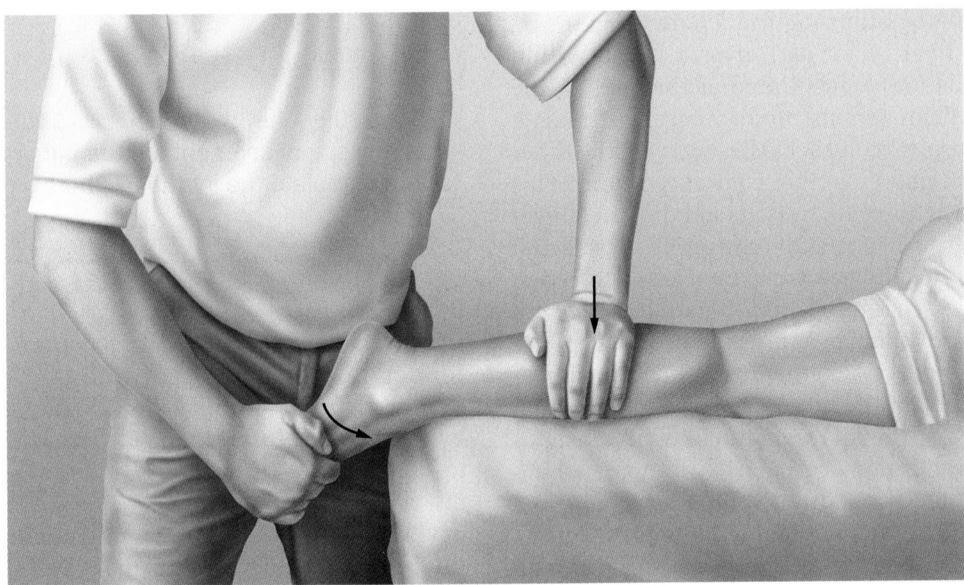

FIGURE 8.1 ▶ Reciprocal inhibition for cramp relief. To relieve a cramp in the gastrocnemius, the therapist employs direct compression and reciprocal inhibition simultaneously. As the therapist firmly presses down on the cramp with one hand, the client is told to pull his foot against the therapist's other hand. This engages the antagonist muscles (dorsiflexors) and inhibits the gastrocnemius.

arc for the stretch reflex. In addition to being part of this reflex arc, the alpha neurons are also part of the sensory and motor pathways used by the primary motor centers of the cerebrum and cerebellum to control and coordinate voluntary movement.

The second reflex arc, called the **gamma loop**, is composed of a ***gamma sensory neuron*** attached toward the ends of the spindle, and the ***gamma motor neuron*** that innervates the intrafusal fibers themselves. The gamma loop regulates the sensitivity of

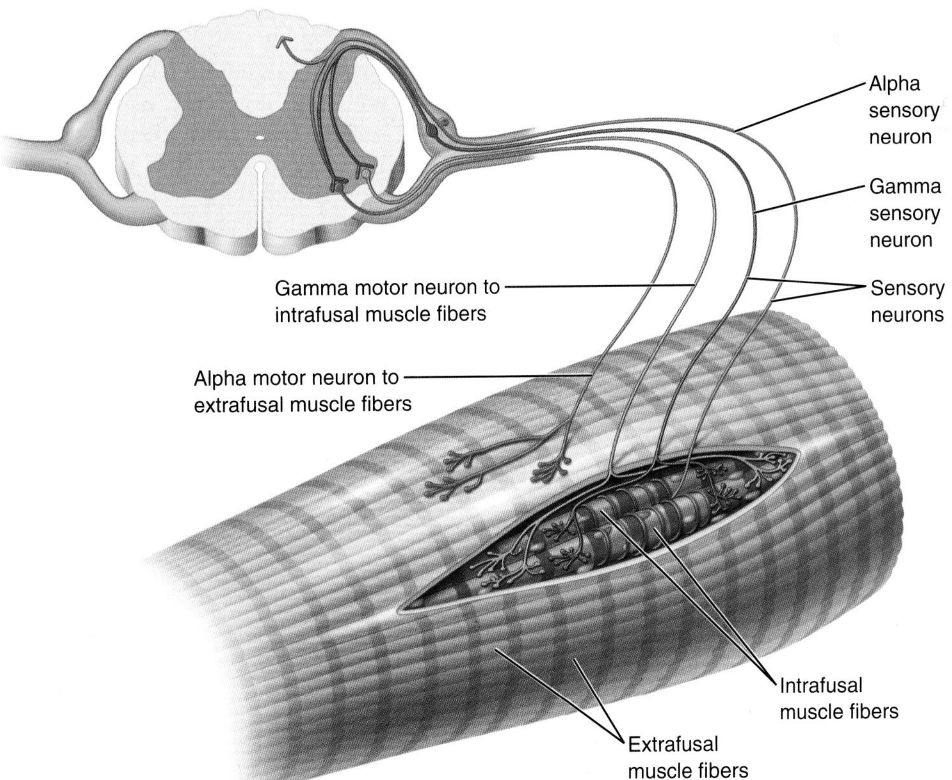

FIGURE 8.2 ▶ Neuromuscular neural loops. The muscle spindle uses two different neuronal loops to monitor muscle length. The alpha loop innervates the extrafusal fibers and manages the stretch reflex, while the gamma loop adjusts the tension in intrafusal fibers to increase muscle spindle sensitivity to rapid lengthening.

the muscle spindle by adjusting the length and tension of the intrafusal fiber. Actin and myosin found within the muscle spindle are concentrated at the ends of the intrafusal fiber. When gamma motor neurons stimulate the intrafusal fiber to contract, the ends of the fiber pull lengthwise, tightening the central portion. This increased tension within the muscle spindle effectively increases its sensitivity so lengthening of the muscle will quickly stimulate the alpha sensory neuron. Referred to as **gamma gain** or gamma loading, this increased tension of the intrafusal fibers increases the proprioceptor's sensitivity to the rate of lengthening. The intensity of the report (signal) from the muscle spindle to the spinal cord is directly proportional to the rate at which lengthening occurs. Gamma gain is especially pronounced if a muscle has been static or shortened for a period of time (▸ FIGURE 8.3). In this case, the high gamma gain causes the muscle spindle to signal an immediate and strong protective contraction of the muscle, not because the muscle is stretched, but because of how quickly its fibers were lengthened. Although slow sustained lengthening of a muscle can also stimulate the stretch reflex, the contraction it stimulates is very mild.

By the Way

A full 30% of *all* motor neurons in the ventral horn of any spinal segment are the gamma efferents that stimulate the intrafusal fibers of the muscle spindle.[1]

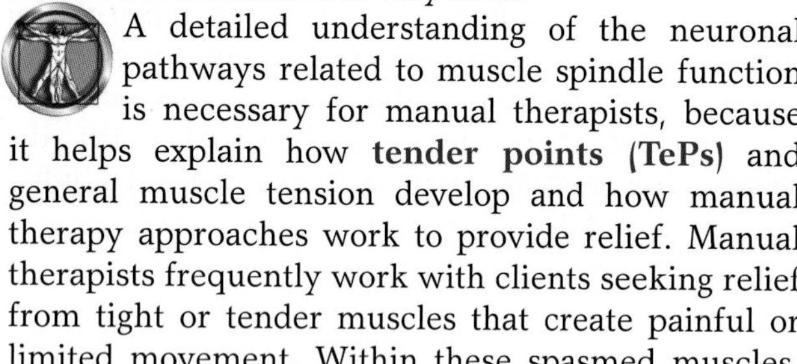

A detailed understanding of the neuronal pathways related to muscle spindle function is necessary for manual therapists, because it helps explain how **tender points (TePs)** and general muscle tension develop and how manual therapy approaches work to provide relief. Manual therapists frequently work with clients seeking relief from tight or tender muscles that create painful or limited movement. Within these spasmed muscles, small nodules or localized areas of tenderness called **tender points** and **trigger points (TrPs)** are commonly detected. While both TePs and TrPs are hypersensitive to mechanical pressure, they appear to be created through very different physiologic mechanisms (▸ TABLE 8-1). TePs are most likely to develop in relationship to gamma gain, which leads to a false stretch reflex signal. When a muscle is held in a shortened position and gamma gain kicks in, sudden lengthening of the muscle stimulates the stretch reflex even though the muscle is not fully stretched. The muscle is then locked into a slight concentric contraction by this neuromuscular reflex, and TePs are likely to develop. Proper treatment must include repositioning the muscle to desensitize and reset the muscle spindle (▸ FIGURE 8.4). Efforts to either actively or passively force the spasmed muscle into a longer position are strongly resisted by the motor tone of the muscle, explaining why stretching often fails to relieve the pain and/or restricted movement experienced by the client. Neuromuscular techniques such as Strain-Counterstrain,

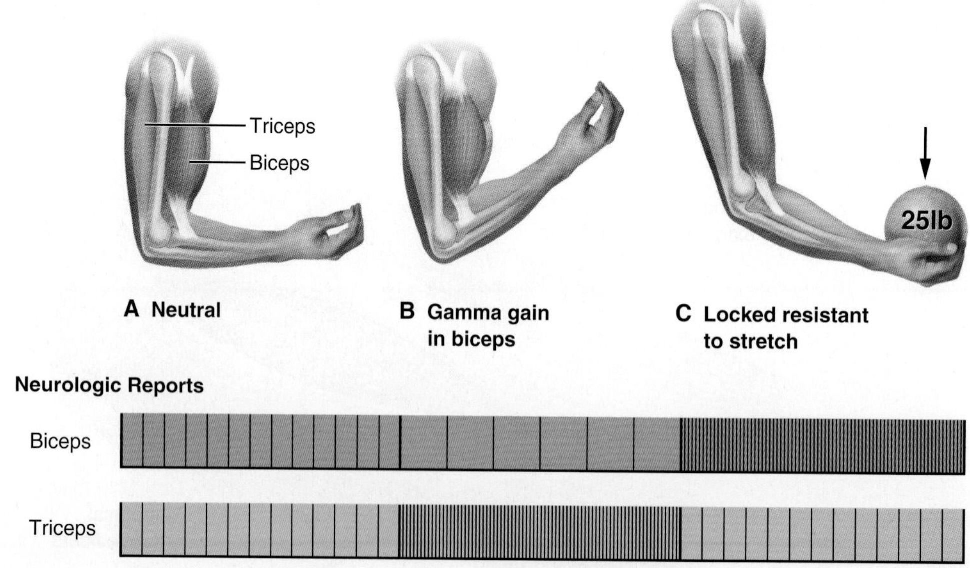

FIGURE 8.3 ▸ Gamma gain. A. When the elbow is held in a neutral position, the neurological report from both the biceps and triceps brachii muscles is low. **B.** When the elbow is flexed, the neurological report from the shorter biceps is decreased, while the report from the lengthened triceps is increased. During this shortened phase, the gamma loops of the spindles in the biceps increase their sensitivity to lengthening (gamma gain). **C.** Due to gamma gain, sudden lengthening of the biceps (weight suddenly dropped into the hand) stimulates a stretch reflex contraction of the muscle even though it is not fully lengthened.

Table 8-1 Characteristics of TrPs and TePs

Type	Pathophysiology	Characteristics
TrPs	Motor end plate dysfunction causes "calcium spill" into sarcomere leading to actin–myosin bonding	Palpable nodule in a taut band of muscle Hypersensitive to mechanical pressure Pressure reproduces pain complaint in a predictable pattern Causes muscle to be hypersensitive to stretch
TePs	High gamma gain in muscle spindle signals contraction in response to sudden lengthening	May or may not be palpable Hypersensitive to mechanical pressure No referred pattern of pain; often silent until compressed Causes muscle to be resistant to stretch

Positional Release, Functional Technique, and Tender Point Release employ repositioning of the muscle or body region to reset the neuromuscular signal and reduce muscle tension and spasm.

In contrast, the pathophysiology of TrPs does not involve neurologic signals. Instead, TrP development seems to be related to a chemical irritation at the motor end plates of several fibers. A variety of factors such as overload stress, overt injury, or underuse of a muscle, can cause acetylcholine to be released at the motor end plate. The acetylcholine stimulates the release of calcium from the sarcoplasmic reticulum, which then allows actin and myosin bonding and the shortening of several sarcomeres in the area. This produces a palpable nodule or ***contraction knot*** that is one of the defining characteristics of TrPs. The sustained contraction of sarcomeres in a TrP creates a maximum energy demand that uses up the stored ATP within the

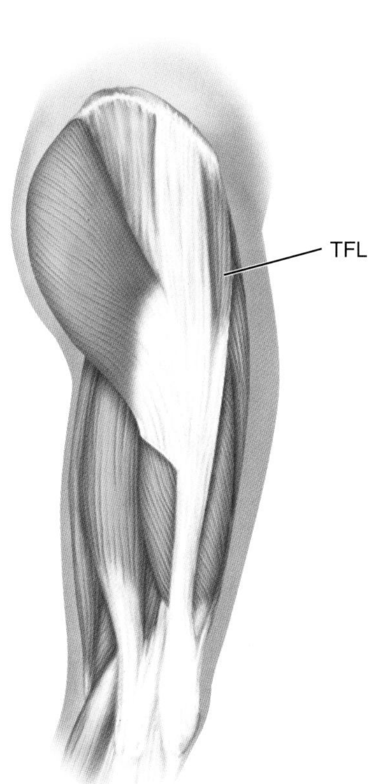

FIGURE 8.4 ▸ Tender point release. A. Location of a common TeP in the tensor fasciae latae. **B.** To release the TeP, the muscle must be returned to a shortened position to desensitize and reset the muscle spindle. For this TeP in the TFL, the primary position is hip flexion, while internal rotation and abduction of the hip are used as fine tuning movements to find the best position of ease.

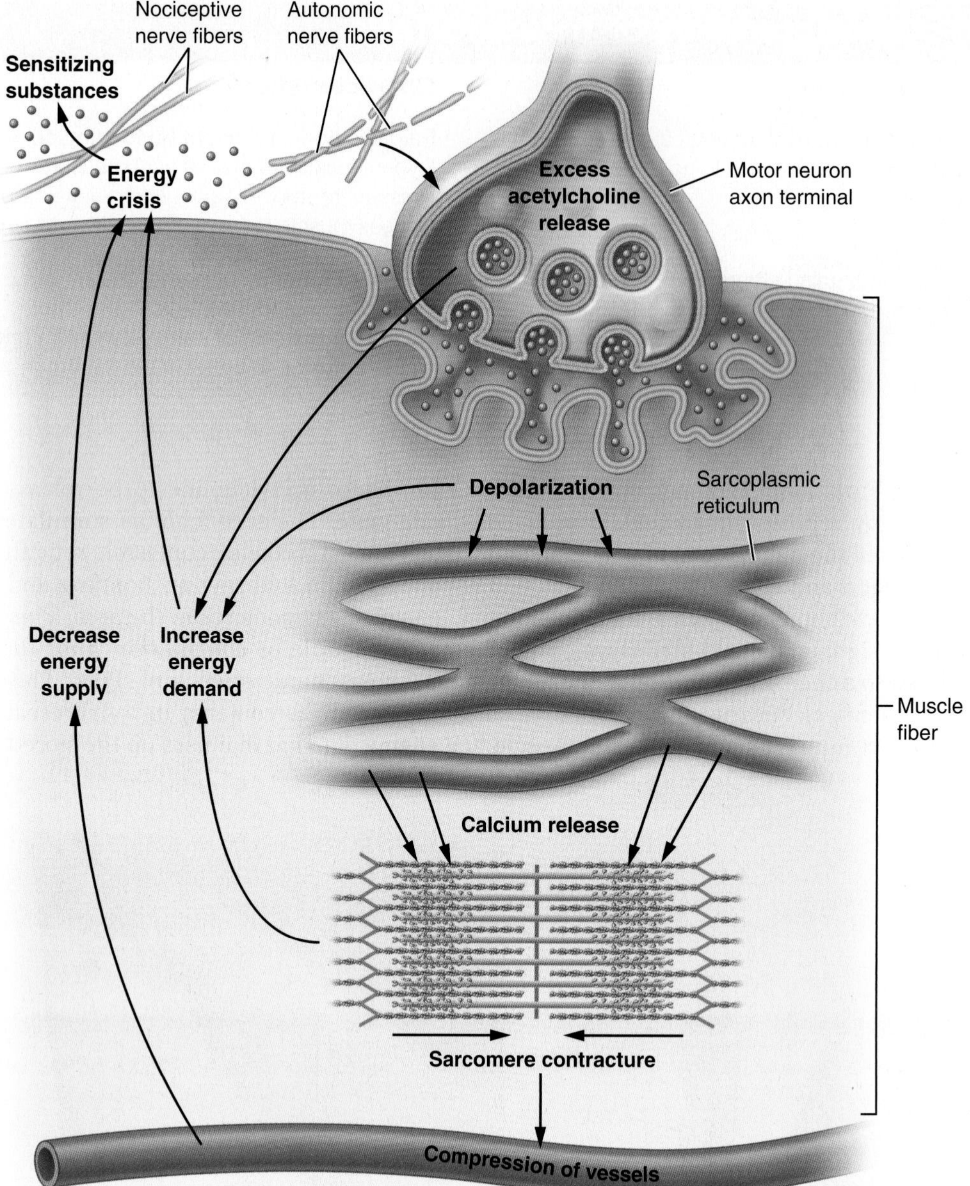

FIGURE 8.5 ▸ Energy crisis related to TrPs is a self-sustaining cycle. Motor end plate dysfunction results in excessive calcium release from the sarcoplasmic reticulum, which allows actin and myosin bonding. The shortening of several sarcomeres in muscle fibers increases the demand for energy to sustain the contraction, while at the same time, the energy supply is decreased due to the restriction of circulation from the contraction.

muscle fibers, while simultaneously constricting blood flow in the area. This creates a metabolic crisis leading to early muscle fatigue[2] (▸ FIGURE 8.5). The combined influences of motor end plate dysfunction, metabolic crisis, and a low-grade inflammatory response in cases of overt injury explains how and why TrPs develop their characteristic taut band, palpable nodule, and hypersensitivity.[2–4] Because the TrP and local muscle spasm are not due to a neurologic signal, repositioning is not absolutely necessary, even though it has been demonstrated to be helpful in some cases. Most TrP treatment protocols focus on mechanically separating the contracture knot of the sarcomeres with moderate compression and direct tissue manipulation (▸ FIGURE 8.6). While it is not essential for manual therapists to make an absolute determination of whether the hypersensitive point they are working with is a TrP or TeP, it can be very helpful in directing and making adjustments to the treatment, especially when the chosen approach doesn't seem to be working. At the very least, the differences between these two types of hypersensitive points support the concept that therapists cannot solve all problems with just one manual therapy form or technique.

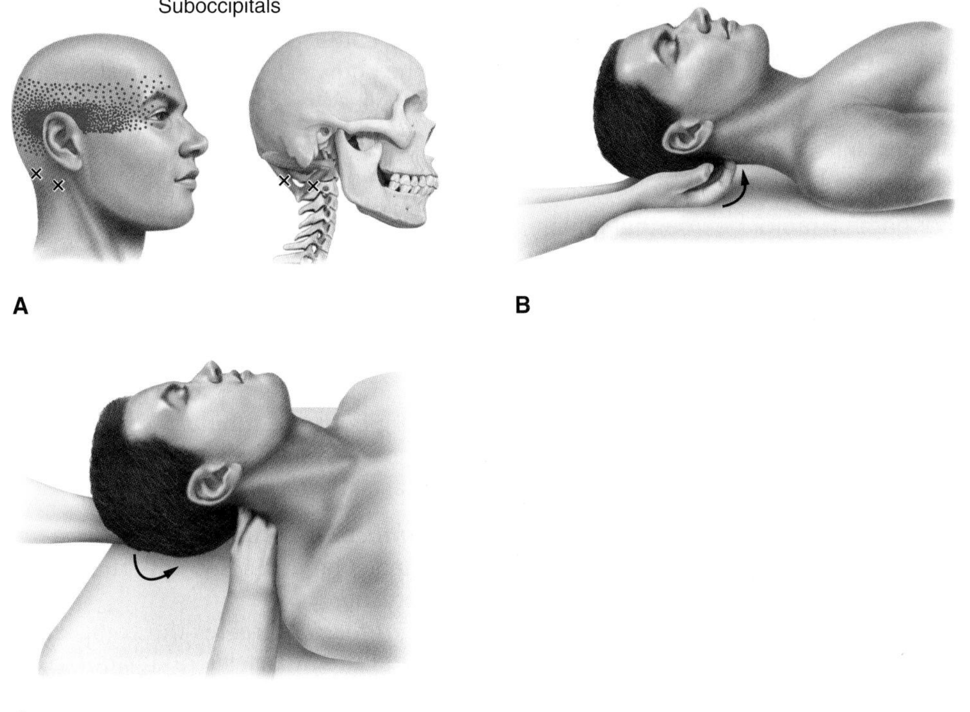

FIGURE 8.6 ▸ Trigger point release. A. Location of common TrPs in the suboccipitals and their associated referred pain pattern. **B.** One method of desensitizing the TrP is to curl the fingers up and under the occiput, placing moderate pressure on the points. **C.** Another method of releasing these TrPs involves placing a thumb over the point then extending and rotating the head over the thumb to increase pressure and desensitize the point.

Inverse Stretch Reflex

Remember that muscles also contain another type of proprioceptor, the Golgi tendon organ (GTO). These proprioceptors are sensitive to tension in the muscle and respond to increased tension by inhibiting contraction. This reflex is called the **inverse stretch reflex**, since the response of the muscle is opposite to that of the stretch reflex. While a few GTOs are scattered throughout the muscle, the highest concentration of these proprioceptors is in tendons and musculotendinous junctions. Since tension is created by all muscle contraction, the GTOs serve as monitors of the force of contraction and protect muscles and their tendons from tearing due to excessive tension. While the GTOs appear to be mildly stimulated when any tension is applied to the muscle or tendon, as might occur with passive stretching or direct compression over the musculotendinous junction, it appears that active muscle contraction provides the most powerful stimulus.

There are several forms of manual therapy that use stimulus to the GTOs as a way to decrease or release muscle tension. The most common are ***active release*** and ***facilitated stretch*** techniques. In active release techniques, the therapist applies manual stretch to the muscles and fascia as the client goes through an active range of motion. These techniques are effective on two levels: they mechanically stretch and loosen the tissue, and the increased tension during contraction stimulates the GTOs, leading to some inhibition of muscle tension.

Manual therapists often employ facilitated stretching as a way to enhance stretch and gain improved range of motion. Different types of facilitated stretch protocols employ a basic contract-relax technique that engages the GTOs and inhibits muscle tension to increase the stretch. In contract-relax, the first step is to place the target muscle into a position of mild stretch. While holding the muscle at that point, the therapist asks the client to engage the muscle in a brief isometric contraction (~10 seconds), and when the client relaxes, the therapist can usually increase the stretch. The isometric contraction of the muscle while in a lengthened position stimulates the GTOs so that muscle tension is inhibited.

Another type of facilitated stretch technique, contract-relax-antagonist-contract, employs two neuromuscular reflex arcs to increase stretch: stimulation of the GTOs and reciprocal inhibition. As the name implies, the target muscle is taken into a light stretch and then contract-relax is applied to engage the GTOs. The client then actively engages the antagonist muscle and moves to a new stretch point (reciprocal inhibition).

What Do You Think? 8.1

- Why is it important to use an isometric contraction when reciprocal inhibition is used to reduce acute muscle cramps?
- Is it the alpha or gamma motor neuron that innervates the motor unit? Explain.
- If there is no increase in stretch after employing contract-relax, what other tissues might be limiting muscle length?

THE FASCIAL SYSTEM

Connective tissue elements have been described as the supportive and protective components within the organs. The location and function of the dermis and hypodermis of the skin, the ligaments and periosteum of the bones, the fascial layers within skeletal muscles, and the meninges of the nervous system have been discussed in prior chapters. In this chapter, these individual connective tissue structures are considered collectively as a unified system: the fascial system. By viewing all the connective tissues as an integrated whole, one begins to appreciate their complexity and the essential roles fascia play in the body.

The fascial system includes *all* fibrous connective tissues, from the disorganized fibrous connective tissue sheets called fascia to the more organized fibrous connective tissue structures such as tendons, aponeuroses, ligaments, and capsules. When viewed as a system, the connective tissues become a unifying structural element for the entire body that extends from the skin to deep within bones and internal organs. This integrated fascial system organizes the growth and development of the embryo; supports proper spatial relationships between vessels, nerves, and organs; protects the body from infection; heals wounds; and allows movement.

In particular, the fascial system defines the structural relationship between the bones and muscles that makes the body strong, flexible, and resilient. The term **tensegrity**, a contraction for ***tension integrity structuring***, was first used by architect R. Buckminster Fuller. It describes how the tension between two opposing forces can be balanced to create structural integrity. Anatomists use *tensegrity* to describe the balance of compression and tension forces of the musculoskeletal system. Gravity creates a compression force on the bones of the skeleton, while the muscles and fascia that connect and functionally integrate the skeleton exert tension on the bones and joints to maintain upright posture. If either of these forces is disrupted, structural integrity is compromised.

In tensegrity systems, structural integrity is maintained by transferring and dispersing stresses throughout the system. However, this also means that trauma or damage to one body area is never confined to a single location but is reflected throughout the system. By viewing the body as a tensegrity system and identifying some of the patterns, layers, and mechanical properties of fascia, we better understand how tension, fibrous restrictions, or damage is easily transmitted through the body, creating pain and dysfunction in distant areas.

Layers

At the 2007 First International Fascia Research Congress, Frank Willard, PhD, from the University of New England shared dissections of the axial body demonstrating both the tubular organization and continuity of the fascial system.[5] These dissections showed four general fascial layers:

- ***Pannicular*** or ***superficial fascia***—made up of subcutaneous fat and some superficial muscles such as the platysma
- ***Axial fascia***—creates the compartments that surround the muscles (epimysium, perimysium, and endomysium)
- ***Meningeal fascia***—surrounds the brain and spinal cord
- ***Visceral fascia***—the areolar connective tissue that holds organs together and the serous membranes of the ventral cavities.

Each layer creates a tube, or organizing sleeve for its structures, and all four sleeves merge together at the lumbosacral aponeurosis (FIGURE 8.7). Willard's finding provides great insight into the structure and connectedness of the various fascial elements of each organ system, and reminds us that all tissues are three-dimensional and connected. However, it may be simpler and more practical for manual therapists to view the fascial system as having only two layers: superficial and deep. In this two-layered perspective, the **superficial fascia** includes the hypodermis that anchors the skin to underlying structures, and a few superficial muscles. The **deep fascia** is considered to be a combination of all other layers that organize and surround individual muscles, bones, and organs. Thick and broad connective tissue sheets such as the lumbosacral aponeurosis and iliotibial (IT) band are difficult to characterize as strictly superficial or deep since they are firmly anchored to the skin, muscles, and bones. Therefore, it is best to consider these structures as part of both the superficial and deep fascial layers.

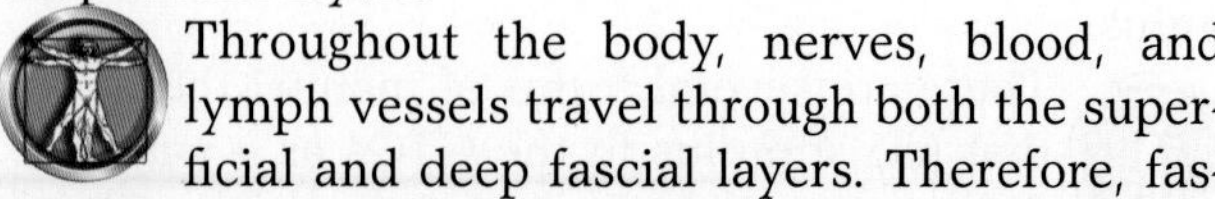

Throughout the body, nerves, blood, and lymph vessels travel through both the superficial and deep fascial layers. Therefore, fascial tension can lead to restriction or diversion of blood and lymph flow as well as stimulation of pain receptors. Dr. Hartmut Heine, a German researcher, identified numerous points throughout the superficial fascia where small neurovascular bundles perforate

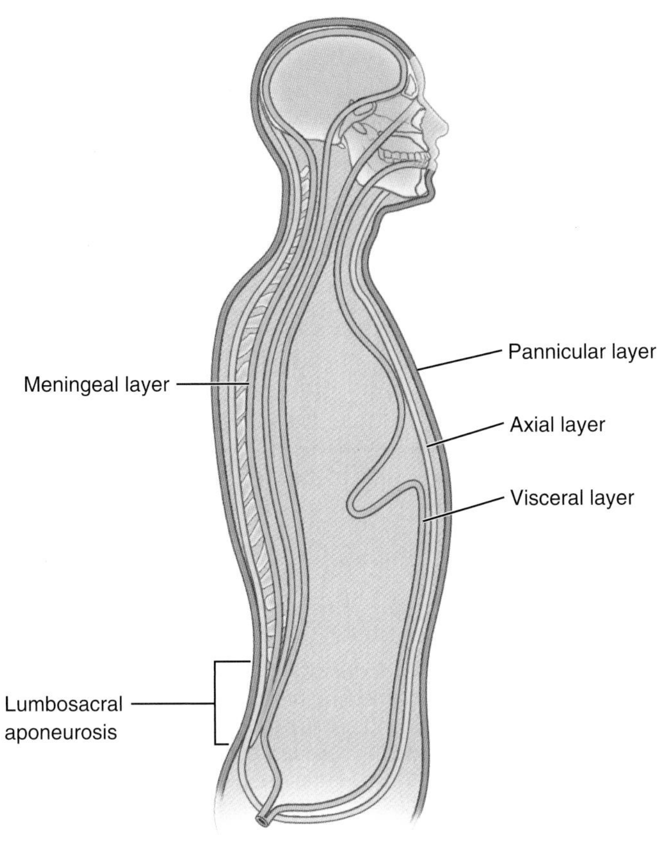

FIGURE 8.7 ▶ Fascial layers. The four fascial layers converge in and around the lumbosacral aponeurosis.

the fascial layer.[6] In a 1998 study of patients suffering from chronic neck-shoulder or shoulder-arm pain, Heine and Günther Bauer found that the neurovascular bundles in these perforations were strangulated by a thickened ring of collagen fibers. When the strangulations were loosened through microsurgery, patients experienced significant improvement. This provides strong evidence to support the importance of stretching, broadening, and loosening the superficial fascia to improve local blood flow and reduce pain. Another interesting finding from Heine's research is that 82% of the "perforation points" in superficial fascia matched up exactly with 361 traditional Chinese acupuncture points.

Fascial Patterns

Both the superficial and deep layers of the fascial system are organized into well-defined vertical, spiral, and horizontal patterns. These patterns form the multiple lines of tension needed to maintain the body's structural integrity and play essential roles in generating and transmitting force for movement. These vertical and spiral patterns have perhaps been best described through the dissections and other studies of anatomist and Rolfer Thomas Myers.[7] He called these structural connections between muscles and bone ***myofascial trains*** or ***meridians***, while others sometimes refer to them as ***myofascial chains***. Researchers have also identified and described several superficial and deep horizontal fascial structures that play different tensegrity roles in the body.

Myofascial Chains

A **myofascial chain** describes the connective tissue links between muscles, bones, and fascial membranes that provide a pathway for the *mechanical* communication of tension and compression throughout the body.[7] Most myofascial chains are vertically aligned. Found superficially and deep, they form straight lines in the coronal or sagittal planes, or they form a spiral vertical line that traverses all three planes (▶ FIGURE 8.8).

Because stress or dysfunction can be reflected or passed through the body along specific myofascial chains, knowledge of these vertical and spiral patterns provides manual therapists with a "map" to guide their assessment procedures and treatment choices. For example, Myer's superficial back line (SBL) maps the fascial connections that link and communicate tension from the sole of the foot to the forehead (▶ FIGURE 8.9). This means that conditions such as plantar fasciitis, Achilles tendonitis, hamstring tendonitis, chronic low-back strain, and even headaches can be related. The therapeutic implication is that treatment along the entire myofascial chain may be required for complete resolution of certain chronic pain and repetitive stress syndromes.

Horizontal Bands and Planes

The superficial layer of fascia has seven flattened horizontal straps or **fascial bands** within the normal subcutaneous contours of the torso. Similar to the retinaculi of the wrists and ankles, these bands serve as "straps" that firmly anchor the softer rounder anterior torso to the vertical, semirigid spine. Though their precise location varies from one individual to another, each band can be correlated to a specific spinal junction.[8,9] ▶ FIGURE 8.10 shows the location and appearance of the bands. They curve slightly, following individual body contours instead of following a straight horizontal line.

Just as a tall bushy plant may be damaged if tied too tightly to a supporting pole, tight or adhered fascial bands can cause problems, including restrictions in breathing, movement, and the body's general sense of ease. Excessive tension or adhesion in the fascial bands tends to increase the compression force translated to the spine, making the bands a type of fulcrum point that leads to stress and pain. These restrictions are often fairly easy to identify in a postural assessment, since they tend to restrict fat deposition. When manual therapists notice a restricted fascial band, they can

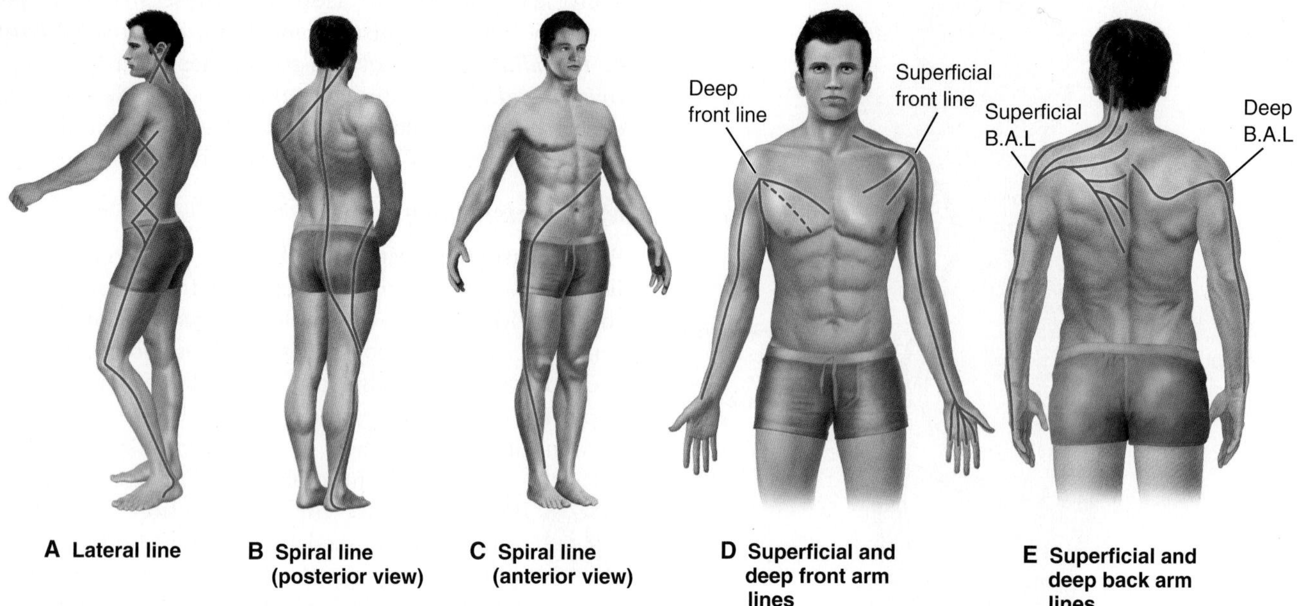

FIGURE 8.8 ▶ Myofascial chains. A. The ***lateral line*** includes the splenius capitis and SCM, external and internal intercostals, lateral abdominal obliques, gluteus maximus, tensor fascia latae, IT band and abductors, anterior ligament at the head of the fibula, and the peroneal muscles with investing fascia. **B.** Posterior view: the ***spiral line*** links the splenius capitis and cervicis to the *opposite side* rhomboids and serratus anterior. **C.** Anterior view: the anterior connections of the spiral line include external obliques, abdominal aponeurosis and *opposite side* internal obliques, tensor fascia latae, IT band, and tibialis anterior. It then passes under the plantar surface of the foot via the fascial connections between anterior tibialis and peroneus longus, up the same side biceps femoris, sacrotuberous ligament, sacrolumbar fascia, and up the erector spinae to the occipital ridge. **D.** The ***superficial front arm line (SFAL)*** connects the pectoralis major, insertion of latissimus dorsi, medial brachial septum, elbow flexor group, flexor retinaculum, and palmar surface of the fingers. The ***deep front arm line (DFAL)*** connects the pectoralis minor, clavipectoral fascia, biceps brachii, radial periosteum, radial collateral ligament, thenar muscles, and the outside of the thumb. **E.** The ***superficial back arm line (SBAL)*** begins at the occipital ridge and thoracic spinous processes and connects the trapezius, deltoid, lateral intermuscular septum, extensor group, and the dorsal surface of the fingers. The ***deep back arm line (DBAL)*** begins at the spinous processes of lower cervical vertebrae and connects the rhomboids, levator scapulae, rotator cuff muscles, triceps brachii, ulnar periosteum, ulnar collateral ligaments, hypothenar muscles, and outside of the little finger. Information based on Chaitow L. *Positional Release Techiques,* 2nd ed. Edinburgh, Scotland: Churchill Livingstone, 2002.

make connections between complaints of posterior pain to structures on the anterior side of the body and vice versa. For example, an individual with chronic pain and restrictions in the lower back might present with a visible umbilical and/or inguinal band. Palpation would likely reveal thickened and restricted connective tissue at the corresponding spinal junctions that would then be addressed. Therapists should also employ myofascial techniques to the anterior portion of the band. General releases throughout the abdominal aponeurosis and some direct work around the pubic symphysis and umbilicus might be beneficial. To minimize the compression fulcrum created by restricted fascial bands, the entire circumference should be assessed and addressed.

The deep fascia has several horizontal structural components called **fascial planes**. These planes are thickened sheets of fascia inside body cavities that provide structural strength to the torso and support the major blood vessels, nerves, and organs within. Both the dorsal and ventral body cavities have horizontal fascial planes that divide and support their internal structures (▶ FIGURE 8.11).[3,10,11] The four horizontal planes are

- ***Cranial base***—separates the spinal and cranial cavities, covers the foramen magnum, and weaves into the meninges.
- ***Thoracic inlet***—crosses the opening at the top of the ribcage and runs from sternum to the cervical thoracic junction of the spine and over the superior aspect of the first ribs.
- ***Diaphragm***—domed skeletal muscle that separates the thoracic and abdominopelvic cavities, which is wrapped in several fascial layers that are attached to the vertebrae and inside of the rib cage.
- ***Pelvic diaphragm***—forms the floor of the pelvic girdle.

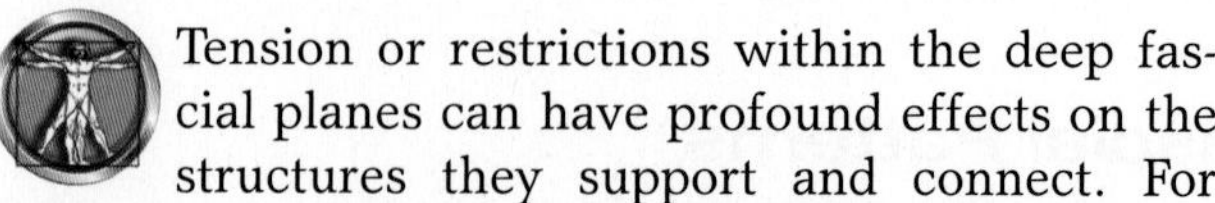

Tension or restrictions within the deep fascial planes can have profound effects on the structures they support and connect. For example, any excess tension or strain from the scalenes can be translated through the thoracic inlet plane into tension and/or compression of the subclavian artery and brachial plexus. Similarly, tension from the hip adductors and external rotators can translate through the pelvic floor and contribute to sciatic nerve irritation. This provides sound rationale for including release of the horizontal fascial planes when treating these kinds of NCTS conditions.

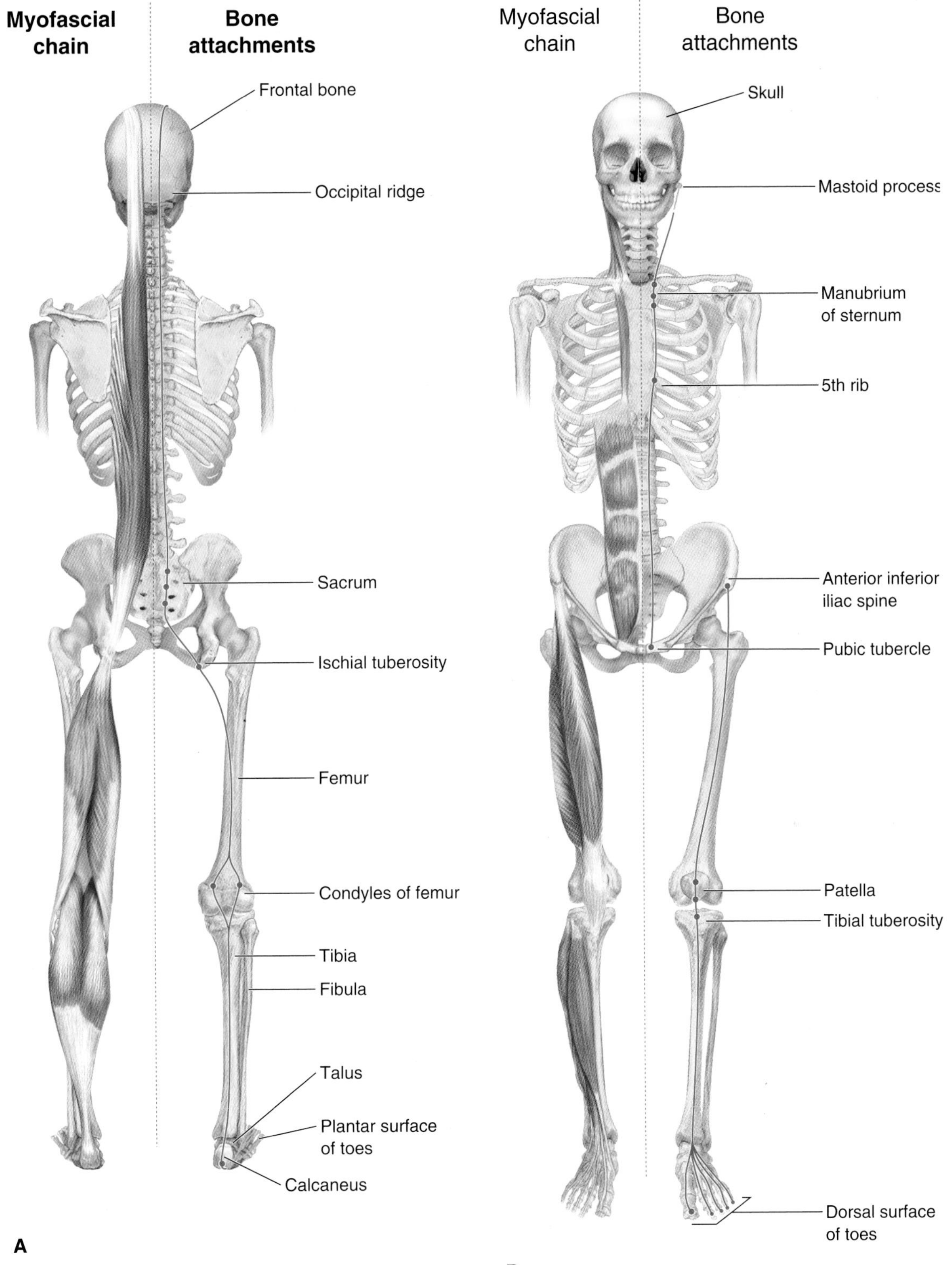

FIGURE 8.9 ▸ Superficial back and front lines. A. The ***superficial back line*** includes the tendons of the toe flexors, plantar fascia, Achilles tendon, investing fascia of the gastrocnemius, hamstrings, lumbosacral aponeurosis, erector spinae, and scalp fascia. **B.** The ***superficial front line*** includes the toe extensors, fascia of the anterior compartment, patellar ligament, quadriceps, rectus abdominis, sternochondral fascia, SCM, and the scalp fascia.

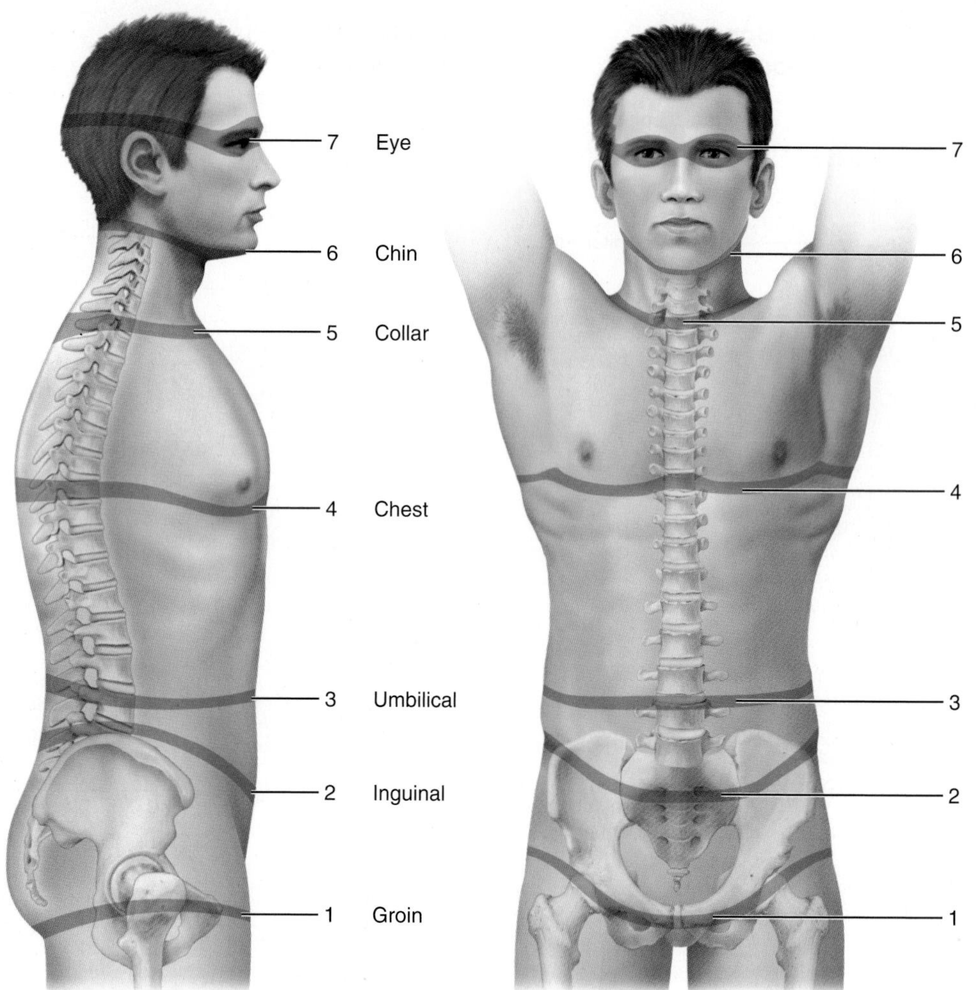

FIGURE 8.10 ▶ Horizontal bands. The body has 11 horizontal bands of superficial fascia that add contour and support to the torso.

Mechanical Properties of Fascia

In addition to its continuous and patterned arrangement in the body, fascia has several mechanical properties that determine how these connective tissue components function, and provides guidance to manual therapists in how to work effectively with the fascial system. The key mechanical properties of fascia are viscoelasticity, thixotropy, and piezoelectricity.

Viscoelasticity

Viscoelasticity describes the ability of tissues to extend and rebound rather than stretch and recoil. This mechanical property is due to the organization and structure of the collagen fibers that make up fibrous connective tissue. Each collagen fiber consists of several bundles of a protein molecule called tropocollagen. This protein's molecular arrangement is similar to DNA, giving it the same unique triple helix appearance (▶ FIGURE 8.12). Several tropocollagen molecules bind together in a parallel arrangement to form one ***collagen fibril***. In turn, multiple fibrils are bundled to form a collagen fiber. Because of the parallel arrangement of molecule to fibril to fiber, the spiral shape of the tropocollagen molecule is reflected and carried through to the collagen fiber itself. When sustained tension is applied, collagen fibers do not stretch. Instead, they extend or slowly lengthen by unwinding these multiple spirals. Once the tension is removed, the fibers slowly return to their normal length, or rebound.

When a collagen fiber is completely unwound and its full length is reached, it cannot lengthen further without structural damage. Visualize a tire swing on a hemp rope suspended from a tree limb. When you jump on the swing, the rope holds you and the tire above the ground as you sway back and forth. Over time, the sustained tension from swinging seems to have stretched the rope, and the tire appears to be closer to the ground. But the rope has not actually stretched; the braided hemp has actually unwound. Once the rope has unwound as much as possible, no further lengthening occurs unless it begins to fray or tear. The same is true for collagen fibers—they unwind and extend when gradual tension is applied but resist the sudden application

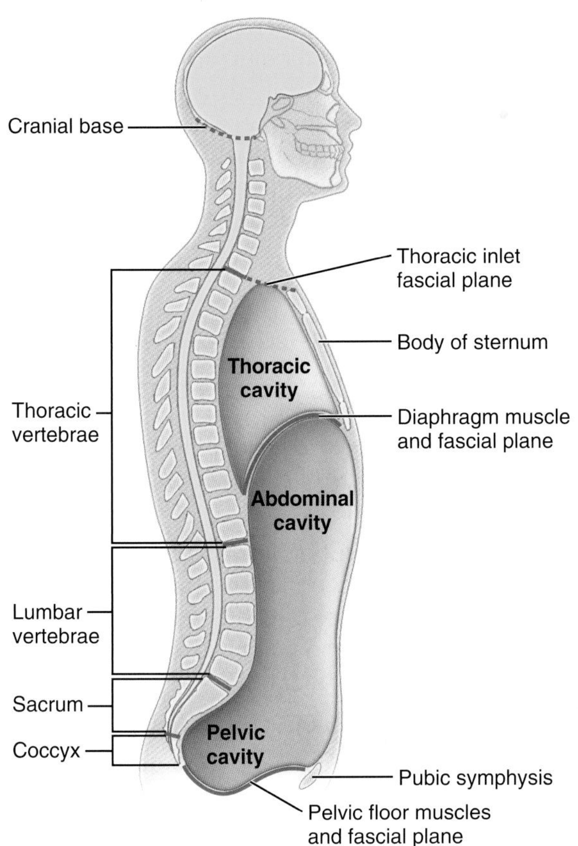

FIGURE 8.11 ▸ Fascial planes. There are four deep fascial planes arranged horizontally across the dorsal and ventral cavities.

of force. When connective tissue is subjected to sustained and repeated tension, collagen fibers do not become more elastic. Instead, more chains of tropocollagen are added to the collagen fibrils to increase their resting length.

An increase in frequency or regularity of exercise increases the demand for more extensibility in the connective tissue, stimulating fibroblasts to increase tropocollagen production. This helps explain why a client's improved range of motion after a single manual therapy session is generally sustained for only a few hours and why frequency of treatments may, at times, be more important than the duration of each session.

Thixotropy

Recall from Chapter 3 that the ground substance of connective tissue is mostly fluid with suspended particles called GAGs (glycosaminoglycans) that serve as water magnets. This keeps the tissue matrix hydrated and gives it the ability to vacillate between a more viscous (gel) state and more liquid (sol) state in response to temperature or movement, a property called **thixotropy**. Of the seven different types of GAGs currently identified, ***hyaluronic acid***, a particularly viscous and slippery substance, is of particular interest. It is present in high quantities in the fascial components of joints and skeletal muscles, where it serves as an essential lubricant for these structures. Because hyaluronic acid is one of the most powerful water magnets, it is essential for maintaining the thixotropic nature of ground substance.

Piezoelectricity

The term **piezoelectricity** comes from the Greek word *piezein*, which means to press or squeeze. When connective tissue is subjected to mechanical pressure, it produces a small electrical charge along its surface. This current has been shown to excite fibroblasts while inhibiting the function of fibroclasts. This effect has been demonstrated in the healing of bones and wounds and is hypothesized to occur in fascia, as well. Theoretically,

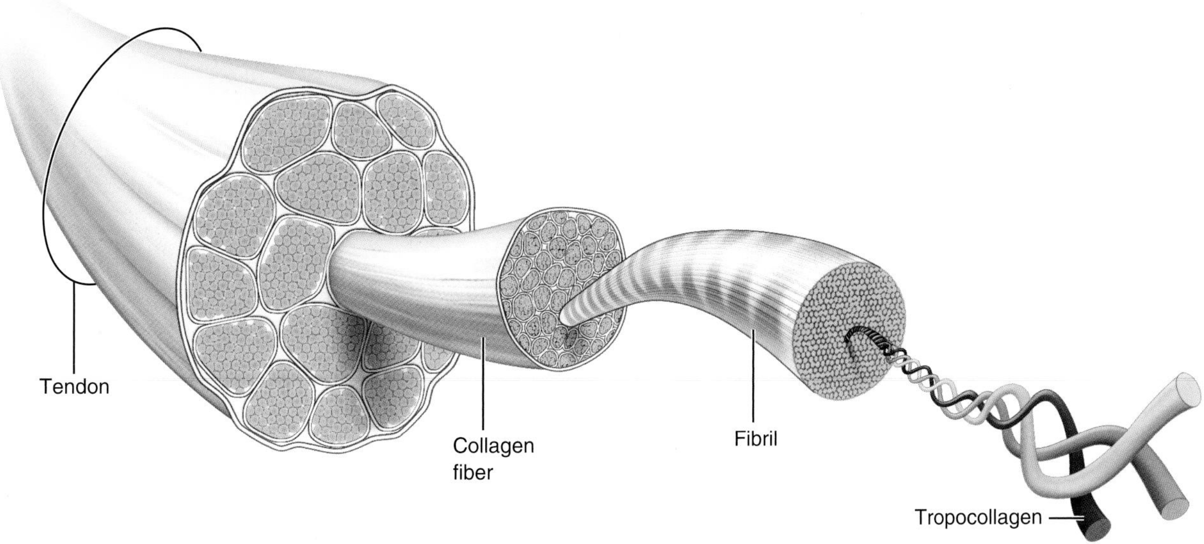

FIGURE 8.12 ▸ Structure of a collagen fiber. Each collagen fiber is made of smaller bundles of fibrils. Fibrils are made up of bundles of microscopic filaments called tropocollagen. The spiral structure of tropocollagen is reflected throughout a collagen fiber, which is why collagen extends or unwinds rather than stretches when tension is applied.

the pressure and movement of manual therapy increases the piezoelectricity of fascia and may contribute not only to healing but also to the softening and loosening sensation felt by clients and therapists during treatment.

Each mechanical property of fascia offers possible physiologic explanations for how and why manual therapy in general, and myofascial techniques specifically, are effective. ▶ FIGURE 8.13 gives an example of a common myofascial release technique called pin-and-stretch. While the application of the sustained moderate pressure shown in the figure is supported by our understanding of the viscoelastic property of fascia, research demonstrates that long-term changes in collagen length require far more force and a longer application than is utilized in manual therapy sessions.[12] Calculations made by A. Joseph Threlkeld showed that at least 60 kg of force and more than an hour of sustained tension were required to create permanent elongation of collagen in the distal portion of the IT band.[13] Therefore, the sensations of tissue unwinding, release, or "creep" described by practitioners cannot be due to viscoelastic changes alone.

Likewise, the thixotropic and piezoelectric properties of fascia help explain the sensation of tissue softening and increased pliability during and immediately after a session. However, these mechanical properties do not fully account for longer-term improvements in range of motion, flexibility, and tissue pliability because they only last a short period of time once heat, pressure, and movement are removed. The piezoelectric property of fascia may provide a rationale for the increased fibroblast activity observed during tissue healing when manual therapy is used in the treatment of conditions such as Achilles, rotator cuff, and patellar tendonitis. Still, the question remains: do these three mechanical properties of fascia provide an adequate explanation for both the immediate and long-term tissue changes observed in response to manual therapy treatments?

What Do You Think? 8.2

- What kinds of musculoskeletal dysfunctions can you think of that might be linked through the superficial front line (SFL)?
- Using your knowledge of myofascial chains and planes, what connections can you make in a client with a history of two C-sections and a fractured clavicle who complains of chronic neck pain and tension headaches?
- What are the therapeutic implications of the above scenario?

FASCIAL PLASTICITY

Plasticity is a term used to describe the ability of something to be molded or changed. The most common usage of the term in anatomy and physiology has been neural

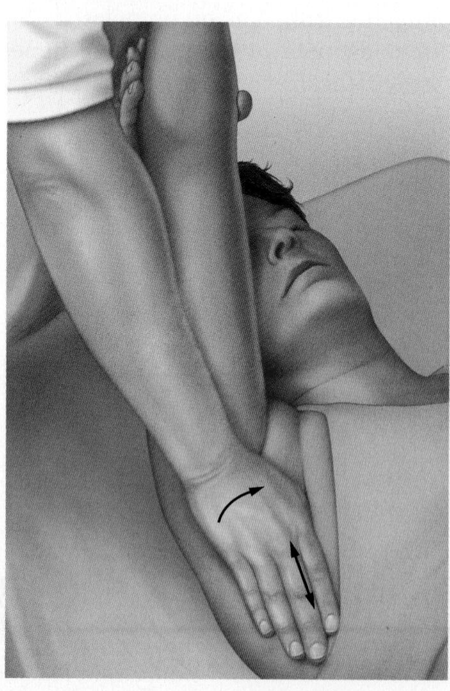

(a)

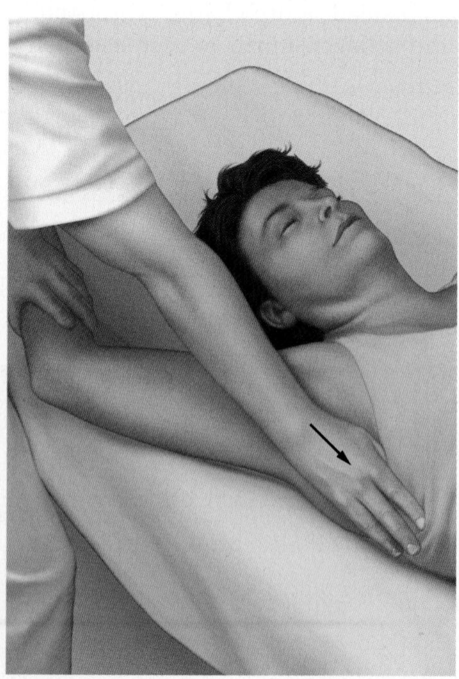

(b)

FIGURE 8.13 ▶ Broad plane fascial release: pin-and-stretch for latissimus dorsi and teres major. A. Starting position. **B.** Finish position.

plasticity, which describes the adaptability of the nervous system. Now that fascia is commonly acknowledged as a fully functioning system, the term ***fascial plasticity*** is used to denote the changeable, responsive, and adaptive nature of fascia. However, the mechanical properties of fascia cannot fully explain its plasticity. So what else is going on? There must be other connections between the fascia, muscles, and nervous system to explain the tissue responses sensed by manual therapists.

Fascia as a Sensory Organ

The role of fascia has been described up to this point based on its function as connective tissue; it surrounds, divides, supports, and connects. Research over the past 20 years has shown that fascia is also extremely rich in sensory receptors. Four types of mechanoreceptors have been identified (TABLE 8-2).[12]

- ***Golgi receptors***—Described in the neuromuscular section as GTOs, these tension receptors are also found throughout fascia and organized fibrous connective tissues such as ligaments and capsules. However, fewer than 10% of the overall numbers of Golgi receptors in the body lie within fascial structures. Most reside on the *muscular* side of the musculotendinous junction to monitor the tension created in the muscle by active contraction.
- ***Pacinian receptors***—These mechanoreceptors are sensitive to vibration and rapid changes in pressure. They are more abundant on the *tendinous* side of the musculotendinous junction, as well as in the deeper portions of joint capsules and thick fascial planes such as the IT band and abdominal and lumbar fascia. These receptors seem to provide important information to the brain to facilitate controlled movement and proprioception.
- ***Ruffini organs***—Unlike pacinian receptors, Ruffini organs are sensitive to *long-term pressure* and are plentiful in all the fascial layers. Because they are particularly sensitive to lateral stretch, they are responsive to slow, deep, multidirectional soft tissue techniques. Stimulation of these receptors seems to be directly linked to a decrease in sympathetic tone.
- ***Interstitial myofascial receptors***—These mechanoreceptors are the most abundant of the fascial sensory receptors. Fascia includes equal numbers of two types of these specialized free nerve endings: ***low-threshold pressure (LTP) units*** that respond to

Table 8-2 Mechanoreceptors in Fascia

Receptor Type	Preferred Location	Responsive to	Known Results of Stimulation
Golgi organs	• Myotendinous junctions • Attachment areas of aponeurosis • Ligaments of peripheral joints • Joint capsules	GTOs at musculotendinous junction to muscle contraction Other Golgi receptors to strong stretch only	Decreased motor tone in related skeletal muscle
Pacinian	• Myotendinous junctions • Deep capsular layers • Spinal ligaments • Epimysium, perimysium, and endomysium	Rapid pressure changes and vibrations	Used as proprioceptive feedback for movement control (kinesthesia)
Ruffini	• Ligaments of peripheral joints • Dura mater • Outer capsular layers • Other tissues associated with regular stretching like fascia	Like pacinian, plus sustained pressure Specifically responsive to lateral and tangential stretch	Inhibition of sympathetic activity; decreased sympathetic tone
Interstitial Type III & IV	• Most abundant receptor type • Found throughout the body, even inside bones • Highest density in periosteum	Rapid and sustained pressure changes. 50% are high-threshold units and 50% are low-threshold units	Decreased sympathetic tone Increased vagal activity that results in global relaxation, a more peaceful mind, and less emotional arousal

Modified from Schleip R. Facial plasticity—a new neurobiological explanation: Part 1. *J Bodyw Mov Ther. 2003;7(1):11–19.*

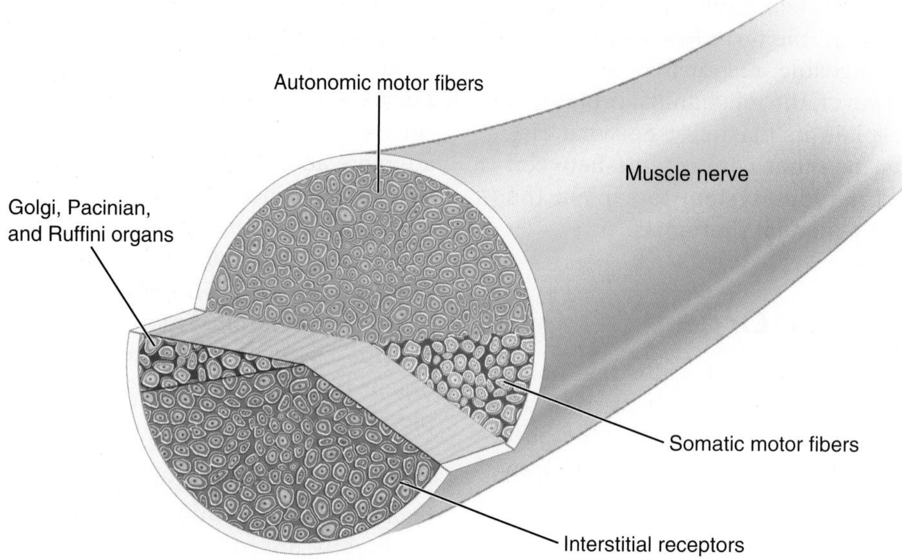

FIGURE 8.14 ▸ Fiber distribution in a typical muscle nerve. A typical skeletal muscle nerve contains only a small number of somatic motor neurons and larger portions of autonomic motor and sensory fibers. Of the sensory fibers, approximately 80% are interstitial myofascial receptors.

light touch and ***high-threshold pressure (HTP) units***, which are activated by stronger pressure stimuli. Changes in heart rate, blood pressure, vasodilation, and respiration have been directly linked to stimulation of **interstitial myofascial tissue receptors**. These changes suggest a strong link between fascia and the autonomic nervous system (ANS). In fact, evidence shows that stimulation of these interstitial receptors leads to a decrease in sympathetic tone as well as an increase in vagal nerve activity.

The discovery of these sensory receptors confirms that the greatest amount of sensory information sent to the central nervous system (▸ CNS) is more likely to come from the myofascial tissues than the skin. If we look at a nerve innervating a skeletal muscle such as the brachial nerve, we find that 40% to 45% of the neurons are sensory fibers (▸ FIGURE 8.14). Of these sensory fibers, 80% are interstitial myofascial receptors, and the other 20% are from the other types of mechanoreceptors, including muscle spindles, Golgi organs, pacinian, and Ruffini organs. This means that the majority of sensory information provided by muscles and fascia is directed to the ANS, rather than musculoskeletal coordination, and creates a new understanding of fascia as a primary sensory organ.

Interestingly, HTP units also function as pain receptors. In the presence of pain, they adapt their sensitivity so that normal pressure changes cause them to fire constantly. This phenomenon has been hypothesized as an explanation for chronic pain syndromes such as fibromyalgia and chronic low back pain when there is no nerve root compression. Therefore, regardless of the specific technique, manual therapy should be considered a

Pathology Alert Fibromyalgia

Fibromyalgia is a debilitating chronic pain syndrome characterized by widespread musculoskeletal pain, disturbed sleep patterns, persistent fatigue, and the development of multiple myofascial TePs in all four quadrants of the body. The diagnosis of fibromyalgia is often only made after other chronic pain conditions, such as lupus, multiple sclerosis, Lyme disease, and rheumatoid arthritis have been ruled out. The origin of fibromyalgia pain has been a bit of a mystery; to a large extent, the presumption has been that pain receptors in the muscles and joints were being stimulated. The discovery of high numbers of pain receptors in fascia offers another possible explanation: that HTP fascial receptors are involved either in addition to, or rather than, muscle and joint receptors. Additionally, the link between these fascial receptors and the ANS also helps explain why stress and emotional ups and downs have such a direct correlation to the intensity of a fibromyalgia patient's pain.

primary treatment for any chronic pain syndrome, since fascial manipulation reduces sympathetic tone, which reduces pain.[6] Fascial plasticity also suggests that a full-body session focused on relaxation and stress reduction is as much of a therapeutic intervention for chronic pain patients as eliminating TePs, and may indeed be preferred. In a 1999 study by Gunilla Brattberg, connective tissue massage (CTM) was used to treat fibromyalgia.[14] CTM, also known as bindegewebsmassage, is a reflexive therapy of the connective tissue developed by Elisabeth Dicke that employs a specific pattern of strokes along the spinal dermatomes to both loosen the fascial layers and bring about ANS changes. In this double-blind study of 48 individuals diagnosed with fibromyalgia, each was treated 15 times during a 10-week period. Levels of average pain, disability, sleep disturbance, anxiety and depression, and quality of life were evaluated three times (immediately after the 10-week treatment period, three months after treatment, and six months after treatment). Immediately following the treatments, those receiving CTM showed a 37% reduction in pain, reduced use of analgesics, decreased anxiety and depression, and improvements in quality of life. All these benefits diminished gradually over the following six months, with pain returning to about 90% of its original rating at the 6-month point.[14]

Intrafascial Smooth Muscle Cells

In 1993, a research team led by L'Hocine Yahia noticed that when fascia was stretched and held at a constant length, it began to slowly resist the tension.[15] They hypothesized that this spontaneous contraction could be due to the presence of smooth muscle tissue in fascia. Just a few years later in 1996, J. Staubesand and Y. Li confirmed that fascia does indeed contain smooth muscle cells (FIGURE 8.15).[6] Taken together, these studies allow us to hypothesize that smooth muscle cells invested within the fascia and controlled by the ANS produce a **fascial tone** that is independent of the motor tone in skeletal muscle.

The direct links between the fascial sensory receptors, ANS, and intrafascial smooth muscle cells provide several new and exciting explanations for how various manual therapies create their physiologic effects. The changes previously attributed to shifts in the thixotropic or viscoelastic properties of connective tissue now appear to be related to the sensory link between the fascia and the ANS. This view of fascial plasticity tells us that the immediate tissue changes therapists sense as they work, as well as general relaxation effects such as decreased blood pressure, heart rate, and respiration, are likely mediated through **neurofascial** loops. Recent research confirms the existence of a large number of fascial receptors that create decreased sympathetic tone in response to direct stimulation. This decrease in sympathetic tone has also been linked to relaxation of intrafascial smooth muscle, causing decreased fascial tension (FIGURE 8.16). Because tension in the axial fascia layers of skeletal muscle is a primary element of muscle tone (not motor tone), decreased fascial tone can be one explanation for

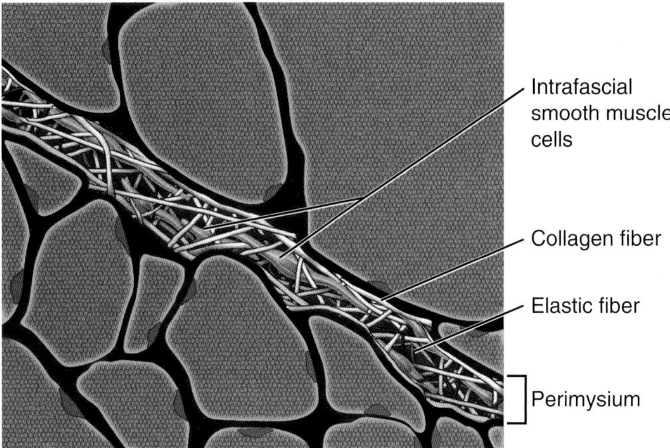

FIGURE 8.15 Intrafascial smooth muscle cells. The perimysium of skeletal muscle contains a high density of intrafascial smooth muscle cells. Like all smooth muscle, the autonomic branch of the nervous system controls these cells.

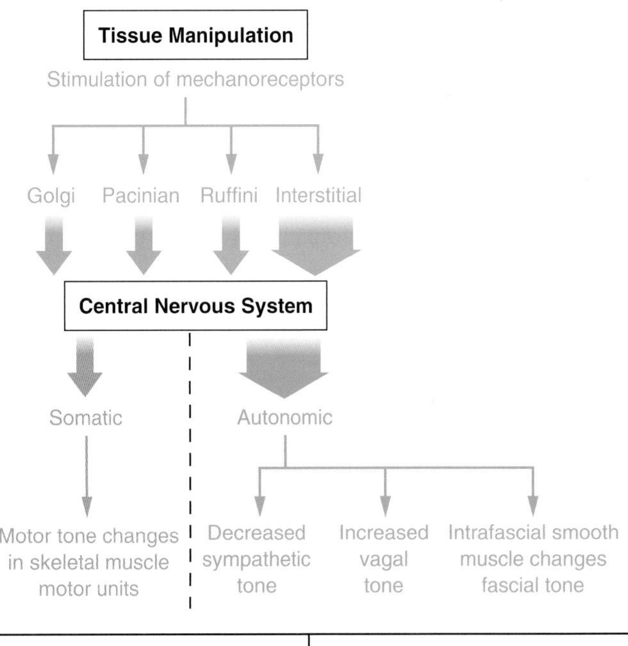

FIGURE 8.16 Tissue manipulation and palpable tissue response. Stimulation provided to mechanoreceptors by tissue manipulation stimulates both somatic and autonomic divisions of the peripheral nervous system. This causes palpable changes in the tissue that manual therapists often refer to as "tissue release."

improvements in flexibility and range of motion. In more practical terms, the new research into fascial plasticity has implications for the choices therapists make about the pressure, pace, pattern, and focus of their work.

By the Way

Physical therapists, athletic trainers, and other physical rehabilitation professionals focus attention on closed kinetic chain exercises once pain-free ROM and sufficient strength has been regained. This type of rehabilitation exercise provides the variety and volume of sensory feedback needed for the cerebellum to retrain neuromuscular reflexes and regain smooth well-coordinated movement.

What Do You Think? 8.3

- How does the information about fascial plasticity shape your thinking about the pressure, pace, pattern, and focus of a massage therapy session?

POSTURE, BALANCE, AND COORDINATED MOVEMENT

The structural tensegrity created by bones, fascia, and skeletal muscle is functionally controlled by the nervous system to maintain posture and balance and to create movement. While the body's structure dictates what movements are possible, the nervous system receives and processes sensory input from millions of proprioceptors to determine body position, plan adjustments, and regulate the sequence and force of muscle contraction. The cerebrum contains the primary motor cortex that plans and commands voluntary muscle activity. The cerebellum works with and through the primary motor center in the constantly shifting effort to maintain both static and dynamic balance and posture. It does so by coordinating and rotating the workload between different motor units in the postural muscles. This process of **motor unit recruitment** regulates the force of contractions (the graded response) and helps control muscle fatigue in postural muscles.

In contrast, the cerebellum creates smooth and efficient movement by coordinating the activation sequence and force adjustments of agonist, synergist, and antagonist. This pattern of stimulation and coactivation between muscle groups is called **muscle recruitment**. The series of muscles used to create each movement form a ***kinetic chain*** that is activated in a specific sequence to create smooth and coordinated motion. For example, the muscles in the arm and shoulder girdle are part of the kinetic chain for throwing, and leg and foot muscles are in the kinetic chain for walking. The key concept is to recognize that the cerebellum coordinates the muscles in a kinetic chain in different patterns based on the demands of a particular motion, and can only do so based on sensory input from proprioceptors. The cerebellum receives more sensory information to guide muscle recruitment when all the muscles in the kinetic chain are engaged. Engaging all the muscles in a particular chain generally requires that the distal joint of an extremity is engaged and weight bearing, which is called ***closing the kinetic chain.***

Voluntary movement and posture can be controlled either consciously or unconsciously depending on the part of the brain from which they originate. Conscious movements involve the alpha motor pathways and originate in areas of the frontal cortex, while areas of the parietal lobe direct most unconscious activities. Interestingly, it seems that areas of the brain involved in unconscious function determine and initiate most movements *before* we consciously decide to make them. Therefore, a conscious decision to move may simply be the recognition of what the unconscious mind has already determined to do. Whether complex movements are conscious or unconscious depends on a person's experience and level of expertise with a specific movement or skill. As movements become more familiar, they become less conscious and more automatic because the neural pathways are "well-worn" through practice. For example, when a teenager is learning to drive a car, the actions of checking mirrors, signaling, and making turns require conscious thought. However, after many repetitions, the same driver can drive any distance without really thinking about those individual actions. Gamma motor pathways in the CNS, which originate in the brain stem, also play a role in unconscious movement by making adjustments in postural tone and muscle tension.

Proprioception

Proprioception is a person's sense of body position. It involves integrating visual information from the eyes and equilibrium information from the ears with all the proprioceptive data from the muscles, joints, and fascia to determine where parts of the body are positioned in space and in relationship to one another. Closely related to proprioception is **kinesthesia**, or the sense of movement. Both of these senses are largely unconscious and are essential to posture and coordinated movement, especially muscle memory and eye–hand coordination.

Considering what is known about the proprioceptors in fascia, we can only imagine how huge its sensory imprint is for proprioception. The muscle spindles and GTOs of muscles, combined with the Golgi and pacinian receptors in multiple fascial layers and joint capsules, provide continuous information about muscle and fascial length and tension and the power needed to create well-coordinated movement. The brain receives this sensory data and sends motor commands to make adjustments in fascial tension, motor tone, and motor unit recruitment to adjust position or stop a motion. This adjusted position or motion produces new sensory input to the nervous system, thereby creating a continuous feedback loop of sensation and adaptation.

Postural and Phasic Muscles

Chapter 6 covered the skeletal muscles and discussed their movement roles as agonist, antagonist, synergist, and stabilizer. In addition to playing these various roles during movement, each major muscle in the body is also designated as either a postural or phasic muscle. A **postural muscle** is essential in maintaining the body's upright position, while a **phasic muscle** is primarily involved in movement. TABLE 8-3 lists the postural and phasic muscles and key characteristics of each group.

By the Way

While a skeletal muscle as a whole is made up of different fiber types, each motor unit only contains one type of fiber. In motor unit recruitment, type I motor units are recruited first, and the units with type II fibers are recruited next. The sequence of motor unit recruitment is also based on the number of fibers in a particular unit, with those containing fewer fibers recruited first.

A muscle's role as postural or phasic is defined somewhat by its location, but more so by slight differences in cellular composition. Differences in the density and number of mitochondria, amount of myoglobin, capillary density, and rate of contraction allows skeletal muscle fibers to be divided into two categories: type I or type II:

- ***Type I fibers*** have a high density of both mitochondria and myoglobin, plus a rich supply of capillaries. This cellular composition gives the type I fibers a

Table 8-3 Postural and Phasic Muscles

Postural Muscles	Phasic Muscles
High endurance type I fibers; Adapt to stress by tightening and weakening	*Fast-twitch, low endurance type II fibers; Adapt to stress by shortening and tightening*
• Upper trapezius • Levator scapula • Latissimus dorsi • Erector spinae • Sacrospinalis • Quadratus lumborum • Piriformis • Sternocleidomastoid • Upper pectoralis major • Pectoralis minor • Iliopsoas • Medial hamstrings • Adductor longus • Oblique abdominals • Tensor fasciae latae • Rectus femoris • Soleus • Gastrocnemius • Tibialis anterior and posterior	• Scalenes • Deltoid • Biceps and triceps brachii • Lower pectoralis major • Middle and lower trapezius • Rhomboids • Serratus anterior • Rectus abdominus • Gluteals • Hamstrings • Vastus muscles • Peroneals • Elbow extensors

reddish appearance and the ability to produce large amounts of ATP to sustain work. They also have a slow rate and velocity of contraction. This combination of characteristics is why type I fibers are also known as ***red*** or ***slow-twitch fibers***.

- ***Type II fibers*** have the opposite cellular composition: few mitochondria, low myoglobin, fewer capillaries, and a higher contraction rate and velocity. For this reason, type II fibers are also known as ***white*** or ***fast-twitch fibers***.

Muscles generally have a 50–50 mix of fiber types; however, these proportions will vary based on genetics, individual training regimes, and whether the muscle functions primarily as a postural or phasic muscle. Postural muscles generally have a higher number of type I slow-twitch fibers due to their higher resistance to fatigue, and the phasic muscles contain more of the fast-twitch type II fibers. The presence of two different fiber types gives skeletal muscle the ability to adapt to differing functional stresses. For example, when stressed by regular aerobic activity such as walking or rowing, type II fibers improve their cellular endurance by gradually increasing myoglobin level, the number of mitochondria, and capillary density. With weight training, fibers thicken to create more powerful contractions. Postural muscles with their high ratio of type I fibers tend to adapt to the long-term stress of gravity by tightening and weakening, while the phasic muscles adapt to their movement responsibility by shortening as they strengthen their fibers. Differences in the adaptation processes of type I and type II fibers remind manual therapists that it is important to employ both neuromuscular and myofascial techniques to relieve "tight" muscles. This is especially true of postural muscles because the tightening and weakening adaptation of type I fibers increases the likelihood of some fibrous build-up.

SUMMARY OF KEY POINTS

- Understanding reciprocal inhibition, the stretch reflex and gamma gain, and the inverse stretch reflex allows manual therapists to utilize these neuromuscular reflexes in an intentional manner to relieve acute muscle cramps, reduce chronic spasms, eradicate TePs or TrPs, and improve muscle lengthening (stretches).
- Muscle spindles contain an alpha neuronal loop that innervates the extrafusal fibers and a gamma loop that innervates the intrafusal fibers inside the spindle itself. The gamma loop increases sensitivity of the spindle, making it more sensitive to rapid lengthening. The process of increasing spindle sensitivity is called gamma gain, which helps explain the origin of TePs in muscles and shows that repositioning can be an essential part of relieving these points and reducing muscle tension.
- TrPs and TePs have different characteristics and etiology. TrPs arise from a motor end-plate dysfunction that causes a localized bonding of actin and myosin, meaning the contraction knot of a TrP is a nonneurologic spasm. TePs arise through the process of gamma gain/hypersensitive muscle spindles, meaning the localized spasm is due to a neuromuscular signal.
- The inverse stretch reflex signaled by GTOs is more fully engaged by tension created with muscle contraction than by passive tension such as static stretch or manual compression.
- When considered together, the connective tissue elements of the human body form the fascial system; the single unifying structural element of the body. These connections between all cells, organs, and systems create a tensegrity system in which compression and tension forces are reflected throughout the body.
- While the fascial system is one fully integrated system, it can be divided into four layers. Moving from superficial to deep, the layers are pannicular, axial, meningeal, and visceral. All four layers are joined at the lumbosacral aponeurosis. For manual therapists, it is easiest to think of the system in two layers: superficial (pannicular) and deep (all others).
- The superficial fascia has seven horizontal bands around the torso that add rigidity and contour to the body. Four fascial planes within the cavities also provide some structural support to the torso as a whole, as well as supporting blood vessels, nerves, and viscera within the cavities.
- The connections between fascia, muscles, and bones can be traced and identified as forming 10 separate myofascial chains. The connections in these myofascial chains provide a "map" of the pathway for mechanical transmission of tension and compression throughout the body.
- The viscoelastic, thixotropic, and piezoelectric properties of fascia may explain some of the changes therapists feel during manual manipulations, but they do not fully explain how all of these "releases" occur or the long-term improvements patients experience from regular manual therapy sessions.
- The high number of mechanoreceptors in fascia provides essential proprioceptive information to the posture and movement centers of the brain.

Additionally, Ruffini and interstitial myofascial receptors appear to have direct links to the ANS, and stimulation of these results in decreased sympathetic tone, relaxation, and peace of mind.

- The presence of smooth muscle cells in fascia (under autonomic control) gives importance to the fascia's role in regulating both muscle and motor tone in the locomotor system.

REVIEW QUESTIONS

Short Answer

1. List and give a brief explanation of the three mechanical properties of fascia.
2. Briefly define and explain the difference between motor unit recruitment and muscle recruitment.
3. List the distinguishing characteristics of a trigger point.
4. Outline how to use reciprocal inhibition to relieve a cramp in the hamstrings.

Multiple Choice

5. Which term describes the process that increases the sensitivity of the muscle spindle?
   a. spindle hypersensitization
   b. gamma gain
   c. stretch reflex
   d. inverse stretch reflex
6. Which neuron loop of the muscle spindle stimulates normal skeletal muscle contraction?
   a. alpha
   b. beta
   c. gamma
   d. delta
7. All of the following manual therapy techniques except ___________ employ some type of repositioning designed to reverse gamma gain and indirectly reduce muscle tension.
   a. positional release
   b. strain counterstrain
   c. functional technique
   d. Rolfing
8. Which action provides the strongest stimulus to the GTOs and the inverse stretch reflex?
   a. sustained compression at the neuromuscular junctions
   b. bouncing types of stretch
   c. active muscle contraction
   d. slow gradual stretching
9. Gamma gain occurs in a muscle spindle when it is held in ___________ for an extended time?
   a. a lengthened position
   b. eccentric contraction
   c. a shortened position
   d. tetanic contraction
10. According to reciprocal inhibition, which muscle is inhibited when the biceps brachii is contracted?
    a. brachioradialis
    b. coracobrachialis
    c. deltoid
    d. triceps brachii
11. What is the pathophysiology behind trigger point development?
    a. calcium leak that causes actin and myosin bonding
    b. false stretch reflex report from the muscle spindle
    c. nerve compression or tension
    d. fibrotic build-up within the sarcomeres
12. The term used to describe the balance of tension and compression forces in the musculoskeletal system is
    a. thixotropic
    b. viscoelastic
    c. piezoelectric
    d. tensegrity
13. What is the functional relevance of the 11 horizontal fascial bands of the body?
    a. attach the appendages to the torso
    b. provide support and some rigidity to the torso
    c. separate the abdominopelvic organs
    d. provide a medium for force transmission between body regions

Continued on page 232

14. What is the purpose of the four horizontal planes of fascia in the body?
    a. transmit force and compression from skeletal muscle contraction to all body regions
    b. serve as the primary spacers and tension strut between organs in the cavities
    c. provide broad and deeply invested attachments for thoracic and abdominal muscles
    d. structurally divide both anterior and posterior cavities and support blood vessels and nerves
15. Which phrase best describes a tender point?
    a. contraction knot
    b. metabolic crisis
    c. always occurs in a taut band of myofascial tissue
    d. local muscle spasm indicating a sensitized muscle spindle
16. According to Willard, the deepest of the four layers of fascia is the
    a. meningeal
    b. visceral
    c. pannicular
    d. axial
17. Which of Myers' myofascial chains connects the plantar fascia of the foot to the fascia on the forehead?
    a. spiral line
    b. superficial back line
    c. superficial front line
    d. lateral line
18. The four horizontal fascial planes include the diaphragm, cranial base, pelvic floor, and
    a. abdominal aponeurosis
    b. thoracolumbar aponeurosis
    c. thoracic inlet
    d. linea alba
19. Which mechanical property of fascia gives it the ability to extend and slowly rebound?
    a. viscoelastic
    b. thixotropic
    c. piezoelectric
    d. stretchability
20. Which of the four types of fascial sensory receptors are particularly sensitive to lateral stretch, and stimulation leads to decreased sympathetic tone?
    a. Golgi organs
    b. pacinian receptors
    c. Ruffini receptors
    d. interstitial myofascial receptors
21. The most abundant type of fascial sensory receptor is the ____________ receptors.
    a. Golgi
    b. pacinian
    c. Ruffini
    d. interstitial myofascial
22. Since 40% to 45% of the neurons in a typical skeletal muscle nerve are sensory, and 80% of these are the interstitial myofascial receptors, it means that the majority of sensory information provided by muscle and fascia is directed to the ____________
    a. coordination of movement
    b. autonomic nervous system
    c. maintenance of posture
    d. regulation of muscle tone
23. The presence of smooth muscle cells in fascia leads us to believe that manipulation of fascia plays a primary role in the regulation of
    a. muscle tone
    b. motor tone
    c. heart rate
    d. blood pressure
24. The ability to sense movement is called
    a. proprioception
    b. equilibrium
    c. kinesthesia
    d. movement awareness
25. Which type of skeletal muscle fiber is characterized as a red or slow-twitch fiber?
    a. phasic fibers
    b. collagen fibers
    c. type II fibers
    d. type I fibers

References

1. Guyton AC, Hall JE. *Textbook of Medical Physiology*. 9th ed. Philadelphia, PA: W.B. Saunders; 1996.
2. Simons DG, Travell JG, Simons LS. *Myofascial Pain and Dysfunction: The Trigger Point Manual*, Vol. I. Upper Half of Body. 2nd ed. Philadelphia, PA: Lippincott Williams & Wilkins; 1999.
3. Chaitow L, DeLany JW. *Clinical Applications of Neuromuscular Techniques, Vol. I: The Upper Body*. Edinburgh, Scotland: Churchill Livingstone; 2000.
4. Kusunose RS. Strain and counterstrain for the upper quarter. Course syllabus, 1990. Presented at NWATA Conference Tacoma, WA.
5. Willard F. *Fascial Continuity: Four Fascial Layers of the Body*. Boston, MA: Disc 3 of First International Fascia Research Congress; 2007.
6. Schleip R. Fascial plasticity—a new neurobiological explanation: Part 2. *J Bodyw Mov Ther*. 2003;7(2):104–116.
7. Myers TW. *Anatomy Trains: Myofascial Meridians for Manual and Movement Therapists*. Edinburgh, London, New York: Churchill Livingstone; 2001.
8. Schultz RL, Feitis R. *The Endless Web: Fascial Anatomy and Physical Reality*. Berkeley, CA: North Atlantic Books; 1996.
9. Juhan D. *Job's Body: A Handbook for Bodywork*. Expanded edition. Barrytown, NY: Station Hill; 1998.
10. Barnes JF. *Myofascial Release I: Seminar Workbook*. Paoli, PA: MFR Seminars; 1989.
11. Mannheim C. *The Myofascial Release Manual*. 3rd ed. Thorofare, NJ: Slack; 2001.
12. Schleip R. Fascial plasticity—a new neurobiological explanation: Part 1. *J Bodyw Mov Ther*. 2003;7(1):11–19.
13. Threlkeld AJ. The effects of manual therapy on connective tissue. *Phys Ther*. 1992;72(12):893–902.
14. Brattberg G. Connective tissue massage in the treatment of fibromyalgia. *Eur J Pain*. 1999;3:235–245.
15. Yahia L et al. Sensory innervation of human thoracolumbar fascia. *Acta Orthop Scand*. 1992;63(2):195–197.

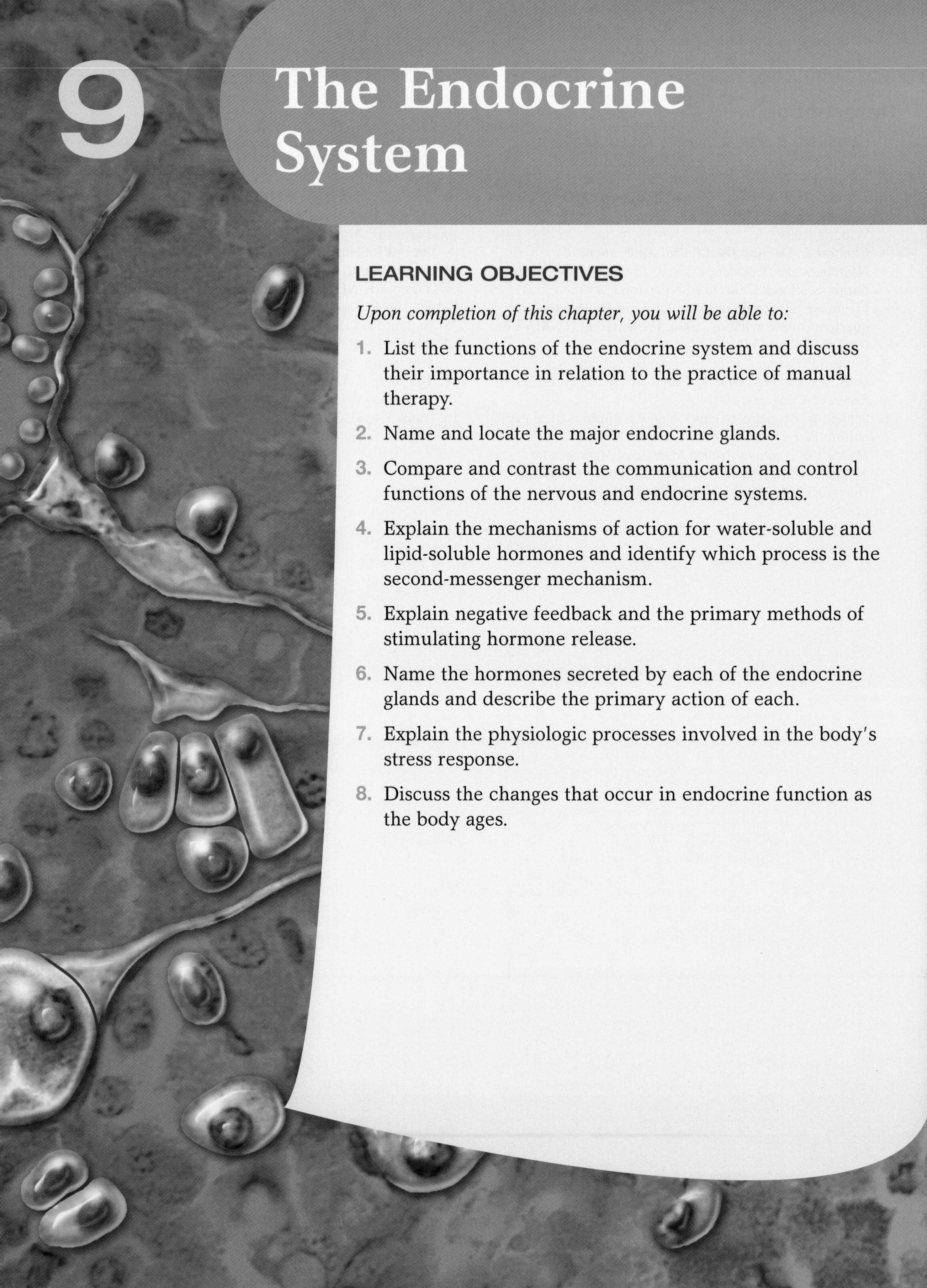

9 The Endocrine System

LEARNING OBJECTIVES

Upon completion of this chapter, you will be able to:

1. List the functions of the endocrine system and discuss their importance in relation to the practice of manual therapy.
2. Name and locate the major endocrine glands.
3. Compare and contrast the communication and control functions of the nervous and endocrine systems.
4. Explain the mechanisms of action for water-soluble and lipid-soluble hormones and identify which process is the second-messenger mechanism.
5. Explain negative feedback and the primary methods of stimulating hormone release.
6. Name the hormones secreted by each of the endocrine glands and describe the primary action of each.
7. Explain the physiologic processes involved in the body's stress response.
8. Discuss the changes that occur in endocrine function as the body ages.

> "What we feel and think and are is to a great extent determined by the state of our ductless glands and viscera."
>
> ALDOUS HUXLEY

KEY TERMS

androgen (AN-dro-jen)
catecholamine (kat-ah-KO-lah-meen)
corticoid (KOR-tih-koyd)
endocrine (EN-do-krin) **gland**
exocrine (EKS-o-krin) **gland**
glucocorticoid (GLU-ko-KOR-tih-koyd)
hormone (HOR-mone)
lipid-soluble (LIP-id SOL-u-bel) **hormone**
mineralocorticoid (MIN-eh-rahl-o-KOR-tih-koyd)
negative-feedback mechanism
second-messenger mechanism
secrete (seh-KREET)
secretion (seh-KREE-shun)
stress response
water-soluble (WAH-ter SOL-u-bel) **hormone**

Primary System Components

- **Hypothalamus**
- **Pituitary gland**
- **Pineal gland**
- **Thyroid gland**
- **Parathyroid glands**
- **Thymus**
- **Pancreas**
- **Adrenal glands**
- **Ovaries**
- **Testes**

Primary System Functions

- **Body-wide communication and control of growth and development**
- **Body-wide communication and control of metabolism**
- **Homeostatic balance of the blood**

The endocrine system, together with the nervous system, works to control and coordinate the functions of the body's cells and tissues. Consisting of a seemingly unrelated group of organs and tissues (FIGURE 9.1), this system communicates by producing chemicals known as hormones that circulate through the bloodstream. This chemical messaging process seems almost archaic when compared to the sophisticated neuronal network of the nervous system. Physiologic responses are initiated in a few seconds, hours, or even days instead of in milliseconds, as occurs with nerve impulses. However, the endocrine system is well designed to regulate key functions such as growth, development, aging, and a wide spectrum of homeostatic mechanisms. Through slower but consistent feedback loops, the endocrine system reads, adjusts, and readjusts various reactions to balance fluids, electrolytes, pH level, metabolic activities, and energy supplies—a huge number of chemical processes every second.

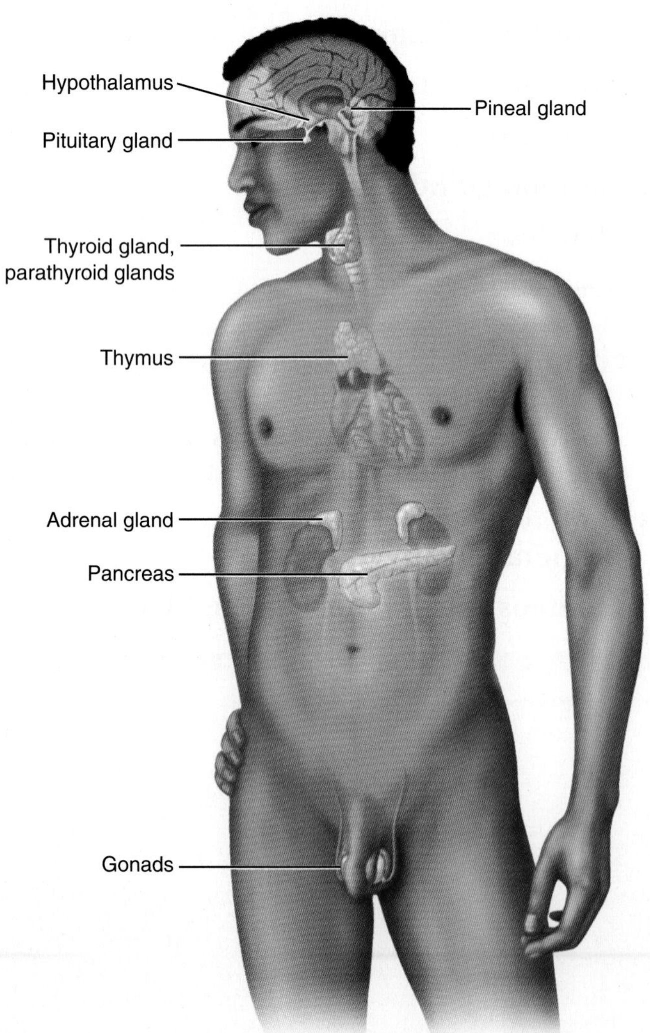

FIGURE 9.1 Glands of the endocrine system.

In most cases, manual therapies only have an indirect effect on endocrine functions due to their stress-reducing benefits. However, some zone and meridian therapies are specifically focused on stimulating major glands and organs, including the endocrine glands. For example, in foot reflexology, key organs and glands are represented as specific points or regions on the sole of the foot (FIGURE 9.2). Manual pressure applied to these reflex points is believed to stimulate change in the related organ or gland to help normalize its function. However, the exact mechanism of effect for foot reflexology has not been clearly defined, even though ample anecdotal evidence of its effectiveness exists.

In this chapter, the endocrine glands and tissues, the hormones they secrete, and the physiologic responses they cause are discussed. You will discover what stimulates hormone secretions, how their levels are regulated, and the mechanism by which hormones create changes in specific target cells. In addition, a section on the body's stress response explores important links between the endocrine and nervous systems.

HORMONES

Endocrine glands and tissues secrete specialized chemical messengers called **hormones** that act on target cells to initiate and regulate multiple physiologic responses (FIGURE 9.3). The word *hormone* originates from a Greek word meaning "to set in motion" or "spur on." The majority of these powerful chemical mediators are called ***circulating hormones***, because they are secreted into the bloodstream and circulated throughout the body. At some point, hormones come in contact with specific target cells whose receptors are sensitive to their particular chemical make-up. These receptors are specialized proteins located either on the plasma membrane or within the target cell. When a hormone is circulated through the body, only target cells with appropriate receptors can bond with the hormone to produce a physiologic response. While hormones are target-specific, any single hormone may stimulate several different target cells to produce a variety of physiologic responses. These

By the Way

Target cells have anywhere from 2,000 to 100,000 receptors for a single hormone. According to Candace Pert in *Molecules of Emotion*, "If you were to assign a different color to each of the receptors that scientists have identified, the average cell surface would appear as a multicolored mosaic of at least seventy different hues."

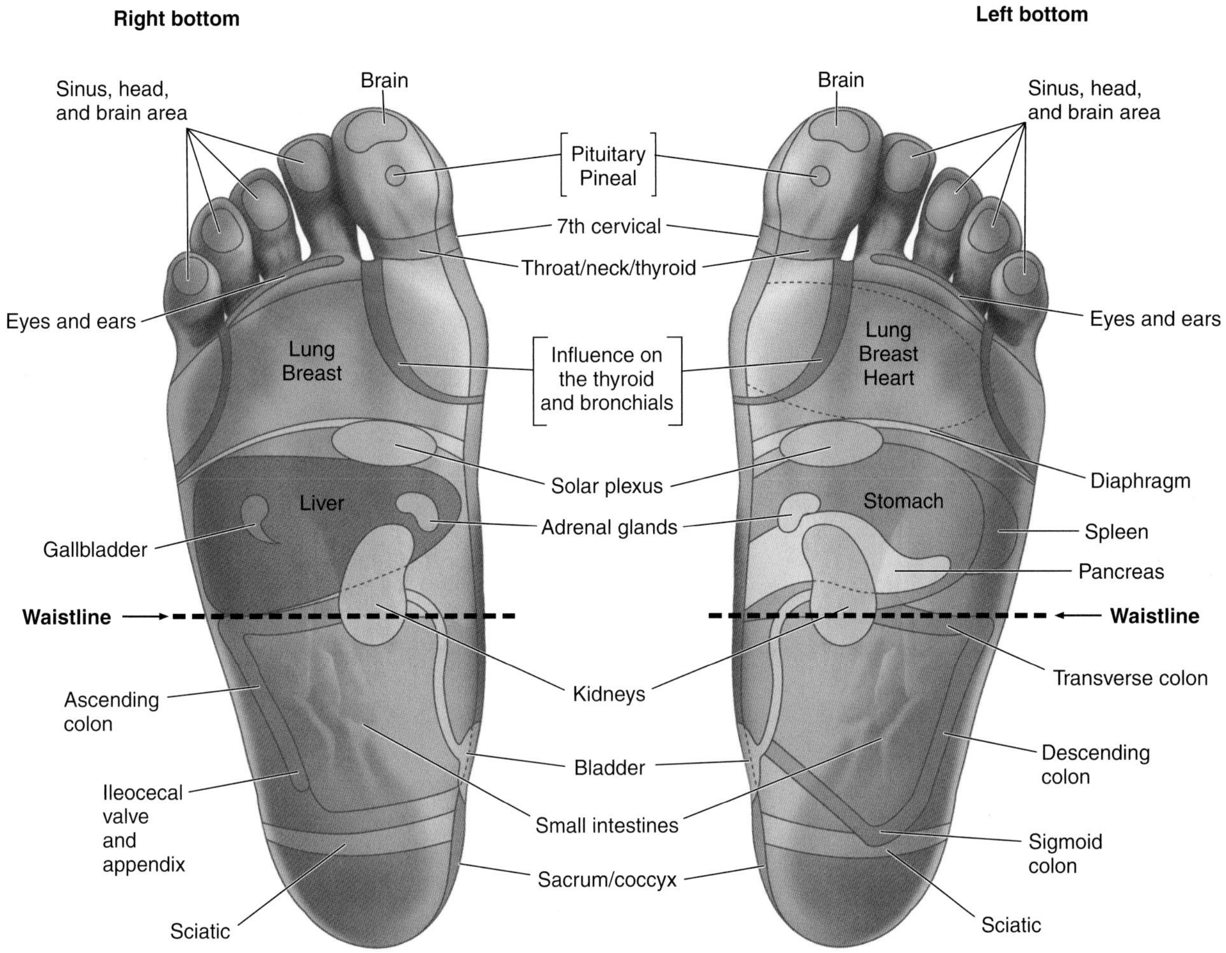

FIGURE 9.2 ▸ Foot reflexology map. Major glands and organs of the body are represented as specific zones or points on the feet. In foot reflexology, digital pressure is used over these points to stimulate change in the correlating organ.

responses can include protein synthesis, increased cellular transport, changes in plasma membrane permeability, or changes in metabolic reactions and rates.

Chemically speaking, there are two broad categories of hormones: water-soluble and lipid-soluble. Most hormones are **water-soluble hormones** made up of amino acids. Some are very simple molecules, while others are made of chains of 3 to 49 amino acids called **peptides** or even longer chains of 50 to 200 amino acids called **proteins. Lipid-soluble hormones** primarily include **steroid hormones**, which are made from cholesterol, and thyroid hormones that are composed of a specific amino acid and iodine. Whether a hormone is water-soluble or lipid-soluble affects its ability to travel through the bloodstream and determines its mechanism of action on the target cells.

Mechanisms of Hormone Action

While water-soluble hormones are easily transported through the blood, they are unable to pass through the plasma membrane of target cells due to the phospholipid

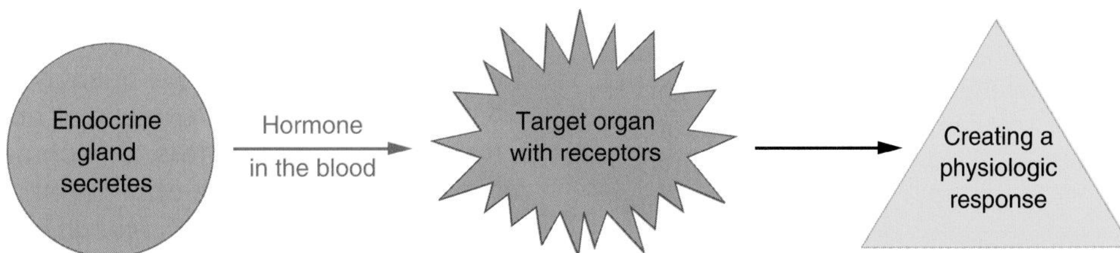

FIGURE 9.3 ▸ Endocrine function. Endocrine glands stimulate metabolic changes by secreting hormones into the bloodstream to be circulated throughout the body. Target organs have hormone-specific receptor cells that are stimulated by that hormone to produce cellular changes (the physiologic response). To view an animation on hormonal control, visit http://thePoint.lww.com/Archer-Nelson.

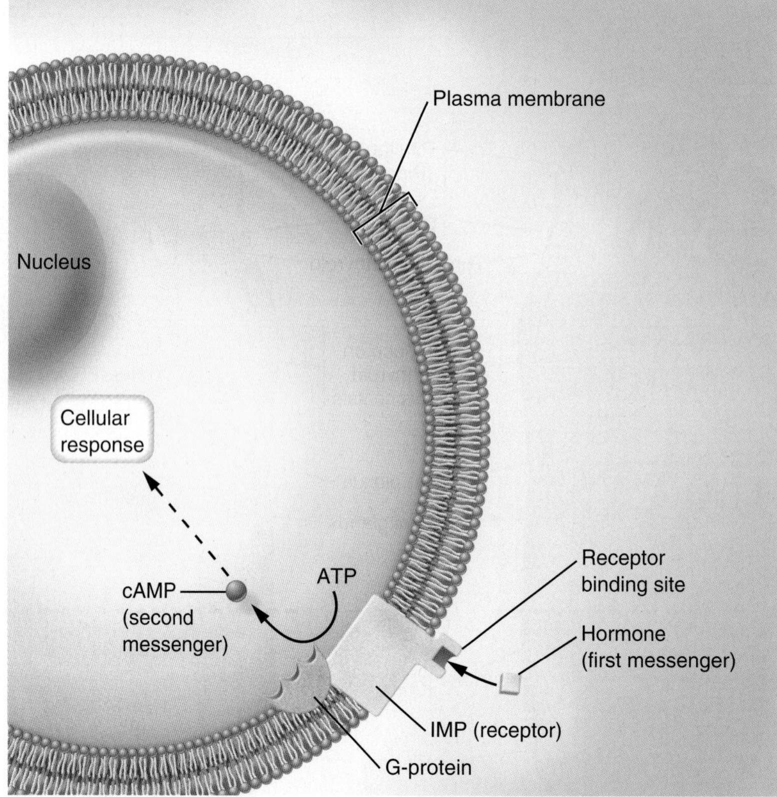

FIGURE 9.4 ▶ Second messenger mechanism. Water-soluble hormones act as the *first messenger*, delivering a message to an IMP in the target cell's plasma membrane. This creates cAMP, the *second messenger*, which stimulates the cellular change (physiologic response) in the target organ, gland, or tissue.

structure of the membrane. Instead, water-soluble hormones are recognized by receptors on the surface of the cell membrane (integral membrane proteins or IMPs), which bond with the hormone. This connection between hormone and receptor activates a complex intracellular protein, a ***G protein***, which initiates the conversion of ATP in the cytoplasm into ***cyclic adenosine monophosphate***, or ***cyclicAMP*** (cAMP). The presence of cAMP in turn activates proteins within the cell that produce a designated physiologic change. This sequence of events is called the **second-messenger mechanism** (▶ Figure 9.4). The hormone acts as the *first messenger* delivering a chemical message to the receptor IMP, similar to a bicycle messenger delivering a package to the front desk of a large office. The receptionist signs for the package (activation of the G protein) and then hands it off to an assistant, the *second messenger*, who delivers the package to the employee for whom it is intended. That employee is the person who will act on or respond to the contents of the package. In the cell, cAMP serves as a second messenger, activating the specific proteins within the cell that will produce the target's physiologic response. Since the majority of hormones are water-soluble, the second-messenger mechanism is the most prevalent method of hormone action in the body.

In contrast to water-soluble hormones, the chemical make-up of lipid-soluble hormones requires them to be transported through the bloodstream by special carrier molecules. Their method of target cell activation is more direct because they are able to pass through the plasma membrane to bond with receptors in the cytoplasm or nucleus (▶ Figure 9.5). This hormone–receptor complex activates a process in which specific genes (segments of DNA) are copied onto an RNA molecule that then directs the production of a new protein by the ribosomes. This new protein alters the metabolic activity of the cell.

The intensity of a target cell's response depends on the concentration of hormone and the number of

By the Way

Cells constantly adapt to their environment by increasing or decreasing the numbers of receptors on their plasma membranes. If a hormone is available in larger-than-normal quantities for a period of time, target cells will decrease or ***down-regulate*** their number of receptors to become less sensitive to the high volume of hormone. In contrast, ***up-regulation*** occurs (the number of receptors is increased) to make a target cell more sensitive to a hormone when levels are abnormally low.

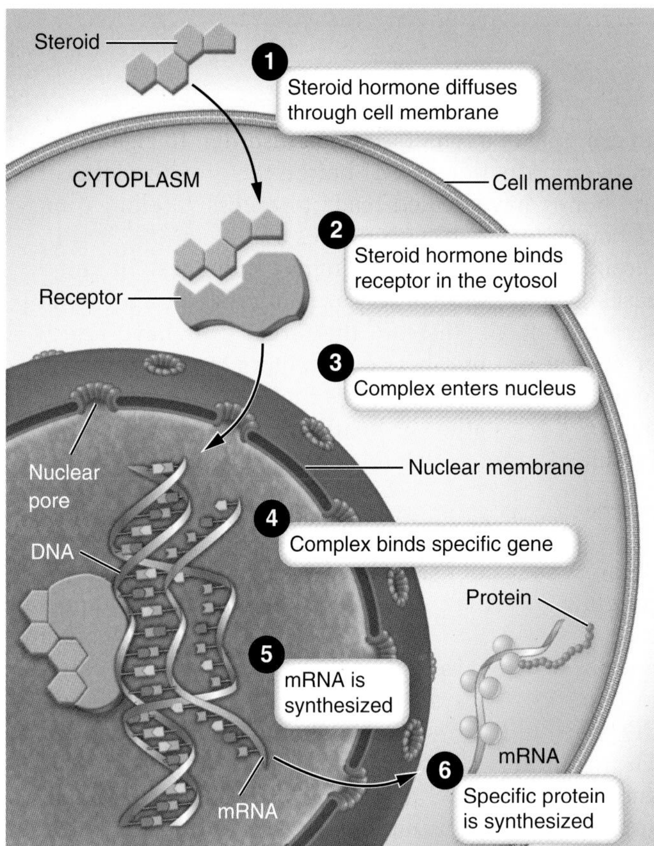

FIGURE 9.5 ▸ Mechanism of action for lipid-soluble hormones. Lipid-soluble hormones, such as steroid hormones, diffuse through the cell membrane and bind to receptors inside the cell. This hormone-receptor complex causes the synthesis of mRNA, which directs the production of a new protein that alters the metabolic activity of the target cell.

receptors available. If either the level of hormone or number of receptors on the target cell is high, the cell responds more intensely. The opposite is also true: low hormone concentrations or fewer receptors create a weak target cell response. Additionally, the intensity of some hormone actions are increased or decreased by the presence of other hormones. For example, two hormones may act together in a **synergistic effect** to enhance or intensify a target cell's response. Other hormones oppose the actions of one another, having an **antagonistic effect** on the target cells, while some hormones have a **permissive effect** that increases a target cell's sensitivity to another hormone.

Control of Hormone Secretion

When an endocrine gland is stimulated, it releases a small amount of hormone in quick bursts to increase the blood level of that hormone. The frequency of these bursts and therefore, the hormone level, is regulated through the **negative-feedback mechanism** described in Chapter 1. With hormones, the physiologic response that occurs in target cells ultimately counteracts the stimulus that caused the original release of hormone from the gland (▸ FIGURE 9.6). For example, when blood contains a high concentration of calcium ions, the thyroid is stimulated to release calcitonin. This hormone causes calcium to be removed from the blood and deposited into bones, thus lowering the amount of calcium in the bloodstream. As blood calcium levels drop, the thyroid is no longer stimulated to release calcitonin. In this way, negative feedback prevents either over- or under-secretion of a hormone by its gland. While most hormones are regulated through negative feedback, recall from Chapter 1 that oxytocin, a pituitary hormone, is a rare example of a hormone regulated via a **positive feedback mechanism**. In positive feedback, the physiologic response reinforces the original stimulus to continue the release of the hormone.

Three types of stimuli activate the endocrine glands:

- ***Hormonal stimulus***—Many glands are stimulated by hormones from other endocrine glands or tissues. The relationship between the hypothalamus and the

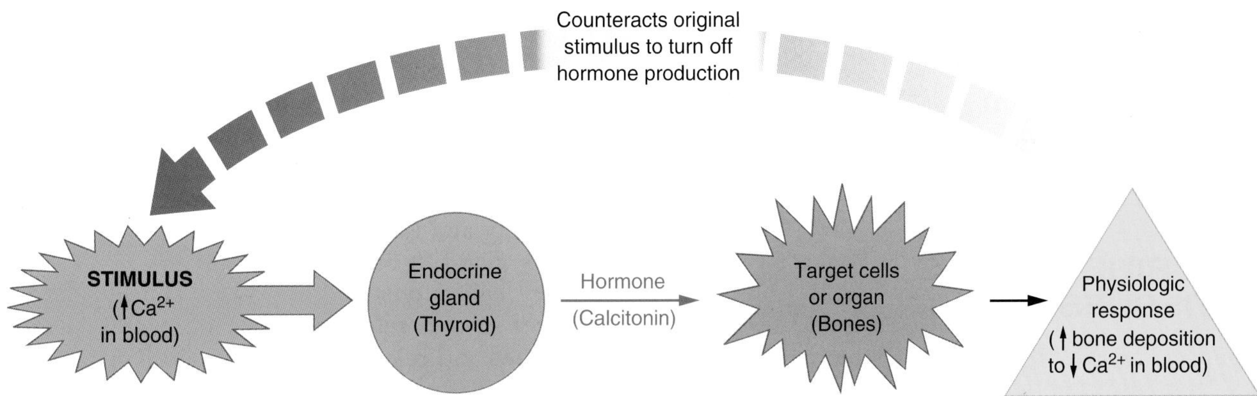

FIGURE 9.6 ▸ Negative feedback mechanism. A stimulus causes the endocrine gland to release a hormone. The physiologic response of the target cells ultimately counteracts the stimulus that caused the hormone's initial release.

Pathology Alert **Endocrine Disorders**

The most common pathologies of the endocrine system involve either ***hyposecretion*** (too little) or ***hypersecretion*** (too much) of a hormone. The causes of these disorders can be faulty stimulation mechanisms or poor regulatory feedback loops caused by genetic predispositions, dietary insufficiencies, or even malignancies. Pituitary dwarfism is a simple example of a genetically caused hyposecretion disorder in which underproduction of human growth hormone (GH) before puberty leads to small stature and associated physical disabilities. In contrast, gigantism is an example of a hypersecretion disorder. In gigantism, an overproduction of GH before the growth plates of the bones have sealed leads to excessive elongation of bones during growth phases and a completely different set of pathophysiologic challenges. In some cases, endocrine disorders are not related to hormone secretion at all. Instead, the target cells have faulty or insufficient numbers of receptors for the hormone to act upon. Type 2 diabetes mellitus is often caused by down-regulation of insulin receptors. Diabetes is discussed in more detail later in this chapter.

pituitary is a prime demonstration of this phenomenon. The hypothalamus releases stimulating and inhibiting hormones that control the function of the pituitary.

- ***Changes in blood concentrations***—In other endocrine glands, a change in the level of a specific ion or nutrient in the blood stimulates hormone secretion. Calcitonin release from the thyroid gland is an example of this sort of stimulus.
- ***Neurologic stimulus***—In a few cases, neurons stimulate specific glands to release their hormones. The relationship between the hypothalamus and posterior pituitary and the sympathetic stimulation of the adrenal glands are examples of this type of endocrine gland activation.

What Do You Think? 9.1

- In addition to the endocrine system, what other system uses the terms *antagonist* and *synergist* to describe structures involved in certain functions? Is there a difference in their meaning between the two systems?
- What would happen if hormones were not target-specific?
- Give an analogy for the negative feedback mechanism using everyday objects or activities.

ENDOCRINE GLANDS AND TISSUES

As depicted in FIGURE 9.1, the major endocrine glands include the pituitary, pineal, thyroid, parathyroids, thymus, adrenals, gonads (ovaries and testes), and the pancreas. Unlike other body systems, the endocrine organs are not structurally linked. Instead, they function together as individual glands and specialized tissues that secrete hormones into blood and lymph. **Endocrine glands** are ductless, in contrast to **exocrine glands** such as the sebaceous glands in the skin and the salivary glands, which have ducts. The pituitary, pineal, thyroid, parathyroids, thymus, and adrenals are endocrine glands. The pancreas and gonads have dual functions as both endocrine and exocrine glands.

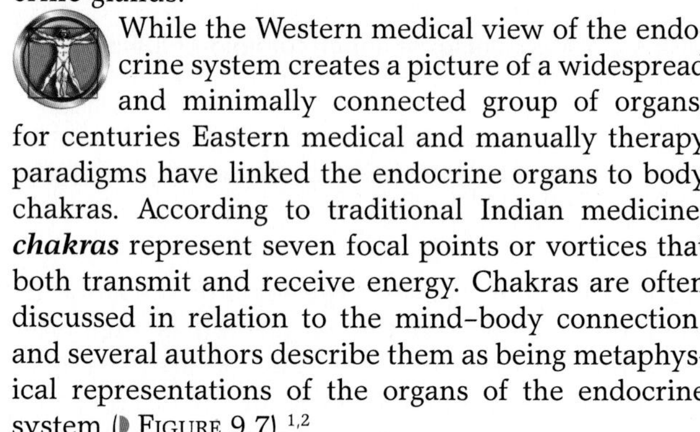

While the Western medical view of the endocrine system creates a picture of a widespread and minimally connected group of organs, for centuries Eastern medical and manually therapy paradigms have linked the endocrine organs to body chakras. According to traditional Indian medicine, ***chakras*** represent seven focal points or vortices that both transmit and receive energy. Chakras are often discussed in relation to the mind–body connection, and several authors describe them as being metaphysical representations of the organs of the endocrine system (FIGURE 9.7).[1,2]

The Brain: Hypothalamus, Pituitary, and Pineal Glands

In Chapter 7, the **hypothalamus** was introduced as a structural component of the brain that functions as the control center for autonomic functions and as a key link between the nervous and endocrine systems (FIGURE 9.8). The inferior aspect of the hypothalamus is structurally and functionally attached to the dual-lobed pituitary gland. It has a circulatory link to the anterior pituitary called ***portal circulation***. The hypothalamus produces several **releasing hormones** and **inhibiting hormones** and releases them into portal circulation. In this way, the hypothalamus controls and regulates all anterior pituitary secretions. The hypothalamus has a neuronal link with the posterior lobe of the pituitary. It

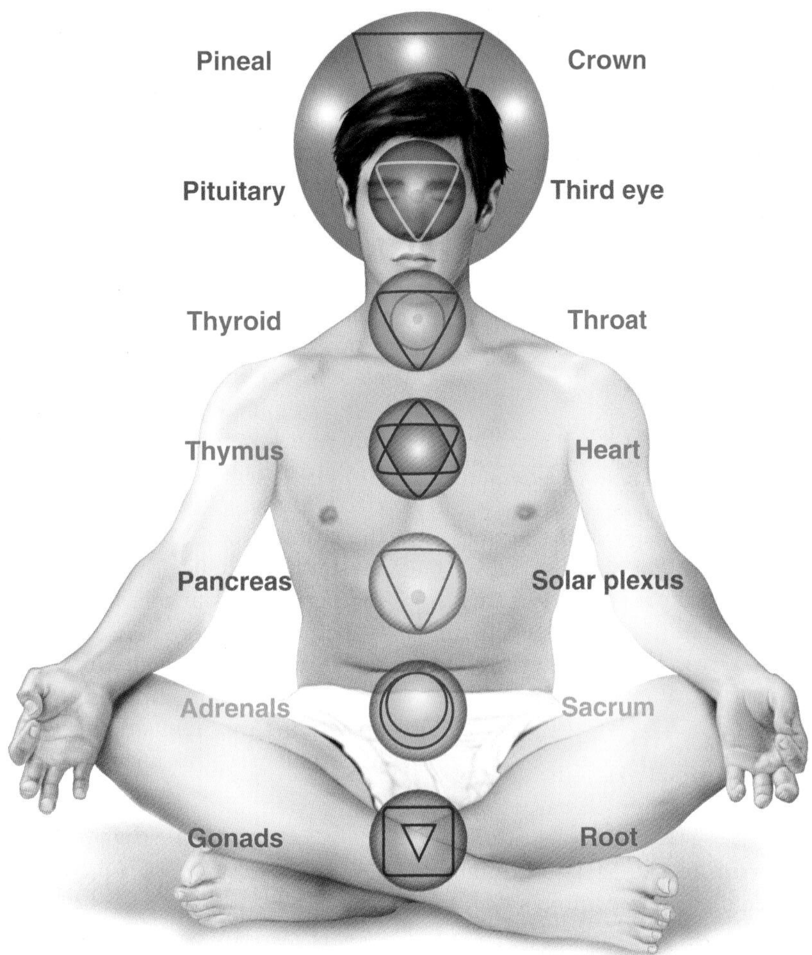

FIGURE 9.7 ▸ Chakras and the endocrine system. Each chakra, or focal point of energy, correlates to a specific endocrine gland.

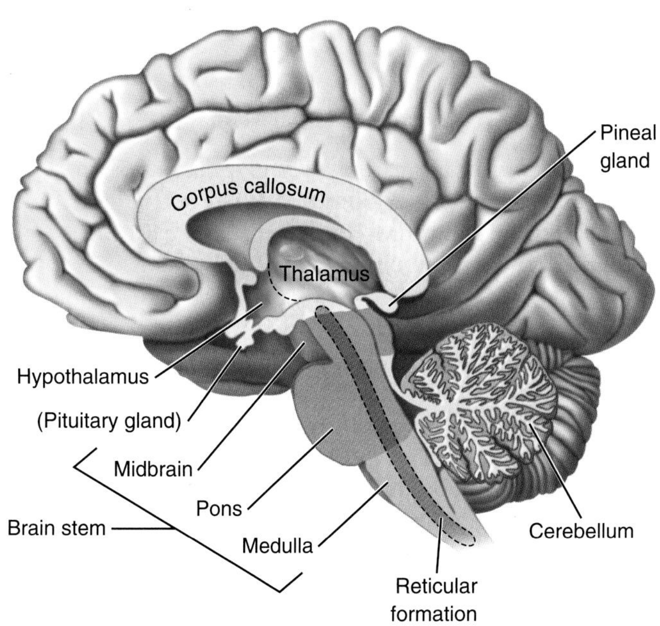

FIGURE 9.8 ▸ The hypothalamus, pituitary and pineal glands. These three endocrine organs are located in the brain, superior to the brain stem.

produces oxytocin and antidiuretic hormone (ADH) and transports them to the posterior pituitary lobe along special neurons known as ***neurosecretory cells.*** The stored hormones in the posterior pituitary are released when signaled by nerve impulses from the hypothalamus.

The **pituitary gland** is only the size of a pea, yet it is considered the keystone, or ***master gland***, for the entire endocrine system because the hormones it releases stimulate many other endocrine glands. While the pituitary exerts significant influence on the other endocrine glands, it is important to remember that its master is the hypothalamus. The pituitary is suspended from the inferior aspect of the hypothalamus via the ***pituitary stalk*** and rests in the small depression of the sphenoid bone called the ***sella turcica*** (▸ FIGURE 9.9). Structurally, it has an ***anterior lobe***, called the ***adenohypophysis*** because it is comprised of glandular tissue, and a ***posterior lobe*** that is mostly nerve tissue, making it the ***neurohypophysis***. Because the two lobes are stimulated through different mechanisms and secrete different types of hormones, it is easiest to describe the function of each lobe as a separate gland.

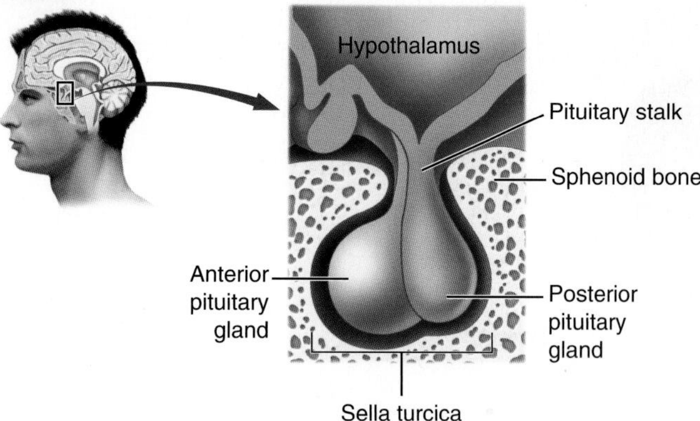

FIGURE 9.9 ▶ Hypothalamus and pituitary links. Located in the sella turcica of the sphenoid bone, the pituitary gland is suspended from the hypothalamus by the pituitary stalk. The anterior lobe of the pituitary is controlled by hormone releases from the hypothalamus, while the posterior lobe is under neural control.

The anterior pituitary produces and secretes a wide variety of hormones that influence growth and homeostatic processes. The hormones that stimulate other endocrine glands are collectively referred to as ***tropic hormones.*** They include:

- **Thyroid-stimulating hormone (TSH)**—stimulates and controls secretion of specific thyroid hormones (T_3 and T_4).
- **Adrenocorticotropic hormone (ACTH)**—stimulates secretions of specific adrenal cortex hormones (***glucocorticoids***).
- **Luteinizing hormone (LH)**—stimulates hormone secretions of the male and female gonads. This includes estrogen and progesterone from the ovaries in females and testosterone from the testes in males.
- **Follicle-stimulating hormone (FSH)**—works with LH to stimulate hormone release from the gonads. For this reason, FSH and LH are collectively referred to as ***gonadotropic*** hormones or ***gonadotrophs***. The gonadotrophs are also responsible for the production and maturation of reproductive cells (eggs and sperm).

Although it is not a tropic hormone, **growth hormone**, also called **somatotropin**, is the most abundant hormone produced and released by the anterior pituitary. GH stimulates and regulates growth in all tissues and plays an important role in some aspects of metabolism. Actions of GH include stimulating the release of various growth factors that promote protein synthesis and tissue repair, inhibit protein breakdown, and stimulate fat breakdown. Two other hormones released by the anterior pituitary are **prolactin**, which initiates milk production in women after giving birth, and **melanocyte-stimulating hormone (MSH)** that affects skin pigmentation. In contrast to the anterior lobe that produces and releases its own hormones, the posterior pituitary is simply a storage area for two hypothalamic hormones. **Oxytocin** is released during childbirth to stimulate powerful contractions of the uterine muscles and weaker contractions in nursing women to help eject breast milk. The second posterior pituitary hormone, **antidiuretic hormone**, functions to inhibit or prevent urine production in the kidneys, as its name implies. When the body becomes dehydrated, the hypothalamus stimulates the posterior pituitary to release ADH, which causes the kidneys to reabsorb more fluid and decrease urine production. This increases blood volume and prevents severe dehydration. When released in large amounts, ADH also creates constriction of the smaller arteries in the body, which tends to raise blood pressure. This explains why ADH is also known as **vasopressin**.

The pineal gland is also found within the brain. This tiny pine-cone–shaped structure secretes the hormone **melatonin**, which plays a role in regulating the body's ***circadian rhythm*** (sleep-wake cycle). While much about the stimulation and function of the pineal gland is still a mystery, it is clear that melatonin levels are higher during periods of darkness and sleep than in strong light or wakefulness. There may also be some correlation between melatonin and reproductive functions, but this has only been measured in animals with specific breeding seasons. ▶ TABLE 9-1 provides an overview of the hormones of the hypothalamus, pituitary, and pineal glands.

What Do You Think? 9.2

- How does your understanding of the relationship between the hypothalamus and pituitary support your understanding of the mind–body concept of health and well-being?
- Do you agree with the title of "master gland" for the pituitary? Why or why not?

Thyroid and Parathyroid Glands

The **thyroid gland** is a highly vascular tissue mass located in the anterior neck (▶ FIGURE 9.10). The soft lateral lobes of the thyroid gland are connected by the ***isthmus*** and are easily palpated on either side of the trachea, just inferior to the voice box (larynx). This important endocrine gland secretes **thyroid hormones** that are essential for regulating cellular metabolism, and **calcitonin**, which plays a key role in the regulation of blood calcium levels.

There are two types of thyroid hormones, each identified by the number of iodine atoms: **triiodothyronine** (T_3) contains three iodine atoms, and **thyroxine** (T_4) contains four iodine atoms. Both hormones increase cellular metabolism and are secreted in response to TSH released from the anterior pituitary. The anterior pituitary secretes TSH in response to ***thyrotropin-releasing hormone*** from the hypothalamus when it senses low levels of either T_3 or T_4 or a low metabolic rate in the body. The **basal metabolic rate (BMR)** is the rate of oxygen consumption

Table 9-1 Hormones of the Hypothalamus, Pituitary, and Pineal Glands

Gland		Stimulated by	Hormone Released	Target	Physiologic Response(s)
Hypothalamus		Stimulated by sensory input plus emotions and perceptions, the hypothalamus makes releasing and inhibiting hormones that regulate the anterior pituitary and produces the hormones released by the posterior pituitary.			
Pituitary	Anterior	Hypothalamic hormones; blood levels of T_3	TSH	Thyroid	Increased secretion of T_3 and T_4
		Hypothalamic hormones; glucocorticoid levels	ACTH	Adrenal cortex	Increased secretion of glucocorticoids
		Hypothalamic hormones	LH	Ovaries and testes	Increased secretion of sex hormones
		Hypothalamic hormones	FSH	Ovaries and testes	Stimulates production of ova and sperm
		Hypothalamic hormones	Prolactin	Breast tissue	Stimulates milk production
		Hypothalamic hormones	GH	Body cells particularly skeletal muscle and bone	Stimulates growth and increases metabolism
	Posterior	Hypothalamus in response to water balance	ADH	Kidneys	Promotes retention of water
		Uterine stretching and breast suckling	Oxytocin	Uterus Breast tissue	Stimulates contraction Stimulates release of milk
Pineal		Light and dark cycles	Melatonin	Brain	Triggers sleep

by the body at rest. T_3 and T_4 increase the rate at which cells utilize oxygen to produce ATP, thus increasing the rate at which they metabolize carbohydrates, lipids, and proteins. In addition to their role in metabolism, thyroid hormones work synergistically with GH and insulin to increase the rate of growth and development, especially of the skeletal and nervous systems. Because all body functions depend on maintaining an adequate energy supply, it is important that a person's daily diet provides sufficient levels of iodine, so that the body can produce these essential thyroid hormones.

Calcitonin released by the thyroid plays an important role in regulating blood calcium (Ca^{2+}) levels. When levels are high, calcitonin is released to inhibit the breakdown of bone and increase calcium uptake into the bone matrix. This leads to a decrease in Ca^{2+} in the blood, which shuts down the release of calcitonin through negative feedback (FIGURE 9.11).

The **parathyroid glands** on the posterior aspect of the thyroid (FIGURE 9.10) secrete **parathyroid hormone (PTH)**, which works as an antagonist to calcitonin. In contrast to calcitonin, PTH release is stimulated when blood calcium levels are too low (FIGURE 9.11). PTH stimulates bone resorption by increasing the numbers and activity of osteoclasts. This increases the breakdown of bone matrix to release Ca^{2+} from the bone and increase blood calcium levels. A secondary effect of PTH is to stimulate the kidneys to release ***calcitriol***, the active form of vitamin D, which accelerates absorption of calcium in the digestive tract to further increase blood calcium levels.

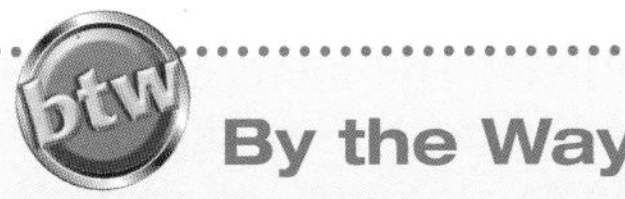

By the Way

The thyroid gland weighs about 1 oz, and it is estimated that it receives 80 to 120 ml of blood per minute.

Thymus

The **thymus** is centrally located in the thoracic cavity, anterior to the aorta and deep to the body of the sternum (see FIGURE 9.1). It produces and secretes **thymosin**, a hormone that plays an essential role in the development and maturation of specialized immune cells called T-cells. In childhood, the thymus can extend from mid-sternum up to the thyroid gland and weigh about 2.3 oz. After

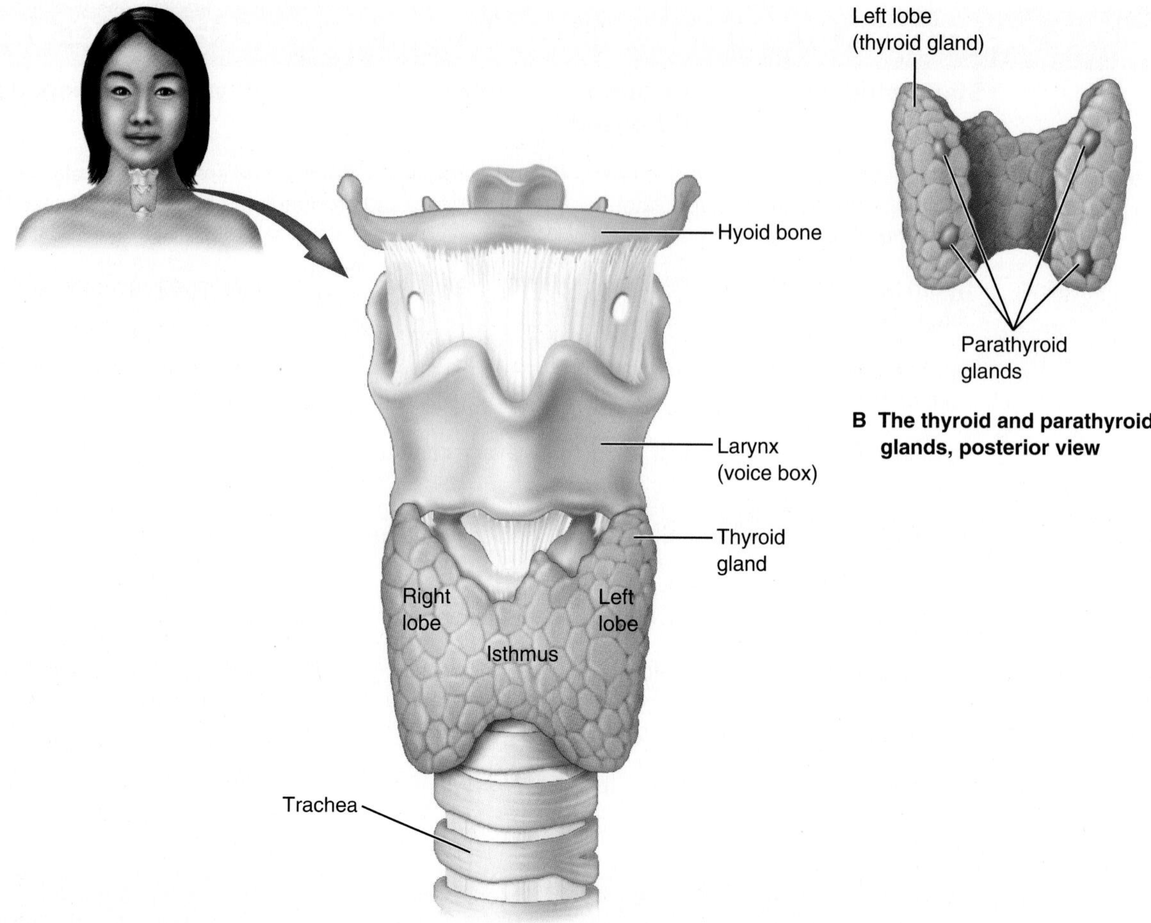

FIGURE 9.10 ▸ The thyroid and parathyroid glands. A. The thyroid gland is anterior to the trachea and just inferior to the voice box (larynx). **B.** The parathyroids are nodes located on the lateral-posterior aspect of the thyroid.

puberty, it begins to shrink and may be less than half this size by adulthood. In the elderly, it may weigh as little as 0.1 oz. The diminishing size of the thymus is due to its decreasing functional role throughout adulthood. More detailed information on the role of the thymus and the immune response is discussed in Chapter 12.

Pancreas

The **pancreas** is located behind the stomach in the medial aspect of the upper left abdominal quadrant (see FIGURE 9.1). While its dominant function is as an exocrine gland in the digestive system, it plays an essential endocrine role as well. The endocrine functions of the pancreas originate in a group of specialized cells called **pancreatic islets**, or **islets of Langerhans**. These islets are interspersed throughout the organ and function like tiny, individualized endocrine organs (▸ FIGURE 9.12). Each islet includes four types of hormone-producing cells, two of which are of primary importance: ***beta cells***, which produce **insulin**, and ***alpha cells***, which produce **glucagon**. Insulin and glucagon work as antagonists in the regulation of blood glucose levels, with insulin *decreasing* blood glucose levels and glucagon *increasing* them.

Insulin is unique because it is the only hormone that specifically decreases blood glucose levels. It does so by improving the rate at which glucose is transported out of the bloodstream and into the cells, thereby increasing cellular metabolism of glucose (▸ FIGURE 9.13). Insulin also promotes fat and protein synthesis. In opposition to insulin, glucagon *increases* the level of glucose in the blood by stimulating the release of stored glucose, or glycogen, from the liver. The increased release of glycogen, called ***glycogenolysis***, raises the level of glucose in blood. In addition to glucagon, several other hormones such as GH and the glucocorticoids from the adrenal glands also act to increase blood glucose levels. ▸ TABLE 9.2 provides an overview of the hormones of the thyroid, parathyroids, thymus, and pancreas.

Pathology Alert Diabetes Mellitus

Diabetes mellitus is a disorder of insulin action and/or secretion that results in high blood glucose (hyperglycemia). It is classified as either type 1 or type 2. ***Type 1***, sometimes referred to as juvenile diabetes, involves insufficient production of insulin from the beta cells in the pancreas. This type represents about 10% of all diabetes cases. Believed to be an autoimmune disorder, type 1 diabetes can be caused by several factors, including genetics or certain drug and chemical exposures. Type 1 diabetes is managed through daily insulin injections and diet. ***Type 2*** represents the remaining 90% of cases in which insulin production may or may not be an issue. In some peoples, a long-term diet high in simple sugars and carbohydrates has overwhelmed the pancreas, causing the beta cells to become less efficient at producing insulin. In other peoples, insulin production remains normal but there is simply too much glucose in the blood to be processed. For others, the insulin production may be normal or even high, but target cells have down-regulated, providing few insulin receptor sites. Type 2 diabetes is managed through changes in diet and exercise, and sometimes oral drugs and/or insulin are prescribed. Stress management can often be helpful.

People with diabetes are susceptible to a wide range of long-term complications such as edema, kidney disease, impaired vision, cardiovascular disease, ulcers, and neuropathy. Individuals with diabetes can receive most forms of manual therapy as long as their tissue has good circulation, no significant neuropathy, and is free of ulcers. Some preliminary research shows an average 20- to 40-point drop in blood sugar levels after massage.[3] Therefore, diabetic clients should be advised to take blood glucose readings before and after therapy to monitor and regulate their response.

Adrenal Glands

The **adrenal glands**, also called the **suprarenal glands**, sit directly on top of the kidneys (FIGURE 9.14). Each gland has two regions that function as independent endocrine glands. The outer region, the **adrenal cortex**, secretes steroid hormones called **corticoids**. The inner or central portion of each gland, the **adrenal medulla**, secretes a group of hormones called catecholamines.

Hormones of the Adrenal Cortex

The adrenal cortex has three layers of cells that secrete different corticoid hormones (FIGURE 9.14C). The outermost layer secretes **mineralocorticoids** that help regulate the levels of specific minerals in the blood. The primary mineralocorticoid is **aldosterone**, which controls sodium (Na^+) and potassium (K^+) ion levels in the blood and helps to regulate blood volume and pressure. Aldosterone stimulates reabsorption of Na^+ from urine into the bloodstream and secretion of K^+ from the blood into the urine. This results in increased Na^+ levels and decreased K^+ levels in the blood. Since water tends to follow sodium ions, aldosterone also causes water retention, which helps increase the fluid volume of blood and increases blood pressure. Because of its multiple physiological effects, aldosterone release is stimulated by several mechanisms. It is directly stimulated by *low Na^+* or *high K^+* levels in the blood. It is also stimulated by decreased blood volume and pressure through the ***renin-angiotensin-aldosterone (RAA) pathway*** (FIGURE 9.15). When blood volume and thus blood pressure drops, special cells in the kidney are stimulated to release an enzyme called ***renin***. Renin causes a cascade of actions that leads to the creation of ***angiotensin II***, which stimulates the adrenal cortex to release aldosterone. As previously explained, aldosterone indirectly increases blood volume and pressure by increasing sodium and water retention. Once blood pressure is normalized, negative feedback shuts down the RAA pathway.

The middle zone of the cortex secretes **glucocorticoids** that help the body regulate metabolism and the stress response (FIGURE 9.14C). Ninety-five percent of glucocorticoid activity is handled by the hormone **cortisol**, which is released when ACTH from the anterior pituitary stimulates the adrenal cortex. ACTH is released in response to ***corticotropin-releasing hormone*** from the hypothalamus whenever it senses low blood cortisol levels. Cortisol has several effects that help to increase nutrients in the blood so the cells have a ready supply of molecules from which to produce ATP. It increases the rate of protein breakdown from muscle fibers, causes the liver to increase its conversion of amino acids and fats into glucose, and increases the breakdown of adipose tissue to release fatty acids. Cortisol also has an anti-inflammatory and mild pain reduction effect because it inhibits production of prostaglandins (local hormones that support inflammation) and plays a major role in helping the body respond to stress.

The innermost layer of the adrenal cortex secretes small amounts of **androgens**, steroids that act as male sex hormones, controlling the development of masculine characteristics in both genders. The primary androgen is ***dehydroepiandrosterone (DHEA)*** whose

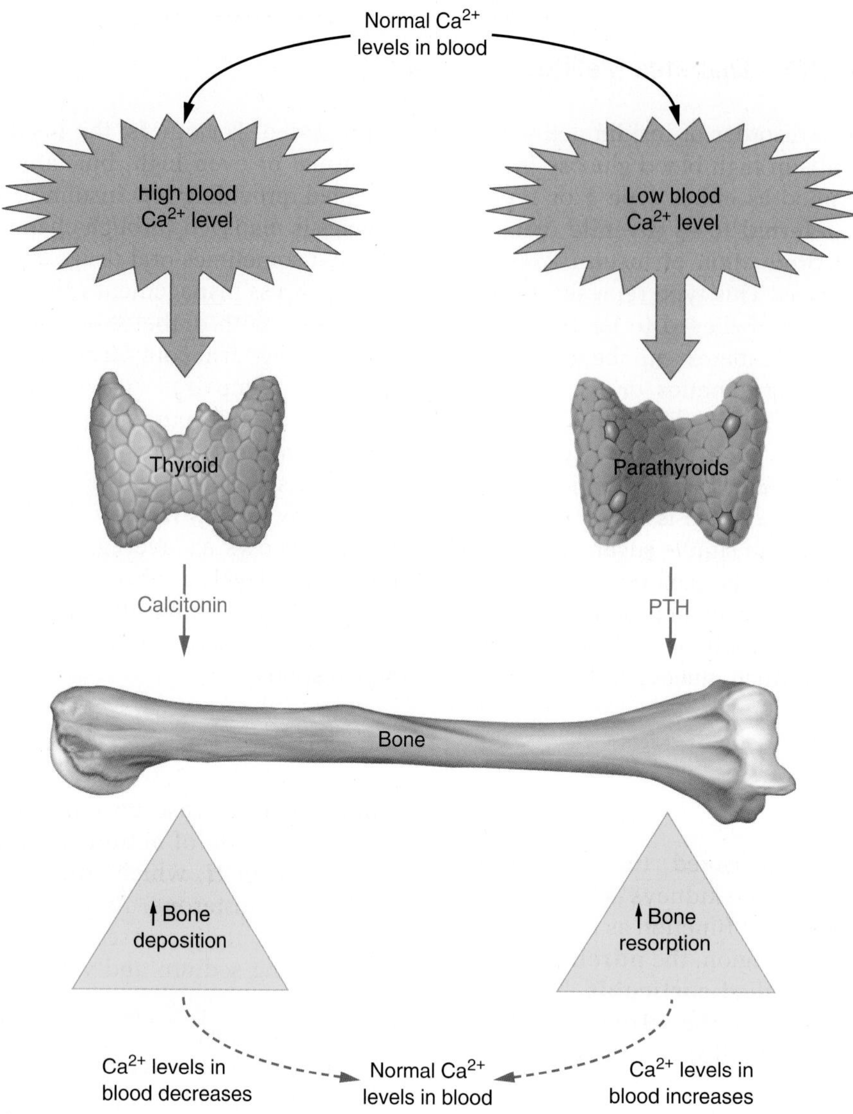

FIGURE 9.11 ▸ Calcitonin and PTH balance. Calcitonin and PTH have antagonistic effects that ensure homeostasis of calcium (Ca^{2+}) blood levels.

release is stimulated by ACTH. In males, DHEA has little effect because of the high amount of testosterone produced by the testes. In females, DHEA promotes libido (sex drive) and is converted into estrogens by other body tissues. In fact, most of a woman's estrogen after menopause comes from the conversion of DHEA.

Hormones of the Adrenal Medulla

The adrenal medulla, the small central region of the adrenal gland, produces and secretes the hormones **adrenaline** and **noradrenaline**, collectively referred to as **catecholamines** (see FIGURE 9.14). Like glucocorticoids, these two hormones play a major role in the body's response to stress, but they create short-term adaptations rather than long-term metabolic shifts. When the adrenals are stimulated by the sympathetic branch of the ANS, these two hormones are released into the bloodstream as part of the fight-or-flight response to sustain increases in heart rate, blood pressure, and blood flow to the heart, liver, and skeletal muscles, and the dilation (opening) of small airways inside the lungs. Adrenaline and noradrenaline also increase the levels of glucose and fatty acids in the blood. All these changes result in faster circulation and better oxygen and glucose supplies for cells to help the body deal with stressful situations.

Chemically, the hormones adrenaline and noradrenaline are exactly the same substances as the neurotransmitters epinephrine and norepinephrine. The distinction between these identical chemicals is simply where they are secreted and what they stimulate. Epinephrine and norepinephrine are released from neurons and stimulate other neurons or effectors. Adrenaline

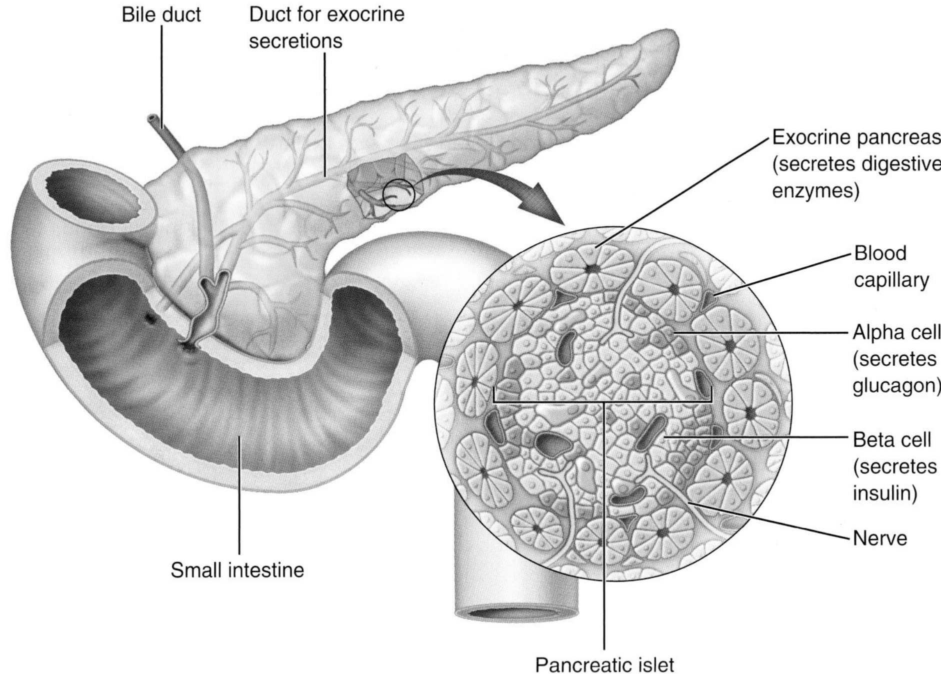

FIGURE 9.12 ▸ Pancreatic islet cells. Specialized alpha cells secrete glucagon, and beta cells secrete insulin.

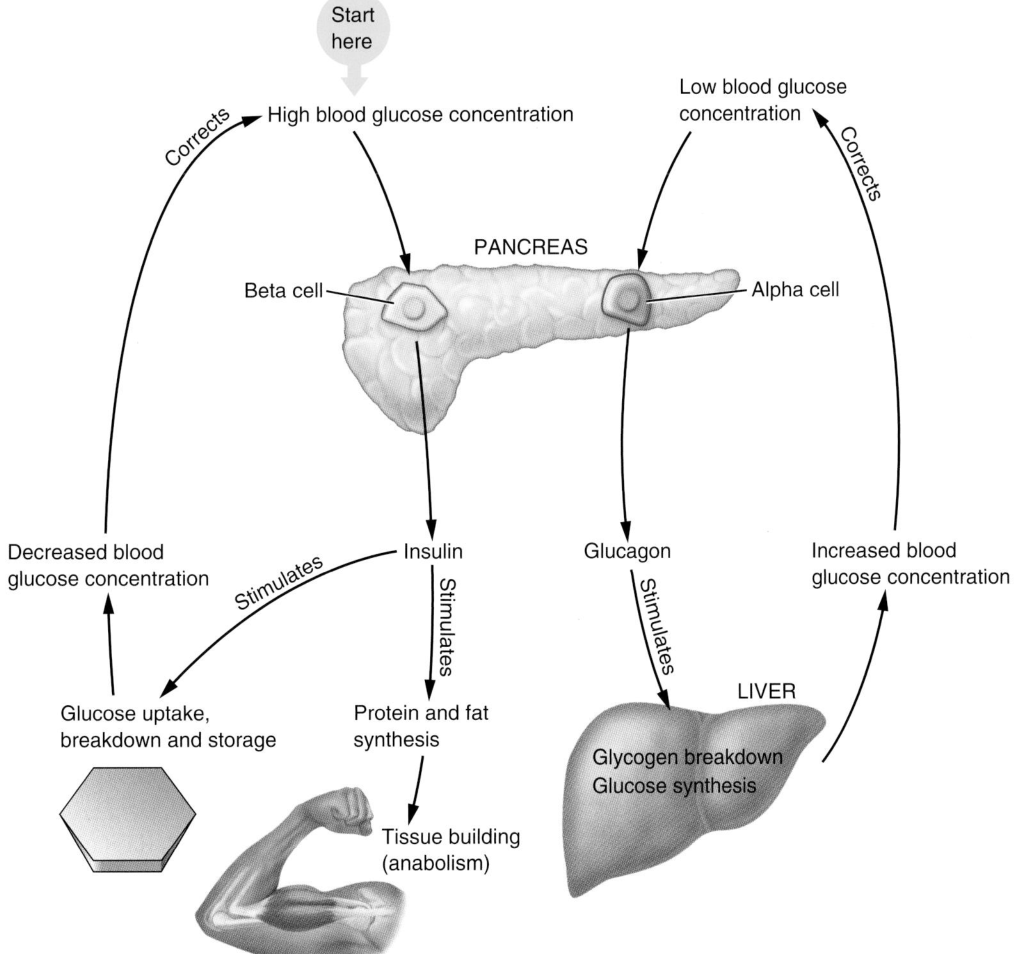

FIGURE 9.13 ▸ Glucose homeostasis. Blood glucose levels are managed through the antagonistic action of insulin and glucagon. Insulin increases glucose utilization by the cells to decrease blood glucose levels, while glucagon stimulates the synthesis of glucose from glycogen stored in the liver to elevate blood glucose.

Table 9-2 Hormones of the Thyroid, Parathyroid, Thymus, and Pancreas

Gland	Stimulated by	Hormone Released	Target	Physiologic Response(s)
Thyroid	TSH from the anterior pituitary	Triiodothyronine (T_3) and thyroxine (T_4)	Body cells	Increases metabolism
	Elevated calcium level in blood	Calcitonin	Bone	Inhibits bone resorption and accelerates calcium uptake to decrease calcium level in blood
Parathyroid	Low calcium level in blood	PTH	Bone	Stimulates bone resorption to increase calcium level in blood
			Kidneys	Stimulates production of calcitriol to increase Ca^{2+} absorption in digestive tract
Thymus	Unknown	Thymosin	T lymphocytes (T-cells)	Promotes maturation of T-cells
Pancreas	Elevated glucose in blood	Insulin	Body cells	Accelerates transport of glucose into cells to decrease glucose level in blood
	Low glucose in blood	Glucagon	Body cells	Accelerates conversion of glycogen and other nutrients to glucose to increase glucose level in blood

and noradrenaline are released into circulation by the adrenal medulla and bind to target cells in various organs. ▶ TABLE 9.3 provides a summary of the hormones secreted by the adrenal cortex and medulla.

Ovaries and Testes

Gonads are the male and female sex glands: the testes and ovaries. Hormones produced in these glands are responsible for the development of reproductive organs and the physical traits that differentiate the genders (▶ FIGURE 9.16). In females, the ovaries contain sex cells called ***ova*** (eggs) each surrounded by a protective casing called a ***follicle***. The follicles produce hormones called **estrogens**, which are responsible for the development of breasts and external genitals, initiating the menstrual cycle, and supporting maturation of ova. When the ovary releases an egg, the follicle collapses and becomes the ***corpus luteum***. This tissue secretes **progesterone**, which supports the growth and preparation of the uterine lining for pregnancy and decreases uterine contractions. In males, the **testes** produce **testosterone**, which regulates the production of ***sperm*** and stimulates the development of masculine characteristics such as beard growth, muscle definition, and deepening of the voice. (▶ TABLE 9.4) A detailed discussion of the sex glands and hormones may be found in Chapter 16.

Prostaglandins and Other Local Hormones

Some hormones do not need to circulate through the blood to find their target. These ***local hormones*** are released directly into neighboring tissues to affect cells in their immediate vicinity. For example, ***interleukin 2*** is a local hormone released by specific immune cells to activate other nearby immune cells. The chemical substances collectively referred to as **prostaglandins** are another example. These local hormones are released by virtually all cells except red blood cells. Prostaglandins alter fat metabolism, intensify pain, promote fever, and enhance the inflammatory response. These local hormones are discussed in more detail in Chapter 12.

What Do You Think? 9.3

- What effect would it have on the body if the thyroid gland were removed?
- How are calcitonin and PTH similar to insulin and glucagon?
- Based on your understanding of the hormone cortisol, what do you think is the purpose of cortisone injections and creams?

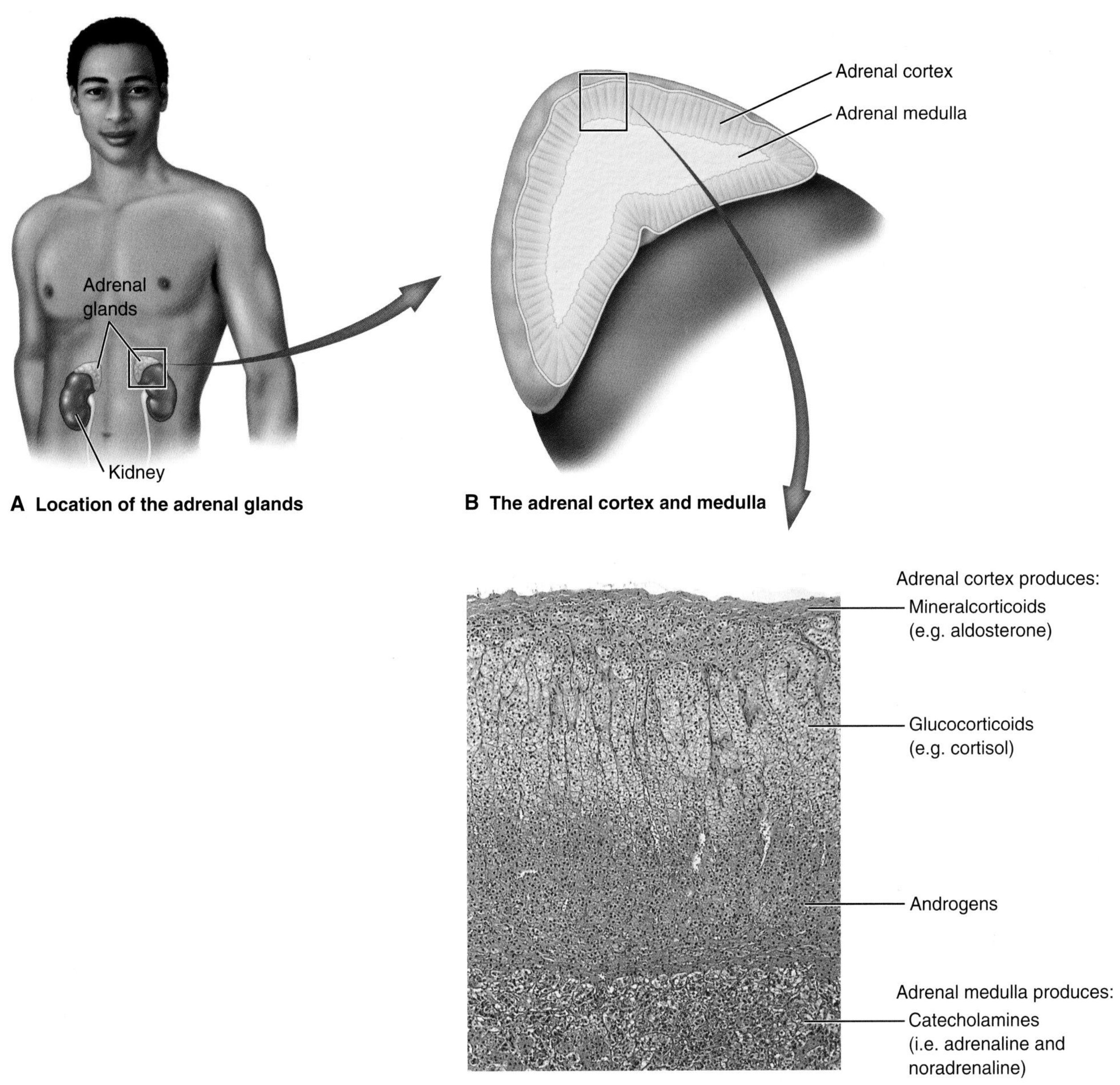

FIGURE 9.14 ▸ Adrenal glands. A. The adrenal glands sit atop each kidney. **B.** Each gland has an outer region, the adrenal cortex, that surrounds the inner adrenal medulla. **C.** The cortex has three regions that produce different corticoid hormones, while the medulla produces the catecholamines adrenaline and noradrenaline.

THE STRESS RESPONSE: LINKING THE NERVOUS AND ENDOCRINE SYSTEMS

Our bodies have to respond to billions of forms of stimuli on a daily basis. Stimuli such as fluctuations in temperature, hydration, blood pH, or respiration are managed through normal homeostatic mechanisms. However, when the stimulus is unusual or perceived as excessive, it prompts the hypothalamus to initiate a group of physiologic responses collectively referred to as the **stress response**. The stimuli, or ***stressors***, that create this response vary widely from one individual to another and can be initiated by physical events or psychological perceptions. When people talk about being "stressed out" or "under a lot of stress," they are referring to the negative aspect of stress called ***distress***. Distress includes the things that anger or frustrate us and drain our strength, stamina, and creativity. However, positive challenges or stress, ***eustress***, cause the same physiological responses within the body, but can also "psych us up" and increase productivity. Situations such as exercising, receiving a surprise of good news, and getting married are examples of eustress. Whether experiencing distress or eustress,

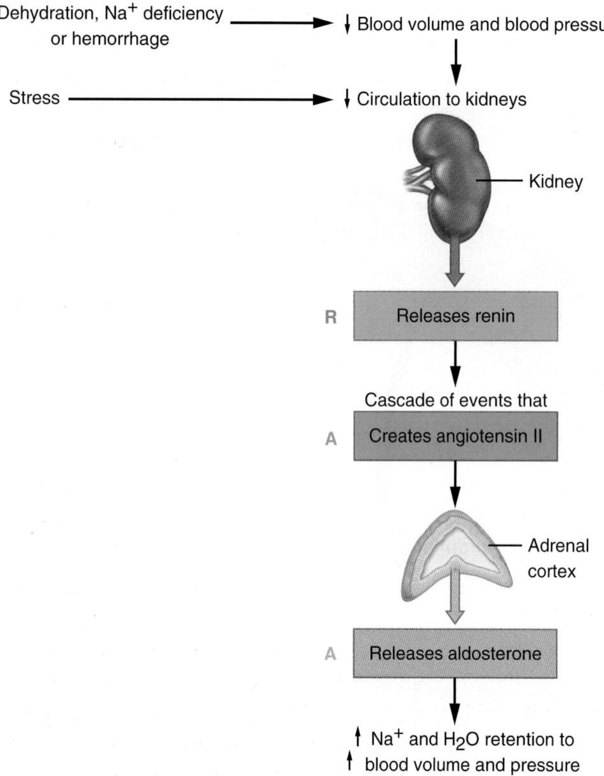

FIGURE 9.15 ▸ RAA pathway. Decreased circulation to the kidneys causes a release of the enzyme renin. Renin stimulates a cascade of chemical events that results in the formation of angiotensin II, which stimulates the adrenal cortex to release aldosterone.

the hypothalamus initiates the stress response by simultaneously engaging the nervous system in the alarm response and the endocrine system in the resistance reaction (▸ FIGURE 9.17).

The Alarm Response

The **alarm response** refers to hypothalamic stimulation of the sympathetic division of the ANS. Sympathetic stimulation initiates immediate, short-lived responses in the visceral effectors to ready the body for quick physical action: fight-or-flight. Recall that these responses include increased heart rate, blood pressure, sweat production, and respiration and increased catabolism for energy production. The sympathetic division also stimulates the adrenal medulla to increase production of adrenaline and noradrenaline to support and prolong the responses of the visceral effectors. These physiologic changes ensure that the body can respond quickly once the determination is made to fight or flee the stressor at hand.

During the alarm response, visceral activities such as digestion and urine production decrease because they are not essential for addressing the stressor. One result of this diminished visceral activity is decreased blood flow to the kidneys, which initiates the RAA pathway and stimulates the release of aldosterone. This mineralocorticoid helps the body manage stress by helping it retain water. ***Shock*** occurs when there is a measurable decrease in circulating blood volume. Therefore, aldosterone helps the body avoid shock by maintaining a normal volume of circulating fluids and stabilizing blood pressure.

The Resistance Reaction

At the same time that the hypothalamus is spurring the sympathetic response, it is also initiating a longer term response. Referred to as the **resistance reaction**, this endocrine response by the hypothalamus includes the secretion of hormones that stimulate the pituitary to release GH, TSH, and ACTH. These hormones help the

Table 9-3 Hormones of the Adrenal Glands

Region	Stimulated by	Hormone Released	Target	Physiologic Response(s)
Adrenal cortex	Elevated K^+ in blood; decreased pressure, volume, or Na^+ level in blood	Mineralocorticoids: aldosterone	Kidneys	Increase Na^+ retention and thus water to increase blood pressure; decrease K^+ in blood
	ACTH from the anterior pituitary	Glucocorticoids: cortisol	Body cells	Increase blood glucose and anti-inflammatory process, especially during stress response
	ACTH from the anterior pituitary	Androgens: DHEA	Body cells	Support libido and source of estrogens after menopause
Adrenal medulla	Sympathetic innervation	Adrenaline and noradrenaline	Body cells	Increase blood glucose; increase metabolism; prolong body changes initiated by alarm response during stress

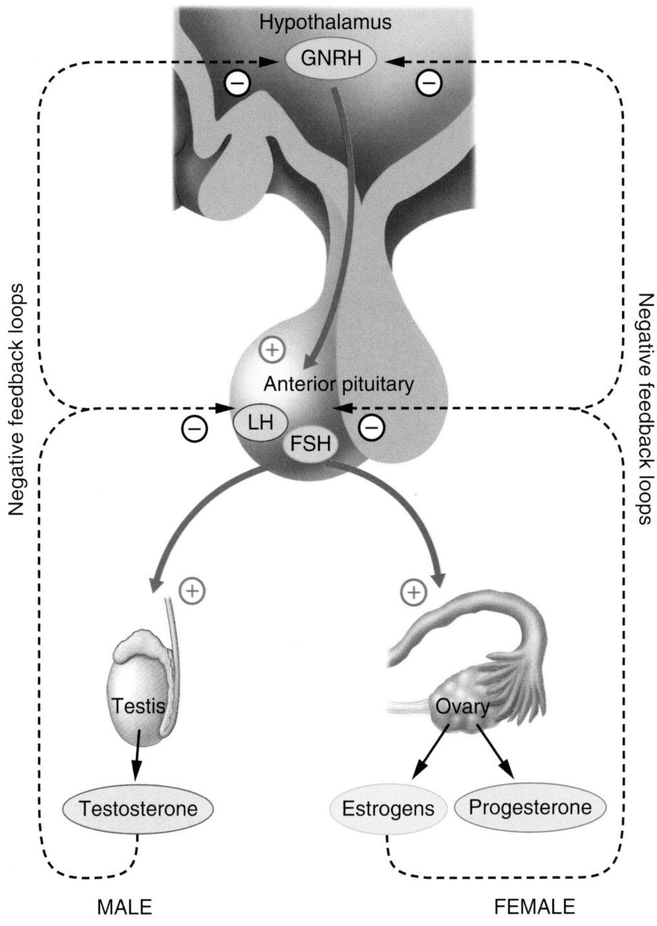

FIGURE 9.16 ▸ Gonads. The hypothalamus stimulates the anterior pituitary to release FSH and LH. These gonadotropins stimulate the ovaries to release estrogens and progesterone and stimulate the testes to release testosterone.

body continue its fight against stressors after the alarm response mediated by the ANS has dissipated.

GH and TSH mediate responses in the body that stimulate metabolism. GH increases catabolism of fats and the conversion of glycogen to glucose in the liver. TSH stimulates the release of thyroid hormones that increase the BMR. Simultaneously, ACTH acts on the adrenal cortex causing the release of cortisol. As previously discussed, this glucocorticoid stimulates the breakdown of proteins to provide amino acids that can be converted to glucose for energy or utilized for the production of enzymes needed for chemical activities throughout the body. Bottom-line, these hormones work together to ensure a ready supply of nutrients for increased ATP production in the cells. Additionally, cortisol plays an important role in reducing inflammation. This is initially helpful, but if the resistance reaction is prolonged, the healing process is slowed and immune responses are compromised.

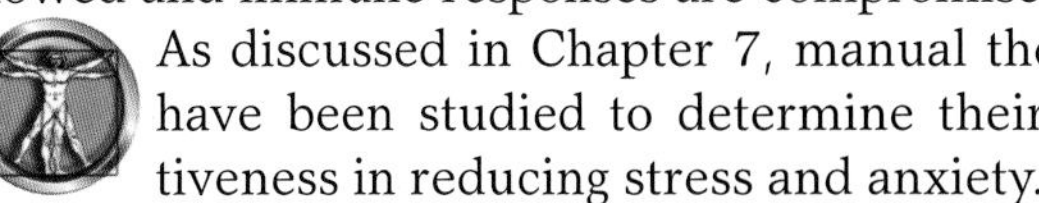
As discussed in Chapter 7, manual therapies have been studied to determine their effectiveness in reducing stress and anxiety. While research consistently supports their benefits, there is insufficient data to provide definitive explanations of exactly how, and to what degree, various manual therapies effect the multiple physiologic changes associated with the stress response. Recent research has measured reductions in salivary cortisol levels, diastolic blood pressure, and heart rate after a single manual therapy session of approximately 30 minutes. While these reductions were consistently measured after a single session, there was no evidence of a sustained reduction in these variables after a series of sessions.[4] Since some studies claim that 80% to 90% of all physical ailments are the direct result of stress, even short-term reductions in the physiological effects associated with stress may offer significant health benefits to clients in today's fast-paced world.[5]

Exhaustion

Generally, the body's alarm response and resistance reaction are able to manage most of life's stressors. However, when either the duration or intensity of a stressor is too great, the body suffers a depletion of its resources to the point of ***exhaustion***. Prolonged exposure to cortisol

Table 9-4 Hormones of the Ovaries and Testes

Gland	Stimulated by	Hormone Released	Target	Physiologic Response(s)
Ovaries	FSH and LH from pituitary	Estrogens	Body cells and ovarian follicles	Maturation of ova; development and maintenance of secondary sex characteristics; prepares uterus for implantation
		Progesterone	Uterus	Promotes growth of uterine lining and decreases uterine contractions to support pregnancy
Testes	FSH and LH from pituitary	Androgens: testosterone	Body cells, especially reproductive organs	Development and maintenance of secondary sex characteristics; sperm production

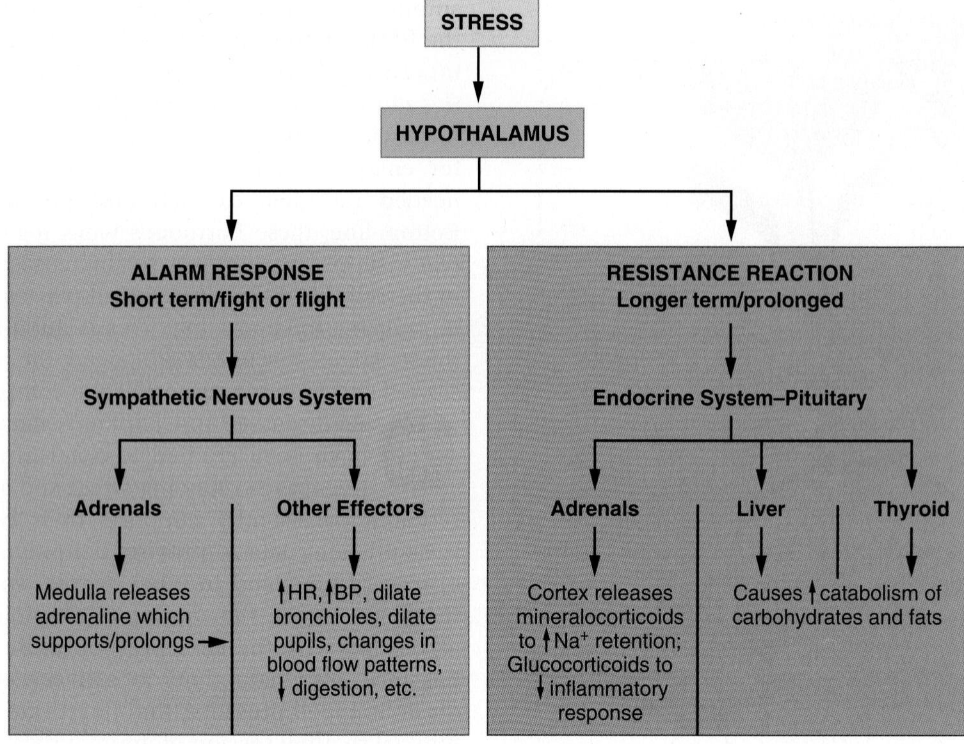

FIGURE 9.17 ▸ The stress response. The hypothalamus initiates the short-term response of the sympathetic nervous system as well as the longer-term resistance reaction of the endocrine system.

and other stress hormones leads to muscle wasting, suppressed immune function, and other organ-system pathologies. Eventually, the body simply gives up and shuts down. In fact, the Japanese have a word for it: *karoshi* means "death caused by overwork."

What Do You Think? 9.4

- What characteristics would you use to describe an individual who calls himself an "adrenaline junkie"?
- What are some possible negative effects of being an adrenaline junkie?
- How would you use your manual therapy skills to maximize stress reduction effects?

AGING AND THE ENDOCRINE SYSTEM

Many of the physical changes associated with aging are related to changes in the endocrine system. These changes include decreases in:

- ***Bone density***—As we age, bone density decreases, leading to osteoporosis and increased risk of fractures. This arises as the thyroid gland produces less calcitonin and the parathyroid glands increase their production of PTH. Women are at additional risk as their ovaries decrease their production of estrogen.
- ***Muscle mass***—A decrease in the production of GH by the pituitary gland contributes to muscle atrophy. This, together with a reduction in resistance-type exercise, decreases overall muscle mass.
- ***Metabolic rate***—The thyroid gland decreases its production of thyroid hormones, which lowers the metabolic rate. When decreased metabolic rate is coupled with decreased muscle mass, it is no wonder that the elderly often have trouble keeping warm.
- ***Resistance to stress***—The adrenal glands decrease the production of aldosterone and cortisol. Together with decreases in GH and thyroid hormones, the body's resistance response is greatly diminished. This explains reduced resistance to stress, including injury and disease.

Other changes in endocrine activity include a decrease in insulin production, combined with decreased receptor sensitivity to insulin in body cells. Therefore, after eating, blood glucose levels climb swiftly and take longer to return to normal, making older adults even more susceptible to type 2 diabetes.

SUMMARY OF KEY POINTS

- The endocrine system and the nervous system are the major regulatory systems of the body. The endocrine system is primarily responsible for communication and control of long-term processes such as growth, maturation, and reproduction, and also plays a key role in maintaining homeostasis.
- Endocrine glands are ductless, spread throughout the body, and not structurally connected like organs of other systems.
- Endocrine glands create their effects by producing and releasing hormones. Most hormones are circulating hormones that travel through the bloodstream to their target. Some hormones are local hormones that are released directly into surrounding tissue to stimulate changes in cellular activity in their immediate vicinity.
- Hormones are target-specific and either water soluble or lipid soluble. Most hormones are water-soluble and alter the metabolic activities of target cells or organs via the second-messenger mechanism. Lipid-soluble hormones such as steroid and thyroid hormones do not use a second messenger; they have a direct effect on the DNA of the cell.
- Most hormone levels are regulated via the negative feedback mechanism. Oxytocin is a rare example of a hormone regulated via positive feedback.
- Endocrine glands are stimulated to release their hormone(s) in one of three ways: by other hormones, by changes in blood concentration, or by the nervous system.
- See Tables 9-1 through 9-4 to review specific hormones and the physiologic responses they produce.
- The body's stress response is a series of physiologic changes initiated by the hypothalamus and mediated by the nervous system (alarm response) and endocrine system (resistance reaction).
- Many of the physical changes associated with aging such as decreases in bone density, muscle mass, metabolic rate, and resistance to stress are related to changes in the endocrine system.

REVIEW QUESTIONS

Short Answer

Define the following four terms:

1. Endocrine
2. Exocrine
3. Hormone
4. Corticoid
5. What is a tropic hormone?
6. Name the four key tropic hormones of the anterior pituitary and explain the function of each.
7. List and explain the three ways the body stimulates the release of hormones.

Multiple Choice

8. The two lobes of the pituitary are named for the type of tissue and method of stimulation from the hypothalamus. The anterior lobe is the ___________, and the posterior is called the ___________.

   A. hypothesis; posteriohypothesis
   B. neurohypophysis; glandular-hypophysis
   C. adenohypophysis; neurohypophysis
   D. glandular-hypophysis; neurohypophysis

9. In addition to the tropins, which other hormone is released from the anterior pituitary?

   A. oxytocin
   B. releasing hormone
   C. ACTH
   D. growth hormone

10. Which hormone released by the thyroid increases cellular metabolism?

    A. thyroxine
    B. thymosin
    C. calcitonin
    D. TSH

11. Where are oxytocin and antidiuretic hormone produced?

    A. anterior pituitary
    B. posterior pituitary
    C. hypothalamus
    D. adrenals

12. Which hormone increases blood calcium levels by accelerating bone matrix breakdown?

    A. calcitonin
    B. parathyroid hormone
    C. growth hormone
    D. T_4

Continued on page 254

13. Which of the following is the target for ACTH?
    A. posterior pituitary
    B. kidneys
    C. adrenals
    D. adenohypophysis
14. What is the function of hormones secreted by the hypothalamus?
    A. stimulate or inhibit release of anterior pituitary hormones
    B. stimulate release of oxytocin
    C. inhibit the production of antidiuretic hormone
    D. regulate the autonomic nervous system
15. What is the function of aldosterone?
    A. increase urine production
    B. increase sodium retention
    C. decrease sodium retention
    D. decrease potassium secretion
16. The name of the primary glucocorticoid is ___________ secreted by the ___________.
    A. glucagon; adrenal medulla
    B. insulin; pancreas
    C. aldosterone; adrenal cortex
    D. cortisol; adrenal cortex
17. Which of the pancreatic islet cells produce insulin?
    A. alpha
    B. beta
    C. delta
    D. gamma
18. What is the function of insulin?
    A. increase blood glucose levels
    B. decrease gluconeogenesis
    C. decrease blood glucose levels
    D. increase glucose storage
19. Where is the hormone glucagon produced?
    A. adrenal cortex
    B. adrenal medulla
    C. posterior pituitary
    D. pancreas
20. The hormones that have the strongest effect on blood glucose levels include insulin, glucagon, glucocorticoids, and
    A. growth hormone
    B. aldosterone
    C. thymosin
    D. parathyroid hormone
21. When blood sodium levels are low, the adrenals secrete which hormone?
    A. cortisone
    B. aldosterone
    C. renin
    D. cortisol
22. What is the secondary effect of aldosterone?
    A. increased blood volume and pressure
    B. decreased blood pressure and pH
    C. support of inflammation and healing
    D. increased glycogenolysis
23. Which of the following would be considered a primary stress hormone?
    A. insulin
    B. oxytocin
    C. cortisol
    D. PTH

References

1. Gardiner P, Osborn G. *The Shining Ones*. London, UK: Watkins Publishing; 2006.
2. Judith A. *Wheels of Life*. Woodbury, MN: Llewellyn Publishing; 1999.
3. Werner R. *A Massage Therapist's Guide to Pathology*. 4th ed. Baltimore, MD: Lippincott Williams & Wilkins; 2009: 572.
4. Moraska A, Pollini R, Boulanger K, et al. Physiological adjustments to stress measures following massage therapy: a review of the literature. *eCAM*. 2010 Dec; 7(4):409–18. Epub 2008 May 7.
5. Pert C. *Molecules of Emotion*. New York: Scribner; 1997.

10 The Cardiovascular System

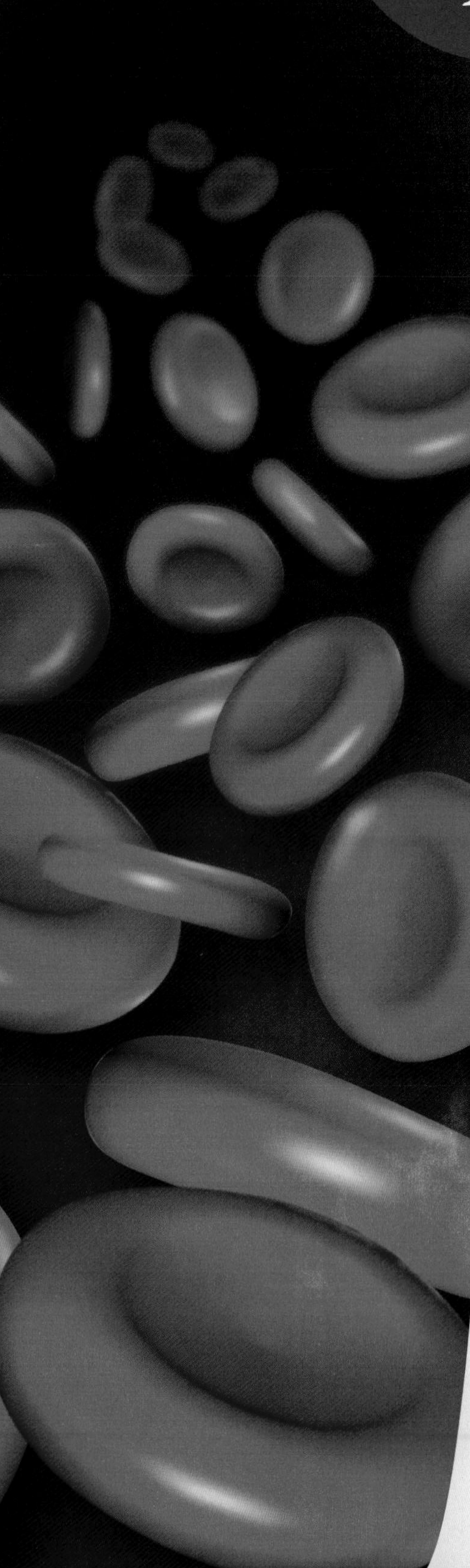

LEARNING OBJECTIVES

Upon completion of this chapter, you will be able to:

1. List the functions of the cardiovascular system and discuss their importance as they relate to the practice of manual therapy.
2. Identify the key components of plasma and explain the functional purpose of each.
3. Name the three types of blood cells and explain the function of each.
4. Name the types of blood vessels and describe the distinguishing characteristics of each.
5. Identify and locate the primary arteries and veins and name those that might be affected by manual therapy.
6. Describe the tissues and functional contribution of each layer of the heart.
7. Identify the chambers, valves, and great vessels of the heart in the order that blood moves through them.
8. Explain the conduction system of the heart, how it creates the cardiac cycle, and mechanisms that regulate heart rate.
9. Name the two divisions of cardiovascular circulation and describe blood flow through each.
10. Explain the primary influences on arterial, capillary, and venous blood flow and how they differ.
11. Define blood pressure and explain how it is regulated and influenced and how the body responds to blood pressure fluctuations.
12. List and describe the Starling forces that create fluid exchange in the capillary beds.
13. List the three stages of tissue healing and describe the key physiologic events of each.
14. Describe changes that occur in the cardiovascular system as the body ages.

> "The heart is not simply suspended in a body, but in a culture, a place, a time."
>
> MIMI GUARNERI, M.D.
> *The Heart Speaks: A Cardiologist Reveals the Secret Language of Healing*

KEY TERMS

atrium (A-tre-um)
capillary (KAP-eh-lar-e) **fluid pressure (CFP)**
cardiac (KAR-de-ak) **cycle**
diastole (di-AS-to-le)
edema (eh-DE-mah)
endocardium (en-do-KAR-de-um)
epicardium (eh-pih-KAR-de-um)
hematopoiesis (he-MAT-o-po-E-sis)
hemoglobin (HE-mah-glo-bin)
hemorrhage (HEM-o-rij)
hemostasis (he-mo-STA-sis)
hyperemia (hi-per-E-me-ah)
interstitial (in-ter-STIH-shul) **fluid pressure (IFP)**
interstitial oncotic (in-ter-STIH-shul on-KAH-tik) **pressure (IOP)**
ischemia (is-KE-me-ah)
lumen (LU-men)
myocardium (mi-o-KAR-de-um)
pericardium (per-eh-KAR-de-um)
peripheral resistance (peh-RIF-er-al re-ZIS-tans)
plasma oncotic (PLAZ-mah on-KAH-tik) **pressure (POP)**
pulmonary (PUL-mah-na-re) **circuit**
pulse (puls)
systemic (sis-TEM-ik) **circuit**
systole (SIS-to-le)
vasoconstriction (va-zo-con-STRIK-shun)
vasodilation (va-zo-di-LA-shun)
ventricle (VEN-trih-kel)

Primary System Components

- **Blood**
  - Plasma
  - Formed elements
- **Vessels**
  - Arteries
  - Arterioles
  - Capillaries
  - Venules
  - Veins
- **Heart**
  - Chambers
  - Great vessels
  - Valves
  - Conduction system

Primary System Functions

- **Transportation of nutrients, waste, hormones, and other substances essential for cellular activity**
- **Regulation of heat, pH, and fluid volumes**
- **Supports protection and immune responses of the body through clotting and tissue healing**

The cardiovascular and lymphatic systems both play a vital role in the movement, distribution, and exchange of fluids. In fact, at one time, these systems were referred to as a single circulatory system. However, the complexities and differences between them make it easier to study them individually. Therefore, this chapter explores the cardiovascular system, while the lymphatic system is covered in Chapter 11.

In some ways, the cardiovascular system seems quite simple and ordinary—just a network of tubes or passageways with a mechanical pump that drives fluids through them. However, each component of the system is really quite complex. Blood vessels create an intricate transportation system capable of instantly adjusting the volume, pressure, and distribution of blood according to tissue needs. The heart is an efficient four-chambered pump due to its thick layer of cardiac muscle tissue. Though the cells of this tissue can self-generate action potentials to initiate their own contractions, they are organized by a conduction system that coordinates their activity and ensures that the heart effectively propels blood through the vascular system. Blood is a complex fluid connective tissue that serves as the medium for nutrient and waste exchange and as a delivery system for chemical messages. It also plays a central role in supporting the body as it heals from injury, fights disease, and maintains homeostasis.

This chapter discusses the specific structure and function of each of the primary components of the cardiovascular system. The dynamics of blood flow, regulation of circulation, and the system's role in inflammation and tissue healing are also reviewed. Additionally, the heart's role in the mind-body-spirit connection is explored, including why factors such as isolation, grief, anger, cynicism, and hostility can be predictors of coronary artery disease in addition to factors such as genetics, smoking, high blood pressure, high cholesterol, and a sedentary lifestyle. Information on the benefits and physiological effects of manual therapy on the cardiovascular system is provided, including current scientific studies that support several positive effects. This chapter also presents information based on research that challenges long-held claims that manual therapy increases circulation, flushes out toxins, improves nutrient and waste exchange, and enhances production of blood cells.

BLOOD

Blood is a fluid connective tissue that consists of two components: a liquid portion, the plasma, and cells and cell fragments called formed elements. Blood acts as the primary transportation medium for nutrients, wastes, and chemicals such as hormones that are needed to support the body's metabolic activities. Nutrients enter the blood from the digestive and respiratory systems or are released from storage sites like the liver. Blood also takes metabolic byproducts away from cells, and carries them to the kidneys, liver, and spleen, where waste is filtered out and usable components recycled. Because it contains specific buffer chemicals, blood helps maintain the slightly alkaline pH of body fluids within the homeostatic range of 7.35 to 7.45. Blood also serves an important thermoregulatory function by transporting heat generated by muscle contraction throughout the body and dissipating that heat through breathing and the skin.

By the Way

Blood makes up about 7% to 8% of a person's total body weight. For example, someone who weighs 150 lb has approximately 5 L (5.2 quarts) of blood in his body.

Plasma

Plasma, the fluid component of blood, accounts for about 55% of blood volume. It is approximately 90% water and 10% dissolved or suspended trace elements and particles, including hormones and electrolytes. The largest portion of these substances consists of **plasma proteins**. Their presence in the blood creates osmotic pressure that allows blood to hold water and remain fluid. In addition to helping to regulate fluids, plasma proteins also play an essential role in blood clotting and immune responses. The primary plasma proteins are:

- ***Albumin***—This is the most abundant type of protein found in plasma. Its primary role is to contribute to osmotic pressure and maintain blood volume. Albumin also acts as a transport or carrier protein for steroid hormones and fatty acids in the blood.
- ***Globulins***—This group of proteins includes ***antibodies*** (***immunoglobulins***) that defend the body against viral and bacterial pathogens and ***complements*** that enhance (complement) certain immune responses. Globulins also include ***clotting factors*** that help in the blood's clotting process and transport proteins that carry iron, lipids, and certain vitamins in the blood.
- ***Fibrinogen***—This protein is converted into ***fibrin*** to form the thin initial threads of a blood clot.

Formed Elements

Three types of cells make up the **formed elements** of blood. These blood cells are produced by stem cells in the red marrow of bones. Through the process of **hematopoiesis**, the stem cells divide and differentiate to produce three types of blood cells: red blood cells (RBCs), white blood cells (WBCs), and platelets (▶ FIGURE 10.1).

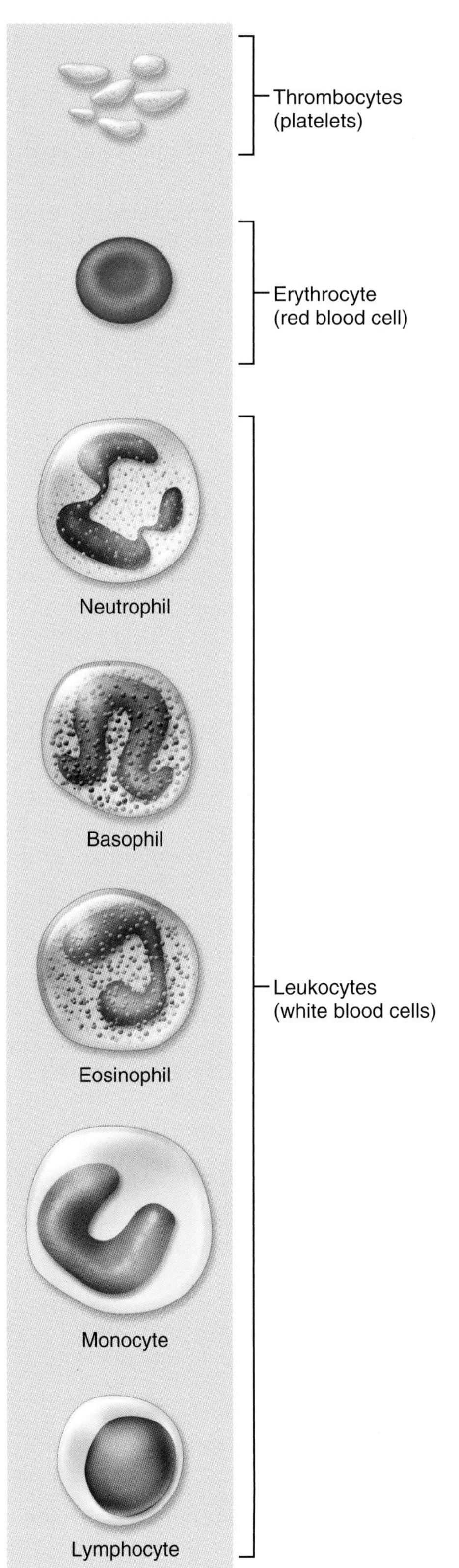

FIGURE 10.1 ▸ Formed elements of blood. Three types of cells make up the formed elements of blood: thrombocytes (platelets), erythrocytes (RBCs), and leukocytes (WBCs).

Red Blood Cells

Red blood cells are also known as **erythrocytes**. They are responsible for transporting the oxygen (O_2) needed for cellular metabolism. **Hemoglobin** is an iron-rich protein in RBCs that binds with O_2 molecules. Blood carrying a lot of oxygen, such as that leaving the lungs, has a bright red hue created by the hemoglobin–oxygen bond. In contrast, blood returning to the heart is a darker red color because the tissues have stripped the oxygen from the hemoglobin molecules of the RBCs. While carrying oxygen is hemoglobin's primary role, it also transports carbon dioxide (CO_2) and hydrogen ions (H^+). Because CO_2 binds to a different part of the hemoglobin molecule, its oxygen-carrying capacity is only slightly reduced. Additionally, H^+ is only picked up once the hemoglobin has given up its oxygen. This ability to bind free-floating hydrogen ions to hemoglobin gives RBCs an important role in maintaining blood pH.

Mature RBCs have a biconcave shape, meaning that they are concave (curved inward) on both sides. They have few organelles and no nucleus (Figure 10.1). The shape and composition of RBCs maximizes their hemoglobin- and oxygen-carrying capacity. However, the lack of a nucleus also means that RBCs are unable to divide. Instead, new cells are produced by red bone marrow through hematopoiesis. The average lifespan of RBCs is approximately 120 days, after which the liver or spleen destroys them. RBC production can also be stimulated in response to low O_2 levels in the blood. In these cases, the kidneys release the hormone ***erythropoietin***, which stimulates hematopoiesis.

Another important feature of RBCs is the presence of receptor integral membrane proteins (receptor IMPs) that serve as protein identity markers called **antigens** (introduced in Chapter 3). Currently, there are more than 100 different antigens that can be identified on the surface of RBCs. These antigens form 24 ***blood groups***, each with two or more ***blood types***. One group of these antigens, the ***ABO blood group***, allows blood to be typed according to the presence or absence of the ***A*** and/or ***B antigen***. Individuals with the A antigen are said to have type A blood, and those with the B antigen have type B.

By the Way

According to the ABO and Rh blood groups, people with type O negative blood are referred to as ***universal donors*** because their RBCs have neither the Rh or AB antigens. Therefore, it is least likely that their RBCs will trigger an immune response in people with other blood types. Conversely, those with type AB positive blood are considered ***universal recipients*** because their RBCs have all three antigens and can receive RBCs from all blood types with the least likelihood of triggering an immune response.

Pathology Alert Anemia

Anemia is a not in itself a disease but an indicator of a deficiency in the amount of hemoglobin or a decreased number or volume of RBCs. It can be caused by iron deficiency, bone marrow dysfunction, or genetic conditions. It can also be a secondary condition induced by complications from cancer, infection, kidney disease, or severe bleeding. Signs and symptoms of anemia include shortness of breath, fatigue, heart palpitations or murmurs, intolerance to cold, as well as pallor of the skin, mucous membranes and nail beds. Manual therapies are generally not contraindicated, but therapists should use caution when anemia is related to bone marrow suppression, bleeding syndromes, and chronic diseases with other side effects and contraindications.

Those without either the A or B antigen have type O blood. In the U.S., Type O is the most common blood type, representing 45% of the population. A very small number of people (4% in the United States) have both protein markers, and their blood is designated as type AB.

Another antigen first discovered in Rhesus monkeys and found in a different area of the plasma membrane of RBCs forms the ***Rh blood group***. Individuals either have or do not have this antigen. Therefore, blood is identified as either Rh positive (Rh^+) or Rh negative (Rh^-). If a person receives a blood transfusion with RBCs carrying antigens different from their own, the antigens will be identified as foreign and produce a defensive immune response. This is why individuals must go through meticulous blood-typing procedures before any type of surgery or transfusion.

White Blood Cells

Leukocytes, the **white blood cells**, play a vital role in healing and the body's immune response. They are outnumbered by a ratio of about 700 to 1 by the RBCs. WBCs have a much shorter lifespan than RBCs, with many circulating for only 6 to 8 hours. However, WBCs that migrate into body tissues can survive much longer, sometimes months or years. Leukocytes are round cells with a prominent nucleus (FIGURE 10.1). There are five types that fall into two broad categories, ***granular*** or ***agranular***, according to the appearance of their cytoplasm. Granular leukocytes are involved in inflammation and tissue healing and include ***neutrophils, basophils***, and ***eosinophils***. The agranular WBCs, ***lymphocytes*** and ***monocytes***, play more specific roles in the immune response. TABLE 10-1 lists and describes the five types of leukocytes.

Platelets

Platelets, or **thrombocytes**, are actually cell fragments that contain mitochondria and a number of active enzymes that are necessary for their function. Like RBCs, platelets cannot reproduce through mitosis because they do not have a nucleus or DNA. Produced in red bone

Table 10-1 White Blood Cells

Category	Name	Percentage of All WBCs	Function	Appearance
Granular	Neutrophils	60–70	Phagocytosis and destruction of bacteria	Multilobular nucleus; cytoplasm has lilac colored granules when stained
	Basophils	0.5–1	Intensify inflammatory response; release histamine and other chemicals in allergic responses	U-shaped or S-shaped nucleus; cytoplasm shows large deep blue granules when stained
	Eosinophils	2–4	Phagocytosis of antibody-antigen complexes; destroy some parasitic worms; combat effects of histamine in allergic reactions	Bilobular nucleus; cytoplasm shows large red-orange granules when stained
Agranular	Lymphocytes	20–25	Mediate specific immune responses	Large round nucleus rimmed by light blue cytoplasm when stained
	Monocytes	3–8	Phagocytosis in immune responses; many migrate into the tissues and transform into macrophages	Kidney-shaped nucleus; blue-gray cytoplasm when stained

By the Way

In each microliter of blood (one-millionth of a liter, equivalent to a cubic millimeter), the average adult has about 5 million RBCs, 7,500 WBCs, and 300,000 platelets.

marrow, platelets are much smaller than red and white blood cells and have a lifespan of about 10 days.

Platelets play a key role in **hemostasis**, the process that stops bleeding or blood flow. Hemostasis is a complex process that is initiated when cells inside the blood vessel release specific chemicals, setting off a chain of events that lead to the formation of a ***clot***, or patch, that prevents further blood loss. As seen in ▶ FIGURE 10.2, hemostasis involves three key steps:

1. ***Vascular spasm***—When blood vessels are damaged, cells of the vessel wall release chemicals that signal the smooth muscle inside the blood vessel wall to constrict and slow the flow of blood through the injured vessel.
2. ***Platelet plug***—The chemicals released from damaged cells cause platelets to become sticky, causing them to clump together and form a soft and temporary plug over the damaged area of the vessel.
3. ***Clot formation***—In this final step, also called **coagulation**, the cells in the platelet plug release clotting factors. These combine with ***prothrombin*** in the plasma to form ***thrombin***. Thrombin reacts with the ***fibrinogen*** in plasma to form ***fibrin***, thin protein threads that form a net around the platelet plug. This netting traps RBCs and platelets to form a clot.

What Do You Think? 10.1

- Explain how the role of each formed element is reflected in the number of each type of cell.
- Given your understanding of clotting and platelet function, what is the danger posed by diseases that create rough spots in blood vessels?

By the Way

A **thrombus** is a stationary blood clot, and an **embolus** is a traveling mass. Since emboli can be made up of bacteria, tissue fragments, air bubbles, or even foreign bodies, a traveling blood clot is referred to as a **thromboembolism.**

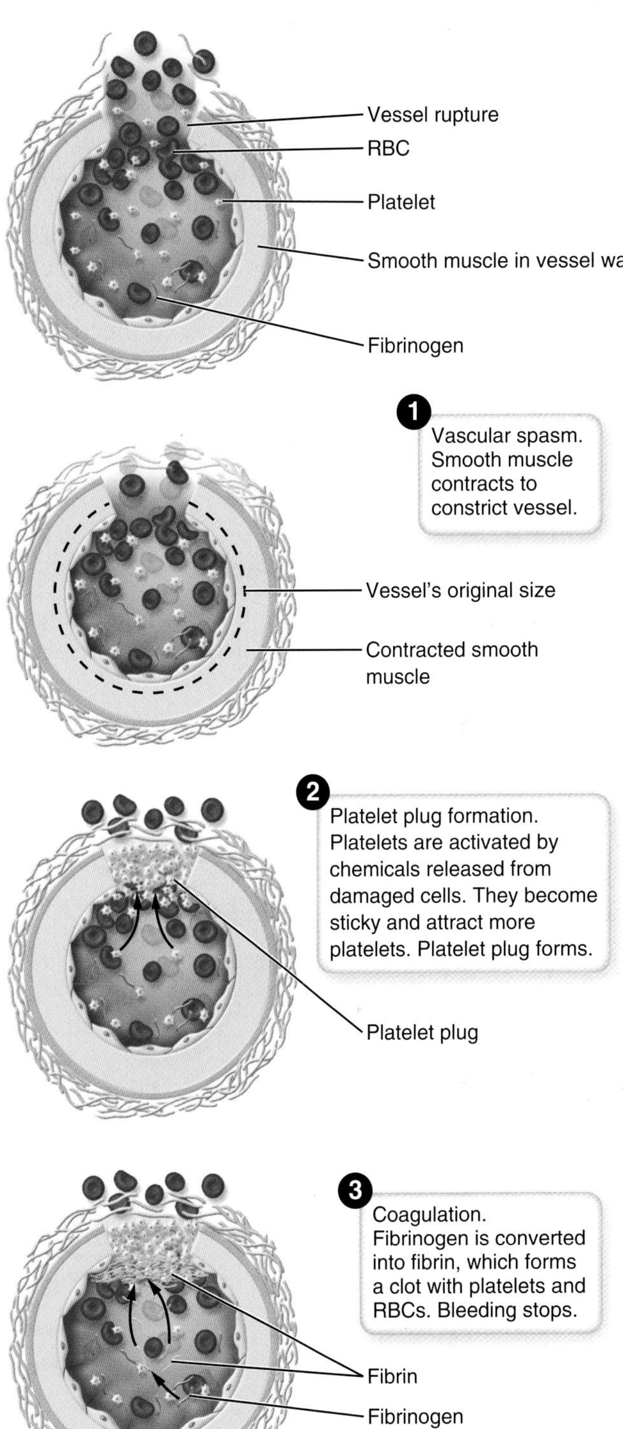

FIGURE 10.2 ▶ Hemostasis. The process of hemostasis stops bleeding or blood flow. To view an animation on hemostasis, visit http://thePoint.lww.com/Archer-Nelson.

BLOOD VESSELS

The network of blood vessels in the cardiovascular system consists of three types of vessels:

- ***Arterial vessels*** that carry blood away from the heart
- ***Capillaries*** that serve as the site of nutrient and waste exchange
- ***Venous vessels*** that return blood to the heart

The arterial vessels include large and small **arteries** and even smaller vessels called **arterioles**. Similarly, the venous vessels include the **veins** (larger) and smaller vessels called **venules**. Structurally, all blood vessels are tubular structures with a cavity or channel in the middle called a **lumen**. The walls of arteries and veins share the same three-layered structure: a smooth inner layer of epithelial cells (***endothelium***) attached to a basement membrane, a middle layer of smooth muscle, and an outer layer of elastic connective tissue (▶ FIGURE 10.3). These layers thin significantly in both arterioles and venules, and capillaries have only one layer. ▶ TABLE 10-2 summarizes the functional role and key structural characteristics of each type of blood vessel.

By the Way

Blood vessels have their own system of tiny vessels called ***vasa vasorum***, meaning vessels of vessels. These little vessels arise from the external layer of the arteries and veins to provide nutrient–waste exchange for the tissues of the vessel wall.

Arteries and Arterioles

In arteries, the middle muscular layer is much thicker than the same layer in veins. This thicker and more elastic middle layer gives arteries the ability to expand and recoil to withstand the pressure and volume of blood that is pumped from the heart. Additionally, smooth muscle contractions control the size of the vessel's lumen, which is essential for regulation of both blood pressure and distribution. **Vasoconstriction** is the narrowing of a blood vessel when the smooth muscle contracts, causing blood flow through the vessel to be diminished

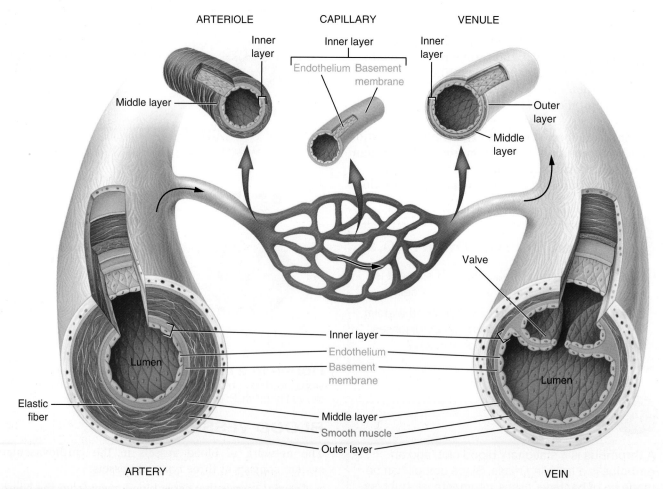

FIGURE 10.3 ▶ Structure of blood vessels. The walls of arteries and veins are made up of the same three layers. However, arteries have a thicker middle muscular layer, while veins have a thicker outer elastic layer and one-way valves to prevent backflow of blood. Capillaries are made of a single layer of epithelial cells attached to a thin basement membrane.

Table 10-2 Characteristics of Blood Vessels

Vessel	Functional Role	Key Structural Characteristics
Artery	Carries blood away from the heart	Thick middle smooth muscle layer with elastic fibers
Arteriole	Controls distribution of blood to the capillary bed	Prominent smooth muscle layer
Capillary	Site of nutrient and waste exchange	Only one cell layer thick
Venule	Collects blood from the capillary beds	Prominent elastic layer
Vein	Returns blood to the heart	Thick outer elastic layer and one-way valves

and the pressure elevated. In contrast, **vasodilation** is the widening of a vessel due to smooth muscle relaxation; this increases blood flow and can decrease blood pressure.

While arterioles are much thinner than arteries, they still have enough smooth muscle to maintain blood pressure and help control blood distribution to the capillaries. At the end of each arteriole, there is an extra ring of smooth muscle called the **precapillary sphincter**, which regulates the volume and rate of blood flow into the capillaries.

The blood distribution pathway formed by the arterial vessels is a continuous network branching out from the heart to the capillaries. To distinguish one section of the network from another, major arteries are named according to their location/body region or the organ they supply with blood (FIGURE 10.4). The ***aorta*** is the great artery that connects to the heart. Almost immediately after it ascends from the heart, three major arteries branch off the aorta: the ***brachiocephalic trunk*** (which branches again to become the right ***carotid*** artery and ***subclavian*** artery), the left ***carotid*** artery, and the left ***subclavian*** artery. The carotids ascend to the neck and head, while the subclavians travel to the upper extremities. In each armpit, the subclavian artery becomes the ***axillary*** artery. The axillary artery becomes the ***brachial***, which then branches into the ***radial*** and ***ulnar*** arteries and finally, the ***palmar*** and ***digital*** arteries of the hand and fingers.

The aorta travels behind the heart and downward through the thoracic cavity as the ***descending aorta***. It continues through the abdominopelvic cavity, where it is called the ***abdominal aorta***. The celiac trunk is the first branch of the abdominal aorta; it further branches into the ***common hepatic*** (liver), left ***gastric*** (stomach), and ***splenic*** (spleen) arteries. The ***mesenteric*** arteries supply the intestines, and the ***renal*** arteries carry blood to the kidneys. At the level of the fourth lumbar vertebrae, the abdominal aorta splits into the right and left ***common iliac*** arteries that deliver blood to the lower extremities. Each common iliac artery has two major branches that supply the tissue and organs in the pelvis, the ***internal*** and ***external iliac*** arteries. From this point, the major arteries of the lower extremity are the ***femoral, popliteal, anterior*** and ***posterior tibial***, and the ***dorsalis pedis***.

Capillaries

The **capillaries**, situated between arterioles and venules, form an intricate network of microscopic vessels called a **capillary bed** (FIGURE 10.5). This web of interconnected vessels is found throughout the tissues and serves as the site for nutrient and waste exchange between the blood and interstitium. Capillary walls are made of a single layer of epithelial cells wrapped in a very thin external connective tissue membrane. The density of capillary beds varies according to the metabolic demand of the particular tissue or organ. Tissues and organs with high metabolic demands, such as the brain, liver, skeletal muscles, and kidneys are densely packed with capillaries. In contrast, ligaments, tendons, and cartilage have very few capillaries, since their metabolic needs are extremely low.

Veins and Venules

In contrast to arteries, the muscular layer in veins is much thinner. This makes them more distensible and allows high volumes of blood to be transported as needed. A unique structural feature of veins is the series of one-way valves interspersed throughout their length (see FIGURE 10.3). These valves prevent backflow and create shorter segments for the blood to travel as it returns to the heart.

With a few exceptions, most of the major veins have the same names as the artery in the same region. In the upper extremity are the ***digital, palmar, radial, ulnar, brachial,*** and ***axillary*** veins (FIGURE 10.6). Additionally, in the antecubital region, three superficial veins also return blood from the arm. The ***basilic*** vein is the most medial, the ***median cubital*** is in the middle, and the ***cephalic*** vein is the most lateral. Its accessible, superficial location makes the median cubital vein the preferred site for drawing blood or for intravenous injections of drugs or fluids. The median cubital joins with

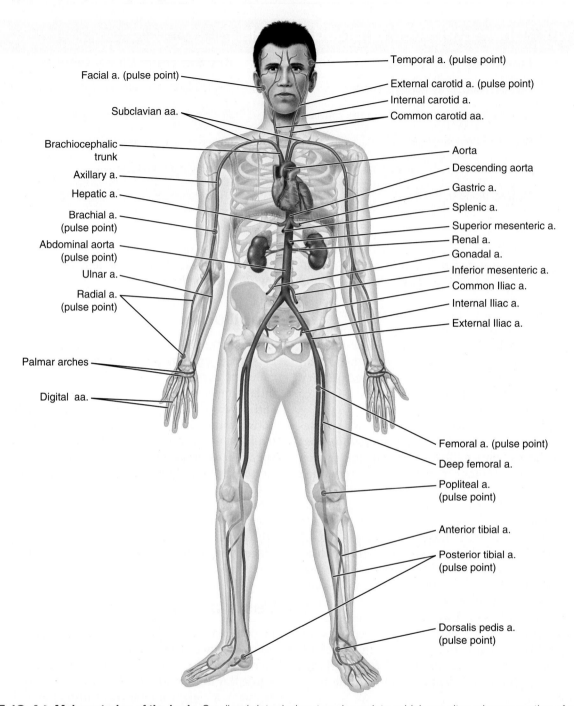

Temporal a. (pulse point)
Facial a. (pulse point)
External carotid a. (pulse point)
Internal carotid a.
Subclavian aa.
Common carotid aa.
Brachiocephalic trunk
Aorta
Axillary a.
Descending aorta
Gastric a.
Hepatic a.
Splenic a.
Brachial a. (pulse point)
Superior mesenteric a.
Renal a.
Abdominal aorta (pulse point)
Gonadal a.
Ulnar a.
Inferior mesenteric a.
Common Iliac a.
Radial a. (pulse point)
Internal Iliac a.
External Iliac a.
Palmar arches
Digital aa.
Femoral a. (pulse point)
Deep femoral a.
Popliteal a. (pulse point)
Anterior tibial a.
Posterior tibial a. (pulse point)
Dorsalis pedis a. (pulse point)

FIGURE 10.4 ▸ Major arteries of the body. Small red dots designate pulse points, which are sites where a portion of an artery becomes superficial arteries. "a." indicates artery; "aa.," arteries.

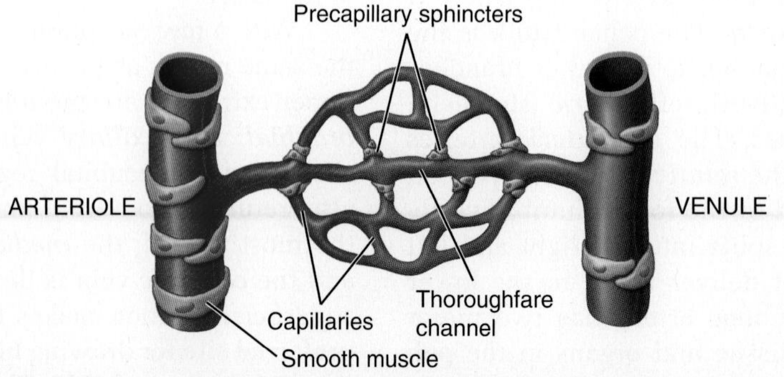

FIGURE 10.5 ▸ Capillary bed. Capillary beds sit between the arterioles and venules and serve as the site for nutrient-waste exchange. The flow of blood through the capillary bed is regulated by precapillary sphincters.

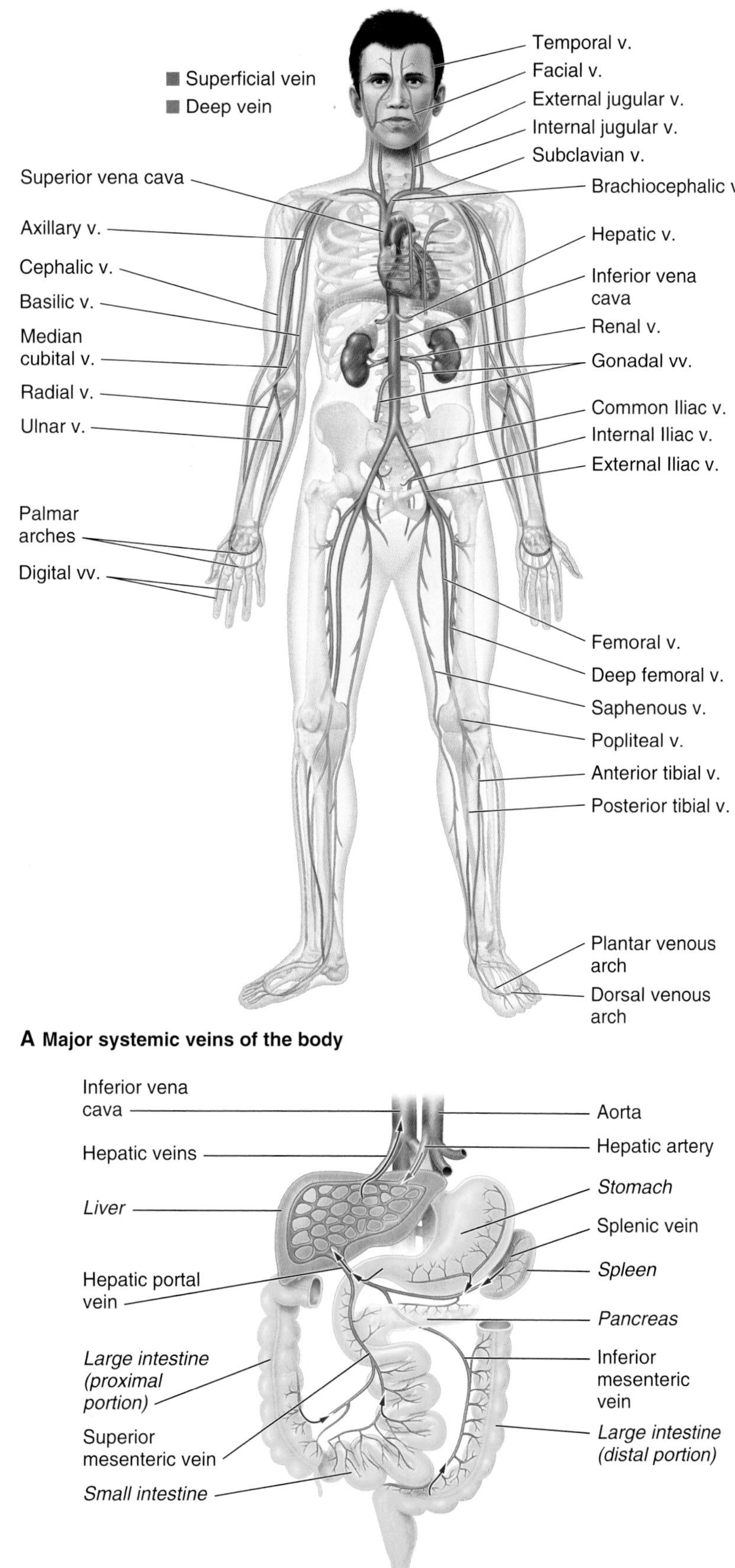

FIGURE 10.6 ▸ Major veins of the body. A. Superficial and deep veins. **Major veins of the body. B.** Veins leaving the stomach, spleen, pancreas, and intestines merge to form the portal vein, which carries blood to the liver for cleansing. The hepatic vein carries blood from the liver to the inferior vena cava. v. indicates vein; vv., veins.

the basilic and cephalic veins in the elbow region. The basilic and cephalic veins merge into the axillary vein, which then becomes the ***subclavian*** vein. The major veins that return blood from the head are the ***jugular*** veins. The right and left subclavian and jugular veins merge into the ***brachiocephalic*** veins before connecting to the ***superior vena cava***.

In the lower extremity, the major veins are the same as the arteries, with one major addition. The ***saphenous*** vein runs the full medial length of the leg before merging with the ***femoral*** vein. The saphenous vein is the longest vein in the body and the one most prone to the condition called varicose veins. All the major veins in the pelvic region and abdominopelvic cavity are named for the organs they serve, and they eventually merge with the ***inferior vena cava*** that connects to the heart. However, the ***splenic, gastric, pancreatic***, and ***mesenteric*** veins all merge with the ***portal*** vein that enters the inferior aspect of the liver. In this case, venous blood is carried to the liver to be cleansed, and blood leaving the liver exits via the ***hepatic*** vein.

What Do You Think? 10.2

- Which everyday analogy uses the term *artery* to describe major thoroughfares?
- What type of tissue do you think the venous valves are made of, and why?

THE HEART

Located center and slightly left in the mediastinum of the thoracic cavity, the heart is a four-chambered pump shaped like an upside-down pear. Its broader superior border, or ***base***, sits under the sternum at the level of the first rib, while its smaller inferior region, or ***apex***, rests atop the diaphragm (FIGURE 10.7). The base is relatively fixed in position, while the apex shifts and slides as the heart pumps.

Base
Sternum
Diaphragm
Apex
Midline

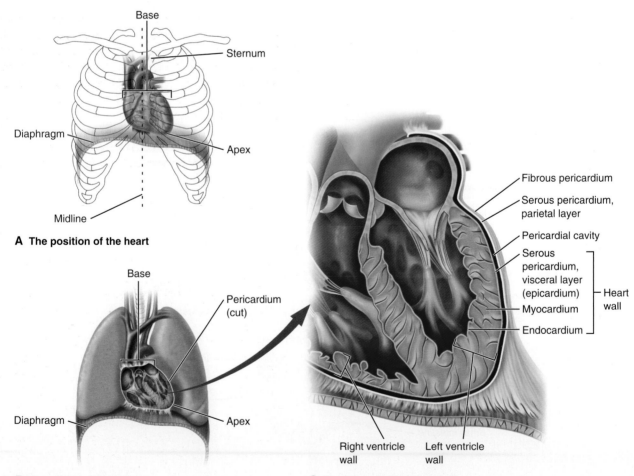

A The position of the heart

Base
Pericardium (cut)
Diaphragm
Apex

B Coverings of the heart

Fibrous pericardium
Serous pericardium, parietal layer
Pericardial cavity
Serous pericardium, visceral layer (epicardium)
Myocardium
Endocardium
Heart wall
Right ventricle wall
Left ventricle wall

C The layers of the heart wall

FIGURE 10.7 Heart. A. The heart is located in the mediastinum of the thoracic cavity. The superior portion of the heart is called the base, while the apex is the inferior point formed by the left ventricle. **B.** Coverings of the heart; note the position of the heart in relation to the lungs and diaphragm. **C.** The heart wall has three layers of tissue.

The heart is surrounded by a protective sac called the **pericardium**. The outermost layer of this sac, the ***fibrous pericardium***, is made of tough disorganized fibrous connective tissue that protects the heart, anchors it in place, and keeps it from overstretching. The inner portion, or ***serous pericardium***, is a two-layered serous membrane that produces ***pericardial fluid***, which acts as a lubricant and decreases friction between the layers when the heart pumps. The outer ***parietal layer*** of the serous pericardium is fused to the inside surface of the fibrous pericardium, while the inner ***visceral layer*** is firmly attached to the heart itself.

The heart walls are made of three layers of tissue. The outermost layer, the **epicardium**, is synonymous with the visceral layer of the serous pericardium (▶ FIGURE 10.7C). The middle stratum is a thick layer of cardiac muscle tissue called the **myocardium**. This muscle tissue produces the coordinated contractions that pump blood throughout the cardiovascular system. A thin epithelial tissue layer, the **endocardium**, forms the inner layer of the heart wall and provides a smooth inner surface for each chamber.

Heart Chambers, the Great Vessels, and Valves

The heart is divided into right and left sides by a muscular partition called the **septum**. Each side has an upper chamber, or **atrium**, and a lower chamber called a **ventricle** (▶ FIGURE 10.8A). Blood returns to the heart via veins entering the atria (receiving chambers). The left atrium receives blood from the ***pulmonary veins*** that has been oxygenated in the lungs. Deoxygenated blood from the body is returned to the right atrium via the ***superior vena cava*** and ***inferior vena cava***. The ventricles are discharging chambers that forcefully pump blood from the heart into the arteries. The right ventricle pumps blood into the ***pulmonary artery***, while the left ventricle pumps blood into the ***aorta***.

The heart has four valves at important structural junctions that control the direction of blood flow through the heart. Two valves, the ***atrioventricular (AV) valves*** are located between the atria and ventricles on each side of the heart. The valves are named based on the number of flaps, or **cusps**, they have. The right AV valve is the **tricuspid valve**, while the left AV valve is the **bicuspid valve**, also known as the **mitral valve**. The other two valves are called ***semilunar valves*** because of their curved half-moon shape. The semilunar valves are located between the ventricles and the major arteries to which they direct blood. The left semilunar valve is the **aortic valve**, while the **pulmonary valve** is the semilunar valve located between the right ventricle and pulmonary artery. Operating like one-way swinging doors, each valve keeps blood from flowing back into the wrong chamber as the heart contracts and relaxes.

Blood Flow through the Heart

The organization of the great vessels, chambers, and valves of the heart provide the means for organized blood flow (see FIGURE 10.8). Remember that a septum separates the right and left sides of the heart, so that each side creates its own circuit. Blood flows from the pulmonary veins into the left atrium. From the left atrium, the blood is pushed through the mitral (bicuspid) valve into the left ventricle. The ventricle then contracts to pump blood through the aortic valve into the aorta to be distributed throughout the body. As blood returns from the body, it ends up in either

Pathology Alert Heart Attacks

Heart attacks, described clinically as ***myocardial infarctions (MI)***, occur when the coronary blood vessels that nourish the myocardial tissue are abruptly closed off. This disruption in blood flow leads to a large area of tissue death that inhibits the function of the heart. Heart attacks can be caused by vascular disease such as ***atherosclerosis*** (build-up of fatty deposits inside arteries), ***arteriosclerosis*** (hardening of the arteries due to mineral deposits), a thromboembolism, or even acute vasospasm. Well-known symptoms of a heart attack include heaviness or crushing pain in the chest, aching of the left arm, and/or neck or jaw pain. However, these symptoms are most common in men. In contrast, the symptoms described by women are much less obvious. Women complain of anxiety, indigestion, shortness of breath, unusual fatigue, and/or difficulty sleeping. This is often dismissed as the flu, an upset stomach, or general tiredness.

A variety of factors contribute to the development of coronary artery disease that leads to heart attacks. These include genetics, high cholesterol, hypertension, diabetes, inactivity, obesity, and stress. Therefore, successful treatment includes a combination of dietary changes, exercise, and stress-reduction techniques. While many people make changes in diet and exercise, research shows that without reducing stress, the risk of coronary artery disease and heart attack is often not significantly diminished.[1–3] Manual therapists need to carefully review their client's past medical history because heart attack survivors are frequently prescribed medications that thin the blood and decrease blood pressure, which requires therapeutic precautions.

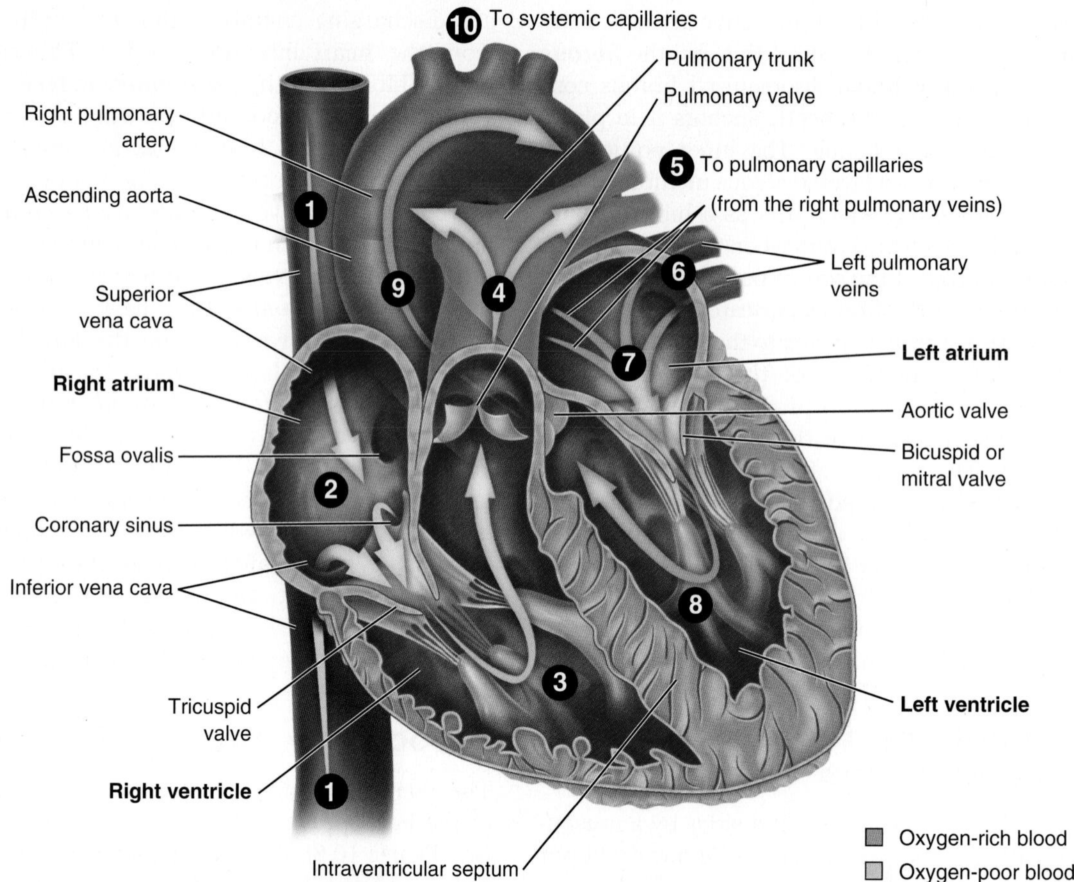

A Anatomical view of cardiac blood flow

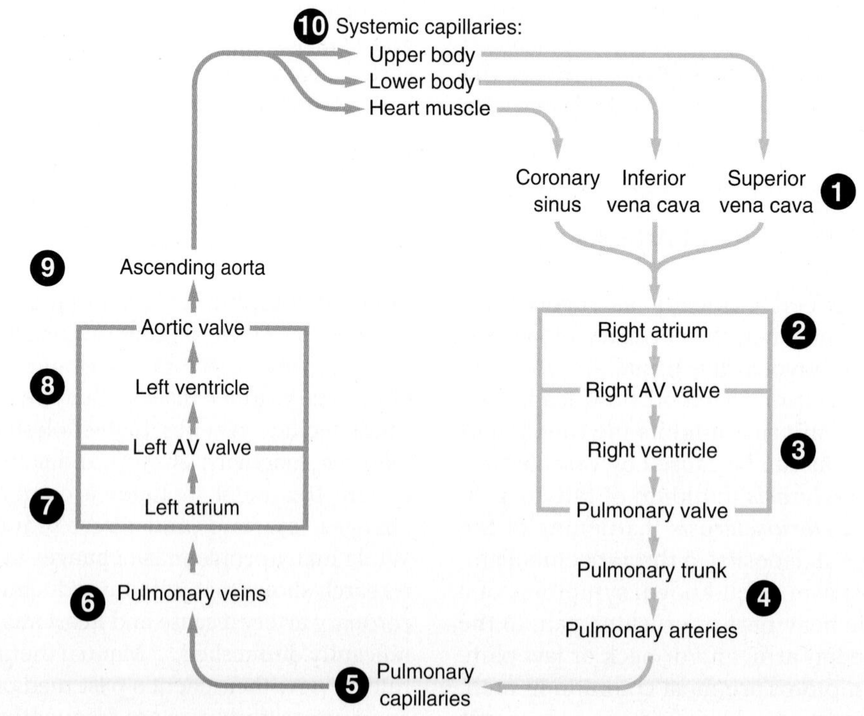

B Schematic view of cardiac blood flow

FIGURE 10.8 ▸ Blood flow through the heart. A. The direction of blood flow through the heart is controlled by the opening and closing of four cardiac valves. **B.** Schematic view of cardiac blood flow.

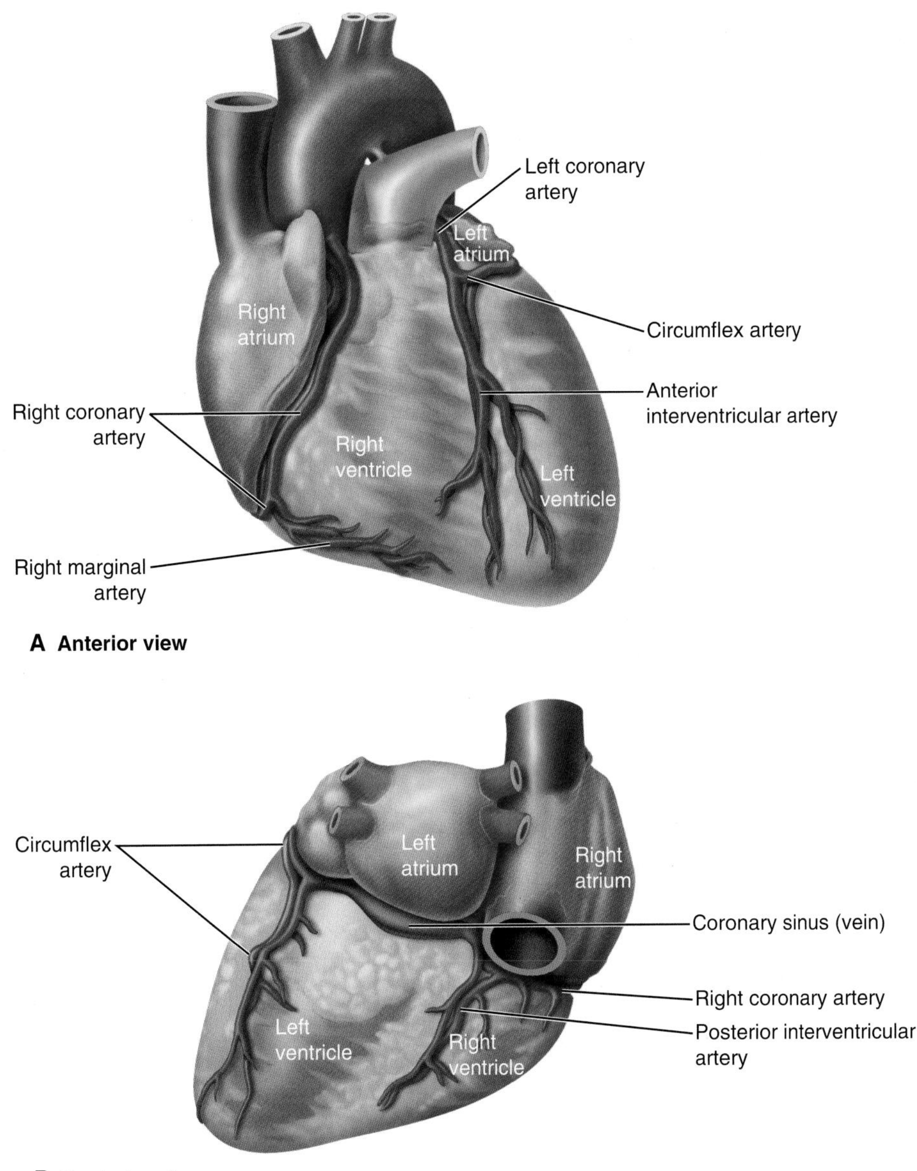

FIGURE 10.9 ▸ Coronary circulation. Circulation is provided to the heart muscle via the coronary vessels. The major vessels of the coronary circulatory loop are shown.

the inferior or superior vena cava and flows through these veins into the right atrium. It then flows through the tricuspid valve into the right ventricle. Next, the ventricle contracts, forcing blood through the pulmonary valve and into the pulmonary artery. A short distance from the heart, the pulmonary artery splits to deliver blood to both lungs.

Like all tissues, the heart wall must have a blood supply to provide nutrients and remove waste products. However, the blood that flows through the heart moves too quickly to enable this exchange with the thick myocardium. Instead, there is a system of vessels throughout the myocardium that takes care of the metabolic needs of the cardiac muscle tissue (▸ FIGURE 10.9). The blood that flows through this system is called ***coronary circulation***.

Conduction System

The pumping action of the heart is produced through the rhythmic contraction of the myocardium. As you recall, cardiac muscle is striated, similar to skeletal muscle tissue. However, cardiac muscle fibers are shorter, branched, and connected to neighboring fibers by thickened areas of the sarcolemma (▸ FIGURE 10.10A). Transverse connections called **intercalated discs** hold the muscle fibers together and form specialized ***gap junctions*** that allow action potentials to be carried from one fiber to the next. This allows large sections of the myocardium to contract as a single unit without individual innervation of each fiber. The entire myocardium is

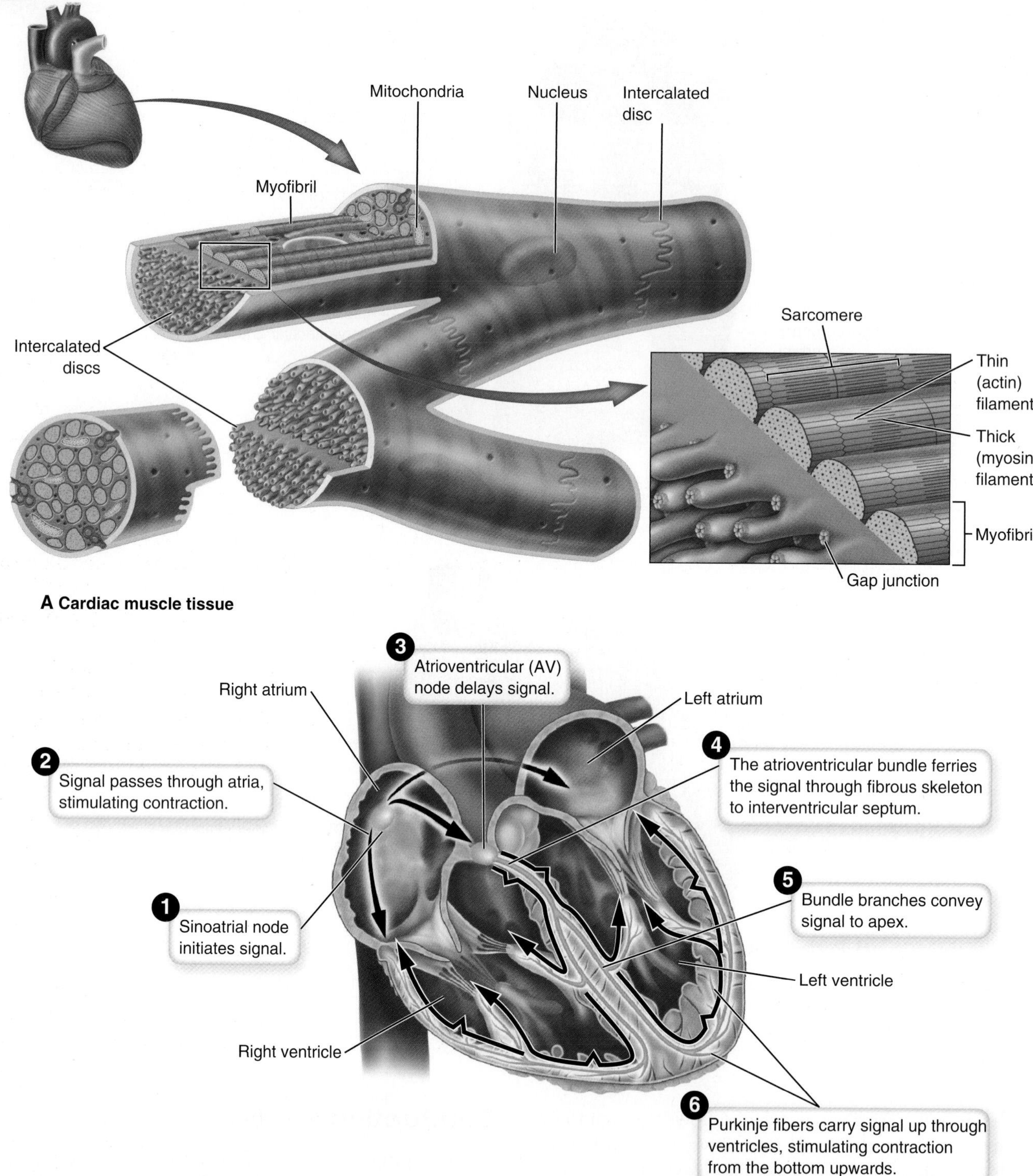
Mitochondria
Nucleus
Intercalated disc
Myofibril
Intercalated discs
Sarcomere
Thin (actin) filament
Thick (myosin) filament
Myofibril
Gap junction
A Cardiac muscle tissue
Atrioventricular (AV) node delays signal.
Right atrium
Left atrium
Signal passes through atria, stimulating contraction.
The atrioventricular bundle ferries the signal through fibrous skeleton to interventricular septum.
Bundle branches convey signal to apex.
Sinoatrial node initiates signal.
Left ventricle
Right ventricle
Purkinje fibers carry signal up through ventricles, stimulating contraction from the bottom upwards.
B The cardiac conducting system

FIGURE 10.10 ▸ Cardiac muscle tissue. A. Cardiac muscle fibers are striated, branched, and connected by intercalated discs. **B.** Contractions of the heart are controlled by its conduction system.

divided into two functional networks of interconnected cells: an atrial and ventricular network.

Approximately 1% of all cardiac muscle cells have an inherent rhythmicity, which means they are self-excitable, able to generate action potentials on their own. Since cardiac fibers can pass action potentials from one to another, these ***autorhythmic fibers*** are responsible for both setting the pace and coordinating the contraction of all cardiac muscle to create an efficient pump. These specialized fibers are organized into a **conduction system** made up of the following components (▸ FIGURE 10.10B):

- **Sinoatrial (SA) node**—Located in the wall of the right atrium, just inferior to the opening of the superior vena cava, this collection of autorhythmic cells serves as the heart's "pacemaker." The SA node self-generates action potentials at the rate of about 100 per minute, signaling the rest of the atrial cardiac muscle tissue to contract.
- **AV node**—The AV node is located in the inferior portion of the septum between the atria. The cardiac fibers of the AV node do not contract as often as the cells of the SA node. Therefore, this structure acts as an area of delay, holding back the conduction of action potentials until both atria have fully contracted.
- **AV bundle**—From the AV node, the action potential passes into the AV bundle, also known as the **bundle of His**, within the interventricular septum. In a healthy person, this is the only region where action potentials pass from the atria to the ventricles.
- **Right and left bundle branches**—The AV bundle branches left and right to carry the action potential down the interventricular septum to the apex of the heart.
- **Purkinje fibers**—These fibers, named for the Czech physiologist who discovered them in 1839, run from the apex of the heart upward around the ventricles. Purkinje fibers carry action potentials through the ventricular myocardial network to produce contraction of the ventricles, resulting in a wringing action.

These components of the conduction system create an organized pathway for action potentials to be initiated and passed across the myocardium. In short, the ***SA node*** initiates the contraction sequence by generating an action potential that quickly spreads across both atria. The ***AV node*** assures that the entire atrial network of cardiac cells has contracted before passing the action potential on to the AV bundle. The coordinated contraction of the ventricular network occurs as the ***AV bundle*** passes the impulse down to the apex of the heart via the ***bundle branches*** and finally, up and across the rest of the ventricular myocardial tissue via the ***Purkinje fibers*** (FIGURE 10.10B).

Cardiac Cycle

A **cardiac cycle** refers to the ordered sequence of atrial and ventricular contraction and relaxation that make up one heartbeat. Simply put, a single heartbeat involves contraction of the two atria while the ventricles relax, followed by ventricular contraction as the atria relax (FIGURE 10.11). The term **systole** refers to a contraction state of the heart chambers, while **diastole** refers to a relaxation state in which the chambers dilate as they fill with blood. The cardiac cycle involves three phases:

1. ***Relaxation phase***—The cardiac cycle starts when the ventricles begin to relax and join the atria in a brief phase of mutual diastole. As the ventricles relax, the pressure within them eventually drops below the pressure in the blood-filled atrial chambers. This allows the AV valves to open and blood to flow into the ventricles. About 75% of the total blood volume within the atria enters the ventricles during this phase.
2. ***Atrial systole***—The SA node generates an action potential that causes the atria to contract while the ventricles continue in diastole. This atrial systole forces the of blood volume left in the atria through the open AV valves into the ventricles. During ventricular filling, the semilunar valves are closed.
3. ***Ventricular systole***—As the impulse is passed through the bundle of His to the bundle branches and Purkinje fibers, the ventricles contract. The AV valves close and the pressure within the ventricles rises. Eventually, the pressure in the ventricles forces the semilunar valves to open, and blood flows from the right ventricle into the pulmonary artery and into the aorta from the left ventricle. During ventricular systole, the atria are experiencing diastole, which allows them to fill with blood from the great veins. Once the ventricles eject their blood and begin to relax, the cycle starts all over again.

By the Way

An ***electrocardiogram*** (abbreviated ECG or EKG) measures the electrical activity of the heart and transfers it to a graph that shows the various stages of the cardiac cycle. It is used to identify the causes of abnormal heart rhythms and heart attacks.

The entire cardiac cycle takes less than one second in a healthy adult when at rest. This translates to an average resting heart rate of 72 to 84 beats per minute. The "lub-dub-lub-dub" sound we call the heartbeat, is actually the sound created by the closing of the heart valves. The "lub" part of the sound occurs when the AV valves shut at the start of ventricular systole, and the "dub" sound is the closing of the semilunar valves. The pause between heartbeats correlates to the relaxation phase. When heart rate increases, the relaxation phase shortens, and there is no real pause between heart sounds.

Scientific studies of the effect of manual therapy on the cardiovascular system have often focused on changes in heart rate, blood pressure, and circulation. It has generally never been considered that manual therapy may impact the heart directly. However, new research in neurocardiology shows that with every beat of the heart, signals are sent to the brain and other organs, causing them to act in synchronicity with the heart's rhythm.[4] While

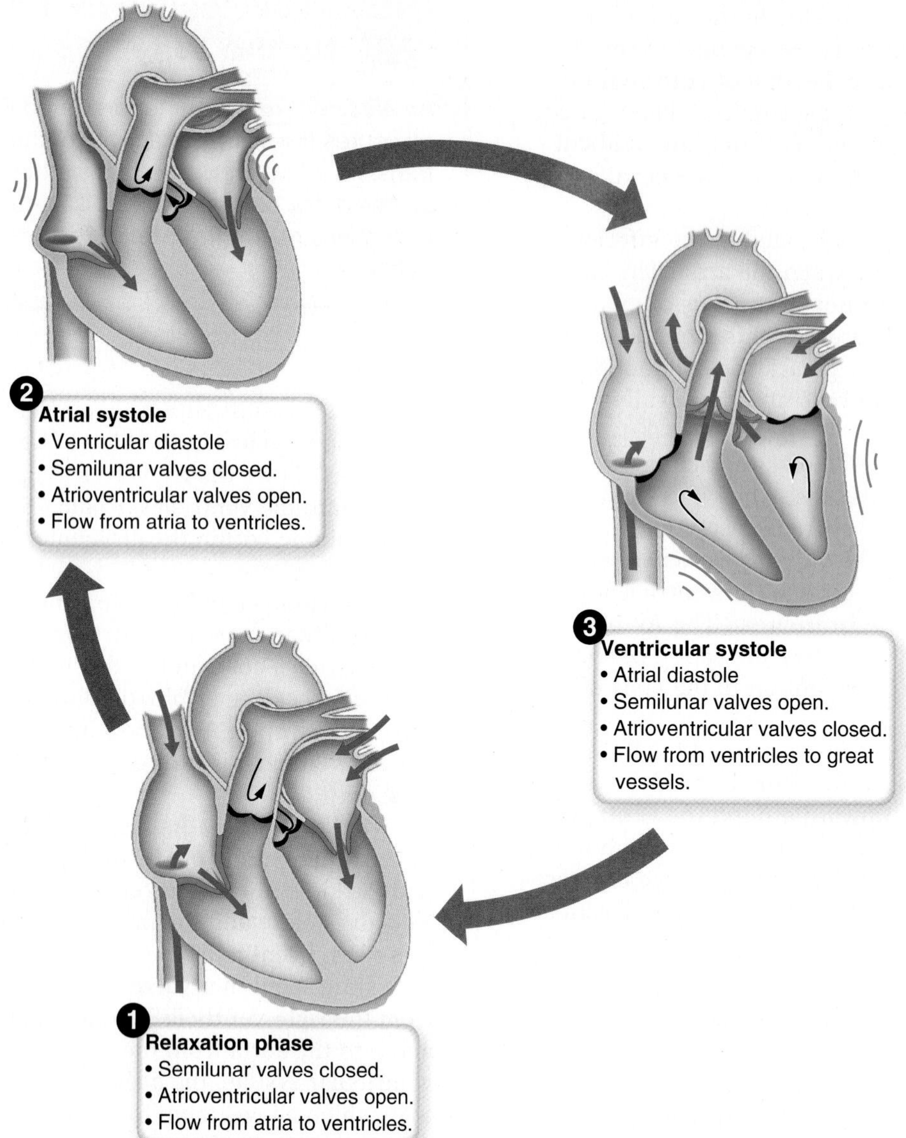

2
Atrial systole
• Ventricular diastole
• Semilunar valves closed.
• Atrioventricular valves open.
• Flow from atria to ventricles.
3
Ventricular systole
• Atrial diastole
• Semilunar valves open.
• Atrioventricular valves closed.
• Flow from ventricles to great vessels.
1
Relaxation phase
• Semilunar valves closed.
• Atrioventricular valves open.
• Flow from atria to ventricles.

FIGURE 10.11 Cardiac cycle. The three phases of the cardiac cycle are the relaxation phase, atrial systole, and ventricular systole. To view an animation on the cardiac cycle, visit http://thePoint.lww.com/Archer-Nelson.

research has proven that manual therapy can decrease stress and anxiety levels, the physiologic mechanisms that produce this change are unclear. In a good deal of the literature, it is hypothesized that the mechanisms for decreasing stress and anxiety are related to increasing parasympathetic stimulus or decreasing sympathetic tone.

Studies on coronary disease have repeatedly shown that factors such as loneliness, grief, cynicism, hostility, and anger are often stronger indicators of heart disease than family history, cholesterol, smoking, or obesity.[1-3,5] The practice of actively listening to a client's history and stories supports a therapeutic relationship that can help a client heal from emotional trauma.[6] It seems that holistically connecting with clients can help to dispel loneliness or create the space and opportunity for an emotional release that allows them to let go of heart-destructive emotions such as grief, anger, and hostility.[7] Could it be that our greatest effect on the cardiovascular system comes not from changes mediated by the brain and nervous system, but by "touching" the heart with positive intentions such as gratitude, compassion, and love?

Regulation of Heart Rate

Factors including age, gender, fitness level, emotional state, temperature, and exercise all affect heart rate, which can vary considerably over the course of a day. Key physiological mechanisms that monitor and regulate heart rate include:

- ***Autonomic nervous system (ANS)***—The ***cardiovascular center*** of the medulla oblongata adjusts heart

By the Way

Research demonstrates that the heart often diverges from the commands made by the ANS. According to Mimi Guarneri, M.D., in *The Heart Speaks: A Cardiologist Reveals the Secret Language of Healing*, it seems that the heart's conduction system acts as an intrinsic nervous system that is able "to act independently of the cranial brain—to learn, remember, even sense and feel."

rate based on sensory input from a variety of receptors. Chief among these are baroreceptors in the sinus at the fork of the left and right common carotid arteries that monitor blood pressure, and chemoreceptors that monitor changes in blood chemistry. The cardiovascular center communicates with the SA and AV nodes via autonomic nerve fibers.

- ***Hormones***—The heart rate alters in response to hormones, including adrenaline and noradrenaline, released from the adrenal medulla during the alarm response. Thyroid hormones that increase metabolic rate also stimulate an increase in heart rate.
- ***Ion levels***—Elevated blood levels of sodium and potassium ions decrease heart rate and contractility, while elevated calcium levels tend to increase the rate and strength of the heartbeat.

What Do You Think? 10.3

- Why is the word *atrium* a good term for a receiving chamber of the heart?
- Ninety-eight percent of all veins carry deoxygenated blood. What are the names of the veins that are the exception to this rule?
- Name situations in which your heart rate has been elevated or has decreased. What mechanisms do you think were responsible for mediating these changes?

CIRCULATION AND DYNAMICS OF BLOOD FLOW

The cardiovascular system is a closed network of blood vessels with the heart as the physiologic center. This network is divided into two circuits, or divisions. The **pulmonary circuit** carries blood between the heart and lungs and the **systemic circuit** carries blood between the heart and the rest of the body (▶ FIGURE 10.12). Blood flow through the pulmonary circuit is controlled by the *right* side of the heart and begins with deoxygenated blood from the body entering the right atrium. Once the blood enters the right ventricle, it is pumped into the pulmonary artery and carried to the lungs. Newly oxygenated blood from the lungs returns to the heart via the pulmonary veins. The *left* atrium receives the oxygenated blood, which begins the systemic circuit. The left ventricle contracts pushing blood into the aorta for arterial distribution throughout the body. The systemic circuit is complete when the inferior vena cava and superior vena cava return deoxygenated blood to the right atrium.

Within each circuit, the vascular routes are always the same: blood travels away from the heart passing through a major artery and into smaller arteries, arterioles, and into capillary beds. Venules carry blood out of the capillaries to small veins and then into major veins for return to the heart. The key physiologic mechanisms that drive circulation are the basic hydrodynamics of all fluid movement:

- Fluid flows from an area of high pressure to an area of lower pressure. Ventricular contractions create the initial driving force for circulation as they forcefully push blood into the arteries. Additionally, blood flow is enhanced any time the difference between high and low pressure is increased.
- Fluid flow is decreased when any resistance is applied. In the cardiovascular system, **peripheral resistance** to blood flow is created by several factors, including blood viscosity, elasticity of blood vessels, and the size of lumens.

Circulation is affected by the structure and location of various vessels. Therefore, arteries, veins, and capillaries require different physiologic mechanisms to maintain blood flow.

Arterial Flow

Arterial blood flow is driven by the combined forces of ventricular contraction and arterial recoil. Ventricular systole forcefully pumps a large volume of blood into the major arteries leaving the heart. Arteries can usually accommodate only about one-third of the blood forced into them. The remaining blood being pushed through causes arterial walls to stretch and bulge. During ventricular diastole, the artery recoils, creating a secondary push that propels the remaining blood further down the arterial network (▶ FIGURE 10.13). This bulge-and-recoil reaction is propagated throughout the entire arterial network, although the force it generates is diminished in the distal arteries. Arterial expansion and recoil can be palpated as a **pulse** in a few superficial arteries, and because it is generated by ventricular contraction, it provides an accurate assessment of heart rate. The major pulse points of the body are indicated by red dots in Figure 10.4.

Another key influence on blood flow through the arteries is blood pressure. **Blood pressure** is a measurement of the hydrostatic pressure generated by blood against the vascular wall. This measurement consists of two numbers,

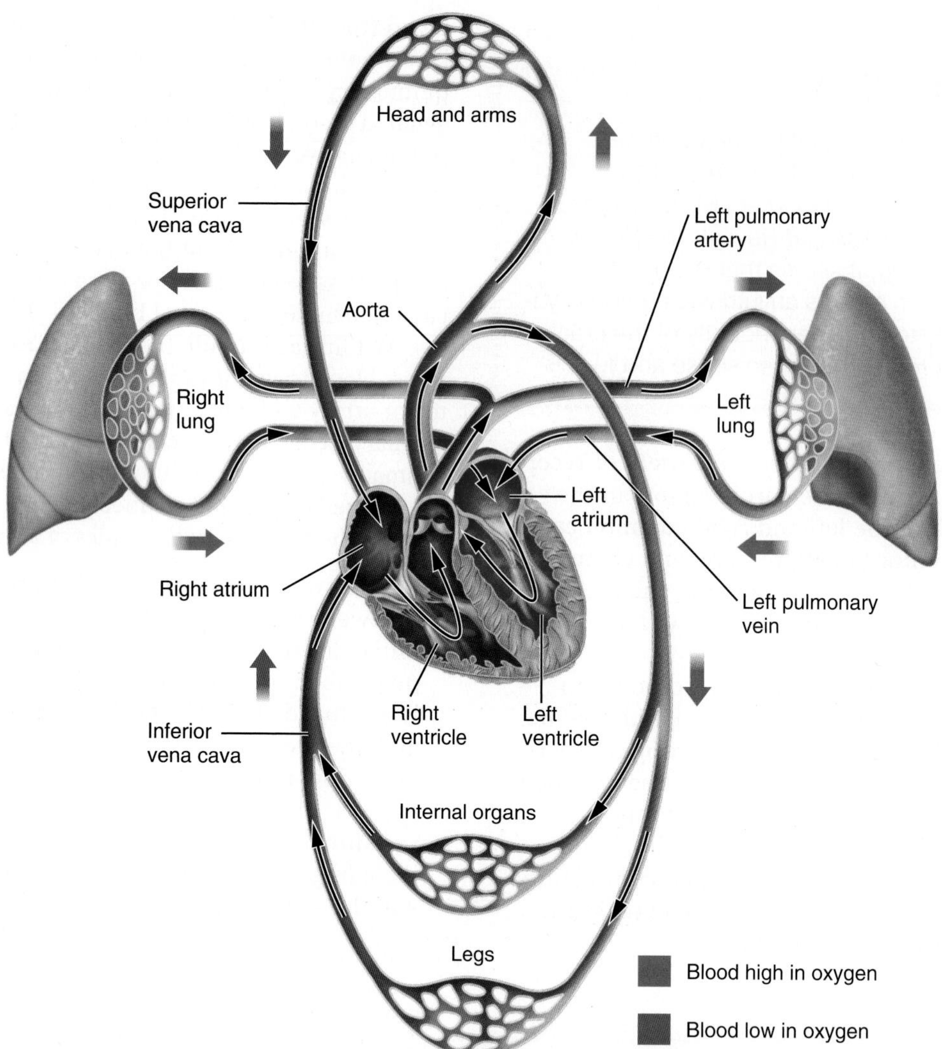

FIGURE 10.12 ▸ Cardiovascular circulation involves the pulmonary and systemic circuits. The pulmonary circuit circulates blood between the heart and the lungs, while the systemic circuit is the blood flow between the heart and the rest of the body.

Pathology Alert Hypertension

A systolic pressure of 140 mm Hg or greater and/or a diastolic reading of more than 90 mm Hg indicate high blood pressure, or hypertension. With no consistent symptoms, **hypertension** can only be diagnosed by taking blood pressure readings over a period of time. For this reason, hypertension is sometimes called the "silent killer." Hypertension can be a sign or a cause of vascular or heart disease. For example, in a patient with arteriosclerosis (hardening of the arteries), the stiffness of the arterial wall increases resistance, elevating blood pressure. In contrast, chronic stress and anxiety lead to elevated blood pressure because of ongoing stimulation of the sympathetic response and long-term exposure to the stress hormones cortisol and adrenaline. This increases cholesterol and glucose levels and makes platelets stickier, which increases the risk of vascular damage and thrombus formation. In these cases, hypertension is not a sign but a cause of vascular and/or heart disease. When hypertension is a sign of cardiovascular disease, manual therapies may be contraindicated, depending on the technique. In stress-related hypertension, manual therapies are valuable complementary therapies and are generally indicated. Clients with hypertension should always check with their primary health care practitioner prior to receiving manual therapy.

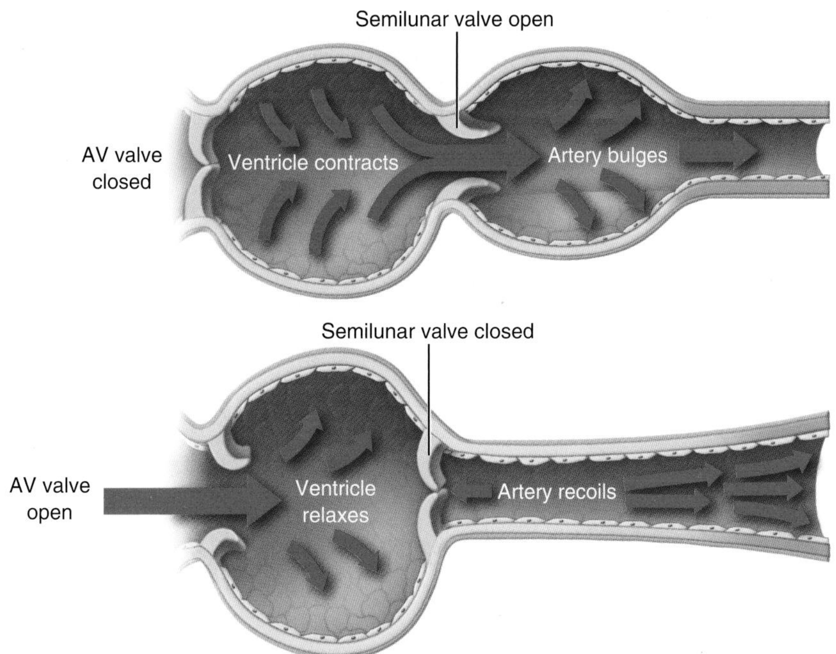

FIGURE 10.13 ▸ Arterial flow. The force of ventricular contraction ejects blood into arteries, creating a bulge-and-recoil response that is a major influence on arterial blood flow.

stated in terms of millimeters of mercury (mm Hg). The higher number is the ***systolic pressure***, which represents the pressure generated during ventricular contraction. The ***diastolic pressure*** is a lower number that represents the pressure in the arteries when the ventricle relaxes. According to the American Heart Association, a reading 120/80 mm Hg is considered the optimal blood pressure for a healthy adult. Blood pressure changes occur throughout the day to meet circulatory demands. For example, standing upright poses more of a challenge to maintain blood return to the heart than does lying down, so blood pressure must increase to meet this demand, as happens upon arising each morning. Numerous factors affect blood pressure, and some of the most common are blood viscosity, blood volume, elasticity of the vessel walls, and efficiency of the heart.

Capillary Flow and Exchange

Blood flow into the capillaries is controlled by the precapillary sphincters. These smooth muscle sphincters constrict and dilate to control both the volume and rate of blood flow into the capillaries. A central pathway called a ***thoroughfare channel*** in each capillary bed allows blood to pass straight through when metabolic need is low (see FIGURE 10.5). However, when demand increases, the precapillary sphincters open to allow the entire capillary bed to fill, so that nutrient–waste exchange can take place.

Capillary exchange between blood and the interstitium involves the movement of particles or solutes, as well as fluids. Solutes including oxygen, carbon dioxide, and glucose move by simple diffusion through the capillary wall. Fluid is exchanged between the blood and interstitium through a combination of hydrostatic and oncotic pressures (osmotic pressures on the capillary wall produced by proteins) (▸ FIGURE 10.14). These pressures are collectively referred to as ***Starling forces***.

- **Capillary fluid pressure (CFP)** is hydrostatic pressure created by blood inside the capillary pushing outward against its wall.
- **Interstitial fluid pressure (IFP)** is created when the interstitial fluid presses inward against the capillary wall.
- **Plasma oncotic pressure (POP)** is an osmotic pressure created inside the capillary by the protein content of the blood.
- **Interstitial oncotic pressure (IOP)** is an osmotic pressure on the outside of the capillary created by the protein content of interstitial fluid.

The movement of plasma and its dissolved solutes out of the blood and into the interstitium is called **filtration. Reabsorption** refers to the movement of fluid and substances back into the capillary bed. Both CFP and IOP create filtration, whereas IFP and POP are the forces involved in reabsorption. In healthy tissue, the dominant Starling force at the arteriole end of the capillary bed is CFP due to residual arterial blood pressure. This is why filtration occurs at the arteriole side of the capillary bed. Once the fluid filters out of the capillary, the concentration of plasma proteins is increased within the vessel. This increases the POP at the venule end of the capillary bed, so that reabsorption can occur. In a healthy body, net filtration is always slightly higher than net reabsorption, producing a net outflow of fluid and solutes from the capillaries. This volume of fluid is regularly absorbed

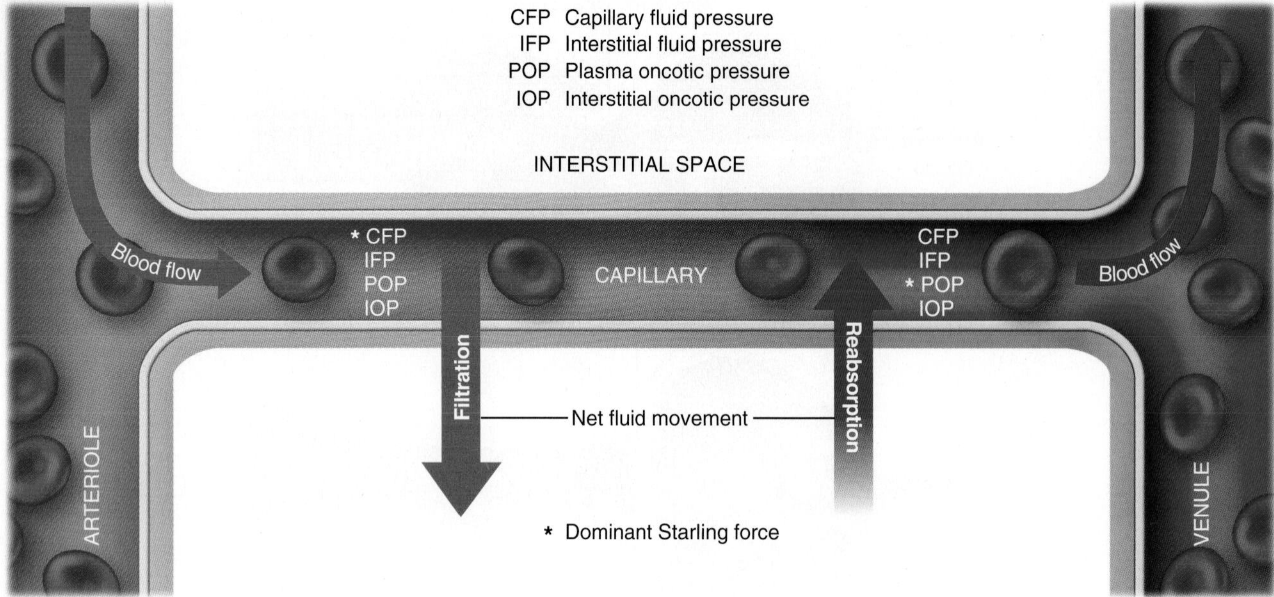

FIGURE 10.14 ▸ Capillary pressures and fluid movement. Capillary exchange is controlled by four Starling forces that create a net filtration pressure at the arterial end of the capillary and a net reabsorption pressure at the venule end.

from the interstitium and returned to circulation via the lymphatic system. If net filtration greatly exceeds net reabsorption, extra fluid accumulates in the interstitium, creating **edema**. Conditions that can lead to edema include hypertension, pregnancy, obesity, diabetes, musculoskeletal trauma, and cardiac and kidney dysfunction.

Venous Flow

Because the pressure created from heart contraction has dissipated by the time blood enters the venules and veins, the physiologic mechanisms for venous flow are more external than those of arterial flow. The location of veins offers a significant clue about the strongest influence: veins are situated more superficially in the tissues than arteries and are surrounded by skeletal muscle. Skeletal muscle contractions apply external compression to the veins, which helps to propel blood back toward the heart (▸ FIGURE 10.15). One-way valves assist venous flow by creating shorter segments for blood to travel. In the ventral cavities where the large veins are not surrounded by skeletal muscle, the driving force for venous flow is the ***respiratory pump.*** This pump is created by the positive and negative pressure changes inside the thoracic and abdominopelvic cavities that occur during breathing. Venous flow is also affected by gravity, which unfortunately works against it most of the time.

Regulation of Circulation

Every day, the cardiovascular system is challenged to maintain adequate blood flow to all body tissues and make circulatory shifts to adapt to the body's changing demands. Changes in aspects such as temperature, activity levels, and emotions can require changes in the circulation affecting organs and tissues throughout the body. Local shifts in the volume and flow of blood are known as **ischemia**, a decrease in blood volume or flow, and **hyperemia**, which indicates an increase.

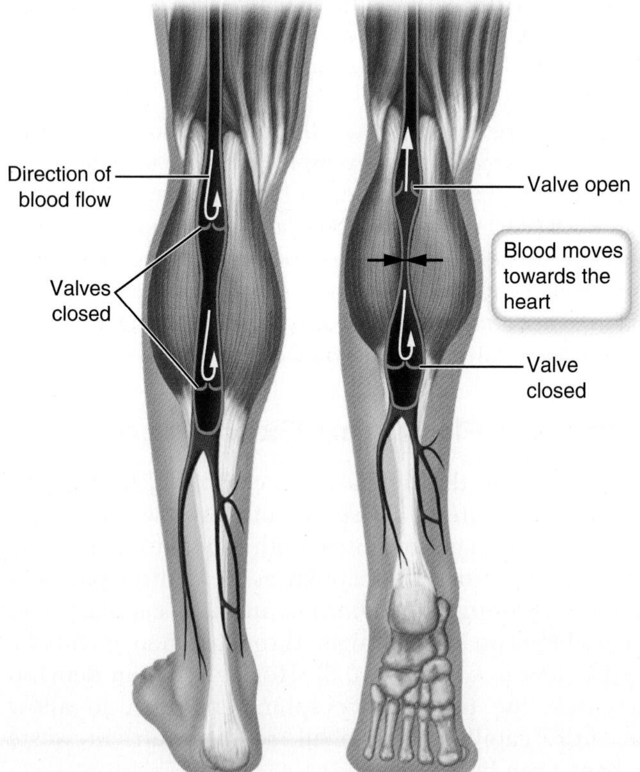

FIGURE 10.15 ▸ Skeletal muscle pump. Skeletal muscle contraction plays an essential role in venous blood flow. The contraction creates external compression of the veins, which assists in moving blood back to the heart.

Pathology Alert Vein Disorders

Varicose veins, thrombophlebitis, and deep vein thrombosis (DVT) are three venous conditions manual therapists are likely to see at some point in their career. While all present some contraindications or cautions to treatment, the least problematic is ***varicose veins***. This condition is identified by the presence of several enlarged and gnarled veins in the legs, which are caused by the failure of the one-way valves. In some cases, there may be associated "spider veins," indicating that capillary beds are also distended. The blood in varicose veins has stagnated and thickened, creating the risk of thrombus formation. Direct manipulation of the varicosities is contraindicated because it could dislodge a thrombus, as could deep stroking in the tissue distal to the varicosity.

When venous flow is obstructed by blood clots in the superficial veins of the leg and inflammation is present, the condition is called ***thrombophlebitis***. Obstruction in the deeper veins (popliteal, femoral, and iliac) is known as ***deep vein thrombosis***. The signs and symptoms of thrombophlebitis include pain, heat, and redness in the calf that cannot be related to musculoskeletal trauma. DVT may not display any of these signs or symptoms but can be indicated by discoloration, pitting edema, intermittent pain that increases with activity, and may also cause the affected limb to feel cool or "clammy." If passive dorsiflexion of the affected leg (knee straight) causes pain, therapists should suspect the possibility of one of these conditions or some other pathology. In any case, manual therapy is contraindicated, and therapists should refer the client to a qualified medical professional.

In general, there are two guiding principles for the regulation of circulation. First, circulation is increased in active tissues and decreased in those that are less active. Second, blood flow to the vital organs must always be maintained. For example, in cases of extreme cold or blood loss, blood flow to the brain, lungs, and heart is prioritized over that of other body parts. The body has three key physiological mechanisms that monitor and regulate circulation:

- ***Autoregulation***—Most body tissues have some ability to signal and create local short-term changes in circulation. For example, during hemostasis, injured cells release chemicals to signal an immediate vasospasm to decrease blood flow in a damaged vessel. Another example is tissue response to local applications of heat or cold; a hot pack creates vasodilation, while a cold pack causes vasoconstriction.
- ***Nervous system regulation***—As described in Chapter 7, the smooth muscle in blood vessels is innervated by sympathetic neurons and controlled by the ***vasomotor center*** in the medulla oblongata. When a sympathetic response is stimulated, vasoconstriction occurs in most organs and tissues, while vasodilation occurs in the skeletal muscles. The activity of the vasomotor center is altered by several factors including emotional state, blood pressure, pain, and oxygen and carbon dioxide levels. Additionally, dramatic shifts in blood flow accompany changes in heart rate. Recall that the cardiovascular center of the medulla oblongata moderates heart rate in response to stimulus from various sensory receptors.
- ***Hormones***—Local hormones such as the prostaglandins promote vasodilation in local blood vessels. Prostaglandins are released when tissue is damaged. Adrenaline and noradrenaline cause the same vascular responses as the sympathetic neurons: vasoconstriction in general but vasodilation in the skeletal muscles and heart.

Increased circulation is one of the most widely claimed physiologic effects of manual therapies, especially massage. However, it is difficult to find current and well-designed research that supports this claim. Circulation involves arterial blood flow away from the heart, capillary flow and exchange, and venous return of blood to the heart—the completion of a circuit from and back to the heart. Since circulation is regulated by several complex physiological processes, it is important not to confuse local changes in blood flow with systemic circulatory changes. For example, hyperemia and ischemia indicate local vasodilation or vasoconstriction. They are not reflective of systemic changes in blood flow and circulation. In fact, local changes in blood flow and pressure demand homeostatic adjustments that maintain adequate circulation throughout the body.

While it has been demonstrated that massage is effective in moving blood through superficial veins and out of local capillary beds, as well as in producing histamine release that leads to vasodilation and local hyperemia, there is no scientific evidence that points to systemic enhancement of circulation.[8–13] Even well-supported systemic physiologic effects of massage such as decreased heart rate and blood pressure are only short-term changes that could actually decrease overall systemic circulation.[14–19] This is not to say that manual therapies have no positive effect on the cardiovascular system, simply that the claim of increased systemic circulation is not supported by current research. Even the most recent research on fascia and its connections with the ANS refers only to changes in blood pressure and *local* circulation. While we may hypothesize that movement therapies will enhance local venous flow due to muscle contraction or that energy techniques that increase or balance energy would be beneficial to circulation, there is a shortage of rigorous scientific research to support these hypotheses. Perhaps ongoing research in neurocardiology will offer new insights into possible systemic effects of massage on the cardiovascular system.

What Do You Think? 10.4

- In cases of hypertension, would you expect a pulse to be strong and pounding or thready and weak? Explain your answer.
- When you are frightened, in what tissues and organs would vasoconstriction occur?
- Explain how both hypertension and pregnancy can create swelling in the ankles.
- How do you think manual therapy may affect local venous flow?

By the Way

Changes in skin coloration can serve as external indicators of changes in local or systemic circulation. The term for reddened skin is **erythema,** which indicates vasodilation. Erythema can be caused by inflammation or elevation in body or tissue temperature. If tissue is not receiving enough oxygen, it has a blue appearance and is described as **cyanotic**. A variety of causes can result in cyanosis, including vasoconstriction, poor circulation related to heart failure, or respiratory dysfunction. **Pallor** is the term used to describe white or pale skin, which generally indicates vasoconstriction and/or poor circulation.

INFLAMMATION AND TISSUE HEALING

When tissue damage occurs, the hemostasis process stops any active bleeding, or **hemorrhage**, in a fairly short time frame. Simultaneously, **inflammation**, an additional protective response, is initiated to help stabilize the damaged tissue and prepare it for repair. Since tissue healing requires the delivery of important repair materials and the clean-up of damaged cells, the cardiovascular system plays a vital role in both the chemical and cellular processes of the inflammatory response. The key physiologic processes of inflammation include vasodilation, changes in capillary permeability, clot formation, and phagocytosis.

Vasodilation and capillary permeability are both enhanced when **histamine** is released by mast cells, basophils, and platelets. These tissue changes allow both the cells and chemical mediators required for clean-up and repair to flood the affected area. Platelets and clotting factors for hemostasis, neutrophils, and other phagocytes for clean-up, and fibroblasts and proteins for tissue repair are some of the major components delivered via blood. Other chemicals are released that support both vasodilation and increased capillary permeability; these include local tissue hormones such as ***prostaglandins*** and ***leukotrienes*** and ***complement factors*** in the blood.

Hemostasis and inflammation are the *initial* tissue responses to trauma. The entire healing and repair process is divided into three phases: acute, subacute, and maturation stages of repair. The rate at which the entire process occurs and the time it takes the body to move through each phase depends on a variety of factors, including the type and amount of tissue damaged, the tissue's vascularity, and the individual's overall nutritional health and fitness status at the time of injury.

Acute Stage

The first phase of the healing process is the **acute stage**, also known as the **inflammatory stage**. It begins when tissue is damaged, leading to hemorrhage and inflammation. These responses result in **primary edema**, a collection of fluids and cellular components in the interstitial space. The amount of primary edema depends on the vascularity of the tissue and the severity of tissue damage and indicates the amount of repair that needs to occur. Other key physiologic events in the acute phase are:

- ***Pain and muscle spasm***—The pressure created by edema and the presence of prostaglandins and other inflammatory chemicals stimulate nociceptors and other free nerve endings in the injured tissue. The resulting pain initiates a reflex contraction of the muscles surrounding the injury. This protective response called **muscle splinting** protects the traumatized region from further damage and allows clotting, clean-up, and hematoma organization to go on undisturbed.
- ***Secondary edema formation***—The extra cells and fluids in the interstitial space create an unusually high concentration of proteins that increases oncotic pressure. This increased IOP creates a siphoning effect that draws fluid out of previously healthy and undamaged cells and capillaries in surrounding

tissues. This process of **secondary edema** formation explains why an area of trauma can continue to swell for 24 to 48 hours after initial injury.

- ***Hematoma organization***—Once secondary edema formation is controlled and all cellular and chemical components are activated, the edema begins to organize. During this process, the leukocytes and phagocytes finish their clean-up and migrate out of the area to be reabsorbed via the lymphatic and venous capillaries. The limited number of fibroblasts initially involved in blood clot formation is increased as phagocytosis is completed, and they begin to move to the perimeter of the edema to establish a loose-knit net around the damaged site. This movement of phagocytes and fibroblasts is called **margination**; it provides a boundary, or margin, for the repair work that needs to be done. The body has now formed a **hematoma**, a localized collection of blood and fluids in tissue, which is the last step in the acute stage of the healing process (FIGURE 10.16A).

Since therapists cannot look inside damaged tissue and see the ongoing physiologic processes, we must rely on a set of external indicators to determine the presence of tissue damage and the most likely stage of healing and repair. Although a multitude of different signs and symptoms must be considered, the key external indicators of musculoskeletal trauma and the tissue's repair status can be summarized by the mnemonic SHARP, which stands for:

- ***Swelling***—Swelling of the area is a result of both primary and secondary edema.
- ***Heat***—An increase in tissue temperature is a clear indicator that the body is actively involved in hemorrhage and/or inflammation.
- ***Asymmetry** and/or **A loss of function***—A visible or palpable change in a tissue's contour, borders, texture, or shape is a good indicator of injury. A decrease in tissue function (range of motion and/or strength) is also a general indicator of injury. Because both can be apparent at any stage of tissue healing, they are better indicators of the severity of a trauma (mild, moderate, or severe) than the stage of healing.
- ***Redness***—Erythema over a site of injury is a clear indicator of vasodilation and chemical activation in the inflammatory process.
- ***Pain***—As previously discussed, pain often accompanies trauma and injury. The quality of the pain and where it occurs during the range of motion help therapists determine the stage of healing. Stabbing pain that limits movement or occurs in multiple points of active motion are good indicators of the acute stage.

All of the SHARP indicators are present to some degree in the acute phase and diminish as healing progresses. Recognizing that observable signs reflect physiologic processes provides therapists with a sound rationale for their therapeutic choices. For example, any redness or heat in the tissue indicates that the injury is still in the inflammatory phase and must be treated accordingly, regardless of how long ago the initial trauma occurred or how much swelling or pain is present. General manual therapy goals for the acute phase include controlling primary and secondary edema and interrupting the pain-spasm-pain cycle. The standard care formula used to achieve these goals is rest, ice, compression, and elevation (RICE). Manual lymphatic techniques that stimulate edema removal can be added to the standard care protocol to help decrease pain and control secondary edema formation.

Subacute Stage

During the **subacute stage**, also called the **proliferative stage**, the actual tissue repair begins. Reabsorption of leukocytes and phagocytes into the lymphatic and cardiovascular capillaries is completed, and the number of fibroblasts is increased. While muscle splinting is an important protective response that allows hematoma

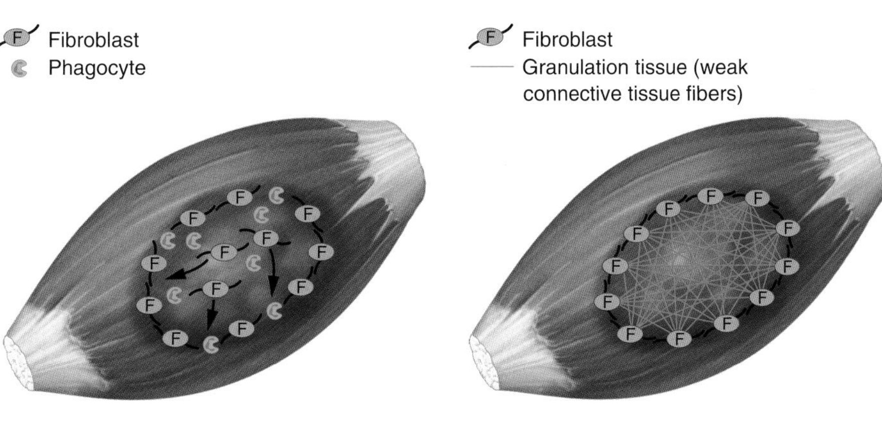

FIGURE 10.16 Hematoma organization. A. Toward the end of the acute stage of tissue healing, the numbers of phagocytes and leukocytes decrease as fibroblasts increase. The fibroblasts migrate to the perimeter of the edema (margination) to form a hematoma. **B.** Fibroblasts begin the repair process by forming a disorganized web of fragile granulation fibers throughout the hematoma.

organization in the acute stage, continued spasm and splinting can create ischemia in the tissues that limits the clean-up and repair process in the subacute stage. With sufficient circulation, fibroblasts proliferate and begin to synthesize fragile thread-like fibers called **granulation tissue** throughout the hematoma (FIGURE 10.16B). These fibers are laid down in a haphazard multidirectional web across the entire hematoma. Once granulation tissue has fully inundated the area, the process of thickening and strengthening the fragile granulation fibers into full strength collagen can begin. During this process of **collagen remodeling**, some granulation threads are dissolved and recycled to make other fibers stronger. Fibers that are in good alignment with the normal lines of stress of the tissue become thickened, while those in poor alignment are dissolved. For example, in muscles with a vertical fiber alignment, the vertical granulation fibers are thickened, while those with a horizontal or oblique fiber direction will be dissolved.

Proper alignment of collagen repair fibers cannot occur without using movement to "show" the body the normal lines of stress in the tissue. This prevents tissue from developing thick, poorly aligned, and matted repair patches, which is especially important in tissues with highly organized fiber arrangements such as muscles, ligaments, and tendons. Disorganized repair patches can disrupt normal tissue function and irritate surrounding healthy tissue. Movement during the subacute phase is also important because it facilitates lymphatic and venous fluid return. In the later part of the subacute stage, therapists may also initiate some form of cross-fiber tissue manipulation to break apart the fibers that interfere with proper fiber alignment in the tissue. However, great care must be taken to not further damage the tissue and return it to the inflammatory stage.

Maturation Stage

The key physiologic event in the **maturation stage** is the continuation and completion of the collagen remodeling process. Again, sufficient circulation in the damaged tissue is essential for rapid and complete tissue repair. During the first half of this stage, the granulation tissue is completely dissolved and recycled, and many of the new collagen fibers are still fragile. As maturation progresses, the fibrocytes generate new collagen until full tensile strength is returned to the tissue. Manual therapy treatments during the early maturation stage continue to focus on creating flexible, well-aligned repair tissue and improving pain-free range of motion. In the latter half of this stage, more vigorous soft tissue manipulation and exercise can be utilized, since the collagen fibers have matured and they are now able to withstand normal tensile stress. The duration of the maturation stage can vary widely, ranging from three months to a full year before full pain-free function is restored to the area.

AGING AND THE CARDIOVASCULAR SYSTEM

Some of the most significant physiologic declines associated with aging occur in the cardiovascular system. In the heart, cardiac muscle fibers decrease in size, and the number of collagen fibers increases, causing a thickening of the valves. Together, these changes create a progressive loss in the size and strength of the heart that leads to diminished blood supply for organs and tissues, including the brain and kidneys. In fact, blood flow to the brain is estimated to be 20% less at age 80 than at age 30. Blood vessels also thicken and become less elastic, and the baroreceptors decrease in sensitivity, which contributes to a general rise in blood pressure. The rate of nutrient–waste exchange is slowed due to the thickening of the connective tissue membrane around capillaries. Total blood cholesterol increases with age, resulting in a high incidence of heart disease, atherosclerosis, and arteriosclerosis. However, there is solid evidence demonstrating that regular physical and social activity, along with good nutritional habits, can delay this decline and greatly reduce the incidence of cardiovascular disease.

Manual therapists who work with older adults must be aware of these changes in cardiovascular function. Due to the high incidence of heart disease, high blood pressure, arteriosclerosis, and atherosclerosis in this population, clients are often on multiple medications that thin the blood, lower blood pressure, and reduce cholesterol. These medications increase the risk of dislodging blood clots or bruising tissue during a session and also increase the likelihood of positional hypotension (low blood pressure) that can cause dizziness when sitting up or getting off the therapy table.

What Do You Think? 10.5

- Why is it important to try to minimize the amount of primary edema formation in the acute phase?
- How does muscle splinting manifest in the case of a sprained ankle?
- Are arteriosclerosis and atherosclerosis only seen in the elderly population? Why or why not?

SUMMARY OF KEY POINTS

- The cardiovascular system consists of the heart, blood vessels, and blood. Together, they transport nutrients and waste; help regulate body temperature, pH, and fluid balance; and play major roles in the body's immune responses.
- Plasma, the liquid component of blood, carries a group of plasma proteins essential for maintaining the osmotic pressure of blood, clotting, and some immune responses.
- The formed elements in blood are:
  - Erythrocytes, the RBCs that transport oxygen via the specific protein hemoglobin, which also transports CO_2.
  - Leukocytes, the WBCs that carry out important immune responses such as phagocytosis and antibody production.
  - Platelets, or thrombocytes, the small cell fragments responsible for blood clotting.
- There are three types of blood vessels:
  - Arterial vessels, the arteries and arterioles that carry blood away from the heart.
  - Capillaries that serve as the site of nutrient-waste exchange.
  - Venous vessels, the veins and venules that carry blood back to the heart.
- Structurally, both arteries and veins have an inner epithelial layer, a middle smooth muscle layer, and an external connective tissue layer. The muscular layer is thicker in arteries, so they can withstand the force of blood propelled into them by the heart. Veins have an internal system of one-way valves that help prevent backflow of blood as it is returned to the heart. Capillaries are a single layer of epithelial cells with a thin connective tissue membrane covering.
- The heart is a pump that provides the primary force to circulate blood. It is located in the mediastinum and surrounded by the two-layered pericardium. The heart is divided into right and left sides by a muscular partition called the septum, and each side has an atrium and a ventricle. There are two AV valves between these chambers; the right AV valve is the tricuspid and the left side is the bicuspid or mitral valve.
- Major arteries that carry blood from the heart are the pulmonary artery that exits from the right ventricle (pulmonary valve between the chamber and artery) and the aorta that exits the left ventricle (aortic valve between the chamber and artery). Major veins that return blood to the heart are the pulmonary veins from the right and left lungs that enter the left atrium and the inferior vena cava and superior vena cava, which both enter the right atrium. There are no valves between the major veins and the atria.
- Blood flow through the heart (beginning on the right side of the heart) starts with blood returning to the heart via the vena cavae to the right atrium. It passes through the tricuspid valve to the right ventricle then through the pulmonary valve into the pulmonary artery to the lungs. Blood returns to the heart through the pulmonary veins to the left atrium, passes through the bicuspid (mitral) valve to the left ventricle, then through the aortic valve into the aortic artery that carries blood to the body.
- A group of autorhythmic fibers in the heart make up the cardiac conduction system, which is responsible for setting the pace and coordinating contractions of the heart. Primary components of this conduction system are the SA node, the AV node, AV bundle (bundle of His), right and left branch bundles, and Purkinje fibers.
- A cardiac cycle refers to the ordered sequence of atrial and ventricular contraction and relaxation that make up one heartbeat. The entire cycle takes less than one second and translates to a normal resting heart rate of about 72 to 84 beats per minute.
- The three regulatory mechanisms for heart rate are:
  - The cardiovascular center of the medulla oblongata; part of the ANS.
  - Hormones that increase general metabolic rate.
  - Ion levels in the blood, such as elevated Na^+ and K^+ that decrease heart rate and elevated Ca^+ that increases the rate and strength of heartbeats.
- The cardiovascular system incorporates a network of blood vessels that create two circuits or divisions: the pulmonary circuit between the heart and lungs and the systemic circuit between the heart and the rest of the body. The pulmonary circuit carries deoxygenated blood from the right ventricle to the lungs via the pulmonary artery and returns freshly oxygenated blood to the left atrium via the pulmonary veins. The systemic circuit carries freshly oxygenated blood out of the left ventricle to the rest of the body via the aorta and returns deoxygenated blood to the right atrium via the superior and inferior vena cavae.
- The physiologic mechanisms that control blood flow through blood vessels vary according to the type of the vessel. Blood flow in arteries is influenced by intrinsic factors: ventricular contraction, blood pressure, and arterial recoil. Venous flow relies on external compression from skeletal muscle contraction, the one-way valves, and the respiratory pump. Capillary exchange is regulated and controlled by four diffusion pressures called Starling forces.
- Capillary exchange has two parts: filtration, which refers to the flow of fluid and nutrients *out* of the

bed at the arteriole side of the capillary bed, and reabsorption, which occurs at the venule side of the bed and refers to the flow of fluids and waste back *into* the capillaries. The dominant Starling force that creates filtration is the higher CFP inside the capillary, while the dominant Starling force for reabsorption is the higher POP created by the plasma proteins.

- The three physiologic mechanisms that monitor and regulate circulation by changing blood pressure, vasodilation, or vasoconstriction are:
  - Autoregulation mechanisms of most body tissues
  - Nervous system regulation from the vasomotor center of the ANS
  - Hormone secretions, including prostaglandins, adrenaline, and noradrenaline.
- The process of tissue repair and healing involves three stages.
  - Acute (inflammatory) stage—The cardiovascular system works to stop the initial hemorrhage via clotting of the blood and begins the inflammation process. The phagocytes that clean up cellular debris and the fibrocytes that organize the hematoma are delivered via the blood.
  - Subacute (proliferative) stage—Actual tissue repair begins as the phagocytes and cellular debris are dissolved and reabsorbed, and an increased number of fibrocytes are delivered to lay down the fragile granulation tissue.
  - Maturation stage—Collagen remodeling is continued to complete tissue repair, which requires dissolving and re-fabricating collagen fibers.

REVIEW QUESTIONS

Short Answer

1. What are the functions of the cardiovascular system?
2. Name the three formed elements in blood and explain the primary function of each.
3. Name the three types of blood vessels and explain the role each plays in circulation.
4. What are the names of the two circulatory divisions (circuits)?
5. Define and explain the cardiac cycle.

Multiple Choice

6. What is the correct anatomic term for a red blood cell?
   a. leukocyte
   b. thrombocyte
   c. erythrocyte
   d. fibrocyte
7. Why is the muscular layer in arteries thicker than that in veins?
   a. because arteries must contract to propel blood forward and veins don't
   b. it makes them more resilient and creates a recoil that is important for arterial flow
   c. to help protect the arteries from trauma
   d. the thinner muscle in veins can contract more rapidly than the thicker layer in arteries
8. Which major artery delivers blood to the upper extremities?
   a. carotid
   b. brachial
   c. subclavian
   d. descending aorta
9. Where is the saphenous vein located?
   a. middle of the cubital fossa
   b. the abdominopelvic cavity
   c. along the lateral aspect of the upper extremity
   d. along the medial aspect of the lower extremity
10. The baroreceptors that sense blood pressure are located in which major blood vessel?
    a. carotid artery
    b. abdominal aorta
    c. jugular vein
    d. pulmonary artery
11. What is the function of the pericardium?
    a. provides nutrient–waste exchange for the heart
    b. applies an external compression to the heart that strengthens the contractions
    c. firmly anchors the heart to the primary vessels and thoracic cavity
    d. acts as a protective covering that secretes serous fluid to decrease friction over heart as it contracts

12. Which blood vessel carries blood from the heart to the lungs?
    a. pulmonary vein
    b. pulmonary artery
    c. respiratory aorta
    d. thoracic artery
13. What is the name of the heart valve located between the right atrium and ventricle?
    a. mitral
    b. bicuspid
    c. tricuspid
    d. pulmonary
14. Which chamber of the heart pumps blood into the systemic circuit of circulation?
    a. left ventrical
    b. right ventrical
    c. left atrium
    d. right atrium
15. Which chamber of the heart receives blood from the inferior vena cava?
    a. left ventrical
    b. right ventrical
    c. left atrium
    d. right atrium
16. Which of the Starling forces is the dominant force behind capillary filtration?
    a. interstitial fluid pressure
    b. capillary fluid pressure
    c. plasma oncotic pressure
    d. interstitial oncotic pressure
17. The amount of pressure exerted by blood against the walls of the arteries is called
    a. peripheral resistance
    b. arterial pressure
    c. blood pressure
    d. essential pressure
18. Which portion of the cardiac conduction system serves as the pacemaker?
    a. left bundle branch
    b. sinoatrial node
    c. AV node
    d. Purkinje fibers
19. The primary influences over venous flow are the one-way valves and
    a. skeletal muscle contraction
    b. ventricular systole
    c. the pulse
    d. myocardial recoil
20. What is the main cause of secondary edema?
    a. muscle splinting
    b. granulation tissue formation
    c. fibrocyte migration
    d. increased interstitial oncotic pressure

References

1. Niaura R, Todaro JF, Stroud L, et al. Hostility, the metabolic syndrome, and incident of coronary heart disease. *Health Psychol.* 2002;21(6):588–593.
2. Suarez EC. Joint Effects of Hostility and severity of depressive symptoms on plasma interleukin-6 concentration. *Psychosom Med.* 2003;65:523–527.
3. Iribarren C, Sidney S, Bild DE, et al. Association of hostility predict the development of atrial defibrillation in men in the Framingham Offspring study. *Circulation.* 2004;109:1267–1271.
4. Armour JA, Ardell J. *Neurocardiology*. New York: Oxford University Press; 1994.
5. Dawber TR. *The Framingham Study: The Epidemiology of Atherosclerotic Disease*. Cambridge, MA: Harvard University Press; 1980.
6. Pennebaker J, Seagal J. Forming a story: the health benefits of narrative. *J Clin Psychol.* 1999;55(10):1243–1254.
7. Emmons R, McCullough, M. *The Psychology of Gratitude*. New York: Oxford University Press; 2004.
8. Braverman DL, Schulman RA. Massage techniques in rehabilitation medicine. *Phys Med Rehabil Clin N Am.* 1999;10(3):631–648.
9. Prentice WE. *Therapeutic Modalities for Physical Therapists*. 2nd ed. New York: McGraw-Hill, 2002.
10. Pornratshanee W, Hume PA, Kolt GS. The mechanisms of massage and effects on performance, muscle recovery, and injury prevention. *Sports Med.* 2005;35(3):235–256.
11. Tiidus PM. Manual massage and recovery of muscle function following exercise: A literature review. *J Orthop Sports Phys Ther.* 1997;25(2):107–112.
12. Hinds T, McEwan I, Perkes J, et al. Effects of massage on limb and skin blood flow after quadriceps exercise. *Med Sci Sports Exerc.* 2004;36(8):1308–1313.

13. Shoemaker JK, Tiidus PM, Mader R. Failure of manual massage to alter limb blood flow: measures by Doppler ultrasound. *Med Sci Sports Exerc.* 1997;1:610–614.
14. Mein EA, Richards DG, McMillin DL, et al. Physiological regulation through manual therapy. Physical Medicine and Rehabilitation: a state of the art review. *Phys Med Rehabil Clin N Am.* 2000;14(1):27–42.
15. Ashton J. In your hands. *Nurs Times.* 1984;80(19):54.
16. Hernandez-Reif M, Field T, Krasnegor J, et al. High Blood pressure and associated symptoms were reduced by massage therapy. *J Bodyw Mov Ther.* 2000;4:31–38.
17. Aourell M, Skoog M, Carleson J. Effects of Swedish massage on blood pressure. *Complement Ther Clin Pract.* 2005;11:242–246.
18. Cady SH, Jones GE. Massage therapy as a workplace intervention for reduction of stress. *Percept Mot Skills.* 1997;84:157–158.
19. Delaney J, Leong KS, Watkins A, et al. The short-term effects of myofascial trigger point massage therapy on cardiac autonomic tone in healthy subjects. *J Adv Nurs.* 2002;37:364–371.

11

The Lymphatic System

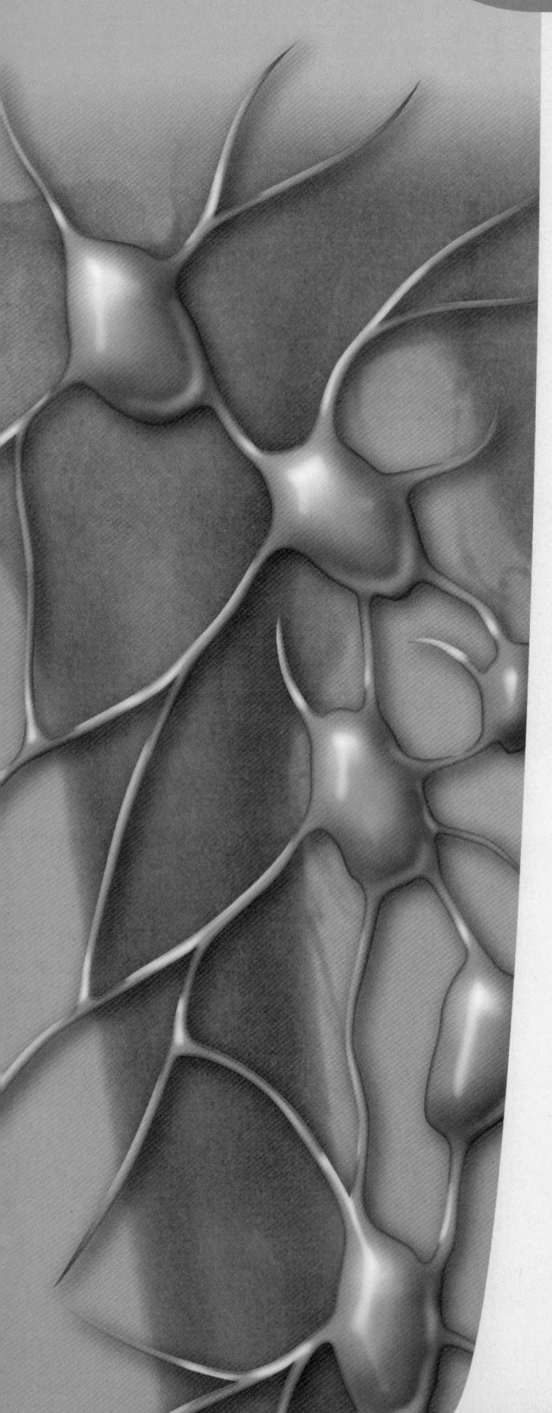

LEARNING OBJECTIVES

Upon completion of this chapter, you will be able to:

1. Discuss the importance of understanding fluid return and immune response as separate roles of the lymphatic system and how this relates to manual therapy practice.
2. Name the primary components of lymph.
3. Identify the five types of lymph vessels and describe the key structural features of each.
4. Explain the general structure and functions of a lymph node.
5. Explain the process of interstitial fluid uptake and distinguish this process from lymph flow.
6. Describe the key internal and external mechanisms that create and influence lymph flow.
7. Name and locate the primary lymphatic catchments and watersheds of the body.
8. Explain the different routes for lymph flow back into cardiovascular circulation for the torso and the upper and lower extremities.
9. Compare and contrast the three major categories of edema.

We are, in fact, mostly water. As terrestrial organisms we may live on solid ground and breathe air, but as a collection of individual cells we still live within the same liquid medium from which we emerged. Every organ and system in the body supports in some way the containment, the renewal, and the circulation of this internal sea."

DEANE JUHAN

Job's Body: A Handbook for Bodyworkers

KEY TERMS

anastomosis (ah-NAS-to-mo-sis)
anchor filaments
angion (AN-ge-on)
angulus venosus (AN-gu-lus ven-O-sus)
catchment (KACH-ment)
dynamic edema
edema uptake
fluid uptake
lymphatic terminus (lim-FAT-ik TERM-en-us)
lymphedema (LIMF-ah-de-mah)
lymphotome (LIMF-ah-tome)
obligatory (o-BLIG-ah-tor-e) **load**
pre-lymphatic channel
traumatic edema
watershed

Primary System Components

- **Lymph**
- **Lymph vessel network**
  - Lymph capillaries (initial vessels and collecting capillaries)
  - Lymphangia
  - Lymphatic trunks (collecting trunks)
  - Deep ducts (right lymphatic duct and thoracic duct)
- **Cisterna chyli**
- **Lymph nodes and lymph node beds (catchments)**

Primary System Functions

- **Returns fluid and proteins to blood**
- **Absorbs fats from the digestive tract**
- **Immune responses via specialized lymphocyte activity (discussed in detail in Chapter 12)**

Fluid makes up about 60% of an adult's body weight. While most fluid resides within the cytoplasm of the cells, the remainder is transported by the cardiovascular and lymphatic systems and exchanged between the blood, lymph, and interstitial fluid. The cardiovascular system acts as a delivery and return system, providing nutrients and removing wastes through capillary exchange. In contrast, the lymphatic system acts only as a fluid return system, picking up extra interstitial fluid, large protein molecules, and cellular debris and returning it to cardiovascular circulation. In fact, the lymphatic system returns approximately 3 liters (just over 3 quarts) of fluid to the cardiovascular system every day. Because the lymphatic system is a one-way return system, its vessels are not connected in a full circuit with the heart like the cardiovascular network. Instead, the lymphatic network is an open system that begins with tiny lymphatic capillaries that absorb interstitial fluid, and ends at large lymphatic ducts that return this fluid to the subclavian veins (▶ FIGURE 11.1).

This chapter focuses on exploring the vascular network of the lymphatic system and the unique structural features and roles of each part. It also describes the mechanisms of fluid pick up from the interstitium and differentiates it from lymph flow and fluid return. Since several forms of manual therapy cite increased fluid flow and/or reduction of edema as primary physiologic effects, understanding the lymphatic system's role in these processes is essential. A clear picture of the key structural features of the lymphatic network and the physiologic mechanisms of fluid movement and flow provides insight into the importance and value of superficial strokes, simple deep breathing techniques, and movement therapies. Additionally, the causes of edema are discussed to help therapists better understand and identify cases of complex systemic edemas for which many manual therapies are contraindicated.

By the Way

While there is still much to learn, researchers, most notably Casley-Smith, Vodder, Foldi, and Chikly, have clearly explained and demonstrated the key systemic features and mechanisms of fluid uptake and movement throughout the lymphatic network.[1–8]

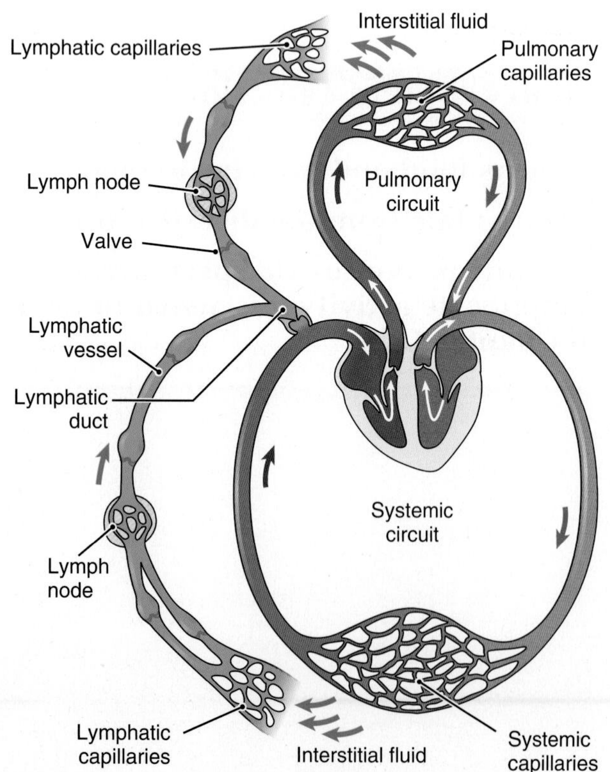

FIGURE 11.1 ▶ Lymphatic system. The lymphatic system absorbs interstitial fluid, filters it through the lymph nodes, and returns it to cardiovascular circulation.

FLUID DYNAMICS OF THE BODY

Approximately two-thirds of the body's total fluid volume is **intracellular fluid**, meaning fluid that is found *inside* the cells. The remaining one-third, found in the interstitium, blood, and lymph, is collectively referred to as **extracellular fluid**. For cells to survive, grow, and perform their special functions, constant exchange of fluid and solutes must occur between the intracellular and extracellular environments. This movement of substances across the cells' plasma membranes is supported by the dynamic exchanges that occur between the interstitium, cardiovascular, and lymphatic systems. While the blood delivers nutrients, both the cardiovascular and lymphatic systems remove waste products and fluid to refresh the interstitium and keep it from stagnating.

To create a simple visual analogy of the body's fluid delivery and return systems, imagine a modified bathtub and shower system (▶ FIGURE 11.2). The shower, tub, and drains represent circulation pathways throughout the body. The shower head represents the arterial side of the cardiovascular system pouring fluid (capillary filtrate) into the tub, or interstitium. The drain at the bottom of the tub represents the venous side of the cardiovascular system taking fluid out of the interstitium (capillary reabsorption). The overflow drain in the wall of the tub represents the lymphatic system as a whole. Water leaves the tub via both drains. Most of the capillary filtrate enters the venous drain via capillary reabsorption, while any remaining fluid is the responsibility of the lymphatic system. Recognize that this analogy is limited because in real life, your bathtub and shower are

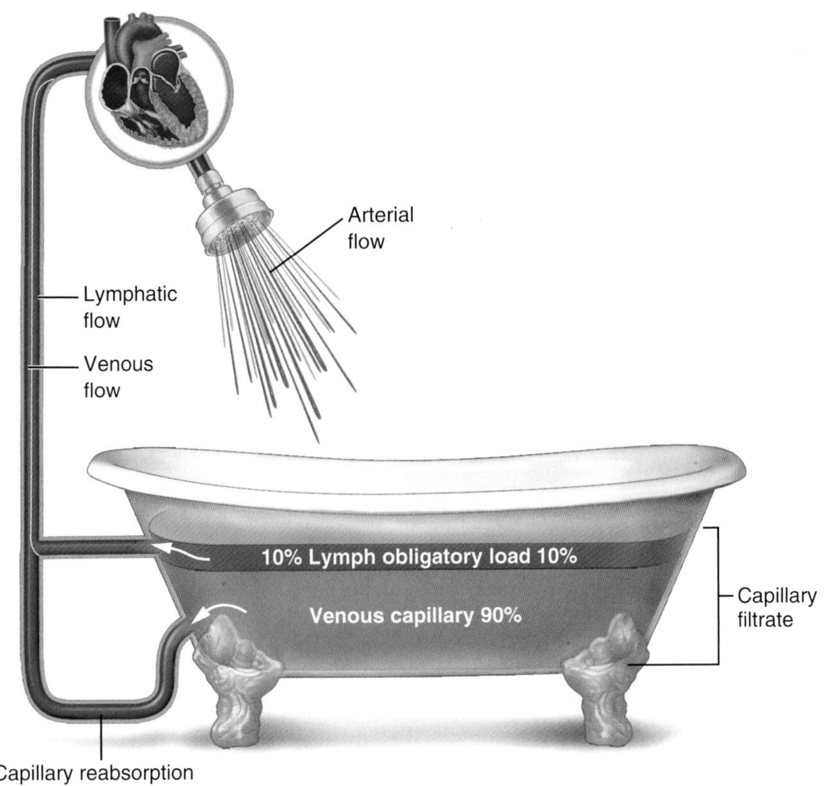

FIGURE 11.2 ▶ Bathtub analogy. This depiction of a modified shower and bathtub with two drains is a good analogy for the interdependent roles of fluid return between the cardiovascular and lymphatic systems.

not a closed system. Thankfully, the water comes from a clean outside source rather than being recirculated from the tub. The take-home point of the analogy is that the lymphatic system serves as an important drain for interstitial fluid; it is an essential fluid return pathway within the body.

LYMPH

Recall from Chapter 10 that the Starling forces create a net filtration on the arterial side of the capillary bed and a net reabsorption on the venous side. In healthy tissue, approximately 10% of the overall volume of capillary filtrate cannot be reabsorbed back into the capillaries due to shifts in concentration gradients and Starling forces. The 10% of filtrate that remains in the interstitial fluid must be returned to blood via the lymphatic system. Because this fluid, together with many proteins, cells, and other substances, cannot return to general circulation via capillary reabsorption, it is called the lymph **obligatory load** (FIGURE 11.2). This means that the lymphatic system is "obliged" to carry out the task of returning 10% of capillary filtrate back to the cardiovascular system.

Once the interstitial fluid is absorbed into lymphatic vessels, it is called **lymph**. Like plasma, lymph is mostly water and electrolytes, but it contains a higher load of metabolic byproducts, cellular debris, and large proteins. Lymph also has a high concentration of **lymphocytes**, specialized white blood cells that are added to the fluid as it passes through the lymph nodes. In addition to water, the primary components of lymph are

- ***Proteins***—The protein load of lymph includes lipoproteins and plasma proteins, as well as hormones and enzymes bound to them. These proteins are too large for reabsorption into blood capillaries and can only be removed from the interstitium via the lymphatic system. Because the nutritional and metabolic demands of tissue in different regions of the body vary, the protein content of lymph also varies from one body region to another.
- ***Cells***—The cellular load of lymph includes lymphocytes, macrophages, dead or damaged cells, and other cellular debris.
- ***Foreign substances***—Substances such as dust, pollen, bacteria, and other environmental particulates are commonly found in lymph.
- ***Long-chain fatty acids***—During digestion, large lipid molecules are absorbed via specialized lymph capillaries in the intestines. These capillaries are called **lacteals**, meaning relating to or resembling milk, because the high fat content gives the lymph a milky white appearance. Lacteals and their function are discussed in Chapter 14.

What Do You Think? 11.1

- How would you summarize the cardiovascular and lymphatic systems' interdependent roles in circulating fluids throughout the body?
- Why do you think it is important to know that the lymph obligatory load is more than just water?

LYMPH VESSEL NETWORK

In the closed circuit of the cardiovascular system, small capillaries are in the middle between the large arteries and veins attached to the heart. In contrast, the vascular network of the lymphatic system begins with capillaries and progresses to the larger vessels of the network (▸ FIGURE 11.3). Interstitial fluid enters the network through microscopic lymph capillaries enmeshed with the cardiovascular capillaries and progresses into slightly

Submandibular node
Right lymphatic duct
Axillary node
Lymph vessels
Cervical node
Thoracic duct
Cisterna chyli
Inguinal node

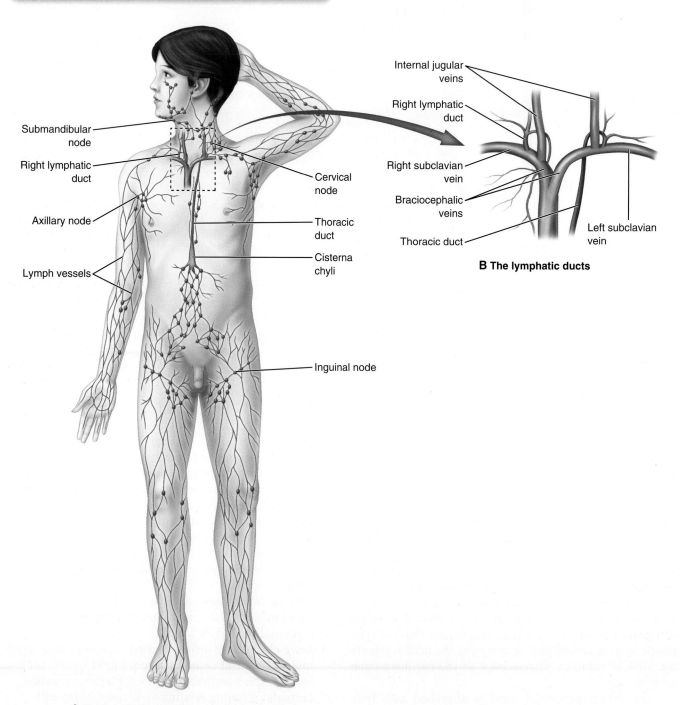

A The lymphatic network

Internal jugular veins
Right lymphatic duct
Right subclavian vein
Braciocephalic veins
Thoracic duct
Left subclavian vein

B The lymphatic ducts

FIGURE 11.3 ▸ Lymph vessel network. A. The lymph system has its own network of vessels that begins with lymphatic capillaries and progresses to larger vessels. Lymph nodes are interspersed along the pathway. **B.** The largest vessels are the thoracic duct and right lymphatic duct that return lymph fluid to cardiovascular circulation at the subclavian veins.

larger capillaries. These capillaries carry lymph into the primary lymphatic vessels that link to larger lymphatic vessels called trunks and ducts. These large vessels eventually join with the subclavian veins to return the lymph to cardiovascular flow.

Lymph Capillaries

There are two types of **lymph capillaries**: initial vessels and collecting capillaries. While authors and researchers acknowledge the slight differences in structure and function between these vessels, very few use distinguishing terminology, referring to both as lymphatic capillaries. For clarity, this text uses the term ***initial vessel*** to denote the first capillary in the network that is entwined with the cardiovascular capillary bed (▶ FIGURE 11.4). This vessel is described as "blind ended" because it begins as a snub-nosed tube in the interstitial spaces of the subepidermis.

Like cardiovascular capillaries, the walls of initial lymphatic vessels are formed by a single layer of epithelial cells. However, the lumen of an initial vessel is four to six times larger than that of a cardiovascular capillary. The ***basement membrane***, also called the ***basal membrane***, that forms the external covering of the initial vessel has a much looser weave, making it only a partial membrane; in several areas, there is no outer membrane at all. This makes initial vessels much more permeable than cardiovascular capillaries and functionally more absorbent. The epithelial cells of the initial vessel overlap each other by approximately one-third, like fish scales (▶ FIGURE 11.5). They have microscopic fibers called **anchor filaments** that hold the initial vessel in

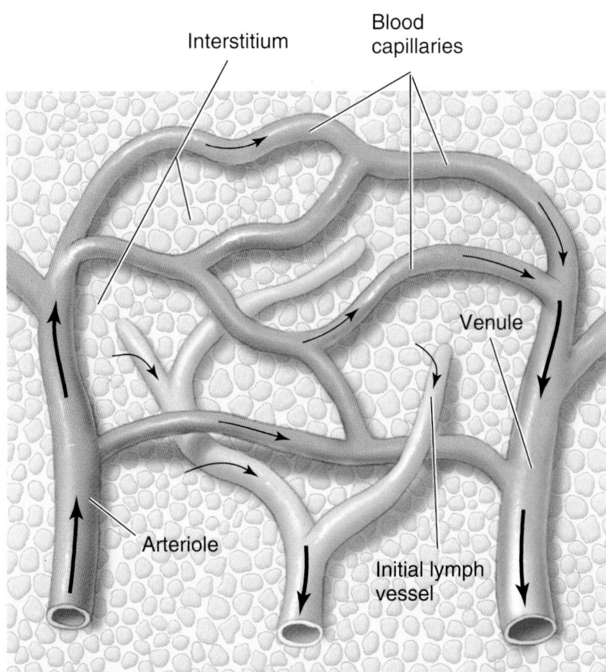

FIGURE 11.4 ▶ Initial vessels. The initial vessels of the lymphatic system are snub-nosed vessels enmeshed with the capillary beds of the cardiovascular system.

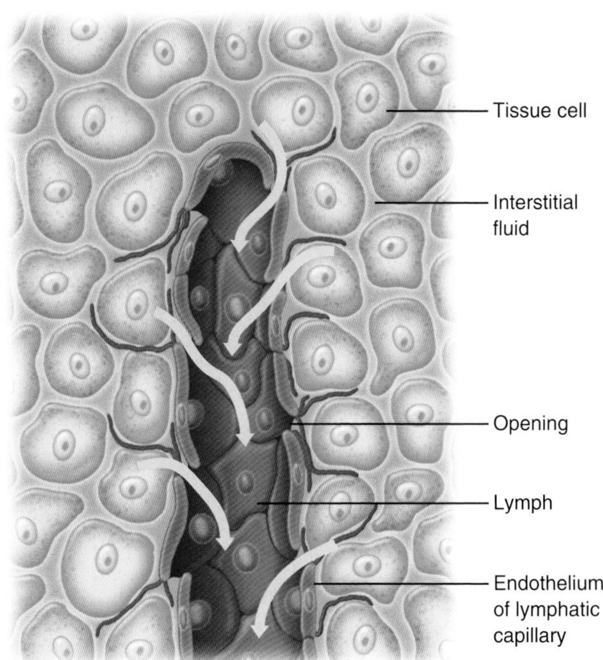

FIGURE 11.5 ▶ Initial vessel, magnified view. The wall of an initial vessel is formed by a single layer of overlapping epithelial cells (endothelium). Each cell has a thin anchoring filament extending from the flap into the interstitium.

the interstitial space and open the vessel by pulling the overlapping flaps of the epithelial cells outward.

Several microscopic initial vessels converge into a slightly larger lymph capillary called a ***collecting capillary***. These vessels are located in the superficial zones of the dermal layer of skin (▶ FIGURE 11.6). Unlike initial vessels, collecting capillaries are less absorbent because

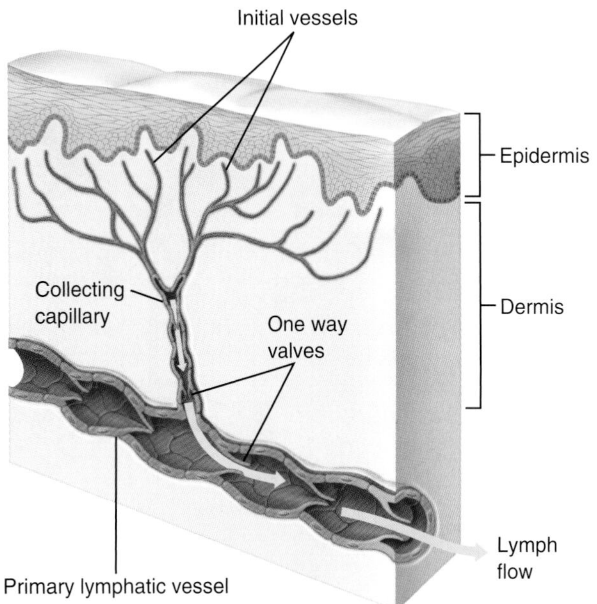

FIGURE 11.6 ▶ Initial vessels and collecting capillaries. Initial vessels are found in the subepidermal layer of skin, while collecting capillaries are deeper, within the upper dermis.

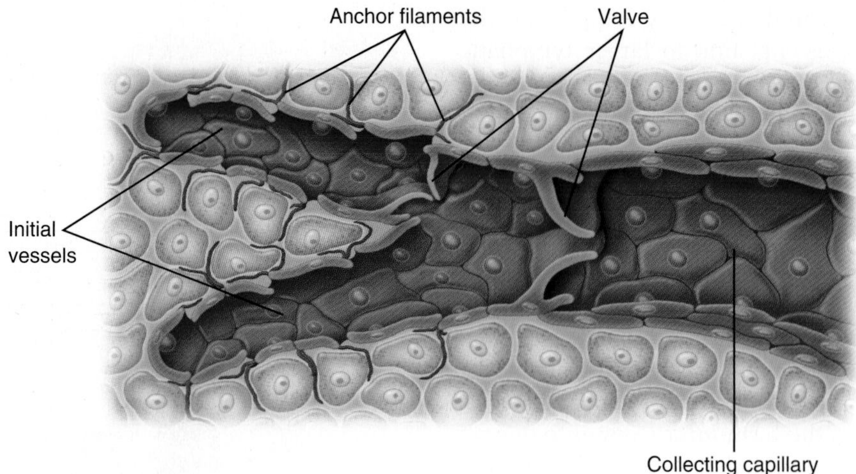

FIGURE 11.7 ▶ Initial vessels converge into a collecting capillary. Several initial vessels converge into a single collecting capillary passing through a weak one-way valve.

their walls have several layers of epithelial cells, no anchor filaments, and a less porous basal membrane around the outside. Additionally, there is generally a single one-way valve at the junction between the initial vessels and collecting capillary, although it is not as strong or rigid as the valves found in veins (▶ FIGURE 11.7). As collecting capillaries get closer to primary lymph vessels, they thicken as the number of epithelial layers increases and the connective tissue sheath around the outside becomes complete. In several regions of the body, multiple collecting capillaries are arranged in an end-to-end manner, similar to a capillary bed in the cardiovascular system. This end-to-end arrangement forms a structural feature called an **anastomosis**. The collecting capillaries in anastomoses are essentially without valves, meaning that lymph can flow in either direction (▶ FIGURE 11.8). The importance of this feature is discussed later in this chapter.

All manual lymphatic techniques include superficial strokes that gently stretch and then release the skin.[1-10] This light stretch of the epidermis opens the initial vessel by pulling on the anchor filaments attached to the overlapping epithelial cells of the vessel wall. The opening created by lifting the epithelial flaps is large enough to allow the passage of proteins, cellular debris, and other substances that are too large to be reabsorbed into blood capillaries. When the tissue stretch is released, the cell flaps of the initial vessels snap back into place. Like slamming a door, this sudden closing of the cell flaps helps propel lymph from initial vessels into collecting capillaries.

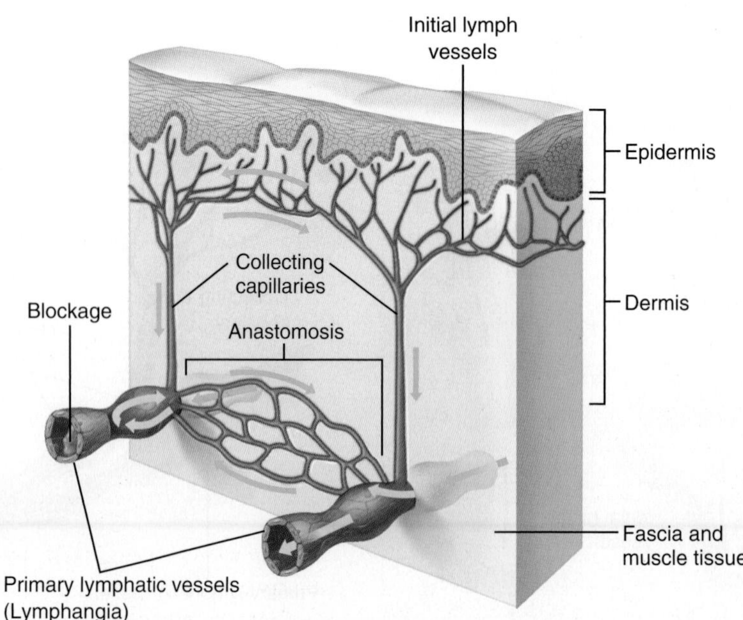

FIGURE 11.8 ▶ Anastomosis connecting two primary lymphatic vessels. If a primary lymphatic vessel is blocked, the anastomosis provides a crossover route for lymph to flow into a different primary vessel.

Primary Vessels: Lymphangia

Several collecting capillaries converge into larger primary lymphatic vessels. These vessels contain a series of one-way ***intralymphatic valves*** along their length, making them structurally similar to veins. The valves, located every 0.6 to 2 cm (~1/4 to 3/4 in.), divide each vessel into a series of smaller segments so lymph only has to be moved from one short segment to the next instead of traveling the full length of the vessel at one time. Each segment forms its own functional unit called an **angion**. Since a primary lymph vessel is actually a series of smaller angions, another term for a primary vessel is **lymphangia**.

Lymphangia have a smooth muscle layer in the vessel wall that spirals around each individual angion, similar to a stripe on a barber pole. Like all smooth muscle, the spiral muscle in the angion is under autonomic control, which stimulates regular rhythmic contractions. These contractions create an internal pump for lymph movement through the vessels (FIGURE 11.9). Superficial lymphangia are located between the dermis and epidermis, and deep lymphangia are found in the subdermal layer of skin as well as in the deeper fascia (see FIGURES. 11.6 and 11.8).

Lymphatic Trunks and Deep Ducts

Several lymphangia converge into larger **lymphatic trunks** that collect lymph from a specific body region or organ. Because of their function, the term **collecting trunk** is also commonly used in manual lymphatic theory and practice. These collecting trunks lie deep within the body tissues and are generally situated alongside major arteries. Similar to lymphangia, collecting trunks have intralymphatic valves that are spaced at wider intervals, approximately every 6 to 10 cm (2.4 to 4 in.). FIGURE 11.10 identifies the major collecting trunks and their location.

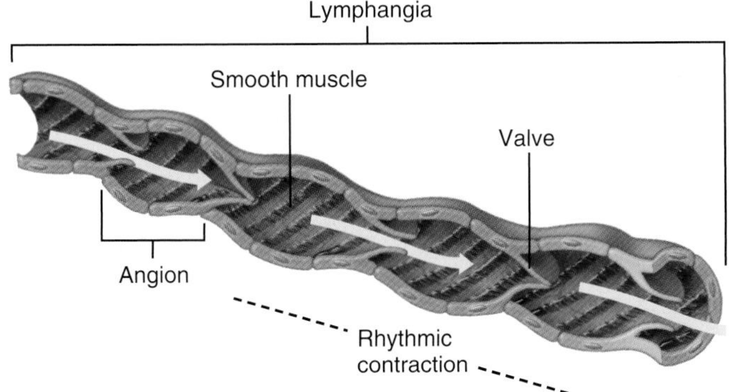

FIGURE 11.9 Lymphangia, cross section. An angion is a segment between the one-way valves of lymphangia. The walls of lymphangia contain a smooth muscle band that spirals around the vessel. This muscle rhythmically contracts to move lymph through the vessel.

The collecting trunks empty lymph into one of the two largest vessels in the lymphatic network, the **right lymphatic duct** or the **thoracic duct** (FIGURE 11.10). Like the collecting trunks, these ducts lay parallel to major arteries and are situated deep within the tissue or ventral cavities. Both the right lymphatic duct and the thoracic duct return lymph to circulation at the subclavian veins close to where they converge with the jugular veins. The lymphovenous junctions between the lymphatic and cardiovascular divisions of circulation are called the right or left **lymphatic terminus**, or **angulus venosus** (FIGURE 11.11), located just posterolateral to the clavicular head of the sternocleidomastoid muscle on each side of the body.

The right lymphatic duct is a very short vessel, ranging from a few millimeters to 1.5 cm (~1/8 to 2/3 in.) in length; it collects lymph from the upper right quadrant of the body (FIGURE 11.12). This deep duct is actually only present in 5% to 10% of the population. In most cases, the three collecting trunks from the upper right quadrant (jugular, subclavian, and bronchomediastinal trunks) connect individually to the subclavian vein at the lymphatic terminus.[1-5] The thoracic duct is centrally located in the thoracic and abdominopelvic cavities, parallel and just to the left of the abdominal aorta. It is the deep duct responsible for collecting lymph from the other three-quarters of the body (upper left, lower left, and right quadrants). A small bulge at the base of the thoracic duct, the **cisterna chyli**, is situated just below the diaphragm at the level of the second lumbar vertebrae (FIGURES 11.10 and 11.12). It functions mainly as a passive collecting well for lymph from the lower extremities (via the lumbar trunks) and some of the abdominal viscera.

What Do You Think? 11.2

- Why is it important that the lymphangia and larger lymph vessels have valves while the initial vessels and collecting capillaries do not?
- What is the importance of both deep ducts returning lymph to circulation at the subclavian veins rather than the subclavian arteries?
- Why is there a structure like the cisterna chyli with the thoracic duct but not with the right lymphatic duct?

LYMPH NODES

Lymph nodes are small specialized lymphoid organs interspersed along the length of the lymphangia, making them a structural part of the network of lymph vessels. Approximately 600 nodes are scattered both superficially and deep along the lymphatic pathway and

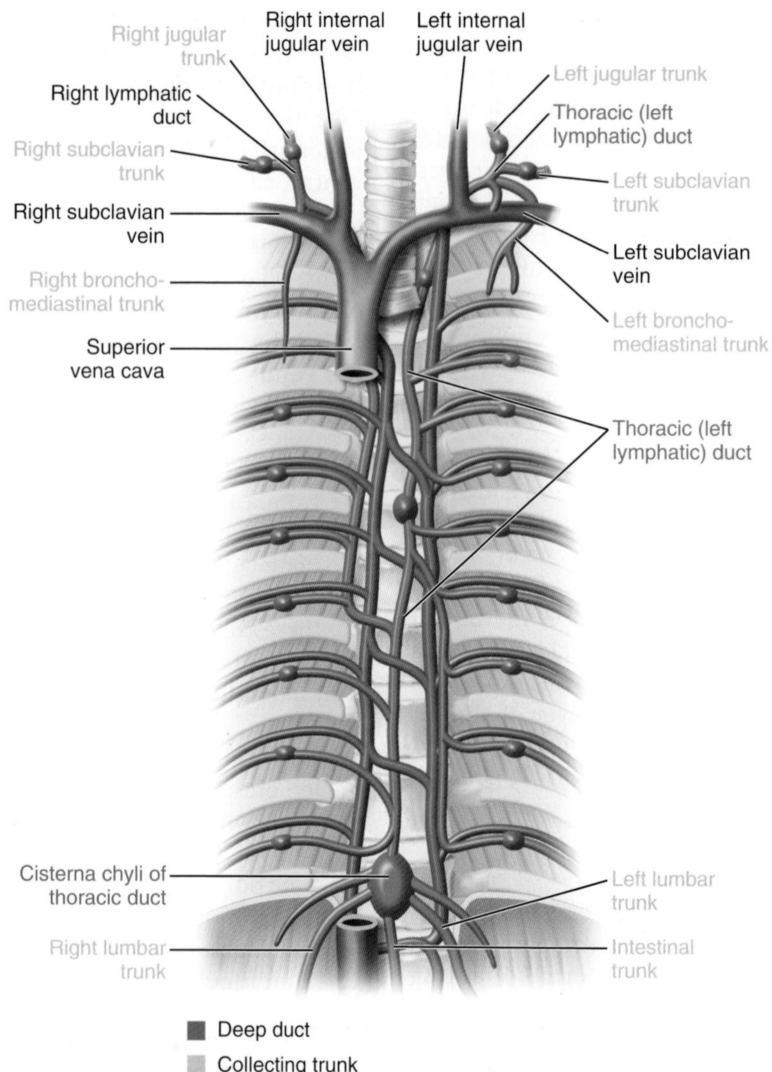

FIGURE 11.10 ▶ Major collecting trunks and deep ducts. The collecting trunks carry lymph from the lymphangia to the two deep ducts of the lymphatic system.

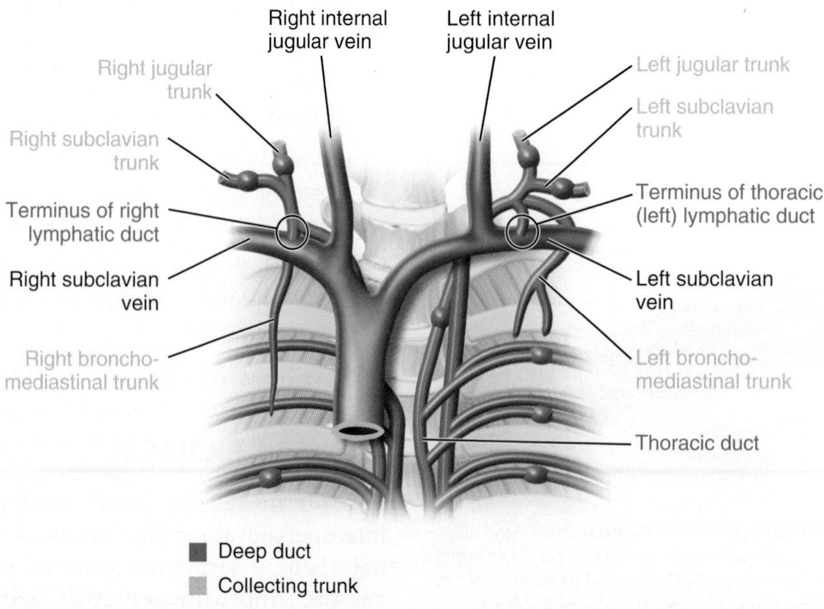

FIGURE 11.11 ▶ Lymphatic terminus. The lymphatic terminus is the point at which fluid from the lymphatic vessel network is returned to the cardiovascular system at the subclavian veins, just lateral to their junction with the jugular veins.

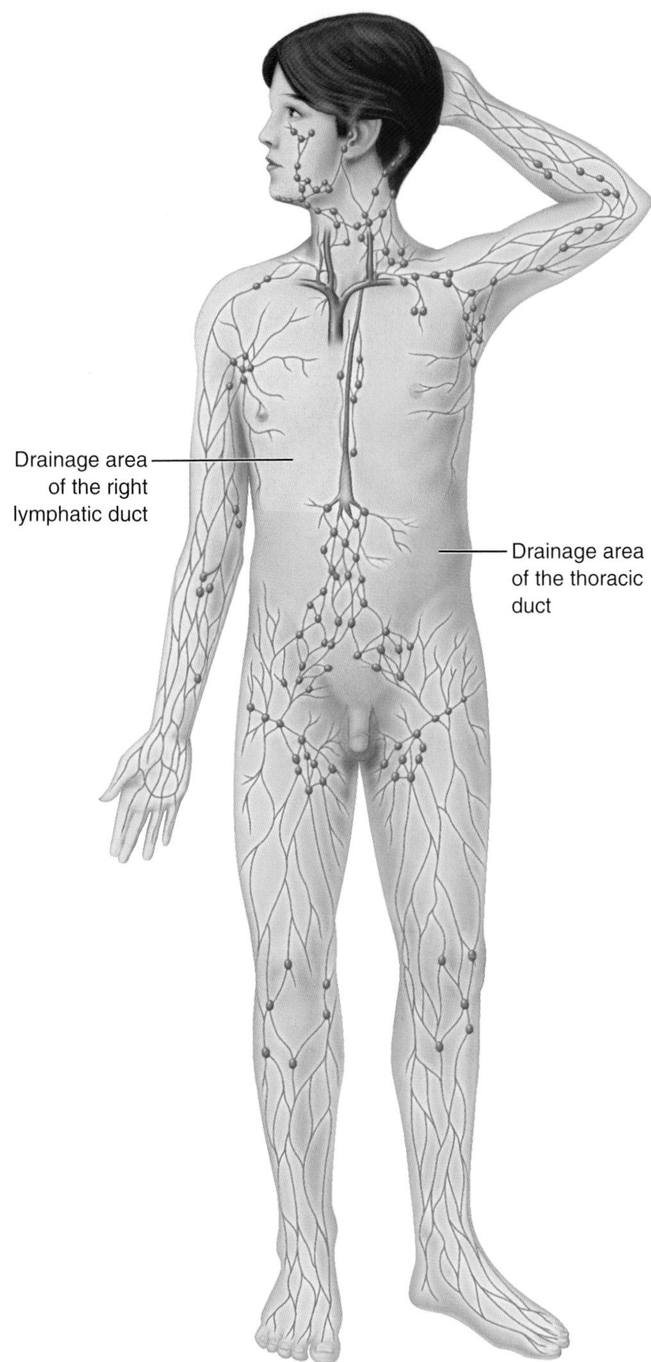

FIGURE 11.12 ▸ Drainage fields for the deep lymphatic ducts. The right lymphatic duct collects lymph from the upper right quadrant of the body, while the thoracic duct receives lymph from the remaining three-quarters of the body.

in clusters called **lymph node beds**. They function as filters, removing particulate matter including dust, pollen, and bacteria, along with damaged cells and other cellular debris, before the fluid is returned to venous flow. Additionally, lymph nodes contain numerous specialized immune cells making them important sites for the specific immune responses discussed in Chapter 12.

Structure of the Lymph Nodes

Lymph nodes can be round, elongated, or bean shaped, and are 1 to 2 cm (~2/5 to 4/5 in.) in size. Each node has several afferent (inward-flowing) vessels with one-way valves that carry lymph into the nodes, and fewer efferent (outward-flowing) vessels that carry lymph out of the node (▸ FIGURE 11.13). The efferent vessels also have one-way valves to prevent backflow into the node. Having more afferent than efferent vessels creates a pressure differential inside the lymph nodes that facilitates the filtration process.

A tough connective tissue covering called a **capsule** surrounds each node, and projections called **trabeculae** extend from the capsule to the inside of the node, dividing it into compartments known as **sinuses**. Within the sinuses there are small islands, or **nodules**, made up of clusters of macrophages and other specialized immune cells (FIGURE 11.13). Similar to a river whose flow is impeded by boulders and numerous small islands, lymph flow through the node is slowed by the irregular network of sinuses and nodules.

Lymph Node Beds: Catchments

As previously described, lymph nodes are interspersed along the length of each lymphangia and clustered together in a few areas to form lymph node beds, also known in manual lymphatic therapies as **catchments.** This term is used because each lymph node bed literally "catches" (collects) fluid from a specific region of the body and slows the rate of flow. Functionally, the decreased rate of lymph flow is needed for the nodes to carry out their filtering and immune system processes. It is estimated that catchments create 100 times more resistance to flow than that of the entire vascular network of the lymphatic system put together. Catchments are located at hinge areas such as the anterior neck, axilla, inguinal, and popliteal regions (▸ FIGURE 11.14), and each catchment has both superficial and deep lymph nodes. Superficial nodes are situated in the subcutaneous layer of the skin, while deep nodes lie within the muscle, fascia, or body cavities. The primary catchments of the body are

- ***Axillary nodes*** receive lymph from the entire upper extremity and the thoracic quadrant of the trunk.
- ***Cervical lymph nodes*** receive lymph from the head and face.
- ***Inguinal catchment*** receives lymph from the thigh and hip, perineum, and lower abdominal regions.
- ***Popliteal catchment*** receives lymph from the foot and leg, and has its own set of deep femoral lymphangia to carry lymph directly into a small group of dedicated deep nodes in the inguinal catchment, which allows lymph to flow into the cisterna chyli at a faster rate. This deep route from the popliteal to inguinal catchment acts as a sort of "express lane" for lymph in the lower extremities.

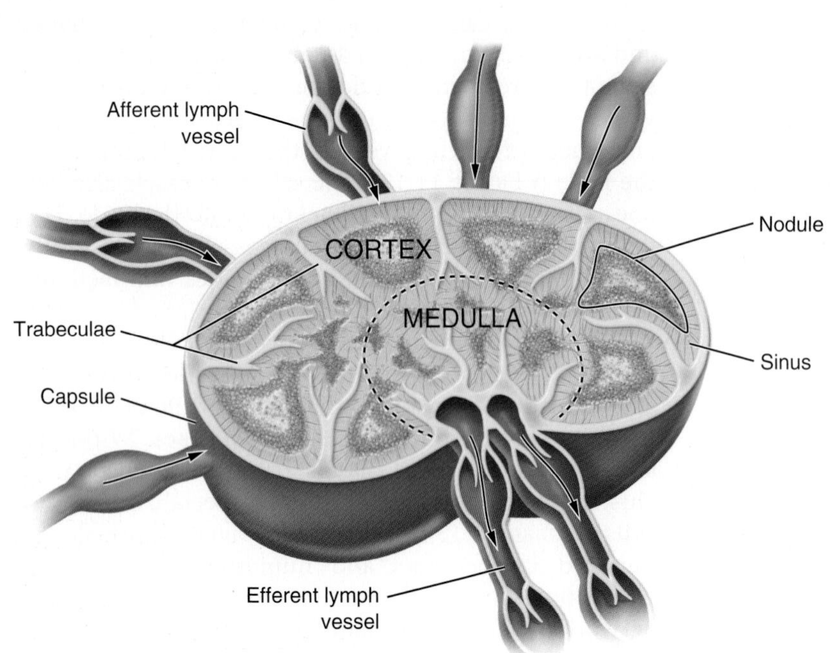

A Cross section of inguinal lymph node

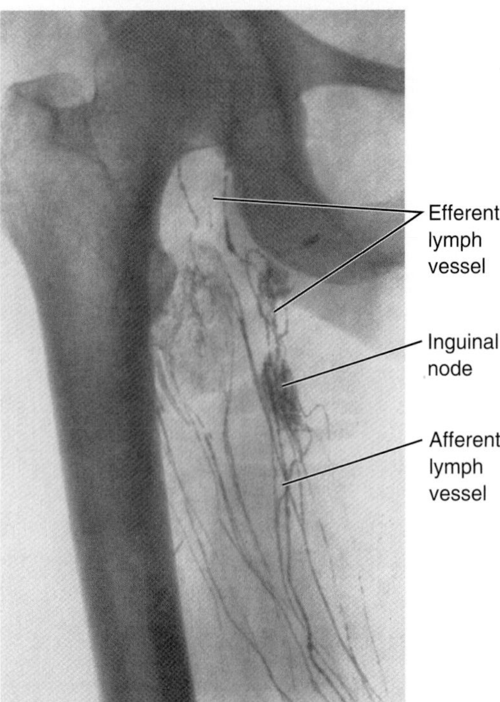

B Scan of inguinal lymph node

FIGURE 11.13 ▸ Lymph node. A. Cross section. **B.** Major inguinal lymph node.

Because catchments add resistance to lymph flow, all manual lymphatic techniques emphasize the importance of clearing these areas to increase lymph flow and reduce edema. Most forms of lymphatic therapy use strokes described as stationary circles or pumping strokes over the catchments to improve lymph flow through these areas. Simply stated, increasing lymph flow through the catchments is necessary for effective edema removal. Furthermore, in cases of lower extremity edema, it is important to address both the popliteal and inguinal catchments to utilize the "express lane" connection to the cisterna chyli. Therapists can increase the rate of edema removal from the lower extremity by focusing on emptying the popliteal and inguinal catchments.[1–3,5]

What Do You Think? 11.3

- Why do lymph nodes have more afferent vessels than efferent vessels?
- Why is it helpful for manual therapists to refer to the clusters of lymph nodes as *catchments* rather than as lymph node beds?
- Why are most of the primary catchments located at hinge points of the body?

In contrast to lymph nodes, which are structurally linked to the lymph vessel network and play a role in fluid return, there are a number of other lymph system structures throughout the body. These **specialized lymphoid tissues** are not connected to the lymph vessel network; instead, they are organs of the cardiovascular, endocrine, digestive, and respiratory systems. Because they are involved with the body's immune response rather than fluid return, they are discussed in Chapter 12.

LYMPH FORMATION AND MOVEMENT

Lymph formation begins when interstitial fluid enters the initial vessels of the lymphatic system. The physiologic mechanisms for this fluid uptake are related to both the unique anatomical features of the lymphatic capillaries and the fluid dynamics between the blood, interstitial fluid, and lymph. Once inside the network of vessels, lymph moves in a specific predictable pattern based on how the vessels are organized. Lymph is moved through a combination of physiologic mechanisms, some similar to venous blood flow and others unique to the lymphatic system. While fluid uptake and lymph flow are linked, this section describes the separate physiologic mechanisms required for each process.

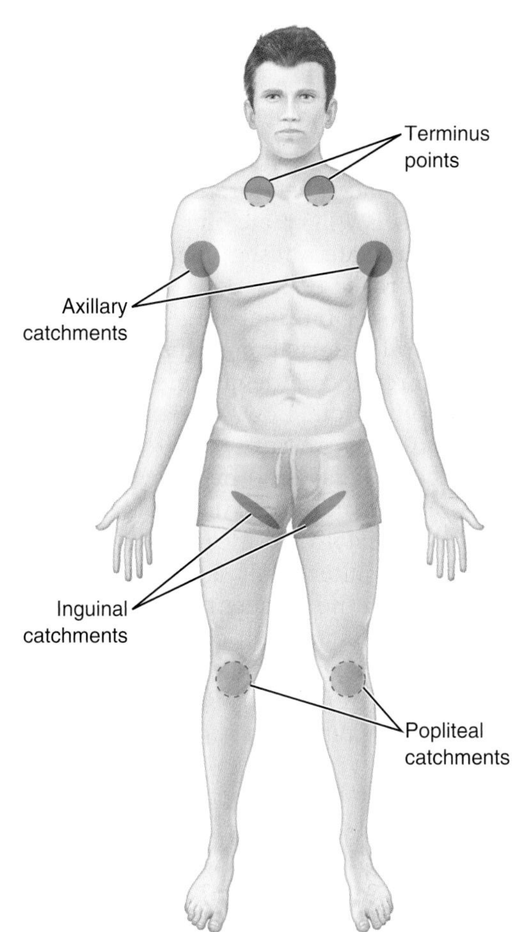

FIGURE 11.14 ▸ Major catchments. Lymph node beds, or catchments, are found at major hinge points of the body.

Pre-Lymphatic Channels and Fluid Uptake

In most body tissues, fluid from the interstitial space drains directly into the initial vessels entwined with the cardiovascular capillaries. There are a few exceptions to this rule, such as the epidermis of the skin, the central nervous system, deeper portions of the peripheral nervous system, fibrocartilage, and the endomysium covering skeletal muscle fibers. These tissues do not contain rich networks of blood capillaries, so the interstitial fluid in these areas must flow between and around the cells to reach the initial vessels of the lymphatic network. In these cases, interstitial fluid flow is facilitated by the presence of minute nonstructural preferred pathways known as **pre-lymphatic channels.**

Each individual has a unique pattern of pre-lymphatic channels, but the purpose and function of this interstitial network is the same for all. Imagine hiking through the woods and encountering a large blackberry bramble. Your first impression may be that this thicket is an impenetrable barrier. However, if you bend down and look at the base of the brambles, you'll see several tiny but distinct trails that provide a way through. The trails are preferred pathways created as mice, rabbits, and other small creatures regularly pass through the brambles. Similarly, pre-lymphatic channels are formed by repeated flow of interstitial fluid toward the initial vessels. Think of how rivulets of water carve a preferred channel of flow into the earth after repetitive watering of a garden.

The movement of interstitial fluid into the initial lymphatic vessels is based on two factors:

- The fluid pressure differentials between the interstitium at the initial lymphatic vessels
- The opening and closing of the epithelial flaps by the anchor filaments that project into the interstitium

In healthy tissue, the interstitial fluid pressure (IFP) outside the initial vessels is slightly higher than the lymph fluid pressure (LFP) inside the vessels. The higher IFP creates a slight stretch that pulls on the anchor filaments of the initial vessels to open the epithelial flaps that will allow interstitial fluid and large proteins to enter. The lower LFP inside the vessels provides the pressure differential needed to draw the interstitial fluid into the opened initial vessel. This process, called **fluid uptake,** or **edema uptake** (in cases of edema), creates a temporary shift in the pressure differentials between the inside and outside of the initial vessels (increased LFP and decreased IFP) that pulls the flaps of the initial vessel closed and holds the fluid (now lymph) inside the initial capillary. Because edema creates a dramatic increase in the IFP, fluid flow into the initial vessels is also increased. However, when IFP increases too much, interstitial fluid movement into the lymph vessels is actually *diminished* because the level of pressure compresses the fragile initial vessels rather than opening their epithelial flaps. This limits the volume of fluid, proteins, and cellular debris that can be pulled into the system.

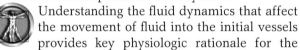

Understanding the fluid dynamics that affect the movement of fluid into the initial vessels provides key physiologic rationale for the importance of using *light* pressure with all manual lymphatic techniques. Light pressure is necessary to avoid compressing and closing off the initial lymph vessels. It is also necessary in directing and supporting interstitial fluid flow through the pre-lymphatic channels.[1-12]

General Lymph Flow

There are several variables that influence lymph flow throughout the lymphatic system. Some are internal systemic forces, while others are external pumping mechanisms.

A key internal force is the siphon effect created by the emptying of lymph back into circulation at the terminus. When lymph is emptied out of the lymphatic network into the subclavian veins, it creates a hydrostatic pressure differential that draws fluid from the lymphatic system into the subclavian veins. Like sucking on a straw, the emptying of lymph into venous flow creates the necessary negative pressure inside the lymphatic vessel network that facilitates pulling fluid into the initial vessels. Once the siphon effect has pulled lymph from the initial vessels and collecting capillaries into the lymphangia, the key internal mechanism influencing lymph flow is contraction of the angions.

By the Way

All forms of manual lymphatic therapy acknowledge the importance of stimulating the siphon effect and generally begin edema removal with strokes in and around the neck and terminus.

Lymph moves through the lymphangia via rhythmic contractions of the spiraled smooth muscle in the wall of each angion. Like all smooth muscle, the contractions are stimulated and regulated by the autonomic nervous system (ANS). In healthy tissue, a normal resting rate of 5 to 7 contractions per minute creates a slow but constant flow of lymph through the primary vessels. In addition to regular autonomic contraction, the smooth muscles inside lymphangia have the ability to boost flow via a stretch reflex similar to that of skeletal muscles. In contrast to the simple recoil response of arteries, when fluid enters the angions and stretches the walls of the lymphangia, it stimulates an actual reflexive contraction of its smooth muscles. This contraction pushes the lymph through that segment of the lymphangia into the next, and so on throughout the entire length of the primary vessel. If the volume of lymph entering the lymphangia increases, the ANS increases the rate of angion contraction to manage the elevated demand. Therefore, increased fluid uptake at the initial vessels will create higher volumes of lymph that, in turn, stimulate a faster contraction rate of lymphangia and improved lymph flow.

In addition to the internal influences just described, there are several external mechanisms (pumps) that move lymph through the deep lymphangia. Similar to the influence on venous flow, skeletal muscle contraction applies external pressure to squeeze fluid through the lymphangia. The arterial pulse of the cardiovascular system exerts an additional influence on lymph flow. As the arteries next to the deep lymphangia expand and rebound, they alternately apply and release external compression to the adjacent lymphangia. So when the artery extends outward, it compresses and empties the parallel lymphangia; when it rebounds, the lymphangia refills. Therefore, any increase or decrease in arterial pulse rate creates an accompanying change in lymph flow.

The respiratory pump created by deep breathing exerts an important external influence on lymph flow through the collecting trunks and deep ducts. Deep inhalation causes a decrease in intrathoracic pressure that enhances the siphon effect that drives lymph movement throughout the system. Additionally, because of the cisterna chyli's proximity to the diaphragm, deep breathing applies direct external compression to the cisterna, adding an extra boost that propels lymph into and through the deep thoracic duct.

Pathology Alert **Pitting Edema**

When pressure from a finger causes a depression or indentation that remains in the skin and soft tissue after pressure is removed, it indicates the condition known as ***pitting edema***. This objective finding demonstrates that the buildup of extracellular fluid (edema) is not being removed effectively by the lymphatic system. There are many causes of pitting edema, including immobilization of an area after injury, deep vein thrombosis, and serious lymphatic, kidney, liver, or heart dysfunctions. In each of these cases, pitting edema is simply a sign of inefficient fluid return. Therefore, the underlying cause of pitting edema must be determined before treatment of any kind is attempted.

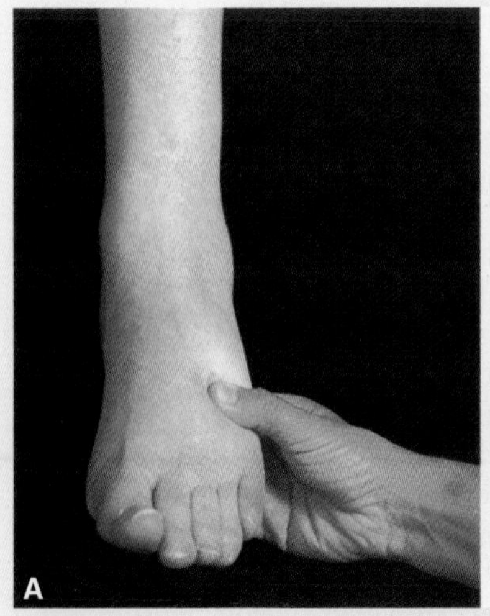

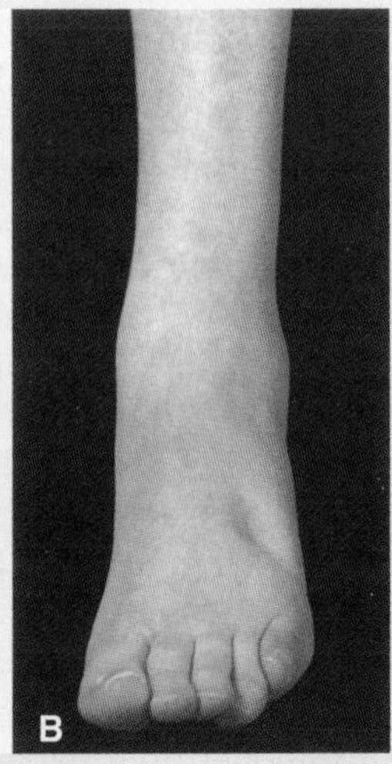

When improved fluid uptake, arterial pulse, skeletal muscle contractions, and diaphragmatic breathing are combined, the rate of angion contraction and lymph flow can be significantly increased.[8,11-17] Therefore, edema removal is greatly enhanced when therapists utilize manual lymphatic techniques and encourage clients to add simple forms of exercise and deep breathing to their daily activities. Light exercise such as walking at a moderate pace is one of the best ways to improve lymph flow. The combination of skeletal muscle contraction, full arterial expansion and recoil, and increased pulse and respiratory rates produced by such exercise leads to an estimated 20% to 30% improvement in lymph flow. By adding manual lymphatic techniques to a program of moderate exercise, the rate of lymph flow and edema removal can be increased by as much as 50%.[1-3,10,11] In cases of lower extremity edema, deep diaphragmatic breathing is of particular importance. Teaching clients to self-stimulate the cisterna chyli with a technique called *exhale-crunch* helps extend improved edema removal well beyond the length of the session. ▶ FIGURE 11.15 depicts the exhale-crunch method of stimulating the respiratory pump and describes how it is explained to clients.

Lymphotomes and Watersheds

The vessels of the lymphatic system are arranged in regional flow patterns called **lymphotomes**. Each lymphotome has its own group of initial vessels, collecting capillaries, and lymphangia that form a predetermined pathway for lymph flow into a particular catchment and often to specific nodes. For example, lymphotomes in the medial and central tissue regions of the arm carry lymph into the axillary catchment. However, lymphotomes on the lateral aspect of the arm bypass the axillary catchment and carry lymph posteriorly around the shoulder and directly into the collecting trunks and ducts.

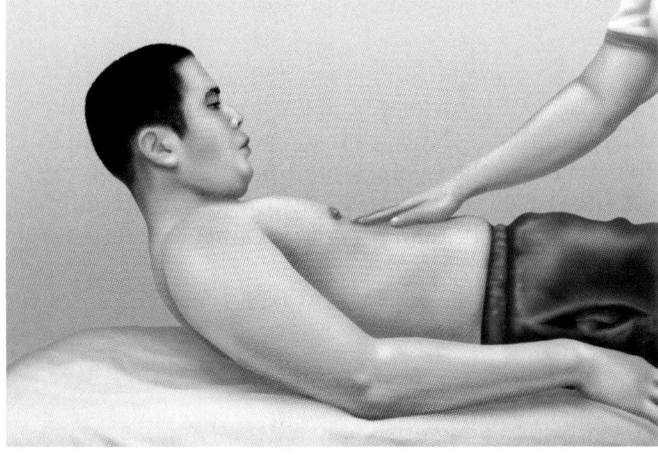

FIGURE 11.15 ▶ Exhale-crunch to empty the cisterna chyli. The exhale-crunch stimulates the respiratory pump and facilitates emptying of the cisterna chyli. In this technique, manual therapists instruct their client to belly breathe. On the exhale, the client must "huff" the air out like blowing out birthday candles and suck in the belly toward the table. A simultaneous crunch that raises the head and shoulders off the table can help exaggerate the squeezing action over the cisterna chyli. Note: The therapist's hand placed over the client's stomach is there only to assure appropriate rise and fall of the belly, *not* to add external compression.

By the Way

Lymphotomes can be seen by injecting radioactive dye into the system and tracking its progression through the vessels via specialized photography using a scanner or probe. This process, known as lymphoscintigraphy, was pioneered in the 1940s and is now used to check the lymph system for disease.

In several regions of the body, thin areas of tissue that contain a high concentration of anastomoses connect lymphangia from one lymphotome to lymphangia in an adjacent region. This zone between lymphotomes is called a **watershed** because the presence of anastomoses allows lymph to flow in either direction through the zone. Watersheds are functional boundaries rather than rigid physical structures. They allow lymph to be transferred between different drainage zones. In the torso, there are two horizontal watershed lines: one at the level of the clavicles and scapular spines and the second at the umbilicus. There is also a vertical watershed boundary at the midsagittal line. These watersheds form large drainage zones, each with multiple lymphotomes (▶ FIGURE 11.16).

- The ***supraclavicular watershed regions*** have lymphotomes that carry lymph from the head and neck directly to the terminus.
- The ***thoracic watershed regions***, located between the clavicular and umbilical lines, drain lymph from these regions to the axillary catchments.
- The ***abdominal watershed regions*** direct superficial lymph from the lower abdomen into the inguinal catchments. However, lymph from the abdominopelvic viscera follows a more direct path into the cisterna chyli or thoracic duct.

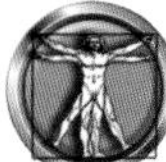

A helpful analogy for understanding watersheds is to visualize the peak of a roof on a house. On each side of the peak is an entire region that empties rain into the gutter on that side of the roof. In this analogy, rain that falls on the north side of the roof flows to gutters on the north side, and rain falling on the south side flows to the south

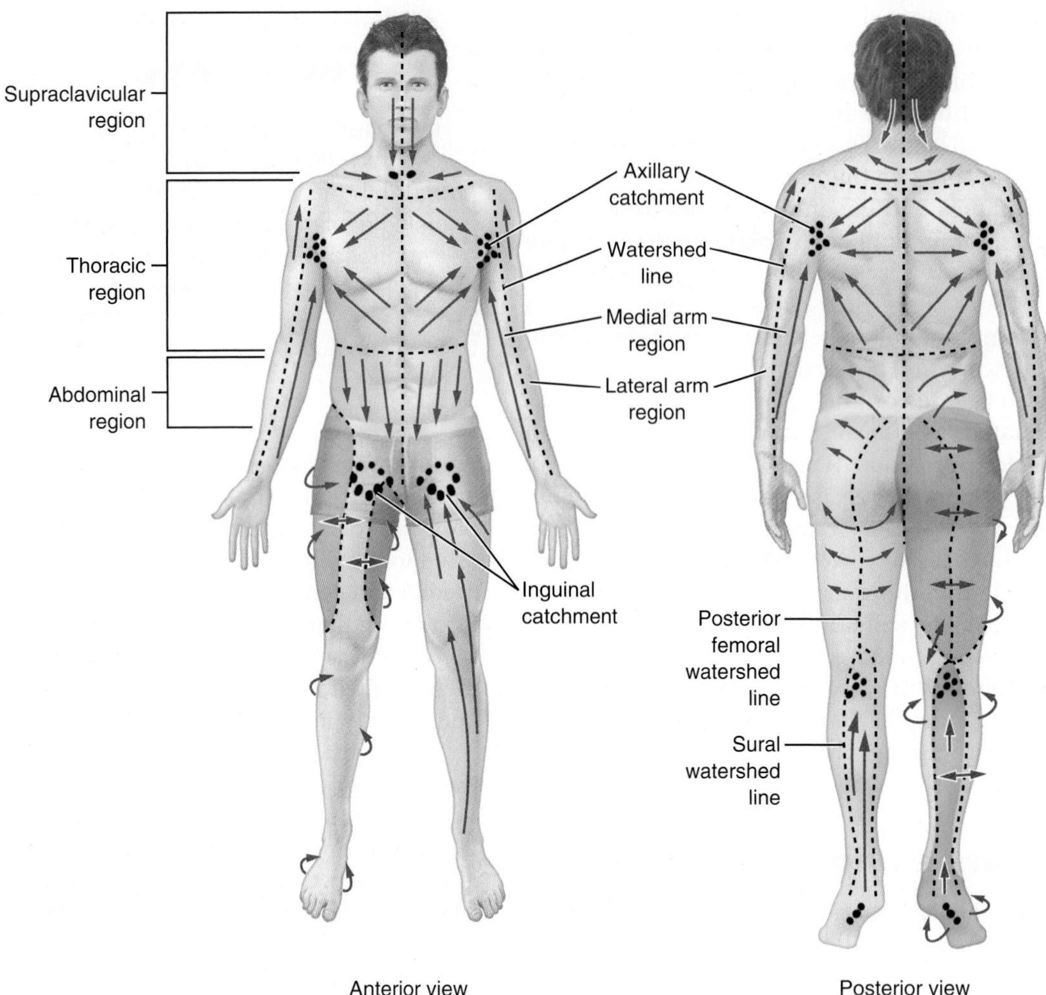

FIGURE 11.16 ▸ Major watersheds. Primary watersheds of the body are designated by dotted lines. *Arrows* indicate the direction of lymph flow toward the catchments within each watershed. While there are multiple lymphotomes within each watershed, only those of the right leg are shown.

gutter. Rain falling directly on the peak of the roof can go either direction, just as the high number of anastomoses at a watershed allows lymph to flow in either direction. In this analogy, it is simply a matter of chance in regard to how much rain hits each side of the roof that determines how much rain flows into the north or south gutter. However, manual therapists who know the location of watersheds can purposefully direct edema to less-congested lymphotomes and catchments to improve its removal. For example, with a shoulder injury or surgery, a portion of edema from the affected side can be taken across the back to the sagittal watershed. That portion of edema now flows into the unaffected and less-congested side's lymphotomes and catchments.

If manual therapists are unaware of lymphotome flow patterns and watershed locations, they may inadvertently move edema away from the most efficient drainage zones. For example, all fluid in the anterior portion of the leg moves into the inguinal catchment. However, it is important to note that there is a watershed between the lymphotomes of the thigh and those of the groin and genitals. If a manual therapist uses the wrong stroke direction through the anterior thigh, fluid may be pushed into the scrotum or genitals rather than into the inguinal catchment.

What Do You Think? 11.4

- In your own words, describe pre-lymphatic channels and how fluid flows through them to the initial vessels.
- Why is the flow through lymphangia subject to ANS control rather than relying solely on respiratory and skeletal muscle pumps?
- Most manual lymphatic techniques begin with deep breathing and strokes around the neck and terminus. Why do you think this is?

TYPES OF EDEMA

Recall that edema is an accumulation of excess fluid in the interstitium, which means that it is not a pathology but a sign of underlying dysfunction or disease. There are many different causes, but ultimately edema occurs because fluids are not being reabsorbed or returned efficiently. When the capillary reabsorption rate is diminished, the lymphatic system's obligatory load increases, and if the lymphatic system is unable to increase its carrying capacity, either temporary or long-term edema occurs. Causes of edema usually originate from one of three general areas of dysfunction:

- Cardiovascular challenges
- Lymphatic system dysfunctions
- Temporary soft tissue inflammation and hemorrhage

In this text, we classify and discuss edema according to these three categories. However, each form of manual lymphatic therapy tends to rely on its own set of descriptive terms for various types of edema. This can create confusion because the terms and their usage vary widely. ▶ TABLE 11-1 outlines the classifications of edema used in this text, along with terminology used by the four major schools of thought in manual lymphatic therapy.

Edema from Cardiovascular Dysfunction

Edema related to dysfunction or disease in the cardiovascular system, **dynamic edema,** is the result of an imbalance between the dynamic forces of capillary filtration and reabsorption. In these cases, the function of the lymphatic system is presumed to be sufficient, although temporarily overwhelmed by the excess fluid in the interstitium. The cause of the imbalance between capillary filtration and reabsorption must be identified and addressed to fully relieve this type of edema. Some common causes are

- Hypertension
- Diabetic-related complications
- Circulatory strain due to pregnancy or obesity
- Venous insufficiency (low ratio of veins per cubic centimeter of tissue)
- Malnutrition

These conditions will lead to increased capillary filtration due to an increase in capillary fluid pressure, decreased capillary reabsorption related to decreased plasma oncotic pressure, or the anatomic limitation created by venous insufficiency. Physicians often treat edemas related to cardiovascular challenges by prescribing blood pressure and/or diuretic medications and recommending dietary changes for weight loss, if obesity is an issue. These measures are generally successful in reducing edema by allowing the lymphatic system to catch up with the increased demand for fluid uptake and return.

In general, manual lymphatic techniques are presumed to be safe and appropriate for use with edemas related to cardiovascular challenges. This is because the lymphatic system is healthy and functioning properly. However, it is always important to identify the underlying etiology of any edema because in some situations,

Table 11-1 Classifications of Edema

Classification	Vodder Terminology	Foldi Terminology	Casley-Smith Terminology	Chikly Terminology
Cardiovascular edema	Dynamic edema	High-volume or dynamic insufficiency	High-flow, low protein	Dynamic
		Hemodynamic insufficiency	High-flow, high-protein	
			Low-flow, low protein	
Lymphedema	Lymphostatic or protein-rich edema	Low-volume or mechanical insufficiency	Low-flow, high-protein, either structural or functional	Lymphostatic
Traumatic edema	Included in discussions of different types of cardiovascular edema or lymphedema as a safety-valve insufficiency; generally defined as a combination of increased obligatory load and decreased lymphatic transport capacity.			

Note: Safety valve insufficiency can be long-term, as in lymphedema, or temporary, as with traumatic edema or cardiovascular edema (e.g., swollen ankles due to standing all day).

it could be dangerous to enhance fluid flow. For example, edema related to kidney disease should not be treated with lymph facilitation techniques because the enhanced fluid return could add stress to already distressed renal functions. Likewise, with congestive heart failure, enhanced edema uptake might help to resolve edema in the ankles initially, but the increased fluid movement and overall return could eventually overtax the heart. This is due to the dynamic relationship between the lymphatic and cardiovascular systems. Even though manual lymphatic techniques are designed to enhance edema uptake and lymph flow, they will affect extracellular fluid circulation throughout the entire body. So, while manual lymphatic techniques can be used in cases of edema related to pregnancy, obesity, or venous insufficiency, they are contraindicated in other cases, such as those due to thrombosis or phlebitis.

Edema from Lymphatic Dysfunction

Edema caused by dysfunction or failure in the lymphatic system is called **lymphedema**. In such cases, the cardiovascular system may be functioning well initially, but it will eventually be compromised. There are two types of lymphedema: primary and secondary.

In ***primary lymphedema***, there is a congenital or genetic defect in lymphatic development, resulting in an insufficient fluid return function. For example, like venous insufficiency, there may be a low ratio of lymphatic vessels per cubic centimeter of tissue, or the valves within the lymphangia may be weak or too sparse. This usually becomes evident in early childhood and most often begins as swelling in the legs (▸ FIGURE 11.17). This systemic insufficiency is a life-threatening condition that eventually develops into full-body lymphedema.

Secondary lymphedema develops when the nodes or vessels of the lymphatic system are damaged or destroyed so that edema uptake and lymph flow are significantly compromised. Common causes of damage leading to secondary lymphedema include surgery, radiation, chemotherapy, infection, or repeated restrictions or injury to the catchments. For example, a high rate of secondary lymphedema is associated with mastectomies, axillary lymph node removal, and radiation treatments. While research over the years has found postmastectomy lymphedema occurrences to be anywhere from 6% to 70%, recent studies show the average to be between 25% and 30%.[18-20] This decrease in occurrences may be due to current treatment protocols that emphasize the removal of only the sentinel axillary nodes and immediate use of therapeutic exercise and medications.

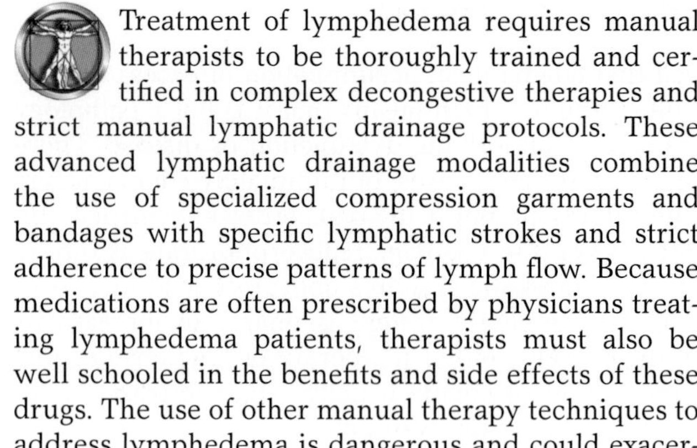
Treatment of lymphedema requires manual therapists to be thoroughly trained and certified in complex decongestive therapies and strict manual lymphatic drainage protocols. These advanced lymphatic drainage modalities combine the use of specialized compression garments and bandages with specific lymphatic strokes and strict adherence to precise patterns of lymph flow. Because medications are often prescribed by physicians treating lymphedema patients, therapists must also be well schooled in the benefits and side effects of these drugs. The use of other manual therapy techniques to address lymphedema is dangerous and could exacerbate the condition.

Edema from Soft Tissue Damage

The localized and temporary swelling associated with soft tissue injury is known as **traumatic edema.** This is a protein-rich edema due to plasma and intercellular proteins that spill into the interstitium from the damaged tissue. As described in Chapter 10, increased interstitial

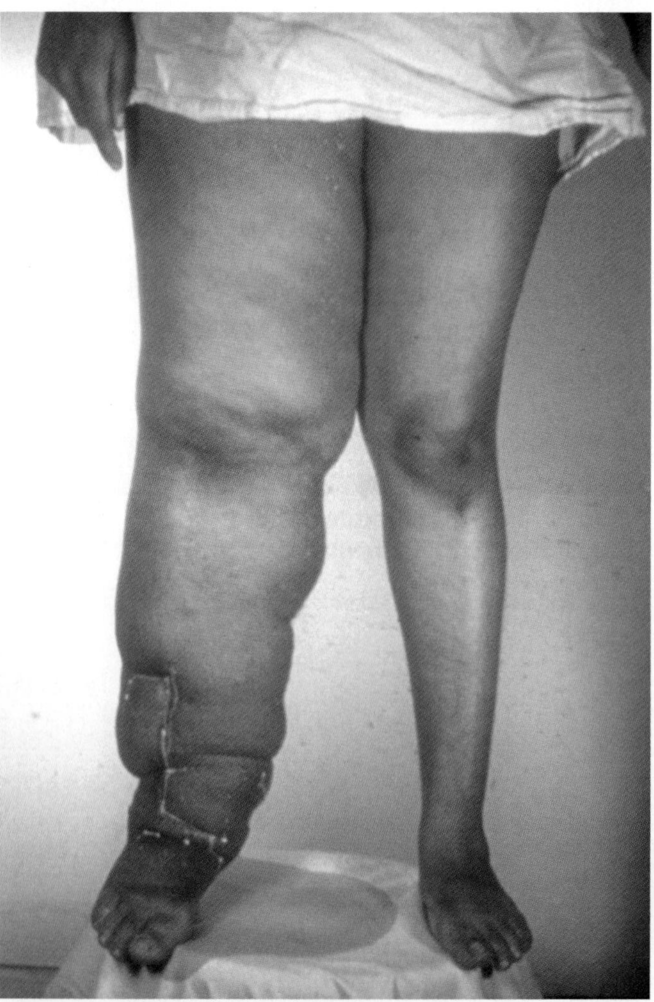

FIGURE 11.17 ▸ Primary lymphedema of the lower body.

By the Way

Another cause of secondary lymphedema is infestation by filaria, thread-like parasitic round worms and their larvae, that are transmitted to humans through insect bites. In this condition, known as lymphatic filariasis, the parasite makes its home in the lymphatic system, blocking lymph vessels, destroying nodes, and causing inflammation. It can progress to elephantiasis, which involves extreme enlargement of the limbs and genitals. The various forms of filariasis occur primarily among people in tropical and subtropical environments; it is rarely seen in North America or Europe.

oncotic pressure (IOP) causes more fluid to be drawn into the area in a process known as secondary edema formation. Since the large protein molecules causing the increased IOP cannot be reabsorbed into the capillaries, they become the sole responsibility of the lymphatic system. Therefore, proper healing of tissue and resolution of traumatic edema relies on normal functioning of both the cardiovascular and lymphatic systems.

Traumatic edema cannot be fully resolved with the standard combination of effleurage and petrissage, because the pressure from these strokes most likely closes the initial lymphatic vessels. However, the effleurage–petrissage combination of strokes may help reduce swelling by shifting interstitial fluid into areas of undamaged tissue where the number of lymph and blood capillaries available for fluid reabsorption is higher. In addition, knowledge of the primary watersheds allows therapists to shift edema and interstitial fluid to entirely different catchments to improve the rate of edema removal. If traumatic edema is not fully resolved in a timely manner, it can become pitting edema.

What Do You Think? 11.5

- Why is it important for manual therapists to recognize that there are different types of edema?
- What health history or objective measurements might a therapist use to determine if it is safe to use manual therapy techniques on a client who presents with edema?

SUMMARY OF KEY POINTS

- The lymphatic system is an open system that returns 10% of capillary filtrate (obligatory load) to cardiovascular circulation.
- Lymph includes fluid plus proteins, cells, foreign substances, and long-chain fatty acids.
- The lymph vessel network (listed smallest to largest) and the key characteristics of each vessel are as follows:
  - Initial vessels are located in the subepidermis entwined with the cardiovascular capillary beds. They have walls formed by a single layer of overlapping epithelial cells with anchor filaments that extend into the interstitium.
  - Collecting capillaries collect lymph from several initial vessels and carry it to the primary lymph vessel or lymphangia.
  - Lymphangia are segmented by one-way valves into multiple units called angions. Their walls contain spiraled smooth muscle under the control of the ANS.
  - Collecting trunks receive lymph from several lymphangia and also contain intralymphatic valves. They are situated alongside major arteries and veins.
  - The right lymphatic duct and thoracic duct are the two deep ducts that return lymph to circulation at the subclavian veins. The cisterna chyli is a small bulge at the base of the thoracic duct. The right lymphatic duct collects lymph from the right upper quadrant, while the thoracic duct is responsible for collecting lymph from the other three-quarters of the body.
- Lymph nodes are specialized lymphoid organs spaced along the length of lymphangia. They remove particulate matter, bacteria, and damaged cells from lymph and serve as primary sites for specific immune responses.
- Lymph formation, or fluid uptake, is a separate process from lymph flow (movement of lymph through the lymphatic vessel network). The primary influences on fluid uptake are the siphon principle that maintains negative pressure within the lymphatic system and changes in IFP due to shifting fluid volume or makeup.
- Lymph flow is influenced by several key internal and external mechanisms. Primary internal influences are the siphon effect and autonomic contractile rate

of lymphangia. External factors affecting lymph flow include the pulse rate, the respiratory and skeletal muscle pumps, and application of manual lymphatic techniques.

- Lymph flows in regional patterns called lymphotomes; these are specific lymphatic drainage regions that contain their own group of capillaries, and lymphangia that carry lymph to a specific catchment.
- Watersheds are functional (nonstructural) zones between the lymphotomes and body regions marked by a high concentration of anastomoses. Watersheds allow lymph to flow in either direction across lymphotomes and regions.
- Edema can be divided into three major categories: dynamic edema, lymphedema, and traumatic edema. Manual lymphatic techniques are most appropriate in the treatment of traumatic edema and may be used with caution to treat some dynamic edemas. Since lymphedema is a specific pathology in which there is a failure of the lymphatic system, its treatment requires that therapists have advanced training and certification in full decongestive manual lymphatic techniques.

REVIEW QUESTIONS

Short Answer

1. List the five lymphatic vessels from the smallest to largest.

Define the following terms:

2. Pre-lymphatic channel
3. Catchment
4. Watershed
5. Angion
6. Anastomosis
7. List the five primary components that make up the lymph obligatory load.

Multiple Choice

8. What is the purpose of the one-way valves in veins and lymphangia?
   a. increase the force propulsion of fluid through the vessels
   b. prevent backflow of fluid and divide the vessels into shorter segments
   c. provide extra surface area in the vessel for nutrient and waste exchange
   d. create extra resistance to fluid flow to mediate blood pressure
9. What is the name for the lymphovenous junction where fluid is returned to the cardiovascular system?
   a. lymphatic jugular junction
   b. cardio-lymphatic junction
   c. cisterna chyli
   d. terminus
10. What percentage of total capillary filtrate is reabsorbed into the cardiovascular capillaries?
    a. 100%
    b. 50%
    c. 90%
    d. 10%
11. Which of the following statements best describes the functions of a lymph node?
    a. filters lymph of impurities and acts as primary site for immune responses
    b. filters blood and produces white blood cells
    c. absorbs edema and filters it of impurities
    d. stores lymph in case of severe trauma
12. Which of these statements best describe the function of the cisterna chyli?
    a. serves as the largest lymph node bed in the body
    b. a collecting well for lymph that propels fluid through the thoracic duct when squeezed
    c. the heart of the lymphatic system that pumps lymph through the entire system
    d. the junction where lymph is returned to the cardiovascular system
13. One-way valves are important structural features in several lymph vessels, but are not found in
    a. lymphangia and deep trunks
    b. collecting trunks and primary vessels
    c. initial and collecting capillaries
    d. thoracic and right lymphatic ducts
14. A group of lymphatic vessels that drain lymph from a specific region of the body into a specific catchment are
    a. lymphangia
    b. angion
    c. collecting trunks
    d. lymphotomes

15. What is the catchment for lymph from the anterior leg and the anterior and posterior thigh?
    a. inguinal
    b. popliteal
    c. perineal
    d. patellar
16. Which of the following are considered the two primary internal forces for creating and maintaining lymph flow?
    a. respiratory and skeletal muscle pumps
    b. arterial pulse and angion contraction
    c. siphon effect and autonomic angion contraction
    d. soft tissue stretch and release
17. Arterial flow and pulse have a major influence on lymph flow through which portion of the lymph system?
    a. anastomosis
    b. collecting trunks
    c. collecting capillaries
    d. catchments
18. The fastest route for lymph from the lower leg to the inguinal catchment is via the
    a. anterior leg lymphotomes
    b. posterior thigh lymphotomes
    c. superficial nodes in the inguinal catchment
    d. popliteal catchment
19. Swelling related to hypertension and obesity is called
    a. primary lymphedema
    b. traumatic edema
    c. secondary lymphedema
    d. dynamic edema
20. A dysfunction in the lymphatic system that results in swelling of an entire body area is called
    a. edema
    b. traumatic edema
    c. lymphedema
    d. circulatory edema

References

1. Casley-Smith JR, Casley-Smith JR. *Modern Treatment of Lymphoedema*. Adelaide, Australia: Henry Thomas Laboratory, Lymphoedema Association of Australia; 1994.
2. Foldi M, Strobenreuther R. *Foundations of Manual Lymph Drainage*. 3rd ed. St. Louis, MO: Elsevier Mosby; 2005.
3. Foldi E, Foldi M. *Textbook of Foldi School*. English translation by Heida Brenneke. Self-published; 1999.
4. Chikly B. *Lymph Drainage Therapy: Study Guide for Level 1*. France: UI and Self-published; 1996, revised 1999.
5. Kasseroller R. *Compendium of Dr. Vodder's Manual Lymph Drainage*. Heidelberg, Germany: Karl F. Haug Publishers; 1998.
6. Wittlinger H., Wittlinger G. *Textbook of Dr. Vodder's Manual Lymph Drainage, Vol. 1: Basic Course*. 6th English translation revised and edited by Robert H. Harris. Heidelberg, Germany: Karl F. Haug Publishers; 1998.
7. Kurz I. *Textbook of Dr. Vodder's Manual Lymph Drainage, Vol. 2: Therapy*. 3rd ed. Heidelberg, Germany: Karl F. Haug Publishers; 1986.
8. Kurz I. *Textbook of Dr. Vodder's Manual Lymphatic Drainage, Vol. 3: Treatment Manual*. Heidelberg, Germany: Karl F. Haug Publishers; 1986.
9. Badger C. Treating lymphoedema. *Nurs Times*. 1996;92(11):84–88.
10. Kolb P., Denegar C. Traumatic edema and the lymphatic system. *Athl Train*. 1983;Winter:339–341.
11. Wallace E, McPartland JM, Jones JM, et al. Lymphatic system: lymphatic manipulative techniques. In: Ward RC, ed. *Foundations for Osteopathic Medicine*. Baltimore, MD: Williams and Wilkins; 1997.
12. Mortimer PS. Managing lymphoedema. *Clin Exp Dermatol*. 1995;20:98–106.
13. Shields JW. Central lymph propulsion. *Lymphology*.1980;13:9–17.
14. Wang G-Y, Zhong S-Z. Experimental study of lymphatic contractility and its clinical importance. *Ann Plast Surg*. 1985;15(4):278–284,.
15. Olszewski WL, Engeset A. Peripheral lymph dynamics. In: *Proceedings of the XIIth International Congress of Lymphology*, Tokyo, August 27 to September 2. Amsterdam, The Netherlands: Excerpta Medica; 1989:213–214.
16. Olszewski W, et al. Flow and composition of leg lymph in normal men during venous stasis, muscular activity and local hyperthermia. *Acta Physiol Scand*. 1977;99:149–155.
17. Guyton AC, Granger HJ, Taylor AE. Interstitial fluid pressure. *Physiol Rev*. 1971;51(3):527–563.
18. Kavanaugh M. Patient information: lymphedema after breast cancer surgery, October 2008, www.uptodate.com/patients/content/topic.do?topicKey = ~ uuz.l9jrtmrR&selectedTitle = 1 ~ 150&source = search_result
19. Harris SR, Hugi MR, Olivotto IA, et al. Clinical practice guidelines for the care and treatment of breast cancer: 11. Lymphedema. *Can Med Assoc J*. 2001;164(2):191–199.
20. Bani HA, Fasching PA, Lux MM, et al. Lymphedema in breast cancer survivors: assessment and information provision in a specialized breast unit. *Patient Educ Couns*. 2007;66(3):311–318.

12 Immunity and Healing

LEARNING OBJECTIVES

Upon completion of this chapter, you will be able to:

1. Explain the function of the immune system and discuss its relationship with and importance to manual therapy practices.
2. Explain the difference between primary and secondary lymphoid tissues.
3. Name, locate, and describe the general function of the primary and secondary lymphoid tissues.
4. Explain the difference between nonspecific defenses and specific immune responses.
5. List the nonspecific immune defenses of the body and explain how each mechanism works.
6. Name and describe the roles of the primary lymphocytes involved in antibody-mediated and cell-mediated immune responses.
7. Discuss the difference between naturally and artificially acquired immunity and give examples of active and passive forms of each.
8. Discuss the immune system changes that commonly occur with aging.
9. Define the field of study known as psychoneuroimmunology and explain what implications this discipline may bring to the practice of manual therapy.

> "Unconditional love is the most powerful stimulant of the immune system."
>
> BERNIE SIEGEL, M.D.

KEY TERMS

antibody (AN-tih-bod-e)
antibody-mediated immunity (AN-tih-bod-e ME-de-a-ted im-MU-nih-te)
antigen (AN-tih-jen)
antimicrobial (AN-tee-mi-KRO-be-al)
cell-mediated immunity (sel ME-de-a-ted im-MU-nih-te)
immunity (im-MU-nih-te)
memory cell
microbe (MI-krobe)
natural killer cell
nonspecific immune defense
pathogen (PATH-o-jen)
psychoneuroimmunology (PSI-ko-NUR-o-im-mu-NOL-o-je)
specific immune response

Primary System Components

- **Lymphocytes**
  - B cells
  - T cells
- **Thymus**
- **Lymph nodes**
- **Spleen**
- **Mucosa-associated lymphoid tissue (MALT)**
  - Tonsils
  - Peyer patches

Primary System Functions

- **Protects and defends the body from foreign substances via general and specific immune responses**

The body's cells, tissues, and organs work together to protect and defend us from a continuous stream of disease-causing agents, or **pathogens**, that challenge our health and well-being. Like a standing army on alert, the body has a multifaceted defense system to provide **immunity**, protection from or resistance to infection and disease. Some of the body's protective mechanisms are generalized, designed to protect us from any and all foreign invaders. These generalized defense mechanisms range from the simple physical barrier created by the skin to complex chemical and cellular processes like the inflammatory response. Other defense mechanisms target specific pathogens and destroy or neutralize them. Most of the time, we are unaware of the body's ongoing defensive responses, but sometimes, such as when we feel fatigued and ache with fever, we are quite aware of the battle being fought in our body. This chapter provides an overview of the organs and tissues of the immune system and discusses their specific functions and the protective mechanisms the body employs to fight immune challenges.

Once an immune challenge is past and its physical symptoms have subsided, the body goes through a recovery phase that returns cells and systems to their normal metabolic status. Like the immune challenge itself, this recovery phase can be short or long. Most of us can remember occasions in which we bounced back quickly from an injury or illness. We can also recall times when we struggled for weeks or even months to return to feeling healthy and vital. But can all the factors that influenced our healing be identified? Did we recover quickly because of a healthy diet, regular exercise, and adequate rest? Or was recovery slowed due to chronic stress or depression? In all cases, a multitude of factors play a role in healing, and complete recovery from any disease or injury involves more than the simple healing of tissues.

While for centuries, Western medical science has approached immunity and healing as purely physical processes, it now generally recognizes that physical healing is profoundly affected by stress, emotions, beliefs, and attitudes. Manual therapists and other holistic practitioners will appreciate this chapter's discussion on psychoneuroimmunology, a science that studies the connections between psychological processes and the immune and nervous systems. The science of PNI challenges health care professionals to think about healing in a more holistic way and provides insight into how touch, love, and laughter are simply "good medicine."

ORGANS AND TISSUES OF THE IMMUNE SYSTEM

Organs and tissues of the immune system are not physically connected. Instead, they are scattered throughout the body, and some are structural components of the cardiovascular, lymphatic, or endocrine systems.

By the Way

Traditional Chinese medicine and several forms of Eastern bodywork do not share Western medicine's views on health and disease. These Eastern practices are based on a holistic belief system that all living things are connected and animated by an energy or life force called *qi* (pronounced "chee") that flows in, around, and through everything. According to this perspective, health is a product of balanced and free flowing qi, and diseases occur when qi is blocked or out of balance.

The structures of the immune system are divided into two broad groups: primary and secondary lymphoid organs and tissues. Recall that **lymphocytes**, a specialized group of white blood cells, are considered the *primary* immune cells. The ***primary lymphoid*** organs and tissues are those in which mature lymphocytes are produced. ***Secondary lymphoid*** organs and tissues serve as the sites where most of the body's immune responses occur (FIGURE 12.1).

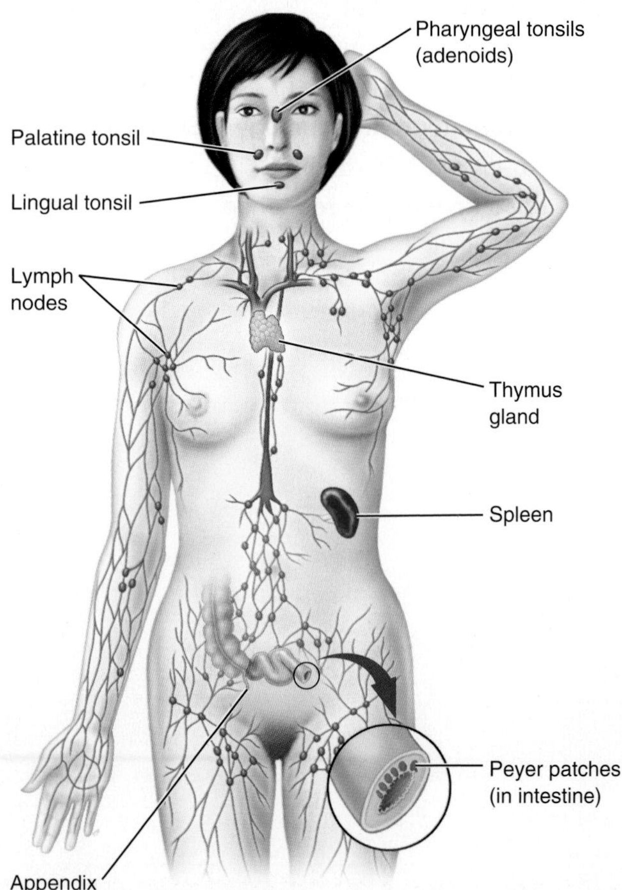

FIGURE 12.1 Organs and tissues of the immune system.

Primary Lymphoid Organs and Tissues

The primary lymphoid organs and tissues that produce lymphocytes include red bone marrow and the thymus. Recall that red bone marrow contains stem cells that divide and differentiate into different types of blood cells, including the agranular leukocytes known as lymphocytes (▶ FIGURE 12.2). Lymphocytes are involved in the specific immune responses, so named because these cells target specific pathogens. There are two types of lymphocytes: **B cells**, B lymphocytes, that are produced and mature in red bone marrow, and **T cells**, T lymphocytes, that are initially produced in red bone marrow and subsequently migrate to the thymus for maturation. In the thymus, immature lymphocytes are exposed to the hormone thymosine, which triggers their proliferation and maturation into T cells. As discussed in Chapter 9, the thymus begins to shrink after puberty, and the organ is gradually replaced with adipose tissue by the time we reach old age. As the thymus atrophies, its ability to produce mature T cells is greatly reduced, and this may reduce the strength of immune responses.

Secondary Lymphoid Organs and Tissues

The secondary lymphoid organs and tissues are sites where the majority of specific immune responses occur. These organs include the lymph nodes, spleen, and unencapsulated masses of lymphoid tissue such as the tonsils and Peyer patches (see FIGURE 12.1). These masses, known as **mucosa-associated lymphoid tissue (MALT)**, are strategically located in the connective tissues of the mucous membranes that line the respiratory, digestive, and reproductive tracts. MALT has a high concentration of lymphocytes, giving these tissues the ability to quickly respond to foreign substances trapped within the mucus.

Lymph Nodes

Recall from Chapter 11 that the movement of lymph through the lymph nodes and lymph node beds is dramatically slowed to enable immune responses. The nodes serve as small filtering stations that trap foreign particles, which can then be destroyed by phagocytes or lymphocytes stored inside the node. As pathogens pass through the node, the specific immune response that destroys or neutralizes the pathogen is set into motion. When the immune response is in full swing, increased phagocyte and lymphocyte activity causes the lymph nodes to swell, which can make us aware that an immune response is occurring and sometimes helps to determine the location of an infection. For example, swollen axillary nodes indicate an infection somewhere in the upper extremity because lymph flow in the arm passes into the axillary catchment. Likewise, swollen submandibular nodes are indicative of an infection of the mouth, teeth, gums, or throat.

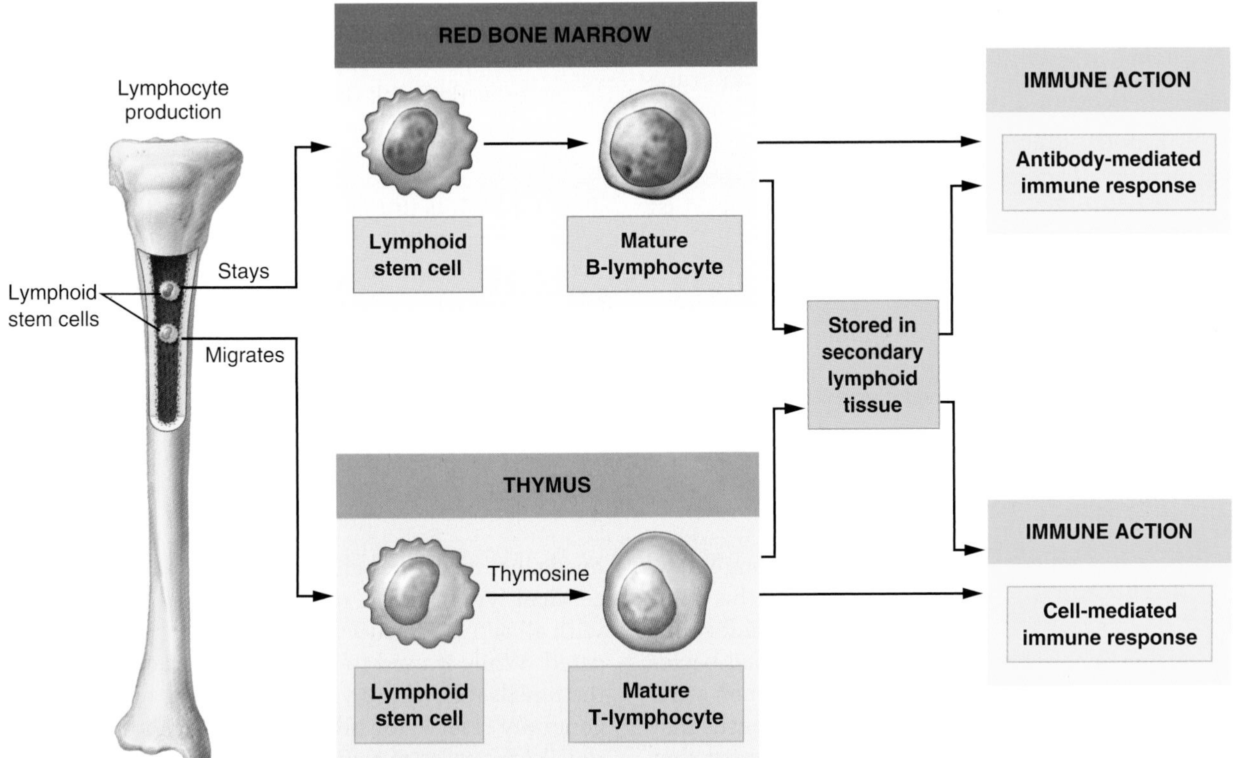

FIGURE 12.2 ▶ Development of lymphocytes. Lymphocytes originate in red bone marrow as lymphoid stem cells. Some remain and mature into B lymphocytes, while others migrate to the thymus to become T lymphocytes. B and T cells are then stored in secondary lymphoid tissues such as the lymph nodes, spleen, and MALTs.

Spleen

The **spleen** is an encapsulated organ located in the upper left quadrant of the abdomen between the stomach and diaphragm at the level of ribs 9 to 11 (see FIGURE 12.1). Similar to lymph nodes, the spleen stores lymphocytes and filters foreign particles. However, the spleen filters blood, not lymph. Foreign particles and pathogens trapped in the spleen initiate the specific immune responses of the lymphocytes or are destroyed by phagocytosis. In addition to its immune functions, the spleen filters out damaged or defective blood cells, stores platelets, and serves as a reserve blood bank holding approximately one pint of blood for emergency use. In cases of severe hemorrhage or infections that destroy red blood cells (such as typhoid fever and malaria), the spleen releases its reserved blood into circulation. Because the spleen is vascular and holds so much blood, trauma that damages or ruptures the organ leads to a life-threatening internal hemorrhage. Fortunately, humans can survive without a spleen if necessary, since its blood-filtering functions can be shifted to the liver and kidneys, and immune responses can be handled by the lymph nodes.

It is not unusual for manual therapists to be asked, "Do my lymph nodes feel swollen?" or, "What do you think this lump is?" For this reason, it is important for therapists to know how to palpate the major lymph node beds and identify a single swollen node that might occur along a lymph vessel. Although idiopathic enlargement of a single lymph node can occur, generally, lymph nodes are not palpable unless there is an active infection. Swollen lymph nodes are about the size of a small grape or kidney bean and feel firm but not hard. Similar to lymph nodes, the spleen should not be palpable when healthy. Therefore, if a client reports tenderness along the left costal border and the therapist can palpate a firm spongy mass in that region, it can indicate an enlarged spleen and/or a severe infection. In both cases, massage is contraindicated and clients should be advised to seek immediate medical attention.

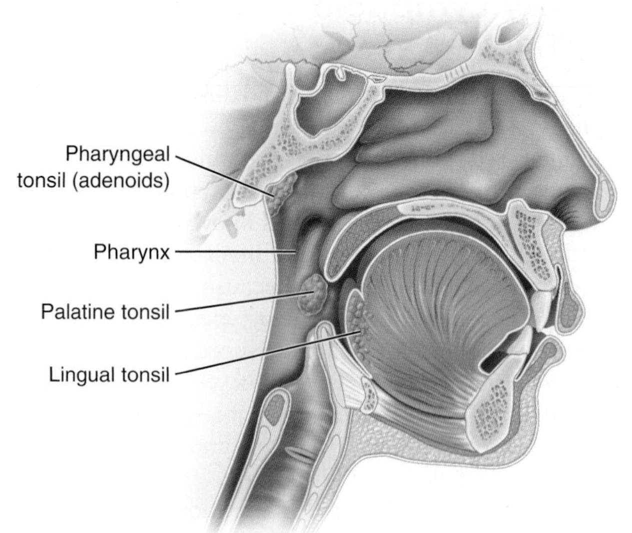

FIGURE 12.3 ▸ Tonsils. Tonsils are a type of MALT located at the back of the nasal cavity, mouth, and throat.

Mucosa-Associated Lymphoid Tissue

The **tonsils** are small masses of lymphoid tissue located in the mucous membranes of the mouth and throat. They trap bacteria and other pathogens that enter through the nose and mouth. Like lymph nodes, the tonsils swell and become inflamed when the immune response is active. There are three groups (▸ FIGURE 12.3):

- ***Lingual tonsils***—at the base of the tongue
- ***Pharyngeal tonsils (adenoids)***—posterior to the nasal cavity at the back of the throat
- ***Palatine tonsils***—at the back of the mouth, on either side of the opening to the throat

Peyer patches are MALTs located in the lower portion of the small intestine (see FIGURE 12.1). They trap and attack pathogens that have entered the body via the digestive tract. The **appendix**, a small twisted tube at the junction between the small and large intestines, also contains MALT nodules. This lymphoid tissue assists the

Pathology Alert Tonsillitis

As the *-itis* suffix indicates, ***tonsillitis*** is an inflammation of the tonsils, usually the palatine tonsils. In this case, the infection is due to either viral or bacterial infection. If the infection is caused by *group A streptococcus* bacteria, it is known as strep throat. The most common signs and symptoms of tonsillitis are red, swollen tonsils that may have white patches (due to pus); sore throat that may refer to the ears; pain or difficulty in swallowing; headache, fever, and chills. The submandibular and/or cervical lymph nodes may be enlarged and tender. As with all active infections, manual therapy is contraindicated. While it used to be common practice to remove the tonsils following an initial infection, physicians now generally reserve tonsillectomy (surgical removal of the tonsils) for cases of recurrent or chronic infections that interfere with daily functioning.

Peyer patches in filtering and trapping pathogens. Like the tonsils, the appendix can become infected and may require surgical removal.

What Do You Think? 12.1

- While many texts and medical professionals use the terms *immune*, *lymphatic*, and *lymphoid* interchangeably, this text distinguishes between *lymphatic* organs, such as lymphangia and the cisterna chyli, and *lymphoid* organs and tissues of the immune system. What are these key distinctions? Why do you think they are helpful in the practice of manual therapy?
- Which of the specialized lymphoid tissues is structurally connected to the lymphatic vessel network?
- What important similarities do you recognize among the secondary lymphoid organs and tissues? And what are important differences?

NONSPECIFIC IMMUNE DEFENSES

The **nonspecific immune defenses** of the body are a group of immune responses that exist in all humans. Also referred to as **innate immune defenses**, these generic and universal responses (nonspecific), are neither stimulated by nor directed toward any single type of pathogen or foreign invader. Nonspecific defenses include the physical and chemical barriers designed to keep foreign substances out of the body, as well as a few more complex internal defense mechanisms. When these general defenses are functioning normally, we spend most of our days in good health.

Physical Barriers

Perhaps the most obvious example of a nonspecific immune defense is the skin. Along with the mucous membranes that line the respiratory, digestive, and reproductive tracts, the skin creates external and internal physical barriers (mechanical obstacles) that block foreign particles. The tough, tightly packed, keratinized cells of the epidermal layer make it difficult for tiny organisms, or **microbes**, to pass through and enter the body's internal environment. If this physical barrier is broken by a cut or scrape, pathogens can enter, and the risk of infection increases. Besides serving as a simple barrier, the continuous shedding of the epidermal layer removes foreign particles from the skin's surface on a regular basis. In a similar fashion, the mucous membranes provide an additional physical barrier to microbes. Sticky mucus secreted by membranes traps foreign particles. In the case of membranes in the respiratory tract, coughing, sneezing, or the movement of cilia propels the trapped microbes outside of the body.

Chemical Barriers

The chemical secretions and excretions of the skin and mucous membranes support the mechanical resistance of the physical barriers. Besides keeping the skin pliable and resilient, sebum secreted by the sebaceous glands creates a slightly acidic layer of protection that inhibits the growth of microbes. The pH of sweat, in combination with some of the enzymes it contains, creates an additional chemical barrier that destroys certain types of bacteria. Tears produced by the lacrimal glands wash the surface of the eyes, and saliva cleans microbes from the teeth, gums, and mucous membranes in the mouth. Any ingested bacteria are subjected to a chemical bath in the mouth and then assaulted by ***gastric juices*** in the stomach. Gastric juice, which is a potent mixture of mucus, hydrochloric acid, and enzymes, has a chemical composition that kills several types of microbes during normal digestive processes. The flow of urine and its slightly acidic pH cleanses the urethra and discourages bacterial growth in the urinary system. Similarly, vaginal secretions also have an acidic pH that discourages bacterial growth as they help flush microbes out of the female reproductive tract.

Internal Antimicrobial Proteins

If a pathogen gets past the body's physical and chemical barriers, it meets with a number of internal general defense mechanisms. Four types of **antimicrobial** proteins found within various body fluids provide the first line of nonspecific internal resistance:

- **Interferons**—Viruses cause disease by first invading cells of the body and then replicating within those cells. Viral-infected fibroblasts, macrophages, or lymphocytes produce chemicals called interferons, which diffuse out of infected cells into nearby cells. Interferons induce the production of specific proteins that shut down the ability of a virus to replicate. In this fashion, interferon production limits the spread of a viral infection once it has occurred.

By the Way

Tears, sweat, saliva, and nasal secretions all contain an antimicrobial enzyme called ***lysozyme*** that breaks down the cell membrane of certain bacteria. Lysozyme is also present in small amounts in some interstitial fluids.

- **Complement proteins**—This particular group of proteins includes about 30 normally inactive proteins found in blood plasma and body tissues. When activated by a pathogen, they initiate a cascade of reactions called the ***complement system*** that enhances the immune response by supporting several other nonspecific immune processes such as phagocytosis, direct microbe destruction, and inflammation.
- **Transferrins**—Most bacterial pathogens require iron to support their metabolism. Transferrins are proteins normally present in blood and other body fluids. When activated by pathogens, they bind to iron, so it is not available to bacteria.
- **Antimicrobial peptides**—These short-chain amino acids are produced by macrophages and mucus-producing epithelial cells. When they come in contact with a pathogen, they destroy it by disrupting its plasma membrane.

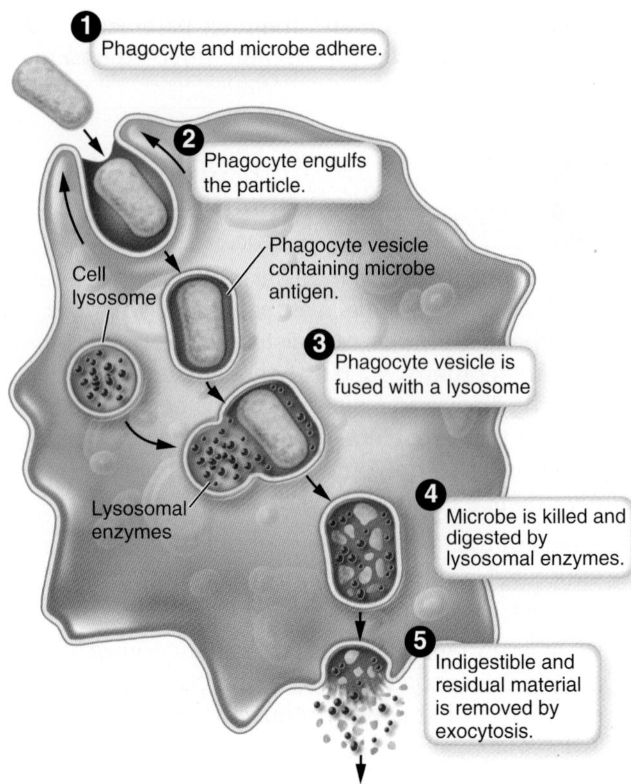
1 Phagocyte and microbe adhere.
2 Phagocyte engulfs the particle.
Cell lysosome
Phagocyte vesicle containing microbe antigen.
3 Phagocyte vesicle is fused with a lysosome
Lysosomal enzymes
4 Microbe is killed and digested by lysosomal enzymes.
5 Indigestible and residual material is removed by exocytosis.

FIGURE 12.4 Key events of phagocytosis.

Phagocytes and Natural Killer Cells

If a microbe survives its journey past physical and chemical barriers, and the body's antimicrobial proteins haven't destroyed it, the next line of internal defense is provided by phagocytes and natural killer cells. Recall that **phagocytes** are a group of cells that destroy microbes and other cellular debris by ingesting them, a process called **phagocytosis** (FIGURE 12.4). White blood cells, specifically the neutrophils and monocytes, are the primary phagocytes in the body. In response to chemicals released by cells at the site of infection, both types of phagocytes quickly migrate to the site, a process known as **chemotaxis**. During their migration through the bloodstream, monocytes transform into macrophages or "big eaters." Macrophages have been previously discussed as primary cellular components of connective tissues. The stationary macrophages found in skin, superficial fascia, liver, spleen, lymph nodes, and red bone marrow are called ***fixed macrophages***. In contrast, macrophages that travel through the bloodstream and migrate into infected tissues are called ***wandering macrophages***. Both types of macrophages, as well as neutrophils, play key roles in the general immune defenses.

Another key player in nonspecific immune responses is the **natural killer (NK) cell**, a type of lymphocyte found in blood. While phagocytes ingest any and all pathogens they encounter, NK cells attach to the surface of an *infected cell* and secrete a toxic substance that destroys its plasma membrane. This direct method of destroying infected cells is called **cytolysis**. In some cases, the cell is destroyed and the microbe or microbe particles are released into the bloodstream. If this occurs, the free-floating microbes are then destroyed through phagocytosis. NK cells make up 5% to 10% of all lymphocytes in the blood and are also present in the red bone marrow, spleen, and lymph nodes. While NK cells have the ability to destroy a wide variety of microbes, they are particularly sensitized to tumor cells.

btw

By the Way

Some microbes are resistant to phagocytosis. For example, one of the organisms that causes pneumonia has an extracellular capsule that makes it difficult for phagocytes to engulf the microbe. The tuberculosis bacterium has a waxy covering over its plasma membrane that prevents it from being exposed to cytolytic enzymes.

Inflammation and Fever

Whether tissue is infected or damaged through traumatic injury, the body defends itself through the process of **inflammation**. As described in Chapter 10, inflammation involves the activation of chemical and cellular processes that enhance vasodilation, capillary permeability, clot formation, and phagocytosis to stabilize the damaged tissue and prepare the body for repair. When this process is initiated by pathogens rather than tissue trauma, the primary goal is to fight and contain the infection rather than to stabilize tissue for repair. In cases of infection, the classic symptoms of inflammation may shift slightly, so there is less localized swelling and pain but an increase in heat and redness. In addition, a local infection may be accompanied by the presence of a thick, whitish fluid called **pus**, which is a mixture of interstitial fluid, destroyed pathogens, phagocytes, and living or dead white blood cells. In cases of systemic infection, or when a local infection spreads, lymph nodes may become enlarged and the core body temperature elevated, resulting in a **fever**. Elevated body temperature speeds up actions involved in enhancing tissue repair, killing or inhibiting the growth of bacteria, and increasing interferons' effect. While it is important to control fevers of 104°F or more, it is not always helpful to decrease a low-grade or moderate fever because it simply indicates an active immune response. ▸ TABLE 12-1 summarizes the nonspecific immune responses.

What Do You Think? 12.2

- Drawing from your everyday life, what analogy would you create to describe the nonspecific immune defenses and responses?
- Have you heard the term antimicrobial before? In what context? What kinds of antimicrobials do we use in our everyday lives?
- How would you describe the difference between the ways that phagocytes and NK cells protect us?

Table 12-1 Nonspecific Immune Defenses

Defense	Components	Functions
Physical barriers	Epidermis of the skin	Form barriers that physically block microbes from entering the body
	Mucous membranes	
Chemical barriers	Sebum	Forms a protective acidic barrier on the surface of the skin
	Sweat	Enzymes and pH discourage microbial growth
	Tears Saliva	Bathes and washes away microbes and contains the antimicrobial protein lysozyme
	Gastric juice	Mix of HCl and enzymes destroys bacteria and other pathogens that reach the stomach
	Urine	Flushes the urethra, and pH inhibits microbial growth
	Vaginal secretions	Flushes the vagina, and acidic pH inhibits bacterial growth
Internal antimicrobial proteins	Interferons	Protect surrounding uninfected cells from viral infection
	Complements	Support phagocytosis, microbe destruction, and inflammation
	Transferrins	Bind with iron to inhibit bacterial growth
	Antimicrobial peptides	Cause cytolysis of microbes
Phagocytes	Neutrophils	Eat microbes and cellular debris (phagocytosis)
	Macrophages	
NK cells		Kill infected cells through cytolysis
Inflammation		Fights and contains infectious agents by increasing vasodilation, capillary permeability, and phagocytosis; prepares tissue for repair
Fever		Speeds up metabolism to facilitate tissue repair; kills or inhibits the growth of certain bacteria; increases the effect of interferons

Pathology Alert Allergies

An ***allergic reaction*** is an abnormal or excessive immune response to an antigen to which the body has been previously exposed and sensitized. Any antigen that provokes an allergic response is called an **allergen**. Allergens can be inhaled, ingested, injected, implanted, or can come in contact with the skin. An allergic reaction is marked by a massive histamine and heparin release, which magnifies vasodilation and vascular permeability. This leads to swelling, increased mucus production, and dramatic increases in phagocytosis and other white blood cell activity that may also damage healthy cells. The intensity and duration of an allergic response depends on the sensitivity of the individual, the quantity of allergen to which they are exposed, and the chemical properties of the allergen. Most allergic responses are localized, such as occurs with hay fever, contact dermatitis, eczema, or gastrointestinal upset. But when certain allergens are ingested or enter the bloodstream, they quickly spread throughout the body and may cause an immediate and widespread response called ***anaphylaxis***, which is an extreme sensitivity to the allergen. In these cases, massive vasodilation and leakage of fluids can cause ***anaphylactic shock***, in which blood pressure drops and breathing becomes difficult. If not treated quickly, it can be fatal.

Some of the most common allergens include dust, pollen, mold, animal dander or hair, detergents and soaps, and certain aromatics. Common food allergens are nuts, gluten, dairy, and shellfish. Manual therapists should carefully question clients who indicate any allergies in their health history to determine if they might have sensitivities to components of lubricants, cleansers, or essential oils. It is generally recommended that manual therapists avoid using strong perfumes, colognes, detergents, oils, or incense to avoid stimulating an allergic response.

SPECIFIC IMMUNE RESPONSES

In contrast to the nonspecific (innate) immune defenses, **specific immune responses**, also known as **adaptive immune responses**, are acquired over time through exposure to specific pathogens. These immune defenses involve the B and T lymphocyte responses to *specific* pathogenic agents. These pathogens may be bacteria, pollen, viruses, parasites, or toxins, but each has a specific chemical marker called an **antigen**. Recall that an antigen, generally a protein on the cell membrane, identifies a cell as a foreign particle and stimulates an immune response. Since B cells and T cells have distinctive ***antigen receptors*** on their surfaces, they are able to recognize and react to antigens. For each unique antigen, there is a lymphocyte with a corresponding antigen receptor. There are millions of B cells and T cells in the body, each ready to go to battle with a particular pathogen. B cells and T cells produce different types of immune responses. The response initiated by B cells is an antibody-mediated response, while that initiated by T cells is a cell-mediated response. In most cases, both specific immune responses are stimulated by any particular pathogen.

Antibody-Mediated Immune Responses

Antibody-mediated immunity, also referred to as ***humoral immunity***, is initiated when a B cell recognizes a specific antigen in the blood, lymph, or interstitial fluid. The antigen binds to receptors on the B cell, which causes the cell to divide and multiply into two types of daughter cells: plasma cells and memory B cells (FIGURE 12.5). **Plasma cells** produce and release **antibodies**, also known as **immunoglobulins**, which are specific plasma proteins that travel through the lymph and/or blood to the site of infection. Once there, they attach themselves to the offending antigens to form ***antigen–antibody complexes***. This binding changes the chemical makeup of the antigen so that it is either neutralized or the invading cell is destroyed through cytolysis. In some cases, antigen–antibody complexes send a chemical signal that attracts additional macrophages to clean up the area. **Memory B cells** are cells in the lymph nodes, spleen,

By the Way

Remember that the term *antigen* describes the protein markers on red blood cells that identify blood types. They are called antigens because they initiate a specific immune response if transfused into an individual with a different blood type.

Pathology Alert Autoimmune Disease

In rare cases, the immune system attacks a person's own cells, reacting as if they were foreign invaders. This situation results in an ***autoimmune disorder*** or ***disease***. Common autoimmune disorders include multiple sclerosis (MS), rheumatic fever, rheumatoid arthritis, systemic lupus erythematosus, scleroderma, ulcerative colitis, type 1 diabetes, and myasthenia gravis. The exact cause of these disorders is not completely understood, but science has identified several possible explanations. For example, MS may be a case of mistaken identity. Because the protein markers in human myelin are very similar to the antigens for measles and the Epstein-Barr virus, antibodies produced to attack these organisms may get confused and also attack the myelin of an affected individual. Rheumatic fever seems to be caused by a type of cross-contamination. The antibodies produced to fight off *Streptococcus pyogenes* bacteria cross-react with the antigens in the heart muscle and end up damaging the myocardium and valves. Other autoimmune diseases may begin when there is a spontaneous appearance of a new type of protein in the blood due to genetic mutations or through alterations that occur during particularly aggressive forms of bacterial or viral infections.

and other lymphoid tissues that "remember" the initial exposure and activation caused by a particular antigen; they stand by ready to respond upon a subsequent exposure.

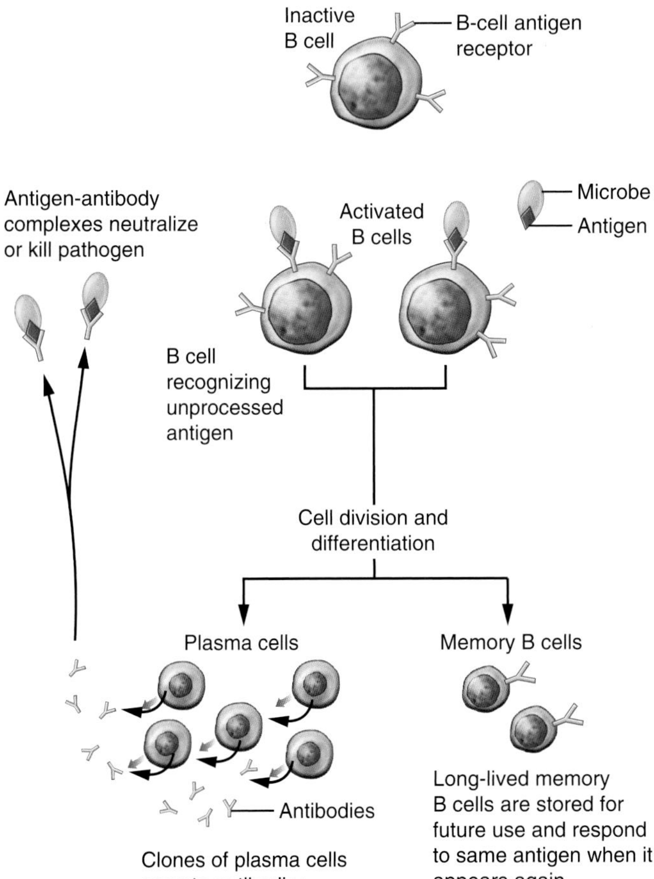

FIGURE 12.5 ▸ Antibody-mediated immunity. Activated B cells divide and multiply to create antibody-producing plasma cells and memory B cells that are stored for future immune challenges.

Cell-Mediated Immune Responses

When a T cell recognizes and responds to a specific antigen, the process of **cell-mediated immunity** occurs, producing cells (instead of antibodies) that travel to the site of infection to destroy the pathogens. Like B cells, T cells must first recognize an antigen. However, while B cells can readily recognize free-floating antigens wherever they are encountered, T cells require an "introduction" to the antigen by an ***antigen-presenting cell (APC)***. Macrophages, the most common type of APC, can ingest and process a pathogen, then insert a recognizable portion of the antigen into their own cell membrane. The macrophage then presents the antigen on its cell membrane to T and B cells for recognition. T cells can only recognize and bind to antigens presented in this manner. Once this binding occurs, the T cell is activated and divides and differentiates to produce cytotoxic T cells and memory T cells (▸ FIGURE 12.6). **Cytotoxic T cells** travel to the site of infection and destroy the foreign invaders. **Memory T cells** function in the same way as memory B cells, "remembering" a particular antigen from an initial exposure and accelerating the immune response upon future exposure.

A unique characteristic of cell-mediated immunity is that it produces several classes of T cells. In addition to cytotoxic T cells and memory T cells, **helper T cells** assist in several key immune responses by releasing special cell-signaling molecules called **cytokines**. These chemicals include interleukin-2 that stimulates the proliferation of cytotoxic T cells and activates NK cells. Interleukin-4 is another cytokine released by helper T cells; it promotes the growth of all T cells and stimulates plasma cell production of antibodies. **Suppressor T cells** function as their name implies. These cells inhibit or shut down both B cell and T cell activity once an immune response is no longer needed.

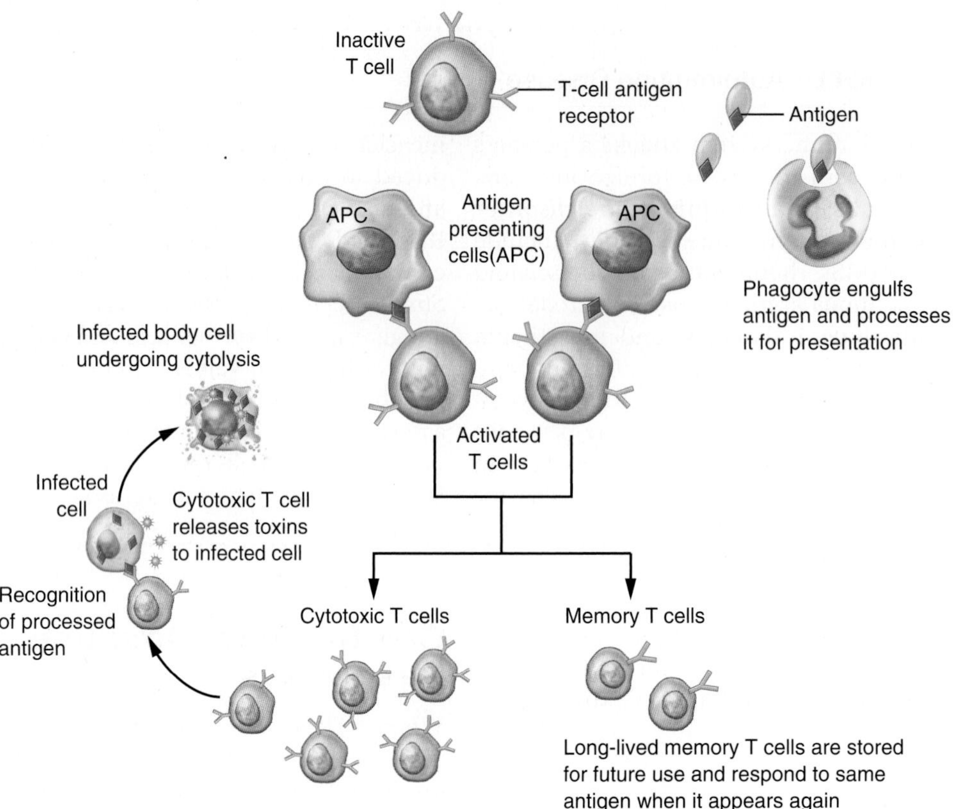

FIGURE 12.6 ▸ Cell-mediated immunity. Phagocytes engulf, process, and present antigens to inactive T cells. Activated T cells divide and multiply to create cytotoxic T cells, which directly battle the antigen, and memory T cells, which are stored for future immune challenges.

Manual therapists must consider extra precautions when working with immune-compromised clients such as the elderly; people with HIV or AIDS; those undergoing dialysis, chemotherapy, or radiation; or anyone taking immune-suppressing medications. For immune-compromised clients, exposure to infection or illness can be dangerous. Therefore, it is advisable for therapists to reschedule appointments if suffering from a cold, flu, or infection. Maintaining universal precautions, including frequent hand washing, as well as sanitizing and proper cleaning of linens and equipment is essential. Additionally, it is recommended that manual therapists consider vaccinations for hepatitis, measles, mumps, diphtheria, and whooping cough, especially if they are working in hospitals or other health care settings where there is a high risk for exposure to these highly contagious conditions.

ACQUISITION OF IMMUNITY

There are two key characteristics that set specific immune responses apart from the nonspecific defenses:

- A specific immune response occurs only in reaction to a particular antigen.
- Lymphocytes have the ability to "remember" exposure to a particular antigen.

Our bodies maintain an ***immunological memory*** because of the **memory cells** produced with each episode of a specific immune response. These memory cells have the same antigen receptors as the original B or T cell that produced them and can live for decades. Upon subsequent exposure to a particular antigen, the memory cells quickly divide and multiply, just like activated B or T cells. Therefore, memory B cells produce plasma cells and more memory B cells, while memory T cells produce cytotoxic and memory T cells.

When one considers the number of memory cells produced during initial exposure, it is not surprising that subsequent responses to a particular antigen occur more quickly than those elicited by the first exposure. During initial exposure, there are only a few lymphocytes with the correct antigen receptors to initiate a response. During a second or subsequent exposure, there are thousands of memory cells available, which increases the possibility of the antigen activating one of these specific lymphocytes. In addition to occurring more rapidly, the second response is also more intense. The larger number of memory cells activated during the subsequent response produce greater numbers of cytotoxic T cells and/or antibody-producing plasma cells. Another reason for the rapid response and increased intensity of subsequent exposures is that antibodies produced from the first exposure are still available, and they are more

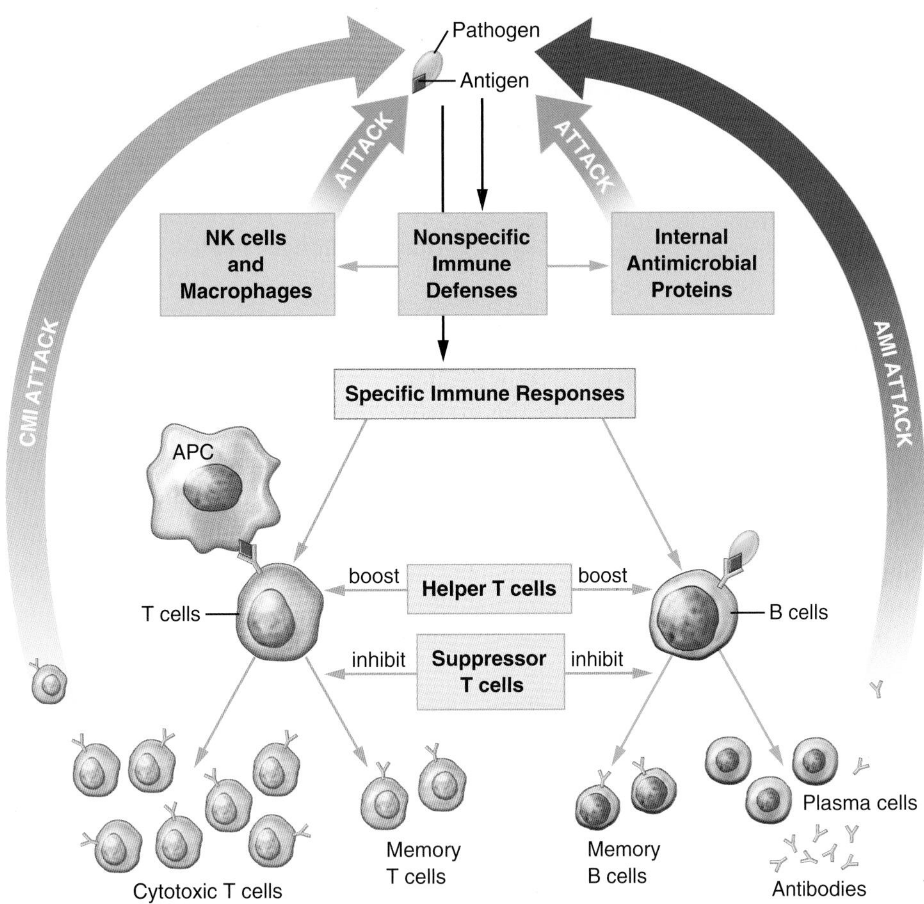

FIGURE 12.7 ▸ Summary of immune responses. Nonspecific immune defenses (green) protect the body from all types of pathogens. Specific immune responses include the recognition of specific antigens by B and T cells. Activation of B cells produces an antibody-mediated immune (AMI) response (blue), and activation of T cells leads to a cell-mediated immune (CMI) response (orange). To see an animation on the immune response, visit http://thePoint.lww.com/Archer-Nelson.

effective in fighting the antigen. ▸ FIGURE 12.7 illustrates the body's immune responses, including the actions of the various types of T cells and B cells.

Immunological memory accounts for how individuals acquire immunity to many diseases. While resistance to certain pathogens can be present from birth, individuals must develop or acquire their immunity to many other pathogens such as the flu or measles. Immunity can be acquired either through natural means or through artificial interventions.

Naturally Acquired Immunity

Naturally acquired immunity is resistance that is obtained naturally, without the intervention of modern medicine. Natural immunity is developed in two ways: actively or passively. ***Naturally acquired active immunity*** is developed through exposure to a pathogen in the course of daily life. Specific immune mechanisms respond and develop the body's resistance. For example, exposure to the varicella-zoster virus leads to chicken pox. Once a person has had chicken pox, it usually results in lifelong immunity. In fact, to develop a child's immunity, it used to be common practice for parents to intentionally expose their children to a sibling or friend with chicken pox. With this exposure, the body *actively* engages in a specific immune response, which develops memory cells and antibodies that provide future resistance.

Immunity can also be naturally acquired through the transfer of antibodies from mother to baby during pregnancy or through breast-feeding. However, since the mother's specific immune responses are engaged in these processes and not the baby's, these are examples of ***naturally acquired passive immunity***; the baby *passively* obtains immunity by receiving antibodies instead of developing them independently.

Artificially Acquired Immunity

Artificially acquired immunity is resistance developed through immunizations. Like natural immunity, it can be actively or passively acquired. ***Artificially acquired active immunity*** is achieved through ***vaccinations***. Often, these injections introduce either a weakened or dead pathogen into the body. While these pathogens are altered, they still retain the antigen and thus, they initiate a specific immune response. In other cases, the vaccination contains a synthesized form of a specific antigen rather than a pathogen. This allows for greater specificity in stimulating the immune response and reduces the risk of introducing an unknown pathogen. Following vaccination, some people will have a low-grade fever and feel achy or fatigued as their body actively engages in the immune response. However, there is a small risk, especially when the pathogen is used, that some recipients can develop the disease in response to a vaccine, particularly if their immune system is compromised or suppressed. Because some vaccinations do not create permanent resistance, ***booster shots*** have been developed to provide a small dose of antigen to boost the levels of antibodies in the bloodstream.

When antibodies are administered directly, it is categorized as ***artificially acquired passive immunity***. This form of immunization is used when there is not time for the recipient to develop their own antibodies, such as during an active infection. For example, antibody or immunoglobulin injections are often used to treat hepatitis A and B, diphtheria, and tetanus. They can also be used to treat conditions such as botulism (a type of food poisoning), to treat venomous bites from spiders or snakes, or used preventively to help protect immune-compromised individuals. Immunity provided through these immunization methods is temporary and generally lasts no more than 3 to 6 months. ▸ TABLE 12-2 summarizes the various ways in which humans acquire immunity.

What Do You Think? 12.3

- What analogy or story can you create to highlight the differences between antibody-mediated immunity and cell-mediated immunity?
- What types of artificially acquired immunity have you received? Are they examples of active or passive methods?

AGING AND THE IMMUNE SYSTEM

Many people live long and healthy lives despite the fact that resistance to disease and infection gradually decreases with age. Overall, the immune system simply becomes less responsive to the foreign antigens that challenge health. As people grow older, most become more susceptible to several types of infection as B and T cells begin to respond more slowly. The overall number of T cells also tends to drop as the thymus shrinks and produces less thymosine. Decreased production of helper T cells diminishes the specific immune responses, the development of NK cells, and phagocytic activity. This decrease in the number and activity of NK cells helps explain why there is a higher incidence of cancer and other malignancies among older populations. Because the immune system of an elderly person is less responsive to vaccinations, older adults are generally encouraged to get flu vaccines every year. Additionally, the incidence of autoimmune dysfunctions increases with age, presumably due to some combination of the autoimmune triggers previously described.

PSYCHONEUROIMMUNOLOGY

In the 17th century, French philosopher, scientist, and mathematician René Descartes inadvertently laid the groundwork for the future direction of Western medicine when he described the mind and body as having radically different natures. Descartes contrasted the mind, a rational thinking entity that could not be teased apart, to the body, which he characterized as a simple physical machine that can be broken apart into smaller functional pieces. While Descartes acknowledged that the mind and body clearly interact, he reasoned that if they are essentially different things, they could be studied and understood separately. In fact, to

Table 12-2 Immunity Acquisition

Type	Method of Acquisition
Naturally acquired active immunity	Exposure to antigen initiates a specific immune response resulting in the formation of plasma cells, cytotoxic T cells, and B and T memory cells. Once recovered, the body retains resistance through immunological memory.
Naturally acquired passive immunity	Antibody transfer from mother to baby across the placenta during pregnancy and through breast-feeding.
Artificially acquired active immunity	Vaccination delivers dead or altered pathogens or synthesized antigens to initiate a specific immune response. This ultimately results in the formation of antibodies and long-lasting B and T memory cells that provide resistance through immunological memory.
Artificially acquired passive immunity	Immunization delivers antibodies to provide immediate but short-term immunity.

pursue his study of the human body, Descartes had to assure the Pope that he would leave the study of the mind and soul to the Church. He pursued the study of the body from a purely physical and mechanical perspective and provided the framework for modern allopathic medicine.

In many ways, this view of mind and body as separate has served mankind and led to remarkable medical advances. The ability to focus on and understand body systems and individual organs, tissues, and even a single cell has exponentially increased the understanding of the structures and functions of the human body. However, it has also narrowed the focus of many Western medical practitioners to the extent that it is common practice to identify a specific systemic or structural problem and treat that issue alone. This tendency in diagnosis and treatment not only overlooks the interdependence between body systems but often fails to consider the impact of emotions, attitudes, and beliefs on health and disease.

Over the past 30 to 40 years, the emerging field of **psychoneuroimmunology (PNI)** has provided an abundance of scientific documentation related to the physiologic connections between the mind and body. This field of study focuses on connections between psychological processes and the functioning of the immune and nervous systems.

However, PNI research has also shown that the term *psychoneuroimmunology*, coined by Dr. Robert Ader in 1975, appears to be incomplete because it fails to include the impact of the endocrine and digestive systems.[1,2] By exploring the chemical and energetic links between the brain, digestive system, endocrine glands, and immune cells, PNI researchers have clearly described two body-wide communication networks. One network is the autonomic nervous system, discussed in Chapter 7, and the other is a vast chemical network of "information molecules."[1]

Peptide Communication Network

The chemical communication system described by PNI researchers is a network created by **peptides** that travel throughout the body. These short amino acid chains include hundreds of chemicals categorized as neurotransmitters, endorphins, hormones, digestive peptides, or cytokines. Since all these substances communicate and stimulate cells by binding with receptors on their plasma membranes, these peptides are collectively referred to as **ligands**, a term taken from the Latin *ligare*, which means "to bind." Familiar examples of ligands are epinephrine, a neurotransmitter, and adrenaline, a hormone. Recall that epinephrine stimulates postsynaptic neurons whose dendrites are sensitive to this neurotransmitter, while adrenaline acts on target cells with adrenaline receptors. While these ligands have different names and functions, chemically, they are exactly the same peptides. PNI research has demonstrated that regardless of how a peptide molecule is categorized (neurotransmitter, hormone, cytokine, etc.) or which system, organ, or tissue releases it, the ligand is capable of stimulating *any* cell that possesses a receptor for it. For example, ligands released by immune cells can circulate to the brain, cross the blood-brain barrier, and stimulate neurons with receptors for that peptide. In a similar manner, neuropeptides released by the brain circulate throughout the body and stimulate specific receptors in the plasma membranes of immune cells. In fact, most peptide receptors found in the brain are also found on the surface of monocytes.[1]

While ligand receptors reside on cells throughout the body, they are most concentrated in regions of the brain associated with the emotions, particularly the limbic system. Scientifically speaking, this suggests that the communication web created by these peptides and their receptors provides a strong functional link between immune responses and emotional states. Our feelings are reflected in what we believe, and what we believe is reflected in the chemistry of our body: a mind–body–spirit connection. In addition to neuroimmune links, recent fascial research has demonstrated neuroendocrine changes that are initiated through stimulation of mechanoreceptors in the abdomen.[3] These changes include shifts in the production of the neurotransmitter serotonin, as well as in the histamines that are key elements of the inflammatory response. These and other discoveries provide physiological rationale for "gut instincts" that tell us something before the mind consciously perceives it and for the boost the immune system receives from laughter and unconditional love.

By the Way

Placebos are often used in clinical research to test the effectiveness of a new therapy. These inert substances (sugar pills) or false treatments have often proven to provide participants with as much relief as the drug or treatment protocol being tested, apparently due to the belief on the part of participants that the drug or treatment would work. Called the *placebo effect*, statistics consistently confirm the power of belief and its impact on healing and provide a clear demonstration of the connection between the mind and body.

THOUGHTS ON HOLISTIC HEALING

The words *health*, *healing*, and *wholeness* all share the common root *hal* (of Germanic origin) that refers to a state of being complete, undivided, and integrated. The insights provided by PNI challenge all health care practitioners to consciously consider their beliefs and mental constructs about health and disease. If health is about wholeness, then disease cannot be limited to its biomedical definition. As discussed in Chapter 1, wholeness and a sense of well-being encompasses the integrated and interdependent nature of the mind, body, and spirit. Therefore, disease is not simply a condition that impairs normal physiological function, but anything that disintegrates us. It is any disruption to the dynamic balance between body, mind, and spirit to the point of causing physical, mental, or emotional trauma.

It is also important to remember that there are times when it is both helpful and necessary to think and speak of disease from a purely biomedical perspective. Without this perspective, there would be no cures or life-saving treatments for conditions such as fractured bones, appendicitis, asthma, malaria, or certain cancers. However, curing a disease or repairing an injury does not always result in complete healing. We sometimes forget that healing is a process of returning to vitality and balance. It requires us to consciously and consistently make good choices in regard to nutrition, exercise, rest, work, play, and relationships. Healing is about moving toward health, and health is not simply the absence of disease but a soundness of body, mind, and spirit that reminds us that we are wonderfully alive.

The practice of manual therapy may include massage, movement, listening, specific kinds of breathing, energy balancing, and other tools and techniques. Many of these techniques have been dismissed by mainstream medical practitioners because the physiologic mechanisms for their effectiveness are not easy to identify, and treatment results are difficult to objectively measure. However, if healing is seen as an ongoing process of moving toward health, the client's belief in the value of a particular treatment can have powerful benefits, even if medical science cannot explain the physiologic mechanisms of its action. To quote Norman Cousins, "It is reasonable ... that science may not have all the answers to problems of health and healing."

Some of the benefits of manual therapy are subjective and difficult to measure. Benefits such as improved mental focus, a greater sense of ease in movement, decreased stress or anxiety, reconnecting or grounding, and an overall feeling of well-being have a profound impact on health and healing. Ultimately, the greatest therapeutic impact of manual therapy may simply be the facilitation of each client's individual healing process.

What Do You Think? 12.4

- What examples of PNI at work can you give from your own life experiences?
- How might you integrate information about the mind–body connection to benefit your clients?

SUMMARY OF KEY POINTS

- The immune system consists of a group of organs and tissues throughout the body that are also structural components of the cardiovascular, lymphatic, and endocrine systems.
- The primary lymphoid organs are sites where lymphocytes are produced: the bone marrow and thymus. Secondary lymphoid organs and tissues serve as sites for immune responses and include the lymph nodes, spleen, and MALT.
- Nonspecific immune responses are general defense mechanisms that are not directed at a particular antigen. See TABLE 12-1 to review.
- The key characteristics of specific immune responses are that they are triggered by specific pathogens, involve B and T lymphocytes, and create an immunological memory.
- Antibody-mediated immunity involves the activation of B lymphocytes to produce antibodies that render antigens harmless. Cell-mediated immunity activates several categories of T lymphocytes, including cytotoxic T cells that directly destroy the antigen.
- Immunity can be acquired through natural or artificial means. Both means include active and passive mechanisms for acquiring immunity. See TABLE 12-2 to review the methods of immunity acquisition.
- The body's immune defenses generally decline as we age, and the incidence of cancer and autoimmune disorders tends to increase.
- PNI is the study of the links between the nervous, endocrine, digestive, and immune systems. Research in this field has provided a good deal of evidence in support of the mind–body connection.

REVIEW QUESTIONS

Short Answer

Define the terms listed in numbers 1–4.

1. Antigen
2. Cytokine
3. Antibody
4. Allergen
5. List the body's nonspecific immune defenses.
6. Describe the difference between how phagocytes and NK cells function.
7. What are the three defining characteristics of the body's specific immune responses?

Multiple Choice

8. Chemical barriers of the body include the pH of the skin and
   a. acidity of gastric juices
   b. aldosterone release
   c. peptide secretion
   d. chemotaxis
9. Mucosa-associated lymphoid tissue (MALT) includes the tonsils and
   a. thymus
   b. spleen
   c. Peyer patches
   d. red bone marrow
10. Where are lymphocytes initially produced?
    a. red bone marrow
    b. thymus
    c. lymph nodes
    d. spleen
11. What is the function of antimicrobial proteins?
    a. maturation of T cells in thymus
    b. first line of internal nonspecific defense
    c. activates killer and helper T cells
    d. specific immune enzyme that creates cytolysis of microbes
12. Which of these lymphocytes produce antibodies?
    a. memory T cells
    b. suppressor B cells
    c. NK cells
    d. plasma cells
13. What is another term for antibody?
    a. antigen
    b. allergen
    c. immunoglobulin
    d. interferon
14. Which of the following are important processes of inflammation, a general immune defense?
    a. vasoconstriction and chemotaxis of NK cells
    b. cytolysis and margination
    c. vasodilation and phagocytosis
    d. activation of B and T cells
15. Which of these antimicrobial proteins prevents bacterial proliferation by binding with iron in cells so it is not available to the microbe?
    a. interleukins
    b. interferons
    c. complements
    d. transferrins
16. Which lymphocytes are involved in antibody-mediated immune responses?
    a. NK cells
    b. B lymphocytes
    c. helper T cells
    d. T lymphocytes
17. What is the function of suppressor T cells?
    a. inhibit and stop the immune process once the challenge has passed
    b. stimulate NK cells to release antigen suppressing enzymes
    c. inhibit the production of viral toxins
    d. monitor and regulate all T lymphocyte activity
18. What happens to antigens when antibodies attach to them?
    a. they are expelled via ciliary action
    b. they are rendered ineffective
    c. they get clumped together for NK cell destruction
    d. their nucleus is ruptured

Continued on page 322

19. Which of these is an example of artificially acquired *active* immunity?
    a. transmission of antibodies from mother to child
    b. being exposed to a disease in childhood
    c. receiving injections of immunoglobulin
    d. vaccinations

20. Psychoneuroimmunology studies the links between the immune system and which other body systems?
    a. digestive, respiratory, and endocrine
    b. digestive, endocrine, and nervous
    c. nervous, lymphatic, and cardiovascular
    d. cardiovascular, respiratory, and nervous

References

1. Pert C. *Molecules of Emotion*. New York: Scribner; 1997.
2. Davis CM, ed. *Complimentary Therapies in Rehabilitation: Holistic Approaches for Prevention and Wellness*. Thorofare, NJ: SLACK, Inc; 1997.
3. Schleip R. Fascial plasticity—a new neurobiological explanation: part 1. *J Bodyw Mov Ther*. 2003;7(1):11–19.

13 The Respiratory System

LEARNING OBJECTIVES

Upon completion of this chapter, you will be able to:

1. List the functions of the respiratory system and discuss their importance as they relate to the practice of manual therapy.
2. Name the major organs of the respiratory system and describe the functions of each.
3. List the organs that make up the upper and lower respiratory tracts and describe the structure of each.
4. Explain the processes of ventilation and respiration.
5. Name and locate the skeletal muscles involved in ventilation and describe how each contributes to this action.
6. Explain the key physiologic processes involved in internal and external respiration.
7. Explain how oxygen and carbon dioxide are transported in blood.
8. Describe the physiologic processes involved in respiratory control and regulation.

Life is not measured by the number of breaths we take, but by the moments that take our breath away."

ANONYMOUS

KEY TERMS

exhalation (ex-hah-LA-shun)
external respiration (EKS-ter-nal res-pih-RA-shun)
inhalation (in-hah-LA-shun)
internal respiration (IN-ter-nal res-pih-RA-shun)
respiration (res-pih-RA-shun)
respiratory membrane (res-PI-rah-tor-ee MEM-brain)
ventilation (ven-teh-LA-shun)

Primary System Components

- **Nose and nasal cavity**
- **Pharynx**
- **Larynx**
- **Trachea**
- **Lungs**
  - Pleura
- **Bronchial tree**
  - Bronchi
  - Bronchioles
  - Alveoli

Primary System Functions

- **Provides oxygen for cellular activity**
- **Removes carbon dioxide, which helps to maintain normal blood pH**
- **Eliminates heat and water through ventilation to help regulate temperature and fluid balance**

Breathing is a continuous process that reflects both the quality and tempo of life. From the first breath of air drawn into a newborn's wet lungs to the last sigh expelled before death, the rhythmic exchange of air provides cells with vital oxygen and eliminates carbon dioxide, the byproduct of respiration. Most of the time, breathing is not consciously directed. Yet, if we do pay conscious attention, the rate and depth of each breath can serve as a reliable indicator of our physical, emotional, and mental state. For example, breathing is slow and even when we are relaxed or resting, but when we are working or exercising strenuously, breathing is fast, even, and deep. The rate of breathing can also increase due to anxiety or excitement; however, in these situations, breaths are often shallow and uneven.

This chapter explores the structure and functions of each major component of the respiratory system. The process of ventilation created through skeletal muscle contraction is explained and distinguished from respiration. Control and regulation of breathing rate and depth is also discussed. Manual therapists are reminded that breathing supports the movement of blood and lymph throughout the body. It can also reflect a client's physical, mental, or emotional status: the pace and quality of breathing can be an important signal of a client's discomfort, tension, or relaxation.

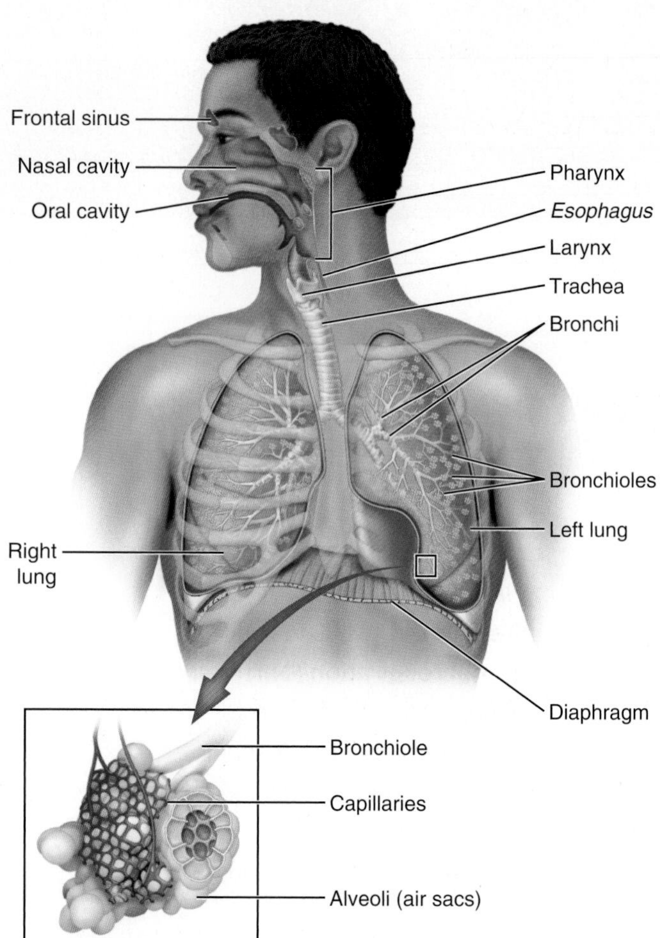

FIGURE 13.1 ▸ Structures of the respiratory system.

STRUCTURES OF THE RESPIRATORY SYSTEM

Anatomically, the respiratory system is fairly simple, consisting of a network of large and small passageways that move air to and from the lungs, where oxygen and carbon dioxide exchange occurs. The system is divided into the upper and lower respiratory tracts (▸ FIGURE 13.1). The ***upper respiratory tract*** consists of the nose, sinuses, pharynx, and larynx. The trachea, lungs, and bronchial tree are the primary organs of the ***lower respiratory tract***. Both tracts are lined with a highly vascular mucous membrane, the **respiratory mucosa**, which consists of pseudostratified columnar epithelial cells interspersed with mucus-producing ***goblet cells*** (▸ FIGURE 13.2). This membrane warms and humidifies the air and traps small inhaled particles. The movement of **cilia** (small hair-like projections) on the mucosal cells moves mucus to the pharynx (throat), where it can be swallowed or expelled.

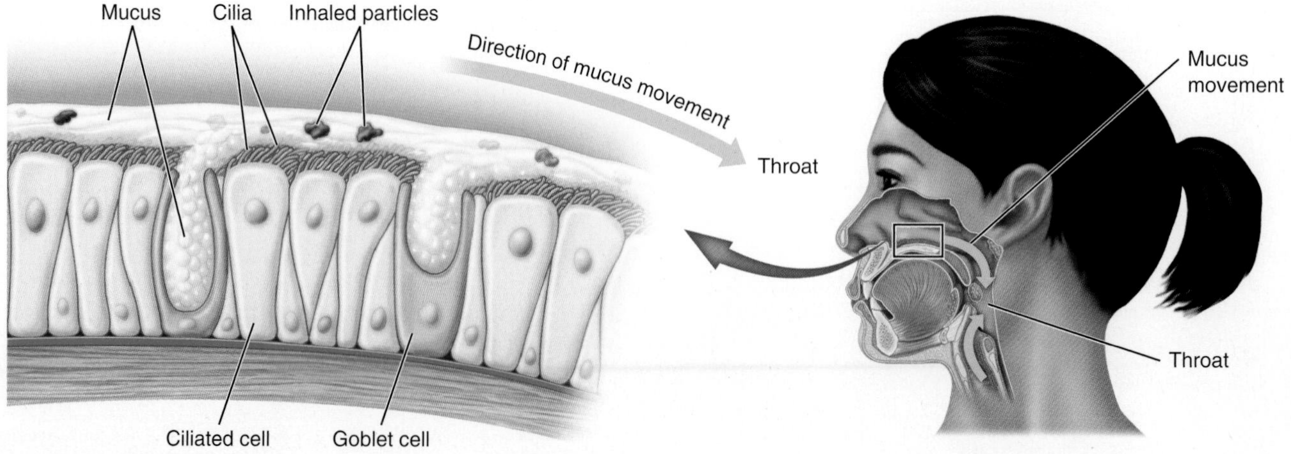

FIGURE 13.2 ▸ Respiratory mucosa. The respiratory tract is lined with a mucous membrane that consists of pseudostratified ciliated, columnar epithelium interspersed with mucus-producing goblet cells. The mucosa warms and humidifies air and traps inhaled particles in its sticky mucus. Cilia move mucus toward the pharynx (throat) to be swallowed or expelled.

By the Way

The olfactory nerve carries information from the olfactory receptors in the nasal cavity directly to the limbic system of the brain. This creates a strong link between odors, memory, and emotion.

Nose and Nasal Cavity

Air enters the respiratory system through the nostrils, or **nares**. Each nostril has fine hairs that trap particles such as dust or pollen and prevent them from moving further along the respiratory tract. The nares open into the **nasal cavity**, a hollow space divided into halves by the **nasal septum** created by the vomer and ethmoid bones and a hyaline cartilage extension (FIGURE 13.3). The floor of the nasal cavity is created by the ***hard palate***, the anterior bony portion of the roof of the mouth formed by the maxillary and palatine bones. Within the nasal cavity are three scroll-shaped bony shelves, the ***nasal conchae***, which slow the flow of air through the cavity and direct air upward to the olfactory receptors (FIGURE 13.3B). Like the rest of the nasal cavity, the conchae are also covered with respiratory mucosa for filtering, warming, and humidifying air.

Recall from Chapter 5 that the paranasal **sinuses** in the frontal, maxillary, and ethmoid bones are cavities that lighten the weight of the skull (see Fig. 5.8). The sinuses open into the nasal cavity and are lined by the same mucous membrane. Therefore, the sinuses help moisten and warm air as it passes through. They also serve as sound resonating chambers for the voice.

Pharynx

The posterior region of the nasal cavity opens into the **pharynx**, commonly called the throat. The pharynx can be divided into three sections (FIGURE 13.4). The ***nasopharynx*** is the superior portion, located between the nasal cavity and ***soft palate*** (the soft tissue partition posterior to the hard palate). The eustachian tubes connect the middle ear to the nasopharynx. The ***oropharynx*** is the middle portion, located at the back of the mouth. The superior and middle sections contain the tonsils. The inferior portion, the ***laryngopharynx***, attaches anteriorly to the larynx and posteriorly to the esophagus. The oropharynx and laryngopharynx act as passageways for both air and food.

Larynx

The **larynx**, also called the voice box, is a cartilaginous structure situated between the pharynx and the trachea. Its location in the anterior neck is easily palpated. The protruding structure known as the "Adam's apple" in males is a protective ring of hyaline cartilage called the ***thyroid cartilage*** (FIGURE 13.5). If you place your index finger in the soft tissue just inferior to the mandible and slowly slide it down the center of the anterior neck, the first protrusion you encounter is the thyroid cartilage, which can be recognized by the tiny notch in its superior rim.

Within the larynx, a portion of the respiratory mucosa folds over two short bands of fibrous connective tissue to create the vocal folds, or **vocal cords** (FIGURE 13.5). When air is forced over the vocal cords, they vibrate like strings on a violin or guitar to produce sound. To raise the pitch of the voice, the laryngeal muscles contract to stretch or tighten the cords, while relaxing these same muscles lowers the pitch. The opening between the vocal cords is the **glottis**, which must be open to allow air to pass from the pharynx into the larynx and trachea. The **epiglottis** is a small cartilage flap attached to the base of the tongue and hyoid bone. During swallowing, it is pulled over the glottis, acting as a trapdoor to prevent food or fluids from entering the trachea.

Pathology Alert Upper Respiratory Tract Infections

Upper respiratory tract infections (***URTIs***) include any number of acute viral or bacterial infections of structures in the upper respiratory tract. Viral infections such as the common cold and influenza, as well as ***sinusitis***, an inflammation of the mucous membranes of the sinuses due to allergies, bacterial or viral infection, or an obstruction, can all be categorized as URTIs. While all three conditions can be accompanied by a runny nose, sneezing, headache, and fever, there are some key differences. The symptoms of the common cold are limited to the upper respiratory tract, generate only a low-grade fever, and generally last less than a week. In contrast, the flu can last up to 2 weeks and cause muscle aches, fatigue, and swollen lymph nodes, as well as a high fever. Influenza can also lead to more serious conditions involving the lower respiratory tract, such as bronchitis or pneumonia. Sinusitis usually presents with a headache, pressure or tenderness over the sinuses, and a runny nose; it can also involve tooth or facial pain. Because massage and other manual therapies are always contraindicated during acute infection, clients with a cold, influenza, and some forms of sinusitis should be rescheduled. Therapists with a cold or flu should always reschedule appointments to avoid spreading infections.

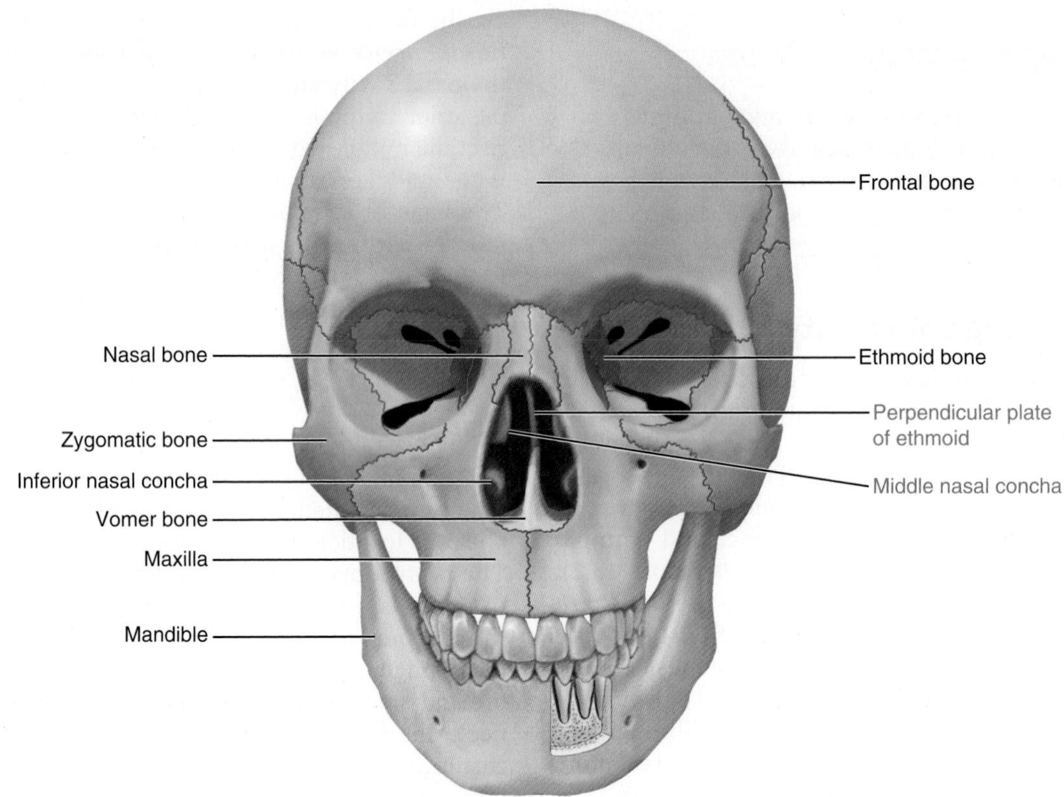

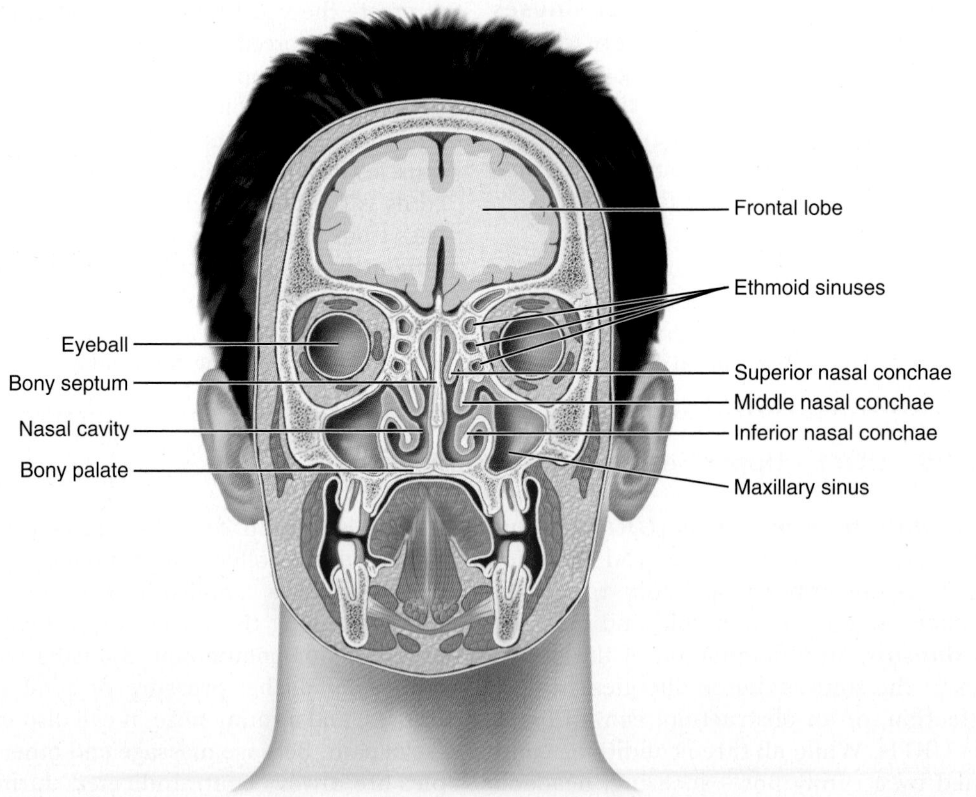

FIGURE 13.3 ▸ Nasal cavity. A. The nasal cavity is divided by the nasal septum created by the vomer and ethmoid bones with a hyaline cartilage extension. **B.** Nasal cavity, frontal section. The superior and middle nasal conchae are projections of the ethmoid bone, while the inferior nasal conchae are separate facial bones. The ethmoid and maxillary sinuses are also shown.

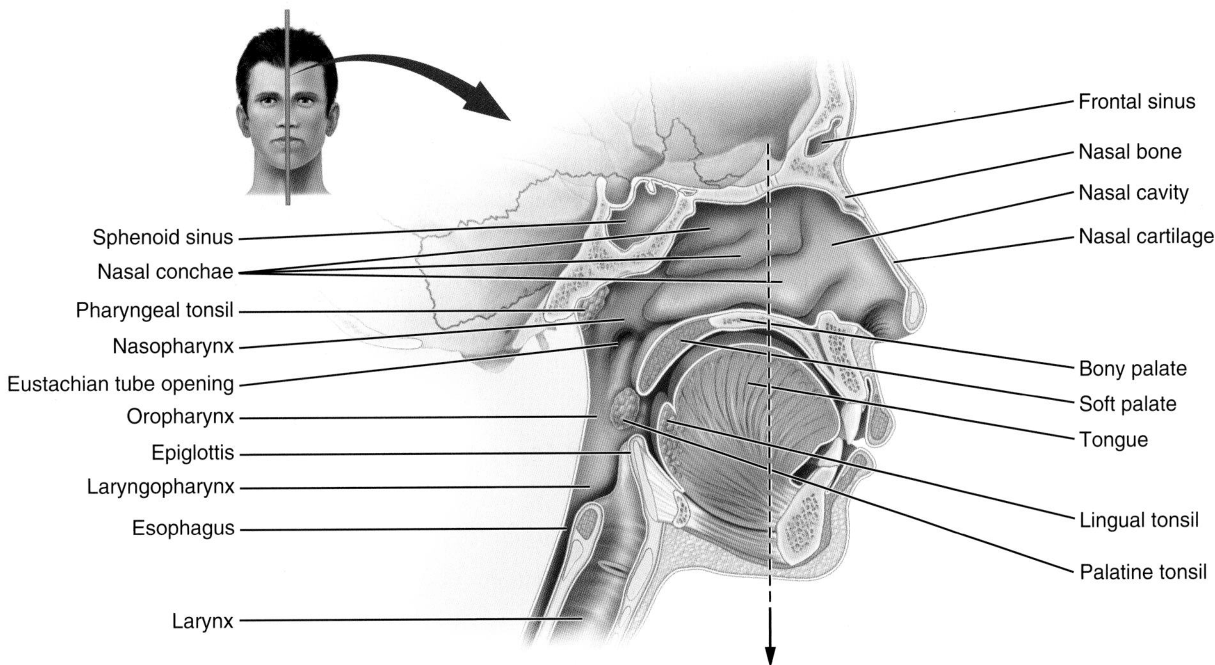

FIGURE 13.4 ▸ Upper respiratory tract. A lateral section of the upper respiratory tract system shows the three segments of the pharynx: the nasopharynx, oropharynx, and laryngopharynx. The tonsils and openings for the eustachian tubes are also found within the pharynx.

By the Way

If a person is eating and talking at the same time, food particles can inadvertently enter the lower respiratory tract, but the body has a built-in remedy—the tissue of the larynx contains many hypersensitive nerves that can trigger a cough reflex. Coughing forcefully blasts air from the lungs through the trachea and larynx, dislodging any foreign particles.

Trachea

The **trachea**, commonly called the windpipe, is a tube that connects the upper respiratory tract to the lungs. The trachea begins at the ***cricoid cartilage***, which is palpable just inferior to the thyroid cartilage, and extends to the superior portion of the thoracic cavity (▸ Figure 13.5). The passageway is protected and kept open by a series of C-shaped hyaline cartilage rings. The opening to these rings faces posteriorly creating a gap for the esophagus, the tube that carries food from the mouth to the stomach (▸ Figure 13.5C). This allows the esophagus to expand anteriorly to pass a large piece of swallowed food, if necessary. The esophagus provides some support to the posterior side of the trachea, together with a band of smooth muscle that connects the ends of each cartilage ring.

The waving action of the mucosal cilia in the trachea creates an "escalator" that transports foreign particles trapped in mucus up and out of the trachea to the pharynx. However, smoking paralyzes these cilia, and longtime smokers develop a chronic "smoker's hack." In this case, coughing becomes the only way for a smoker to draw up or clear mucus and trapped particles from the lower respiratory tract.

Lungs and Bronchial Tree

The **lungs** occupy the majority of the thoracic cavity and are situated on either side of the heart (▸ Figure 13.6). The superior narrow portion of each lung, the ***apex***, is located just deep to the clavicle, while the broader ***base*** of each lung rests on top of the diaphragm. Deep fissures in the external tissue divide each lung into smaller regions, or **lobes**. The right lung has three lobes and is slightly larger than the left lung, which has two lobes. The left lung is a bit smaller to allow room for the heart on the same side of the thoracic cavity. Similar to the heart, the lungs are surrounded by a double-layered serous membrane called the **pleura**. The ***visceral pleura*** covers the outer surface of the lungs, while the ***parietal pleura*** lines the walls of the thoracic cavity. The microscopic space between the layers is filled with ***pleural fluid***, a serous fluid that reduces friction as the lungs expand and contract during breathing. ***Pleurisy***, an

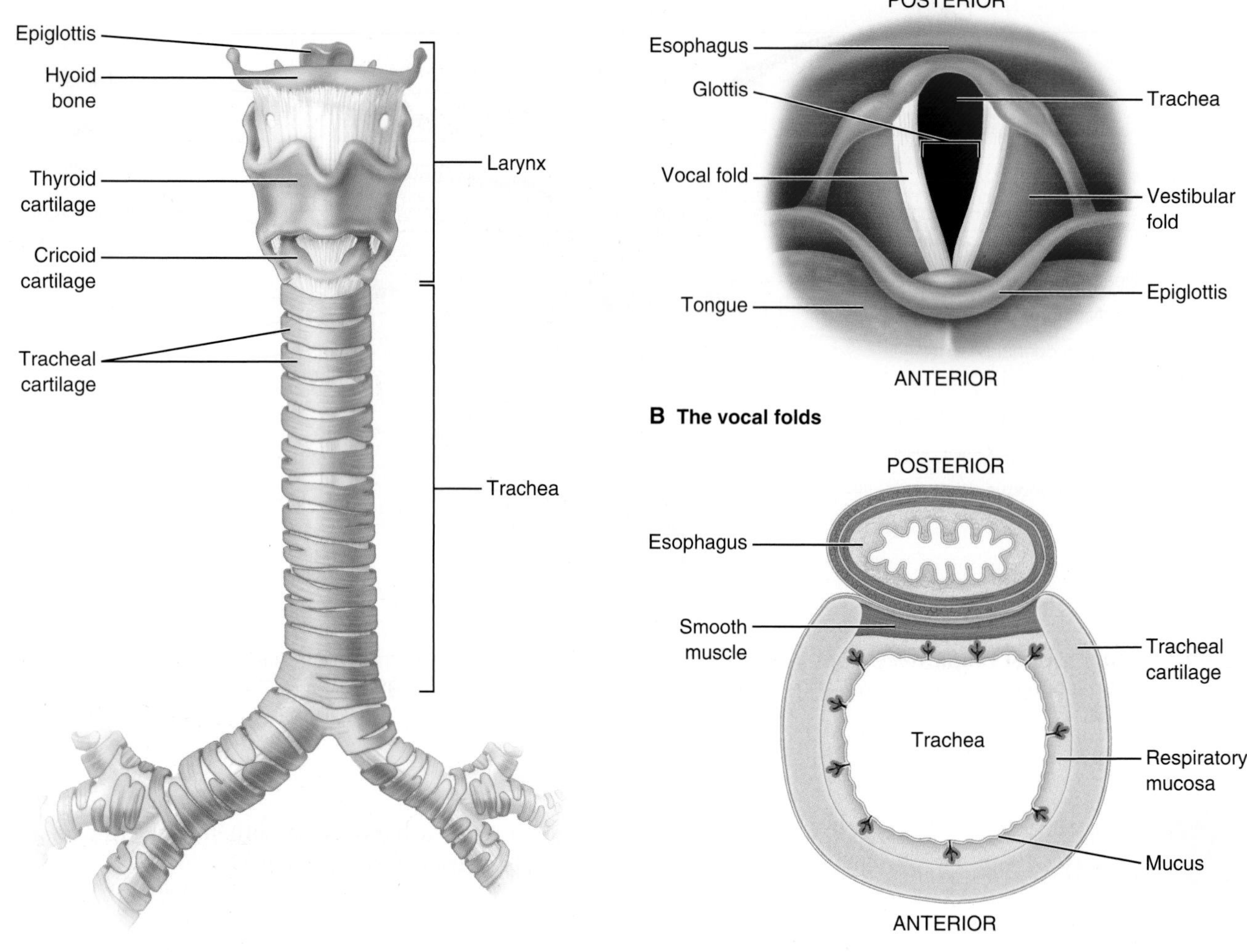

FIGURE 13.5 ▸ Larynx and trachea. A. Anterior view shows the thyroid cartilage that protects the larynx, as well as the cricoid cartilage and cartilage rings of the trachea. **B.** Superior view shows the vocal folds and the glottis. **C.** Trachea cross section shows the C-shaped tracheal cartilage ring and the smooth muscle and esophagus that provide support on the posterior side.

infection or inflammation of the pleura, increases the friction between the membranes, making every breath quite painful.

The **bronchial tree** is a network of passageways inside the lungs (▸ FIGURES 13.6 and 13.7). The right and left primary **bronchi** are the largest branches that divide from the trachea at about the T-5 vertebral level. Once inside the lungs, each primary bronchus branches into several smaller secondary bronchi, which then divide into even smaller passageways. Similar to the trachea, the bronchi have incomplete rings or plates of cartilage to help keep them open. The smallest passageways are the **bronchioles**, which are small tubes with no cartilage in their walls; instead, the walls are made of smooth muscle. The respiratory mucosa in the bronchioles is comprised of cuboidal instead of columnar epithelium, with fewer cilia and goblet cells, because most foreign particles have already been filtered by the mucosa of the bronchi.

Alveoli and the Respiratory Membrane

The bronchioles subdivide into smaller ***alveolar ducts*** surrounded by numerous **alveoli**. These microscopic air sacs cluster around the alveolar ducts; their shape resembles a bunch of grapes (▸ FIGURE 13.8). Alveoli are comprised of a single layer of epithelial cells, just like the capillaries that surround them. Some of these cells produce a slippery substance called ***surfactant*** that reduces surface tension within the alveoli to keep them open during breathing. Macrophages are also found in the alveoli to take care of any particles or microbes that were not trapped and expelled by the respiratory mucosa.

Oxygen and carbon dioxide are exchanged between air in the alveoli and blood in the surrounding capillaries. This exchange occurs via diffusion through the extremely thin barrier called the **respiratory membrane** formed by the alveolar and capillary

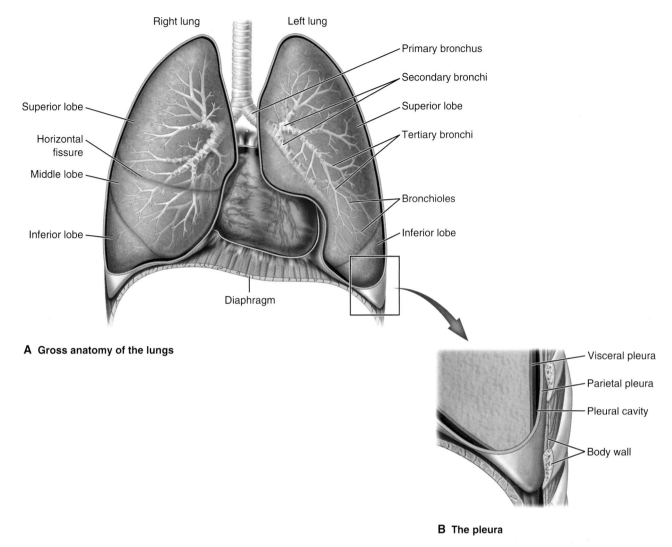

A Gross anatomy of the lungs

B The pleura

FIGURE 13.6 ▶ Lungs and pleura. A. The lungs sit on either side of the heart and are divided into lobes. **B.** The lungs are surrounded by the pleura, a double-layered serous membrane.

walls (▶ FIGURE 13.8C). Since there are literally millions of alveoli in the lungs, there is a large surface area for gas exchange. In fact, if spread out flat, the alveoli would cover an area of approximately 70 m^2 (~750 ft^2), which is a much larger surface area than that covered by the skin.

What Do You Think? 13.1

- What everyday household items could you use to construct a model of the respiratory system? List each item and the anatomical structure it would represent.

VENTILATION AND RESPIRATION

Breathing includes two interdependent processes: ventilation and respiration. **Ventilation** refers to the process of drawing air into and expelling air from the lungs. While respiration is reliant on this mechanical process, it is not the same thing. The term **respiration** refers to the passive diffusion of oxygen and carbon dioxide. Although the primary purpose of breathing is gas exchange, ventilation and respiration also support several important homeostatic functions, including temperature regulation and the maintenance of pH and fluid balances.

Ventilation

Pulmonary ventilation has two phases: **inhalation**, also called **inspiration**, which draws air into the lungs, and **exhalation**, also called **expiration**, which expels air, moisture, and heat from the lungs (▶ FIGURE 13.9). Both phases result from changes in the shape and size of the thoracic cavity. These changes, produced through the contraction and relaxation of respiratory muscles, cause variances in pressure between the inside of the

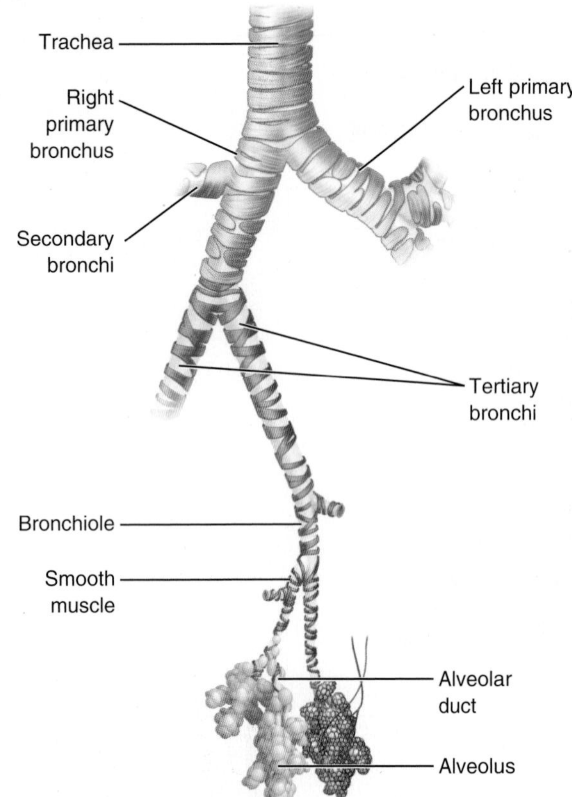

FIGURE 13.7 ▸ Bronchial tree. The bronchial tree has 25 orders of branches. Shown are several major branches from the right primary bronchus to the alveolar ducts.

lungs and the external environment. Muscles contract to enlarge the size of the thoracic cavity, thus decreasing the air pressure in the lungs. This causes air to rush from the external environment (an area of higher pressure) into the lungs, resulting in inhalation. Conversely, when these respiratory muscles relax, the size of the thoracic cavity decreases, raising the air pressure in the lungs, and exhalation occurs. Inhalation is always an active process created through the contraction of ***inspiratory muscles***, while exhalation is generally caused by the passive relaxation of the same muscles and the elastic recoil of the lungs. However, exhalation can also be actively driven through the contraction of ***expiratory muscles***, such as happens when we forcefully cough.

As seen in ▸ FIGURE 13.10, muscles that lift and expand the rib cage facilitate inhalation, while those that depress and compress it are muscles of exhalation. However, contraction and relaxation of the diaphragm, the dome-shaped muscular partition that separates the thoracic and abdominal cavities, has the most effect on the size of the thoracic cavity. When the diaphragm contracts, the center of the muscle is pulled inferiorly into a flattened position. This increases the size of the cavity, producing inhalation by decreasing the air pressure within the lungs. While the diaphragm is considered

By the Way

Pulmonary ventilation is measured by an instrument called a ***spirometer***. This simple instrument has a mouthpiece and a tube connected to a measuring device; it can quantify both the volume and rate of air flow into and out of the lungs.

the primary respiratory muscle, there are several other muscles of ventilation, listed according to their relative importance:

- Internal and external intercostals
- Scalenes
- External and internal oblique abdominals
- Sternocleidomastoid
- Rectus abdominus
- Pectoralis major and minor
- Serratus anterior

Several muscles of ventilation are key contributors to neck and upper back pain. Therapists often observe a forward head posture with rounded shoulders in those who suffer from chronic neck and back pain. Since the muscles in the upper back and posterior neck are strained and locked into eccentric contraction, this posture is often a primary cause of the client's pain. In addition to general strain, this posture has a negative impact on ventilation. The scalenes and pectoralis minor can shift roles, so instead of assisting with ventilation by elevating the ribs, they become postural muscles that feel fibrous and tense. Without the ability to efficiently expand and elevate the rib cage, breathing becomes shallow, and other muscles must work harder to ventilate the lungs. This may lead to an overall decrease in oxygen levels, which can accelerate muscle fatigue. Therefore, relieving muscle tension and fibrous buildups in the muscles of ventilation plays a key role in relieving chronic neck and back pain and in stress reduction. Additionally, since breathing is an important contributor to lymph and blood flow in the large vessels of the thoracic cavity, releasing muscle tension to improve ventialtion also assists in reducing edema, especially in the lower extremities.

Respiration

Respiration is the physiologic process of oxygen and carbon dioxide exchange. This gaseous exchange takes place in two locations:

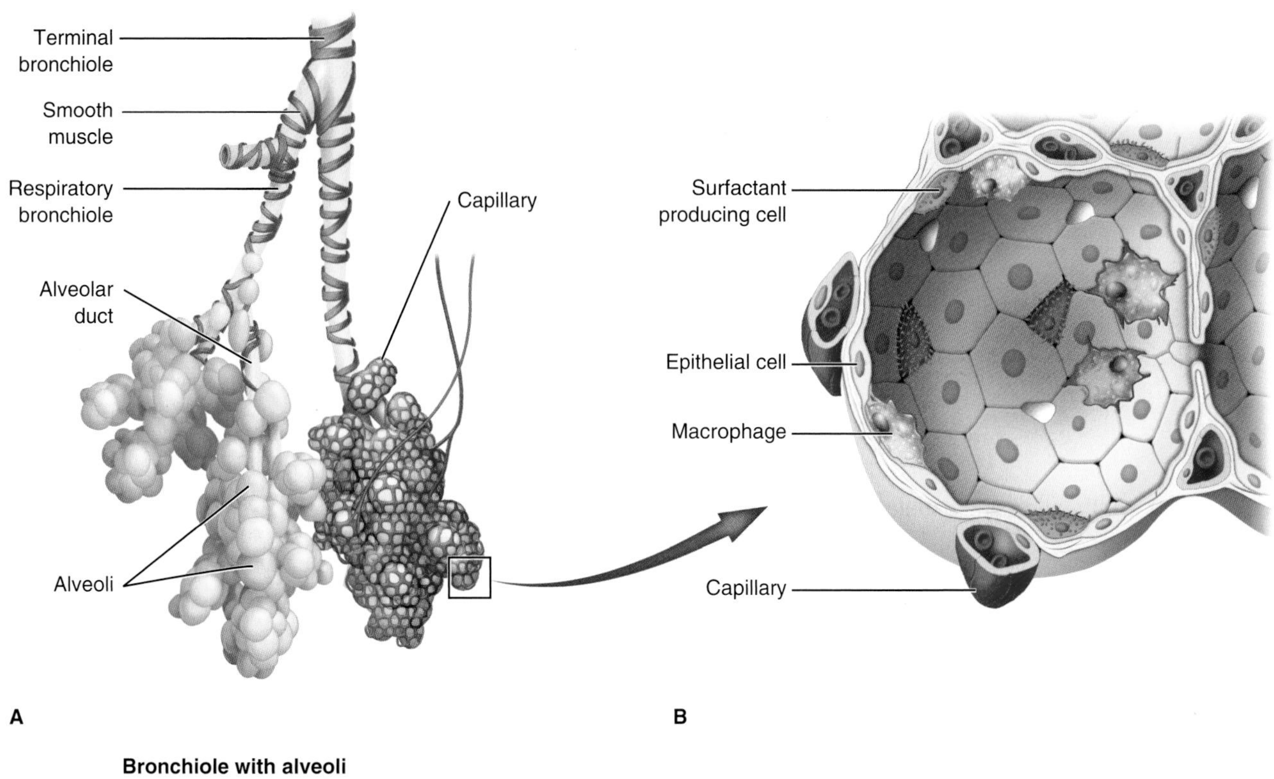

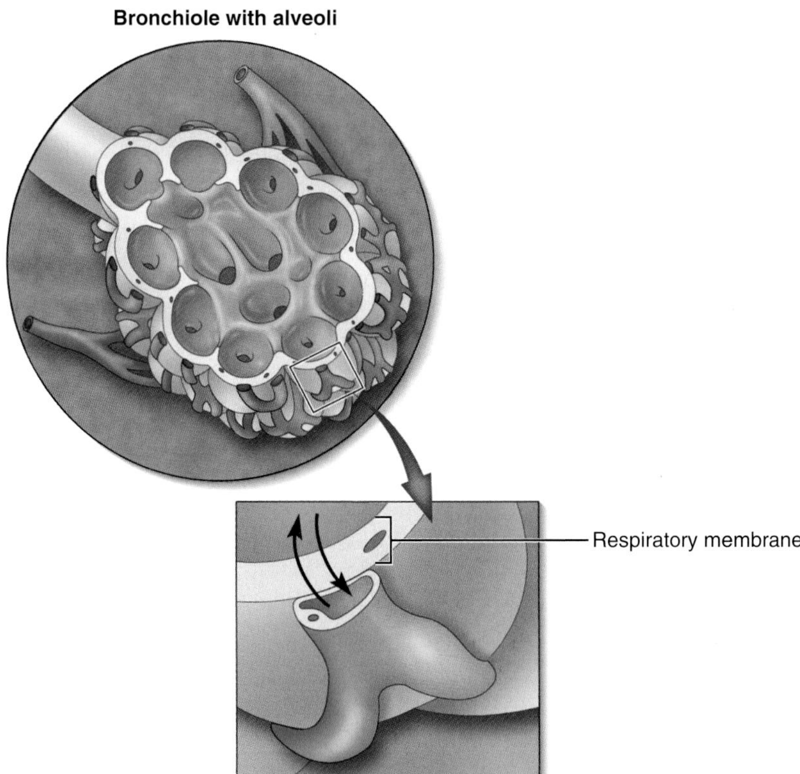

FIGURE 13.8 ▸ Alveoli and respiratory membrane. A. Alveoli are tiny air sacs at the distal end of the bronchial tree surrounded by capillaries. **B.** The wall of each alveolus consists of epithelial cells, some of which are specialized surfactant-producing cells. Macrophages are also present in alveoli. **C.** Both the alveolar and capillary walls form the respiratory membrane through which oxygen and carbon dioxide are exchanged.

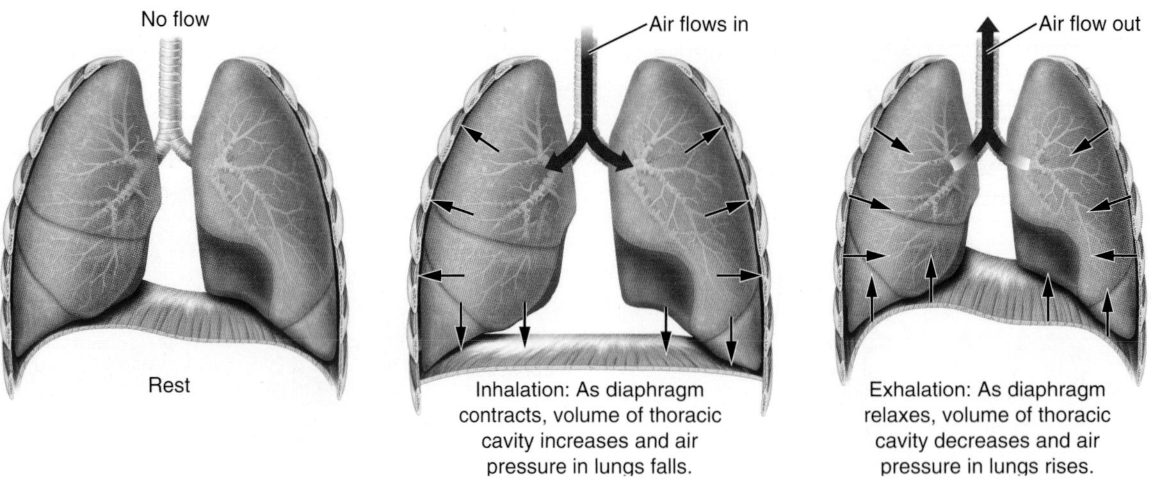
No flow
Rest
Air flows in
Inhalation: As diaphragm contracts, volume of thoracic cavity increases and air pressure in lungs falls.
Air flow out
Exhalation: As diaphragm relaxes, volume of thoracic cavity decreases and air pressure in lungs rises.

FIGURE 13.9 ▶ Inhalation and exhalation. During inhalation, the diaphragm contracts and flattens to enlarge the thoracic cavity. This decreases the internal pressure in the lungs, and air flows in. During exhalation, the diaphragm relaxes and returns to its original domed shape, decreasing the size of the thoracic cavity and expelling air from the lungs. To view a video on the diaphragm, visit http://thePoint.lww.com/Archer-Nelson.

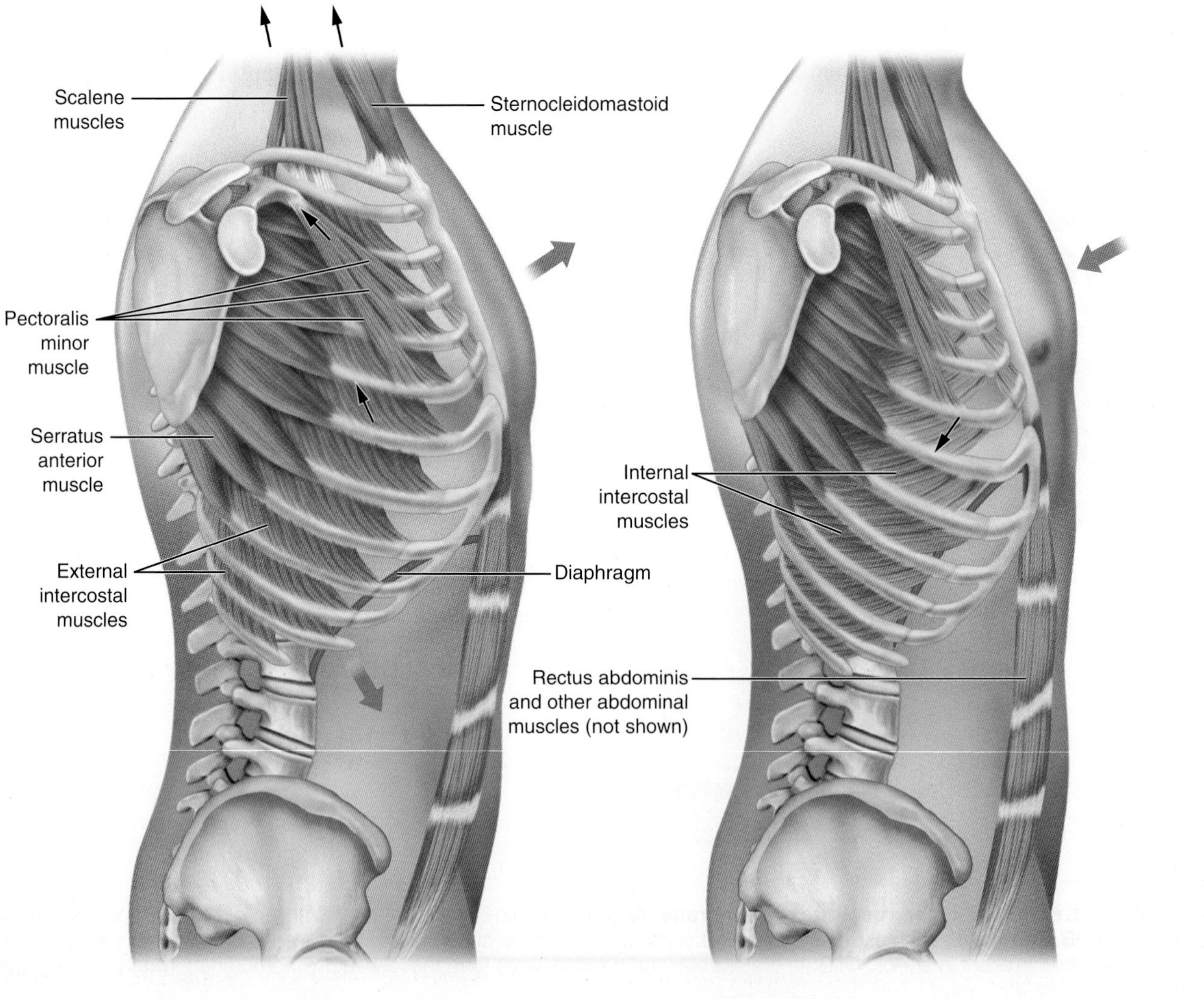

Scalene muscles
Sternocleidomastoid muscle
Pectoralis minor muscle
Serratus anterior muscle
External intercostal muscles
Diaphragm
Internal intercostal muscles
Rectus abdominis and other abdominal muscles (not shown)

A Inspiratory muscles **B Expiratory muscles**

FIGURE 13.10 ▶ Muscles of ventilation. A. When inspiratory muscles contract, they elevate and expand the rib cage, enlarging the thoracic cavity. **B.** Exhalation is generally a passive process of relaxing the inspiratory muscles. However, forced exhalation occurs when the expiratory muscles are contracted.

- Inside the lungs between alveoli and the blood
- In the capillary beds throughout the body between blood and the interstitium.

In both cases, the method of exchange is simple diffusion, and the gases are transported by binding to the hemoglobin (Hb) molecules on red blood cells or as a dissolved substance in plasma (▶ FIGURE 13.11). The exchange of oxygen (O_2) and carbon dioxide (CO_2) between the alveoli and blood is called **external respiration** because the air inside the lungs comes from the external environment. In normal circumstances, the concentration of oxygen inside the lungs is higher than the concentration inside the blood, so diffusion down the concentration gradient will move O_2 out of the alveoli and into the blood where it binds with Hb to create ***oxyhemoglobin*** (HbO_2). Simultaneously, carbon dioxide diffuses out of the blood and into the alveoli to be exhaled from the lungs.

Internal respiration is the gas exchange that occurs between the blood and body tissues. During internal respiration, O_2 needed to fuel cellular metabolism is released from the HbO_2, and the metabolic byproduct CO_2 diffuses into the bloodstream. Carbon dioxide is transported through the blood in three ways: as a dissolved gas in plasma, as bicarbonate ions (HCO_3^-) formed from CO_2 and H_2O, or as ***carbaminohemoglobin*** ($HbCO_2$) formed when carbon dioxide binds with Hb. Approximately 80% of CO_2 is transported by converting it to bicarbonate, while the other 20% is transported either as $HbCO_2$ or as a dissolved gas. While CO_2 is essential for regulating the pH of body fluids, excess levels interfere with normal cellular metabolism. Therefore, eliminating excess CO_2 is a key homeostatic function of the respiratory system.

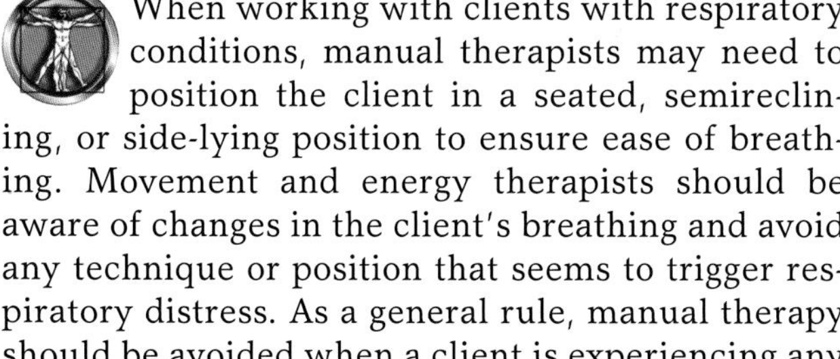

When working with clients with respiratory conditions, manual therapists may need to position the client in a seated, semireclining, or side-lying position to ensure ease of breathing. Movement and energy therapists should be aware of changes in the client's breathing and avoid any technique or position that seems to trigger respiratory distress. As a general rule, manual therapy should be avoided when a client is experiencing any type of respiratory infection or breathing difficulty.

Advanced training is available in the technique known as ***postural drainage***, a type of positioning massage that helps to relieve lung congestion. When respiratory conditions such as chronic bronchitis or asthma cause mucus congestion in the lungs, therapists trained in postural drainage utilize a series of specialized positions to methodically loosen and drain the lobes of the lungs. ▶ Figure 13.12 shows two positions that are used, along with others in a specific sequence, to drain the lungs. Therapists often use a combination of tapotement and vibration over the back and chest to loosen mucus buildup before positioning for drainage.

What Do You Think? 13.2

- Why do you think some muscles, such as the serratus anterior and pectoralis minor, do not contribute as much to ventilation as do the scalenes and intercostals?
- What could you do to observe a client's breathing during a manual therapy session? How do you think you might use that information?
- How would you alter a standard manual therapy session for clients with breathing issues

CONTROL AND REGULATION OF BREATHING

Regular and rhythmic breathing is so subconscious that the complexities of the process are rarely, if ever, considered. While inhaling is a simple mechanical process of contracting the muscles of respiration, the rhythm, rate, and depth of breathing is controlled and regulated through several physiologic mechanisms.

- ***Respiratory centers***—As discussed in Chapter 7, two regions of the brain stem contain the respiratory reflex centers: the medulla oblongata and pons. Two small groups of self-excitatory neurons in the medulla oblongata are responsible for setting the basic rate and rhythm of breathing. A region of the pons, the ***pneumotaxic area***, regulates the breathing rate by coordinating the stimulation of the inspiration center in the medulla (▶ FIGURE 13.13). In adults, a normal resting respiratory rate is 12 to 18 breaths per minute.
- ***Respiratory nerves***—The respiratory centers in the brain stem send commands to the inspiratory muscles via two key nerves. A primary role is played by the ***phrenic nerve***, which innervates the diaphragm. The ***intercostal nerves*** send signals to both groups of intercostal muscles.
- ***Stretch receptors***—There are pulmonary stretch receptors located throughout the lungs, mostly in the walls of the bronchioles and alveoli. They are specialized mechanoreceptors that send signals that inhibit the inspiration areas of the brain stem when the lungs are full and stretched. By inhibiting the inspiration centers, exhalation can occur.
- ***Chemoreceptors***—The strongest influence on the respiratory centers comes from chemoreceptors located within the walls of the carotid and aortic arteries. These receptors, known as the ***carotid*** and ***aortic bodies***, stimulate the respiratory centers via cranial nerves IX (glossopharyngeal) and X (vagus) and are sensitive to changes in blood levels of oxygen and carbon dioxide. The respiratory centers are

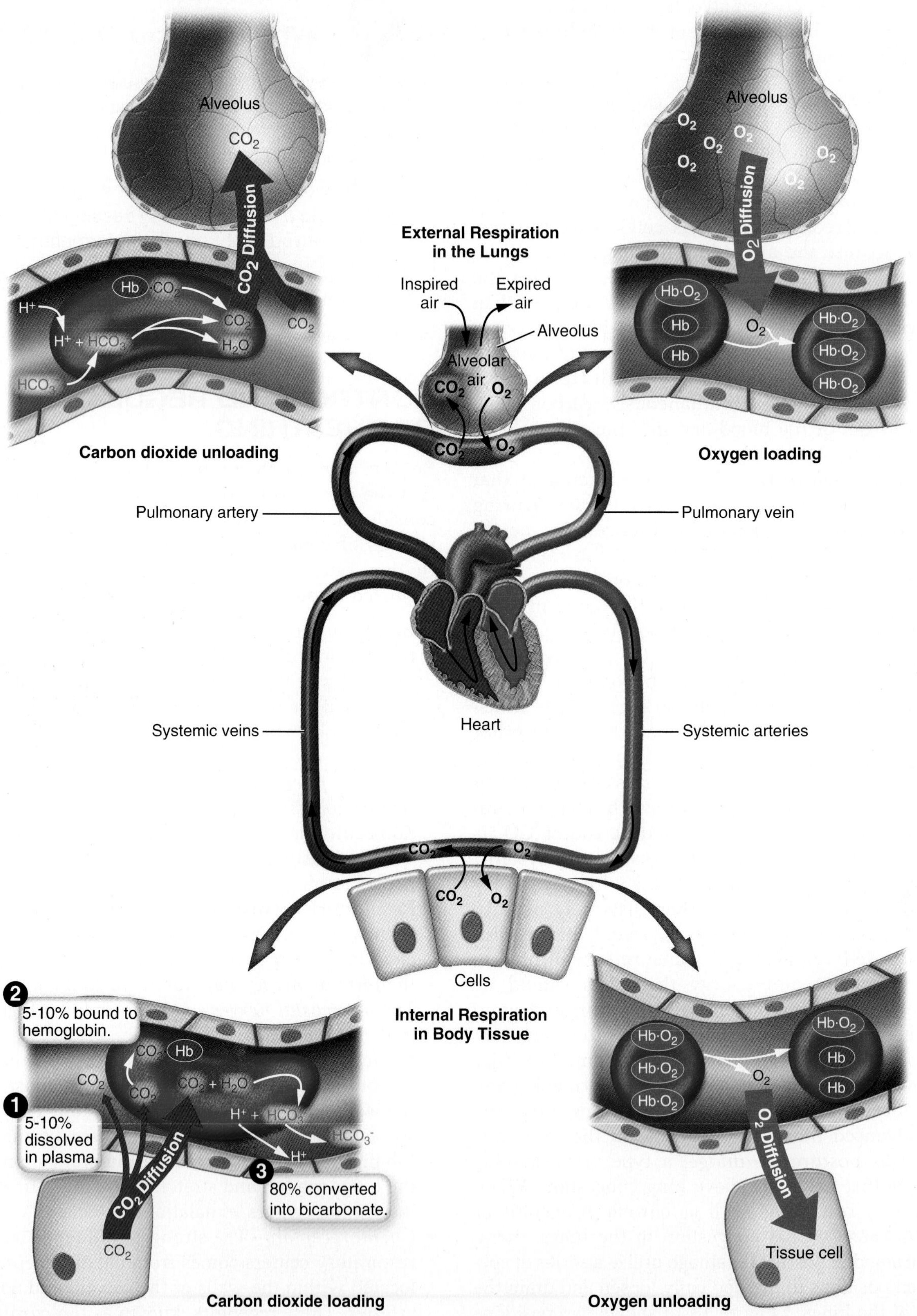

FIGURE 13.11 ▸ Internal and external respiration. External respiration is the exchange of gases between the alveoli and surrounding capillaries. Internal respiration occurs between the blood and tissues of the body. Oxygen and carbon dioxide are exchanged via diffusion during both processes. To view an animation on oxygen transport, visit http://thePoint.lww.com/Archer-Nelson.

Pathology Alert Chronic Obstructive Pulmonary Disease

There are many chronic conditions involving the lungs and/or bronchial tree that cause respiratory discomfort and distress. These include asthma, chronic bronchitis, and emphysema, which are categorized as types of ***chronic obstructive pulmonary disease* (*COPD*)** because they are associated with structural changes that block air flow or the exchange of air in the lungs.

Asthma is characterized by attacks of shortness of breath, wheezing, and coughing caused by reversible inflammation and secondary spasm of the smooth muscle of the bronchioles. The bronchioles of asthma patients are chronically inflamed and hypersensitive to allergens and other irritants. When irritated, the bronchioles sympathetically dilate and then reflexively spasm, making it difficult to breathe. Anxiety and stress can exacerbate the symptoms of an attack or serve as an additional trigger. Chronic irritation of the bronchial tubes due to asthma, allergies, multiple infections, smoking, or environmental pollutants can lead to ***chronic bronchitis***. This long-term inflammation of the lining of air passageways is characterized by a chronic cough and increased susceptibility to respiratory tract infections. Chronic bronchitis can lead to or occur together with ***emphysema***, a condition in which the walls of the alveoli first become less elastic then eventually break down, fuse together, and thicken into fibrous connective tissue. The breakdown of alveoli decreases the overall surface area for gas exchange and requires great effort to obtain enough oxygen and to expel air from the lungs. Therapists working with clients with COPD may need to adjust the duration of the session and position the client to ensure that they can breathe comfortably.

stimulated by low O_2 levels and high levels of CO_2. The inspiration respiratory centers are far more sensitive to high carbon dioxide levels, and thus, the signal for inspiration is more vigorous. This explains why we generally can't hold our breath beyond a minute or two.

While the mechanisms of breathing are all reflexive, communication between the brain stem and cerebral cortex enables the conscious control of breathing. Thoughts and emotions can create shifts in the depth, rate, and rhythm of breathing. Other influences include pain, extreme shifts in blood pressure, and irritation of the airway.

Since there is such a strong link between emotional state and breathing, in many manual and movement therapies and in some spiritual practices, conscious breathing is used to promote relaxation. Slow, deep, rhythmic breathing is extremely calming and nourishes the mind and spirit as it provides the body with vital oxygen and relieves it of carbon dioxide. Integrating some type of deep breathing practice into daily life can provide an opportunity to calm and renew the body, mind, and spirit. We are literally *inspired* and more creative when we make time for conscious deep breathing.

AGING AND THE RESPIRATORY SYSTEM

As the body ages, lung tissue becomes less elastic, the number of capillary networks in alveoli decrease, and the respiratory centers become less responsive

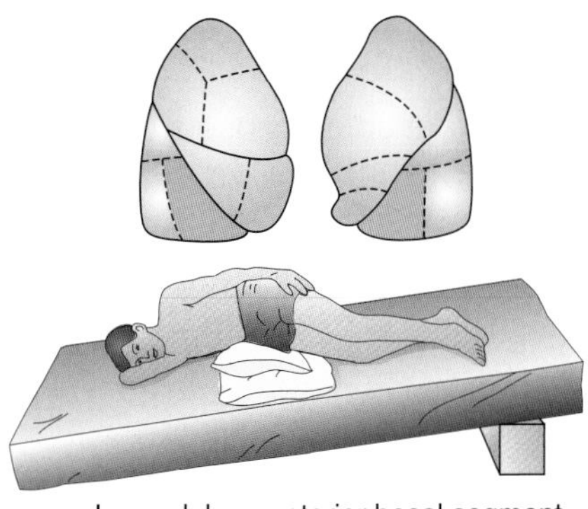

A Lower lobes, anterior basal segment

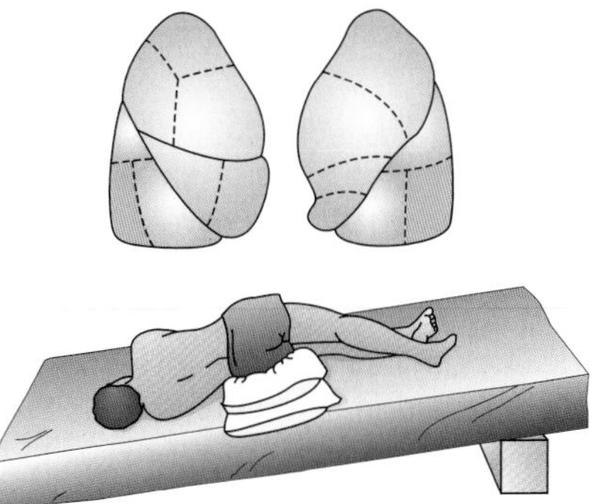

B Upper lobes, lateral basal segment

FIGURE 13.12 ▸ Postural drainage techniques. Special training is required in the sequence and positions of postural drainage, a technique used to relieve mucus congestion in the lungs. **A** and **B** show just two of the positions used for drainage of particular lung regions. To view an animation on hemostasis, visit http://thePoint.lww.com/Archer-Nelson.

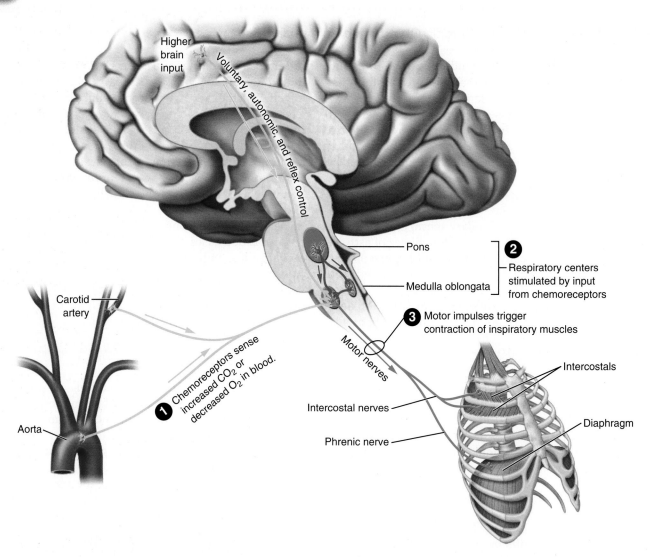

FIGURE 13.13 ▸ Regulation of breathing. The rate, rhythm, and depth of breathing is controlled by respiratory centers in the pons and medulla oblongata based on input from chemoreceptors in the carotid and aortic arteries and stretch receptors in the lungs. These brain stem centers stimulate muscles of inhalation via the phrenic and intercostal nerves.

By the Way

When a person becomes highly excited or distressed, the respiratory rate and force of exhalation often increases. This ***hyperventilation*** (overbreathing) leads to a reduction in the level of CO_2 in the bloodstream as large quantities of carbon dioxide are exhaled. The diminished CO_2 levels mean the stimulus for inhalation is minimal and if not corrected, the individual will pass out from lack of O_2. Treatment for hyperventilation requires reinhaling exhaled air by placing a paper bag over the nose and mouth and breathing into it. This restores CO_2 to normal levels, which stimulates inhalation and returns breathing to its normal rate and depth.

What Do You Think? 13.3

- What could cause a client's breathing to become rapid and shallow during a manual therapy session?
- Why do you think the strongest stimulus for inhalation is related to CO_2 levels?

to changes in oxygen and carbon dioxide levels. Age-related changes in the skeletal system such as osteoporosis and calcification of the costal cartilage make the thoracic cavity more rigid, which decreases ventilation capacity. This rigidity of the chest wall may cause the respiratory muscles to weaken, further diminishing respiratory function. Older adults also become more susceptible to respiratory diseases, including pneumonia and bronchitis, because macrophage activity in the lungs decreases and the number and activity of cilia declines. This makes it difficult to expel pathogens trapped in the mucous membrane. Age-related decline in respiratory function can be slowed or minimized through positive choices such as not smoking, maintaining a healthy body weight, and engaging in an active lifestyle.

SUMMARY OF KEY POINTS

- The primary function of the respiratory system is to deliver oxygen to and remove carbon dioxide from body tissues. Breathing also helps regulate blood pH, fluid balance, and temperature.
- The respiratory system has two divisions: the upper respiratory tract formed by the nose and nasal cavity, pharynx, and larynx and the lower respiratory tract, which consists of the trachea, lungs, and bronchial tree (including alveoli).
- The respiratory mucosa serves to warm and humidify air before it enters the lungs and also traps foreign particles. Cilia embedded in the mucosa move the particle-laden mucus to the pharynx, where it is swallowed or expelled.
- The trachea splits into the right and left primary bronchi that carry air into the lungs. Each bronchus has multiple smaller branches and bronchioles that terminate in clusters of tiny air sacs called alveoli.
- Alveolar and capillary walls form the respiratory membrane, the point where the exchange of oxygen and carbon dioxide occurs.
- The pleura, a continuous serous membrane, covers the lungs and lines the thoracic cavity. The parietal pleura lines the inner surface of the thoracic cavity, while the visceral pleura surrounds the lungs.
- Ventilation is the mechanical process of breathing: drawing air into and expelling it from the lungs. Inhalation is an active process created by skeletal muscle contractions that elevate the ribs and expand the thoracic cavity. The muscles involved include the diaphragm, external intercostal, and scalene muscles. Exhalation is usually a passive process of relaxing the muscles of ventilation and reducing the size of the thoracic cavity. Other muscles of ventilation include the obliques abdominals, sternocleidomastoid, pectoralis minor and major, rectus abdominus, and serratus anterior.
- Respiration is the exchange of oxygen and carbon dioxide. External respiration occurs in the lungs, where oxygen moves from alveoli into the blood, and carbon dioxide moves from the blood into the alveoli. Internal respiration occurs between the capillaries and tissues throughout the body. During internal respiration, O_2 moves from blood into the tissue, while CO_2 moves from tissue into the blood.
- During external respiration, oxygen bonds with the Hb molecules in red blood cells to form HbO_2. During internal respiration, a small amount of CO_2 binds with Hb to form $HbCO_2$, while most CO_2 is converted into bicarbonate ions for transport back to the lungs.
- Breathing rate and rhythm is controlled and regulated by the respiratory centers in the medulla oblongata and pons. The medulla oblongata centers are responsible for setting the basic rhythm of breathing, while those in the pons assist with rhythm and set the rate. Two types of receptors provide sensory information to these centers: stretch receptors in the walls of the bronchioles and alveoli, and chemoreceptors in the carotid and aortic bodies of those arteries.

REVIEW QUESTIONS

Short Answer

1. Name the skeletal muscles responsible for breathing and briefly describe the role of each.
2. What structures form the upper respiratory tract?
3. Describe the location and function of the respiratory mucosa, the respiratory membrane, and the two types of pleura.
4. Describe the location and function of the two respiratory centers of the brain.

Multiple Choice

5. What is a functional purpose of the nose and sinuses?
   a. trap pathogens and control air volume
   b. warm and moisten air before it enters lungs
   c. balance the pH of the air
   d. provide an easy method of discharging mucus from the system
6. Which respiratory structure houses the structures of voice production in addition to functioning as a passageway for air?
   a. pharynx
   b. sinus
   c. larynx
   d. trachea
7. What is the anatomic name for the structure commonly called the windpipe?
   a. nasopharynx
   b. laryngopharynx
   c. larynx
   d. trachea
8. Which term describes the process of moving air into and out of the lungs?
   a. ventilation
   b. respiration
   c. inspiration
   d. pulmonary process
9. The exchange of oxygen and carbon dioxide that occurs between the air and capillaries inside the lungs is called
   a. ventilation
   b. internal respiration
   c. external respiration
   d. exhalation
10. Which transport mechanism is used for the exchange of oxygen and carbon dioxide between the lungs, blood, and body tissues?
    a. osmosis
    b. active transport
    c. filtration
    d. diffusion
11. Which of these statements correctly describes the gas exchange of internal respiration?
    a. oxygen moves out of the blood and carbon dioxide moves into it
    b. carbon dioxide and oxygen move into the blood
    c. carbon dioxide moves into the lungs and oxygen into the blood
    d. active transport moves oxygen into the tissues and carbon dioxide into blood
12. What change occurs in the thoracic cavity when the diaphragm contracts?
    a. decreases in size
    b. increases in size
    c. the rib cage elevates
    d. the rib cage is depressed
13. Which of these skeletal muscles is considered *the* primary muscle of inspiration?
    a. diaphragm
    b. internal intercostals
    c. pectoralis minor
    d. scalenes
14. How is oxygen transported in the blood?
    a. as a dissolved gas in the plasma
    b. O_2 binds with plasma proteins
    c. as oxyhemoglobin on red blood cells
    d. as part of bicarbonate ions
15. The respiratory centers of the brain respond to stimulus from which types of receptors?
    a. nociceptors and mechanoreceptors
    b. stretch receptors and Golgi bodies
    c. thermoreceptors and chemoreceptors
    d. chemoreceptors and stretch receptors

16. Which condition provides the strongest stimulus to inhale?
    a. low levels of oxygen in the blood
    b. filling the lungs with air
    c. high levels of oxygen in the lungs
    d. high blood levels of carbon dioxide
17. Where are the chemoreceptors for breathing located?
    a. inside the walls of bronchioles and alveoli
    b. on outer surface of the pleural and respiratory membrane
    c. inside the walls of the carotid and aortic arteries
    d. on the cell membrane of red blood cells
18. In healthy adults at rest, what is the average number of breaths per minute?
    a. 12 to 18
    b. 6 to 10
    c. 20 to 30
    d. 30 to 35
19. Carbaminohemaglobin is formed during
    a. ventilation
    b. external respiration
    c. internal respiration
    d. pulmonary obstruction
20. Which nerves are the primary innervations for the muscles of ventilation?
    a. vagus and thoracic
    b. intercostal and vagus
    c. hepatic and phrenic
    d. phrenic and intercostal

14 The Digestive System

LEARNING OBJECTIVES

Upon completion of this chapter you will be able to:

1. Discuss the importance of the digestive system as it relates to the practice of manual therapy.
2. List and explain the functions of the digestive system.
3. List the four layers of the gastrointestinal (GI) tract and describe the functional characteristics of each.
4. Name and locate the major organs of the GI tract and describe the functions of each.
5. Name and locate the accessory organs of the digestive system and describe the general functions of each.
6. Explain the basic metabolic processes of anabolism and catabolism in regard to macronutrients.
7. Discuss digestive system changes that commonly occur as the body ages.

Although there is a great deal of controversy among scientists about the effects of ingested food on the brain, no one denies that you can change your cognition and mood by what you eat."

ARTHUR WINTER
Smart Food

KEY TERMS

digestion (di-JES-chun)
enteric nervous system (ENS) (en-TER-ik NUR-vus SIS-tem)
ingestion (in-JES-chun)
mastication (mas-tih-KA-shun)
metabolic rate (meh-tah-BAL-ik)
motility (mo-TIL-ih-te)
mucosa (mu-KO-sah)
muscularis (mus-ku-LAR-is)
peristalsis (per-eh-STAHL-sis)
peritoneum (per-ih-tah-NE-um)
segmentation (seg-men-TA-shun)
serosa (seh-RO-sa)
submucosa (SUB- mu-ko-sah)

Primary System Components

- **Gastrointestinal tract**
  - Mouth
  - Pharynx
  - Esophagus
  - Stomach
  - Small intestine
  - Large intestine
- **Accessory digestive organs**
  - Teeth and tongue
  - Salivary glands
  - Liver
  - Gallbladder
  - Pancreas

Primary System Functions

- Digests food to provide nutrients for cellular metabolism
- Eliminates solid wastes from the body
- Assists in regulating body temperature by generating heat as a byproduct of metabolism

The digestive system consists of a convoluted pathway of chambers and tubes called the gastrointestinal (GI) tract plus several accessory digestive organs. This amazing system processes the nutrients in food and makes them available to the body for immediate use, storage, or elimination. Like the skin, the GI tract serves as both a barrier and an interface between the external and internal environment, filtering out unwanted substances and absorbing essential nutrients. The digestive system is populated with millions of neurons that sense and regulate its functions, and it produces numerous peptides that can stimulate responses in the brain and spleen. Perhaps this is why the gut seems to have its own wisdom; an intelligence based on peptide communication between the nervous and digestive systems. This "gut instinct" can create a physical awareness of emotions such as fear, anxiety, or excitement before there is conscious recognition by the brain.

In this chapter, the general functions of the digestive system and the specific anatomy and physiologic roles of its components are explained. The body's metabolic processes and nutritional needs are also discussed. Although most manual therapies have limited impact on the digestive system, several zone or meridian therapies such as reflexology and shiatsu have demonstrated their ability to stimulate change in specific digestive organ functions or systemic processes. Manual therapists will appreciate the discussions on general nutritional considerations, digestive pathologies, and possible physiologic effects of other therapeutic interventions.

FUNCTIONS OF THE DIGESTIVE SYSTEM

The digestive system is designed to break down food into particles that can be used or eliminated from the body. Its involuntary functions are controlled by an intrinsic set of neurons that form the **enteric nervous system (ENS)** or "brain of the gut" (FIGURE 14.1). Made up of approximately 100 million neurons, the ENS senses the nature of the contents and degree of distension of the GI tract. This sensory information is communicated to motor neurons that reflexively activate or inhibit the secretion of glandular epithelium and regulate the strength and frequency of smooth muscle contractions in the GI tract. These actions occur independently from the functioning of the rest of the nervous system. Although not technically part of the autonomic nervous system, the ENS can also be regulated through sympathetic or parasympathetic motor stimuli.

The six functions of the digestive system are

- **Ingestion**—The process of taking in foods and liquids (eating and drinking) is known as ingestion.
- **Secretion**—The inner lining of the digestive tract and accessory organs produce and secrete mucus and a variety of enzymes, acids, and buffer chemicals. These substances facilitate movement, protect the GI tract, and break down foods into usable basic nutrients.
- **Digestion**—Food is broken down into usable molecules through the process of digestion. Actions such as chewing and stomach churning are referred to as ***mechanical digestion*** because they physically grind or break apart food particles. In contrast, ***chemical digestion*** refers to the processes that use enzymes and acids to break down substances into smaller molecules.
- **Motility**—Motility refers to movement. In the GI tract, this action is a combination of mixing and forward movement. The contraction of smooth muscle within the walls of the GI tract mixes food particles with digestive secretions and propels it forward.
- **Absorption**—Once food has been reduced to its smallest molecular form, nutrients are then absorbed through the wall of the GI tract into the bloodstream or lymph.

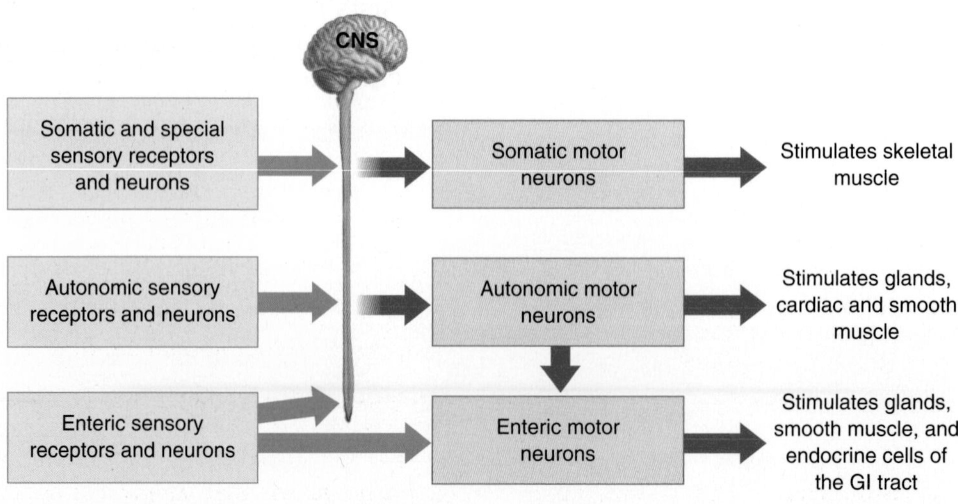

FIGURE 14.1 ▸ Enteric nervous system. This "brain of the gut" can function independent of the central nervous system (CNS), as well as in response to autonomic nervous system regulation.

- **Elimination**—Unabsorbed substances, along with bacteria, dead cells from the lining of the GI tract, and indigestible materials are eliminated from the body. Compacted solid waste, or ***feces***, is eliminated through the process of **defecation**.

What Do You Think? 14.1

- Why do you think people tend to eat more during the winter months than in the summer?
- What everyday analogy would you use to demonstrate each process of the digestive system?

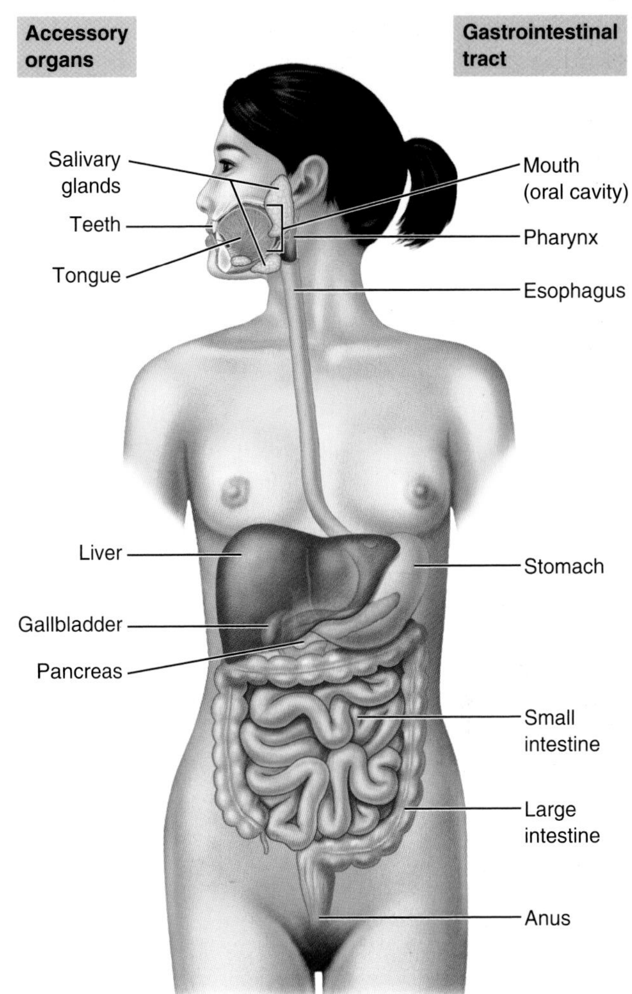

FIGURE 14.2 ▶ Structures of the digestive system. The digestive system includes the organs of the GI tract and several accessory organs.

GASTROINTESTINAL TRACT

The digestive organs involved in the transport and processing of food make up the **gastrointestinal (GI) tract**. The GI tract begins at the mouth and continues through the pharynx, esophagus, stomach, small intestine (SI), large intestine, and ends at the anus, where waste products are excreted from the body (▶ FIGURE 14.2). Each organ carries out specialized digestive or absorption processes to provide nutrients necessary for cellular metabolism.

Layers of the Digestive Tract

Even though a number of organs make up the GI tract, each has the same four layers of tissue forming the wall of the tract (▶ FIGURE 14.3); they are listed below from the innermost to the outer layer.

1. **Mucosa**—The innermost layer is a mucous membrane whose secretions coat food particles to help propel them through the tract and protect the wall from irritation or damage from enzymes and acids. This layer has millions of chemoreceptors and stretch receptors that sense the contents and distension of the GI tract to stimulate secretion and motility. Recall that the Peyer patches (a type of mucosa-associated lymphoid tissue [MALT]) are located within the mucosa of the small intestine.
2. **Submucosa**—This areolar connective tissue layer binds the mucosa to the muscular layer below. It contains blood and lymphatic vessels, as well as secretory glands formed by deep pockets of mucosal epithelium. This layer also contains motor neurons that form the ***submucosal nerve plexus*** of the ENS that help regulate mucosal function.
3. **Muscularis**—Two distinct types of smooth muscle fibers make up the third layer of the GI tract. The inner fibers are arranged circularly, while the outer fibers are longitudinally aligned. Contraction of these muscle fibers mechanically digests, mixes, and propels food particles along the GI tract. Action of the muscularis is controlled by the motor neurons of the ***myenteric nerve plexus*** of the ENS. In addition to smooth muscle, skeletal muscle is also present in organs such as the mouth, pharynx, and anus, where voluntary control of movement is important.
4. **Serosa**—Also called the ***visceral peritoneum***, this outer layer of the digestive tract is a serous membrane. Together with the ***parietal peritoneum*** that lines the abdominal cavity, the serosa secretes serous fluid into the ***peritoneal space*** between the layers, which allows the GI tract to slide and glide against the other abdominal organs.

The **peritoneum** is a thin, translucent serous membrane that lines the walls of the abdominal cavity and encloses the internal organs (▶ FIGURE 14.4). The nerves, blood, and lymphatic vessels that supply the abdominal organs are found within the peritoneum. The **mesentery** is a double

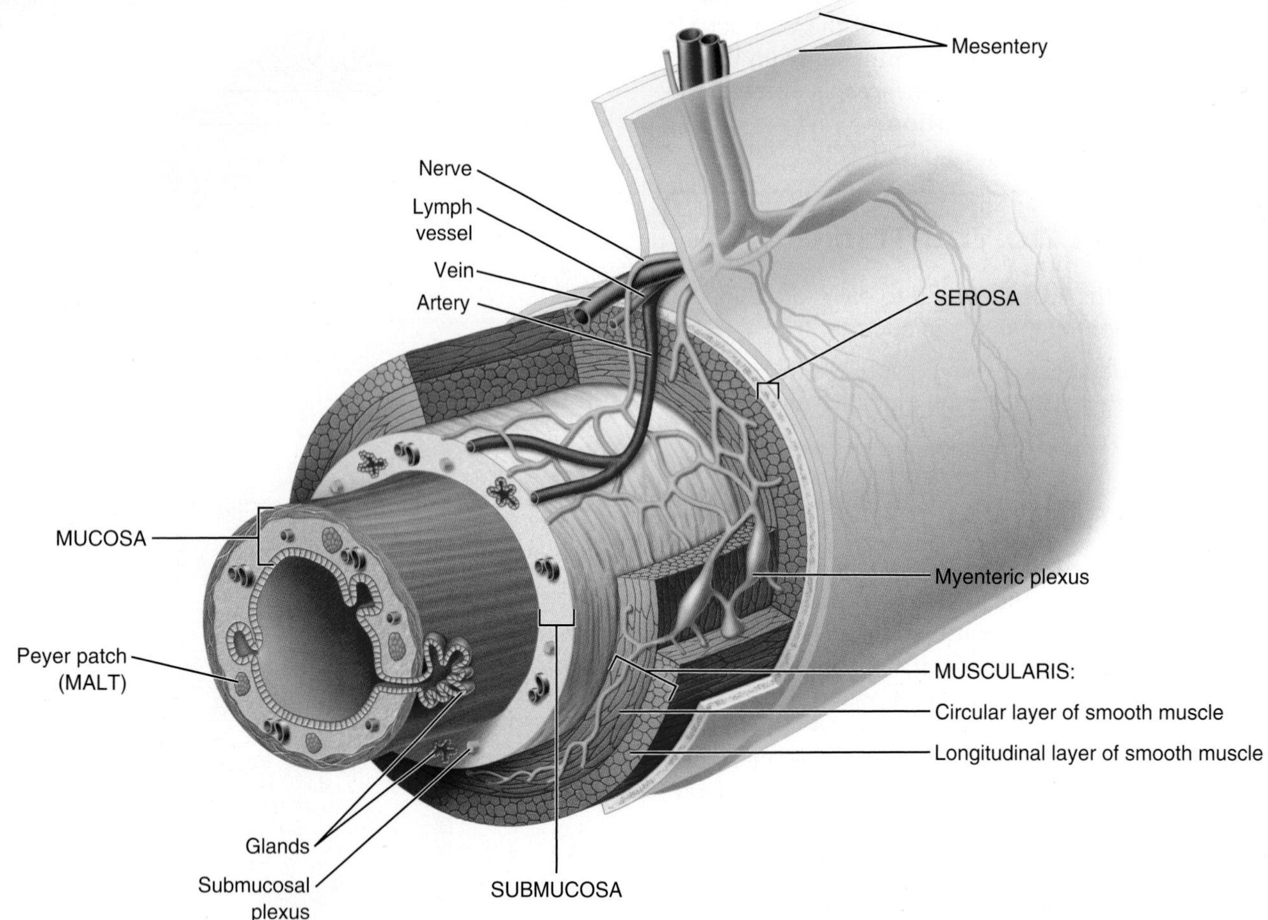

FIGURE 14.3 ▸ Layers of the GI tract. The four layers of the GI tract wall include the mucosa, submucosa, muscularis, and serosa. The neuronal plexuses of the ENS are located within the submucosa and muscularis.

fold of peritoneum attached to the posterior side of the abdominal cavity that supports most of the small intestine. An extension of the peritoneum, the **greater omentum**, is a fatty apron that hangs down over the front of the abdominal cavity, forming a protective layer that cushions and insulates the organs within. The greater omentum extends from the inferior portion of the stomach downward into the pelvic portion of the cavity, then loops back upward and attaches to the transverse colon. As one of the body's preferred storage sites for fat, the greater omentum thickens and enlarges as an individual gains weight, producing a protruding belly. The **lesser omentum** is a small extension of this tissue that connects the stomach to the underside of the liver and supports the hepatic vessels.

Mouth and Pharynx

Food enters the GI tract through the mouth, where it is chewed. **Mastication** is the mechanical process of chewing, in which pieces of food are smashed or broken into smaller pieces. While the teeth play the primary role, the cheek and lip muscles and the tongue move and position food so it can be thoroughly chewed. For this reason, the teeth and tongue are considered accessory organs of the digestive system. During mastication, food is also mixed with saliva secreted by the salivary glands (▸ FIGURE 14.5). **Saliva** serves as a lubricant to help food pass through the GI tract and begins the process of chemical digestion. It contains the enzyme **amylase** that breaks down starches, a type of carbohydrate. You can do a quick experiment to demonstrate this process by holding a small cube of bread in your mouth for a few minutes. Do not chew or swallow. You'll notice that saliva begins to chemically digest the bread, breaking it down into a slightly sweet-tasting basic starch. This experiment also demonstrates that the taste buds on the tongue and the olfactory receptors in the roof of the mouth are stimulated by the presence of food. Recall that there is a direct link between olfaction and the limbic system, which is why eating can either *stimulate* an

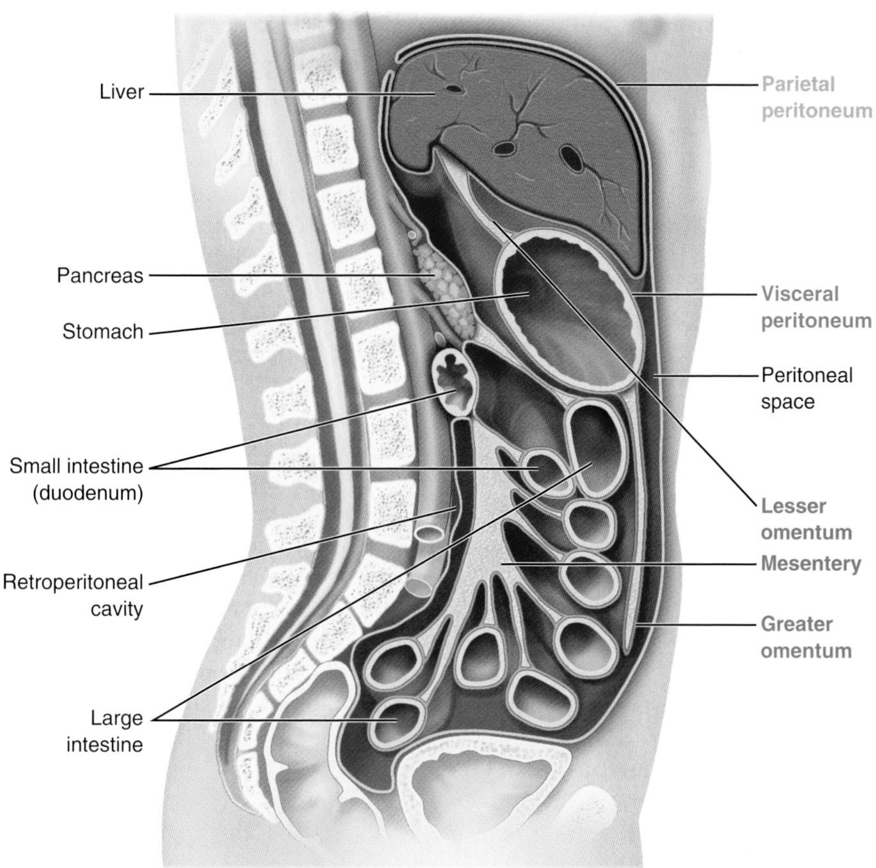

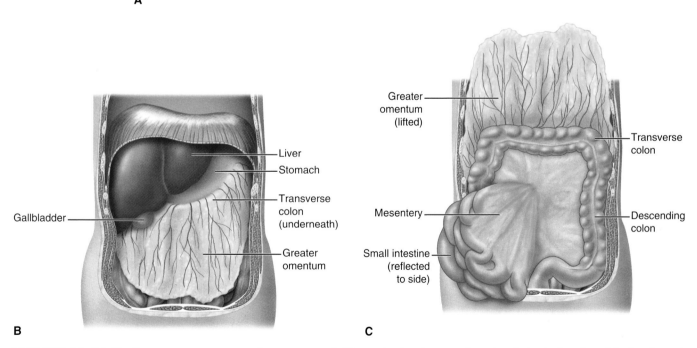

FIGURE 14.4 ▸ Peritoneum, omentum, and mesentery. A. The parietal peritoneum lines the abdominal cavity, while the visceral peritoneum surrounds the abdominal organs. Notice that the duodenum and pancreas lie posterior and outside the peritoneum (retroperitoneal). Double folds of peritoneum form additional supportive and protective membranes: the mesentery and greater and lesser omentum. **B.** Anterior view. The greater omentum is a fatty extension of the peritoneum. **C.** With the greater omentum lifted, the mesentery that supports the small intestine can be seen. To view a video on the abdominal cavity, visit http://thePoint.lww.com/Archer-Nelson.

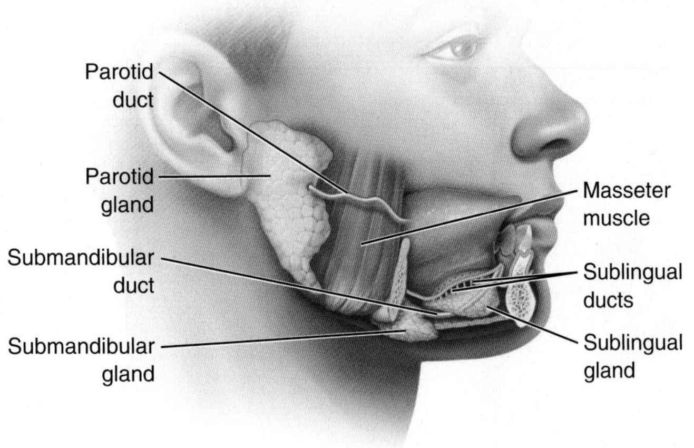

FIGURE 14.5 ▶ Salivary glands. There are three salivary glands: the sublingual gland beneath the tongue, the submandibular gland under the mandible, and the parotid gland anterior and inferior to the ear, superficial to the masseter muscle.

emotional response or *be* the active response to powerful emotions and feelings.

Once food has been chewed, the mixture of food and saliva, now called a ***bolus***, is swallowed. Swallowing, or ***deglutition***, is an involuntary reflex that moves the bolus through the pharynx and into the esophagus. During swallowing, the soft palate and uvula (the small mass of tissue hanging from the soft palate at the back of the mouth) are raised to prevent food from entering the nasopharynx, and the epiglottis is pulled over the opening of the trachea.

Esophagus

The esophagus is the muscular tube just posterior to the trachea that connects the pharynx to the stomach. It serves as a passageway for food, and no digestion takes place here. Lubricated by mucus secretions, the bolus moves through the esophagus via **peristalsis**, a series of wave-like contractions and relaxation in the smooth muscle layer, which propels food along the GI tract (▶ FIGURE 14.6). Although peristalsis begins in the esophagus and continues throughout the GI tract, it is most apparent in the lower half of the stomach and throughout the small intestine. Vomiting is the nontechnical term for reverse peristalsis, a protective reflex of the GI tract. If the lining of the stomach becomes irritated, as happens with bacterial food poisoning or viral infections like the flu, receptors in the stomach wall signal the ***emetic reflex center*** in the medulla oblongata. The medulla signals smooth muscles in the stomach to contract in a reverse peristalsis sequence, while simultaneously signaling involuntary contractions of the diaphragm and abdominal muscles that forcefully eject the stomach contents.

Stomach

Food passes through the esophagus and enters the **stomach**, the saclike muscular organ located in the upper left quadrant of the abdominal cavity (▶ FIGURE 14.7). The stomach has three regions: a large proximal portion called the **fundus**, a central **body**, and a narrow distal portion, the **pylorus**. While an empty stomach is smaller than a closed fist, its inner mucosal

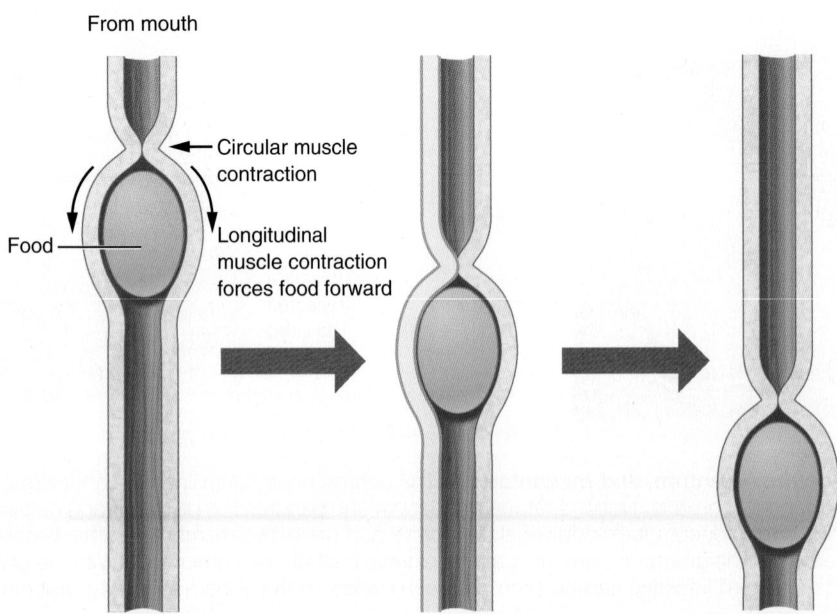

FIGURE 14.6 ▶ Peristalsis. Coordinated, wave-like contractions of both the circular and longitudinal muscle layers propel food forward along the GI tract.

Pathology Alert Heartburn

When acidic gastric juices back up in the lower end of the esophagus and cause irritation, it can produce a painful burning sensation in the chest under the sternum. Often called ***heartburn***, this condition is clinically known as cardialgia pyrosis and more commonly referred to as ***acid reflux***. Pain is often accompanied by gas, gastric bloating, or a bitter taste in the mouth. While heartburn can be caused by too much fat in the diet, consuming spicy foods, obesity, or even pregnancy, it can also indicate the presence of an ulcer or a ***hiatal hernia***, a condition that occurs when a superior portion of the stomach gets irritated or trapped at the esophageal opening in the diaphragm. Frequent bouts of esophageal irritation can lead to damage and structural changes in the esophagus, a condition known as ***gastroesophageal reflux disease (GERD)***. When working with clients who suffer from frequent heartburn or GERD, manual therapists may need to position them in a semireclined position, since lying flat can exacerbate heartburn symptoms. Direct manual pressure over any tender areas in the diaphragm and abdomen should also be avoided.

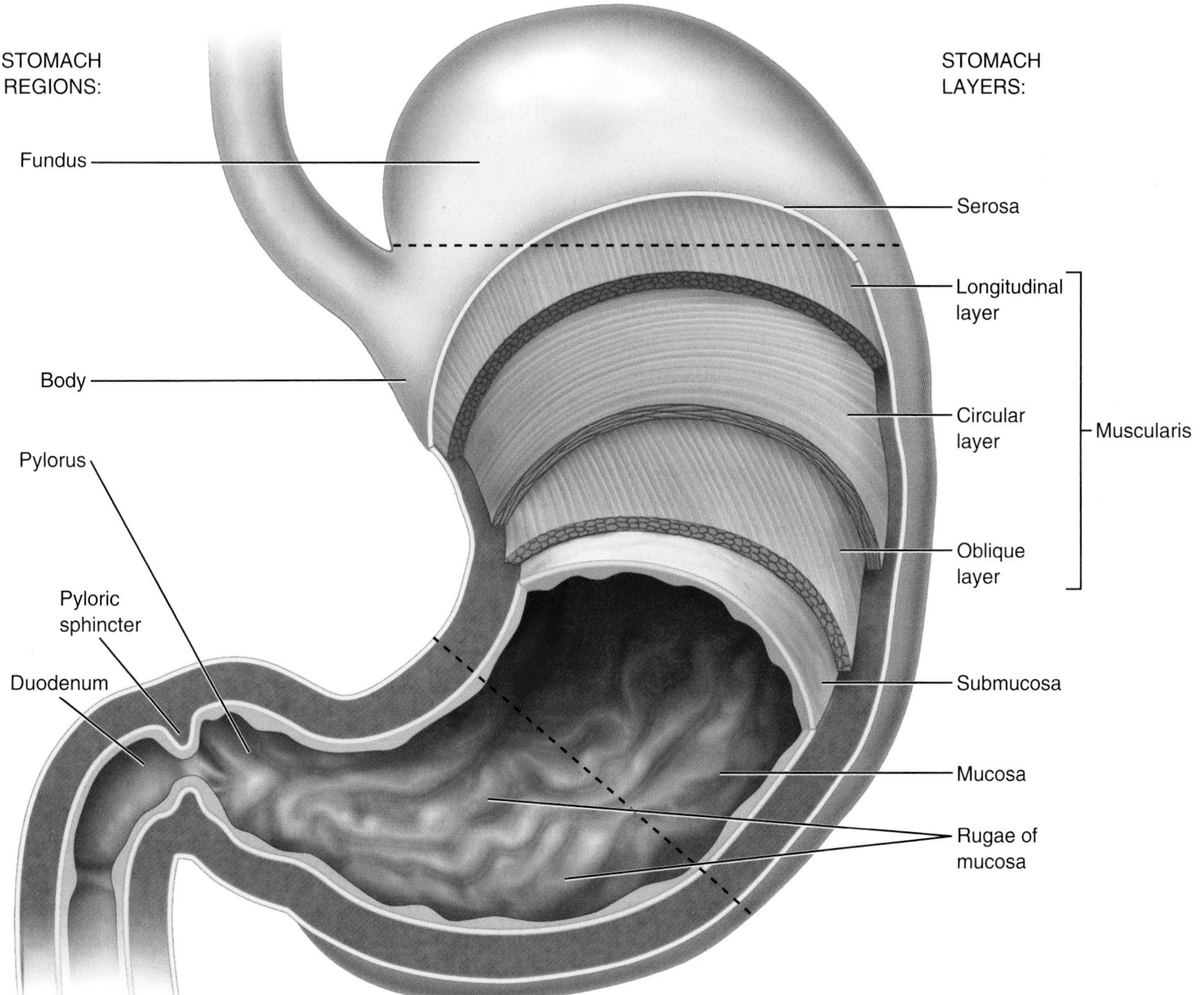

FIGURE 14.7 ▸ Stomach. The three regions of the stomach are the fundus, body, and pylorus. The muscularis layer has an additional oblique smooth muscle layer that helps produce strong contractions to thoroughly mix the stomach contents.

By the Way

The terms *pepsin* and *peptic* are derived from the Greek term *pepsis*, which means digestion.

layer has hundreds of folds called **rugae** that allow it to expand to accommodate large quantities of food. The thick muscularis layer is strengthened by an additional inner layer of smooth muscle with an oblique fiber arrangement. Smooth muscle rings called **sphincters** are located at the proximal and distal openings to the stomach. These ring-shaped muscles contract and relax to control the flow of food into and out of the stomach. The **esophageal sphincter** lies between the esophagus and stomach and is also referred to as the ***lower esophageal***, ***gastroesophageal***, or ***cardiac sphincter***. The sphincter that separates the stomach from the small intestine is called the **pyloric sphincter**.

The stomach plays a role in both chemical and mechanical digestion. When food enters the stomach, ***hydrochloric acid*** is secreted to begin the breakdown of proteins and destroy pathogens that might enter with the food. Simultaneously, mucus production increases to protect the stomach lining from this strong acid. The stomach also secretes **pepsin**, a protein-digesting enzyme. Initially secreted in an inactive form (***pepsinogen***) to protect the stomach lining, pepsin is activated by the secretion of hydrochloric acid. Strong smooth muscle contractions churn and mix the stomach contents with gastric juices to form a semiliquid substance called **chyme**. Via peristalsis, chyme is forced against the pyloric sphincter, stimulating it to open. Each time it opens, only a small amount of chyme is allowed to push through into the small intestine. When it closes, the remaining chyme is forced back into the body of the stomach, where it is subjected to further mixing. Because the pyloric sphincter regulates the process of ***gastric emptying***, it normally takes about four hours for the stomach to empty after a meal; it can take up to 6 hours if the meal is high in fats.

Small Intestine

Most of the digestive and absorption functions of the digestive system occur in the **small intestine (SI)**. Named for its small diameter of approximately 1 in (2.5 cm), the majority of this coiled tube is situated in the gastric region of the abdominal cavity (FIGURE 14.8). If fully relaxed and extended, the SI is approximately 20 ft (6 m) in length. Divided into three segments, the SI begins at the pyloric sphincter with the **duodenum** that extends approximately 10 in. The 3-ft central segment is the **jejunum**, while the 6-ft-long **ileum** is the longest and most distal portion of the SI. While most of the SI resides within the peritoneum, the duodenum is located in the retroperitoneal space.

Because chyme entering the duodenum from the stomach is highly acidic, the SI secretes a large amount of mucus to protect its lining. While the SI also secretes small amounts of digestive enzymes, the majority of digestion that occurs there is due to enzymes secreted by the liver and pancreas. These enzymes enter the duodenum through a common duct; they are released through the secretion of two hormones in the duodenum:

- ***Secretin*** stimulates the release of bicarbonates and water from the pancreas; this inactivates pepsin and raises the pH of chyme by neutralizing the hydrochloric acid.
- ***Cholecystokinin (CCK)*** stimulates the release of digestive enzymes from the pancreas as well as bile from the gallbladder and liver. The brain and spleen also have receptors for CCK, which clearly links the digestive system to both the nervous and immune systems.

Pathology Alert **Peptic Ulcers**

Peptic ulcers are sores or lesions in the mucosa of the GI tract that do not heal, leading to long-term inflammation and tissue loss. Most commonly found in the lower esophagus, stomach, or duodenum, these lesions can be caused by long-term nonsteroidal anti-inflammatory drug use or stress, but most often are due to a bacterial infection, specifically Helicobacter pylori bacterium. These factors damage the mucosal lining, making it unable to effectively protect the tract wall from the strong acid and pepsin produced by the stomach. Signs and symptoms of peptic ulcers include heartburn or a gnawing abdominal pain, bloating, burping, nausea, or vomiting. If left untreated, ulcers can penetrate through the epithelial lining into the submucosa and begin to bleed. They are considered a local contraindication for any form of direct manual therapy such as massage.

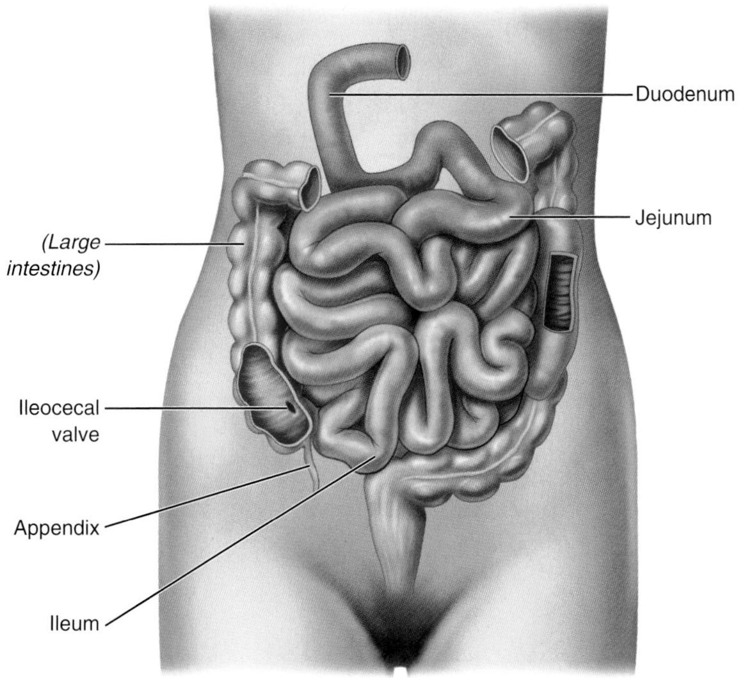

FIGURE 14.8 ▸ Small intestine. The SI has three segments: the duodenum, jejunum, and ileum.

In addition to chemical digestion, **segmentation**, a form of mechanical digestion, also occurs in the SI. Segmentation involves alternating contractions of single segments of the circular smooth muscle, which mixes and churns the chyme (▸ FIGURE 14.9).

Most nutrient absorption occurs in the SI. **Plicae**, circular folds in the mucosa and submucosa, increase the surface area and cause chyme to spiral as it moves through. Small mucosal projections called **villi** further increase the surface area for absorption (▸ FIGURE 14.10). Each villus contains a lacteal surrounded by a rich network of cardiovascular capillaries. Lacteals are specialized lymphatic capillaries responsible for the absorption of fats. Microscopic brush-like extensions from the surface of each epithelial cell of villi, the ***microvilli***, also increase surface area to speed up nutrient absorption.

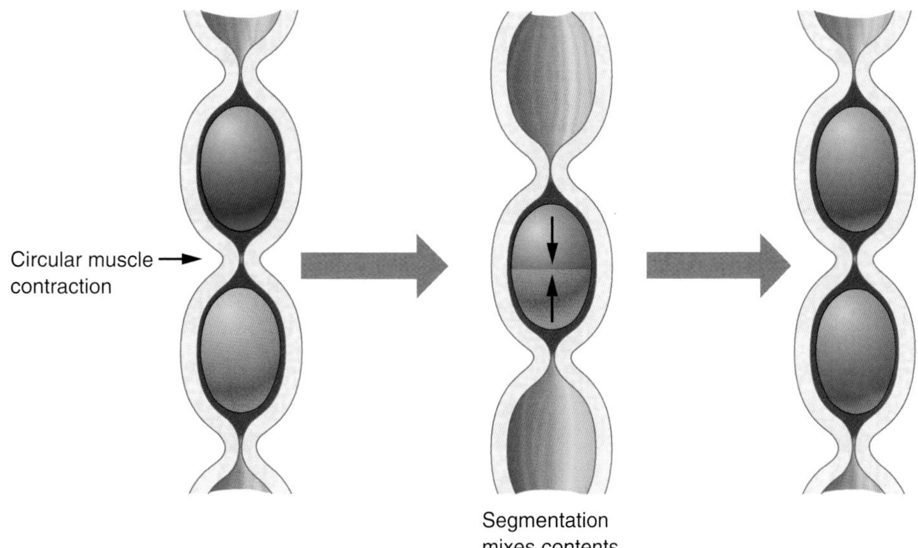

FIGURE 14.9 ▸ Segmentation. The circular layer of the muscularis contracts to create segments that mix the contents of the SI.

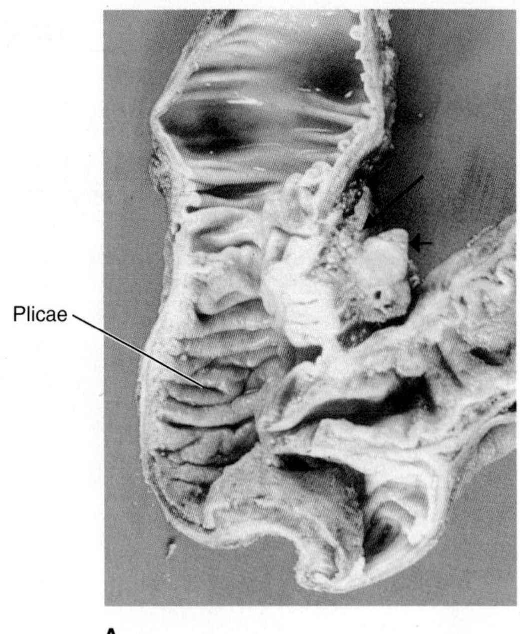

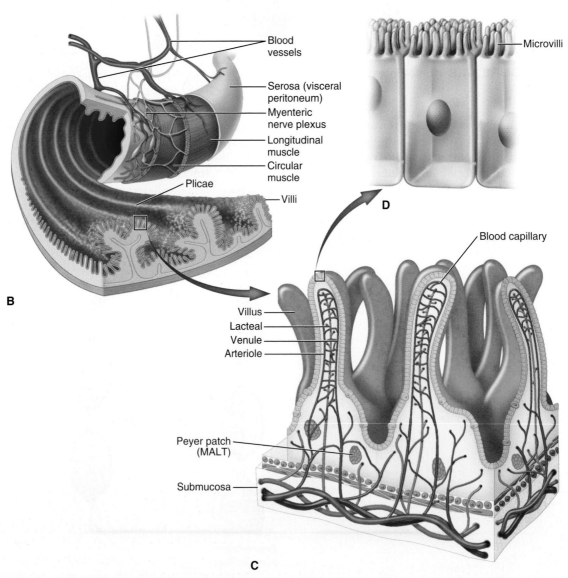

FIGURE 14.10 ▸ Internal anatomy of the SI. A. Plicae are circular folds in the SI. **B.** Plicae are covered with thousands of tiny projections called villi. **C.** Each villus has a lacteal surrounded by a capillary bed for nutrient absorption. **D.** Villi are covered by epithelial cells with microvilli.

By the Way

Remember that long-chain fatty acids absorbed by the lacteals are a primary component of lymph. *Chyle* is the cloudy white or pale yellow liquid product of digestion that consists mainly of emulsified fats. It is taken up by the lacteals during digestion and carried by the lymphatic system through the thoracic duct and into the circulation at the subclavian veins, ultimately reaching the liver for processing.

Large Intestine

The **large intestine** extends from the ileum to the anus, forming an arch around the SI. It begins at the **cecum**, a short pouch-like segment separated from the ileum by the ***ileocecal valve***, and includes the colon, rectum, and anal canal. Its longest portion is the **colon**, which has several segments (FIGURE 14.11). A vertical segment called the **ascending colon** is located on the right side of the abdominal cavity. It connects to the **transverse colon**, which crosses the upper abdominal cavity just below the stomach and connects to the **descending colon**, which extends downward on the left side. The intersection points between the three segments of the colon are named according to the closest organs. The **hepatic flexure** is the bend between the ascending and transverse colon (near the liver), while the bend between the transverse and descending segments (near the spleen) is called the **splenic flexure**. At the distal end of the descending colon, there is a short S-shaped segment called the **sigmoid colon**, which ends at the **rectum**. The last few centimeters of the rectum make up the **anal canal**, which ends at an opening, the **anus**.

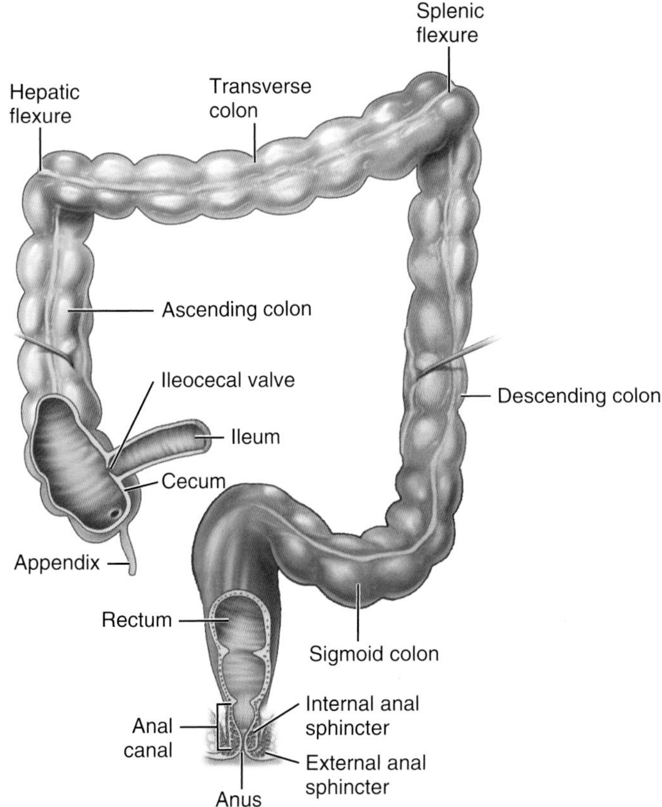

FIGURE 14.11 ▸ Segments and flexures of the large intestine.

No digestive processes take place in the large intestine. Instead, it serves as a primary site for reabsorption of water and the compacting of wastes into solid matter. Feces are temporarily stored in the colon, where bacteria act on it to produce vitamin K and some varieties of vitamin B. If these helpful bacteria are destroyed by antibiotic medications, diarrhea and other unpleasant side effects occur. ***Defecation*** is the elimination of fecal matter through the anus. The smooth muscles of the colon methodically move feces toward the rectum and anal canal. As fecal matter builds up and stretches the rectum, it creates pressure against the anus, a sphincter made up of an inner ring of smooth muscle and an outer skeletal muscle ring. This pressure causes the inner involuntary sphincter to open, which creates pressure against the external skeletal muscle ring. When directed by the cerebral cortex, this outer muscle ring voluntarily relaxes and defecation occurs.

Direct soft tissue manipulation to the abdomen is indicated in some digestive system conditions and contraindicated in others. Chronic constipation is one of the most frequently cited indications for abdominal massage, and its effectiveness is supported by anecdotal evidence. However, this effectiveness is not likely due to mechanical loosening or movement of fecal matter through the colon. Instead, abdominal massage seems to relieve constipation by decreasing stress and sympathetic tone to enhance motility. Therapists must always determine the probable cause of constipation, since it may be a sign of a more serious metabolic condition or mechanical blockage. With conditions such as irritable bowel syndrome, Crohn disease, and ulcers, general massage may be prescribed and can be helpful, but abdominal massage may be contraindicated.

By the Way

The appendix is a worm-like organ that extends from the colon. It attaches at the cecum and contains a small amount of lymphoid tissue. ***Appendicitis***, inflammation of the appendix, usually occurs due to blockage or infection, sometimes caused by trapped food particles. Surgical removal of the appendix is the preferred treatment for acute appendicitis.

What Do You Think? 14.2

- What household items would you choose to construct a model of the GI tract?
- In a manual therapy session, what changes or precautions should the therapist take after learning that a client has just consumed a big meal?

ACCESSORY DIGESTIVE ORGANS

With the exception of the teeth and tongue, the accessory digestive organs are glands located outside the GI tract that supply the tract with important chemicals and enzymes that aid in the digestive process. In addition to the salivary glands already discussed, the accessory organs include the liver, gallbladder, and pancreas.

Liver and Gallbladder

The **liver** is a large reddish-brown organ with a large right and smaller left lobe. It is situated between the diaphragm and the lower margin of the rib cage in the upper right quadrant of the abdominal cavity, suspended by the peritoneum and several specialized ligaments. Its lobes are divided into functional units called **lobules** made up of **hepatocytes**, which are specialized epithelial cells surrounding a central vein. Hepatocytes secrete **bile** into small intercellular channels that drain into ***bile ducts*** on the outside of the lobules. The bile ducts merge and form the right and left **hepatic ducts** that eventually drain into the **common hepatic duct**, the tube that carries bile from the liver to the gallbladder, where it is concentrated and stored until needed for digestion. The **gallbladder** is a small pear-shaped pouch located under the right lobe of the liver that delivers bile to the duodenum through the **common bile duct** (FIGURE 14.12).

Bile is a yellowish-brown or greenish fluid that **emulsifies**, or breaks apart large fat globules into smaller ones, so that fat enzymes can more easily digest and process lipids in the SI. The presence of bile also facilitates the absorption of fats. In addition to producing bile, the liver is also important as a nutrient storage facility, acting as a "warehouse" for fat-soluble vitamins, minerals such as iron and copper, as well as glycogen, the stored form of glucose. When blood sugar levels are low, the liver converts stored glycogen into glucose and releases it into the blood to return levels to normal. Although

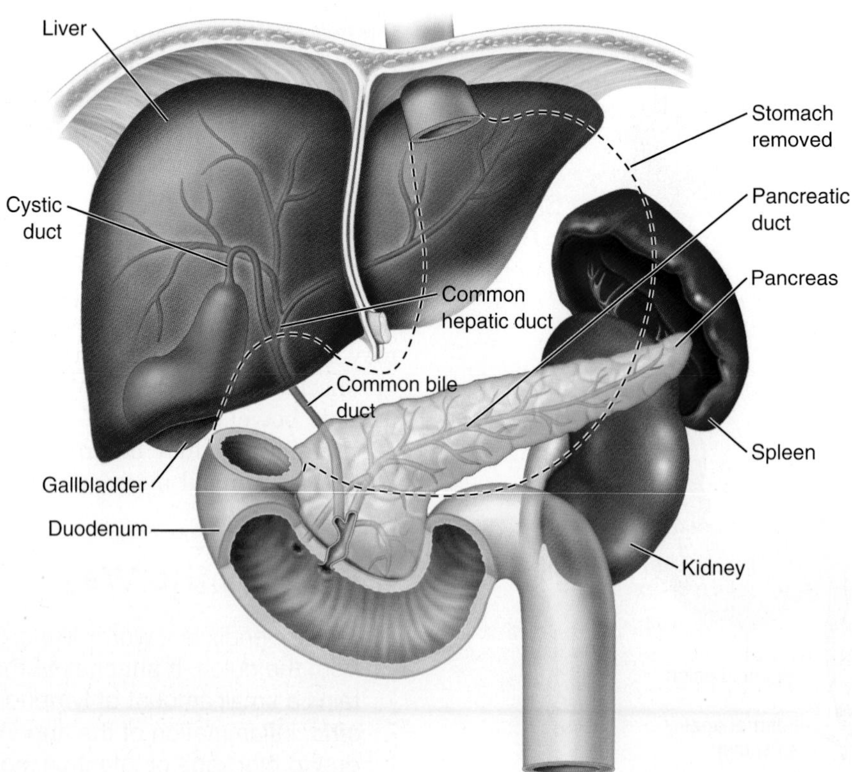

FIGURE 14.12 ▸ Liver, gallbladder, and pancreas. The digestive secretions of the liver, gallbladder, and pancreas pass through ducts that merge and share a common entrance to the duodenum.

Pathology Alert Irritable Bowel Syndrome

Irritable bowel syndrome (IBS) is a functional condition rather than a structural disorder of the colon. It is characterized by abdominal bloating, distension, and pain, as well as constipation or diarrhea. Its symptoms are often aggravated by stress and diet. Little is understood about IBS, and there is no definitive diagnostic test for the condition. While commonly diagnosed in adolescence or early adulthood, its symptoms can mimic those of parasitic infestation, celiac disease, food allergies, or chronic GI tract infections, making diagnosis difficult. IBS affects more women than men, especially those who have suffered psychological trauma or abuse and is often associated with chronic fatigue or fibromyalgia. Because of its stress-reducing and relaxation benefit, manual therapy is generally indicated for people with IBS. However, if treatment exacerbates the client's symptoms, manual therapy should be modified or discontinued.

proteins are a less desirable source for energy, hepatocytes can assist in the breakdown of amino acids to produce adenosine triphosphate (ATP). If amino acids are used for energy, ammonia is produced, and the liver must convert this toxin into a substance called **urea** that is then excreted through urine.

Besides its digestive, storage, and metabolic roles, the liver forms certain plasma proteins, including albumin and clotting factors. It also filters and detoxifies the blood of harmful products, including alcohol and toxins, as well as common medications such as ibuprofen, naproxen, and chondroitin. Similar to the spleen, the liver destroys old red blood cells, recycling or eliminating the byproducts, including ***bilirubin*** a yellow-orange pigment released from the breakdown of hemoglobin. Most bilirubin produced in the liver is secreted into the SI as a component of bile, where it is metabolized. It is eventually eliminated in feces. However, if excess amounts of bilirubin enter the bloodstream, it produces a yellow discoloration of the skin, mucous membranes, and whites of the eyes known as jaundice.

It is important for manual therapists to remember that organ dysfunctions are a common source of referred pain. Right shoulder pain may actually indicate liver dysfunction, and back pain is a common indicator of gallstones. If viscera are the source of pain, there are usually associated signs such as nausea, sweating, and pale skin. Another consideration in treating clients with suspected or diagnosed organ dysfunction is that direct manipulation in the abdominal area may be contraindicated. In these cases, therapists trained in foot reflexology or meridian therapies can consider these alternatives, as they have been shown to be effective at stimulating digestive organ functions.

Pancreas

The **pancreas** is 5 to 6 in (12 to 15 cm) long and is situated behind the stomach in the upper left abdominal quadrant between the duodenum and spleen (FIGURE 14.12). Like the duodenum, the pancreas is retroperitoneal. While the pancreas is an important endocrine gland, 99% of its cells are engaged in exocrine functions, making it a vital accessory digestive organ. Enzymes from the pancreas are transported through the **pancreatic duct** to the duodenum, where they are active in the chemical digestion of proteins, fats, and carbohydrates. These digestive enzymes are mixed and transported in an alkaline fluid called ***pancreatic juice*** that protects the lining of the SI by neutralizing the acidic chyme. Enzymes in pancreatic juice and their digestive actions include:

- ***Amylase***—Similar to the starch-digesting enzyme in saliva, amylase is a primary component of pancreatic juice.
- ***Lipase***—Proper digestion of fats is dependent on this highly active enzyme. Lipase breaks down fat into its more absorbable components: glycerol and fatty acids.
- ***Trypsin***—This enzyme is a ***protease***, meaning that it digests protein into amino acids.

These enzymes, together with small amounts of ***lactase***, ***maltase***, ***sucrase*** (enzymes that break down sugars), and other proteases secreted by the duodenum, are responsible for the majority of chemical digestion that occurs in the GI tract (TABLE 14-1).

What Do You Think? 14.3

- When a client reports nagging right shoulder pain, what questions and/or objective assessments would be important to help identify the possible cause of the pain?
- Why do you think the pancreas and duodenum are located in the retroperitoneal space?

Table 14-1 Summary of Digestive Organ Functions

Organ	Ingestion	Secretion	Digestion	Motility	Absorption	Elimination
Mouth and pharynx	Eating	Mucus	M: Chewing C: Salivary amylase	Swallowing	Sublingual medications	
Salivary glands		Saliva				
Esophagus		Mucus		Peristalsis		
Stomach		Mucus, hydrochloric acid, pepsin	M: mixing, grinding C: breakdown of all foods via HCl and proteins by pepsin	Peristalsis Gastric emptying	Alcohol and some medications such as aspirin	
Small intestine		Mucus, sucrase, lactase, and maltase plus proteases Local hormones secretin and CCK	M: mixing via segmentation C: breakdown of all nutrients with enzymes from the SI and pancreas, and fats with bile from the liver and gallbladder	Segmentation Peristalsis	Glucose, triglycerides, and amino acids as well as water and medications	
Large intestine		Mucus		Peristalsis	Water and vitamin K	Defecation
Liver		Bile				
Gall bladder		Concentrated bile made by the liver				
Pancreas		Amylase, lipase, and trypsin				

M, mechanical; C, chemical.

METABOLISM AND BASIC NUTRITION

Recall from Chapter 3 that the term ***metabolism*** refers to all the chemical reactions that occur within the body. Metabolism includes ***anabolic*** processes, in which complex molecules are formed from simpler ones, tissue is created or repaired, and molecules such as hormones, neurotransmitters, and ligands are synthesized. Metabolism also includes ***catabolic*** processes such as glycolysis and the Krebs cycle, which break down complex molecules to produce the energy molecule ATP. Both anabolism and catabolism rely on the body's intake and assimilation of nutrients through the digestive system (FIGURE. 14.13). The nutrients required for health can be divided into two broad categories. ***Macronutrients*** refer to the carbohydrates, lipids, and proteins needed in large quantities to meet the body's energy demands. Other substances, such as vitamins and minerals, are categorized as ***micronutrients*** because the body only requires small amounts to maintain homeostasis and health. A summary of the body's nutrient usage is shown in TABLE 14-2 at the end of this section.

Carbohydrates

Carbohydrates, commonly called sugars or starches, are organic compounds formed by rings of carbon, oxygen, and hydrogen linked together to make up larger molecules. Sugars come from a variety of sources and are generally named according to their source. For example, ***fructose*** is the sugar found in fruit, ***lactose*** is a milk sugar, and table sugar, or ***sucrose***, is usually derived from sugar cane or sugar beets. Starchy food sources include root vegetables such as potatoes, carrots, and turnips, and many varieties of squash and beans. All sugars and starches are broken down into **glucose**, which is then absorbed into the bloodstream through the SI. As discussed in Chapter 6, glucose molecules are a primary source of cellular energy because the mitochondria readily convert it into ATP molecules, water, and carbon

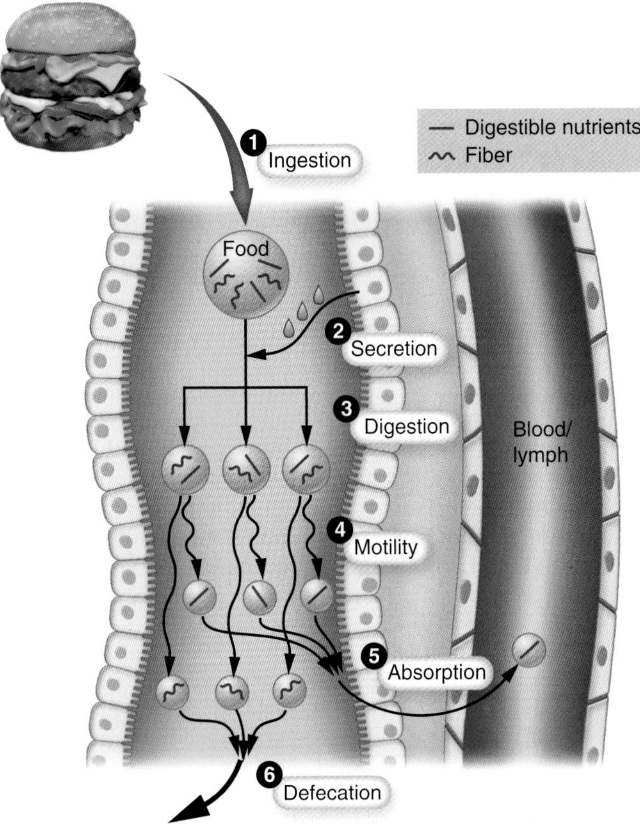

FIGURE 14.13 ▸ Intake and assimilation of nutrients. Ingested food is broken down into usable nutrients and fiber. Nutrients are absorbed through the wall of the digestive tract for use in metabolic processes, while fiber is eliminated via defecation.

dioxide (▸ FIGURE 14.14). All body cells can utilize glucose for energy production. Some tissues, primarily the brain, use glucose exclusively except during extreme starvation, when glucose is unavailable. Another type of carbohydrate, ***fiber***, is also found in foods (especially fruits, vegetables, and whole grains) but does not serve as an energy source. Instead, its purpose is to help form undigested particles of food into feces and facilitate elimination of solid waste (see FIGURE 14.13).

Sometimes more glucose is supplied than is needed for immediate energy, and since it is a highly osmotic particle, it cannot be stored as is. In fact, if a cell were to hold on to all of its glucose, it would continue to absorb water until it exploded. Therefore, to store glucose, the body converts it into **glycogen** through an anabolic process. Glycogen can be stored as a ready source of glucose in muscles or in the liver for later use. For longer-term storage, excess glucose is generally converted and stored as fat.

Lipids

Like carbohydrates, lipids can be catabolized to produce ATP. Fats are provided for metabolism either through the diet or from adipose stored in body tissues and the liver. Dietary fat sources include meat, poultry, fish,

By the Way

The rate at which a food elevates the blood glucose level is called its ***glycemic index***. Most sugars and some starches swiftly raise the level of glucose in the blood, causing a large surge of insulin to be released by the pancreas. This in turn leads to a large increase in glucose utilization by the cells, especially skeletal muscle cells, followed by a huge drop in the blood glucose level, which leads to fatigue. This may account for the "energy crash" some people experience shortly after eating a carbohydrate-heavy meal.

dairy, and other animal products; oils; nuts; seeds; and avocados. Dietary fats are broken down through digestion into large molecules called **triglycerides**. Once absorbed, triglycerides can be broken down into smaller component molecules of **glycerol** and **fatty acids**. Many body cells can convert glycerol into a substance that can either enter the Krebs cycle or be converted back into glucose. Fatty acids can be converted into another substance that can directly enter the Krebs cycle to produce ATP. However, since converting triglycerides for this purpose requires some work, fats are the body's second choice for energy and are generally only utilized when glucose reserves have been depleted.

Unused lipids are stored as triglycerides in adipose tissues and the liver. Since fats are not the body's first choice for energy production, excess dietary consumption of lipids leads to fat accumulation in tissue, as well as increased lipids in the blood. While the body requires some level of fat stores for insulation, cushioning of vital organs, hormone production, construction of cellular organelles, and future sources of energy, excess fat and obesity can lead to a variety of health challenges.

Proteins

Proteins are large molecules formed by chains of nitrogen-containing molecules known as **amino acids.**

By the Way

The fatty plaques that collect inside arteries with atherosclerosis were once believed to be the result of a diet rich in highly saturated fatty acids, such as those in animal products. However, recent research indicates that most plaques consist predominantly of polyunsaturated fats, such as those found in safflower, sesame, corn, and soy oils.[1]

Table 14-2 Nutrient Usage

Nutrient	Digested and Absorbed as	Physiologic Use
Carbohydrates	Glucose	*Catabolism*: Glucose catabolized for energy via glycolysis and the Krebs cycle *Anabolism*: Converted into glycogen to be stored in skeletal muscle and the liver for future use. Glucose can also be converted into fat and stored as adipose tissue.
Lipids	Triglycerides	*Catabolism*: Triglyceride molecules are broken into glycerol and fatty acids that can be catabolized for energy. *Anabolism*: Triglycerides are stored as adipose tissue. Lipids are necessary for absorption and storage of fat-soluble vitamins, hormone production, maintenance of plasma membranes, and cushioning and insulation of the body and vital organs.
Proteins	Amino acids	*Catabolism:* Excess amino acids can be stripped of their amino group by the liver and catabolized to produce ATP, but they are the body's last choice for energy production, after carbohydrates and fats. *Anabolism:* Amino acids are used for all types of protein synthesis. Amino acids are necessary to make and repair tissue, form enzymes, receptors, antibodies, hormones, and ligands.
Vitamins		Support normal metabolic function by serving as coenzymes in chemical reactions.
Minerals		Support homeostatic mechanisms including pH, fluid, and energy balance. Important components of blood and bone tissues. Many serve as coenzymes that help regulate chemical reactions in the body.
Water		Necessary for cellular health and maintaining homeostasis. Primary component of all body fluids and essential for the transportation of enzymes, hormones, and chemicals throughout the body. Facilitates the action of enzymes that create chemical digestion, digestive motility, and absorption.

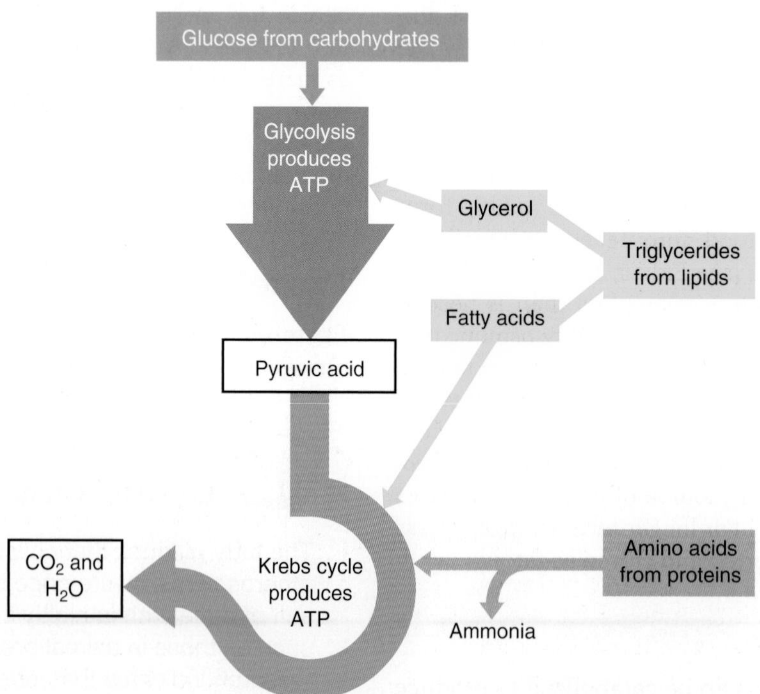

FIGURE 14.14 ▸ Catabolism of nutrients. Usable nutrients are catabolized to provide the body with energy. The glucose from carbohydrates can be converted to CO_2 and H_2O through glycolysis and the Krebs cycle to produce ATP. Fatty acids and glycerol from triglycerides and the amino acids from protein must be converted into intermediary substances before they can enter these catabolic processes.

By the Way

There are 20 different amino acids, 10 of which are ***essential amino acids***, so named because they must be supplied through dietary sources on an ongoing basis to sustain healthy metabolism.

When proteins are digested, they are broken down into their amino acid components, which are then absorbed into the bloodstream through the wall of the SI. Most amino acids are used as building blocks to synthesize a wide array of peptides and proteins. Some examples of important proteins are actin, myosin, keratin, collagen, and elastin, as well as enzymes, receptors, and hundreds of peptide molecules. Amino acids not needed for anabolism can be converted into glucose or triglycerides or transported in blood to the liver, where they are catabolized into a substance that can enter the Krebs cycle. This catabolic process produces urea, the ammonia byproduct excreted in urine.

Vitamins and Minerals

Vitamins are organic micronutrients required by the body for proper growth and normal metabolism. While most vitamins are provided through dietary intake, some are synthesized by the body. Vitamins fall into two broad categories, fat-soluble or water-soluble. The ***fat-soluble*** group includes vitamins A, D, E, and K. These vitamins are absorbed in the SI along with lipids, and can be stored in various cells, especially in the liver. ***Water-soluble*** vitamins include the B and C vitamins, which dissolve in body fluids and cannot be stored. Excess amounts of water-soluble vitamins are excreted in urine. Most vitamins act as coenzymes in various chemical reactions.

Inorganic elements such as calcium, phosphorus, potassium, sodium, iron, magnesium, and others account for about 4% of a person's body weight. Collectively known as **minerals**, they are important to homeostatic mechanisms, including pH, fluid, and energy balance. Many minerals are found in bones and similar to vitamins, many serve as coenzymes that help regulate chemical reactions in the body.

Water

While water is not a nutrient, it is imperative for cellular health and maintaining homeostasis. As the primary component of all body fluids, it is essential for the transportation of enzymes, hormones, and chemicals throughout the body. In the digestive tract, water facilitates motility, transports vitamins and minerals, and facilitates the action of the enzymes involved in chemical digestion. To accomplish this, the body secretes nearly 7 l (slightly <2 gal) of water into the digestive tract each day in addition to the 2 liters (just over 2 quarts) that should be consumed as drinking water and other liquids.

For decades, manual therapists have been reminding clients to increase their water intake after a session in order to "flush out toxins" that therapy has "released" into the tissues. However, there is no real scientific evidence to support this flushing concept, since water must be processed and absorbed into blood and lymph through the small and large intestines just like any other nutrient. Additionally, our understanding of metabolism should lead us to question the idea that catabolic processes create toxins that get stuck in muscle, fascia, and skin that external manipulation then releases into the bloodstream. However, suggesting an increase in water intake to clients may still be helpful, but for different reasons. An increase in water intake will assist in the rehydration of tissues (especially fascia), support nutrient absorption, and facilitate the elimination of waste products.

DIET AND HEALTH

Food choices are influenced by a broad spectrum of factors. Clearly, the body has nutritional requirements, but what and how much is ingested can be affected by a person's emotional state, daily habits, socioeconomic influences, cultural norms, memories, and beliefs. Physiologically, eating is regulated by homeostatic mechanisms that balance energy consumption with energy expenditures. ***Hunger***, the sensation of discomfort or weakness that signifies the body's need for food, is regulated by centers in the hypothalamus that respond to changing levels of nutrients in the blood. When nutrient levels drop, the hypothalamus causes us to feel hungry, and when nutrient levels rise, we feel full or satisfied. Since catabolism produces most of the body's heat, stimulus to eat is also closely related to the mechanisms that control body temperature. **Metabolic rate** is the measure of the amount of heat produced by the body. While the body's metabolic rate shifts throughout the day in response to food consumption, physical activity, sympathetic responses, or hormones, each person has a baseline energy requirement called the **basal metabolic rate (BMR)**. The BMR, represented in ***calories*** of heat, is the amount of energy a body needs to support basic life functions such as breathing, heartbeat, and kidney function. Individual BMRs differ due to factors such as lean mass, stress levels, thyroxine production, age, gender, and even climate. This explains why people tend to eat more during cold weather and less when temperatures are warmer.

In addition to the homeostatic mechanisms that mediate hunger, eating is affected by a powerful reward-driven response.[2] Our senses, particularly taste, are linked to pleasure centers in the brain, so when we eat something with a taste that appeals to us, we are driven to continue eating it in pursuit of the associated pleasure sensation. Some people recognize and control this desire, or ***appetite***, understanding that they do not need the food, they simply want it. However, in others, this pursuit of pleasurable sensation becomes an addiction, driving them to eat more and more, even when they are physically full. This reward response can override the body's homeostatic mechanisms and make us feel a type of hunger even when we are completely full. This equation is further influenced by the availability of foods, memories, social environment, attitudes, and mood. As in all issues related to health, eating must be considered holistically. Our diet and food choices ideally work together to feed the mind, emotions, and spirit as well as meet the nutritional needs of the body.

As allied healthcare practitioners, most manual therapists are aware that dietary change can serve as an important therapeutic tool. For example, some therapists may know that the type of dietary fats an individual consumes can have a positive or negative influence on inflammatory conditions such as arthritis or lupus. Other therapists may have experienced health benefits from increasing intake of vitamin D or taking certain herbal supplements. However, it is imperative to recognize that *giving nutritional or dietary advice is clearly outside the scope of practice* for manual therapists, unless they also have training and licensing as a registered dietician. Every individual body is a unique chemical environment, and there are often large variations in responses to identical dietary therapies. Clients may misinterpret a casual conversation on supplements or food choices as professional advice, so it is important that you either avoid these topics altogether or if they are discussed, make it clear that you are not providing professional advice.

What Do You Think? 14.4

- Why does the body choose to first catabolize carbohydrates, then fats, then proteins for energy production?
- Describe the difference between hunger and appetite. Which do you think is a bigger challenge for people in your community?

AGING AND THE DIGESTIVE SYSTEM

Age-related changes in the digestive system include decreased secretion, motility, and tone of the muscularis layer of the GI tract. Changes in the ENS include decreased feedback on enzyme and hormone release, as well as a decrease in internal sensations overall. For these reasons, many elderly people experience more digestive discomfort such as increased belching and flatulence. Some common conditions associated with increased age include periodontal disease, difficulty swallowing, gastritis, gallbladder problems, constipation, diverticulosis, and hemorrhoids.

SUMMARY OF KEY POINTS

- The functions of the digestive system are ingestion, secretion, digestion, motility, absorption, and elimination.
- Ingestion, digestion, absorption, and elimination are the physiologic processes through which the digestive system provides energy for cellular metabolism, removal of solid waste, and generates heat to help regulate body temperature.
- Organs of the digestive system are divided into those in the GI tract, where digestive processes take place, and accessory organs that do not transport food but contribute several important enzymes to the digestive process.
- Organs in the GI tract include the mouth, pharynx, esophagus, stomach, and small and large intestines. The teeth, tongue, and salivary glands, plus the liver, gallbladder, and pancreas, are the accessory organs of the system. See TABLE 14-1 to review the digestive processes for each organ.
- Digestion, breaking down food into usable components, occurs through mechanical and chemical processes. In mechanical digestion food is smashed or broken into smaller pieces; chewing and segmentation are examples of mechanical digestion. In chemical digestion specific enzymes work to dissolve food into smaller nutritional components.
- Metabolism encompasses all chemical processes in the body that are necessary to maintain life. Anabolic metabolism is a building-up process in which complex molecules are formed from simpler ones, tissue is created or repaired, and hormones, neurotransmitters, and ligands are synthesized. Metabolic processes that break down complex molecules to produce energy are catabolic.
- Carbohydrates are broken down into different forms of sugar. The simple sugar glucose is either catabolized for energy or anabolized as glycogen for storage in the liver and skeletal muscle.

- Lipids are digested and absorbed as triglycerides. These molecules are stored as adipose tissue, used in the production of hormones and enzymes, or broken into glycerol and fatty acid molecules that can be catabolized for energy.
- Proteins are digested and absorbed as amino acids. These molecules are most often used in anabolic processes to build and repair tissue and to form ligands of all kinds. Amino acids can be catabolized for energy, but this process produces ammonia, which must be converted to urea and excreted.
- With age there is decreased secretion, motility, and smooth muscle tone in the digestive system. Some common digestive difficulties associated with age are gastritis, gallbladder problems, constipation, diverticulosis, and hemorrhoids.

REVIEW QUESTIONS

Short Answer

1. List and give a brief explanation of the six functions of the digestive system.
2. Which organs of the GI tract do not play a role in either chemical or mechanical digestion?
3. What is the difference between segmentation and peristalsis?
4. List the layers of the gastrointestinal tract starting with the external layer and progressing inward.

Multiple Choice

5. Where does the chemical digestion of starch begin?
   a. mouth
   b. esophagus
   c. small intestine
   d. stomach
6. Where does the process of peristalsis begin?
   a. pharynx
   b. stomach
   c. esophagus
   d. small intestine
7. Where does the chemical digestion of protein begin?
   a. mouth
   b. stomach
   c. small intestine
   d. pancreas
8. Which organ is an accessory organ to digestion?
   a. salivary glands
   b. stomach
   c. esophagus
   d. small intestine
9. Besides elimination, what is the other function of the large intestine?
   a. protein digestion
   b. complete carbohydrate digestion
   c. secretion of bicarbonates to neutralize acidic chyme
   d. absorption of water
10. Which valve is between the large and small intestine?
    a. duodenal
    b. esophageal
    c. pyloric
    d. ileocecal
11. Which two organs are retroperitoneal?
    a. stomach and esophagus
    b. pancreas and duodenum
    c. jejunum and transverse colon
    d. pancreas and gallbladder
12. What is the function of the greater omentum?
    a. cover and protect the abdominal viscera
    b. lubricate the peritoneum
    c. insulate and cushion the abdominopelvic cavity
    d. hold the blood supply for all organs in GI tract
13. What is the point of intersection between the transverse and descending colon?
    a. splenic flexure
    b. transverso-inferior flexure
    c. hepatic flexure
    d. sigmoid flexure

Continued on page 362

14. What is the function of the enteric nervous system?
    a. signaling the external anal sphincter for defecation
    b. regulation of bile production and secretion
    c. inhibition of ingestion
    d. regulation of digestive secretions and motility
15. Digestion of fats results in the production of which absorbable molecule?
    a. amino acid
    b. glycogen
    c. glucose
    d. triglyceride
16. What is the function of the lacteals?
    a. secrete lipase for fate digestion
    b. absorb fats from the small intestine
    c. remove water from the colon to compact particles into feces
    d. secrete amylase for carbohydrate digestion
17. The majority of digestion and absorption occurs in which organ?
    a. stomach
    b. pancreas
    c. small intestine
    d. large intestine
18. Which process builds actin, myosin, and collagen?
    a. anabolism of amino acids
    b. catabolism of fatty acids
    c. anabolism of glycerol
    d. catabolism of lipids
19. Which of the following is not part of the digestive juice secreted by the pancreas?
    a. peptidases
    b. bicarbonate
    c. lipase
    d. secretin
20. Which of the following is not a function of the liver?
    a. secretion of bile
    b. storage of glycogen
    c. secretion of protein digestive enzymes
    d. filtering toxins from the blood

References

1. Felton CV, Crook D, Davies MJ, Oliver MF. Dietary polyunsaturated fatty acids and composition of human aortic plaques. *Lancet*. 1994;344(8931):1195–1196.
2. Kessler DA. *The End of Overeating: Taking Control of the Insatiable American Appetite*. New York: Rodale, Inc.; 2009.

15 The Urinary System

LEARNING OBJECTIVES

Upon completion of this chapter, you will be able to:

1. Explain the function of the urinary system and discuss its relationship and importance to manual therapy practices.
2. Name, locate, and explain the general function of urinary system organs.
3. Name the key structural components of the kidneys and describe the general function of each.
4. Describe the role of the urinary system in fluid management.
5. Name the key structural components of a nephron and describe the processes that occur in each.
6. List and explain the three processes involved in urine formation.
7. Discuss urinary system changes that commonly occur as the body ages.

> "The length of a film should be directly related to the endurance of the human bladder."
>
> ALFRED HITCHCOCK

KEY TERMS

filtration (fil-TRA-shun)
micturition (mik-chur-IH-shun)
reabsorption (re-ab-SORP-shun)
renal (RE-nal)
renal corpuscle (RE-nal KOR-pus-al)
renal tubule (RE-nal TU-bu-el)
secretion (seh-KRE-shun)
urination (yur-ih-NA-shun)
urine (YUR-in)

Primary System Components

- **Kidneys**
  - Nephrons
- **Ureters**
  - Bladder
  - Urethra

Primary System Functions

- **Filters fluid from the blood to regulate fluid volume, blood pressure, and electrolyte and pH levels**
- **Eliminates liquid waste through excretion of urine**
- **Secretes hormones**

As the primary organs of the urinary system, the kidneys continually cleanse and refresh about 40 gal of blood per day, ridding it of toxins and wastes. In a process similar to panning for gold, the kidneys filter (sift) the blood to collect and conserve valuable "nuggets" of essential nutrients and ions, while unwanted substances are passed on for elimination. At the same time, the kidneys make adjustments to maintain blood volume and pressure and to keep pH and electrolyte levels within their homeostatic ranges. This chapter explains the primary components of the urinary system and their functions. The process of urine formation and the importance of each step in maintaining fluid balance are also discussed. For manual therapists, understanding the impact of stress on kidney function emphasizes the vital role that general relaxation and stress reduction play in maintaining overall health and well-being.

COMPONENTS OF THE URINARY SYSTEM

The urinary system has four major components. The two kidneys are the primary organs responsible for the majority of system functions. The term **renal** means pertaining to the kidneys. The ureters, bladder, and urethra form the urinary tract. The ureters carry urine from each kidney to the bladder, which serves as a temporary storage site before urine is excreted through the urethra (FIGURE 15.1).

Kidneys and Ureters

The **kidneys** are responsible for filtering blood and forming **urine**, or liquid waste. Each kidney (FIGURE 15.2) is a bean-shaped organ about the size of a fist, situated parallel to the spine and posterior to the peritoneum lining the abdominopelvic cavity (retroperitoneal). The right kidney sits slightly lower than the left kidney to accommodate the liver. The superior portion of each kidney is under the 12th rib, and the inferior edges are at the level of the third lumbar vertebra. The kidneys are anchored to the parietal peritoneum and protected by a three-layered envelope of fascia and other connective tissue called the **renal capsule**. The outermost anchoring layer of the capsule is called the ***renal fascia***, while the middle fatty layer is known as the ***adipose capsule***. The innermost layer consists of more fibrous connective tissue. Along the medial border of each kidney, there is an indentation called the **hilum**, where the ureters, blood vessels, and nerves enter and exit the organ.

Since the superior one-third of the kidneys are situated just anterior to the floating ribs, manual therapists need to be cautious with the amount and direction of pressure applied over the lower rib cage when working on the back. If deep or even moderate pressure is forcefully or abruptly applied in a lateral to medial direction, it could push a floating rib into the kidney and cause damage. To reduce this risk, it is important for deep or heavy percussive strokes such as tapotement to be focused over the erector spinae muscles or sacrum.

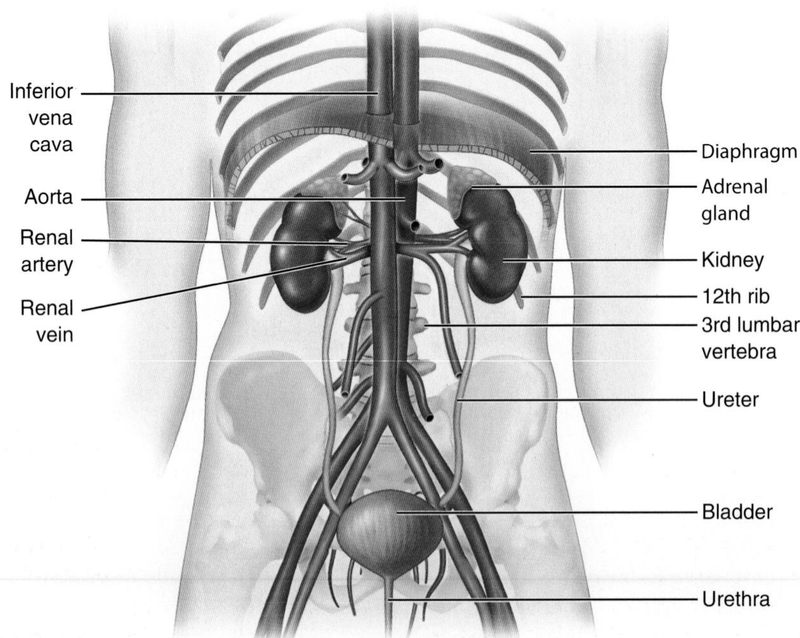

FIGURE 15.1 Structures of the urinary system. The urinary system consists of the kidneys, ureters, urinary bladder, and urethra.

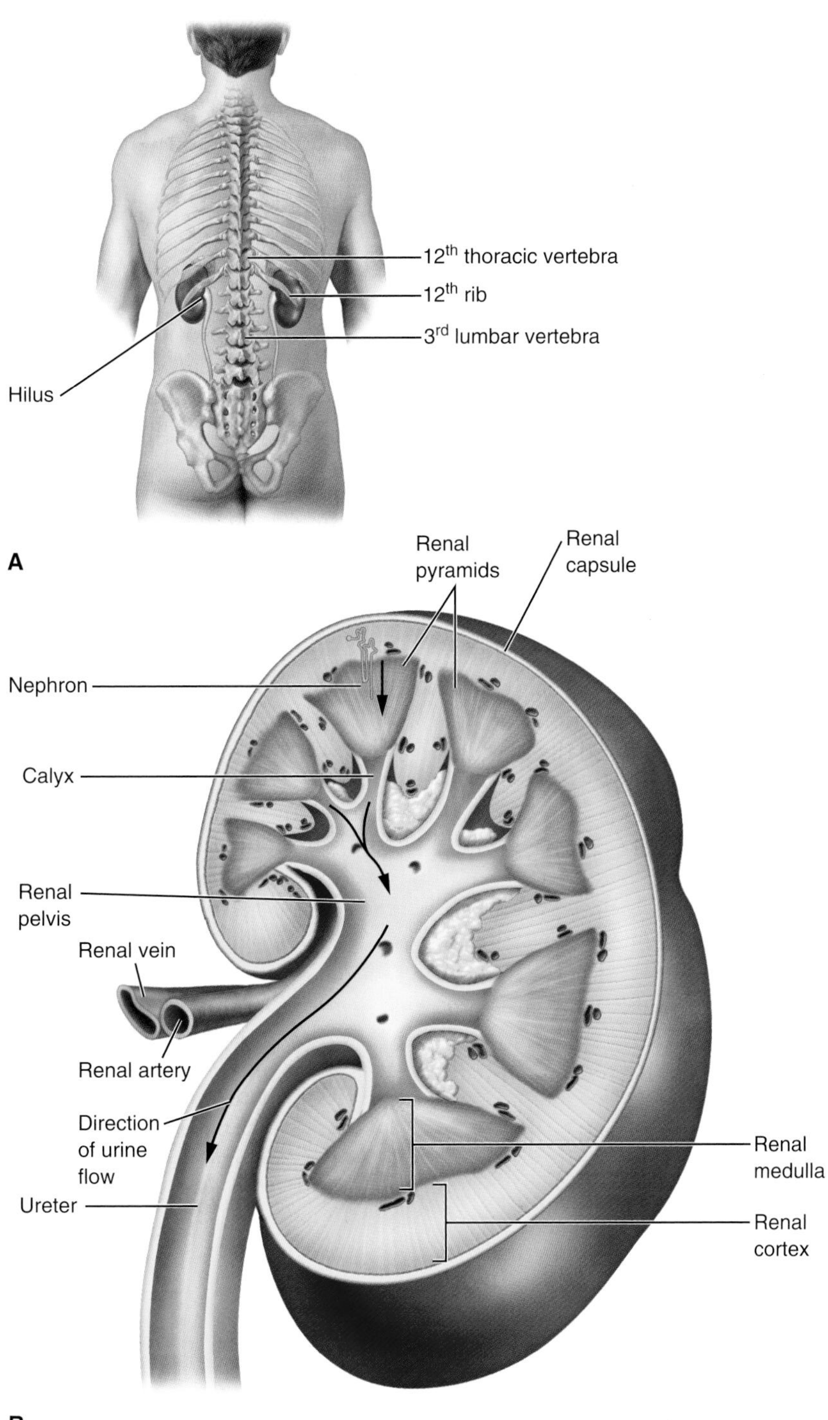

FIGURE 15.2 ▸ Kidneys. A. The kidneys are situated parallel to the lumbar spine; the superior third of each kidney is under the floating ribs. **B.** Each kidney has an outer region, the cortex, and an inner region, the medulla. The medulla is organized into renal pyramids that drain urine into a calyx. The calices pass the urine into the renal pelvis, which drains into a ureter.

The kidneys are also a common source of nonspecific low back pain. Therapists should be alert to associated signs such as sweating or fever, and the client's description of the onset of pain to distinguish the referred pain of kidney dysfunction from structural low back pain. Kidney stones that are lodged or passing through the ureters can create severe abdominal pain, as well as back pain. In fact, kidney stones are the most common cause of sudden onset, left-side abdominal pain.

Internally, the kidneys have two distinctive sections of tissue: a relatively thin outer ring, the **renal cortex**, and a thicker inner **renal medulla** (▸ FIGURE 15.2B). **Nephrons**, the microscopic functional units of the

By the Way

As relatively small organs that make up only one-twentieth of the body's weight, the kidneys have a huge functional impact; they filter and cleanse 1,200 ml (about five cups) of blood per minute, which represents 20% to 25% of total cardiac output.

kidneys, lie predominantly within the cortex. The renal medulla contains **renal pyramids**, triangular bundles of microscopic tubes that collect urine from a specific group of nephrons. At the inferior apex of each pyramid, a small cup, or **calyx**, collects urine and transfers it through the **renal pelvis** into the ureters.

Ureters are muscular tubes lined with a mucous membrane that carry urine from the kidneys to the bladder. Like the kidneys, they are retroperitoneal and attach to the posterior wall of the abdominopelvic cavity via fibrous connective tissue. Peristaltic contractions of the smooth muscles in ureters move urine toward the bladder as it is produced. The ***ureteral opening*** between ureter and bladder is a small slit, so it is easily closed by bladder contractions during urine excretion to prevent backflow.

Bladder and Urethra

The urinary **bladder** functions as the temporary holding site for urine before it is excreted. The superior aspect of the bladder is attached to the peritoneum. In males, the bladder is anterior to the rectum, but in females it is positioned just inferior to the uterus and anterior to the vagina. The bladder is similar in structure to the stomach, with multiple folds in its mucous membrane. It also has a three-layered arrangement of smooth muscle in its wall (FIGURE 15.3). This structure allows the bladder to distend as it fills to hold approximately 750 ml (25 oz or slightly more than 3 cups) of urine. Because the posterior aspect of the bladder is attached to the uterus, the average bladder capacity for women is approximately 500 ml (17 oz or about 2 cups) and even less during pregnancy.

The narrow tube that passes urine from the bladder to the external environment is the **urethra**. The process of eliminating urine is **urination**, or **micturition**. Similar to defecation, urination is regulated by two muscular sphincters. The ***internal urethral sphincter***, an involuntary smooth muscle ring under the control of the autonomic nervous system, is situated at the junction between the bladder and urethra. At the distal opening of the urethra, a circular arrangement of voluntary skeletal muscle forms the ***external urethral sphincter***. As the bladder fills, stretch receptors signal a ***micturition reflex*** that causes relaxation of the internal sphincter. The stretch receptors also send this information to the cerebral cortex, creating the urge to urinate. Simultaneously, the brain sends a motor signal to the external sphincter of the urethra to contract. This inhibits urination until an appropriate time when through conscious control, the external sphincter is relaxed and urine is excreted. This ability to voluntarily control micturition (and defecation) is a learned behavior developed during early childhood through the process of "potty training."

What Do You Think? 15.1

- What structural similarities and differences do you notice between the urinary, respiratory, and digestive systems?

Pathology Alert Kidney Stones

When the body is dehydrated, calcium salts or uric acid found in urine can crystallize in the renal pelvis to form deposits referred to as ***kidney stones*** or ***renal calculi***. Small stones may be no larger than a grain of sand and will move through the urinary tract without creating any symptoms. But kidney stones can grow as large as an inch in diameter and cause considerable pain as they pass through or lodge in the ureters. The pain caused by large kidney stones is due to smooth muscle spasms stimulated by the irritation to the mucosal lining of the ureters. Although the onset of pain is usually sudden, it comes and goes in waves and is frequently accompanied by nausea or vomiting. It can be experienced as back or abdominal pain and may also refer into the groin. Kidney stones are most often caused by dehydration but may also be caused by parathyroid dysfunction, too much vitamin D, a variety of metabolic dysfunctions, or chronic urinary tract infections. Manual therapy is contraindicated only for people suffering an acute attack; otherwise, all forms of manual therapy are appropriate once clients are asymptomatic.

Pathology Alert Urinary Tract Infection

The urinary tract is a sterile environment protected by both the slightly acidic pH of urine and the mucus-producing inner lining. While this environment is generally devoid of microorganisms, bacteria can be introduced through the urethra and is the most common cause of infections. ***Urinary tract infection (UTI)*** is the term used to describe an infection in one or more of the structures that make up the urinary system. These infections generally affect the lower urinary tract, but since the mucosal lining of the urinary tract is continuous, they can spread up the ureters and even into the kidneys, resulting in a dangerous inflammation of the kidney called ***nephritis***. ***Cystitis*** is an inflammation of the urinary bladder, and ***urethritis*** is an inflammation of the urethra. These conditions can be caused by an infection, or they can result from some other condition, injury, or disease that impairs urinary function.

Common UTI symptoms include frequent urination, pain or a burning sensation while urinating, and cloudy or blood-tinged urine. Other symptoms include abdominal, pelvic, or low back pain, as well as fever and general fatigue. Women are more susceptible to UTIs because their urethra is shorter and closer to the anus, which is populated with many types of bacteria. The risk of suffering from a UTI is increased in catheter users, people with diabetes, and in women who use spermicides and/or a diaphragm for birth control. While many forms of manual therapy are systemically contraindicated during acute infection, zone and meridian therapies are often viewed as safe and indicated. These reflexive techniques are intended to stimulate organ functions without direct manipulation of the inflamed structures. Note that these techniques should be used with caution, since even indirect stimulation of the urinary tract has been known to exacerbate or spread infection.

ROLE OF THE URINARY SYSTEM IN FLUID MANAGEMENT

Recall that water is a major component in all body tissues and is critical for optimal functioning and health. Blood, lymph, and the cerebrospinal and interstitial fluids that make up the extracellular fluids transport dissolved nutrients, wastes, and ligands. These fluids also distribute heat throughout the body and together with the intracellular fluid, serve as the primary environment for all chemical reactions. Fluid is continuously lost through the skin via evaporation, through respiration, and eliminated with urine and feces (▶ FIGURE 15.4). The majority of water is replaced by ingested fluids and foods, even though some is generated in the tissues during ATP production via aerobic catabolism.

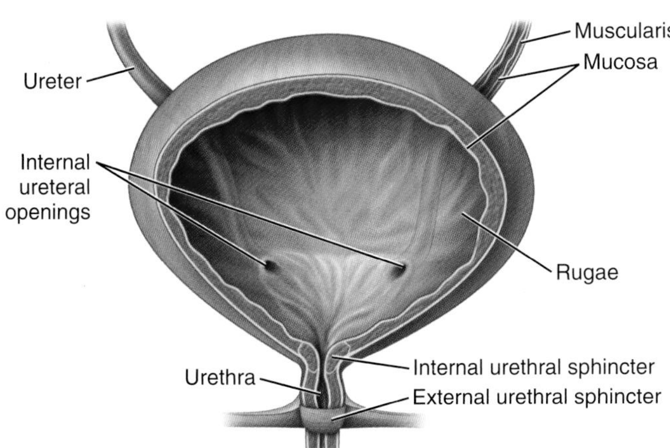

FIGURE 15.3 ▶ Bladder and urethra. The urinary bladder is a temporary holding tank for urine before it is excreted through the urethra. Like the stomach, the bladder has rugae that allow for expansion.

Balancing fluid intake and output is a function of the ***thirst center*** within the hypothalamus. When water output exceeds intake, dehydration begins, saliva production decreases (leading to dry mouth), and the thirst center is stimulated. This stimulation creates the sensation of thirst, which signals a person to drink water or other beverages. While the signal to increase fluid intake is important in keeping tissues well hydrated, the homeostatic mechanisms that regulate output are far more important to maintaining fluid level balance and composition throughout the body. The kidneys play a central role in these fluid management processes. Specifically, the kidneys:

- ***Regulate blood volume***—As the kidneys filter blood, they adjust the body's overall fluid volume by returning fluid to the plasma or eliminating it in urine. This balance of fluid retention versus excretion is directed by hormones, especially antidiuretic hormone (ADH) and aldosterone, in response to decreased blood volume and pressure, or low blood sodium levels. Both ADH and aldosterone increase water retention in the kidneys, which increases blood volume and blood pressure.
- ***Regulate blood pressure***—Recall from Chapter 9 that a decrease in blood volume or pressure, as

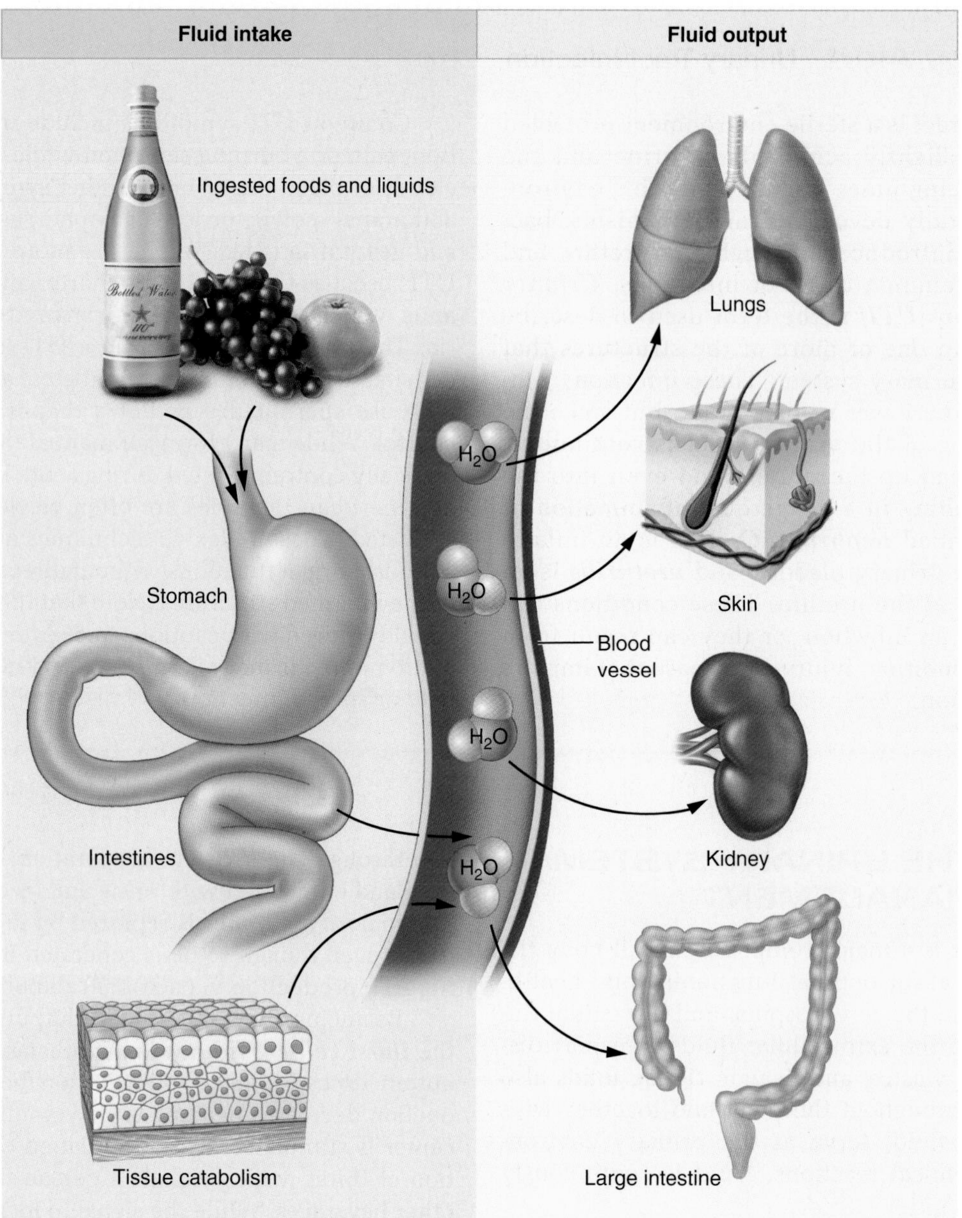

FIGURE 15.4 ▸ Sources of fluid intake and output. Ingested foods and liquids, as well as nutrient catabolism in the tissues, provide fluid to the body. Water is eliminated through the skin, respiration, and excretion.

well as any decrease in renal circulation, stimulates the release of renin from the kidneys. This enzyme initiates the renin-angiotensin II-aldosterone (RAA) pathway. The aldosterone released via this process causes retention of sodium ions by the kidneys. These ions act as osmotic particles that increase water reabsorption. The net result of the RAA pathway is an increase in both blood volume and pressure. Stress commonly causes a decrease in renal circulation, which leads to and stimulation of the RAA pathway. In contrast, the kidneys respond to high blood pressure by increasing water excretion through urine.

- ***Manage blood electrolytes***—The blood levels of essential ions such as sodium (Na^+), potassium (K^+), calcium (Ca^{2+}), and phosphate (HPO_4^{2-}) are all regulated by the kidneys during urine formation.
- ***Maintain blood pH***—During urine formation, the kidneys regulate the concentration of hydrogen ions (H^+) to be excreted in urine and the retention of bicarbonate ions (HCO_3^-). Maintaining a normal blood pH depends on the kidneys' ability to balance the acidic H^+ and alkaline HCO_3^-.
- ***Eliminate wastes***—Waste products of cellular metabolism are filtered out of the blood by the kidneys. These include urea and ammonia from amino acid catabolism, uric acid produced by nucleic acid breakdown, and bilirubin from the destruction of old red blood cells. Additionally, many drugs, toxins, and excess amounts of vitamins, minerals, and other

nutritional supplements are filtered out and excreted in urine.
- ***Produce hormones***—Although the kidneys are not primary endocrine glands, they do secrete two hormones. ***Calcitriol*** is the active form of vitamin D, which is essential for calcium absorption, and ***erythropoietin*** stimulates red blood cell production in the bone marrow.

What Do You Think? 15.2

- How does stress stimulate the RAA pathway?
- How does increased urine output help to lower blood pressure?

URINE FORMATION AND COMPOSITION

The kidneys cleanse the blood of metabolic waste products through urine formation. In this process, nephrons exchange fluids and dissolved substances between the blood and their own small tubules to either retain or eliminate water, specific ions, and molecules from the body. Urine formation involves three distinct steps that occur in different portions of the nephron. Filtration occurs in the **renal corpuscle**, which is the bulbous initial portion of the nephron (FIGURE 15.5). The twisted hollow **renal tubule** that extends from the corpuscle is the site of reabsorption and secretion, the other two steps in urine formation. All three processes (summarized in TABLE 15-1 at the end of this section) occur through the passive or active transport mechanisms outlined in Chapter 3.

Filtration

As the first step in urine formation, filtration occurs in the renal corpuscle of the nephron. The bulbous renal corpuscle is made up of a tangled network of capillaries called a ***glomerulus*** surrounded by a cup called the ***glomerular*** or ***Bowman capsule*** (FIGURE 15.5). Blood enters the glomerulus through an afferent arteriole and leaves through a smaller efferent arteriole. The narrow outgoing vessel combined with the knotted arrangement of capillaries significantly slows blood flow and increases pressure within the glomerulus. Just as an espresso machine forces pressurized water through tiny holes to make cappuccino, fluid is forced through the capillary walls into the waiting glomerular capsule via **filtration** (FIGURE 15.6). This passive transport mechanism forms a ***glomerular filtrate*** of water and dissolved substances, including electrolytes, glucose, amino acids, hydrogen and bicarbonate ions, and metabolic waste products such as urea, ammonia, uric acid, and bilirubin. Changes in blood volume, arterial blood pressure, or blood composition all affect the rate of filtration in the renal corpuscle.

By the Way

People with hypertension tend to urinate more frequently as the kidneys attempt to lower blood pressure by increasing fluid output. Patients being treated for hypertension are often prescribed diuretic medications to further increase urinary output.

Reabsorption

Glomerular filtrate leaves the renal corpuscle through the Bowman capsule and enters the renal tubule, a passageway made of a single layer of epithelium. The tubule is divided into three regions, each named for its general shape and location (FIGURE 15.5). The initial coiled portion that attaches to the glomerular capsule is the ***proximal convoluted tubule***. It connects to a prominent hairpin bend, the ***loop of Henle***, which in turn attaches to another coiled portion called the ***distal convoluted tubule***. All three sections of the renal tubule are surrounded by a network of ***peritubular capillaries*** formed when the efferent arteriole that exits the glomerulus divides. The thin walls in both the capillaries and tubule allow for easy exchange between blood in the capillaries and filtrate inside the renal tubule.

Together the nephrons of both kidneys produce about 125 ml (just over 4 oz) of glomerular filtrate per minute, which translates to 180 l (47.5 gal) every day. Based on these amounts, it is obvious that most of this filtrate is not eliminated as urine; 97% to 99% of the glomerular filtrate is passed from the renal tubule back into the capillary and retained in the bloodstream through **reabsorption** (FIGURE 15.6). A majority of reabsorption occurs at the proximal convoluted tubules, including all of the glucose and amino acids initially filtered, about two-thirds of the water, and varying quantities of electrolytes. At this point, fluid that remains inside the renal tubule is referred to as ***tubular fluid***. The reabsorption process continues along the entire length of the tubule to allow the balancing of electrolytes, pH, and fluid volumes.

Secretion

The third and final process of urine formation is **secretion**, which is movement of substances from the blood in the peritubular capillaries back into the renal tubule (FIGURE 15.6). Secretion helps the blood eliminate metabolic wastes such as ammonia and urea, excess hydrogen and potassium ions, as well as many medications. Similar

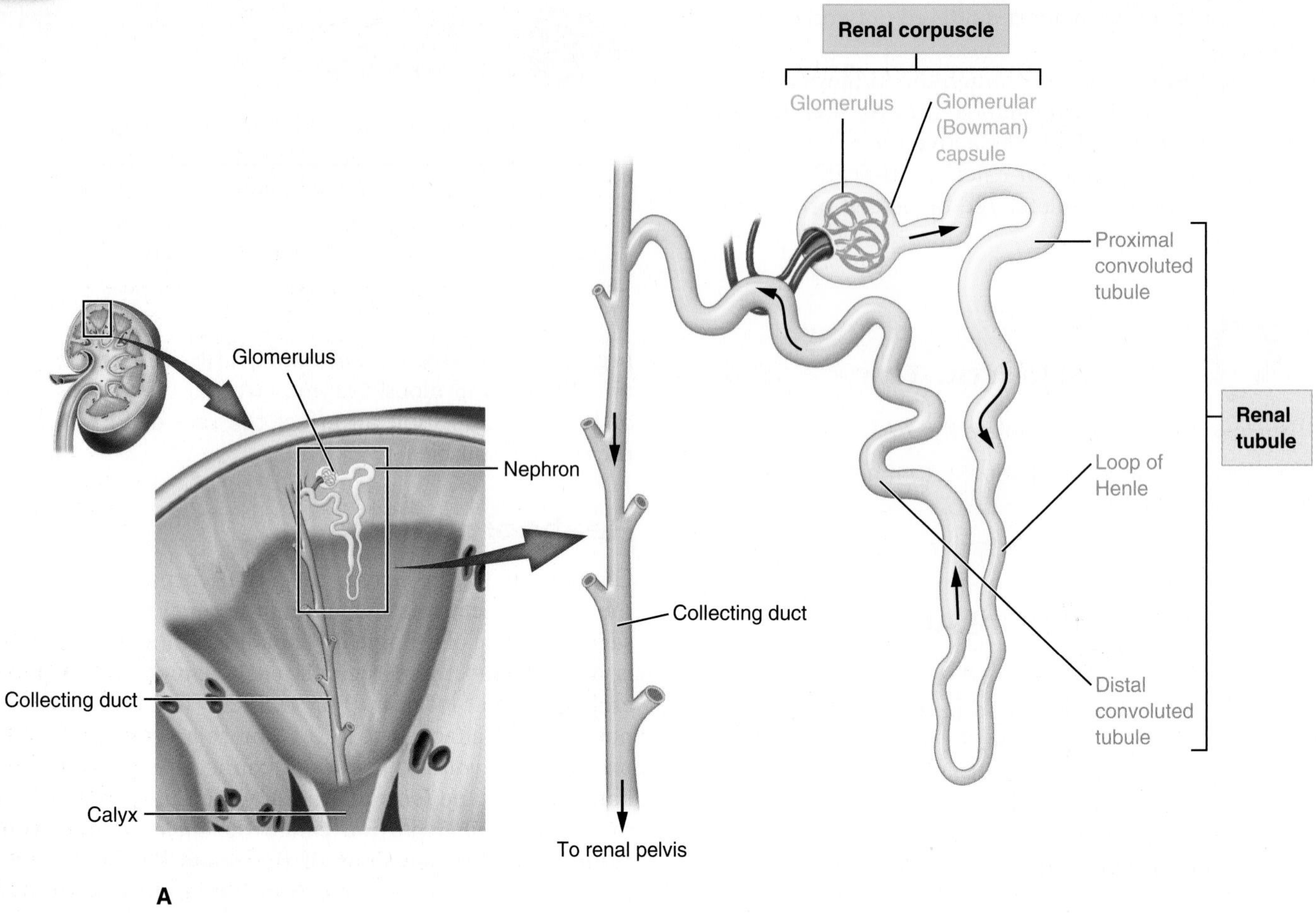

Renal corpuscle
Glomerulus
Glomerular (Bowman) capsule
Proximal convoluted tubule
Renal tubule
Loop of Henle
Distal convoluted tubule
Glomerulus
Nephron
Collecting duct
Collecting duct
Calyx
To renal pelvis

A

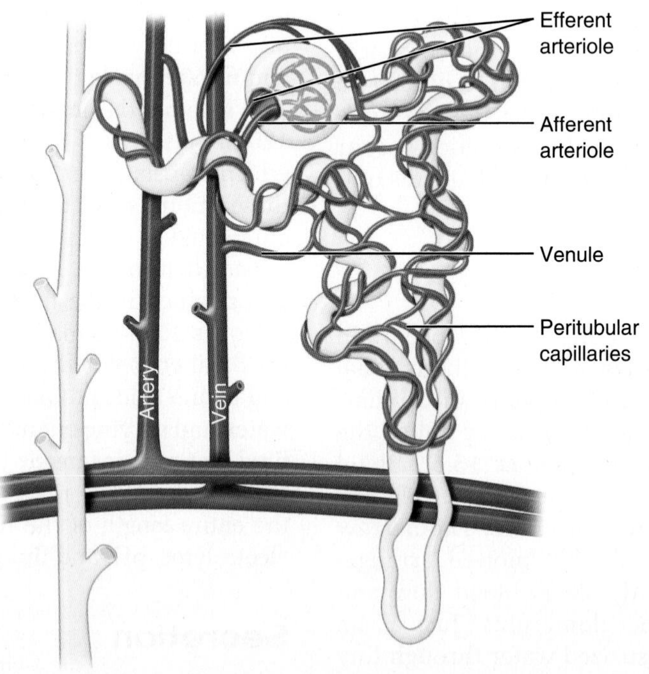

B

FIGURE 15.5 ▶ Nephron. A. Each nephron consists of a renal corpuscle (which includes the glomerulus and the Bowman capsule) and the renal tubule. **B.** An afferent arteriole carries blood into the glomerulus, and it exits the capsule via a smaller efferent arteriole. This arrangement increases the pressure inside the capsule for filtration. The tubule is surrounded by a network of peritubular capillaries so that fluid and dissolved substances can be exchanged between the blood and tubular filtrate. To view an animation on kidney function, visit http://thePoint.lww.com/Archer-Nelson.

Table 15-1 Summary of Urine Formation

Process	Nephron Location	Direction of Fluid and Particle Movement	Description and Purpose
Filtration	Renal corpuscle	From blood to renal tubule	Fluid is forced out of the blood via capillary pressure through the walls of the glomerulus and into the glomerular (Bowman) capsule to begin the process of urine formation.
Reabsorption	Along entire course of the renal tubule and in the collecting ducts	From renal tubule to blood	Fluid and substances move from the renal tubule back into the blood of the peritubular capillaries. This conserves water and vital substances, including glucose, amino acids, and various electrolytes and buffers.
Secretion	Along entire course of the renal tubule and in the collecting ducts	From blood to renal tubule	Fluid and substances move from the blood into the renal tubule to be excreted with the urine. This includes H^+, K^+, ammonia, and urea, as well as many medications.

to reabsorption, secretion occurs throughout the length of the renal tubule, although the majority occurs in the distal convoluted tubule. Because reabsorption and secretion both occur along the entire length of the renal tubule, regulation of pH and electrolytes can be readily maintained within homeostatic limits.

Tubular fluid left within the distal convoluted tubules drains into ***collecting ducts*** (FIGURE 15.5). Several nephrons converge into one collecting duct where reabsorption and secretion continue. Water reabsorption in the collecting ducts concentrates tubular fluid into urine, which passes into the renal pelvis of the kidney, where it drains into the ureter.

Composition of Urine

Since the kidneys are responsible for cleansing waste and metabolic byproducts from blood, analysis of urine composition provides as much information about general health as it does about kidney function. In healthy individuals, urine is a straw-colored or transparent yellow fluid that is 95% water, plus pigments and variable levels of nitrogenous waste and electrolytes. The pigments that color urine come from the breakdown of bile and hemoglobin. Urine normally contains high levels of nitrogenous waste such as urea and ammonia from amino acid catabolism, uric acid from the breakdown of nucleic acids, and creatinine (the byproduct formed when creatine phosphate is used by muscle fibers for energy). Smaller levels of sodium, potassium, chloride, sulfates, and phosphates are excreted in urine to maintain homeostatic ranges of these important electrolytes. The average pH of urine is slightly acidic at 6.0, but its normal range is between 4.5 and 8.0. Changes in either pH or concentrations of discarded substances are often related to diet. For example, high-protein diets tend to make urine more acidic, while a vegetarian diet increases alkalinity. Eating beets gives urine a reddish

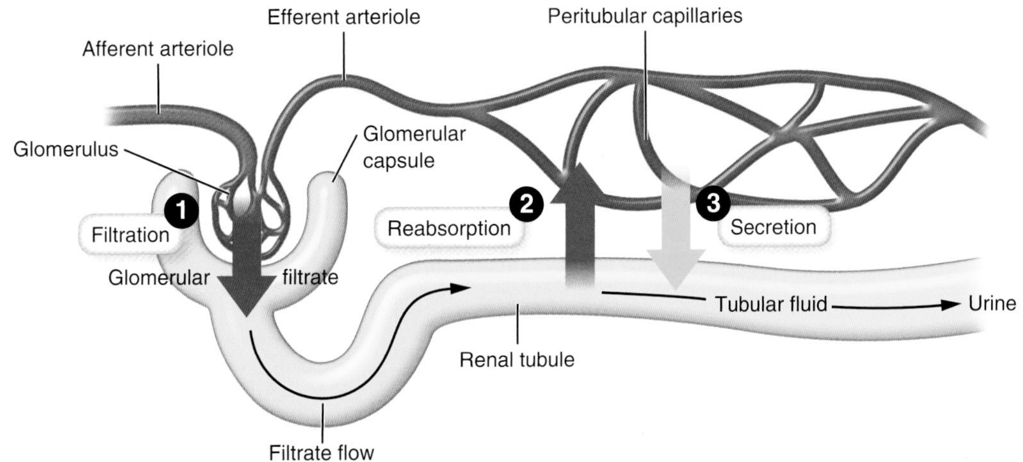

FIGURE 15.6 ▸ Urine formation. Urine formation involves three processes: filtration, reabsorption, and secretion. Each process occurs in a different portion of the nephron.

By the Way

It is estimated that each kidney contains around one million nephrons.

tinge, and asparagus can give it a distinct odor. However, changes in the composition of urine can also be a sign of disease, so urinalysis is used to diagnose various pathologies. For example, the presence of white blood cells in urine can indicate infection. Since urine should never contain glucose or proteins, their presence can indicate several conditions such as diabetes, malnutrition, or nephritis. Urinalysis can also be used to test for the presence of drugs or hormones. Home pregnancy tests check urine for the hormone ***human chorionic gonadotropin (HCG)***, which is produced during pregnancy.

What Do You Think? 15.3

- What changes in blood composition and blood pressure would increase filtration in the nephrons?
- Why do you think so much reabsorption occurs in the proximal convoluted tubule?

AGING AND THE URINARY SYSTEM

Several structural and physiologic changes associated with aging result in diminished urinary function, especially when accompanied by decreased circulation due to cardiovascular disease or diabetes. For example, the number of nephrons decreases with age, resulting in a slower rate of filtration and reduced ability to cleanse the blood. The kidneys' renin production and sensitivity to ADH also decreases, which hinders their efficiency in regulating fluid volume and blood pressure. Many older adults experience a general reduction in thirst and subsequently reduce their water intake. All these combined factors predispose the elderly to dehydration. Additional renal changes include a decline in the production of the hormones erythropoietin and calcitriol, which contributes to anemia and poor calcium reabsorption from the small intestines.

This general decline in kidney function, as well as increased incidence of conditions such as diabetes, hypertension, and congestive heart failure, can lead to chronic edema in the lower extremities. In these cases, improving edema uptake and local fluid flow may overtax the already struggling kidneys and/or heart. Since it is not uncommon for geriatric clients to have swollen feet or ankles, it is important for therapists to avoid direct manipulation of the edematous tissue or vigorous reflexive stimulation of the urinary system until kidney and cardiac dysfunctions have been ruled out as causes of the edema. In clients already diagnosed with these conditions, regular manual therapy sessions administered before serious edema develops can be a helpful preventive or early management method. However, these sessions should still avoid heavy soft tissue manipulation and focus more on gentle stroking, passive movement, and stretching.

SUMMARY OF KEY POINTS

- Urinary system functions include regulation of fluid volumes, blood pressure, and electrolyte and pH levels; the excretion of liquid waste products in urine; and secretion of the hormones calcitriol and erythropoietin.
- The organs of the urinary system are the kidneys, ureters, bladder, and urethra. All processing of fluids occurs in the kidneys. The ureters and urethra act as transportation corridors, and the bladder serves as a temporary storage tank for urine before it is excreted.
- Nephrons are the functional units of the kidney where urine formation occurs. These microscopic structures have two divisions: the renal corpuscle and renal tubule. The renal corpuscle consists of a knotted network of capillaries called the glomerulus, surrounded by the cup-shaped glomerular (Bowman) capsule. The renal tubule has three primary sections: the proximal convoluted tubule, loop of Henle, and distal convoluted tubule.
- Urine formation in the nephron occurs through three processes. Filtration, which takes place in the renal corpuscle, moves substances from the blood into the renal tubule. Absorption occurs primarily in the proximal tubule, returning water and other needed substances back to the blood. Secretion is the final process, in which waste products and unneeded substances move out of blood and into urine for excretion.
- Micturition, or urination, is the excretion of liquid waste as urine. This process is controlled by two sphincters in the urethra. The internal sphincter at the junction between the bladder and urethra is involuntary muscle controlled by the autonomic nervous system, and the external sphincter is voluntary muscle. Therefore, the ability to control urination is a learned behavior.

REVIEW QUESTIONS

Short Answer

1. List the four components of the urinary system.
2. What is the difference between the ureters and the urethra?
3. List the key steps in urine formation in the order that they occur in the kidneys.
4. Name the two hormones (one secreted by the posterior pituitary and the other by the adrenal medulla) that affect kidney functions and explain the effect of each.

Multiple Choice

5. Which organ serves as the site for urine storage?
   a. kidneys
   b. bladder
   c. ureters
   d. urethra
6. Which of the following is considered the functional unit of the urinary system?
   a. glomerulus
   b. nephron
   c. renal medulla
   d. renal cortex
7. What is the name of the outer section of tissue in the kidney?
   a. renal cortex
   b. renal medulla
   c. renal hilum
   d. renal tubule
8. What is the name for the bulbous portion of the nephron?
   a. hilum
   b. renal cortex
   c. renal circularis
   d. renal corpuscle
9. The expansion at the upper end of the ureter inside the kidney is called the
   a. calyx
   b. pyramid
   c. pelvis
   d. internal ureter
10. Which two structures make up the renal corpuscle?
    a. proximal and distal corpuscle
    b. glomerulus and proximal tubule
    c. renal and adipose capsule
    d. Bowman capsule and glomerulus
11. The majority of secretion processes occur in which portion of the nephron?
    a. proximal convoluted tubule
    b. distal convoluted tubule
    c. loop of Henle
    d. collecting duct
12. Which urine formation process moves substances such as glucose and water out of the renal tubule and into the blood?
    a. filtration
    b. reabsorption
    c. secretion
    d. urine concentration
13. What is the name for the bundle of capillaries inside the Bowman capsule?
    a. glomerulus
    b. peritubular capillaries
    c. Bowman capillaries
    d. renal corpuscle
14. What is the normal concentration of water in urine?
    a. 10%
    b. 50%
    c. 95%
    d. 75%
15. Urine generally contains high concentrations of nitrogenous waste products such as urea, ammonia, and
    a. phosphates
    b. uric acid
    c. calcium bicarbonate
    d. amino acid

16 The Reproductive System

LEARNING OBJECTIVES

Upon completion of this chapter, you will be able to:

1. Explain the function of the male and female reproductive systems and discuss their relationship and importance to manual therapy practices.
2. List the structural features and hormonal control processes that the male and female reproductive systems have in common.
3. Name, locate, and explain the general function of the male genitalia in terms of the primary and accessory reproductive organs.
4. Name, locate, and explain the general function of the female genitalia in terms of the primary and accessory reproductive organs.
5. Describe the stages and hormonal regulation of the female reproductive cycle.
6. Explain the key physiologic processes that occur during each stage of pregnancy and childbirth.
7. Discuss the reproductive system changes that occur as the body ages.

> "The reproduction of mankind is a great marvel and mystery."
>
> MARTIN LUTHER
> *16th Century Theologian*

KEY TERMS

amniotic (am-ne-OH-tik) **sac**
blastocyst (BLAS-to-sist)
conception (kon-SEP-shun)
ejaculation (e-jak-u-LA-shun)
embryo (EM-bre-o)
erection (e-REK-shun)
estrogen (ES-tro-jen)
fertilization (fer-tih-li-ZA-shun)
fetus (FE-tus)
gamete (GAM-eet)
gestation (jes-TA-shun)
implantation (im-plan-TA-shun)
labor (LA-ber)
menstrual (MEN-stru-al) **cycle**
ovulation (ah-vu-LA-shun)
ovum (O-vum)
placenta (plah-SEN-tah)
progesterone (pro-JES-teh-rone)
semen (SE-men)
sexual reproduction (SEKS-u-al re-pro-DUK- shun)
sperm
testosterone (tes-TOS-ter-own)

Primary System Components

Male Reproductive System

- **Testes**
  - Seminiferous tubules
- **Duct system**
  - Epididymis
  - Vas deferens
  - Ejaculatory duct
  - Urethra
- **Accessory glands**
  - Seminal vesicles
  - Prostate
  - Bulbourethral glands
- **Penis**
- **Scrotum**

Female Reproductive System

- **Ovaries**
  - Follicles
  - Corpus luteum
- **Fallopian tubes**
- **Uterus**
  - Endometrium
- **Vagina**
- **Vulva**
- **Clitoris**

Primary System Functions

- **Production of hormones and development of sex cells (sperm and egg)**
- **Producing offspring (reproduction)**

This chapter explores the commonalities between the male and the female reproductive systems along with the organs, functions, and regulatory processes that are unique to each. For manual therapists, the links between reproductive functions, the hypothalamus, and pituitary gland are of particular interest. Understanding these connections leads to an appreciation of how stress and emotions affect menstruation, fertility, and pregnancy. This knowledge encourages the use of manual therapies to support reproductive processes since stress reduction is a key benefit. The information in this chapter will be useful for those students who may ultimately choose to specialize in providing therapeutic support to women during pregnancy and childbirth.

Like all living organisms, humans possess a biological need to reproduce. In humans, this need is driven by specific hormonal changes. **Sexual reproduction** is the process in which one male sex cell and one female sex cell join. The sex cells from each parent, called **gametes**, carry half the genetic code for an offspring. When joined, a complete and unique DNA blueprint is formed. This single cell divides and differentiates to create a new individual who may share some similar qualities but does not duplicate either parent.

In humans, the basic need to reproduce is a powerful and complicated process due to connections between the pleasure centers of the brain and the endocrine structures that control reproduction. It can be driven by a desire for sex (libido) or overpowered by a lack of desire. In the same way that emotions and other factors influence eating patterns, libido is deeply affected by memories, emotions, values, thoughts, and beliefs. This makes sexual desire and reproduction part of the much larger "marvel and mystery" of life.

COMMON STRUCTURES AND PHYSIOLOGY

While the organs and processes of the male and female reproductive systems differ (see TABLE 16-1 for a comparison), they also share a few organizational and functional features, which include:

- ***Primary reproductive organs*—Gonads** are organs that produce the gametes (sex cells). In males, the testes produce sperm, and in females, the ovaries produce eggs.
- *Genitalia*—While all reproductive organs can be called **genitalia**, the term is more commonly used to refer to the male and female external sex organs.
- *Accessory organs*—These organs include the ducts and chambers that make up the tracts connecting the gonads with genitalia, as well as the glands that secrete fluids that protect, transport, or facilitate the movement and joining of gametes.
- ***Hormonal control process*—**Both the male and female reproductive systems are stimulated and controlled by the endocrine system. In a two-step process, the hypothalamus stimulates the anterior pituitary to release two gonadotropins, **follicle stimulating hormone (FSH)** and **luteinizing hormone (LH)**. In turn, these hormones stimulate the gonads to produce gametes and sex hormones. Interestingly, even though the same gonadotropins are released in both males and females, their names describe their functions within the female reproductive system.

By the Way

Gametes are produced through a cell division process called ***meiosis*** in which parent cells go through two successive nuclear divisions, resulting in a gamete (sperm or egg) that contains half the genetic code.

MALE REPRODUCTIVE SYSTEM

The primary reproductive organs in males are the testes. A complex network of ducts and tubules plus a number

Table 16-1 Comparison of the Male and Female Reproductive Systems

Feature	Male	Female
Gonads	Testes	Ovaries
Gametes	Sperm	Eggs (ova)
External genitalia	Penis and scrotum	Vulva
Reproductive tract	Duct system: seminiferous tubules, seminal vesicles, epididymis, vas deferens, ejaculatory ducts, and urethra	Fallopian tubes, uterus, and vagina
Hormones	Testosterone	Estrogen and progesterone
Hormone secreting cells	Leydig cells	Follicles and corpus luteum

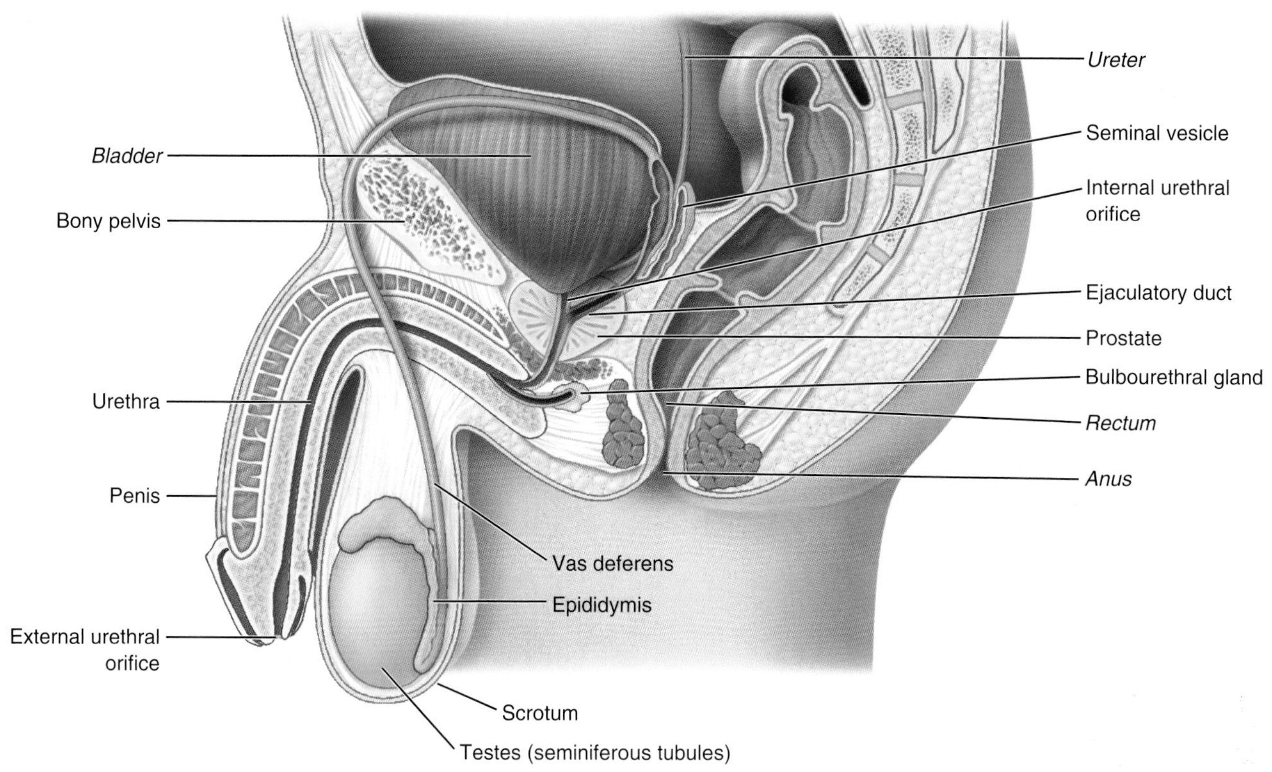

FIGURE 16.1 ▶ Male reproductive system, sagittal section.

of secretory glands make up the accessory organs, and the penis and scrotum are the external genitalia (▶ FIGURE 16.1).

Scrotum and Testes

The **testes**, also called **testicles**, are paired, oval-shaped gonads that lie within a supportive external sac of skin, the **scrotum** (▶ FIGURE 16.2). Each testis is surrounded by a fibrous connective tissue capsule that also extends inward to divide the gonad into 200 to 300 small compartments, or **lobules**. Within each lobule is one to three tightly coiled tubes called **seminiferous tubules** that produce **sperm**, the male gamete. A group of specialized cells called ***Sertoli cells*** line the seminiferous tubules. These cells nourish developing gametes, produce a fluid for the transport of sperm, and play a role in regulating sperm production. The external position of the testes is by design, since the optimal temperature for sperm production and survival is a few degrees below the normal core body temperature of 98.6°F.

By the Way

If straightened out, the seminiferous tubules would measure the length of five football fields.

As discussed in Chapter 9, the testes also function as endocrine glands. When stimulated by LH from the anterior pituitary, **testosterone** is secreted by ***Leydig cells*** located between the seminiferous tubules. Testosterone, along with FSH from the pituitary, stimulates the production of sperm. Testosterone is responsible for the development of the male genitalia at puberty, as well as the development of secondary sex characteristics, such as enhanced muscular and skeletal growth, deepening of the voice, thickening of skin, and the development of facial and body hair.

Accessory Organs

Sperm leaves the seminiferous tubules, and flows into an extensive duct system that stores and transports it. As sperm moves through this network of ducts, various accessory glands secrete fluids that provide nutrition and protection to the gametes. This combination of sperm and secretions forms a thick whitish fluid called **semen.** The spermatic duct system includes the epididymis, vas deferens, ejaculatory duct, and urethra. The accessory organs that produce and secrete the liquid components of semen include the seminal vesicles, prostate, and bulbourethral glands.

Duct System

The production of sperm and fluid within the seminiferous tubules of each testicle eventually creates enough internal pressure to force the contents out into a tightly coiled segment, the **epididymis**,

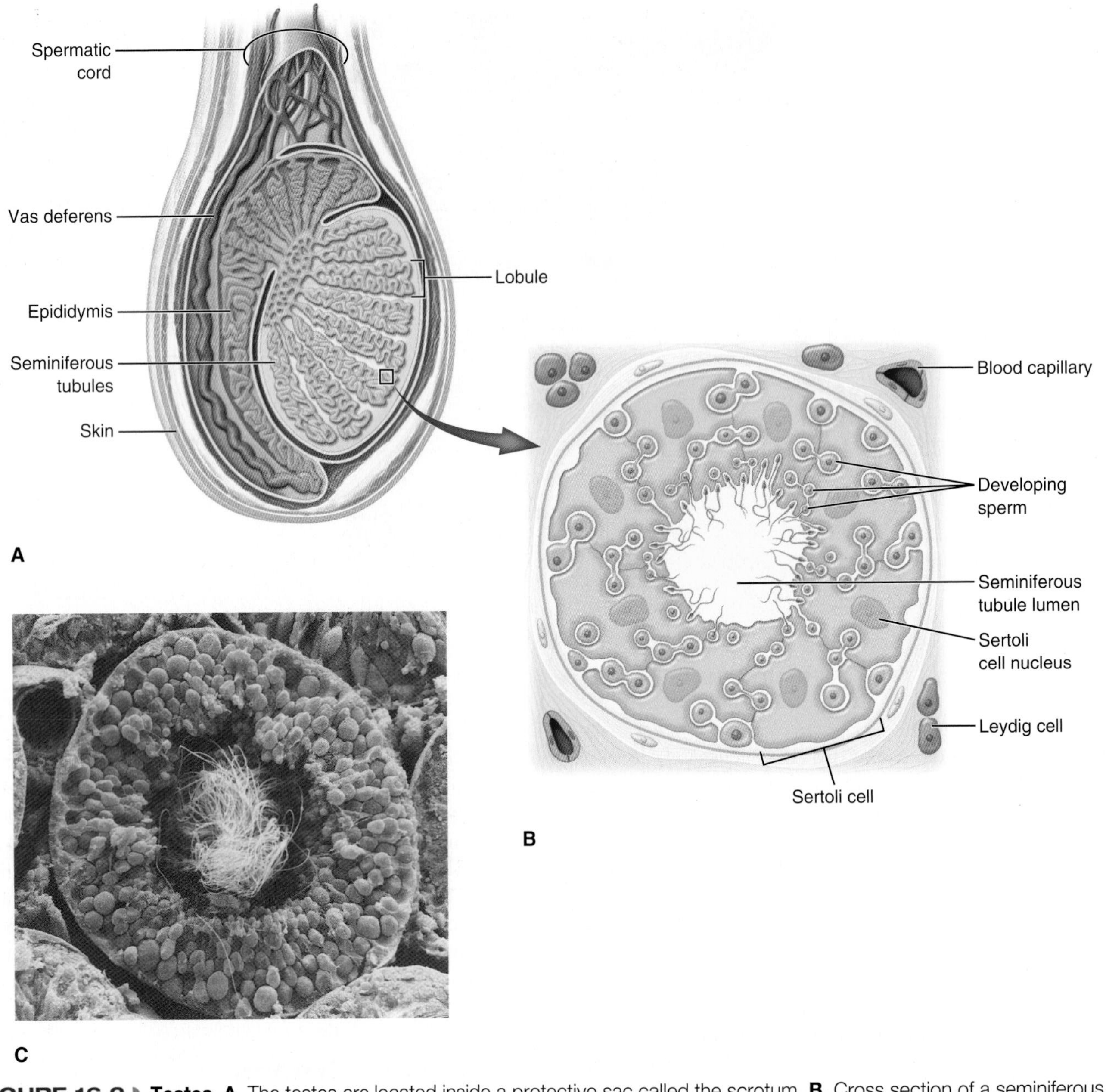

FIGURE 16.2 ▸ Testes. A. The testes are located inside a protective sac called the scrotum. **B.** Cross section of a seminiferous tubule shows sperm, Sertoli and Leydig cells. **C.** Micrograph shows the tails of mature sperm cells projecting into the lumen of the seminiferous tubule.

which hugs the posterolateral side of each testis (▸ FIGURE 16.2A). Sperm mature in the epididymis, developing their motility and ability to fertilize a female egg. While this maturation process takes about 10 to 14 days, sperm can be stored in the epididymis for several months. During sexual arousal, peristaltic contractions of smooth muscle propel sperm into the next duct, the vas deferens.

The **vas deferens** is a larger and less convoluted tube that carries sperm out of the scrotum then up and over the posterior side of the bladder. At this point, connective tissue wraps the vas deferens together with nerves,

By the Way

A *vasectomy* is a form of male contraception in which a portion of the vas deferens within the scrotal sac is surgically removed or cauterized, thereby preventing sperm from being ejaculated during sexual intercourse. It can sometimes be reversed but is generally considered to be a complete sterilization procedure.

blood, and lymph vessels to form a bundled cord known as the ***spermatic cord***. The vas deferens can also store sperm for several months. During sexual arousal, smooth muscle in the walls of the vas deferens contract to forcefully propel sperm through the ejaculatory ducts and urethra. The **ejaculatory ducts** pass through the prostate gland before merging into the **urethra**. **Ejaculation** is the forceful ejection of semen from the urethra.

Accessory Glands

There are several accessory glands that secrete the liquid portion of semen. About 60% of this fluid is produced by the **seminal vesicles** that lie between the bladder and rectum (see FIGURE 16.1). This pair of glands secretes a thick alkaline fluid that helps to neutralize the acidic pH inside the urethra. This fluid also contains fructose, which sperm catabolize for ATP, and prostaglandins, which support sperm motility and viability. The seminal vesicles secrete their fluid into the vas deferens where it joins the ejaculatory duct.

The **prostate** is a donut-shaped gland that encircles the upper portion of the urethra. During ejaculation, the prostate secretes a milky fluid directly into the urethra. This prostatic fluid makes up about 25% to 35% of semen and contains several substances that support sperm mobility and ensure that semen does not clot or congeal. Two additional secretory glands, the **bulbourethral glands**, are situated inferior to the prostate on either side of the urethra. These pea-sized glands produce a clear mucus-like substance that lubricates the lining of the urethra to protect sperm during ejaculation. The secretion from the bulbourethral glands is also quite alkaline, which helps to neutralize any remaining acids from urine.

Penis

The **penis** contains the urethra, which serves as a passageway for both urine and semen. As a copulatory organ, the primary function of the penis is to deliver sperm into the reproductive tract of the female. It consists of three segments: the ***root*** is the point where it attaches to the underside of the bony pelvis, the middle portion is the ***shaft*** (body), and the distal portion is the head, or ***glans penis***. The shaft consists of three fascia-covered cylinders of spongy ***erectile tissue*** divided into multiple compartments by thin walls of smooth muscle and elastic connective tissue. During sexual arousal, parasympathetic stimulation leads to relaxation of the smooth muscle in the erectile tissue, allowing large quantities of blood to flow into the compartments and fill the erectile tissue within the fascial cylinders. When the tissue is engorged with blood, it causes the penis to enlarge and stiffen, creating an **erection**, which helps the penis penetrate the vagina during intercourse.

Because the smooth muscle surrounding the erectile tissue is under parasympathetic control, a man may experience an erection during a manual therapy session as a part of a relaxation response rather than sexual arousal. Understandably, this can be a source of consternation and embarrassment for the client, therapist, or both. Generally, some type of acknowledgement or response by the therapist is required. While there is no single correct way to mediate this situation, therapists using techniques that focus on relaxation must be prepared for this possibility. Therapists should consider their history and communications with the client, as well as their own personal and professional boundaries, to determine how to tactfully and respectfully handle this occurrence.

What Do You Think? 16.1

- What positive or negative impact do the stress reduction and general relaxation effects of manual therapy have on the reproductive system, and why does this occur?
- If a couple is having difficulty getting pregnant, the male is often advised to shift from wearing tight-fitting types of underwear to looser-fitting boxer shorts. Why would this be helpful?

FEMALE REPRODUCTIVE SYSTEM

The female system is specifically designed to conceive, carry, and nourish a developing fetus and deliver offspring. Organs include the ovaries, fallopian tubes, uterus, and vagina, which collectively make up the female reproductive tract (FIGURE 16.3). In addition to these components, the **mammary glands**, or breasts, are considered accessory reproductive organs. While breasts develop at puberty, they only produce milk after childbirth when **prolactin** from the anterior pituitary stimulates milk production, and **oxytocin** (from the posterior pituitary) enables milk release.

Ovaries

Each female gonad, or **ovary**, is about the size of a large peach pit. Situated on either side of the pelvic cavity, each ovary is held in place by ligaments that attach it to the abdominal wall and uterus. Similar to the testes, the primary function of the ovaries is production of gametes: the **eggs** or ***ova*** (the singular form is **ovum**, which is Latin for *egg*). There are thousands of eggs in the ovaries, each held within a protective ring of specialized cells called a **follicle**. Approximately once a month, FSH from the pituitary stimulates several follicles to ripen or mature together with the ovum contained within. Generally, only one of these follicles migrates to the edge of the ovary and ruptures to release its egg, a process known as **ovulation.** Once the egg has been discharged, the remaining follicle is transformed into a glandular mass known as the **corpus luteum**.

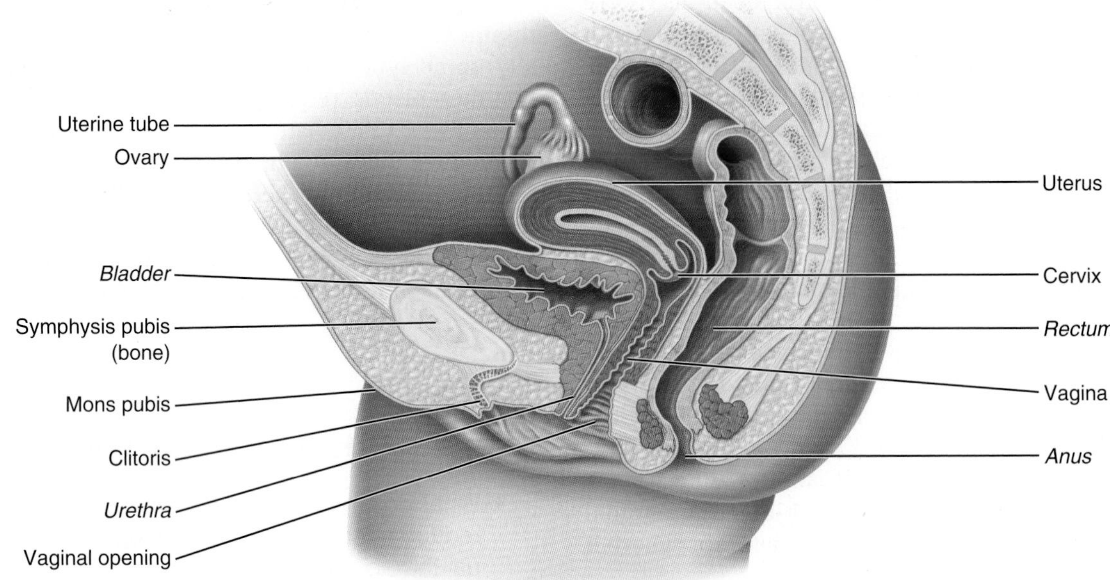

FIGURE 16.3 ▸ Female reproductive system, sagittal section.

The mature follicles and eggs that don't get released disintegrate inside the ovary, and the cycle begins again.

The ovaries also produce the sex hormones **estrogen** and **progesterone** when stimulated by gonadotropins from the anterior pituitary. Like testosterone in men, estrogen stimulates the development of the reproductive organs and the secondary sex characteristics at puberty. In women, these secondary characteristics include enlargement of the breasts, hair growth in the axilla and pubis, widening of the pelvis, and increased subcutaneous fat deposits in the hips, thighs, buttocks, and breasts. Estrogen is also necessary for the onset of the menstrual cycle.

Female Genitalia

The external female sex organs are collectively referred to as the **vulva** (▸ FIGURE 16.4). Just superior to the vulva, there is a soft mound of fatty tissue over the pubic bone, the ***mons pubis***, which literally translates as "pubic mountain." After puberty, the mons pubis is covered with pubic hair. The region between the mons pubis and anus is called the **perineum**. (In males, the perineum is the region between the scrotum and the anus.)

Soft tissue folds called **labia** (lips) surround the ***vaginal*** and ***urethral orifices***. The larger folds are the ***labia major***, while the ***labia minor*** are the smaller folds immediately medial to the major. The labia's protective folds over the vaginal opening help keep microbes out but easily separate for intercourse. The ***vestibular glands*** are two mucus-producing glands located in the rims of the vaginal orifice. They secrete large amounts of mucus during sexual arousal to assist in penile penetration. At the anterior junction between the two sides of the labia minor, there is a small bud of highly sensitive tissue, the **clitoris**. The clitoris can be compared to the shaft of the penis because during sexual arousal, it too becomes engorged with blood, making it firm and erect.

Fallopian Tubes, Uterus, and Vagina

Females have three major accessory sex organs, beginning with the **fallopian tubes**, or **oviducts**, that carry the eggs from the ovaries to the uterus. Attached to the superior aspect of the uterus, these two short muscular tubes extend laterally and end at the ovaries in a cup-shaped

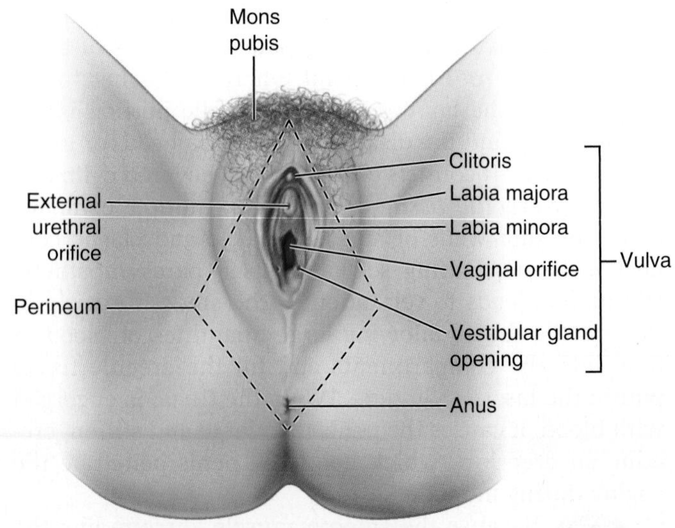

FIGURE 16.4 ▸ Female genitalia. The vulva is located in the perineum, the region from the mons pubis to the anus.

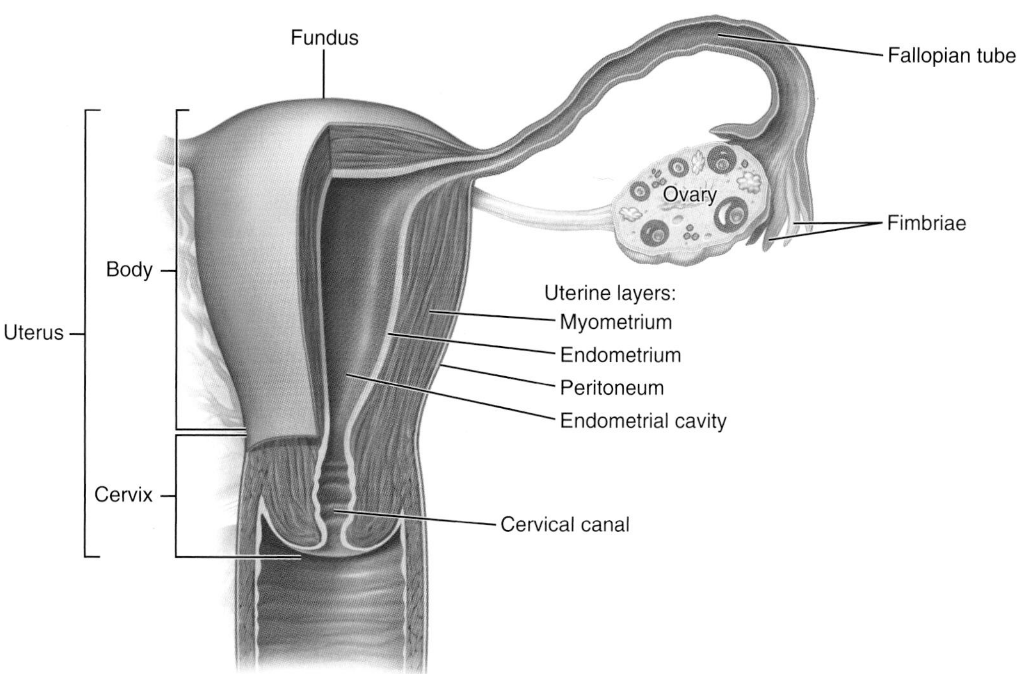

FIGURE 16.5 ▶ Fallopian tubes, uterus, and vagina, anterior view. Notice the different sections and layers of the uterus.

terminal portion with multiple projections called ***fimbriae*** (▶ FIGURE 16.5). These projections surround the ovary to catch an egg as soon as it is released, and they wave rhythmically to direct the captured ovum into the body of the fallopian tube. Once inside, the ovum moves toward the uterus via the peristaltic action created by smooth muscles in the fallopian tube walls. Ciliated cells in the lining create a waving motion with each muscle contraction that gently guides the ovum toward the uterus, a trip that takes approximately 5 days. The lining of the fallopian tube also has a large number of nonciliated cells that secrete a fluid that further facilitates movement of the egg and keeps it moist and healthy.

Uterus

The **uterus**, or ***womb***, is a pear-shaped organ with thick muscular walls and a small internal cavity (FIGURE 16.5). It is normally situated superior and posterior to the bladder in the pelvic cavity, but during pregnancy when it shelters a growing fetus, it expands to rest over the top of the bladder. Structurally it has three sections: the superior domed aspect, or ***fundus***, the broad central region called the ***body***, and the narrow inferior section, or ***cervix***. The lining of the uterus is a vascular mucous membrane called the **endometrium**. During the female reproductive cycle, hormonal signals cause the endometrium to thicken in preparation for implantation of a fertilized egg. If this doesn't happen, the lining dies, forming a mixture of dead cells and blood that is sloughed away as menstrual flow. At the time of childbirth, the muscular contractions of the uterus provide the force needed to push the fetus through the birth canal.

Pathology Alert Endometriosis

Sometimes, endometrial cells implant and grow outside the uterus in the peritoneal cavity, a condition called ***endometriosis.*** These cells grow and decay in response to the hormones that control the menstrual cycle. Because decayed cells outside the uterus cannot be shed through menstruation, they stimulate an inflammatory response and eventually become surrounded by fibrous scar tissue. Women with endometriosis may experience no symptoms. However, these growths, most commonly found on the fallopian tubes, ovaries, bladder, and colon, can cause dysmenorrhea (painful menstruation), pelvic and abdominal pain, painful intercourse, or difficulty with urination or defecation. The most common complaint associated with endometriosis is infertility. This usually results from excessive scar tissue, which causes adhesions in the fallopian tubes or ovaries. Because many women with endometriosis find this condition to be frustrating and stressful, manual therapy for relaxation and stress reduction is indicated. However, direct manipulation of the abdominal soft tissues is generally contraindicated, especially during menses.

Vagina

The **vagina** is a wide muscular tube approximately four inches long situated between the bladder and rectum that connects the cervix of the uterus with the exterior environment (FIGURES 16.3 and 16.5). Sperm is ejaculated into the vagina during sexual intercourse, and babies are delivered through it, which is why it is also called the birth canal. Because the vagina opens to the exterior environment, it has a mucous membrane lining contiguous with the linings of the uterus and fallopian tubes. This factor explains why vaginal infections and sexually transmitted diseases such as gonorrhea can be spread into the pelvic cavity if untreated.

1	2	3	4	5	6	7
8	9	10	11	12	13	14
15	16	17	18	19	20	21
22	23	24	25	26	27	28

Menses

Ovulation

Proliferative phase

Secretory phase

FIGURE 16.6 ▸ Female reproductive cycle. The 28-day menstrual cycle can be divided into three phases: the menses, the proliferative phase, and the secretory phase.

Menstrual Cycle

Although the reproductive systems of men and women are controlled by the same hypothalamic and pituitary hormones, the cyclic nature of these processes is unique to females. Throughout a woman's reproductive years, from puberty to menopause, the reproductive system goes through a series of predictable patterns of changing hormone levels, production of ova, and shifts in the vascularity and thickness of the uterine lining. The ***female reproductive cycle***, or **menstrual cycle**, represents the sequencing of these combined changes. Although it varies, the average menstrual cycle is 28 days long and is divided into three phases: menses, proliferation, and secretion (▸ FIGURE 16.6).

- ***Menses***—The first day of the cycle is the onset of **menstruation,** the bleeding caused by the sloughing away of the uterine lining. If the ovum has not been fertilized, both the egg and corpus luteum die, leading to decreased levels of estrogen and progesterone. Without the support of secretions from the egg and corpus luteum, the endometrium cannot be maintained and menstrual flow begins. In most women, menstruation lasts 4 to 5 days, but this can vary based on health condition and other factors.
- ***Proliferative Phase***—In days 6 through 14 of the cycle, the endometrium is signaled by increased levels of estrogen to grow, or proliferate. Estrogen is secreted from the follicles as they grow and mature inside the ovary. High levels of estrogen during this phase lead to a thickening of the endometrium and an increase in its vascularity. Estrogen also facilitates the passage of sperm by thinning the cervical mucus.
- ***Secretion Phase***—The secretion phase (days 15 to 28) begins with ovulation and ends with the onset of menses. While the exact timing of ovulation can vary, it generally occurs on day 14. The corpus luteum continues to secrete small levels of estrogen after ovulation in addition to large amounts of progesterone, which stimulate the endometrial changes needed for egg implantation. In addition to a general thickening and

Pathology Alert Common Menstrual Disorders

Painful menstruation is called ***dysmenorrhea.*** While many women experience some general discomfort or mild cramping, the symptoms associated with dysmenorrhea are severe enough to keep a woman from engaging in regular daily activities. Dysmenorrhea is characterized by deep aching pain in the abdomen and lower back, and/or sharp pain and cramping in the abdominopelvic region. Common associated symptoms include gastrointestinal disturbances such as nausea, vomiting, diarrhea or constipation, as well as headaches. Dysmenorrhea can be either a primary condition or a sign of an underlying pathology such as fibroid tumors, pelvic inflammatory disease (PID), endometriosis, or any number of sexually transmitted diseases. Manual therapies are generally indicated for reducing pain and stress, but direct abdominal work is contraindicated during menstruation.

Amenorrhea is the absence of menstruation. Common causes include insufficient estrogen synthesis due to low body weight or a body fat percentage below 9%. Amenorrhea is common among female athletes and dancers. Because the menstrual cycle is controlled by the hypothalamus, emotions, stress, and other psychological factors can also lead to amenorrhea.

increased vascularity of the endometrium, high levels of progesterone stimulate the endometrial glands to produce a nutrient-rich fluid that supports the growth of a fertilized egg. If the egg is not fertilized, menstruation occurs and the cycle begins again.

What Do You Think? 16.2

- How are the uterus and stomach similar in terms of their structure and function?
- Which structure in the digestive system would you consider comparable to the fallopian tubes (oviducts)? Why?

PREGNANCY AND CHILDBIRTH

The female reproductive cycle repeats approximately every 28 days unless **fertilization**, the joining of sperm and egg also known as **conception**, occurs. An egg is viable for up to 24 hours after it is expelled from the ovary, and sperm can survive for up to 5 days in the female reproductive tract. Due to this time frame, couples who want to get pregnant ideally would have sexual intercourse within a window of 24 hours before and no more than 3 to 4 days after ovulation. The period from conception to birth is the **gestation** period, commonly referred to as pregnancy or the **prenatal** period.

Fertilization and Implantation

In addition to timing issues, fertilization is dependent on the number of sperm as well as their health and motility. During intercourse, healthy males ejaculate millions of sperm, which travel up the female reproductive tract into the fallopian tubes. Vaginal and uterine contractions assist sperm in their upward movement. If intercourse occurs at the time of ovulation, chemicals emitted from the ovum help to attract the sperm. Even with this assistance, it is estimated that only 100 or so of the millions of ejaculated sperm contact the ovum, which usually occurs in the upper third of the fallopian tube. When sperm bump against the ovum, they release enzymes that break down the egg's protective outer covering. Eventually, one is successful, completely breaking through and entering the ovum. At this point, the fertilized egg or **zygote** undergoes changes in its cell membrane that prevent any other sperm from entering, ensuring that only one complete set of DNA is created. This means that conception is more complex than sperm simply meeting egg; it is the joining of genetic material between male and female gametes.

By the Way

Several forms of manual therapy, particularly reflexive and zone therapies such as foot or hand reflexology, acupressure, shiatsu, and bindegewebsmassage (connective tissue massage [CTM]), have been known to help in cases of infertility. This may be related to the relaxation response or specific stimulation of the reproductive organs.

After conception, the zygote continues its trip along the fallopian tube toward the uterus. As it travels, it undergoes rapid mitosis to form a multicelled mass called a **blastocyst**. The blastocyst has an outer ring of cells surrounding an inner cavity. This inner cavity contains a mass of cells clustered to one side (FIGURE 16.7). This rapidly growing mass of cells secretes a hormone similar to LH called ***human chorionic gonadotropin (hCG)***. This is an essential hormone because it stimulates the corpus luteum to continue production of estrogen and progesterone to maintain the uterine lining and prevent menses. Between the fourth and seventh day after conception, the blastocyst burrows its way into the endometrial lining, a process called **implantation.**

By the Way

In some instances, the blastocyst implants in the fallopian tubes or elsewhere in the pelvic cavity. This condition, known as an ***ectopic pregnancy***, is relatively uncommon but its incidence has increased in the last several decades. It presents a serious health risk to both mother and embryo, since a fetus cannot develop outside of the uterus.

Pregnancy

While pregnancy begins with fertilization and implantation, the entire prenatal period is 38 to 40 weeks. Development begins with continued mitosis and adaptation of the blastocyst. The inner mass of cells develops into the embryonic layers of the **ectoderm**, **mesoderm**, and **endoderm** that differentiate to produce the various cells, tissues, and systems in the body (FIGURE 16.8). As discussed in Chapter 3, the ectoderm gives rise to the integumentary and nervous systems, the mesoderm to

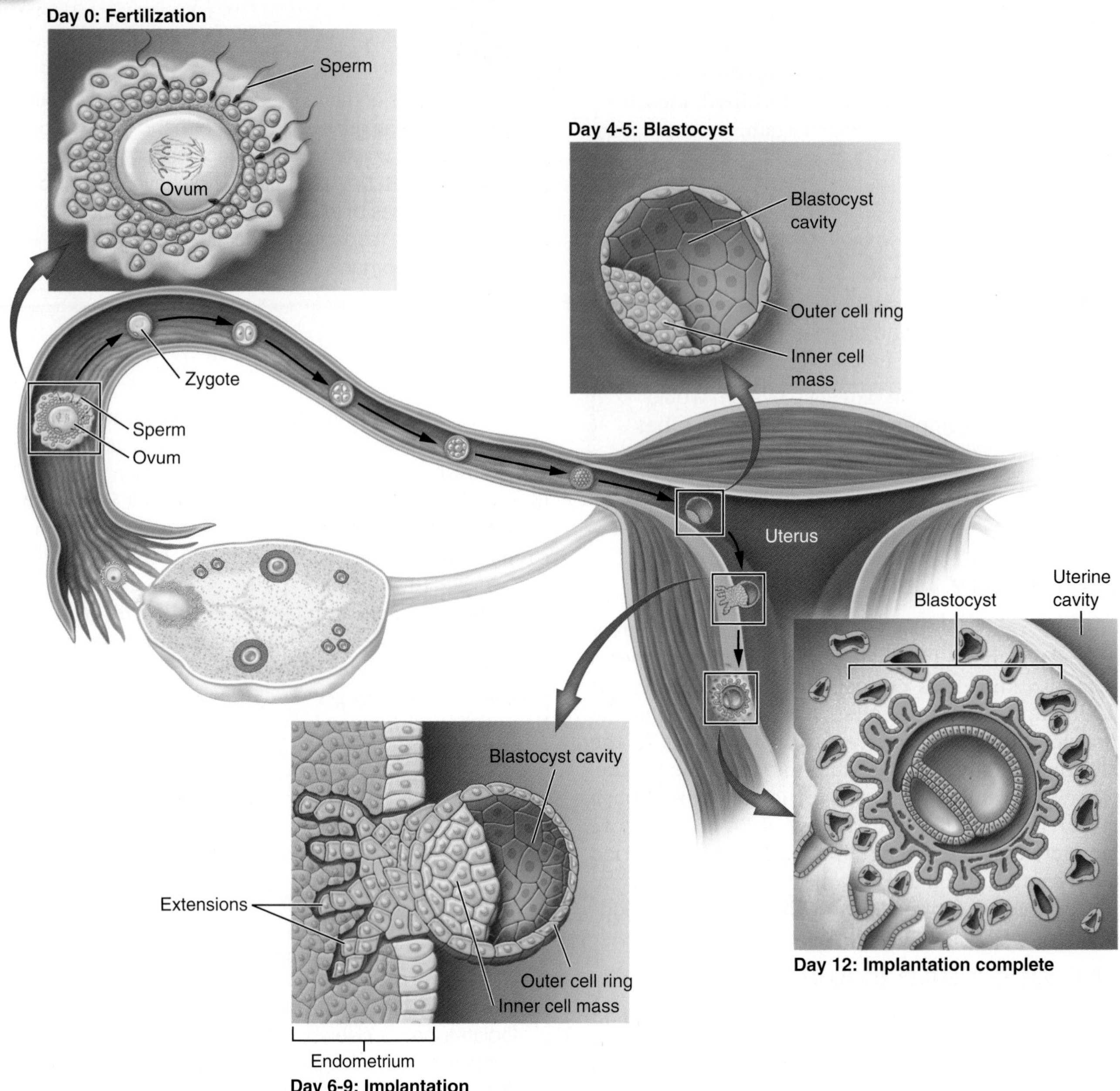

FIGURE 16.7 ▸ Fertilization and implantation. After fertilization, the zygote develops into a blastocyst that implants into the endometrium. Notice the two cellular regions of the blastocyst: the inner cell mass and outer cellular ring.

the muscles and connective tissues, and the endoderm to the internal organ systems comprised of mucosa and glandular epithelium. The internal cellular mass of the blastocyst also produces the **amniotic sac** that surrounds and protects the embryo. The outer layer of the blastocyst produces extensions that become invested in the endometrium and eventually form the membrane called the **placenta.**

By week 3 of gestation, the developing tissue mass, now called an **embryo**, becomes attached to the placenta via the **umbilical cord**. By the end of week 8, the rudiments of every organ system have developed, and as of week 9, the developing offspring is referred to as a **fetus**. Blood flows through the placenta, which serves as a protective barrier that provides oxygen and nutrients for the fetus and carries away fetal waste products. The placenta begins to secrete its own gonadotropins so that eventually the corpus luteum of the ovary becomes inactive. For the next 30 weeks, the fetus continues to grow and develop.

As the fetus develops, the mother's body undergoes extensive changes to nourish and support the growing child. For example, during gestation, the mother's blood and fluid volume increases by 25%

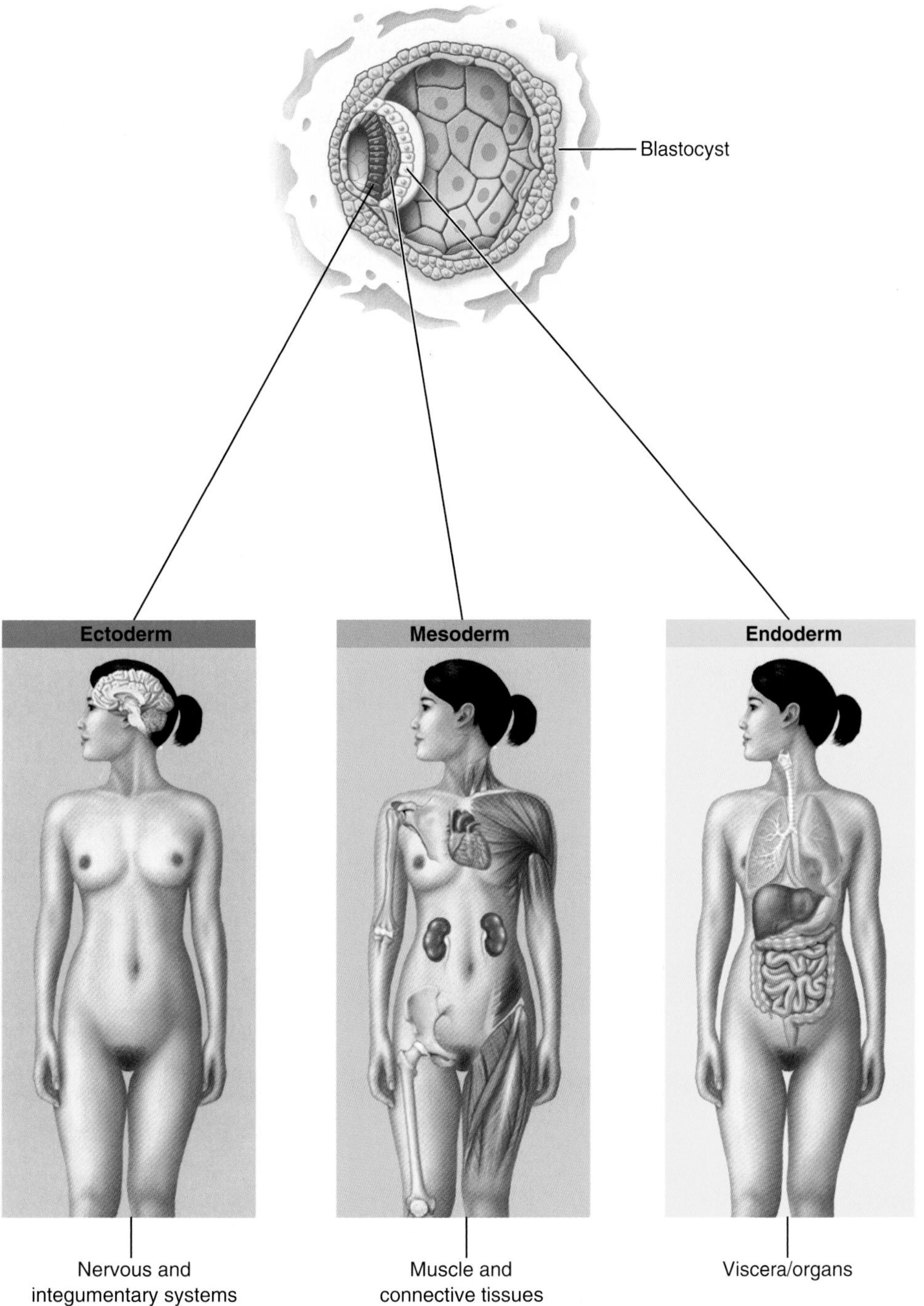

FIGURE 16.8 ▸ Embryonic layers. The inner cell mass of the blastocyst develops into three embryonic layers that grow and differentiate into various tissues and organs.

to 40%. Maternal heart rate and blood pressure must also increase to circulate this additional fluid load. To meet the increased demand for oxygen for the fetus, the mother's respiratory rate increases. Kidney function also increases to process the additional fluid and fetal wastes, and pressure on the bladder from the growing fetus increases the frequency and urgency of urination.

The excessive levels of reproductive hormones needed to sustain pregnancy and prepare the mother's body for birth often cause bouts of nausea or vomiting known as ***morning sickness***. While pregnancy-related changes are natural and positive, they create a tremendous amount of structural and systemic stress on the mother's body. At the end of a full-term pregnancy, the fetus has grown to occupy a space from the pubic bone to the xiphoid process, compressing the liver, spleen, and intestines up against the diaphragm. This compression reduces the mother's lung volume, creates constipation, and may result in heartburn due to acid reflux. The mother's center of gravity has shifted forward, placing stress on the lumbar spine. The skin and connective tissue over the abdomen are distended, causing stretch marks to appear (▶ FIGURE 16.9). Additionally, the breasts grow larger and prepare to produce milk for the newborn baby.

Most manual therapists will work with a pregnant client sometime during their professional career. Therefore, it is important that all manual therapists are familiar with the special considerations, indications, and contraindication for therapy when working with a pregnant client. Pregnant clients need to be cleared by their physician or midwife before the initial manual therapy session. In addition, it is important that therapists carefully screen the client and identify high-risk situations in which therapy might be contraindicated or require specialized advance training. Some of these high-risk factors include

- A mother younger than age 16 or over age 35
- A mother with diabetes or heart disease
- A history of miscarriage or multiple abortions
- A pregnancy with twins, triplets, or multiples

If the mother is healthy enough to receive manual therapy, adaptations in normal bolstering, positioning, and technique are required. Generally,

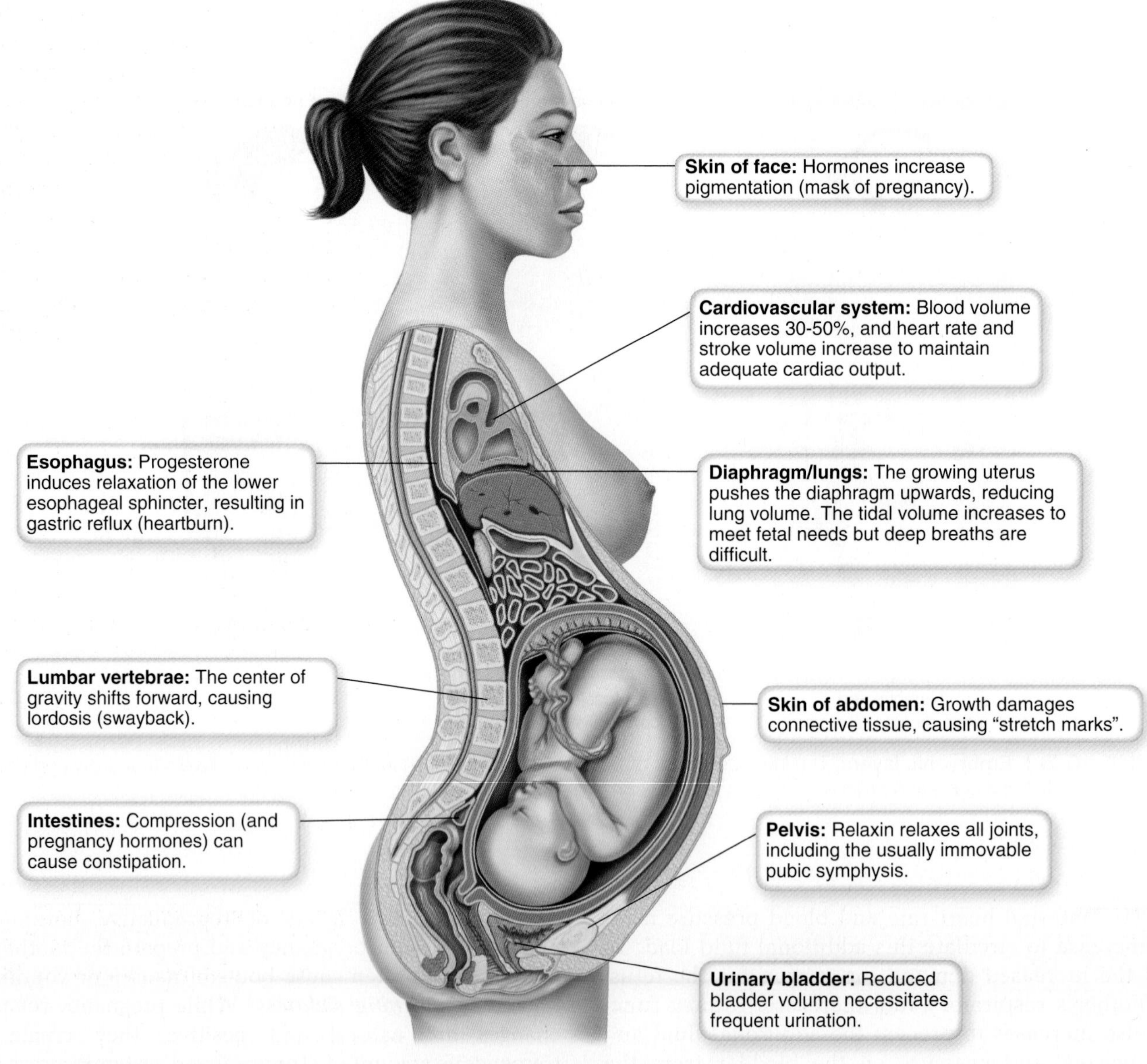

Skin of face: Hormones increase pigmentation (mask of pregnancy).
Cardiovascular system: Blood volume increases 30-50%, and heart rate and stroke volume increase to maintain adequate cardiac output.
Esophagus: Progesterone induces relaxation of the lower esophageal sphincter, resulting in gastric reflux (heartburn).
Diaphragm/lungs: The growing uterus pushes the diaphragm upwards, reducing lung volume. The tidal volume increases to meet fetal needs but deep breaths are difficult.
Lumbar vertebrae: The center of gravity shifts forward, causing lordosis (swayback).
Skin of abdomen: Growth damages connective tissue, causing "stretch marks".
Intestines: Compression (and pregnancy hormones) can cause constipation.
Pelvis: Relaxin relaxes all joints, including the usually immovable pubic symphysis.
Urinary bladder: Reduced bladder volume necessitates frequent urination.

FIGURE 16.9 ▶ Structural and systemic stressors of pregnancy. The impact of a full-term pregnancy on the structures and systems of the mother are shown.

side-lying, semireclining, or seated positions are required for safety and comfort. Pregnant clients may need to be assisted on and off the table, and unscented oils and lotions are recommended. Therapists must be watchful for excessive edema and varicose veins and treat these as local contraindications. In the third trimester, the risk of blood clots in the lower extremities is high, so deep work, particularly over the medial aspect of the legs (greater saphenous vein), should be avoided. Throughout the pregnancy, deep abdominal work and aggressive or heavy pressure over joints and tendons is contraindicated.

Childbirth

Birth most often occurs within 2 weeks of the estimated due date. As this date approaches, the fetus shifts into a head-downward position, and the uterus becomes more sensitive to stimulus. **Labor**, the sequence of actions that push the baby through the birth canal, begins with regular contractions of the uterus (stimulated by the hormone oxytocin) and rupturing of the amniotic sac. The birth process can be divided into three stages:

1. ***Cervical dilation***—This stage begins with the onset of uterine contractions. Simultaneously, the cervix begins to ***efface*** (thin and shorten) and dilate to permit the baby to pass through the birth canal. The stage is complete when the cervix is fully dilated, at approximately 10 cm.
2. ***Delivery of the fetus***—Beginning with the complete dilation of the cervix, this phase is completed when the baby emerges from the birth canal. At this point, the baby takes its first breaths, and the umbilical cord is cut to physically separate the mother and child.
3. ***Expulsion of the afterbirth***—Within minutes of the baby's birth, the placenta separates from the wall of the uterus and is expelled, along with the remaining fetal membranes.

What Do You Think? 16.3

- Even though an expectant mother is "eating for two," why do you think pregnant women generally prefer to eat small meals?
- At this stage of your education, what are your chief concerns about working with pregnant clients? How can you best resolve these concerns?

AGING AND THE REPRODUCTIVE SYSTEM

In both genders, reproductive development starts sometime after age 10 and becomes fully active at puberty, when sexual reproduction becomes possible. In females, puberty is marked by ***menarche***, the first menstrual period and the beginning of the reproductive cycle. Between the ages of 45 and 55, the number of follicles in the ovary is exhausted, and the ovaries become less responsive to the gonadotropic hormones. This leads to a dramatic decrease in estrogen levels and the eventual cessation of the menstrual cycle. **Menopause** is the permanent cessation of menses (defined as the point when a woman has had no menstrual periods for 12 consecutive months) and the end of fertility. ***Perimenopause*** is the interval (3 to 5 years or more) during which a woman's body experiences hormonal changes and makes the transition to menopause. Fluctuating hormone levels and declining estrogen cause irregularities in the normally predictable menstrual cycle and changes in the reproductive organs. While menopause is a natural process, some common symptoms may occur, including hot flashes, night sweats, vaginal dryness, weight gain, insomnia, mood swings, and hair loss. Women may experience some or all of these symptoms, and for some, they can be intense. However, for most women, symptoms of menopause are manageable. Since estrogen plays an important role in calcium absorption, the risk of osteoporosis is significantly higher for postmenopausal women. For this reason, many healthcare practitioners encourage women to have bone-density scans in their early forties to establish a baseline and again at the onset of menopause.

Age-related changes occur later and are generally less noticeable for men. While gonadotropin levels remain high, testosterone levels slowly decrease between the ages of 50 and 60. This decline in testosterone leads to decreased desire for sexual activity, smaller numbers of viable sperm, and a decrease in the mass and strength of skeletal muscles. A common condition for men in this age range is an enlargement of the prostate gland called ***benign prostatic hyperplasia*** or ***BPH***. In BPH, the prostate can grow up to four times its normal size and obstruct the urethra. This increases the urgency and frequency of urination, decreases the force of the urine stream, and creates a sensation of incomplete emptying of the bladder. For this reason, physicians often recommend including a prostate exam in every annual physical beginning at age 50. Men experiencing any type of urinary changes should see their doctor, regardless of age.

SUMMARY OF KEY POINTS

- The reproductive system is responsible for the production and release of gametes and the secretion of sex hormones; its primary purpose is to produce offspring.
- A hormonal process involving the hypothalamus, anterior pituitary, and gonads controls the male and female reproductive systems. Reproductive organs are organized into three categories: gonads, genitalia, and accessory organs. See Table 16-1 for a comparison of the features of the male and female reproductive systems.
- In men, the testes produce male gametes, or sperm, and secrete testosterone, which is responsible for the development of the reproductive organs and secondary sex characteristics. Sperm are produced in the seminiferous tubules, and Leydig cells produce testosterone.
- After leaving the testes, sperm pass through a series of ducts: the epididymis, vas deferens, ejaculatory duct, and urethra, before exiting the penis. The seminal vesicles, bulbourethral glands, and prostate produce and secrete various fluids into the ducts to make semen. These fluids provide nourishment to sperm and assist its movement through the duct system.
- In women, the ovaries produce female gametes, called ova or eggs, and secrete estrogen and progesterone. These hormones are responsible for the development of the reproductive organs and secondary sex characteristics, as well as the reproductive cycle and preparation of the uterus for childbirth. Ovarian cells called follicles produce eggs and secrete estrogen. Progesterone is produced and secreted by the corpus luteum, the glandular remains of a follicle once the egg has been released.
- Once an ovum is released from an ovary, it is scooped into a fallopian tube, where peristaltic contractions and ciliary waves move it toward the uterus. An unfertilized egg disintegrates and is sloughed off with the endometrial lining of the uterus during menstruation.
- The successful fertilization of an egg is dependent on sexual intercourse occurring within a window of 24 hours before and no more than 3 to 4 days after ovulation.
- A fertilized egg is a zygote. The zygote passes through the fallopian tube as it develops into a multicellular mass called a blastocyst that implants in the endometrial lining of the uterus. The blastocyst grows and about week 3 of pregnancy becomes an embryo; after week 8, it is called a fetus.
- During pregnancy, a mother's body changes to provide support and nourishment for the developing baby. Changes include increased blood and fluid volumes, increased blood pressure and kidney function, and increased heart and respiratory rates. The growing fetus compresses the abdominal organs into the diaphragm, creates pressure on the bladder, and moves the mother's center of gravity forward, placing strain on the lumbar spine.
- Labor is the series of actions that push the baby through the birth canal. Labor begins with regular uterine contractions and the rupturing of the amniotic sac. The three stages in the birth process are cervical dilation, delivery of the fetus, and expulsion of the afterbirth.
- Between the ages of 45 and 55, women experience menopause, the permanent end of menses and fertility. Hormonal fluctuations and decreased levels of estrogen can create symptoms, including hot flashes, night sweats, vaginal dryness, hair loss, mood swings, weight gain, and insomnia.
- Age-related changes in males begin between 50 and 60 years of age. Changes include decreased desire for sexual activity, smaller numbers of viable sperm, and decreased mass and strength of the skeletal muscles.

REVIEW QUESTIONS

Short Answer

1. Briefly describe the function of the male and female reproductive systems.
2. List the four structural features and physiologic processes that are common to both the male and female reproductive systems.
3. Name the primary male and female reproductive organs and briefly describe their functions.

Briefly define the following terms:

4. Gamete
5. Ovulation
6. Fertilization
7. Genitalia
8. Menarche
9. Gestation

Multiple Choice

10. Which structure in the testes produces sperm?
    a. epididymis
    b. Leydig cells
    c. seminal vesicle
    d. seminiferous tubules

11. Which of the following structures secrete progesterone?
    a. follicle
    b. ovum
    c. corpus luteum
    d. fimbriae
12. Which of the following is not a part of the duct network that carries sperm?
    a. seminal vesicle
    b. epididymis
    c. vas deferens
    d. urethra
13. Which structure in the female reproductive system is also called the birth canal?
    a. uterus
    b. vagina
    c. oviducts
    d. fallopian tubes
14. At what stage of pregnancy is the term embryo applied?
    a. day one
    b. first week
    c. third week
    d. ninth week
15. The cervix is which region of the uterus?
    a. superior dome
    b. wide central zone
    c. narrow distal zone
    d. the muscular wall
16. Which hormone is responsible for building and maintaining the uterine lining?
    a. progesterone
    b. estrogen
    c. follicle stimulating hormone
    d. luteinizing hormone
17. What is the function of the prostate gland?
    a. secretion of semen
    b. development of sperm
    c. initiating ejaculation
    d. secrete a milky fluid that prevents semen from clotting
18. Where does fertilization of an egg generally occur?
    a. inside the fallopian tube
    b. inside the vagina
    c. at the entrance of the uterus
    d. at the junction between uterus and oviduct
19. During which days of the menstrual cycle does menstruation occur?
    a. 6 through 14
    b. 9 through 20
    c. 1 through 5
    d. 14 through 28
20. What is the name of the protective sac that encloses the testes?
    a. vas deferens
    b. spermatic sac
    c. seminal vesicle
    d. scrotum

Appendix A

Benefits and Physiologic Effects of Swedish Massage

As the manual therapies push for increased acceptance as complementary/alternative health care professions, more and more research is being conducted to evaluate the benefits and effects of various modalities. While the body of evidence for all forms of manual therapy increases each year, much of the research focuses on establishing the efficacy of massage, most often, Swedish massage. This appendix is provided to help students and teachers clearly articulate the benefits and effects of Swedish massage based on valid, current evidence. Our understanding of the effects of massage on the human body is advancing, and new evidence emerges each year. This evidence is primarily provided through three types of supportive data:

- ***Anecdotal evidence***—Subjective evidence that reflects individual or collective experience without regard to controls for variables or rigorous statistical analysis.
- ***Empirical evidence***—Objective evidence gathered through observation and/or experiments with controls and statistical analysis for validity and reliability.
- ***Presumptive evidence***—Logical evidence that has been neither proved nor disproved but reflects a common understanding and current thinking.

SPECIFIC EFFECTS OF SWEDISH MASSAGE STROKES

Swedish massage, also known as relaxation massage, classically includes the blending of five foundational strokes: effleurage, petrissage, friction, vibration, and tapotement. These strokes are most often sequenced according to two general guidelines:

- ***Work superficial to deep***—The superficial tissues (skin and superficial fascia) are addressed before deep muscles or fascial zones. Effleurage and petrissage are used to fulfill this guideline.
- ***Work general to specific***—Once a body region has been generally massaged and the superficial tissues loosened using effleurage and petrissage, friction can be used to address deeper structures more specifically.

Vibration and tapotement can be used superficial or deep, generally or specifically, depending on the technique being used and the desired effect.

While the effects of any form of manual therapy are cumulative (created by a combination of strokes or techniques, and more apparent after several sessions), it is common for teachers and therapists practicing Swedish massage to assign specific physiologic effects to each of the basic massage strokes. The following table briefly defines each of the foundational Swedish massage strokes, describes its purpose, and lists the physiologic effects that are supported by research and/or a volume of practical evidence. Discussion points are included to explain the rationale for the physiologic effects shown.

Purpose and Effects of Individual Swedish Massage Strokes

Stroke Definition and Purpose	Physiological Effects	Discussion
Effleurage is a sliding, gliding stroke used to: • Begin, end, and transition between strokes during a massage. • Introduce touch, pace, and rhythm; define the area to be worked; and spread lubricant. • Provide sensory feedback that reconnects each individual region to the whole. • Warm and assess superficial tissues (skin and fascia).	• Stretching and loosening of superficial tissues • Warming of superficial tissues • Improved local and superficial fluid flow	• Effleurage is not attributed with increasing circulation as a whole. The physiologic mechanisms that regulate cardiovascular circulation and lymphatic flow are complex internal processes that are not easily affected by external manipulation. • The local fluid moved is most likely interstitial fluid, not blood. • Even superficial fluid movement is dependent on the amount of pressure applied with the stroke, the body area being worked, and the duration of application. • There is no evidence to support effleurage having any influence on arterial flow, metabolic rate, nutrient–waste exchange, or removal of lactic acid. In fact, lactic acid is not removed from muscles or the bloodstream but is metabolized during recovery (O_2 debt).

Purpose and Effects of Individual Swedish Massage Strokes *(Continued)*

Stroke Definition and Purpose	Physiological Effects	Discussion
Petrissage is a kneading stroke used to: • Assess tissue pliability and/or stiffness. • Assess restrictions and tender areas within the superficial layers of tissue. • Prepare an area for deeper work.	• Loosening and softening of superficial tissues • Broadening and stretching of superficial tissues • May prevent or decrease rate of muscle atrophy and enhance muscle tone	• The loosening, softening, broadening, and stretching effects are structural effects, which may be related to the thixotropic and viscoelastic qualities of connective tissue. • A secondary response to these structural changes may be reduction of tissue ischemia and decreased resistance to fluid movement. However, there is no evidence to support that petrissage has any measurable effect on fluid flow. • Stimulation of sebaceous and sudoriferous glands is often attributed to petrissage alone. However, any time the skin is manipulated, these glands are presumed to be stimulated. • Decreasing the rate of muscle atrophy or enhancing muscle tone is based on presumptive and anecdotal evidence.
Friction is a compression, stretching, or broadening stroke used to: • Add specificity to structural elements (e.g., posterior edge of the ITB, muscle origins and insertions, or "tight/sore spots"). • Address specific fascial zones or structures (e.g., septum between the gastroc heads) and ligaments and tendons in the area being massaged.	• Loosening and softening of tissue • Broadening and stretching of tissue • Breaking up connective tissue adhesions • Increased tissue pliability • Reduction in muscle tension • Creates superficial hyperemia • Generates superficial heat (when applied with repeated short, rapid strokes)	• Loosening, softening, broadening, and stretching effects are structural effects, likely to be related to the thixotropic and viscoelastic qualities of connective tissue. • Breaking up of adhesions is also structural. This effect is best used when working within the healing process or on specific scars and/or adherent tissue. • Improved tissue pliability is listed here rather than increased range of motion (ROM). Because ROM may be limited due to pain, emotional resistance, or changes in inert joint structures (e.g., ligament, capsule, cartilage), increased flexibility may not improve the ROM. Therefore, increased ROM implies improved tissue flexibility, but improved tissue pliability/flexibility may not mean improved ROM. • Reduction in muscle tension can be related to reduction of specific trigger points, tender points, or spasms. Because the localized contraction knot of a trigger point is not mediated by the nervous system, reduced muscle tension would be considered a structural effect. However, muscle spasm and tender points *are* mediated by the nervous system. In these cases, reducing muscle tension would be considered a systemic effect. Additionally, reduced muscle tension due to the global relaxation response is also a systemic effect. • Hyperemia related to friction is most likely due to the release of histamine from the mast cells rather than heat generated by the stroke. Histamine causes local vasodilation, which gives the tissue a reddened appearance.

(Continued)

Purpose and Effects of Individual Swedish Massage Strokes *(Continued)*

Stroke Definition and Purpose	Physiological Effects	Discussion
Vibration is a rhythmic shaking or jostling of tissue or a body part used to: • Release subconscious tension or holding by the client. • Provide kinesthetic feedback to the client to help with a movement or stretch.	• Decreased muscle guarding • Traction vibration may redistribute synovial fluid within the joint cavity • Stationary vibration to sinuses may reduce local congestion	• Reduced muscle guarding from jostling or shaking is well supported by clinical, presumptive, and anecdotal evidence. • The statements about traction vibration and stationary vibration over the sinuses are presumptive based on anecdotal evidence.
Tapotement is a rhythmic alternate striking of tissues used to: • Bring the client to a more alert or conscious state before concluding the massage. • Create muscle relaxation in well-muscled areas such as the gluteals, thighs, and upper back.	• General muscle relaxation • Increased motor tone that may slow down atrophy • Cupping over the lungs and tapping over the sinuses may relieve respiratory or sinus congestion • Increased hyperemia • General invigoration	• General muscle relaxation has been measured in at least one study, and there is good anecdotal evidence. This is likely related to the pace, rhythm, and duration of the application; moderate pace and rhythm plus long duration seem to create relaxation. There is some anecdotal evidence that tapotement over the sacrum contributes to general relaxation, perhaps by decreasing sympathetic tone by stimulating the intrafascial receptors. • Increased motor tone is presumptive based on anecdotal evidence, and may only occur in healthy but untoned muscles. • Relief of lung congestion with tapotement is an advanced technique that requires a specific sequence of application and special positioning. There is no evidence that standard tapotement over the back or chest during Swedish massage has this effect. • Relief of sinus congestion is presumptive based on anecdotal evidence. • Like friction, the increased hyperemia from tapotement is probably due to release of histamine from mast cells in superficial tissues. • Regardless of pace and depth, the alternate percussion of tapotement must provide a certain level of stimulation. Therefore, therapists must decide when and where to apply this invigoration during the massage. This is generally not appropriate for every body part in a full-body massage.

Appendix B

Innervation of Major Skeletal Muscles

Note that each muscle innervation table is presented in the same order as the muscle tables in Chapter 6.

Muscles of the Head and Face

Group	Muscle	Spinal Segment(s)	Peripheral Nerve
Skull	Occipitalis	None	Cranial nerve VII (facial)
	Suboccipitals	C1	Suboccipital nerve
	Frontalis	None	Cranial nerve VII (facial)
	Temporalis	None	Cranial nerve V (trigeminal)
Face	Masseter	None	Cranial nerve V (trigeminal)
	Medial pterygoid	None	Cranial nerve V (trigeminal)
	Lateral pterygoid	None	Cranial nerve V (trigeminal)
	Buccinator	None	Cranial nerve VII (facial)
	Orbicularis oris	None	Cranial nerve VII (facial)
	Orbicularis oculi	None	Cranial nerve VII (facial)
	Zygomaticus • Major • Minor	None	Cranial nerve VII (facial)

Muscles of the Neck

Group	Muscle	Spinal Segment(s)	Peripheral Nerve
Anterior neck	Platysma		Cranial nerve VII (facial)
	Sternocleidomastoid	C2–C3	Cranial nerve XI (spinal accessory)
	Suprahyoids	C1–C2	Cranial nerves V, VII
	Infrahyoids	C1–C3	Cranial nerve XII (hypoglossal)
	Scalenes • Anterior • Middle • Posterior	• Anterior C4–C6 • Middle C3–C8 • Posterior C6–C8	Cervical spinal nerves
Posterior neck	Splenius capitis	C2–C5	Middle cervical nerves
	Splenius cervicis	C4–C7	Lower cervical nerves
	Levator scapulae	C3–C5	Dorsal scapular nerves

Muscles of the Chest and Abdomen

Group	Muscle	Spinal Segment(s)	Peripheral Nerve
Chest	Pectoralis major	C5–T1	Lateral and medial pectoral nerve
	Pectoralis minor	C8–T1	Medial pectoral nerve
	Subclavius	C5–C6	Subclavian nerve
	Serratus anterior	C5–C7	Long thoracic nerve
	Intercostals	T1–T11	Intercostal nerves
	Diaphragm	C3–C5	Phrenic nerve
Abdomen	Rectus abdominis	T6–T12	Intercostal nerves
	External obliques	T8–T12	Intercostal nerves
	Internal obliques	T7–L1	Intercostal, iliohypogastric, and ilioinguinal nerves
	Transverse abdominis	T7–L1	Intercostal, iliohypogastric, and ilioinguinal nerves
	Psoas	L1–L4	Lumbar plexus

Paraspinal Muscles

Group	Muscle	Spinal Segment(s)	Peripheral Nerve
Erector spinae	Iliocostalis	By associated cervical, thoracic, or lumbar spinal segments	Dorsal primary divisions of spinal nerves
	Longissimus	By associated cervical, thoracic, or lumbar spinal segments	Dorsal primary divisions of spinal nerves
	Spinalis	By associated cervical, thoracic, or lumbar spinal segments	Dorsal primary divisions of spinal nerves
Transversospinales	Semispinalis	By associated cervical, thoracic, or lumbar spinal segments	Dorsal primary divisions of spinal nerves
	Multifidi	By associated cervical, thoracic, or lumbar spinal segments	Dorsal primary divisions of spinal nerves
	Rotatores	By associated cervical, thoracic, or lumbar spinal segments	Dorsal primary divisions of spinal nerves
	Interspinales	By associated cervical, thoracic, or lumbar spinal segments	Dorsal primary divisions of spinal nerves
	Intertransversarii	By associated cervical, thoracic, or lumbar spinal segments	Dorsal primary divisions of spinal nerves

Muscles of the Back

Group	Muscle	Spinal Segment(s)	Peripheral Nerve
Back	Trapezius	C3–C4	Cranial nerve XI (spinal accessory)
	Latissimus dorsi	C6–C8	Thoracodorsal nerve
	Teres major	C5–C7	Lower subscapular nerve
	Rhomboids	C4–C5	Dorsal scapular nerve
	Quadratus lumborum	T12, L1–L3	Lumbar plexus
	Supraspinatus	C5–C6	Suprascapular nerve
	Infraspinatus	C5–C6	Suprascapular nerve
	Teres minor	C5–C6	Axillary nerve
	Subscapularis	C5–C6	Upper and lower subscapular nerves

Muscles of the Brachium

Group	Muscle	Spinal Segment(s)	Peripheral Nerve
	Deltoid	C5–C6	Axillary nerve
Anterior brachium	Biceps brachii	C5–C6	Musculocutaneous nerve
	Coracobrachialis	C5–C7	Musculocutaneous nerve
	Brachialis	C5–C7	Musculocutaneous nerve
Posterior brachium	Triceps brachii	C6–C8	Radial nerve
	Supinator	C6	Radial nerve

Muscles of the Forearm

Group	Muscle	Spinal Segment(s)	Peripheral Nerve
Posterior lateral forearm	Brachioradialis	C5–C6	Radial nerve
	Extensor carpi radialis longus	C6–C7	Radial nerve
	Extensor carpi radialis brevis	C6–C7	Radial nerve
	Extensor digitorum	C6–C8	Radial nerve
	Extensor carpi ulnaris	C6–C8	Radial nerve
Anterior medial forearm	Pronator teres	C6–C7	Median nerve
	Flexor carpi radialis	C6–C7	Median nerve
	Palmaris longus	C7–C8	Median nerve
	Flexor carpi ulnaris	C7–T1	Ulnar nerve
	Flexor digitorum superficialis	C7–T1	Median nerve
	Flexor digitorum profundus	Digits 2–3: C8–T1 Digits 4–5: C8–T1	Median nerve Ulnar nerve

Muscles of the Pelvic Girdle

Group	Muscle	Spinal Segment(s)	Peripheral Nerve
Posterior	Gluteus maximus	L5–S2	Inferior gluteal nerve
	Gluteus medius	L4–S1	Superior gluteal nerve
	Gluteus minimus	L4–S1	Superior gluteal nerve
	Piriformis	S1–S2	Sacral plexus
	GO-GO-Qs • Gemellus superior • Obturator internus • Gemellus inferior • Obturator externus • Quadratus femoris	 • L5–S1 • L5–S1 • L5–S1 • L3–L4 • L4–S1	 • Lumbosacral plexus • Lumbosacral plexus • Lumbosacral plexus • Obturator nerve • Lumbosacral plexus
Anterior	Tensor fascia latae	L4–S1	Superior gluteal nerve
	Iliacus	L2–L3	Femoral nerve

Muscles of the Thigh

Group	Muscle	Spinal Segment(s)	Peripheral Nerve
Anterior	Rectus femoris	L2–L4	Femoral nerve
	Vastus medialis	L2–L4	Femoral nerve
	Vastus intermedius	L2–L4	Femoral nerve
	Vastus lateralis	L2–L4	Femoral nerve
	Sartorius	L2–L3	Femoral nerve
Medial	Pectineus	L2–L3	Femoral and sometimes obturator nerve
	Adductor brevis	L2–L4	Obturator nerve
	Adductor longus	L2–L4	Obturator nerve
	Adductor magnus	Ant. L2–L4 Post. L4–S1	Obturator nerve Sciatic nerve
	Gracilis	L2–L3	Obturator nerve
Posterior	Biceps femoris	Long: S1–S3 Short: L5–S2	Tibial branch of sciatic nerve Peroneal branch of sciatic nerve
	Semimembranosus	L5–S2	Tibial branch of sciatic nerve
	Semitendinosus	L5–S2	Tibial branch of sciatic nerve

Muscles of the Leg

Group	Muscle	Spinal Segment(s)	Peripheral Nerve
Anterior	Tibialis anterior	L4–S1	Deep peroneal nerve
	Extensor hallucis longus	L4–S1	Deep peroneal nerve
	Extensor digitorum longus	L4–S1	Deep peroneal nerve
Lateral	Peroneus longus	L4–S1	Superficial peroneal nerve
	Peroneus brevis	L4–S1	Superficial peroneal nerve
Posterior	Gastrocnemius	S1–S2	Tibial nerve
	Soleus	S1–S2	Tibial nerve
	Tibialis posterior	L5–S1	Tibial nerve
	Flexor hallucis longus	L5–S2	Tibial nerve
	Flexor digitorum longus	L5–S1	Tibial nerve

Chapter 1

What Do You Think?

1.1

- An everyday analogy for the five levels of body organization could be an office building: bricks = cells; wall = tissue; office suite = organ; single building level = system; building = whole organism. Another possible analogy could be words = cells; sentence = tissue; paragraph = organ; chapter sections = organs; chapters = systems; book = whole organism.
- Examples of homeostatic changes can include a number of choices. Think of processes that are shifted and balanced on regular basis, such as blood pressure changes from sitting to standing to moving, heart rate, breathing rate and depth, hormonal shifts, and pH or electrolyte levels of the blood.

1.2

- Your instructor can help you categorize various forms of manual therapy (shown in Table 1-1) or explain why he or she believes a particular form doesn't fit into any of the listed categories.
- Benefits of manual therapy for you may include aspects such as keeps me grounded, helps me focus, supports healthy choices, reminds me to feel as well as think, and others.
- In thinking of the structural and/or systemic effects of manual therapy you have experienced, the goal is to understand the difference between a benefit and an effect. You may also choose to do some research on your own. Note that other sources may use different terminology and discuss other benefits and effects that you can compare with your own experiences and the information in this text.

Review Questions

1. tissue
2. organs
3. stimulus
4. A level of internal stability or balance; a dynamic equilibrium between physiologic processes necessary to sustain life.
5. negative
6. d
7. f
8. a
9. g
10. b
11. i
12. c
13. j
14. e
15. h
16. b
17. c
18. a
19. c
20. b
21. d
22. d
23. c
24. a
25. b

Chapter 2

What Do You Think?

2.1

- Compare your examples of body planes with those of your classmates.
- Examples of everyday activities that illustrate basic movement terms: opening and closing a car trunk lid can be compared to flexion and extension; swinging a door open is like horizontal abduction and adduction; stirring a pot of stew can be either rotation or circumduction.

2.2

- Examples of common words that use the prefixes, suffixes, or word roots in Table 2-3 include circumference = around the whole; premonition = before it happens; contradiction = opposite meanings, Northern or Southern hemispheres = two halves of the globe.
- Knowing body regions and body cavities will be helpful for a wide range of reasons that may include communication with other professionals, locations of organs, understanding the relationship between problems within the same body region, names of muscles, nerves, bones, blood vessels.

2.3

- Examples of illnesses may include the following: Infectious: conjunctivitis (pink eye), measles, mumps, chicken pox. Environmental: Skin rashes, mesothelioma, some forms of asthma. Hereditary: Huntington chorea, diabetes, psychological conditions such as bipolar disorder. Nutritional and lifestyle: diabetes, heart disease, high blood pressure.

Review Questions

1. coronal
2. sagittal
3. palms, face, feet
4. thoracic, abdominopelvic
5. cranial
6. d
7. b
8. a
9. c
10. a
11. d
12. b
13. a
14. c
15. d
16. a
17. c
18. c
19. a
20. d
21. a
22. b
23. c
24. d
25. a

Chapter 3

What Do You Think?

3.1

- Most substances you come in contact with are compounds because they are a combination of different molecules. Some of the elements you might come into contact with could be pure gold or silver in jewelry or lead in pipes or paint.
- Regardless of the methods used in growing vegetables, they all contain carbon molecules and are therefore classified as organic compounds.

3.2

- Examples of organ systems you can associate with the functions of the organelles of a cell include cell membrane = brain, nervous system; cytoplasm = blood, cardiovascular system; nucleus = reproductive system.
- In the nature versus nurture debate, each student will answer this question differently. Options might include the following: Nurture is most important because the membrane is constantly responding to different stimulus to direct all cellular activity, or nature plays the larger role because you must have the proper make-up of IMPs to respond correctly and efficiently.

3.3

- Here are some examples of analogies for diffusion and filtration: making coffee, playing soccer, using toothpaste or a garden hose, closing window screens.
- Cyto- is a prefix that means cell, and the other part of the term is *-kinesis*. Since the term is used to describing the cell separation process during mitosis, *-kinesis* is a word part that means movement.

Review Questions

1. Plasma membrane—defines and encloses the cell, regulates what enters and leaves the cell, contains the identity markers for the individual. Cytoplasm—contains all of the organelles, and carries out all cellular processes except those of the nucleus. Nucleus—contains the genetic code for the cell and organism as whole in the form of DNA, responds to receptor proteins in plasma membrane for mitosis and to replicate specific segments of DNA.
2. Inorganic compounds are made up of molecules that do not contain carbon atoms. Examples of inorganic compounds are water, salts, acids, and bases. Organic compounds always contain carbon molecules and often have hydrogen, as well. Organic compounds that make up all cells and tissues include carbohydrates, lipids, proteins, and nucleic acids.
3. Epithelial cells are named for their shape: squamous = flat; cuboid = square; columnar = cylindrical. A specialized collection of secreting cells called glandular epithelium is found inside glands such as the salivary, sweat, and endocrine glands. When a single layer of cells are present, it is simple epithelium, two or more layers are stratified. Columnar cells arranged in a single layer can appear to be multiple layers, so it is called pseudo-stratified epithelium.
4. See Table 3-3 for types and locations of connective tissue.

5. b
6. a
7. b
8. b
9. c
10. c
11. a
12. b
13. c
14. d
15. b
16. a
17. d
18. a
19. a
20. c

Chapter 4

What Do You Think?

4.1

- The skin's ability to sense makes it more effective at protecting the body. It is like having a wall with motion sensors or video surveillance installed; this high-tech wall is a more effective protective barrier than one without these features. Remember that the skin provides more sensory input to the brain than any other sensory organ, and this input is essential for the healthy development of the nervous system and overall vitality of the body.
- Examples of the skin's excretion and absorption functions include the following: Many ingested substances such as garlic and alcohol, or inhaled substances such as tobacco smoke, are excreted as metabolic byproducts through the skin when we sweat. Substances absorbed through the skin include water, lotions and other skin care products, medications, and toxins from lawn and garden products.

4.2

- Multiple layers in the epidermis create better protection by dividing the protective functions into different layers. Disruption of one layer doesn't affect all processes of the skin.
- Scrapes only damage the epidermis, which is avascular, while cuts damage the dermis where the blood vessels are located.
- Superficial scrapes only affect the epithelial tissues, which repair quickly with identical epithelial tissue. In contrast, deep cuts affect the connective tissue layer of the dermis, hypodermis, and possibly muscle tissue. These tissues repair themselves with fibrous connective tissue.

4.3

- People with disabilities can use the sense of touch to enhance their perceptions: temperature, pressure, and/or pain information from the cutaneous sensory receptors provides enough information to the brain for it to interpret the input with some accuracy. Blind people can feel the shape and texture of objects and form an image or concept in their minds; deaf people can use movement and vibration to sense sounds.
- Examples of thoughts and emotions being displayed through the skin could include a situation when your purse or wallet was stolen and you felt your face pale with fear or flush with anger, or one in which a friend began sweating profusely when you asked him about his readiness for an exam.
- Examples of touch, pressure, or temperature creating certain feelings could include a sharp, abrupt squeeze on a hand that creates feelings of fear, while gently holding someone's hand might show compassion.

4.4

- If a client informs you that he or she has been diagnosed with ringworm, you should continue with your usual hygienic and universal precaution habits: washing your hands thoroughly before and after every session (and after you have properly handled dirty linens); avoiding contact with open skin lesions and undiagnosed outbreaks; using a nonpermeable covering on the table or therapeutic surface; cleaning surfaces on a daily basis; and washing any materials in hot water that have come in contact with clients. After receiving this information from your client, you might re-clean all surfaces and launder all linens and thoroughly review your last session with this client to determine whether you should contact other clients and alert them to possible signs and symptoms and/or cancel appointments for the next day. If you are sure that you were thorough in the proper cleaning of your hands, linens, surfaces, etc., you probably won't need to do that.
- Precautions to take in a session with a client with severe acne include: do not use a lubricant in affected areas (if you are able to work over these regions at all). If the client requests that you do work in these areas, discuss the possibility of irritating

the skin and/or rupturing lesions and the need for wearing protective gloves to protect yourself from infection.

- Some considerations in working with a geriatric client might be to use extra lubricant, lighten up pressure in some strokes to avoid overstretching tissue, or to slow down the pace and rhythm.

Review Questions

1. A membrane is a broad sheet of two or more types of tissue; generally serves as a lining or covering.
2. Synovial—inside lining of fibrous joint capsule in free moving joints; Serous—lining of body cavities that are closed to the external environment (such as the thoracic and abdominopelvic cavities); also covers the organs in these cavities; Mucous—internal lining of system open to external environment (such as the respiratory, digestive, and reproductive tracts); Cutaneous—the outer protective covering of the body; primary organ of the integumentary system.
3. Protection, helps to regulate body temperature and water balance, serves as body's largest sensory organ, absorption of substances through pores, synthesis of vitamin D.
4. The parietal layer of serous membrane lines the cavity (thoracic and abdominopelvic), and the visceral layer covers the outside of the organs within the cavity.
5. Any four of the following: Meissner's corpuscles, hair root plexuses, Merkel discs, Ruffini corpuscles, Pacinian corpuscles, free nerve endings, or nociceptors.
6. Both the integumentary and nervous systems form from the same embryologic layer, the ectoderm.
7. c
8. b
9. a
10. c
11. d
12. b
13. b
14. d
15. a
16. b
17. c
18. c
19. c
20. a
21. N
22. N
23. C
24. C
25. N
26. N
27. C
28. C
29. C
30. N

Chapter 5

What Do You Think?

5.1

- The medullary cavity provides strength to the bone by adding density (because it is filled with yellow marrow) but keeps it light enough to be moveable.
- Growing pains can be a very real and uncomfortable condition, especially in active children and teenagers. Because the epiphysis is already quite active during growth spurts, it is not uncommon for weight-bearing bones such as the femur and tibia to ache a little. If extra stress is added through activities such as intense weight lifting or running, this may increase growing pains or even cause Osgood-Schlatter or epiphyseal fractures.

5.2

- A dislocation is a glenohumeral joint injury in which the head of the humerus has been forced out of the glenoid fossa. Because there are no muscles across the inferior aspect of the joint, most dislocations are inferior and occur when the arm is in an abducted or fully flexed position. A shoulder separation is the term used to describe an AC joint sprain, a separation between the acromion process and the clavicle. In contrast to a dislocation, this injury occurs when the arm is at the side and force is delivered to the tops of the AC joint (such as a fall in which the top of the shoulder hits a hard surface) or driven up through the humerus into the AC joint (such as falling directly on to your elbow).
- In part due to the interosseous membrane, the fibula serves as a shock absorber, stabilizer, and suspension strut for the leg. The space between the tibia and fibula provides a broad base for the calf muscles and a passageway for blood vessels and nerves.

5.3

- Some examples of types of joints might include a kitchen whisk stirring a pot (ball-and-socket), a rider on a bicycle (saddle), a TV wall bracket that tilts up

and down and side-to-side (condyloid), a door handle (pivot), and a pull-out pantry shelf (gliding).
- The hinge joint with the strongest bony arrangement is the elbow because of the bony hook and spool arrangement of the trochlea (humerus) and trochlear notch (ulna). If you were thinking of a different hinge joint, explain your rationale.
- Infants have an opening or "soft spot" called a fontanel covered by cartilage in the front of the cranium. This allows the cranium to be compressed for passage through the birth canal and space for rapid growth of the infant. Gradually, the fontanel closes and the frontal bone fully fuses around 2 years of age.

Review Questions

1. Framework, protection, levers for movement, storage of important mineral salts like calcium and phosphorous, site for the production of blood cells
2. Long, short (cuboid), flat, irregular
3. Diarthrotic (synovial) joints are freely moveable; amphiarthrotic (cartilaginous) joints allow partial movement; synarthrotic (fibrous) joints have little or no movement.
4. Hinge = flexion and extension only; pivot = rotation only; condyloid = flexion, extension, abduction, and adduction; saddle = all except circumduction; gliding = sliding; ball-and-socket = all movements + circumduction
5. X
6. A
7. A
8. X
9. A
10. A
11. A
12. X
13. A
14. X
15. C
16. C
17. F
18. C
19. C
20. F
21. F
22. F
23. C
24. c
25. b
26. a
27. a
28. d
29. b
30. b
31. c
32. b
33. c
34. a
35. d
36. c
37. a
38. c
39. d
40. b
41. b
42. a
43. f
44. a
45. g
46. c
47. f
48. e
49. c
50. j
51. k
52. c
53. e
54. a
55. k
56. f
57. e
58. a
59. f
60. c
61. m
62. k

Chapter 6

What Do You Think?

6.1

- Force generation and tension are directly linked to the organization of a skeletal muscle. The parallel arrangement of fibers in a fascicle helps to generate tension throughout the entire muscle. The arrangement of the fascicles and their relationship to tendons affect the level of force and the types of movements a muscle generates.
- Ways in which knowledge of fascia and tendons can be applied in practice include the following: knowing the fiber direction in the muscle helps to determine

when techniques are across or with the fiber of the soft tissue; the concept that fascia is stretchy, but tendons' unwind is helpful in making choices in regard to the amount of force, duration of strokes, and frequency of treatments; the gel-sol (thixotropic) nature of fascia implies that heat and movement will help to loosen stiff and adhered muscles tissue.

6.2

- Of the three analogies, the best is the eight-man crew rowing because of the action of the oars; they reach forward, exert force against the water, and pull the boat along. This is similar to myosin extensions as they bond to actin and slide the filaments closer together.
- Heat lowers the threshold, or makes it easier for the muscle to contract, and cold raises the threshold, making it harder for contraction to occur. Think about how difficult it is to move when you are in cold water. It makes your muscles feel stiff, and movement is slow.
- If all the fibers in a muscle were controlled by a single motor unit, every movement would use all the muscle fibers in the muscle, so there would be no such thing as a small or subtle motion. Blinking the eyes might produce a grimace, and eating with utensils could be life threatening. Furthermore, there would be no way to rotate the tonic contractions required to maintain posture, so simply standing for several minutes would lead to extreme muscle fatigue.

6.3

- There are different training regimes because short-distance runners need to maximize their anaerobic mechanisms, while long-distance runners need to highly train their aerobic mechanisms. Therefore, repeated bouts of short sprints are used to train sprinters, and distance runners do more mileage than sprints to prepare them for competition.
- Attachment points are areas of stress and strain and therefore, common areas of tissue damage. However, muscles need to be able to broaden in order to generate full force with contraction. Cross-fiber techniques that break up adhesions between fibers in the muscle belly allow for more effective sliding of filaments and shortening (bunching) of the muscle.
- Isometric contractions are not particularly effective in developing overall strength because they only strengthen a muscle at a single point in its range of motion. Muscles need to be strengthened throughout their range. Therefore, isotonic contractions are more effective in increasing one's overall strength and force generation.
- Negative lifting stresses working the muscle during eccentric contraction. Because this involves the muscle working as it lengthens and working as a brake, it requires a great deal of control and is more effective at developing strength than shortening the muscle during concentric contraction. However, it is also been shown that the highest incidence of injury and muscle soreness comes from this type of training.

6.4

- Examples of analogies for the types of parallel and pinnate muscle arrangements can include a broom to represent triangular muscles, a rubber gasket for circular, an umbrella for multipennate, a flag on a pole for unipennate, and a feather for bipennate.
- The prime movers for knee flexion are on the posterior side of the leg, while the muscles of extension are on the anterior side. Muscles produce movement where they cross a joint. Since knee flexion draws the leg posteriorly, the muscle must be on the posterior side of the leg and vice versa.
- The antagonists for elbow flexion must be on the posterior side of the arm, opposite the agonists of elbow flexion (biceps, brachialis, brachioradialis) on the anterior side.
- A sprain involves inert tissue injury and therefore, passive range of motion (PROM) would be painful, especially at end feel. Additionally, there may be some pain with active range of motion (AROM) and resistive range of motion (RROM), since muscles help to stabilize the joint along with the inert tissues. Conversely, strains that are injuries to contractile tissues should be most painful with RROM. They will also cause challenges with AROM, but PROM should be unremarkable except when stretching the strained muscle.

Review Questions

1. The skeletal muscle system creates movement, maintains posture, stabilizes joints, and generates heat.
2. Skeletal muscles are extensible, excitable, elastic, and contractile.
3. Aponeurosis is a broad flat sheet of connective tissue serving as a tendon for groups of muscles. Examples include the abdominal, thoracolumbar,

cranial, and pectoral aponeurosis, as well as the iliotibial tract (band).

4. The all-or-none principle of muscle contraction states that when threshold stimulus is applied, all muscle fibers in the motor unit fully contract.
5. Examples for each category:
   - Size (a and b): gluteus maximus, medius, minimus; adductor magnus, longus, brevis; peroneus longus or brevis; longissimus; palmaris longus; vasti muscles of quadriceps
   - Shape (c and d): deltoid, plantaris, serratus, trapezius
   - Function (e and f): pronator, supinator, extensor carpi, flexor carpi, adductors (magnus, longus, brevis), flexor and extensor pollicus, flexor and extensor hallucis, flexor and extensor digitorums, erector spinae, rectus abdominus, rectus femoris
   - Location (g and h): serratus anterior, tibialis posterior, tibialis anterior, vastus lateralis and medialis, gluteal muscles, suboccipitals, subclavius, subscapularis, supraspinatus, infraspinatus, frontalis, temporalis, extensor/flexor carpi radialis and ulnaris
6. The events of muscle contraction are
   i. Calcium ions stored in the sarcoplasmic reticulum are released into the sarcoplasm.
   ii. The presence of calcium exposes binding sites on the actin filaments.
   iii. The myosin heads bind to these exposed sites, forming cross-bridges that pull the myofilaments across one another.
   iv. ATP is used to detach the myosin heads, so they can flip forward to the next binding site. The net result is continued sliding of the filaments across one another.
   v. When the stimulus is removed, more energy is expended to pump calcium back into the sarcoplasmic reticulum, and chemical bridges can no longer form.
7. d
8. c
9. d
10. b
11. a
12. c
13. d
14. b
15. b
16. d
17. b
18. a
19. b
20. c
21. c
22. c
23. c
24. a
25. d
26. b
27. d
28. c
29. a
30. d

Chapter 7

What Do You Think?

7.1

- Sensory neurons simply transmit data from sensory receptors to the central nervous system (CNS). Therefore, they do not need to have many branching dendrites. They need to relay very specific information from the receptors to be integrated. In contrast, motor neurons need to stimulate responses in effectors in response to various sensory data. Therefore, they need to have a lot of dendrites to receive impulses from many sources and transmit them to specific effectors.
- A common mnemonic for remembering the names and order of the 12 cranial nerves is On Old Olympus's Towering Top, A Finn And German View Some Hops. Notice that this mnemonic does not exactly match the names shown in Table 7-1. Cranial nerve VIII (vestibulocochlear) was formerly called the acoustic nerve, and cranial nerve XI (accessory) was known as the spinal accessory nerve.
- With fictitious names, there are a lot of options. Some ideas for naming a nerve that innervates the lungs could be respiratory nerve, diaphragmatic nerve, pneumabreathalotory, or neurolungo nerve. The point is that nerves are often named for the organ they innervate.
- There is no thoracic nerve plexus because the spinal nerves in the thoracic region do not merge and then re-divide before heading out into the body; they just stretch out and branch.

7.2

- An action potential is conducted along the full length of an axon once a threshold stimulus is received. The

membrane potential shifts, causing the charge on the outside of the neuron's plasma membrane to be more negative, since positive sodium ions rush in during depolarization.
- *Systemic effects* and *reflex effects* are both acceptable terms, so you can form an opinion based on your understanding of the far-reaching effects of manual therapy.
- Examples of systemic effects that occur locally can include local shifts in blood flow to the tissue compared to overall changes in blood pressure, or alleviation of a local trigger point or spasm versus body-wide muscle relaxation.

7.3

- There is a limit to the range of temperatures thermoreceptors are able to sense because tissue damage occurs outside those temperature ranges, and the nociceptors are stimulated at that point to initiate appropriate responses.
- It would be dangerous for all sensory receptors to adapt completely. For example, if pain receptors adapted completely, we would stop sensing and therefore, stop protecting ourselves from harmful stimuli. However, 100% adaptation by smell receptors allows us to remain in odorous environments when necessary.
- Slow stretching keeps us from activating the muscle spindles that initiate contractions of the muscle, and stimulates the GTOs that inhibit contraction. Bouncing will stimulate muscle spindles and the stretch reflex and cause muscles to tighten; it also increases the risk of muscle strain.

7.4

- Because the brain "floats" in a cushion of CSF, it can slam against one side of the skull and then bounce back and hit the other side, causing bruising on both sides.
- Because of the location of the kidneys and bladder, it makes more sense that these organs are innervated by nerves from the lumbar and/or sacral segments of the spinal cord.

7.5

- The thalamus may be compromised when sensory information is confused or poorly prioritized. The limbic system could be creating a challenge by associating specific sensory input with powerful emotions and memories. Otherwise, any associative region of the cerebrum could confuse or misinterpret sensory input.
- Individual experiences, memories, and emotions all influence a person's perception of beauty or value. Compare your views on art, movies, fashion, music, or books with those of your classmates and discuss your criteria for determining their beauty or value.
- Real estate agents often suggest to sellers that they should bake bread or cookies before a showing or open house because the smell often evokes a powerful memory of home, safety, and warmth.

7.6

- The longest axons in the peripheral nervous system (PNS) are those of somatic motor neurons that innervate the distal muscles in the feet and toes. The shortest axons are most likely the sympathetic preganglionic neuron fibers that exit the spinal cord and travel to the paravertebral chain or the somatic neurons that innervate the muscles of the eye.
- In a massage session, an invigorating response is produced by increasing depth, speed, and/or rhythm of strokes to stimulate or enhance sympathetic tone.
- Since the parasympathetic motor neurons originate in the cranial and sacral regions of the body, placing one hand at the base of the client's skull and the other at his or her sacrum may have a soothing and calming effect.

7.7

- Terms used to describe various types of pain include achy, sharp, dull, stabbing, zinging, pounding, buzzing, heavy, and cramping.
- Heat supports relaxation and may reduce pain associated with muscle spasm. Conversely, ice raises muscle threshold and may alleviate cramping in that way. Both methods of hydrotherapy could support pain reduction, according to the gate control theory, by providing a high volume of temperature sensory input.
- When a client requests more pressure and the therapist doesn't think it is advisable, it is appropriate to explain to the client that the tissue will not tolerate the increase. It is a fine balance between hearing a client's request and trusting your assessment skills.

Review Questions

1. The three functions of the nervous system are communication, coordination, and control of virtually all body processes by:
   - Sensing changes in the internal and external environment

- Interpreting and integrating sensory information to determine motor responses
- Stimulating motor responses of muscles, glands, and organs.

2. Sensory and motor neurons are in the peripheral nervous system, while interneurons or integrative neurons are found in the CNS.
3. The six types of glial cells: *Schwann cells* are PNS cells that produce myelin; *astrocytes* are CNS neuroglia that structurally support neurons, maintain the chemical environment required for impulse conduction, and form scar tissue for nerves after an injury; *oligodendrocytes* are CNS cells that produce and maintain the myelin for brain and spinal cord; *microglia* are CNS cells that carry out phagocytosis in the CNS; *ependymal* cells produce cerebrospinal fluid of CNS; and *satellite cells* form ganglia in the PNS.
4. Brain stem, diencephalon, cerebrum, cerebellum
5. Cranial nerves originate from the brain or brain stem; spinal nerves originate from the spinal cord; cranial nerves can be mixed, sensory or motor only while spinal nerves are all mixed; cranial nerves are designated by Roman numerals, and spinal nerves are designated by regular numbers with the associated spinal region (C = cervical, T = thoracic, L = lumbar, S = sacral, Co = coccygeal).
6. The key steps of nerve impulse conduction are
    i. Threshold stimulus is applied to neuron and increases plasma membrane permeability to sodium.
    ii. Sodium enters the neuron depolarizing the plasma membrane.
    iii. Potassium exits the cell to repolarize the membrane as adjacent area depolarizes.
    iv. This self-propagating wave of negativity continues along the axon until the impulse reaches the axon terminals.
    v. Neurotransmitters are released from the axon terminals into the synaptic cleft to stimulate a postsynaptic neuron or effector.
    vi. Energy is expended to remove the neurotransmitter from the synaptic cleft.
7. b
8. c
9. a
10. c
11. d
12. b
13. a
14. c
15. d
16. c
17. a
18. d
19. c
20. a
21. c
22. d
23. b
24. a
25. c
26. d
27. b
28. c
29. d
30. b

Chapter 8

What Do You Think?

8.1

- It is important to use an isometric contraction when reciprocal inhibition is used to reduce acute muscle cramps because there is no change in length with an isometric contraction, which minimizes the stimulus to the muscle spindle. This ensures that relaxation rather than contraction is the overall message to the muscle.
- The alpha motor neuron innervates the motor unit since this neuron stimulates extrafusal fibers, and the gamma loop stimulates intrafusal fibers of the muscle spindle.
- The connective tissue elements are most likely responsible for the decreased flexibility since they do not receive the neuromuscular signals engaged by contract-relax. These tissues include fascia, ligaments, tendons, and joint capsules.

8.2

- There are many examples of connected dysfunctions, since a challenge anywhere in the chain can be linked to other structures in the chain. Some examples are appendectomy or hysterectomy scars that could be related to headaches, neck, or hip pain. Similarly, patellar tendinitis could be associated with a groin strain or inguinal bursitis.
- The scar tissue and fascial restriction associated with the C-sections are most directly related to the headaches because they pull the torso forward placing strain on the neck and head. Likewise, the fractured

clavicle magnifies the strain further up and down this anterior chain.
- Therapeutic implications of the above scenario are that to relieve the client's tension headaches, all the restrictions on the chain must be addressed. In this case, a therapist may choose to begin with work in the head and neck and progress through the pelvic girdle or vice versa.

8.3

- Your opinion about the pressure, pace, pattern, and focus of an MT session will be based on your understanding of this information and your manual therapy experiences thus far. Be sure to consider that the interstitial fascial receptors are part of autonomic nervous system neural pathways that regulate fascial tone.

Review Questions

1. The three mechanical properties of fascia are
   - Viscoelasticity—the ability to stretch and rebound
   - Thixotropy—the quality to become more liquid (sol) or solid (gel) in response to temperature and movement
   - Piezoelectricity—the ability to generate an electrical current in response to pressure
2. Motor unit recruitment is the adding of additional motor units within a muscle to increase the strength of a contraction: a graded response. Muscle recruitment is the pattern of stimulation and coactivation between muscle groups to create smooth and coordinated movement.
3. The distinguishing characteristics of a trigger point include
   - Palpable nodule in a taut band of muscle
   - Hypersensitive to mechanical pressure
   - Pressure reproduces pain complaint in a predictable pattern
   - Causes muscle to be hypersensitive to stretch
4. To relieve a cramp in the hamstrings, first apply direct pressure to the cramp. Then, using the other hand at the ankle to create resistance against knee extension, instruct the client to extend or straighten the leg against the pressure of your hand.
5. b
6. a
7. d
8. c
9. c
10. d
11. a
12. d
13. b
14. d
15. d
16. b
17. b
18. c
19. a
20. c
21. d
22. b
23. a
24. c
25. d

Chapter 9

What Do You Think?

9.1

- The muscular system uses the terms *antagonist* and *synergist* to describe the roles that muscles play in coordinated movement. Their meanings are essentially the same; synergists help and antagonists oppose.
- If hormones were not target specific, all cells would have receptors for all hormones (a lot of receptors on one cell), but likely fewer receptors for any specific hormone. This would decrease each cell's responsiveness to a hormone. Also, every cell in the body would respond to the hormone signals. This would create chaotic activity that would significantly disrupt homeostasis instead of preserving it.
- An analogy for the negative feedback mechanism could be using a punishment such as grounding a child for behavior modification as described in Chapter 1.

9.2

- Consider that the hypothalamus is connected with the autonomic nervous system, limbic system, and endocrine system. Therefore, emotions, autonomic and endocrine functions are anatomically and physiologically linked.
- The differing opinions on the pituitary as "master gland" hinge on the role of the hypothalamus and its connection with the pituitary. While the pituitary does control the function of many other endocrine glands, its function is controlled by the hypothalamus.

9.3

- If the thyroid gland were removed, metabolism would be radically altered, as would calcium homeostasis. In cases of thyroid damage where it must be removed, thyroid hormones are prescribed.
- Calcitonin and PTH are similar to insulin and glucagon because each set of hormones maintains the homeostasis of a particular substance (Ca^{2+} or glucose) by acting as antagonists.
- The purpose of cortisone injections and creams is to isolate the anti-inflammatory effect of cortisol. Therefore, they are used to decrease specific local tissue inflammation.

9.4

- "Adrenaline junkies" are risk takers who often thrive on physical challenges. Activities such as sky diving, bungee jumping, base jumping, hang gliding, and other extreme sports are generally a hallmark of people with this characteristic. They may also be employed in dangerous careers, such as search and rescue or firefighting. Adrenaline junkies seem to be addicted to the "rush," the physiologic changes associated with adrenaline and noradrenaline releases.
- Possible negative effects of being an adrenaline junkie include physiologic exhaustion due to constant engagement of the stress response. Also desensitization to everyday challenges that normally initiate the stress response. This secondary effect can be considered a positive because they don't get stressed out by every little thing. However, it can also be a negative characteristic because their desensitization can make them unaware of common risks and dangers.
- How manual therapy skills are used to maximize stress reduction effects depends on the form of manual therapy being practiced. Techniques that decrease sympathetic tone/stimulation help with stress reduction. Consider how choices in rhythm, pressure, speed, and intention, as well as the environment and setting, can influence the client's stress reduction.

Review Questions

1. Endocrine—refers to a ductless gland.
2. Exocrine—refers to a gland with a duct.
3. Hormone—a substance released by endocrine glands that stimulates physiologic responses in target cells or organs.
4. Corticoid—refers to the steroid hormones such as those released from the adrenal cortex.
5. A tropic hormone is a hormone that affects the function of another endocrine gland.
6. The four key tropic hormones of the anterior pituitary are
   - Thyroid-stimulating hormone (TSH)—stimulates the thyroid to control secretion of T_3 and T_4.
   - Adrenocorticotropic hormone (ACTH)—stimulates secretions of the adrenal cortex glucocorticoids (e.g., cortisol).
   - Luteinizing hormone (LH)—stimulates secretions of estrogen and progesterone from the ovaries and testosterone from the testes.
   - Follicle-stimulating hormone (FSH)—works with LH to stimulate hormone release from the gonads and the production and maturation of eggs and sperm.
7. The three ways the body stimulates hormone releases are
   - Hormonal stimulus from other endocrine glands or tissues.
   - Changes in blood concentrations of a specific ion or nutrient in the blood.
   - Neurologic stimulation of a specific gland.
8. c
9. d
10. a
11. c
12. b
13. c
14. a
15. b
16. d
17. b
18. c
19. d
20. a
21. b
22. a
23. c

Chapter 10

What Do You Think?

10.1

- Red blood cells carry oxygen and carbon dioxide and so their numbers are extremely high. White blood cells are involved in body defense, which means we

need a small number of them but a constant presence. Large numbers of WBCs are only required when fighting a specific immune challenge. Platelets are needed to stop bleeding quickly any time it occurs. Therefore, we need a significant presence available throughout the bloodstream at all times.
- Platelets become trapped and attracted by any damage to a vessel, whether or not there is active bleeding. Roughened areas inside the vessels attract and collect platelets and increase the risk of the formation of a blood clot (thrombus) in the vessel.

10.2

- Freeways are often called major arteries, as are main thoroughfares through cities.
- The venous valves are extensions of the basement membrane and epithelial cells that make up the inner lining of the veins. The basement membrane is a little thicker to allow the valves to close firmly.

10.3

- An atrium is a central court or room in a building with a skylight or that is open to the sky. The heart atria are open (without valves at their entrance) and on the top of the heart.
- The pulmonary veins carry oxygenated blood from the lungs back to the heart.
- Your answer in regard to situations in which your heart rate was elevated will be specific to you. However, increases may be due to autonomic nervous system stimulation due to fear or anger or due to hormones such as adrenaline when under stress. Decreases are generally related to rest and relaxation.

10.4

- In cases of hypertension, the pulse is strong and bounding. Increased blood pressure is due to an increase of force against the arterial walls. Therefore, the pulse would feel stronger.
- When a person is frightened, vasoconstriction occurs in the stomach, kidneys, intestines, pancreas, bladder, and gonads since they are not needed for the flight-or-fight response.
- Hypertension increases capillary fluid pressure (CFP), which increases net filtration and resistance to reabsorption. Pregnancy can cause pressure on the vessels of the lower extremity, which also increases CFP, but on the venous side of the capillary, thus reducing net reabsorption.
- Massage has been shown to improve local venous flow. Movement therapies should also improve venous flow by stimulating the skeletal muscle and respiratory pumps.

10.5

- The amount of primary edema directly links the increases in interstitial oncotic pressure that cause secondary edema formation. Therefore, controlling primary edema helps to minimize the overall amount of edema formation.
- People with a sprained ankle tend to hold the ankle in a neutral 90-degree position. This causes them to limp when they walk.
- Arteriosclerosis and atherosclerosis are seen in people of all ages; they can be caused by numerous factors, including diet, stress, activity level, and genetics.

Review Questions

1. The cardiovascular system is responsible for transporting nutrients and wastes; helping to regulate body temperature, pH, and fluid balance; and playing a major role in the body's immune responses.
2. The three formed elements in blood are *erythrocytes* that carry O_2 and CO_2 in the blood, *leukocytes* that carry out important immune responses, and *platelets* that are responsible for blood clotting.
3. The three types of blood vessels are *arterial vessels* (arteries and arterioles) that carry blood away from the heart and distribute it to the tissues, *capillaries* that serve as the site of nutrient–waste exchange, and *venous vessels* (veins and venules) that collect blood from the tissues and return it to the heart.
4. The two circulatory divisions are the *pulmonary circuit* between the heart and lungs and the *systemic circuit* between the heart and body.
5. The cardiac cycle is the three-step process of contraction and relaxation that moves blood through the heart. It begins with the relaxation phase that allows blood to move from the atria into the ventricles. The second phase is atrial systole, in which the atria contract to push blood through the atrioventricular valves into the ventricles. The semilunar valves are closed during this stage. The third phase is ventricular systole, in which the ventricles contract to pump blood through the semilunar valves into the great arteries. The atria fill during this stage.
6. c
7. b

8. c
9. d
10. a
11. d
12. b
13. c
14. a
15. d
16. b
17. c
18. b
19. a
20. d

Chapter 11

What Do You Think?

11.1

- The cardiovascular system is a closed system that is responsible for delivery of nutrients and return of waste products, while the lymphatic system is an open system. The lymphatic system is a fluid-return system only; it plays an essential role in returning fluid to the cardiovascular system.
- The large proteins that cannot be reabsorbed through the cardiovascular capillaries are essential to maintaining the Starling forces in the capillary beds. The fats absorbed by the lacteals are essential for the production of hormones, enzymes, and cellular components.

11.2

- Similar to veins, the larger lymph vessels must return large volumes of fluid against gravity. Therefore, the smaller segments of vessels created by valves create a "stair-step process" that allows movement of fluids through shorter segments rather than the entire vessel. Valves in the initial vessels and collecting capillaries would increase resistance to fluid uptake from the interstitium.
- Fluid return is a function of the venous system. Dropping lymph into the arteries would simply allow it to be moved back into the capillary beds and filtered right out again.
- The thoracic duct handles a larger volume of fluid and works against gravity most of the time. Therefore, the collection of fluid at the base of the thoracic duct helps move larger volumes through the duct at one time.

11.3

- Lymph nodes have more afferent vessels than efferent vessels, creating an internal pressure within the node that slows the flow of lymph to support filtration.
- The term *catchment* is more descriptive of lymph node bed function: slowing and holding the flow of lymph to support filtration. Therefore, manual therapists must focus attention on speeding lymph flow through the catchment to reduce edema.
- Hinge points are natural points of compression in the body. Therefore, body movement helps to increase the flow of lymph through the catchments.

11.4

- Your individual description of pre-lymphatic channels and how fluid flows through them to the initial vessels is appropriate here.
- The lymphatic system has no central pump, but lymphangia have spiraled smooth muscle. Smooth muscle is innervated by the autonomic nervous system to create a pumping action that maintains lymph flow, even during periods of inactivity.
- Most manual lymphatic techniques begin with deep breathing and strokes around the neck and terminus to create negative pressure within the lymphatic system that enhances the siphon effect.

11.5

- It is important for therapists to recognize that there are different types of edema because it is not safe to treat all forms of edema with many forms of manual therapy. For example, treatment of lymphedema requires advanced specialty certification.
- There are many possible questions and clues related to the cause of edema. Some questions a therapist may ask the client are: When did the edema begin? Is there any kidney or heart dysfunction? Has there been an injury in the area? Does the edema go away at night and develop over the course of the day? Is there any history of cancer, surgeries, radiation, or chemotherapy?

Review Questions

1. The five lymphatic vessels, from smallest to largest, are initial capillary, collecting capillary, lymphangia, collecting trunk, and deep ducts.
2. Pre-lymphatic channels—minute nonstructural preferred pathways for interstitial fluid in tissues without initial lymphatic capillaries.

3. Catchment—lymph node bed; a cluster of lymph nodes.
4. Watershed—functional zone (boundary) between lymphotomes with a high concentration of anastomoses to allow lymph flow in either direction.
5. Angion—segment of a lymphangia between two intralymphatic valves.
6. Anastomosis—end-to-end arrangement of collecting capillaries along a watershed.
7. The five primary components that make up the lymph obligatory load are fluid, proteins, cells, foreign substances, and long-chain fatty acids.
8. b
9. d
10. c
11. a
12. b
13. c
14. d
15. a
16. c
17. b
18. d
19. d
20. c

Chapter 12

What Do You Think?

12.1

- In manual therapy practice, the term *lymphatic* refers to the structures involved in fluid return, while *lymphoid* is reserved for the cells, tissues, and structures involved with immune responses. This helps to clarify the importance of understanding the two different roles of the lymphatic system.
- The lymph nodes are structurally connected to the lymphatic vessel network.
- The secondary lymphoid tissues trap pathogens and contain lymphocytes to fight them off. However, the lymph nodes filter pathogens from lymph, while the spleen filters blood. The mucosa-associated lymphoid tissues trap pathogens that enter the digestive, respiratory, and reproductive tracts.

12.2

- Analogies for nonspecific immune defenses include fences or walls for physical barriers, and treated wood in a fence or mace can represent chemical barriers. Lawn feed that supports the growth of grass and inhibits weed growth could represent antimicrobial proteins. Phagocytosis could be represented by crows or seagulls that engulf and digest just about anything. Natural killer (NK) cells could be represented as a community block watch group who protect a neighborhood.
- Most people have used antimicrobial soaps, detergents, or sanitary wipes. You may even own a garment made from antimicrobial material.
- Phagocytes eat and digest pathogens and other particles, while NK cells destroy infected cells by cytolysis.

12.3

- An example might be the difference between a gang boss and a lower-level member of the group. The boss represents AMI because he sends other members out to do his dirty work, similar to a plasma cell sending out antibodies to deal with an antigen. Lower-level members do their own dirty work, similar to cytotoxic T cells in CMI.
- Your response in regard to artificially acquired immunity is specific to your medical history. If you are unaware of your history of immunizations, it is recommended that you obtain this information.

12.4

- Think of times when you had a "gut feeling" or your emotions affected your thinking, activity, and/or health. Have you ever called in sick at work so you could go play and then ended up feeling sick? Or read a book, seen a movie, or listened to music that lifted your spirit and made your energy rise?
- There are a variety of ways that knowledge of the mind-body-spirit connection can benefit clients. Take into consideration the setting, your tone of voice, intention, mindset and beliefs, and your form of manual therapy practice.

Review Questions

1. Antigen—substance that elicits a specific immune response
2. Cytokine—cell-signaling molecule
3. Antibody—plasma protein released by plasma B cells that can bind with an antigen to neutralize or kill it; immunoglobulin
4. Allergen—antigen that triggers an allergic response of the immune system
5. The body's nonspecific immune defenses include physical barriers, chemical barriers, internal

antimicrobial proteins, phagocytes, natural killer cells, inflammation, and fever.
6. Phagocytes engulf and digest pathogens, while NK cells attach themselves to infected body cells and secrete toxic substances that kill the entire cell through cytolysis.
7. The body's specific immune responses include recognition of specific antigens, the activity of specific B and T lymphocytes, and development of immunological memory.
8. a
9. c
10. a
11. b
12. d
13. c
14. c
15. d
16. b
17. a
18. b
19. d
20. b

Chapter 13

What Do You Think?

13.1

- A garden hose could represent the trachea, or the larynx could be represented by a blade of grass held between your thumbs to create a whistle or buzz.

13.2

- The insertion points of the serratus anterior and pectoralis minor are on the scapula; therefore, their primary action is movement or stabilization of the scapula. Additionally, the fiber direction of the serratus anterior is too horizontal to assist in elevation, depression, or expansion of the thorax. Conversely, the scalenes and intercostals insert on the ribs. Therefore, their primary actions include movement of the ribs.
- Breathing is generally easy to assess visually by watching the rise and fall of the chest. You can also listen to a client's breathing and sometimes feel their breath. This information can indicate their level of relaxation, anxiety, tension, pain, and other sensations or emotions. You may also be able to notice what muscles are being used. For example, shallow breathing often indicates that the scalenes are engaged more than the diaphragm. Discuss you responses with your classmates and teacher.
- To modify a standard manual therapy session for clients with breathing issues, you might consider altering the position to semirecumbent, side-lying, or seated. The duration of the session may also need to be shortened.

13.3

- Many things can shift a client's breathing. For example, a memory or emotional release can be stimulated and shift breathing, or a pain response is stimulated by too much pressure or an abrupt movement.
- The metabolic interference caused by a build-up of carbon dioxide is dangerous to the body. Therefore, it is important that it is quickly recognized and handled.

Review Questions

1. The diaphragm, external intercostals, and scalenes exert the most influence over inhalation. The pectoralis minor and serratus anterior can also assist with inhalation. Exhalation is primarily a passive process of relaxation, but forced exhalation is accomplished through contraction of the internal intercostals, rectus abdominus, and other abdominal muscles.
2. The upper respiratory tract includes the nose and nasal cavities, pharynx, and larynx.
3. The pleura is the double-layered serous membrane that lines the thoracic cavity and surrounds the external surface of the lungs. The respiratory mucosa is the continuous mucous membrane that lines the respiratory tract. The respiratory membrane is formed by the alveolar and capillary walls; this is where oxygen and carbon dioxide is exchanged between the lungs and blood.
4. The respiratory reflex centers are located in the brain stem. Two small groups of self-excitatory neurons in the medulla oblongata are responsible for setting the basic rate and rhythm of breathing, with some assistance from the respiratory centers in the pons. The pneumotaxic area of the pons regulates breathing rate by coordinating the stimulation of the inspiration center in the medulla.
5. b
6. c
7. d
8. a

9. c
10. d
11. a
12. b
13. a
14. c
15. d
16. d
17. c
18. a
19. c
20. d

Chapter 14

What Do You Think?

14.1

- During winter months, the body must work harder to maintain its core temperature. Therefore, increased nutrients are needed and people tend to eat more.
- Cracking a nut could be an analogy for the mechanical digestive process of chewing, and chemical digestion can be equated to the use of vinegar or lemon juice in meat marinades. Squeezing toothpaste or frosting out of a tube could be used to show motility and/or elimination.

14.2

- A garden hose could be used to represent the esophagus, a wine pouch could be used for the stomach, and gasket rings could be sphincters.
- In treating a client who has just consumed a large meal, positioning may need to be altered. For example, the prone position could be uncomfortable. Direct manipulation in the abdomen or diaphragm could also be uncomfortable.

14.3

- When a client reports nagging right shoulder pain, it is important to ask if there has been an injury or change in activity that may have aggravated musculoskeletal structures. Additionally, thorough palpation and range of motion assessment should help to identify structural issues in the shoulder. However, if the shoulder pain cannot be related to the function and structure of the shoulder, therapists must consider the possibility of referred pain from the liver or other organ dysfunction. In this case, clients may have a fever, report nausea or abdominal pain, and appear pale and/or sweaty.
- The pancreas and duodenum contain a lot of chemical enzymes. Therefore, they are located retroperitoneal so if they are damaged, they will not spill their contents into the entire abdominal cavity.

14.4

- Carbohydrates break down into glucose, which provides the most ATP. The byproducts (CO_2 and H_2O) produced by glucose catabolism are easily recycled or excreted by the body. In comparison, both fats and proteins must be converted into intermediary substances before they can be catabolized for energy, and protein catabolism also produces ammonia (a toxin), which must be converted to urea and excreted in the urine.
- Hunger is the physiologic drive to eat, while appetite is a desire for food that can be stimulated by mental, emotional, social, cultural, and other factors. Whether hunger or appetite is a bigger challenge in your community depends on the socioeconomic status of your community and your perception of it.

Review Questions

1. The six processes of the digestive system are
   - Ingestion—taking in food and liquids by mouth; eating and drinking.
   - Secretion—mucus, enzymes, buffers, acids, and hormones are produced and secreted by the digestive organs to assist in the functions of the system.
   - Digestion—breakdown of food into absorbable molecules through mechanical and chemical processes.
   - Motility—movement of substances through the gastrointestinal (GI) tract.
   - Absorption—nutrients are absorbed through the wall of the digestive tract into the bloodstream.
   - Elimination—defecation of solid waste.
2. The pharynx, esophagus, and large intestine do not contribute to chemical or mechanical digestion.
3. Segmentation is the contraction of the circular smooth muscle of the small intestine that assists in mixing its contents with enzymes. Peristalsis involves the coordinated contractions of both the longitudinal and circular smooth muscle that move substances through the GI tract.
4. The layers of the gastrointestinal tract, from external to internal, are the serosa, the muscularis, the submucosa, and mucosa.
5. a
6. c
7. b

8. a
9. d
10. d
11. b
12. c
13. a
14. d
15. d
16. b
17. c
18. a
19. d
20. c

Chapter 15

What Do You Think?

15.1

- All three systems (urinary, respiratory, and digestive) are mucous-lined tracts and holding areas. However, the digestive system has openings to the external environment at both ends, while the respiratory and urinary systems only have one external opening each. Also, the stomach and bladder both have rugae to allow for distension.

15.2

- During the alarm response, sympathetic autonomic nervous system stimulation of the smooth muscle of the renal vessels decreases blood flow to the kidneys. This reduction in blood flow stimulates the RAA pathway.
- Increased urine output decreases the overall volume of fluid within the body. Any decrease in fluid volume decreases the circulating blood volume, which in turn, decreases blood pressure.

15.3

- Anything that increases blood pressure will increase filtration. Examples include arterial disease, diabetes, obesity, and stress.
- Since the overall amount of glomerular filtrate is so high, it is important that the fluid is reabsorbed right away. The body must retain fluid, so it is prioritized at the proximal convoluted tubule.

Review Questions

1. The four components of the urinary system are the kidneys, ureters, bladder, and urethra.
2. The ureters are tubes that carry urine from the kidneys to the bladder. The urethra is the passageway from the bladder to the external environment that allows for urination.
3. The key steps in urine formation (in the order that they occur) are filtration, reabsorption, and secretion.
4. Antiduretic hormone (ADH) and aldosterone act on the kidneys to regulate fluid volume and pressure in the body. ADH secreted by the posterior pituitary causes water retention, while aldosterone secreted by the adrenal medulla causes sodium retention. Since water follows sodium, aldosterone leads to water retention as well. Water retention ensures that circulating blood volume and blood pressure are maintained.
5. b
6. b
7. a
8. d
9. c
10. d
11. b
12. b
13. a
14. c
15. b

Chapter 16

What Do You Think?

16.1

- Stress has a negative impact on libido, sexual performance, and fertility. Therefore, general relaxation and stress reduction from manual therapy can be considered supportive therapy.
- Loose-fitting underwear allows the testes to drop away from the body, thus lowering the temperature to support better sperm production and motility.

16.2

- The uterus and stomach are both highly muscular and designed as holding areas. They also contract to expel their contents.
- The fallopian tubes are similar to the esophagus; both are tubes lined with a mucous membrane whose muscular contractions move contents to a larger organ.

16.3

- A growing fetus creates pressure on all the internal organs, including the stomach. Therefore, the

processing of large meals is difficult and uncomfortable and smaller meals are preferred by most pregnant women.
- Your concerns about working with pregnant clients can be discussed with your classmates and instructor.

Review Questions

1. The function of the male and female reproductive systems is to produce hormones and gametes; they are involved in all the processes related to sexual development and producing offspring.
2. The four structural features and physiologic processes common to both the male and female reproductive systems are gonads and gametes, genitalia, accessory organs, and hormonal production and control.
3. The primary male reproductive organs are the testes that produce sperm and testosterone; the reproductive tract (seminal vesicles, epididymis, vas deferens, ejaculatory ducts, and urethra) that carries sperm and semen out of the body; accessory organs, including the prostate and bulbourethral glands that secrete fluids that make up semen; and the penis that serves as the copulatory organ. Female reproductive organs include the ovaries that store and release eggs and produce estrogen and progesterone; the fallopian tubes, uterus, and vagina that form the reproductive tract; and the vulva is the external genitalia.
4. gamete—a sex cell (sperm or egg)
5. ovulation—expelling of an egg from the ovary
6. fertilization—joining of the genetic material from one sperm and one egg; conception
7. genitalia—external sexual organs (penis and vulva)
8. menarche—the onset of menstruation
9. gestation—pregnancy
10. d
11. c
12. a
13. b
14. c
15. c
16. a
17. d
18. a
19. c
20. d

Glossary

abdominal (ab-DOM-i-nal) Relating to the abdomen (belly).

abdominopelvic cavity (ab-dom-ih-no-PEL-vik KAV-ih-te) One of two ventral cavities of the body; contains all of the abdominopelvic organs.

abduction (ab-DUK-shun) Movement away from the midline of the body.

absorption Uptake or incorporation of a substance such as a gas or liquid or energy (light or heat) into another substance or tissue.

acid (AS-id) Electrolyte compound that releases hydrogen ions (H^+) when dissolved in water; a compound with a pH below 7.

acidic (ah-SID-ik) The quality of having a high number of hydrogen ions (H^+) and with a pH below 7.

acromial (ah-KRO-me-al) Pertaining to the highest point of the shoulder.

actin (AK-tin) Thin protein myofilaments in the muscle sarcomere.

action potential (AK-shun po-TEN-shul) Nerve impulse.

active transport Cellular transport mechanism that requires energy to move substances against a concentration or pressure gradient.

acute (ah-KUTE) Having a sudden or rapid onset; not chronic.

acute (ah-KUTE) **stage** The first phase of the healing process that includes hemorrhage, inflammation, secondary edema formation, spasm, and hematoma organization; also known as the *inflammatory stage*.

adaptation (a-dap-TA-shun) Adjustment in the sensitivity of a sensory receptor to continued or repeated stimulus.

adaptive immune response Defensive body reaction involving B- and T-cell activation by a specific pathogen; also known as a *specific immune response*.

adduction (ad-DUK-shun) Movement toward the midline of the body.

adenosine diphosphate (ADP) (ah-DEN-o-seen di-FOS-fate) Compound formed when one phosphate group is broken off ATP molecules to provide energy for cellular work.

adenosine triphosphate (ah-DEN-o-zeen tri-FOS-fate) **(ATP)** Energy molecule produced by the mitochondria within cells.

adhesion (ad-HE-zhun) Point in connective tissue where fibers are bound together and/or where ground substance is thick and stiff.

adrenal cortex (ah-DREE-nal KOR-teks) Outer section of the adrenal glands; secretes mineralocorticoids, glucocorticoids, and small amounts of androgens.

adrenal (ah-DREE-nal) **gland** Endocrine gland situated on top of each kidney; also known as *suprarenal gland*.

adrenal medulla (ah-DREE-nal meh-DOO-la) Inner section of adrenal glands; secretes adrenaline and noradrenaline.

adrenaline (ah-DREN-ah-len) Hormone secreted by the adrenal medulla; its function is to support the fight-or-flight response of the ANS, sustaining increased heart rate, blood pressure, and blood flow to heart, liver, and skeletal muscles, and dilation of small airways inside the lungs.

adrenocorticotropic (ah-dreen-o-kor-teh-ko-TRO-pik) **hormone (ACTH)** One of the tropic hormones secreted by the anterior pituitary gland; stimulates the release of glucocorticoids from the adrenal cortex.

aerobic (ah-RO-bik) **metabolism** Method of energy production that requires oxygen; includes the Krebs cycle in which pyruvic acid is converted into ATP, water, and carbon dioxide in the mitochondria of cells.

afferent (AF-fer-ent) Conducting toward a center; inflowing.

agonist (AG-on-ist) Muscle that generates most of the power for motion; also known as the *prime mover*.

alarm response The part of the stress response that stimulates the sympathetic division of the ANS, which initiates the fight-or-flight response.

aldosterone (al-DOS-ster-rone) Hormone secreted by the adrenal cortex; causes kidneys to increase sodium retention and secretion of potassium.

alkaline (AL-kah-lin) The quality of having a high number of hydroxide ions (OH^-) and with a pH above 7.

all-or-none response Muscle physiology principle stating that all muscle fibers in a motor unit must contract when threshold stimulus is delivered.

allergen (AL-er-jen) Antigen that triggers an allergic response of the immune system.

alpha (AL-fah) **loop** Neuronal reflex loop that mediates the stretch reflex; created by an alpha sensory neuron originating in the muscle spindle and an alpha motor neuron that stimulates extrafusal muscle fibers.

alveoli (al-VE-o-lus) Microscopic air sacs in the lungs at the end of the bronchial tree that serve as the site of oxygen and carbon dioxide exchange (singular form is alveolus).

amino acid (ah-ME-no AH-sid) Molecular building block of proteins.

amniotic (am-ne-OH-tik) **sac** Fluid-filled sac that surrounds and protects the growing embryo or fetus.

amphiarthrosis (am-fe-ar-THRO-sis) Type of joint that allows partial movement in which there is a cartilage disk between bony surfaces; also known as a *cartilaginous joint*.

amylase (AM-ah-lace) Enzyme that digests starch.

anabolism (ah-NAB-o-lizm) Process that occurs when the body uses nutrients as building blocks, either storing the nutrient for use at a later time or using it to repair and build new tissue.

anaerobic (AN-ah-RO-bik) **metabolism** Method of energy production that occurs without the presence of oxygen; includes glycolysis, in which glucose is converted to pyruvic acid.

anal canal (A-nal kah-NAL) Distal portion of the rectum.

anastomosis (ah-NAS-to-mo-sis) An end-to-end arrangement of nerves, blood, or lymph vessels.

anatomic position (an-ah-TOM-ik po-ZIH-shun) Position in which a person is standing upright with arms at the sides and the face, palms, and feet facing forward; this position is the basis for all regional, directional, and movement terminology.

anatomy (ah-NAT-o-me) The form and structure of an organism, such as the human body; also the branch of science that studies the structure of organisms.

anchor filaments Thin, hair-like fibers attached to the epithelial flap of lymphatic initial vessels; hold or anchor the initial vessel in place and open the vessel when pulled by tissue stretch.

androgen (AN-dro-jen) Steroid that acts as a male sex hormone; controls the development of masculine characteristics in both genders.

anemia (ah-NE-me-ah) Condition that indicates a decreased number of red blood cells or decreased hemoglobin level.

angion (AN-ge-on) Segment of the primary lymphatic vessel (lymphangia) marked by intralymphatic valves at each end.

angulus venosus (AN-gu-lus ven-O-sus) Junction between the lymphatic ducts and the subclavian veins; also known as the *lymphatic terminus*.

antagonist (an-TAG-ah-nist) Muscle that opposes the agonist (prime mover).

antagonistic (an-TAG-ah–NIS-tik) **effect** Two hormones working against each other; one increases and the other decreases blood levels of a chemical.

antebrachial (an-teh-BRA-ke-al) Pertaining to the forearm.

antecubital (an-teh-KU-beh-tal) Pertaining to the anterior elbow.

anterior (an-TE-re-or) Front (ventral).

antibody (AN-teh-bod-e) Plasma protein released by plasma B cells that can bind with an antigen to neutralize or kill it; also known as an *immunoglobulin*.

antibody-mediated immunity (AN-teh-bod-e ME-de-a-ted im-MU-nih-te) **(AMI)** Disease resistance produced through the activation of B cells.

antidiuretic (anti-di-u-REH-tik) **hormone (ADH)** Hormone secreted by the posterior pituitary that inhibits urine production in the kidneys; also known as *vasopressin*.

antigen (AN-teh-jen) Cellular marker that identifies a cell as foreign and causes an immune response.

antimicrobial (AN-tee-mi-KRO-be-al) Any substance that destroys or inhibits the growth and development of microbes, preventing their pathogenic action.

antimicrobial (AN-tee-mi-KRO-be-al) **peptide** Short-chain amino acid produced by macrophages and mucus-producing epithelial cells; destroys pathogens.

anus (A-nus) Opening at the end of the anal canal through which waste is excreted.

aortic (a-OR-tik) **valve** Heart valve located between the left ventricle and the aorta.

apocrine (AP-o-krin) Pertaining to a specialized sweat gland found in the axilla, groin, and areola of the breasts; secretes a milky fluid consisting of sweat and pheromones.

aponeurosis (AP-o-nu-RO-sis) Broad sheet of connective tissue that serves as the attachment point for several muscles in the torso or thigh.

appendicular (ap-pen-DIK-u-lar) Pertaining to the portion of the skeleton made up of the bones of the limbs, including the bones of the pectoral and pelvic girdles.

appendix (ah-PEN-diks) Small twisted tube located at the junction between the small and large intestines that contains lymphoid tissue.

arachnoid mater (ah-RAK-noyd MAH-ter) Middle layer of the meninges, made of a web-like arrangement of connective tissue fibers.

arrector pili (ah-REK-tor PE-li) Small muscle attached to a hair follicle; when it contracts, the hair stands up straight.

arteriole (ar-TE-re-ole) Small artery.

artery (AR-ter-e) Blood vessel that carries blood away from the heart.

articular (ar-TIK-u-lar) **cartilage** Type of hyaline connective tissue that covers the articulating surfaces of bones in a synovial joint.

articulation (ar-tik-u-LA-shun) Joint; the point at which two or more bones meet.

artificially acquired immunity (ar-tih-FISH-ah-le ah-KWI-yurd im-MU-nih-te) Disease resistance developed in response to an initial vaccination or subsequent booster shot.

ascending colon (a-SEN-ding KO-lun) Vertical segment of the large intestine on the right side of the abdominal cavity.

atom Smallest particle (unit) of an element that has all the properties of the element; composed of electrons, neutrons, and protons; atoms make up molecules.

atrioventricular (A-tre-o-ven-TRIK-u-lar) **(AV) bundle** Located in the interventricular septum, this is the only region where action potentials can pass from the atria to the ventricles; also known as *bundle of His*.

atrioventricular (A-tre-o-ven-TRIK-u-lar) **(AV) node** Collection of autorhythmic cardiac muscle cells located in the inferior portion of the septum between the atria that acts as an area of delay.

atrium (A-tre-um) Receiving chamber of the heart.

atrophy (AT- tro-fe) Wasting or shrinking of tissue or organs.

autonomic (ah-to-NAH-mik) **nervous system (ANS)** Motor division of the peripheral nervous system that innervates glands, smooth muscle, and cardiac muscle.

autonomic ganglion (ah-to-NAH-mik GANG-le-on) Collection of cell bodies and dendrites that form a point of synapse for the two motor neurons in an autonomic pathway; a junction box.

avascular (a-VAS-cu-lar) Without blood vessels.

axial (AKS-e-al) Pertaining to the portion of the skeleton made up of the bones in the skull, spine, and torso; includes the cranial, facial, and hyoid bones, the sternum, ribs, and the 33 vertebrae.

axillary (AK-sil-air-e) Located in or near the arm pit.

axon (AKS-on) Fibrous portion of a neuron that carries a nerve impulse from the cell body toward another neuron or effector.

axon hillock (AKS-on HIL-lok) Small "bump" in a neuron where the axon attaches to the cell body.

axon terminal (AKS-on TER-min-al) Small branched end of an axon.

B cell Specialized lymphocyte involved in antibody-mediated immunity (also called B lymphocyte).

ball-and-socket joint Type of synovial joint in which the rounded end of one bone fits into a socket-like depression of another; allows all five basic movements.

baroreceptor (BAR-o-re-SEP-tor) Sensory receptor sensitive to changes in fluid pressure.

basal cell carcinoma (BA-zehl SEL kar-sih-NO-mah) Most common type of skin cancer.

basal metabolic (BA-zal meh-tah-BAL-ik) **rate (BMR)** Rate at which energy is used by the body at rest; determined by the amount of oxygen used by body cells and represented as Calories of heat.

base Substance whose molecule or ion is capable of combining with a proton (hydrogen ion) to form a new substance; alkaline substance having a high number of hydroxide ions (OH^-) and with a pH above 7.

basement membrane Thin connective tissue layer to which epithelium is attached.

bicuspid (bi-KUSS-pid) **valve** Heart valve located between the left atrium and ventricle; also known as the *mitral valve*.

bile A fat emulsifier produced by the liver and stored in the gall bladder.

bladder (BLA-der) Expandable membranous sac; the urinary bladder serves as a holding tank (storage site) for urine before it is excreted from the body.

blastocyst (BLAS-to-sist) Multi-celled mass of tissue that develops into an embryo.

blood pressure Force exerted by blood on the wall of a blood vessel.

body Main portion or mass of any structure.

body planes Planes that divide the body to establish front, back, top, bottom, right, and left sections.

brachial (BRA-ke-al) Pertaining to the upper arm.

brain stem Portion of the brain just above the spinal cord that relays information between the spinal cord and upper regions of the brain; consists of the medulla oblongata, pons, and midbrain.

bronchi (BRON-kus) Tube-like passageways that carry air to and from the lungs (singular form is bronchus).

bronchial tree (BRON-ke-al TRE) Air passageways from the trachea through the bronchi and bronchioles.

bronchiole (BRON-ke-ole) Smallest air passageway of the bronchial tree.

bulbourethral (BUL-bo-yur-RE-thral) **gland** One of two pea-sized glands inferior to the prostate that secretes a mucus-like substance that lubricates the urethra so sperm can easily pass through.

bundle branch A component of the heart's conduction system that diverges left and right from the AV bundle to carry the action potential down the interventricular septum to the apex of the heart.

bundle of His Located in the interventricular septum, this is the only region where action potentials can pass from the atria to the ventricles; also known as *atrioventricular bundle*.

bursa (BUR-sah) Small synovial fluid-filled sac found in some synovial joints; cushions and reduces friction between the bone and ligaments, tendons, and/or muscles.

calcaneal (kal-KA-ne-al) Pertaining to the heel of the foot.

calcify (KAL-sih-fi) To deposit or store calcium salts; hardening of bone due to the deposition of calcium salts.

calcitonin (KAL-seh-TO-nin) Hormone released by the thyroid gland; decreases blood calcium levels by enhancing bone absorption of the mineral.

calyx (KA-liks) Cup-like structure or organ; in the kidney, a small cup at the bottom of a renal pyramid that collects urine and transfers it to the renal pelvis.

canaliculi (kan-ah-LIK-u-li) Intricate network of channels in dense bone tissue that allows for the passage of nutrients and wastes to and from the osteocytes.

cancellous (KAN-sel-us) Pertaining to soft or spongy bone tissue.

capillary One of the tiniest blood vessels; connects arterioles and venules.

capillary (KAP-eh-lar-e) **bed** Dense network of the smallest blood vessels that allows for nutrient and waste exchange with the tissues; located between an arteriole and venule.

capillary (KAP-eh-lar-e) **exchange** Movement of substances through the wall of a capillary.

capillary (KAP-eh-lar-e) **fluid pressure (CFP)** Force created by blood pressing out against the wall of the capillary.

capsule (KAP-sel) Fibrous connective tissue covering that envelops an organ, gland, or joint.

carbohydrate (car-bo-HI-drate) Molecular compound containing a mixture of hydrogen, carbon, and oxygen atoms; breaks down into the simple sugar glucose, which is used by the body to produce energy.

cardiac (KAR-de-ak) **cycle** The ordered sequence of atrial and ventricular contraction and relaxation that makes up one heart beat.

cardiovascular (kar-de-o-VAS-ku-lar) **system** Body system made up of the heart and blood vessels.

carpal (KAR-pal) Pertaining to the wrist.

cartilaginous (kar-tah-LAJ-en-us) **joint** Type of joint that allows partial movement in which there is a cartilage disk between bony surfaces; also known as an *amphiarthrosis*.

catabolism (cah-TAB-o-lizm) Any chemical process the body uses to break down nutrients to release energy.

catchment (KACH-ment) Grouping of lymph nodes; also known as a *lymph node bed*.

catecholamines (kat-ah-KO-lah-meen) Group of hormones secreted by the adrenal medulla; adrenaline and noradrenaline together.

cauda equina (KAW-dah e-KWI-nah) Bundle of spinal nerve roots that fan out at the end of the spinal cord.

caudal (KAW-dal) Below; closer to the feet; inferior.

cecum (SE-kum) Initial short pouch-like segment of the large intestine.

cell (SEL) Basic building block of the body; the least complex level of organization.

cell-mediated immunity (sel ME-de-a-ted im-MU-nih-te) **(CMI)** Disease resistance produced through the activation of T cells.

cell membrane (MEM-brain) Semipermeable boundary surrounding all cells; also known as the *plasma membrane*.

central canal Small opening in the center of the spinal cord that allows for circulation of cerebrospinal fluid through the cord.

central nervous system (CNS) A primary division of the nervous system; includes the brain and spinal cord.

cephalad (SEF-ah-lad) Above; closer to the head; superior.

cephalic (seh-FAL-ik) Pertaining to the head.

cerebellum (ser-ah-BEL-um) Portion of the brain located posterior to the brain stem and inferior to the cerebrum that coordinates voluntary muscle contraction and maintains muscle tone, posture, and balance.

cerebral cortex (seh-RE-brul KOR-teks) Outer layer of the cerebrum; consists of gray matter.

cerebrospinal fluid (CSF) (seh-RE-bro-SPI-nal FLU-id) Clear fluid that lubricates and cushions the brain and spinal cord.

cerebrum (seh-RE-brum) The largest and uppermost portion of the brain that is the center for consciousness, cognition, and motor activity.

cervical (SER-vih-kal) Pertaining to the neck.

chemoreceptor (KE-mo-re-sep-tor) Sensory receptor sensitive to chemical changes.

chemotaxis (ke-mo-TAX-is) Movement of cells in response to chemicals.

chondrocyte (KON-dro-site) Cartilage cell.

choroid plexus (KO-royd PLEKS-us) Collection of specialized capillaries in the ventricles of the cerebrum that secrete cerebrospinal fluid.

chronic (KRAH-nik) Having a gradual or long-term onset or prolonged duration.

chyme (KIME) Semiliquid mixture of food and digestive juices created by the stomach and moved through the small intestine.

cilia (SIL-e-ah) Hair-like projections that extend outward from the plasma membrane of certain cells.

circuit (SIR-kut) Patterned neuronal pathway; convergence or divergence are types of circuits.

circumduction (sur-kum-DUK-shun) Multiaxial movement about a fixed point.

cisterna chyli (sis-TERN-ah KI-le) Enlarged sac at the base of the thoracic duct; a collecting well for lymph from the lower extremities and abdominopelvic organs.

clitoris (KLIH-tor-is) Small bud of highly sensitive tissue at the anterior junction of the labia minor that becomes engorged during sexual arousal.

coagulation (ko-ag-u-LA-shub) Process of blood clotting.

cochlea (KO-kle-ah) Snail shell–shaped structure of the inner ear where the sensory receptors for hearing are located.

cognition (kog-NIH-shun) Higher mental process involved in knowing and perceiving.

collagen (KAHL-ah-jen) Thick extensible but not elastic fiber found in connective tissues.

collagen remodeling Breakdown, recycling, and thickening of granulation tissue to form permanent scar tissue during tissue healing.

collecting trunk Large lymphatic vessel that connects lymphangia from a specific body region to a deep lymphatic duct; also known as a *lymphatic trunk*.

colon (KO-lun) Longest segment of the large intestine.

common bile duct Tube that carries bile from the gall bladder to the duodenum.

common hepatic duct (hah-PAH-tik DUKT) Tube that carries bile out of the liver to the gall bladder.

compact bone Dense (cortical) bone tissue.

complement protein (KOM-pleh-ment PRO-teen) Plasma protein involved in the body's immune response that initiates the complement system when activated by a pathogen.

compound Substance consisting of atoms or ions of different elements bound together.

concentric contraction (kon-SEN-trik) Isotonic contraction in which the muscle shortens; insertion moves closer to the origin.

conception (kon-SEP-shun) Joining of one sperm and one egg; also known as *fertilization*.

conduction Conveyance of energy; process by which a nerve impulse is transmitted.

conduction (kon-DUK-shun) **system** Organized network of autorhythmic cardiac muscle cells that produce the coordinated contractions of the heart.

condyle (KON-dile) Rounded projection at the end of a bone; usually articulates with another bone.

condyloid (KON-deh-loyd) Type of synovial joint characterized by two oval-shaped articular surfaces, one convex and one concave, that fit into one another; allows flexion, extension, abduction, and adduction.

cone Photoreceptor of the eye sensitive to bright light and specific color wavelengths.

congenital (kon-JEN-ih-tal) Disease or defect present from birth.

connective tissue Most abundant and widespread type of tissue in the body; functions to bind, support, protect, insulate, and transport.

contractile (kon-TRAK-tile) **tissue** Tissue involved with contraction; skeletal muscle, tendons, and fascia.

contraindication (kon-trah-in-dih-KA-shun) Any condition or factor that makes a specific course of treatment improper or inadvisable because a negative effect is likely.

contralateral (kon-trah-LAH-ter-al) On the opposite side of the median.

convergence (kon-VER-jents) Neural circuit in which several neurons make a synaptic connection to one neuron.

corpus callosum (KOR-pus ka-LOS-sum) Bridge of nerve fibers that connects the right and left hemispheres of the cerebrum.

corpus luteum (KOR-pus LU-te-um) Glandular mass formed from the follicle once the ovum has been discharged.

cortical (KOR-teh-kal) Pertaining to dense (compact) bone tissue.

corticoid (KOR-tih-koyd) Hormone secreted by the adrenal cortex, including the mineralocorticoids and glucocorticoids.

cortisol (KOR-tih-zol) The most important of the glucocorticoids secreted by the adrenal cortex; increases rate of protein breakdown, the liver's conversion of amino acids and fats into glucose, and the breakdown of adipose to release fatty acids.

coxal (KOKS-al) Pertaining to the hip area.

cranial (KRA-ne-al) Pertaining to the head.

cranial cavity (KRA-ne-al KAV-ih-te) Cavity of the skull that houses the brain and other tissues.

cranial nerve Nerve originating from the brain.

creatine Nitrogenous organic acid found in muscle tissue (usually combined with phosphorus) that supplies energy for contraction.

creatine phosphate Organic compound found in muscle tissue that consists of creatine with an attached phosphate ion; stores and provides energy for muscle contraction.

crural (KRUH-ral) Pertaining to the lower leg.

cubital (KU-bih-tal) Pertaining to the elbow region.

cusp Fold or flap in a heart valve.

cutaneous (ku-TA-ne-us) **membrane** Skin; the outer covering of the body made up of the epidermis and dermis.

cutaneous receptor (ku-TA-ne-us re-SEP-tor) Sensory receptor located in the skin.

cyanotic (si-ah-NAH-tik) Bluish appearance of the skin that results from tissues not receiving enough oxygen.

cytokine (SI-to-kine) Cell-signaling molecule.

cytokinesis (si-to-kin-E-sis) Separation of the cytoplasm during cell division; final stage of cell division.

cytology (si-TOL-o-je) Study of cells.

cytolysis (si-TOL-eh-sis) Dissolution of a cell.

cytoplasm (SI-to-plaz-um) Internal environment of a cell between the plasma membrane and the nucleus; includes the cytosol and organelles.

cytoskeleton (si-to-SKEL-eh-ton) Series of protein filaments and tubules that extend through the cytoplasm forming an internal scaffolding that provides shape, strength, and mobility to cells.

cytosol (SI-to-zahl) Fluid component of cytoplasm.

cytotoxic (si-to-TOKS-ik) **T cell** Specialized T lymphocyte produced during a cell-mediated immune response that attacks and destroys pathogens; killer T cell.

deep Farther from the surface.

deep fascia Connective tissue layers that organize and surround individual muscles, bones, and organs.

defecation (def-eh-KA-shun) Elimination of feces through the anus.

dendrite (DEN-drite) Fibrous portion of a neuron capable of receiving a stimulus.

deoxyribonucleic acid (de-ocks-e-ri-bo-NU-kle-ik AH-sid) **(DNA)** Nucleic acid that carries the human genetic code.

depolarization (DE-po-lah-ri-ZA-shun) The first stage of an action potential when positive sodium ions rush in through open channels in the neuron's cell membrane.

deposition (deh-po-ZIH-shun) Formation of new bone; building up of bone tissue. Literally means "the act of depositing or laying down."

dermal papillae (DER-mal pah-PIL-i) Finger-like projection of the dermis that extends into the germinating layer of the epidermis.

dermatome (DER-mah-tome) Specific region of skin innervated by the branches of a particular spinal nerve.

dermis (DER-mis) Connective tissue layer of the skin just below the epidermis that contains nerve endings, sweat glands, sebaceous glands, and blood and lymph vessels.

descending colon (de-SEN-ding KO-lun) Vertical segment of the large intestine on the left side of the abdominal cavity.

diagnosis (di-og-NO-sis) Determination of the cause and nature of a disease, injury, or disorder.

diaphragm (DI-ah-fram) Large flat skeletal muscle separating the thoracic and abdominopelvic cavities; serves as the primary muscle of inspiration.

diaphysis (di-AF-eh-sis) Shaft of a long bone.

diarthrosis (di-ar-THRO-sis) Freely movable joint; also known as a *synovial joint*.

diastole (di-AS-to-le) Relaxation state in which the chambers of the heart dilate as they fill with blood.

diencephalon (di-en-SEF-ah-lon) Middle portion of the brain between the cerebrum and brain stem consisting of the thalamus, hypothalamus, and pineal gland.

differentiation (dif-fer-en-she-A-shun) Ability of cells to adapt and specialize their functions and create a wide variety of cell types.

diffusion (deh-FU-zhun) Cellular passive transport mechanism in which substances move according to a concentration gradient; movement from an area of higher concentration to lower.

digestion (di-JES-chun) Process of breaking down food into smaller components that can be absorbed and assimilated by the body.

digestive (di-JES-tiv) **system** Body system that ingests food, converts it to a form the body can use, absorbs nutrients into the circulation, and eliminates solid wastes.

digital (DIH-jih-tal) Pertaining to the fingers or toes.

direct phosphorylation (fos-for-ah-LA-shun) Method of energy production in which a phosphate group is broken off creatine phosphate and added to ADP to create ATP.

disk Fibrocartilage pad between vertebrae and the bones of other cartilaginous joints. (Alternate spelling is disc.)

distal (DIS-tal) In extremities, farther from attachment point to the body.

divergence (di-VER-jents) Neural circuit in which the axon of one neuron branches out to make synaptic connections with several other neurons.

dorsal (DOR-sal) Back (posterior).

dorsiflexion (DOR-se-flek-shun) Movement at the ankle in which the dorsum of the foot is pulled upward toward the knee.

dual innervation (DU-al in-er-VA-shun) Referring to glands and organs that are functionally connected with both sympathetic and parasympathetic motor pathways.

duodenum (du-AH-deh-num) Initial segment of the small intestine.

dura mater (DER-ah MAH-ter) Outermost layer of the meninges made of tough fibrous connective tissue.

dynamic edema (ah-DE-mah) Edema related to cardiovascular dysfunctions; caused by an imbalance between capillary filtration and capillary reabsorption or poor venous return.

eccentric contraction (e-SEN-trik) Isotonic muscle contraction in which the muscle lengthens; insertion moves away from the origin.

eccrine (EK-rin) Pertaining to a sweat gland.

ectoderm (EK-to-derm) Outer embryonic layer that differentiates to form the epidermis, nervous tissue, and sense organs.

edema (eh-DE-mah) Accumulation of excess fluid in the interstitium.

edema (ah-DE-mah) **uptake** Movement of excess interstitial fluid (edema) into lymphatic capillaries; fluid uptake in the case of edema.

effector (e-FEK-tor) Target cell, tissue, or organ that responds to a specific stimulus, such as a motor command from the central nervous system.

efferent (E-fer-ent) Conducting away from a center; outflowing.

egg Female sex cell before fertilization; also known as an *ovum*.

ejaculation (e-jak-u-LA-shun) Forceful ejection of sperm from the penis.

ejaculatory duct (e-JAK-u-lah-toh-ree DUKT) Tube between the vas deferens and urethra.

elastic (e-LAS-tik) Having the ability to return to the original shape after being stretched.

elastic fibers (e-LAS-tik) Thin stretchy connective tissue fibers made of the protein elastin.

elastin (e-LAS-tin) Protein that makes up the elastic fibers found in connective tissue.

electrolyte (e-LEK-tro-lite) Compound containing charged particles (ions) that conducts an electrical current.

element Substance formed when like atoms and molecules are bound together.

elimination In general, the process of expelling or removing something; excreting waste from the body.

embolus (EM-bo-lus) Traveling mass of undissolved material in a blood vessel; can be a foreign object or made up of tissue fragments, bacteria, or gas.

embryo (EM-bre-o) Developing human offspring from week 3 through week 8 of gestation.

emulsification (e-MUL-sih-fi-KA-shun) In general, the process of suspending small globules (droplets) of one liquid in a second liquid; in digestion, the breakdown of large fat globules into smaller, evenly distributed particles.

end feel Quality of resistance sensed at the end of a normal range of motion when the joint creates resistance to stop movement; can be assessed during PROM.

endocardium (en-do-KAR-de-um) Thin smooth inner layer of the heart that consists of epithelium.

endocrine (EN-do-krin) **gland** Ductless gland that secretes hormones into the blood.

endocrine (EN-do-krin) **system** Body system composed of glands that secrete hormones; works with the nervous system for communication, coordination, and control of all other body systems.

endoderm (EN-do-derm) Innermost embryonic layer that differentiates into the epithelial linings of organs and glands.

endometrium (en-do-ME-tre-um) Inner lining of the uterus that thickens and sloughs off with each menstrual cycle.

endomysium (en-do-MI-se-um) Fine connective tissue layer surrounding a muscle fiber.

endoneurium (EN-do-NUR-e-um) Connective tissue covering of an individual nerve fiber within a nerve.

endoplasmic reticulum (en-do-PLAZ-mik reh-TIK-u-lum) Cell organelle that creates pathways that allow for intracellular movement of substances; can be rough (with ribosomes) or smooth (without ribosomes).

endosteum (en-DOS-te-um) Layer of cells lining the medullary cavity.

enteric nervous system (en-TER-ik NUR-vus SIS-tem) Complex neuronal network that controls the secretions and smooth muscle contractions of the digestive tract.

enzyme (EN-zime) Protein that functions as a catalyst for biochemical reactions in the body, speeding up the breakdown and synthesis of various substances.

epicardium (ep-ih-KAR-de-um) Outer layer of the heart; visceral pericardium.

epicondyle (ep-eh-KON-dile) Bone projection just superior to the condyle of certain bones.

epidermis (ep-eh-DER-mis) Superficial stratified epithelial tissue layer of the skin.

epididymis (ep-ih-DID-ih-mis) Narrow, coiled tube in the spermatic duct system located along the posterior side of the testes; the location where sperm develop their motility and ability to fertilize eggs.

epiglottis (ep-ih-GLAH-tis) Small cartilage flap at the entrance to the larynx that prevents food and liquids from entering the lower respiratory tract during swallowing.

epimysium (ep-ih-ME-se-um) Fibrous connective tissue layer surrounding an entire muscle.

epineurium (EP-e-NUR-e-um) Outermost connective tissue covering of a nerve.

epiphyseal (eh-pif-eh-SE-al) **plate** Growth plate of bones.

epiphysis (eh-PIF-eh-sis) End portion of a long bone.

epithelial (ep-ih-THE-le-al) **tissue** One of the four basic tissue types of the body; functions to line, cover, secrete, and protect; also known as the *epithelium*.

epithelium (ep-ih-THE-le-um) One of the four basic tissue types of the body; functions to line, cover, secrete, and protect; also known as *epithelial tissue*.

erection (e-REK-shun) Condition in which the penis is stiff, enlarged, and engorged.

erythema (er-ih-THE-mah) Reddened appearance of the skin.

erythrocyte (eh-RITH-ro-site) Formed element of blood responsible for transporting oxygen throughout the body; also known as a *red blood cell*.

esophageal sphincter (eh-sof-ah-GE-al SFINK-ter) Smooth muscle ring at the lower end of the esophagus that controls the flow of food into the stomach and prevents its contents from pushing back into the esophagus.

estrogen (ES-tro-jen) Female sex hormone that stimulates the development of the reproductive organs and the secondary sex characteristics; during the menstrual cycle, promotes an environment suitable for fertilization and implantation of an egg and for sustaining an embryo.

etiology (e-te-AH-lo-je) The cause of a disease; factors involved in the development of a disease.

excitable Having the ability to react to a stimulus; in muscle tissue, the ability to respond quickly to stimulus from nerve impulses.

exhalation (ex-hah-LA-shun) Expelling air from the lungs; also known as *expiration*.

exocrine (ECKS-o-krin) **gland** Gland that secretes into a duct that carries the secretion to a specific location (e.g., sweat, salivary, and sebaceous glands).

exocytosis (ex-o-si-TO-sis) Active transport mechanism in which a substance is carried from inside to outside the cell membrane.

expiration (ex-peh-RA-shun) Expelling air from the lungs; also known as *exhalation*.

extensible (eks-TEN-seh-bel) Having the ability to stretch.

extension (ek-STEN-shun) Posterior movement of a body part in the sagittal plane; increases the anterior angle between two bones (exception: knee joint).

external respiration (EKS-ter-nal res-pih-RA-shun) Gas exchange between the air in the alveoli of the lungs and the bloodstream.

extracellular (EKS-trah-SEL-u-lar) **fluid** Fluid present outside of cells.

extrafusal fiber (EKS-trah-fu-zal FI-ber) Skeletal muscle fiber.

facet (fah-SET) Small, smooth, oval area on a bone.

facilitated diffusion (fah-sil-ih-TA-ted deh-FU-zhun) Passive transport mechanism in which special carrier molecules in the cell membrane assist in moving specific substances across the plasma membrane.

fallopian (fah-LO-pe-an) **tube** One of two short muscular tubes that carry eggs from the ovary to the uterus; also known as *oviduct.*

fascia (FAH- shah) Multiple layers of disorganized fibrous connective tissue that surround and invest all structures of the body down to the cellular level.

fascial (FAH-shul) **band** One of seven flattened horizontal straps of superficial fascia.

fascial (FAH-shul) **plane** One of four horizontal deep fascial components inside the body cavities.

fascial (FAH-shul) **system** A unifying structural and functional system that includes *all* fibrous connective tissues from the disorganized fibrous connective tissue sheets called fascia to the organized fibrous connective tissue structures such as tendons, aponeuroses, ligaments, and capsules.

fascial tone Low-grade tension created by smooth muscle cells within the fascia that is independent of motor tone in surrounding skeletal muscles; regulated by the ANS.

fascicle (FAS-eh-kel) Bundle of muscle fibers.

fatty acid Molecular component of triglycerides.

feedback system Part of the body's homeostatic process in which a stimulus occurs and the body responds.

femoral (FEM-or-al) Pertaining to the thigh.

fertilization (fer-tih-li-ZA-shun) Joining of genetic material from one sperm and one egg; also known as *conception.*

fetus (FE-tus) Developing human offspring from week 9 until birth.

fever Elevated core body temperature.

fibroblast (FI-bro-blast) Specialized connective tissue cell that synthesizes certain protein molecules to create connective tissue fibers.

fibrocartilage (FI-bro-KAR-tih-lej) Dense fibrous connective tissue pad found between the vertebrae and at the pubic symphysis.

fibrous (FI-brus) **joint** Type of joint in which the bone ends are held together with fibrous connective tissue; most but not all fibrous joints are also synarthrotic.

filtration (fil-TRA-shun) In general, the process of passing liquid through a filter; a cellular passive transport mechanism in which substances move according to a pressure gradient, from an area of higher to lower pressure. In the cardiovascular system, movement of fluid out of the capillary and into the interstitium. In the urinary system, the first step in urine formation that occurs in the renal corpuscle in which fluid is forced out of the blood through the glomerulus and into the Bowman capsule.

fissure (FISH-ur) Deep furrow or cleft.

fixator (FIKS-a-tor) Muscle that fixes or stabilizes the origin end of a prime mover so that the movement is more efficient; also known as the *stabilizer.*

flexion (FLEK-shun) Anterior movement of a body part in the sagittal plane; decreases the anterior angle between two bones (exception: knee joint).

fluid (FLU-id) **uptake** Movement of interstitial fluid into lymphatic capillaries.

follicle (FOL-ih-kl) Small sac at the base of a hair shaft that produces new keratinocytes; also the specialized ring of cells surrounding an ovum.

follicle stimulating hormone (FAH-lih-kel SITM-u-la-ting HOR-mone) **(FSH)** Gonadotropin released by the anterior pituitary that stimulates estrogen secretion and the growth of ovarian follicles in females and the formation of sperm in the male testis.

foramen (for-A-men) Opening or hole in a bone.

formed element Cellular component of blood, including red and white blood cells and platelets.

fossa (FOS-sah) Saucer-like depression.

fovea (FO-ve-ah) Small pit or depression in a bone.

free nerve ending Superficial cutaneous receptor that is sensitive to light touch, pressure, temperature, and tissue damage (pain).

frontal (FRONT-al) Pertaining to the forehead or relating to the frontal plane.

frontal plane (FRUN-tal plane) Coronal plane; a vertical plane that divides the body into anterior and posterior sections.

fulcrum (FUL-krum) The point of movement in a system of levers.

fundus (FUN-dus) Upper portion of an organ such as the stomach or uterus.

gall bladder Small, pear-shaped sac under the right lobe of the liver that stores bile secreted by the liver until it is needed for digestion.

gamete (GAM-eet) Male or female sex cell; sperm or egg (ovum).

gamma (GAM-a) **gain** Pre-tensioning of the muscle spindle that increases its sensitivity to rapid lengthening (also called gamma loading).

gamma (GAM-a) **loop** Neuronal reflex loop that regulates the sensitivity of the muscle spindle; created by a gamma sensory neuron originating at the ends of the muscle spindle and a gamma motor neuron that stimulates the intrafusal muscle fiber.

ganglion (GANG-le-on) Collection of cell bodies and dendrites found outside the central nervous system.

gastrointestinal (GI) tract (GAS-tro-in-TES-tih-nal TRAKT) The pathway that transports food and liquids that begins at the mouth and includes the pharynx, esophagus, stomach, small and large intestines, and the anus; site of digestive processes.

gate-control theory A theory to explain the mechanism of pain.

gel (JEHL) Semisolid material.

gene (JEEN) Portion of the DNA molecule that corresponds to a specific inherited trait or characteristic.

general receptor Type of sensory receptor that is sensitive to changes in the environment, located throughout the tissues of the body.

general sense Sensory information generated by cutaneous and other general sense receptors; includes touch, temperature, pain, pressure, vibration, movement, and body position.

genitalia (jeh-neh-TA-le-ah) External sexual organs (penis and vulva).

gestation (jes-TA-shun) Period of fetal development from conception until birth; pregnancy.

gland Organ that secretes specific chemicals (enzymes or hormones) necessary for body processes.

glial cell (GLE-al) Type of cell of the nervous system that functions to support and protect neurons; also known as *neuroglia*.

gliding joint Type of synovial joint with smooth flat articular surfaces (*facets*) that slide or glide across one another; normally found between short bones.

glottis (GLAH-tis) Space between the vocal cords of the larynx.

glucose (GLU-kose) A simple sugar.

gluteal (GLU-te-al) Pertaining to the buttocks.

glucagon (GLU-kah-gon) Hormone secreted by the alpha islet cells of the pancreas; increases blood glucose levels.

glucocorticoid (GLU-ko-kor-tih-koyd) Any of a group of steroid hormones secreted by the middle layer of the adrenal cortex; involved in metabolism of carbohydrates, proteins, and fats and have anti-inflammatory properties.

glycerol (GLIS-er-ol) Molecular component of triglycerides.

glycogen (GLI-co-jen) Storage form of glucose.

glycolysis (gli-COL-eh-sis) An anaerobic metabolic process in which glucose is broken down and converted into pyruvic acid to produce ATP.

Golgi apparatus (GOLE-je ap-ah-RAT-us) Cellular organelle that processes and packages proteins and lipids.

Golgi tendon organ (GOLE-je TEN-dun OR-gan) **(GTO)** Proprioceptor found in skeletal muscle that senses tension.

gonad (GO-nad) Male or female sex gland that produces and stores the sex cells; the testes in males and the ovaries in females.

graded response Increasing or decreasing the number of motor units stimulated to control the force of muscle contraction.

granulation (gran-u-LA-shun) **tissue** Thin thread-like fibers formed by fibroblasts across a hematoma.

gray matter Areas of the brain or spinal cord made primarily of cell bodies and dendrites of neurons rather than myelinated axons; has a grayish color.

greater omentum (o-MEN-tum) Large fatty extension of the peritoneum.

ground substance Fluid component of connective tissues.

growth hormone (GH) Most abundant hormone secreted by the anterior pituitary; stimulates and regulates growth in all tissue and plays important role in some aspects of metabolism.

gustatory (GUS-tah-tor-e) Pertaining to taste.

gyrus (JI-rus) Fold or convolution of the cerebrum.

hair root plexus (PLECKS-us) Deep cutaneous receptor that wraps around the hair follicle and is sensitive to movement of the exposed hair shaft.

hair shaft Exposed portion of a hair that projects above the surface of the skin.

haversian (hah-VER-zhen) **canal** Central canal of the haversian system in compact bone through which blood and lymph vessels and nerves pass.

haversian system (hah-VER-zhen) Basic structural unit of compact bone, made up of the haversian canal and its concentric rings of lamellae; also known as an *osteon*.

helper T cell Specialized lymphocyte that boosts the body's specific immune response.

hematoma (he-mah-TO-mah) Localized collection of blood and other fluids in tissue.

hematopoiesis (he-MAT-o-po-E-sis) Blood cell production in the red bone marrow (also called hemopoiesis).

hemoglobin (HE-mah-glo-bin) Iron-containing protein in red blood cells that transports oxygen from the lungs to the tissues and gives red blood cells their color.

hemorrhage (HEM-o-rij) Active bleeding.

hemostasis (he-mo-STA-sis) Physiological process of stopping vascular bleeding.

hepatic duct (hah-PAH-tik DUKT) One of two tubes (right or left) within the liver that carries bile to the common hepatic duct.

hepatic flexure (heh-PAH-tik FLEK-shur) Point of intersection in the large intestine between the ascending and transverse colon.

hepatocyte (hah-PAT-o-site) Liver cell.

hilum (HI-lum) Indentation or opening through which ducts, nerves, or blood vessels pass in a gland or organ; in the kidney, the opening through which the ureters, blood vessels, and nerves pass.

hinge joint Type of synovial joint in which the rounded end of one bone articulates with a groove or trough in another bone; allows flexion and extension only.

histamine (HISS-tah-meen) Chemical released in damaged tissue that causes vasodilation and increased capillary permeability.

histology (HISS-tahl-o-ge) Study of tissues.

holistic (ho-LIS-tik) **approach** Health and wellness concept guided by the principle that the physical body, cognitive processes (mind), and emotional and spiritual aspects are inseparable parts of a whole and integrated person.

homeostasis (ho-me-o-STA-sis) State of internal stability or balance in the body.

horizontal abduction (hor-ih-ZON-tal ab-DUK-shun) Movement away from the midline within the transverse plane.

horizontal adduction (hor-ih-ZON-tal ad-DUK-shun) Movement toward the midline within the transverse plane.

hormone (HOR-moan) Specialized chemical messenger released by endocrine glands into the blood.

horn Specific region of the gray matter of the spinal cord.

hyaline cartilage (HI-ah-line KAR-tih-lej) Smooth connective tissue found on the articular surface of bones.

hydrostatic (HI-dro-stat-ik) **pressure** Force caused by fluid pushing against a wall or membrane; fluid pressure.

hyperemia (hi-per-E-me-ah) Local increase in blood volume or flow.

hypertension (hi-per-TEN-shun) High arterial blood pressure.

hypertonicity (hi-pur-to-NIS-eh-te) Excessive muscle tension.

hypertrophy (hi-PER-tro-fe) Increase in the size or diameter of tissues or organs.

hypodermis (hi-po-DER-mis) Fat and areolar connective tissue that lies between the dermis and underlying tissues and organs; also known as the *subcutaneous layer* or *superficial fascia*.

hypothalamus (hi-po-THAL-a-mus) Anterior medial portion of the diencephalon that controls the autonomic nervous system and is the primary link to the endocrine system.

idiopathic (ih-de-o-PATH-ik) Disease or condition of unknown origin.

ileum (IL-e-um) Lowest portion of the small intestine.

immunity (im-MU-nih-te) Protection from or resistance to disease.

immunoglobulin (IM-mu-no-GLOB-u-lin) Plasma protein released by plasma B cells that can bind with an antigen to neutralize or kill it; also known as an *antibody*.

implantation (im-plan-TA-shun) Process by which the blastocyst burrows into the uterine lining.

indication (in-di-KA-shun) Sign or circumstance that provides a reason to proceed with a specific treatment because a positive effect is likely.

inert tissue (in-ERT) Noncontractile tissues, including ligaments, joint capsules, and cartilage.

inferior (in-FE-re-or) Below; closer to feet; caudal.

inflammation (in-flah-MA-shun) Defensive response that helps to stabilize damaged tissue and prepare it for repair.

inflammatory (in-FLAM-mah-tor-e) **stage** First phase of the healing process that includes hemorrhage, inflammation, secondary edema formation, spasm, and hematoma organization, also known as the *acute stage*.

ingestion (in-JES-chun) The act of eating or taking in food or liquids.

inguinal (ING-gwih-nal) Pertaining to the groin.

inhalation (in-hah-LA-shun) Drawing air into the lungs; also known as *inspiration*.

inhibiting hormones Group of hormones synthesized and released by the hypothalamus that inhibit hormone releases from the anterior pituitary gland.

innate (in-NATE) **immune defense** Any of a group of general body defenses that are not directed at a particular pathogen; also known as *nonspecific immune defense*.

innervate (IN-er-vate) To supply nerves to an organ or body part; to stimulate an organ or body part via nerve impulses.

inorganic (IN-or-gan-ik *or* in-or-GAN-ik) **compound** Compound made of molecules that do not contain carbon atoms.

insertion (in-SIR-shun) Moving end of a muscle attachment.

inspiration (in-speh-RA-shun) Drawing air into the lungs; also known as *inhalation*.

insulin (IN-suh-lin) Hormone released by the beta cells in the pancreatic islets; cause a decrease in blood glucose levels.

integral membrane proteins (IMPs) Groups of specialized protein molecules in the plasma membrane of cells that monitor the internal and external environment, shuttle nutrients and wastes across the membrane, and direct cellular responses.

integumentary (in-teg-u-MEN-tah-re) **system** Body system made up of the skin and its accessory organs: hair, nails, oil glands, and sweat glands.

intercalated (in-TER-kah-la-ted) **discs** Small connections between cardiac muscle cells made by thickened areas in the sarcoplasmic reticulum.

internal respiration (IN-ter-nal res-pih-RA-shun) Gas exchange between the bloodstream and tissues of the body.

interstitial (in-ter-STIH-shal) **fluid** Fluid between cells in tissues that is not part of the blood or lymph.

interstitial (in-ter-STIH-shul) **fluid pressure (IFP)** Force created by fluid in the interstitial space pressing in against the wall of the capillary.

interstitial myofascial (in-ter-STIH-shul mi-o-FAH-shul) **tissue receptors** Specialized free nerve endings found within fascia that cause autonomic changes when stimulated.

interstitial oncotic (in-ter-STIH-shul on-KAH-tik) **pressure (IOP)** Osmotic pressure created by protein within the interstitial fluid that draws fluid out of the capillary.

interstitium (in-ter-STIH-she-um) The space between cells in a tissue; contains the interstitial fluid.

intracellular fluid (in-trah-SEL-u-lar) Fluid inside cells.

intrafusal fiber (IN-trah-fu-zal FI-ber) Specialized striated muscle fiber within a muscle spindle.

inverse stretch reflex Reflex mediated by the Golgi tendon organs that inhibits muscle contraction, causing a stretched or contracted muscle to relax when the tension level reaches a certain point; opposite of the stretch reflex.

ion (I-on) Charged particle of matter.

ion pump Active transport mechanism that moves charged particles across the plasma membrane; movement against the concentration gradient.

ipsilateral (ip-seh-LAH-ter-al) On the same side of the median.

ischemia (is-KE-me-ah) Local decrease in blood volume or flow.

islets of Langerhans (I-lets of LANG-er-hans) Group of specialized hormone-producing cells scattered throughout the pancreas that produce glucagon and insulin. Also known as *pancreatic islets*.

isometric contraction (i-so-MET-rik) A type of muscle contraction that increases tension but does not create movement.

isotonic contraction (i-so-TON-ik) A type of muscle contraction that creates movement.

interferon (in-ter-FE-ron) Cytokine that inhibits the spread of viral infection from infected to uninfected body cells.

jejunum (je-JUN-um) Central portion of the small intestine.

joint capsule Fibrous connective tissue sleeve around the bone ends in synovial joints.

joint receptor Receptor in the joint capsule and ligaments that monitors the pressure, tension, and movement of a joint.

keratin (KARE-a-tin) Substance secreted by keratinocytes of epidermis to toughen the outer layer and make it water resistant.

keratinocyte (kare-ah-TIN-o-sites) Type of cell in the outer layer of epidermis that secretes keratin.

kidney (KID-ne) One of two bean-shaped organs located in the lumbar region; responsible for cleansing the blood and forming urine.

kinesthesia (kin-eh-STE–zjah) The sense and awareness of movement.

kinetic (kin-EH-tik) **chain** Series of muscles engaged to create a complex movement.

Krebs cycle An aerobic metabolic process in which pyruvic acid is broken down and converted into carbon dioxide and water, producing ATP and heat (also called the citric acid cycle).

kyphotic (ki-FOT-ik) **curve** Convex curve of the thoracic and sacral spine (as seen in a sagittal view).

labia (LA-be-ah) Lateral tissue folds that surround the vaginal and urethral openings.

labor (LA-ber) Series of actions that push an infant through the birth canal and result in expulsion (birth).

lacteal (lak-TEELS) Specialized lymphatic capillary inside the microvilli of the small intestine that absorbs fats.

lactic acid Acid produced in the body during the anaerobic metabolism of glucose, as occurs in muscle tissue during exercise.

lacuna (lah-KOO-nah) Small cavity or chamber that houses an osteocyte in mature bone tissue or a chrondrocyte in cartilage.

lamella (lah-MEL-lah) Concentric bony plate that surrounds the central canal (haversian canal) in an osteon.

Langerhans (LAHN-ger-hans) **cells** Specialized immune cells found in the spiny layer of the epidermis.

large intestine Portion of the intestine that extends from the ileum to the anus and includes the cecum, colon, rectum, and anal canal.

larynx Part of the respiratory tract between the pharynx and trachea that contains the vocal cords; commonly called the voice box.

lateral (LAH-ter-al) Situated or extending away from the midline of the body (to the side).

length-strength ratio Relationship between the length of a muscle and the strength of its contraction.

lesion (LE-zhun) Localized pathological change in an organ or tissue, such as a wound, sore, rash, ulcer, tumor, boil, or other abnormal tissue change.

lesser omentum (o-MEN-tum) Extension of the greater omentum between the stomach and liver.

leukocyte (LU-ko-site) Formed element of blood that plays a vital role in the body's healing and immune responses; includes neutrophils, basophils, eosinophils, lymphocytes, and monocytes; also known as a *white blood cell*.

ligament (LIG-ah-ment) Fibrous connective tissue that connects bone to bone in synovial joints.

ligand (LIH-gand) Peptide molecule that binds to receptor proteins in the plasma membrane of target cells to stimulate cellular activity.

limbic system (LIM-bik SIS-tem) Group of brain structures involved in processing memory and emotion and controlling unconscious aspects of behavior related to survival; called the "emotional brain."

lipid (LIP-id) A fat; an organic compound made of carbon, hydrogen, and oxygen atoms that breaks down to smaller usable fatty acid and glycerol molecules.

lipid-soluble (LIP-id SOL-u-bel) **hormone** Hormone transported in blood via special carrier molecules, which then passes easily into the target cell to directly alter its metabolic activity; no second messenger is involved.

liver Glandular organ that occupies most of the upper right abdominal cavity and is essential to many metabolic processes; secretes bile, stores fat and sugar as energy resources, filters out toxins, and regulates the amount of blood in the body.

lobe Subdivision or region of an organ or body part, such as the lobes of the brain or lungs.

lobule (LO-bule) Smaller functional unit of a lobe, such as the lobule of the liver.

lordotic (lor-DAH-tik) **curve** Concave curve of the cervical and lumbar spine (as seen in a sagittal view).

lumbar (LUM-bar) Pertaining to the lower back area.

lumen (LU-men) Opening in a tube.

lung One of two organs of ventilation and external respiration.

luteinizing (LOO-tee-ni-zing) **hormone (LH)** Tropic hormone secreted by the anterior pituitary; targets the gonads to secrete their gonadotropic hormones.

lymph (LIMF) Fluid inside the lymphatic vessels.

lymph capillary (LIMF KAP-eh-la-re) Smallest lymphatic vessel that is entwined with the cardiovascular capillary bed; includes initial vessels and collecting capillaries.

lymph node Specialized lymph organ interspersed along the lymphangia; functions as a filter for lymph.

lymph node bed Grouping of lymph nodes; also known as a *catchment*.

lymphangia (lim-FAN-ge-ah) Primary lymphatic vessels that carry lymph from the colleting capillaries to the collecting (lymphatic) trunks.

lymphatic (lim-FAT-ik) **system** Body system consisting of the lymph vessels and lymphoid tissue; involved in immunity, nutrient absorption, and fluid return.

lymphatic terminus (lim-FAT-ik TERM-en-us) Junction between the lymphatic ducts and the subclavian veins; also known as the *angulus venosus*.

lymphatic trunk (lim-FAT-ik) Large lymphatic vessel that connects lymphangia from a specific body region to a deep lymphatic duct; also known as a *collecting trunk*.

lymphedema (LIMF-ah-de-mah) Edema caused by dysfunction or failure of the lymphatic system; primary or secondary condition.

lymphocyte (LIMF-o-site) Specialized white blood cell that carries out specific immune responses.

lymphotome (LIMF-o-tomes) Specific lymphatic drainage region that contains its own network of lymphatic vessels; regional network of vessels that carries lymph from one specific region of tissue into a designated catchment.

lysosome (LI-so-zome) Small sac filled with digestive enzymes that break down a variety of molecules within the cell.

macrophage (MAK-ro-fahj) Large phagocyte; the name literally means "big eater."

malignant melanoma (mah-LIG-nant mel-ah-NO-mah) Most serious type of skin cancer that is particularly aggressive and can be fatal.

mammary glands (MAH-meh-re) Breasts.

manual (MAN-u-al) Pertaining to the hand; an action performed by the hands.

manual therapy Patterned and purposeful application of touch and/or movement with therapeutic intent.

margination (mar-jih-NA-shun) Movement of phagocytes and fibroblasts to the periphery of an area of inflammation; occurs during the acute stage of the healing process.

marrow (MAR-ro) Soft gelatinous material found in the spaces of spongy bone (red marrow) and filling the medullary cavity (yellow marrow).

mast cells Specialized connective tissue cells that contain histamine.

mastication (mas-tih-KA-shun) Process of chewing.

matrix (MA-triks) Combination of fibers and ground substance in a connective tissue.

maturation (mah-cher-A-shun) **stage** Final phase of the healing process that involves the final thickening and alignment of scar tissue.

meatus (me-A-tus) Short channel or canal in a bone.

mechanoreceptor (meh-KAN-o-re-SEP-tor) Sensory receptor sensitive to mechanical changes in the environment such as touch, pressure, movement, and vibration.

medial (ME-de-al) Closer to or toward the midline of the body.

mediastinum (me-de-AS-tin-um) Central region in the thoracic cavity between the pleural sacs; contains the heart and all the thoracic viscera except the lungs.

medulla oblongata (meh-DU-la ob-long-GAH-tah) The most inferior portion of the brain stem that is continuous with the spinal cord.

medullary (MED-u-lar-e) **cavity** Hollow channel in the diaphysis of a long bone.

meiosis (mi-O-sis) Cellular division of sex cells.

Meissner corpuscle (MIZE-ner KOR-pus-sel) Superficial cutaneous receptor that is sensitive to vibration and light touch.

melanin (MEL-ah-nin) Dark pigment that determines skin and hair color.

melanocyte (meh-LAN-o-site) Specialized cell of the epidermis that produces melanin.

melanocyte- (meh-LAN-o-site) **stimulating hormone (MSH)** Hormone released by the anterior pituitary that effects skin pigmentation.

melanoma (mel-ah-NO-mah) Most common type of skin cancer.

melatonin (mel-ah-TO-nin) Hormone secreted by the pineal gland that plays a role in regulating the body's circadian rhythm (sleep-wake cycle).

membrane (MEM-brain) Thin layer of tissue that covers a surface, lines a cavity, or separates or connects structures or organs.

memory cell Specialized B cell or T cell produced during a specific immune response that "remembers" the initial exposure to a pathogen; specific types are called memory B cells or memory T cells.

meninges (meh-NIN-jeez) Membranous connective tissue coverings of the brain and spinal cord.

meniscus (meh-NIS-kus) Fibrocartilage pad in the knee joint.

menopause (MEN-o-pahz) Permanent cessation of the menses and of fertility; generally defined as occurring when a woman has gone 12 consecutive months with no menstrual period.

menstrual cycle (MEN-stru-al SI-kel) Women's monthly reproductive cycle, marked by hormonal shifts and recurring physiologic changes in the endometrium; culminates in menstruation.

menstruation (men-stru-A-shun) Bleeding from the vagina during days 1 to 5 of the menstrual cycle, caused by sloughing of the uterine lining.

Merkel disc (MER-kl) Superficial cutaneous receptor that is sensitive to light touch.

mesentery (MEZ-en-ter-e) Double fold of the peritoneum that attaches the small intestine to the posterior side of the abdominal cavity.

mesoderm (MEZ-o-derm) Middle embryonic layer that differentiates into bone, blood, muscle tissue, and certain epithelial tissues in the urinary and endocrine systems.

metabolic (meh-tah-BAL-ik) **rate** The speed of cellular processes measured as the amount of heat produced by the body.

metabolism (mah-TAB-o-lizm) The sum of all biochemical reactions that occur within a cell or organism that are necessary to maintain life; includes building up molecules (anabolism) and breaking down molecules (catabolism).

metaphysis (meh-TAF-eh-sis) In a long bone, the flared area where the diaphysis joins the epiphysis.

microbe (MI-krobe) A minute organism, including many types of pathogens such as bacteria and viruses.

microvilli (mi-kro-VIL-i) Microscopic hair-like projections from the plasma membrane of certain cells; small cilia.

micturition (mik-chur-IH-shun) Passing urine out of the urethra; also known as *urination*.

midbrain Uppermost portion of the brain stem.

mineral (MIN-er-al) Inorganic micronutrient required for maintaining tissues such as bone and blood and homeostatic balances including pH, fluid, and energy balance.

mineralocorticoid (MIN-eh-rahl-o-KOR-tih-koyd) Any of a group of hormones released by the outer layer of the adrenal cortex; aldosterone is the most important.

mitochondria (mi-to-KON-dre-ah) Cellular organelle responsible for the production of ATP (the energy molecule).

mitosis (mi-TO-sis) Process of cell division in which the nucleus divides to produce two nuclei, each containing the same chromosome and DNA content as the original cell.

mitral (MI-tral) **valve** Heart valve located between the left atrium and ventricle; also known as the *bicuspid valve*.

molecule (MAHL- eh-kule) Particle formed by the bonding of two or more atoms; the smallest unit of a substance that exhibits the properties of an element or compound.

motility (mo-TIL-ih-te) Ability to move spontaneously.

motor end plate Region of the neuromuscular junction located on a muscle fiber that is sensitive to neurotransmitters released from the knobs of a motor neuron.

motor neuron Nerve cell that transmits impulses from the central nervous system to muscular or glandular tissue.

motor tone A consistent state of low-grade tension generated through tonic contractions; palpated as firmness in the muscle.

motor unit A motor neuron and the muscle fibers it innervates.

motor unit recruitment Increasing the number of motor units stimulated to increase the force of a muscle contraction.

mucosa (mu-KO-sah) Inner mucous membrane layer, such as the innermost tissue layer of the digestive tract.

mucosa-associated lymphoid tissue (MALT) Small masses of lymphoid tissue found in the mucous membranes of the throat (tonsils) or digestive tract (Peyer patches).

mucous membrane (MU-kus) Epithelial membrane that secretes mucus and lines cavities that open to the outside environment.

mucus (MU-kus) Thick, clear secretion of the mucous membrane.

muscle belly Middle bulky portion of a muscle.

muscle cramp Acute involuntary muscle contraction that can last for several minutes.

muscle fatigue Inability of muscle to contract forcefully after prolonged activity, even when stimulated to do so.

muscle recruitment Organizational pattern of stimulation and coactivation of muscle groups regulated by the cerebellum to produce complex, coordinated movements.

muscle spasm Involuntary muscle contraction sustained over hours, days, weeks, or months; can lead to problems such as postural adaptations, limited or painful movement, and poor circulation.

muscle spindle Proprioceptor found in skeletal muscle that senses length and rate of change in length; consists of intrafusal fibers and nerve endings.

muscle splinting Reflexive contraction of muscles surrounding an injured area to help keep the area still and protected.

muscle tissue One of four major classes of tissue in the body; has the ability to contract.

muscle tone Natural firmness of a muscle created by its fluid and connective tissue elements.

muscular (MUS-ku-lar) **system** Body system made up of muscles that move the skeleton, support and protect internal organs, and maintain posture.

muscularis (mus-ku-LAR-is) Smooth muscular layer within the wall of an organ such as the digestive tract.

musculotendinous junction (MUS-ku-lo-TEN-din-us JUNK-shun) Tissue zone where muscle transitions to tendon.

myelin (MI-eh-lin) Lipid insulating layer found around the axons of many neurons.

myocardium (mi-o-KAR-de-um) Thick middle layer of the heart composed of cardiac muscle tissue.

myofascial (mi-o-FAH-shul) **chain** Connective tissue links between muscles, bones, and fascial membranes that provide a pathway for the mechanical communication of tension and compression throughout the body.

myofibril (mi-o-FI-bril) Small cylindrical organelle within a muscle fiber, made up of myofilaments.

myofilament (mi-o-FIL-ah-ment) Thin protein strand found within a muscle cell; actin or myosin.

myoglobin Oxygen-transporting protein of muscle; provides an immediate source of O_2 to the cell when needed.

myosin (MI-o-sin) Thick protein myofilaments in a muscle sarcomere.

nares (NA-reez) Nostrils.

nasal (NA-sal) Pertaining to the nose.

nasal cavity (NA-zal cah-vih-te) Space on either side of the nasal septum lined with a mucous membrane; the beginning of the respiratory tract.

nasal septum (NA-zal SEP-tum) Cartilage and bony divider that separates the nasal cavity.

natural killer (NK) cell Lymphocyte found in blood that destroys pathogens.

naturally acquired immunity (NAH-chur-ah-le ah-KWI-yurd im-MU-nih-te) Disease resistance obtained without medical intervention.

negative feedback Most common homeostatic control mechanism in which the effector response counteracts the original stimulus; method by which hormone levels are regulated.

nephron (NEH-fron) Microscopic functional unit of the kidney.

nerve Bundle of nerve fibers in the peripheral nervous system.

nerve fiber Process extending from the cell body of a neuron; dendrites and axons.

nerve impulse Electrochemical signal that travels along a neuron; an action potential

nerve root Either the posterior (dorsal) afferent sensory root or the anterior (ventral) efferent motor root of a spinal nerve.

nervous (NER-vus) **system** Body system made up of the brain, spinal cord, and peripheral nerves; works with the endocrine system for communication, coordination, and control of all other body systems.

nervous tissue One of four major classes of tissue in the body; found only in the nervous system.

neurilemma (nur-ih-LEM-mah) Plasma membrane of a Schwann cell (also spelled neurolemma).

neurofascial loops Reflexive neuronal pathways connecting the ANS to smooth muscle cells in the fascia.

neuroglia (nu-RAHG-le-ah) Cells of the nervous system that support and protect neurons; also known as *glial* cells.

neuromuscular junction (NER-o-MUS-ku-lar JUNK-shun) Structural interface between a motor neuron and muscle fibers of a motor unit.

neuromuscular reflex Reflex that controls skeletal muscle contraction, includes simple reflex arcs as well as more complex neuronal pathways involving multiple interneurons in the spinal cord and brain.

neuron (NUR-on) Nerve cell specialized to conduct nerve impulses; consists of a cell body, dendrites, and an axon.

neuronal pool (nu-RON-al pool) Functional grouping of nerve cells in the central nervous system.

neurotransmitter (ner-o-TRANZ-mit-ter) Chemical released by a neuron that transmits an impulse (electrical signal) to another neuron or to an effector cell.

nociceptor (NO-se-sep-tor) Free nerve ending that is sensitive to chemical, thermal, or mechanical damage to tissue; pain receptor.

node of Ranvier (rahn-ve-A) Gap between Schwann cells along a myelinated axon where depolarization and repolarization can occur.

nodule (NOD-jule) Small node.

nonspecific immune defense Any of a group of general body defenses that are not directed at a particular pathogen; also known as *innate immune defense*.

noradrenaline (NOR-ah-DREN-ah-len) One of the two catecholamine hormones released by the adrenal medulla; works with adrenaline to create the fight-or-flight response when stimulated by the autonomic nerves.

nucleic acids (nu-KLE-ik) Organic molecules made of carbon, hydrogen, oxygen, nitrogen, and phosphorous atoms; primary structural component of DNA and RNA molecules.

nucleus (NU-kle-us) Large cellular organelle that contains the DNA.

obligatory (ah-BLIG-ah-tor-e) **load** The 10% of cardiovascular capillary filtrate that must be picked up and returned via the lymphatic system.

occipital (ok-SIP-ih-tal) Pertaining to the back of the head; the posterior region of the skull.

olecranal (o-LEK-crah-nal) Pertaining to the posterior elbow (the point of the elbow).

olfaction (ol-FAK-shun) Sense of smell; perceiving and distinguishing odors.

oral (OR-al) Pertaining to the mouth.

orbital (OR-bih-tal) Pertaining to the eye.

organ (OR-gan) Group of tissues working together to accomplish specific tasks.

organelle (or-gah-NEL) Microscopic organ found in the cytoplasm of cells.

organic (or-GAN-ik) **compound** Compound that contains carbon molecules (and often hydrogen molecules).

organism (OR-gah-nizm) A living thing that functions as a whole and is able to use nutrients, excrete wastes, move, grow, respond or adapt to changes in its environment, and reproduce.

origin (OR-eh-jin) The fixed (nonmoving) muscle attachment.

osmosis (oz-MO-sis) Type of diffusion (passive transport) in which water moves from an area of high to low concentration.

osseous (OS-se-us) Bony or bone-like consistency or structure.

osteoblast (OS-te-o-blast) Bone-building cell.

osteoclast (OS-te-o-klast) Bone cell that breaks down bone tissue.

osteocyte (OS-te-o-site) Bone cell.

osteogenic (OS-te-o-JEN-ik) Producing bone; derived from or composed of bone-producing tissues.

osteon (OS-te-on) Basic structural unit of compact bone, made up of the haversian canal and its concentric rings of lamellae; also known as a *haversian system*.

otic (AH-tik) Pertaining to the ear.

ovary (O-vah-re) One of two female gonads located in the pelvic cavity; produces eggs (ova) and secretes hormones.

oviduct (O-vih-dukt) One of two short muscular tubes that carry eggs from the ovary to the uterus; also known as *fallopian tube*.

ovulation (ah-vu-LA-shun) The expelling of an egg from the ovary.

ovum (O-vum) Female gamete; also known as an *egg* (plural form is ova).

oxygen (OCKS-eh-gin) **debt** Cumulative deficit of oxygen that results from a period of exercise; leads to deep and rapid breathing after exertion has stopped to obtain more oxygen.

oxytocin (ox-e-TO-sin) Hypothalamic hormone stored and released by the posterior pituitary gland; released during labor to stimulate uterine contractions and also facilitates the ejection of milk during breastfeeding.

Pacinian corpuscle (pa-SIN-e-an KOR-pus-sel) Deep cutaneous receptor that is sensitive to vibrations and deep pressure.

pain threshold The amount of stimulus required to produce a pain signal.

pain tolerance The amount of pain a person can withstand.

pallor (PAL-or) Whitened or pale appearance, usually refers to skin or tissue coloration.

palmar (PALM-ar) Pertaining to the palm of the hand.

pancreas (PAN-kree-as) Elongated glandular organ situated behind the stomach that secretes pancreatic juice and the hormones insulin and glucagon; classified as both an exocrine gland (digestive functions) and endocrine gland (islet cells).

pancreatic duct (PAN-kre-AH-tik DUKT) Tube that carries pancreatic juice from the pancreas to the duodenum.

pancreatic islets Group of specialized hormone-producing cells scattered throughout the pancreas that produce glucagon and insulin; also known as *islets of Langerhans*.

papillary (PAP-eh-lar-re) **region** Superficial region of the dermis that contains blood and lymph vessels and cutaneous receptors.

parallel muscle Muscle whose fibers are all the same length and in parallel arrangement.

parasympathetic (par-ah-sim-pah-THET-ik) Pertaining to the division of the autonomic nervous system concerned with normal functioning and conservation of body energy; the "feed-and-breed" division.

parathyroid gland One of four small nodules located on the posterior aspect of the thyroid gland; secretes parathyroid hormone (PTH).

parathyroid hormone (PTH) Hormone secreted by parathyroid glands to help regulate calcium blood levels; antagonist to calcitonin by increasing level of CA+ in blood.

paravertebral (par-ah-VER-te-bral) **chain** Chain of sympathetic ganglia beside the vertebral column; also known as the *sympathetic chain*.

parietal (pah-RI-eh-tal) Pertaining to the outer layer of a serous membrane that lines a body cavity.

passive transport Cellular transport mechanism that does not require the use of energy.

patellar (pa-TEH-lar) Pertaining to the anterior knee.

pathogen (PATH-o-jen) Disease-causing agent such as a virus, parasite, bacteria, fungi, or a chemical or environmental particulate.

pathology (path-OL-o-je) Study of disease.

pectoral (PEK-to-ral) Pertaining to the chest.

pedal (PE-dal) Pertaining to the foot.

pelvic (PEL-vik) Pertaining to the hip girdle.

penis (PE-nis) External male organ of copulation and urination.

pennate muscle Muscle with shorter fibers that run in an oblique line to attach to a central tendon.

pepsin (PEP-sin) Strong protein enzyme produced by the stomach.

peptide (PEP-tide) Compound made up of a chain of 3to 49 amino acids.

perception (per-SEP-shun) Mental process of becoming aware of or recognizing an object or idea; interpretation of external stimuli.

pericardium (per-eh-KAR-de-um) Connective tissue sac that surrounds the heart.

perimysium (per-e-MI-se-um) Thin connective tissue layer surrounding each fascicle within a muscle.

perineum (per-ih-NE-um) Region between the mons pubis and anus in females or between the scrotum and anus in males.

perineurium (per-e-NUR-e-um) Connective tissue covering that binds multiple nerve fibers together to create a fascicle.

periosteum (pair-e-OS-te-um) Connective tissue covering of all bones.

peripheral nervous system (PNS) A primary division of the nervous system consisting of the nerves and ganglia that connect the various parts of the body with the central nervous system.

peripheral resistance (peh-RIF-er-al re-ZIS-tans) Forces within the cardiovascular network that resist blood flow, including the elasticity of blood vessels, the size of lumens, and blood viscosity.

peristalsis (per-eh-STAHL-sis) Wave-like contraction of the smooth muscle layer that propels food forward through the GI tract.

peritoneum (per-ih-tah-NE-um) Serous membrane that covers the abdominal organs and lines the abdominal cavity.

permissive effect Effect of one hormone on a target cell's sensitivity in which sensitivity to another hormone is increased.

peroneal (peh-RO-ne-al) Pertaining to the lateral lower leg.

Peyer patch (PI-yer PACH-ez) Small mass of lymphoid tissue located within the mucous membrane of the lower portion of the small intestine.

pH scale Scale used to measure a compound's acidity or alkalinity.

phagocyte (FAG-o-site) Cell whose primary function is engulfing, ingesting (eating), and absorbing other cells, cell fragments, or debris.

phagocytosis (fag-o-si-TO-sis) Active transport mechanism; the process of engulfing, ingesting (eating), and absorbing other cells, cell fragments, or debris.

pharynx (FAIR-inks) Passageway for air and food between the nose and larynx; commonly called the throat.

phasic (FA-zik) **muscle** Skeletal muscle whose primary role is to create movement.

pheromone (FARE-ah-mones) Type of protein secreted by apocrine glands.

photoreceptor (fo-to-re-SEP-tor) Sensory receptor that is sensitive to light.

physiologic effect (fiz-e-o-LOJ-ik e-FEKT) Specific, objective, quantifiable changes in the body's structure and/or function.

physiology (fiz-e-OL-o-je) Study of the body's functional processes.

pia mater (PE-a MAH-ter) Thin and delicate innermost layer of the meninges.

piezoelectricity (PE-zo-e-lek-TRIS-eh-te) Small electrical charge along a cell or tissue surface created by mechanical pressure.

pineal gland (pi-NE-al) Tiny gland in the diencephalon that produces melatonin.

pinocytosis (pe-no-si-TO-sis) An active transport mechanism by which extracellular fluid is taken into a cell; cell "drinking."

pituitary (pih-TOO-eh-tary) **gland** Master gland of the endocrine system; attached to and regulated by the hypothalamus; has an anterior and posterior lobe.

pivot (PIV-et) **joint** Type of synovial joint in which the round articular surface of one bone fits into the bony ring of another; allows rotation around a single axis.

placenta (plah-SEN-tah) Blood-rich membrane that serves as a protective barrier, providing oxygen and nutrients for the fetus and carrying away fetal waste products.

plantar (PLAN-tar) Pertaining to the sole of the foot.

plantarflexion (plan-tar-FLEK-shun) Ankle movement in which the plantar surface moves downward; pointing the foot down.

plasma (PLAZ-mah) Fluid or liquid component of blood.

plasma (PLAZ-mah) **cell** Specialized B lymphocyte produced during an antibody-mediated immune response that produces and releases antibodies.

plasma membrane (PLAZ-mah MEM-brain) Semipermeable boundary surrounding all cells; also known as the *cell membrane*.

plasma oncotic (PLAZ-mah on-KAH-tik) **pressure (POP)** Osmotic pressure created by plasma proteins within the capillary that draws fluid in from the interstitium.

plasma proteins Protein elements in the liquid component of blood, including albumins, complements, clotting factors, and antibodies.

plasticity (plas-TIS-eh-te) The ability of a substance to be molded or changed.

platelet (PLATE-let) Formed element of blood that is a cellular fragment involved in clotting; also known as a *thrombocyte*.

pleura (PLUR-ah) Serous membrane that lines the thoracic cavity and surrounds the lungs.

plexus (PLEKS-us) In general, a network of nerves, veins, or lymphatic vessels; in the peripheral nervous system, refers to a network of intersecting spinal nerves.

plicae (PLI-ka) Circular folds in the lining of the small intestine that increase the surface area for absorption.

pons (PONZ) Structure of the brain stem that forms a bridge connecting the medulla with the midbrain.

popliteal (pop-leh-TE-al) Pertaining to the posterior knee.

pore Small opening in a membrane.

positive feedback Least common homeostatic control mechanism; response to stimulus is sustained until no longer needed.

posterior (pos-TE-re-or) Back (dorsal).

postural muscle Skeletal muscle that plays an essential role in maintaining the body's upright position.

precapillary sphincter (PRE-CAP-eh-la-re SFINK-ter) Small ring of smooth muscle that encircles the entrance of the capillary at the end of the arteriole; controls the blood volume and rate of flow into the capillary.

pre-lymphatic channel (pre-lim-FAT-ik CHAN-els) Minute functional (nonstructural) preferred pathway for interstitial fluid flow toward the initial vessels.

prenatal (pre-NA-tal) Pertaining to the gestation period between conception and birth when a developing fetus grows inside the uterus.

primary edema (PRI-mar-e eh-DE-mah) Collection of fluids in the interstitial space caused by hemorrhage and the initial inflammatory response.

prime function The strongest movement a muscle creates during concentric contraction.

prime mover Muscle that generates most of the power for motion; also known as the *agonist*.

process (PRAH-ses) Projection from a bone.

progesterone (pro-JES-ter-rone) Female steroid hormone that stimulates the development and maintenance of the endometrium.

prognosis (prog-NO-sis) Expected progression or outcome of a disease or dysfunction.

projection (pro-JEK-shun) Referring of a sensation to the body region producing it.

prolactin (pro-LAK-tin) Hormone secreted by the anterior pituitary gland to initiate milk production after childbirth.

proliferative (pro-LIF-er-ah-tiv) **stage** Second phase of the healing process that involves the creation of granulation tissue and the beginning of collagen remodeling; also known as the *subacute stage*.

propagation (prah-pah-GA-shun) In general, the process of spreading to a larger area or disseminating; in regard to the nervous system, refers to the conduction of an impulse along a nerve fiber.

proprioception (PRO-pre-o-SEP-shun) The sense and awareness of where body parts are positioned in space and in relationship to one another.

proprioceptor (pro-pre-o-SEP-tor) Sensory receptor sensitive to movement and/or changes in length or tension in the muscles, tendons, ligaments, or joint fascia.

prostaglandin (PROS-tah-GLAN-din) Any of a group of locally acting hormone-like compounds derived from fatty acids and released by virtually all cells except red blood cells; effects on surrounding tissue include alteration of fat metabolism and enhancing the inflammatory response.

prostate (PRAH-state) Donut-shaped gland that surrounds the upper portion of the urethra and secretes a milky fluid that is a primary component of semen.

protein (PRO-teen) Organic molecule made of carbon, hydrogen, oxygen, and nitrogen; breaks down into amino acids.

protein synthesis (SIN-theh-sis) Anabolic process in cells in which amino acids are chained together to produce proteins.

proximal (PROKS-ih-mal) In extremities, closer to attachment point to the body.

psychogenic (si-ko-JEN-ik) Arising from the mind.

psychoneuroimmunology (PNI) (PSI-ko-NUR-o-im-mu-NOL-o-je) Study of the communication links between the nervous, endocrine, immune, and digestive systems.

pubic (PU-bik) Pertaining to the genital region.

pulmonary (PUL-mah-na-re) **circuit** Cardiovascular circulatory pathway between the heart and lungs.

pulmonary (PUL-mah-na-re) **valve** Heart valve located between the right ventricle and pulmonary artery.

pulse (puls) The rhythmic expansion of an artery in response to ventricular contraction; can be felt by placing a finger on the skin over the artery.

Purkinje (per-KIN-je) **fibers** Small fibers that are part of the heart's conduction system; these fibers carry the action potential across the ventricles.

pus Whitish fluid mixture of interstitial fluid, destroyed pathogens, phagocytes, and living or dead white blood cells.

pyloric sphincter (pi-LOR-ik SFINK-ter) Smooth muscle ring between the stomach and duodenum that controls the flow of chyme from the stomach into the small intestine.

pylorus (pi-LOR-us) Lower portion of the stomach.

pyruvic acid Organic compound formed as an end product of glycolysis, the anaerobic phase of glucose metabolism; may be converted to lactic acid in muscle tissue.

range of motion Extent or range of movement of a joint, measured in degrees (as in degrees of a circle).

reabsorption (re-ab-SORP-shun) In general, the process of a substance being absorbed again. In the cardiovascular system, applies to the movement of fluid back into the capillary from the interstitium. In the urinary system, the second step in urine formation that occurs in the renal tubule in which water, nutrients, and ions are moved out of the renal tubule back into the blood.

receptor (re-SEP-tor) Sensory organs that are sensitive to specific types of stimuli.

reciprocal inhibition (re-SIP-ro-cal in-hi-BISH-un) Reflex mechanism that coordinates the effort between agonist and antagonist muscles.

rectum (REK-tum) Straight distal portion of the large intestine between the sigmoid colon and the anus.

red blood cell (RBC) Formed element of blood responsible for transporting oxygen throughout the body; also known as an *erythrocyte*.

referred pain Pain felt in body areas that are distant from the location of the affected organ; generally associated with visceral pain.

reflex (RE-fleks) An involuntary and automatic response to a stimulus.

reflex arc (RE-fleks ARK) Simple neuronal pathway involving two or three neurons that produces a predictable motor response to a specific sensory stimulus.

refractory period (re-FRAK-tor-e PE-re-od) The time following an effective stimulus during which the neuron is unable to respond to another threshold stimulus because its membrane has not yet repolarized.

releasing hormones Group of hormones synthesized and released by the hypothalamus that stimulate hormone releases from the anterior pituitary gland.

remodeling (re-MOD-el-ing) Cycle of breaking down and rebuilding bones carried out by osteoclasts and osteoblasts.

renal (RE-nal) Pertaining to the kidneys.

renal capsule (RE-nal KAP-sul) Large fascial envelope that surrounds each kidney and attaches it to the abdominal wall.

renal corpuscle (RE-nal KOR-pus-al) Initial portion of a nephron comprised of the glomerulus and Bowman capsule.

renal cortex (RE-nal KOR-teks) Outer region of the kidney.

renal medulla (RE-nal meh-DU-lah) Inner region of the kidney.

renal pelvis (RE-nal PEL-vis) Region of the kidney where urine is collected before being passed into the ureter.

renal pyramid (RE-nal PEER-ah-mid) Triangle-shaped bundle of microscopic tubes within the renal medulla that collect urine from a specific group of nephrons.

renal tubule (RE-nal TU-bu-el) Second portion of a nephron consisting of the proximal and distal convoluted tubules and the loop of Henle.

repolarization (RE-po-lah-ri-ZA-shun) The second stage of an action potential when positive potassium ions rush out the neuron to return the cell to its original resting membrane potential.

reproductive (re-pro-DUK-tiv) **system** Body system consisting of gonads and accessory organs responsible for male and female sexual characteristics and reproductive processes.

resistance reaction The part of the stress response with longer-term effects; stimulates endocrine activity when stress stimulus is sustained beyond the ability of the alarm response to manage.

resorption (re-ZORP-shun) Process of breaking down bone tissue.

respiration (res-peh-RA-shun) Exchange of oxygen and carbon dioxide across the respiratory or cell membrane.

respiratory membrane (res-PI-rah-tor-ee MEM-brain) Thin barrier formed by the alveolar and capillary walls where external respiration occurs.

respiratory mucosa (mu-KO-sah) Mucous membrane that lines the respiratory tract.

respiratory (RES-peh-rah-tor-e) **system** Body system made up of the lungs and air passages involved in exchange of oxygen and carbon dioxide between the atmosphere and the blood.

resting membrane potential The natural polarized state of a neuron's cell membrane when it is not stimulated or conducting a nerve impulse.

reticular (reh-TIK-u-lar) **fiber** Thin collagen-like fibers made from the protein reticulin.

reticular formation (reh-TIK-u-lar for-MA-shun) Net-like arrangement of small masses of gray and white matter within the brain stem that plays a role in managing autonomic functions, muscle tone, and states of alertness and sleep.

reticular (reh-TIK-u-lar) **region** Deeper region of the dermis.

reticulin (reh-TIK-u-lin) Type of thin, delicate collagen that makes up reticular fibers in connective tissues.

retina (RET-in-ah) Light-sensitive membrane that forms the innermost layer of the wall of the eyeball.

ribonucleic acid (ri-bo-NU-kle ik AH-sid) **(RNA)** Nucleic acid that copies portions of DNA to be translated by ribosomes to make proteins.

ribosome (RI-bah-zomes) Cell organelle that synthesizes proteins.

right lymphatic duct (lim-FAT-ik DUKT) One of the two largest deep lymphatic vessels that collects lymph from the right upper quadrant of the body and returns it to the right subclavian vein.

rod Photoreceptor of the eye sensitive to dim light.

rotation (ro-TA-shun) Movement about a fixed point in a single axis that can occur in any plane.

Ruffini corpuscle (ru-FE-ne KOR-pus-sel) Deep cutaneous receptor that is sensitive to deep touch, pressure, and tissue distortion.

rugae (RU-gi) Specialized folds in the lining of the stomach that allow it to distend to accommodate large quantities of food.

sacral (SA-kral) Pertaining to the tail bone region.

saddle joint Type of synovial joint in which bone surfaces are shaped like a saddle; allows flexion, extension, abduction, and adduction.

sagittal plane (SAJ-ih-tal plane) Vertical plane that divides the body into right and left sections.

saliva (sah-LI-vah) Secretion of the salivary glands.

saltatory conduction (SAL-teh-tor-e con-DUK-shun) Nerve transmission along a myelinated nerve fiber in which the impulse seems to jump from one node of Ranvier to the next.

sarcolemma (sar-ko-LEM-ma) Plasma membrane of a muscle fiber.

sarcomere (SAR-ko-meer) Contractile unit within a muscle fiber made up of a repetitive pattern of actin and myosin filaments.

sarcoplasm Cytoplasm of a muscle fiber.

sarcoplasmic reticulum (sar-ko-PLAZ-mik reh-TIK-u-lum) Endoplasmic reticulum of a muscle fiber.

scapular (SKAP-u-lar) Pertaining to the shoulder blade.

scoliosis (sko-le-O-sis) Abnormal lateral curvature of the spine.

scrotum (SKRO-tum) Pouch-like sac that encloses the testes.

sebaceous gland (sah-BA-shus) Oil gland located in the skin.

sebum (SE-bum) Oily secretion of the sebaceous glands that is released onto the surface of the skin.

secondary edema (SEH-kon-da-re eh-DE-mah) Fluid drawn into the interstitial space in response to an increase in oncotic pressure caused by primary edema.

second-messenger mechanism Method of action for water-soluble hormones; the hormone is the first messenger, and the cAMP produced inside the cell is the second.

secretion (seh-KRE-shun) In general, the process of producing and releasing substances from a cell; in the urinary system, the third step in urine formation that occurs in the renal tubule in which toxins, metabolic wastes, and unused ions and nutrients move out of the blood into the renal tubule.

segmentation (seg-men-TA-shun) Mechanical digestive process of the small intestine that mixes and churns chyme through alternating contractions in single segments of the circular smooth muscle.

semen (SE-men) Thick, whitish fluid consisting of sperm and secretions from several accessory reproductive organs.

semicircular canal (SEM-e-SIR-kyu-lar KAH-nal) Three bony tubes of the inner ear that contain the mechanoreceptors responsible for dynamic equilibrium.

seminal vesicle (SEM-in-al VES-ih-kel) One of two ducts that secrete a thick alkaline fluid that is a primary component of semen.

seminiferous tubules (seh-mih-NIF-er-us TU-buelz) Small coiled tubes within the testes that produce sperm.

sensation (sen-SA-shun) Impression created by the interpretation of a sensory stimulus by the brain, such as hot, cold, yellow, round, etc.

sensory receptor (SEN-sor-e re-SEP-tor) Sensory nerve ending that is sensitive to changes in the environment and responds to various stimuli.

septum (SEP-tum) Dividing wall of tissue.

sequela (seh-KWEL-ah) Lasting results of a specific disease or dysfunction.

serosa (seh-RO-sa) Outermost serous membrane layer of the digestive tract; visceral peritoneum.

serous (SEER-us) **fluid** Fluid secreted by a serous membrane.

serous (SEER-us) **membrane** Epithelial membrane that folds to cover organs and line a cavity without openings to the external environment.

sexual reproduction (SEKS-u-al re-pro-DUK-shun) Process of producing offspring that requires the joining of one male and one female sex cell.

sigmoid colon (SIG-moyd KO-lun) Short S-shaped segment at the distal end of the descending colon.

sign Objective indicator of disease that is usually measurable.

sinoatrial (SI-no-A-tre-al) **(SA) node** Collection of autorhythmic cardiac muscle cells located in the wall of the right atrium just inferior to the opening of the superior vena cava; acts as the heart's pacemaker.

sinus (SI-nus) Cavity inside a structure, such as within a bone.

skeletal (SKEL-eh-tal) **system** Body system consisting of the bones and joints.

skeleton (SKEL-eh-ton) Bones and connective tissues that form the rigid framework that supports the body.

sliding filament mechanism (FIL-ah-ment MEK-ah-niz-em) The physiology of muscle contraction; the bonding between actin and myosin when calcium is present that causes the filaments to slide over one another.

small intestine Narrow, winding upper portion of the intestine, consisting of the duodenum, jejunum, and the ileum, in which digestion is completed and nutrients are absorbed by the blood.

sol (SAWL) A state in which a solid is suspended throughout a liquid.

somatic nervous system (so-MAT-ik) Motor division of the peripheral nervous system that innervates skeletal muscle.

somatogenic (so-mah-to-JEN-ik) Arising from the body.

somatotropin (so-mah-to-TRO-pin) Another name for growth hormone (GH).

special receptor Sensory receptor that is sensitive to changes in the environment and is responsible for the senses of sight, smell, taste, and hearing.

specialized lymphoid tissue A patch of tissue, gland, or organ composed of lymphatic tissue and involved in immune responses; includes the spleen, thymus, tonsils, Peyer patches, lacteals, and appendix.

specific immune response Defensive body reaction involving B and T cell activation by a specific pathogen; also known as an *adaptive immune response*.

sperm Male sex cell; also known as a *gamete*.

sphincter (SFINK-ter) Ring of smooth muscle that controls the opening and closing of a body passage or orifice.

spinal (SPI-nal) Pertaining to the backbone region; vertebral.

spinal cavity (SPI-nal KAV-ih-te) One of the two dorsal cavities; cavity that contains the spinal cord.

spinal nerve Nerve originating from the spinal cord.

spinal segment (SPI-nal SEG-ment) Transverse section of the spinal cord that gives rise to a pair of spinal nerves.

spleen Large vascular lymphoid organ located in the upper left abdominal quadrant; functions include lymphocyte and platelet storage, blood filtering, and serving as an emergency blood reserve.

splenic flexure (SPLEH-nik FLEK-shur) Point of intersection in the large intestine between the transverse and descending colon.

spongy (SPUN-je) **bone** Cancellous bone.

sprain Joint injury in which ligaments are stretched or torn.

stabilizer (STA-bil-i-zer) Muscle that stabilizes or fixes the origin end of a prime mover so that the movement is more efficient; also known as the *fixator*.

sternal (STER-nal) Pertaining to the breastbone region.

steroid (STAIR- oyd) hormone Lipid-soluble hormones made from cholesterol.

stimulus (STIM-u-lus) Any internal or external change in the environment that produces a response.

stomach (STUM-ik) Pouch-shaped muscular organ of digestion located in the upper left abdominal quadrant.

strain Injury that occurs when a muscle or tendon is stretched or torn.

strata (STRAT-ah) Layers (singular form is stratum).

stress response Group of physiologic responses initiated by the hypothalamus when stress is perceived; includes the alarm response of the autonomic nervous division and the resistance reaction of the endocrine system.

stretch reflex Reflexive contraction of a muscle after it is stretched due to stimulation of proprioceptors.

structural (STRUK-chur-al) **effect** Physical changes created by manual therapy, such as stretching, loosening, broadening, or unwinding of muscles and connective tissue.

subacute (SUB-ah-kute) **stage** Second phase of the healing process that involves the creation of granulation tissue and the beginning of collagen remodeling; also known as the *proliferative stage*.

subcutaneous (sub-ku-TA-ne-us) **layer** Fat and areolar connective tissue that lies between the dermis and underlying tissues and organs; also known as the *hypodermis* or *superficial fascia*.

submucosa (SUB- mu-ko-sah) Tissue layer directly beneath the mucosa.

sudoriferous gland (su-do-RIF-ur-us) Sweat gland.

sulcus (SUL-kus) Shallow groove or depression, as in the cerebrum.

superficial (su-per-FISH-al) Closer to the surface.

superficial fascia (su-per-FISH-al FASH-ah) Fat and areolar connective tissue that lies between the dermis and underlying tissues and organs; also known as the *hypodermis* or the *subcutaneous layer*.

superior (su-PE-re-or) Above; closer to head; cephalad.

suppressor (suh-PRES-er) **T cell** Specialized lymphocyte the inhibits or shuts down the body's specific immune response when it is no longer needed.

suprarenal (su-prah-RE-nal) **gland** Endocrine gland situated on top of each kidney; also known as the *adrenal gland*.

sural (SUR-al) Pertaining to the calf; posterior lower leg.

suture (SU-chur) Fibrous line of junction or an immovable joint between two bones, as in the skull. (Also refers to the surgical process of closing a wound or joining tissues using stitches.)

sympathetic (sim-pah-THET-ik) Pertaining to the fight-or-flight (emergency response) division of autonomic nervous system.

sympathetic chain Chain of sympathetic ganglia beside the vertebral column; also known as the *paravertebral chain*.

symptom (SIMP-tom) Subjective indicator of disease as perceived by the patient; not easily measured or quantified.

synapse (SIN-aps) The functional junction point between a neuron and another neuron or an effector.

synaptic bulb (sih-NAP-tik bulb) The small knob at the distal end of each axon terminal where the vesicles containing neurotransmitter are stored.

synaptic cleft (sih-NAP-tik kleft) Tiny gap between the presynaptic and postsynaptic neurons across which the neurotransmitter diffuses.

synarthrosis (SIN-ar-THRO-sis) Type of joint that does not allow movement.

synergist (SIN-er-jist) Muscle that assists the agonist in creating movement.

synergistic effect Effect of two hormones working together to enhance or intensify a target's response beyond that which occurs when each hormone acts singly.

synovial (sin-O-ve-al) Pertaining to synovial membrane or fluid.

synovial (sin-O-ve-al) **fluid** Fluid secreted by a synovial membrane.

synovial (sin-O-ve-al) **joint** Joint with a fibrous capsule and synovial lining that allows free movement; also known as a *diarthrosis*.

synovial (sin-O-ve-al) **membrane** Thin connective tissue lining of the fibrous capsule in diarthrotic joints and bursa.

system (SIS-tem) Group of organs working together to accomplish a specific set of tasks.

systemic (sis-TEM-ik) **circuit** Cardiovascular pathway between the heart and the tissues of the body.

systemic (sis-TEM-ik) **effect** Responses to manual therapy that are cellular, circulatory, and/or nervous system mediated; responses occur regionally or throughout the body.

systole (SIS-to-le) Contraction state of a heart chamber.

T cell Specialized lymphocyte involved in cell-mediated immunity (also called T lymphocyte).

tarsal (TAR-sal) Pertaining to the ankle.

taste bud Cluster of gustatory cells (chemoreceptors for taste) found primarily in the papillae (tiny bumps) on the tongue's surface.

temporal (TEMP-or-al) Pertaining to the temple of the skull.

tender point Localized area of tenderness within a muscle that is painful when compressed.

tendinopathy Condition in which one or more tendons are strained or degenerated.

tendon (TEN-dun) Band of fibrous connective tissue that attaches muscle to bone.

tenoperiosteal junction (TE-no-per-e-OS-te-al JUNK-shun) Tissue zone where a tendon transitions and weaves into the periosteum to attach to a bone.

tensegrity (ten-SEG-rih-te) System in which tension between two opposing forces is balanced to create structural integrity; a contraction for "tension integrity structuring."

testicle (TES-tih-kel) One of two male gonads located within the scrotum that produce sperm and secrete hormones; also known as a testis.

testis (TES-tis) One of two male gonads located within the scrotum that produce sperm and secrete hormones; also known as a *testicle*.

testosterone (tes-TOS-ter-own) Male androgen that stimulates the production of sperm and the development of the reproductive organs and secondary sex characteristics at puberty.

tetanic contraction (teh-TAN-ik kon-TRAK-shun) Nonproductive contraction in which the muscle is bombarded with constant stimuli, causing a sustained contraction of multiple fibers that effectively "locks" the muscle.

thalamus (THAL-ah-mus) Portion of the diencephalon responsible for sorting and prioritizing almost all incoming sensory input and relaying it to appropriate centers of the cerebrum.

thermoreceptor (ther-mo-re-SEP-tor) Sensory receptor that is sensitive to changes in temperature.

thixotropy (THICKS-o-tro-pe) Ability of a substance to shift between a semisolid (gel) state and a more liquid (sol) state.

thoracic (thor-AS-ik) Pertaining to the chest.

thoracic cavity (thor-AS-ik KAV-ih-te) The most superior of the two ventral cavities of the body; chest cavity that contains the lungs and heart.

thoracic duct (thor-AS-ik DUKT) One of the two largest deep lymphatic vessels that collects lymph from the left upper and both lower quadrants of the body and returns it to the left subclavian vein.

threshold (THRESH-hold) **stimulus** Minimum amount of stimulus required to evoke a response, such as producing an action potential or creating a muscle contraction.

thrombocyte (THROM-bo-site) Formed element of blood that is a cellular fragment involved in clotting; also known as a *platelet*.

thromboembolism (throm-bo-EM-bo-liz-um) Traveling blood clot.

thrombus (THROM-bus) Stationary blood clot.

thymosin (thi-MO-sin) Hormone secreted by the thymus; plays an essential role in the development and maturation of T cells for immune response.

thymus (THI-mus) Endocrine gland located deep to the sternum and anterior to the trachea; secretes thymosin.

thyroid (THI-royd) **gland** Endocrine gland in the throat just inferior to the voice box; secretes thyroxine (T_4), T_3, and calcitonin.

thyroid hormone Metabolism boosting lipid-soluble hormone secreted by the thyroid gland; triiodothyronine (T_3) and thyroxine (T_4).

thyroid-stimulating hormone (TSH) Tropic hormone secreted by the anterior pituitary; stimulates the thyroid gland to release the thyroid hormones.

thyroxine (thi-ROKS-in) Hormone produced by the thyroid gland that increases cellular metabolism; contains four iodine atoms so known as T_4.

tissue (TISH-u) Group of like cells working together.

tonic contraction (TAH-nik) Sustained muscle contraction in which low-grade tension is maintained, as occurs in maintaining posture.

tonsil (TON-sils) Small mass of lymphoid tissue located in the mucous membranes of the mouth and throat.

trabecula (trah-BEK-u-la) Small beam or supporting structure; part of a mesh-like structure in spongy bone.

trachea (TRA-ke-ah) Tube that connects the upper respiratory tract to the lungs; commonly called the windpipe.

tract (TRAKT) Bundle of myelinated axons within the spinal cord.

transferrin (tranz-FER-in) Plasma protein that binds iron when activated by a pathogen to make it unavailable to bacteria.

transverse colon (tranz-vers KO-lun) Segment of large intestine that extends across the top of the abdominal cavity between the ascending and descending portions of the colon.

transverse plane (TRANZ-vers plane) Horizontal plane that divides the body into superior and inferior sections.

traumatic edema Edema related to soft tissue damage, such as that resulting from sprains, strains, or contusions.

tricuspid (tri-KUSS-pid) **valve** Heart valve located between the right atrium and ventricle.

trigger point Hyperirritable nodule within a skeletal muscle that produces a pattern of referred pain when moderately compressed.

triglyceride (tri-GLIS-er-ide) Lipid molecule absorbed from the digestive tract for metabolism by the body.

triiodothyronine (TRI-i-O-do-THI-ro-neen) Hormone produced by the thyroid gland that increases cellular metabolism; contains three iodine atoms so known as T_3.

trochanter (TRO-kan-ter) Large condyle on the proximal femur.

tropocollagen (tro-po-KAL-ah-gen) Protein molecule that is the basic structural unit of collagen; consists of three coiled polypeptide chains, giving it a triple-helix appearance similar to DNA.

tubercle (TU-ber-kul) Small mound or elevation; a rounded projection from a bone.

Tuberosity (tu-ber-OS-eh-te) Projection or protuberance, especially at the end of a bone where a muscle or tendon attaches.

twitch A sudden, spasmodic muscle contraction.

ulcer (UL-ser) Localized lesion in the skin or mucous membrane; often an open sore that heals slowly.

umbilical (um-BIL-ih-kal) Pertaining to the umbilicus (navel).

umbilical (um-BIL-ih-kel) **cord** Vascular cord that attaches the developing fetus to the placenta.

urea (yur-E-ah) Byproduct of amino acid catabolism created from ammonia and excreted in urine.

ureter (YUR-eh-ter) Muscular tube that carries urine from the kidney to the urinary bladder.

urethra (yur-RE-thrah) Narrow tube that carries urine from the urinary bladder to the external environment; in males, also carries semen.

urinary (YER-eh-nar-e) **system** Body system involved in elimination of fluid wastes and the regulation of water and electrolyte balance.

urination (yur-ih-NA-shun) Process of passing urine out of the urethra; also known as *micturition*.

urine (YUR-in) Liquid waste formed by the kidneys.

uterus (U-teh-rus) Pear-shaped, hollow, muscular organ of the female reproductive tract that houses the developing fetus; also called womb.

vagina (vah-JI-na) Muscular genital canal between the cervix and exterior environment (also called the birth canal).

vas deferens (vas DEF-er-enz) Tube that carries sperm out of the epididymis and over the posterior side of the bladder to the ejaculatory duct.

vasoconstriction (va-zo-con-STRIK-shun) Decrease in the size of the lumen of a blood vessel caused by contraction of the smooth muscle within the vessel's wall.

vasodilation (va-zo-di-LA-shun) Increase in the size of the lumen of a blood vessel caused by the relaxation of the smooth muscle within the vessel's wall.

vasopressin (VA-zo-pres-sin) Hormone secreted by the posterior pituitary that inhibits urine production in the kidneys; also known as *antidiuretic hormone*.

vein (VAIN) Blood vessel that carries blood toward the heart.

ventilation (ven-teh-LA-shun) Movement of air into and out of the lungs; breathing.

ventral (VEN-tral) Front (anterior).

ventricle (VEN-trih-kel) Chamber or hollow cavity in a structure or organ, such as the chambers of the heart or cavities in the brain filled with cerebrospinal fluid.

venule (VEN-ule) Small vein.

vertebral (ver-TE-bral) Pertaining to the backbone region; spinal.

vesicle (VES-eh-kal) Small sac containing liquid inside cells or tissues.

vestibule (VES-tih-byul) Small space at the base of the semicircular canals of the inner ear that contains mechanoreceptors responsible for static equilibrium.

villi Finger-like projections from the small intestine mucosa that increase the surface area for absorption.

visceral (VIS-er-al) Pertaining to an organ or to the serous membrane layer that surrounds an organ.

visceral (VIS-er-ahl) **effector** Gland or organ with smooth and cardiac muscle innervated via autonomic pathways.

viscoelasticity (VIS-ko-e-las-TIS-ity) Ability of tissues to extend and rebound rather than to stretch and recoil.

vitamin (VI-tah-min) One of various fat-soluble or water-soluble micronutrients necessary for proper growth and normal metabolism.

vocal cord (VO-kal KORD) Small fold in the mucous membrane of the larynx that produces sound as air passes over and causes it to vibrate; vocal fold.

volar (VO-lar) Pertaining to the posterior hand.

Volkmann (VOLK-man) **canal** Small canal (channel) in dense bone that communicates with the haversian canals, allowing for the passage of blood vessels into the bone.

vulva (VUL-vah) Female external genitalia.

watershed (WAH-ter-shed) Zone between lymphotomes.

water-soluble (WAH-ter SOL-u-bel) **hormone** Hormone that is easily transported in blood and requires a second messenger to stimulate metabolic changes in target cells.

white blood cell (WBC) Formed element of blood that plays a vital role in the body's healing and immune responses; includes neutrophils, basophils, eosinophils, lymphocytes, and monocytes; also known as a *leukocyte*.

white matter Areas of the brain or spinal cord made primarily of the myelinated axons of neurons.

zygote (ZI-gote) Fertilized egg.

Figure Credit List

Following are the sources of art that was either borrowed or adapted for *Applied Anatomy & Physiology for Manual Therapists*.

CHAPTER 1

McConnell TH, Hull K. *Human Form, Human Function: Essentials of Anatomy and Physiology*. Philadelphia, PA: Lippincott Williams & Wilkins; 2011:*1, System figures, pp. 10–12.*

CHAPTER 2

Cael C. *Functional Anatomy: Musculoskeletal Anatomy, Kinesiology, and Palpation for Manual Therapists*. Philadelphia, PA: Lippincott Williams & Wilkins; 2011:*3*.

Cohen BJ, Taylor J. *Memmler's Structure & Function of the Human Body*, 8th ed. Philadelphia, PA: Lippincott Williams & Wilkins; 2005:*3–6*.

McConnell TH, Hull K. *Human Form, Human Function: Essentials of Anatomy and Physiology*. Philadelphia, PA: Lippincott Williams & Wilkins; 2011:*2, 7–12*.

Premkumar K. *The Massage Connection: Anatomy and Physiology*, 2nd ed. Philadelphia, PA: Lippincott Williams & Wilkins; 2004:*1*.

CHAPTER 3

Cohen BJ, Taylor J. *Memmler's Structure & Function of the Human Body*, 8th ed. Lippincott Williams & Wilkins; 2005:*1, 8, 9*.

McConnell TH, Hull K. *Human Form, Human Function: Essentials of Anatomy and Physiology*. Philadelphia, PA: Lippincott Williams & Wilkins; 2011:*2–7, 10, 13–15; By the Way p. 36*.

Sadler T. *Langman's Medical Embryology*, 9th ed. Image Bank. Philadelphia, PA: Lippincott Williams & Wilkins; 2003:*11*.

CHAPTER 4

Anatomical Chart Company: *Pathology Alert: Skin Cancer*.

Bear M, Conner B, Paradiso M. *Neuroscience: Exploring the Brain*, 2nd ed. Philadelphia, PA: Lippincott Williams & Wilkins; 2000:*8, 10 (the latter adapted from Penfield and Rasmussen, 1952*.

Goodheart HP. *Goodheart's Photoguide of Common Skin Disorders*, 2nd ed. Philadelphia, PA: Lippincott Williams & Wilkins; 2003:*11A, 11C-E, 12A, 12C, 12D*.

McConnell TH, Hull K. *Human Form, Human Function: Essentials of Anatomy and Physiology*. Philadelphia, PA: Lippincott Williams & Wilkins; 2011:*1, 4A, 5, 6, 7*.

Mills SE. *Histology for Pathologists*, 3rd ed. Philadelphia, PA: Lippincott Williams & Wilkins; 2007:*4B*.

Stedman's Medical Dictionary, 27th ed. Philadelphia, PA: Lippincott Williams & Wilkins; 2000:*9*.

Werner R. *A Massage Therapist's Guide to Pathology*, 4th ed. Philadelphia, PA: Lippincott Williams & Wilkins; 2009:*11B, 12B*.

CHAPTER 5

Cael C. *Functional Anatomy: Musculoskeletal Anatomy, Kinesiology, and Palpation for Manual Therapists*. Philadelphia, PA: Lippincott Williams & Wilkins; 2011:*Box 5.1, Table 5.4*.

McConnell TH, Hull K. *Human Form, Human Function: Essentials of Anatomy and Physiology*. Philadelphia, PA: Lippincott Williams & Wilkins; *2011:1–6, 8–15, 18, 19, 22–26*.

Oatis CA. *Kinesiology: The Mechanics and Pathomechanics of Human Movement*. Philadelphia, PA: Lippincott Williams & Wilkins; 2003:*16A, 20*.

Premkumar K. *The Massage Connection: Anatomy and Physiology*, 2nd ed. Philadelphia, PA: Lippincott Williams & Wilkins; 2004:*16B, 17*.

Rubin R, Strayer DS. *Rubin's Pathology: Clinicopathologic Foundations of Medicine*, 5th ed. Philadelphia, PA: Lippincott Williams & Wilkins; 2008: *Pathology Alert: Osteoporosis*.

Twietmeyer A, McCracken T. *Coloring Guide to Human Anatomy*. Philadelphia, PA: Lippincott Williams & Wilkins; 2001:*21*.

Werner R. *A Massage Therapist's Guide to Pathology*, 4th ed. Philadelphia, PA: Lippincott Williams & Wilkins; 2009:*11B, 12B; Pathology Alert: Spinal Deviations; Pathology Alert: Spinal Disc Disorders*.

CHAPTER 6

Cael C. *Functional Anatomy: Musculoskeletal Anatomy, Kinesiology, and Palpation for Manual Therapists*. Philadelphia, PA: Lippincott Williams & Wilkins; 2011:*6A, 15, 16, 18, 26, 27*.

Clay JH, Pounds DM. *Basic Clinical Massage Therapy*, 2nd ed. Philadelphia, PA: Lippincott Williams & Wilkins; 2006:*Box 6.1 (last four images)*.

Cohen BJ, Taylor J. *Memmler's Structure & Function of the Human Body*, 8th ed. Lippincott Williams & Wilkins; 2005:*3, 5, 6B*.

Hendrickson T. *Massage for Orthopedic Conditions*. Philadelphia, PA: Lippincott Williams & Wilkins; 2002:*14B, 14C*.

Kendall FP, McCreary EK. *Muscle Testing and Function*, 5th ed. Philadelphia, PA: Lippincott Williams & Wilkins; 2005:*177*.

LifeArt: *24B*.

McArdle WD, Katch FI, Katch VL. *Essentials of Exercise Physiology*, 2nd ed. Philadelphia, PA: Lippincott Williams & Wilkins; 2000:*2*.

McConnell TH, Hull K. *Human Form, Human Function: Essentials of Anatomy and Physiology*. Philadelphia, PA: Lippincott Williams & Wilkins; 2011:*1, 4, 9, 11, 12, 14, 22–25, 28*.

Premkumar K. *The Massage Connection: Anatomy and Physiology*, 2nd ed. Philadelphia, PA: Lippincott Williams & Wilkins; 2004:*10, 13, Box 6.1 (first three images)*.

CHAPTER 7

Bear M, Conner B, Paradiso M. *Neuroscience: Exploring the Brain*, 2nd ed. Philadelphia, PA: Lippincott Williams & Wilkins; 2000:*26*.

Clay JH, Pounds DM. *Basic Clinical Massage Therapy*, 2nd ed. Philadelphia, PA: Lippincott Williams & Wilkins; 2006: *Pathology Alert: Neural Compression-Tension Syndromes, left image*.

Cohen BJ, Taylor J. *Memmler's Structure & Function of the Human Body*, 8th ed. Lippincott Williams & Wilkins; 2005:*12A, 16–18*.

McConnell TH, Hull K. *Human Form, Human Function: Essentials of Anatomy and Physiology*. Philadelphia, PA: Lippincott Williams & Wilkins; 2011:*1, 2, 4A, 5, 6, 10, 19–21, 23–25, 27–30*.

Moore KL, Dalley AF. *Clinically Oriented Anatomy*, 4th ed. Philadelphia, PA: Lippincott Williams & Wilkins; 1999:*9*.

Porth CM. *Pathophysiology: Concepts in Altered Health States*, 6th ed. Philadelphia, PA: Lippincott Williams & Wilkins; 2002:*34*.

Snell R. *Clinical Neuroanatomy*. Philadelphia, PA: Lippincott Williams & Wilkins; 2001:*3*.

Werner R. *A Massage Therapist's Guide to Pathology*, 4th ed. Philadelphia, PA: Lippincott Williams & Wilkins; 2009:*7, 8; Pathology Alert: Neural Compression-Tension Syndromes, right image*.

CHAPTER 8

Archer P. *Therapeutic Massage in Athletics*. Philadelphia, PA: Lippincott Williams & Wilkins; 2007:*1, 3–6, 9–13*.

CHAPTER 9

McConnell TH, Hull K. *Human Form, Human Function: Essentials of Anatomy and Physiology*. Philadelphia, PA: Lippincott Williams & Wilkins; 2011:*1, 8–10, 12–14*.

Premkumar K. *The Massage Connection: Anatomy and Physiology*, 2nd ed. Philadelphia, PA: Lippincott Williams & Wilkins; 2004:*2, 16*.

CHAPTER 10

Archer P. *Therapeutic Massage in Athletics*. Philadelphia, PA: Lippincott Williams & Wilkins; 2007:*16*.

Cohen BJ, Taylor J. *Memmler's Structure & Function of the Human Body*, 8th ed. Lippincott Williams & Wilkins; 2005:*12*.

McConnell TH, Hull K. *Human Form, Human Function: Essentials of Anatomy and Physiology*. Philadelphia, PA: Lippincott Williams & Wilkins; 2011:*1–11, 14, 15*.

CHAPTER 11

Archer P. *Therapeutic Massage in Athletics*. Philadelphia, PA: Lippincott Williams & Wilkins; 2007:*2, 5, 6, 8–11, 15, 16*.

McConnell TH, Hull K. *Human Form, Human Function: Essentials of Anatomy and Physiology*. Philadelphia, PA: Lippincott Williams & Wilkins; 2011:*3, 12, 13A*.

Moore KL, Dalley AF. *Clinically Oriented Anatomy*, 4th ed. Philadelphia, PA: Lippincott Williams & Wilkins; 1999:*1*.

Premkumar K. *The Massage Connection: Anatomy and Physiology*, 2nd ed. Philadelphia, PA: Lippincott Williams & Wilkins; 2004:*4*.

Rubin R, Strayer DS. *Rubin's Pathology: Clinicopathologic Foundations of Medicine*, 5th ed. Philadelphia, PA: Lippincott Williams & Wilkins; 2008:*17*.

Stedman's Medical Dictionary, 27th ed. Philadelphia, PA: Lippincott Williams & Wilkins; 2000:*13B*.

CHAPTER 12

Cohen BJ, Taylor J. *Memmler's Structure & Function of the Human Body*, 8th ed. Lippincott Williams & Wilkins; 2005:*1, 3*.

CHAPTER 13

McConnell TH, Hull K. *Human Form, Human Function: Essentials of Anatomy and Physiology*. Philadelphia, PA: Lippincott Williams & Wilkins; 2011:*1–7, 8A, 8B, 9, 11, 13*.

Premkumar K. *The Massage Connection: Anatomy and Physiology*, 2nd ed. Philadelphia, PA: Lippincott Williams & Wilkins; 2004:*12*.

Werner R. *A Massage Therapist's Guide to Pathology*, 4th ed. Philadelphia, PA: Lippincott Williams & Wilkins; 2009:*8C*.

CHAPTER 14

McConnell TH, Hull K. *Human Form, Human Function: Essentials of Anatomy and Physiology*. Philadelphia, PA: Lippincott Williams & Wilkins; 2011:*2, 3, 4A, 5–13*.

Moore KL, Dalley AF. *Clinically Oriented Anatomy*, 4th ed. Philadelphia, PA: Lippincott Williams & Wilkins; 1999:*4B, 4C*.

CHAPTER 15

McConnell TH, Hull K. *Human Form, Human Function: Essentials of Anatomy and Physiology*. Philadelphia, PA: Lippincott Williams & Wilkins; 2011:*1–3, 5, 6*.

CHAPTER 16

McConnell TH, Hull K. *Human Form, Human Function: Essentials of Anatomy and Physiology*. Philadelphia, PA: Lippincott Williams & Wilkins; 2011:*1–5, 7–9*.

Index

Page numbers in *italics* denote figures; those followed by t denote tables; those followed by b denote boxes

G

H

I

J

K

L

N

O